NTRODUCTION TO FOOD SCIENCE & FOOD SYSTEMS

2nd Edition

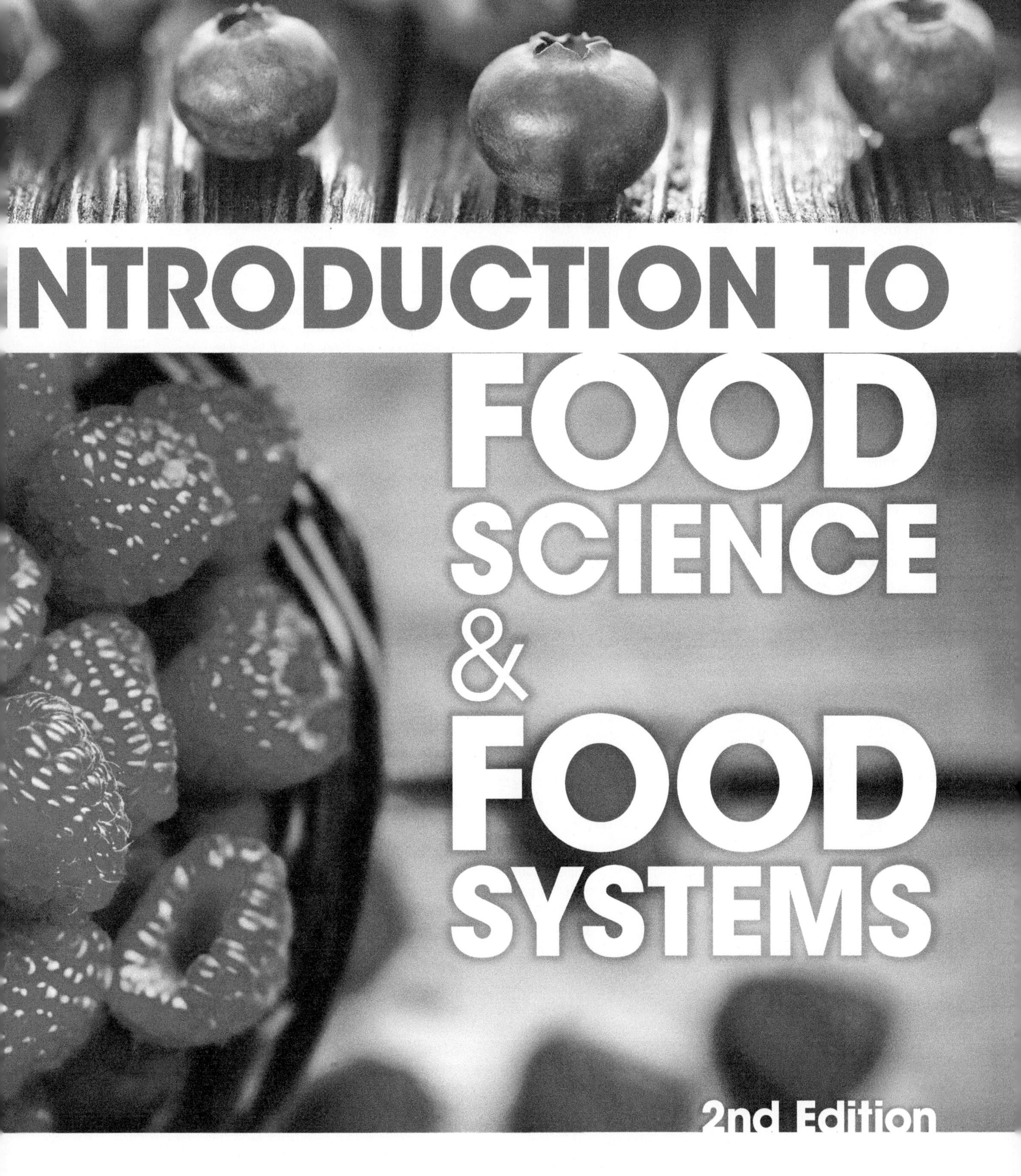

Rick Parker and Miriah Pace

Australia • Brazil • Mexico • Singapore • United Kingdom • United States

Introduction to Food Science & Food Systems, Second Edition
Rick Parker and Miriah Pace

Senior Vice President, GM Skills & Global Product Management: Dawn Gerrain

Product Team Manager: Erin Brennan

Product Manager: Nicole Robinson

Senior Director Development: Marah Bellegarde

Senior Product Development Manager: Larry Main

Senior Content Developer: Jennifer Starr

Product Assistant: Maria Garguilo

Vice President Marketing Services: Jennifer Baker

Market Manager: Jonathan Sheehan

Senior Production Director: Wendy A. Troeger

Senior Content Project Manager: Betsy Hough

Senior Art Director: Benj Gleeksman

Software Development Manager: Pavan Ethakota

Cover Image Credits:
© Zeljko Radojko/Shutterstock
© Zigzag Mountain Art/Shutterstock
© inacio pires/Shutterstock
© Alena Haurylik/Shutterstock
© symbiot/Shutterstock
© Gayvoronskaya_Yana/Shutterstock
© Grisha Bruev/Shutterstock

Interior Design Image Credits:
© Elina Li/Shutterstock
© Sasha_Ivv/Shutterstock

Library of Congress Control Number: 2015943892

[Book Only] ISBN-13: 978-1-4354-8939-4

Cengage Learning
20 Channel Center Street
Boston, MA 02210
USA

Cengage Learning is a leading provider of customized learning solutions with employees residing in nearly 40 different countries and sales in more than 125 countries around the world. Find your local representative at **www.cengage.com**

Cengage Learning products are represented in Canada by Nelson Education, Ltd.

To learn more about Cengage Learning, visit **www.cengage.com**

Purchase any of our products at your local college store or at our preferred online store **www.cengagebrain.com**

Printed in the United States of America
Print Number: 06   Print Year: 2021

To Marilyn, *wife, mother, partner, friend, and one true love for more than 47 years, through good times and bad, helping me enjoy the journey.*

CONTENTS

Preface

Introduction to Food Science & Food Systems, Second Edition, is designed for high school agriscience and consumer science programs and for postsecondary students enrolled in a food science course. It is an excellent overview for anyone interested in attaining a basic understanding of food science.

As the title suggests, science is an important component of the book. Food science as understood by humans represents a specific body of knowledge that approaches and solves problems by the scientific method—a continuous cycle of observations, hypotheses, predictions, experiments, and results. The *science* of food science is emphasized throughout the book.

HOW THIS TEXT IS ORGANIZED

Introduction to Food Science & *Food Systems* makes teaching easy. The information is divided into four basic sections, and chapters are based on a thorough, easy-to-follow outline.

- **Section I: Introduction and Background** provides the necessary background information for understanding the science of foods. This includes an introduction to the industry, a new chapter on food systems and sustainability, the chemistry of foods, nutrition and digestion, food composition and quality, unit operations, and food deterioration. These chapters are the foundation.
- **Section II: Preservation** groups the chapters that relate to methods of food preservation, including heat, cold, drying, radiant and electrical energy, fermentation, microorganisms, biotechnology, chemicals, and packaging. These chapters are the basics of food science.
- **Section III: Foods and Food Products** includes chapters on milk, meat, poultry and eggs, fish and shellfish, cereal grains, legumes and oilseeds, fruits and vegetables, fats and oils, candies and sweets, and beverages. These chapters are the application of food science.

- **Section IV: Related Issues** includes chapters that cover environmental concerns, food safety, regulations and labeling, world food needs, and career opportunities as well as a new chapter on food as it relates to health. These chapters represent the challenges of food science.

FEATURES OF THIS EDITION

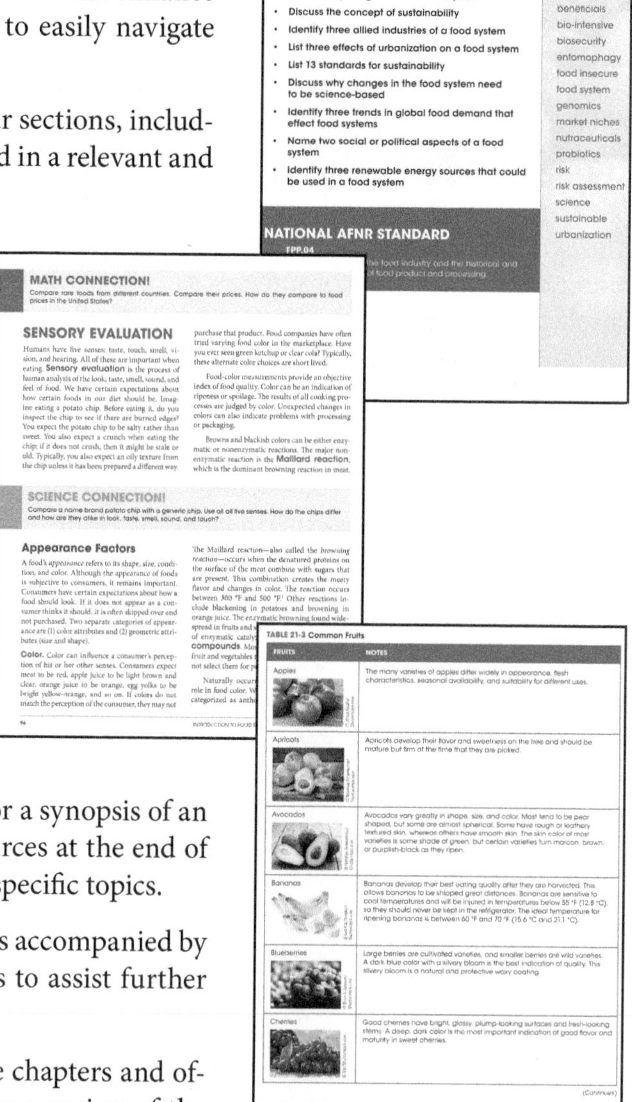

OBJECTIVES

After reading this chapter, you should be able to:

- Describe food systems
- List five major segments of a food system
- Discuss the concept of sustainability
- Identify three allied industries of a food system
- List three effects of urbanization on a food system
- List 13 standards for sustainability
- Discuss why changes in the food system need to be science-based
- Identify three trends in global food demand that effect food systems
- Name two social or political aspects of a food system
- Identify three renewable energy sources that could be used in a food system

KEY TERMS

anaerobic
aquaponics
beneficials
bio-intensive
biosecurity
entomophagy
food insecure
food system
genomics
market niches
nutraceuticals
probiotics
risk
risk assessment
science
sustainable
urbanization

NATIONAL AFNR STANDARD

FPP.04

MATH CONNECTION!

Compare rare foods from different countries. Compare their prices. How do they compare to food prices in the United States?

SENSORY EVALUATION

Humans have five senses: taste, touch, smell, vision, and hearing. All of these are important when eating. **Sensory evaluation** is the process of human analysis of the look, taste, smell, sound, and feel of food. We have certain expectations about how certain foods in our diet should be. Imagine eating a potato chip. Before eating it, do you inspect the chip to see if there are burned edges? You expect the potato chip to be salty rather than sweet. You also expect a crunch when eating the chip; if it does not crush, then it might be stale or old. Typically, you also expect an oily texture from the chip unless it has been prepared a different way.

purchase that product. Food companies have often tried varying food color in the marketplace. Have you ever seen green ketchup or clear cola? Typically, these alternate color choices are short lived.

Food-color measurements provide an objective index of food quality. Color can be an indication of ripeness or spoilage. The results of all cooking processes are judged by color. Unexpected changes in colors can also indicate problems with processing or packaging.

Browns and blackish colors can be either enzymatic or nonenzymatic reactions. The major nonenzymatic reaction is the **Maillard reaction**, which is the dominant browning reaction in meat.

SCIENCE CONNECTION!

Compare a name brand potato chip with a generic chip. Use all all five senses. How do the chips differ and how are they alike in look, taste, smell, sound, and touch?

Appearance Factors

A food's *appearance* refers to its shape, size, condition, and color. Although the appearance of foods is subjective to consumers, it remains important. Consumers have certain expectations about how a food should look. If it does not appear as a consumer thinks it should, it is often skipped over and not purchased. Two separate categories of appearance are (1) color attributes and (2) geometric attributes (size and shape).

Color. Color can influence a consumer's perception of his or her other senses. Consumers expect meat to be red, apple juice to be light brown and clear, orange juice to be orange, egg yolks to be bright yellow-orange, and so on. If colors do not match the perception of the consumer, they may not

The Maillard reaction—also called the *browning reaction*—occurs when the denatured proteins on the surface of the meat combine with sugars that are present. This combination creates the meaty flavor and changes in color. The reaction occurs between 300 °F and 500 °F.[1] Other reactions include blackening in potatoes and browning in orange juice. The enzymatic browning found widespread in fruits and

96 INTRODUCTION TO FOOD

TABLE 21-3 Common Fruits

FRUITS	NOTES
Apples	The many varieties of apples differ widely in appearance, flesh characteristics, seasonal availability, and suitability for different uses.
Apricots	Apricots develop their flavor and sweetness on the tree and should be mature but firm at the time that they are picked.
Avocados	Avocados vary greatly in shape, size, and color. Most tend to be pear shaped, but some are almost spherical. Some have rough or leathery textured skin, whereas others have smooth skin. The skin color of most varieties is some shade of green, but certain varieties turn maroon, brown, or purplish-black as they ripen.
Bananas	Bananas develop their best eating quality after they are harvested. This allows bananas to be shipped great distances. Bananas are sensitive to cool temperatures and will be injured in temperatures below 55 °F (12.8 °C), so they should never be kept in the refrigerator. The ideal temperature for ripening bananas is between 60 °F and 70 °F (15.6 °C and 21.1 °C).
Blueberries	Large berries are cultivated varieties, and smaller berries are wild varieties. A dark blue color with a silvery bloom is the best indication of quality. This silvery bloom is a natural and protective waxy coating.
Cherries	Good cherries have bright, glossy, plump-looking surfaces and fresh-looking stems. A deep, dark color is the most important indication of good flavor and maturity in sweet cherries.

(Continues)

Each chapter is designed to provide students with features that enhance learning and a learning pathway that enables them to easily navigate through food-science topics:

LOGICALLY ORGANIZED, the text is divided into four sections, including content in context so that information is presented in a relevant and meaningful way that reinforces learning.

LEARNING FEATURES such as **Learning Objectives** and **Key Terms** set the stage for the chapter and help learners identify key concepts and information. Also included are **National Agricultural Education Standards Correlations** that highlight the specific core competency that is met through successful completion of each chapter.

APPLICATION is emphasized throughout with engaging **Math** and **Science Connection activities** integrated into each chapter and a set of **Review Questions** and **Student Activities** to conclude each chapter.

FURTHER LEARNING is encouraged, with **engaging articles** highlighting interesting tidbits about the industry—whether historical insights, fun facts, or a synopsis of an emerging food trend—and a list of Additional Resources at the end of each chapter to inspire students to learn more about specific topics.

HIGHLY ILLUSTRATED, and now in full color, the text is accompanied by many tables, charts, graphs, photos, and illustrations to assist further understanding of the topics under discussion.

ROBUST APPENDICES reinforce topics learned in the chapters and offer valuable reference materials. **Appendix A** contains a review of the chemistry behind food science for those students who require additional learning in this area. **Appendix B** includes a multitude of reference tables, in particular the Food Composition Table, which provides specific nutrient information of different foods. It also features another list of Internet resources so that students can expand on their learning of specific food-science topics. **Appendix C** covers the details of harvesting and storing fruits, nuts, and vegetables. Also included is a combined **Glossary and Glosario** that provides terms and definitions in both English and Spanish.

NEW TO THIS EDITION

CURRENT INFORMATION, including new chapters on the food system and sustainability and food and health, along with a new section on genetically engineered foods, will keep students in the know about significant industry trends.

ALL-NEW, FULL-COLOR DESIGN featuring photos, charts, graphs, and illustrations that visually demonstrate food production and process techniques to engage students. **National Agricultural Education Standard Correlations** are highlighted at the start of each chapter to address core competencies, while **MATH** and **SCIENCE CONNECTION ACTIVITIES** throughout the chapters encourage students to demonstrate these skills in the context of food science.

ALIGNED to FFA CAREER DEVELOPMENT EVENTS (CDEs), where applicable, to help students focus on important job skills

ADDITIONAL RESOURCES, including a list of valuable and relevant Web sites in each chapter, encourage students to further explore specific food-industry topics.

GLOSSARY and GLOSARIO provides terms and definitions in English and Spanish.

EXTENSIVE TEACHING AND LEARNING PACKAGE

NEW! COMPANION SITE

Instructor Resources are available on the Companion Site to accompany *Food Science & Food Systems,* Second Edition, to facilitate teaching and learning. This site offers **FREE**, secure access to the following resources:

ANSWERS TO QUESTIONS includes the answers to all end-of-chapter questions to validate learning.

LESSON PLANS that outline the key concepts in each chapter, along with correlations to the corresponding PowerPoint® presentations, provide tools for classroom instruction.

POWERPOINT® PRESENTATIONS map the Lesson Plans and include photos and illustrations to reinforce learning. This feature is only also available to students for important self-review.

COGNERO ONLINE TESTING system includes quiz questions for each chapter, providing the ability to:

- Author, edit, and manage test-bank content from multiple resources
- Create multiple test versions in an instant
- Deliver tests from instructor- or institution-specific LMS or classrooms

IMAGE GALLERY, containing all the images from the book, enables instructors to enhance classroom presentations or review key concepts and information.

NEW! MINDTAP FOR INTRODUCTION TO FOOD SCIENCE & FOOD SYSTEMS, SECOND EDITION

The MindTap for *Introduction to Food Science & Food Systems,* Second Edition, features an integrated course offering a complete digital experience for both students and teachers. This MindTap is highly customizable and combines assignments, videos, interactivities, lab exercises and quizzes along with the enhanced e-book to enable students to directly analyze and apply what they are learning as well as allow teachers to measure skills and outcomes with ease.

- **A Guide:** Relevant interactivities combined with prescribed readings, featured multimedia, and quizzing to evaluate progress will guide students from basic knowledge and comprehension to analysis and application.
- **Personalized Teaching:** Teachers are able to control course content—hiding, rearranging existing content, or adding and creating their own content to meet the needs of their specific programs.
- **Promote Better Outcomes:** Through relevant and engaging content, assignments, and activities, students are able to build the confidence they need to ultimately chart a course to success. Likewise, teachers are able to view analytics and reports that provide a snapshot of class progress, time in course, engagement, and completion rates.

Acknowledgments

Without the support of my wife Marilyn, any of the writing I have done would still be a dream or idea. As I have discovered, writing requires the goodwill and support of an understanding spouse. Our marriage has thrived for 47 years, and it continues strong. Marilyn is a friend who critiques ideas, types parts of the manuscripts, writes questions and answers, organizes artwork, takes photographs, and checks format. She is a partner in the production of a text and in all other aspects of my life.

Finally, I appreciate the support, understanding, help, and encouragement of Nicole Robinson, Jennifer Starr, and the rest of the Cengage team.

As always, we wish to express our sincere appreciation to those who have contributed to the development of this and past editions:

Daniel Andrews
Wauneta-Palisade High School
Wauneta, Nebraska

Roy Crawford
Lancaster High School
Lancaster, Texas

Diane Ryberg
Eau Claire North High School
Eau Claire, Wisconsin

Dr. Janelle Walter
Baylor University
Waco, Texas

About the Authors

RICK PARKER

R. O. (Rick) Parker grew up on an irrigated farm in southern Idaho. His love of agriculture guided his education. Starting at Brigham Young University, he received his bachelor's degree and then moved to Ames, Iowa, where he finished his PhD in animal physiology at Iowa State University. After completing his PhD, he and his wife, Marilyn, and their children moved to Edmonton, Alberta, Canada, where he completed a postdoctorate at the University of Alberta. His next move was to Laramie, Wyoming, where he was a research and teaching associate at the University of Wyoming. After Wyoming, he moved to Clovis, California, where he wrote with Dr. M. E. Ensminger, author of numerous early animal science textbooks.

Returning to Idaho, the author served as division director and instructor at the College of Southern Idaho for 19 years. He then worked as director for AgrowKnowledge, the National Center for Agriscience and Technology Education, a project funded by the National Science Foundation. Currently, he is president of the National Agricultural Institute and the director of the North American Colleges and Teachers of Agriculture (NACTA). In addition, he is the editor of the peer-reviewed *NACTA Journal*, which focuses on the scholarship of teaching and learning, and he teaches biology, food science, and animal science for the College of Southern Idaho. Dr. Parker is also the author of the following Cengage Learning texts: *Aquaculture Science*, *Introduction to Plant Science*, *Fundamentals of Plant and Soil Science*, and *Equine Science*. He is also the co-author of *Fundamentals of Plant Science*.

MIRIAH PACE

Miriah Pace is currently a board member and assistant editor at the National Agricultural Institute. Her role includes developing and maintaining Web sites, developing curriculum and editing the peer-reviewed *NACTA Journal* for the North American Colleges and Teachers of Agriculture. She also serves as a teaching assistant for a College of Southern Idaho class, Food Systems and

Science. Miriah joined the staff of the National Agricultural Institute in July 2012 and is currently working toward her bachelor's degree.

CONTRIBUTING AUTHORS

For the second edition, we were fortunate to find two talented individuals to assist in revising the content. We gratefully acknowledge their contributions.

FARRAH JOHNSON

Farrah Johnson is the Agriscience Educator at Deltona High School in Deltona, Florida. Ms. Johnson earned her BS in Agricultural Education from the University of Florida and her MS from Mississippi State University also in Agricultural Education. Ms. Johnson began working with food-science curriculum through a grant project with Cornell University, the University of Florida, and the University of California—Davis early in her teaching career. She now teaches food-science courses as part of the agriculture program at Deltona High School. Farrah is active in the professional organizations for agricultural education and works with new agriculture teachers in Florida. She served as president of the National Association of Agricultural Educators in 2012–2013 and served a 2-year term on the National FFA Board of Directors as a stakeholder. She has served in numerous roles for the Florida Association of Agricultural Educators during her teaching career.

LEVI CAHAN

Levi Cahan has a distinguished background in livestock production from years of training and educating others as well as being self-employed as a farmer. He is the lead Agriculture Educator at Schuylerville High School in upstate New York, where he instructs and manages student learning in several agricultural subjects with a focus on animal science. He received his BS in Animal Science and his MS in Agricultural Education from Cornell University. He also studied abroad in New Zealand at Lincoln University, specializing in animal science and rotational grazing practices. Mr. Cahan stays active in agriculture and education as an FFA advisor; he has served on the New York State FFA Governing Board as a trustee, as a trustee for the NYAAE, and currently as the chairman of the NYS FFA Foundation board, and he is a member of the NYS Beef Council. He has also contributed to other Cengage titles, including serving as author of the *Modern Livestock and Poultry Production Lab Manual*, Ninth Edition.

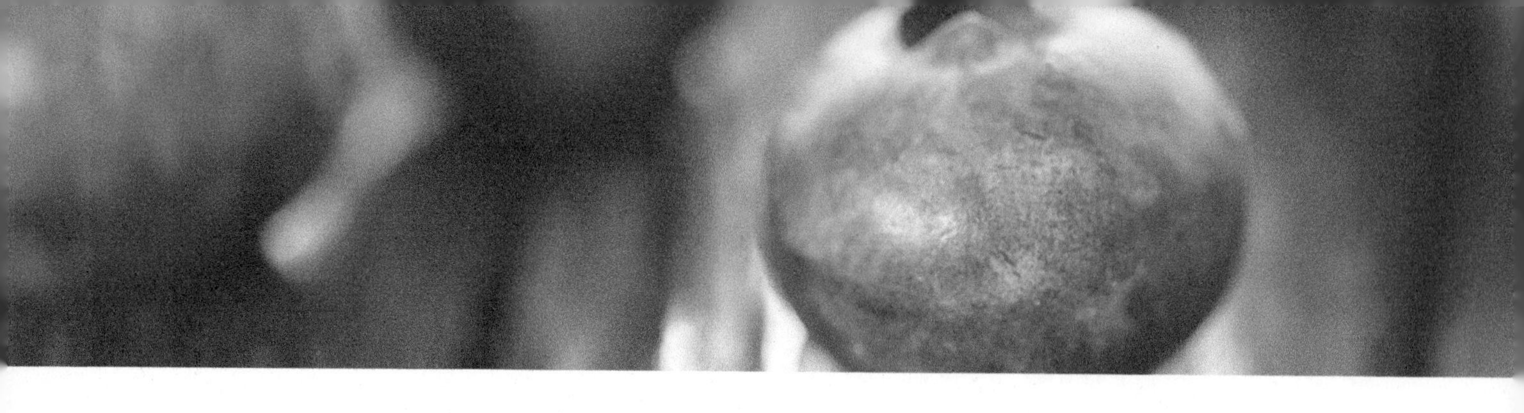

SECTION *One*

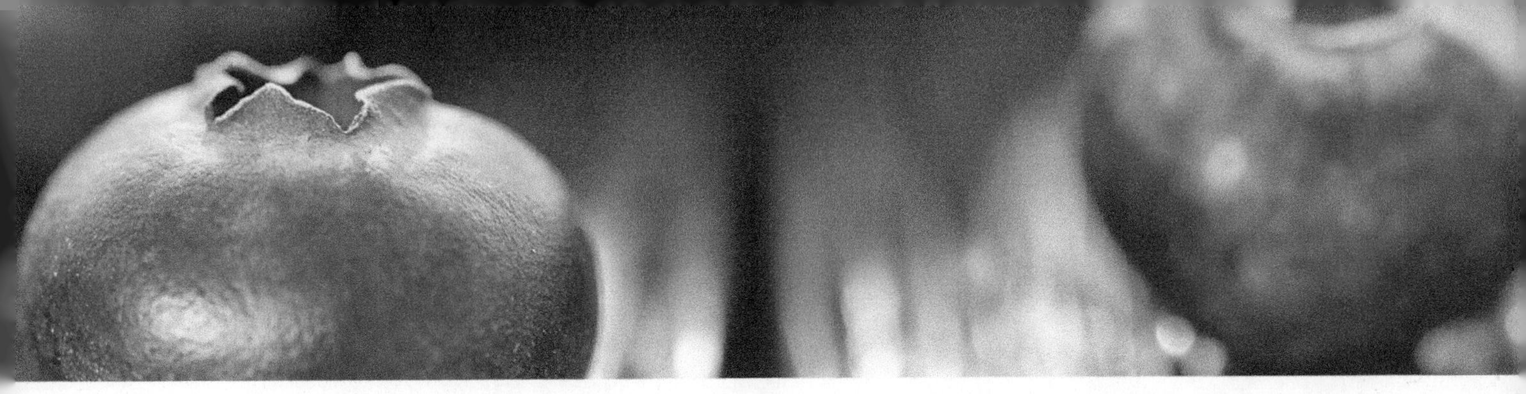

Introduction and Background

CHAPTER 1

Overview of Food Science

OBJECTIVES

After reading this chapter, you should be able to:

- Name the four parts of the food industry
- Describe consumer food buying trends
- Divide the food industry by major product lines
- Compare spending for food in the United States to that in other countries
- List four consumption trends
- Discuss trends in consumer meal purchases
- Identify allied industries
- Explain the international scope of the food industry

KEY TERMS

allied industry
consumer
distribution
expenditures
manufacturing
marketing
per capita
production
tariffs
trends

NATIONAL AFNR STANDARD

FPP.04

Explain the scope of the food industry and the historical and current developments of food product and processing.

No matter where people live or what they do, they are food consumers. We consume food on a daily basis so we make choices every day about what foods to purchase and consume based on a variety of needs and wants. Consumers vote every day in the marketplace with their dollars, and the market listens carefully to their votes. A continuous feedback exists from consumers responding to offerings by marketers who are trying to meet the perceived wants of consumers. Price, availability, health, and convenience are all factors that affect personal purchasing choices. Changes in the makeup of the population, lifestyles, incomes, and attitudes on food safety, health, and convenience have drastically altered the conditions facing the producers and marketers of food products. Food manufacturers and distributors work hard to meet changing consumer demands.

PARTS OF THE FOOD INDUSTRY

The food industry is divided into four major segments:

1. Production
2. Manufacturing and processing
3. Distribution
4. Marketing

Production is the raising or growing of plant and animal products for food consumption. It includes such industries as farming, ranching, orchard management, fishing, and aquaculture. Technologies involved in the production of raw materials include the selection of plant and animal varieties; their cultivation, growth, harvest, and slaughter; and the storage and handling of raw materials. **Manufacturing** converts raw agricultural products to more refined or finished products. For example, peanuts are manufactured into hundreds of different products, only one of which is peanut butter. Manufacturing requires many unit operations and processes that are at the core of food technology. As many foods are being processed into ready-to-eat products, more specialized manufacturing is required. **Distribution** deals with those aspects conducive to product sales, including product formation, weighing and bulk, transportation, and storage requirements and stability. **Marketing** is the selling of foods and involves wholesale, retail, institutional, and restaurant sales (Figure 1-1). Marketing is the segment that consumers are most involved with. Advertising through television, print, and social media is a multibillion-dollar effort that food companies use to reach consumers.

These four divisions are not clear-cut and often overlap one another. For example, when farmers take their crops to a farmer's market, distribution and marketing merge into the same category. Nevertheless, the food industry requires planning

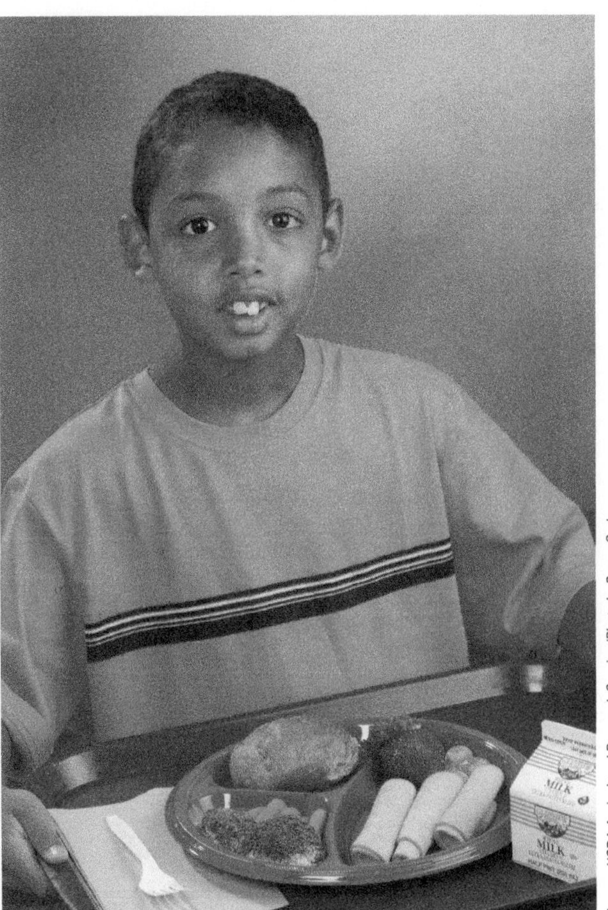

Source: USDA Agricultural Research Service/Photo by Peggy Greb

FIGURE 1-1 School meals are just another example of consumer products that are influenced by food industry marketers.

and synchronization in all its divisions to be successful. When the entire food system is analyzed, additional divisions are often included. These are discussed in more detail in Chapter 2.

Another way to divide the food industry is along major product lines:

- Cereals and bakery products
- Meats, fish, and poultry
- Dairy products
- Fruits and vegetables
- Sugars and other sweets
- Fats and oils
- Nonalcoholic beverages
- Alcoholic beverages

These divisions are typically where **consumer** consumption is measured and reported.

TRENDS

Although consumers' food spending has increased considerably over the years, the increase has not matched the gain in disposable income. As a result, the percentage of income spent for food has declined. The decline is the direct result of the income-inelastic nature of the aggregate demand for food: As income rises, the proportion spent for food declines (Figure 1-2). The **expenditures** for food require a large share of income when income is relatively low—in any country.

Americans spent only about 6% of their personal consumption expenditures for food eaten at home (Figure 1-3). This compares with 10% for Canada and 11% for Switzerland. In less developed countries, such as Kenya and the Philippines, at-home food expenditures often account for more than 40% of a household's budget (Table 1-1).

Americans do not have the highest **per capita** income (the average Swiss income is higher). In relation to total per capita personal consumption expenditures, however, Americans spend the lowest percentage on food. Factors other than income alone influence food expenditures in developed nations. Thanks to abundant arable land and a varied climate, Americans do not have to rely as heavily on imported foods as do some other nations. The American farm-to-consumer distribution system is highly successful at moving large amounts of perishable food over long distances with a minimum of spoilage or delay. Finally, American farmers use a tremendous wealth of agricultural information and state-of-the-art farming equipment. This allows them to produce food more efficiently.

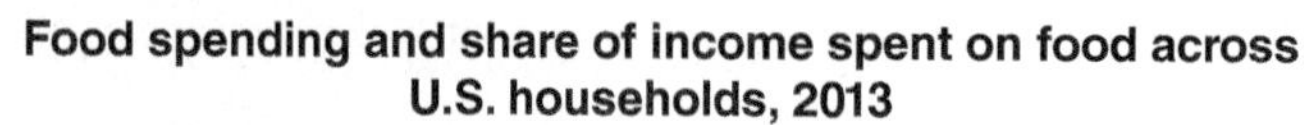

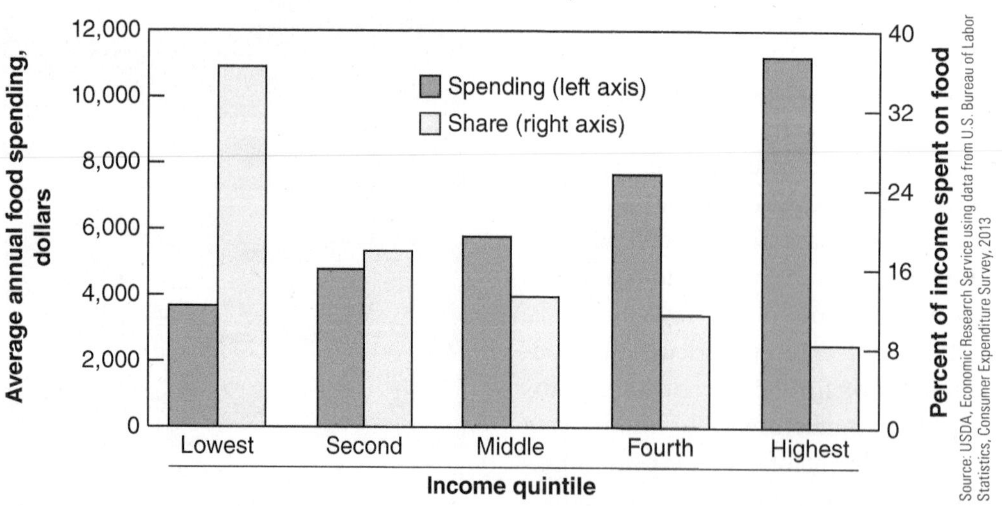

FIGURE 1-2 As income rises the proportion spent on food declines.

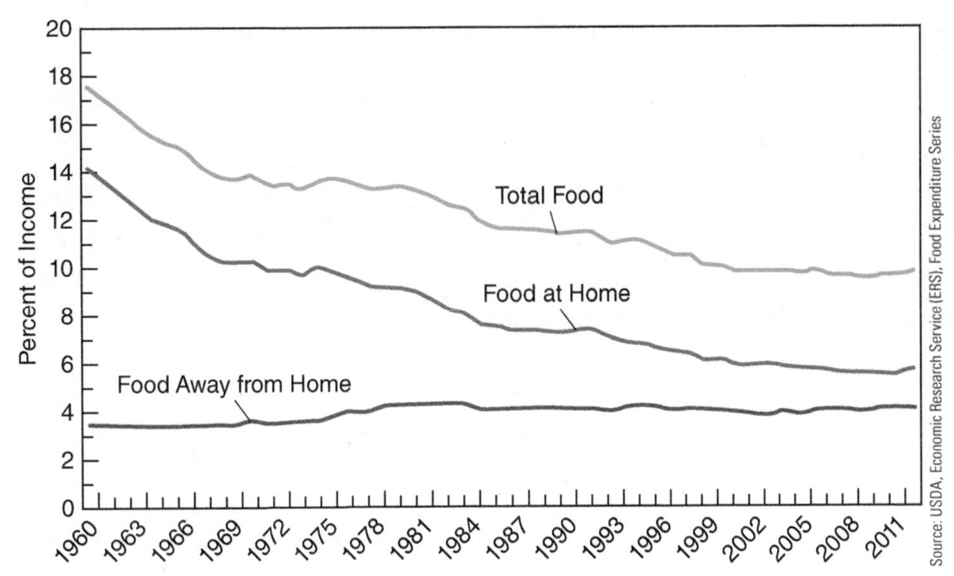

FIGURE 1-3 Percent of Disposable Income Spent on Food, 1960–2011. The percentage of income spent on food at home has steadily declined in recent years.

MATH CONNECTION!

Track your food consumption for a day. What is the average cost of your consumption in the United States? Research and compare those costs with another county. How much would it cost for the same food in that other country?

TABLE 1-1 Percent of Consumer Expenditures Spent on Food, Alcoholic Beverages, and Tobacco Consumed at Home, by Selected Countries, 2012[1]

	SHARE OF CONSUMER EXPENDITURES			
	FOOD[2]	ALCOHOLIC BEVERAGES AND TOBACCO	CONSUMER EXPENDITURES[3]	EXPENDITURE ON FOOD[2]
COUNTRY/TERRITORY	PERCENT		U.S. DOLLARS PER PERSON	
United States	6.6	1.9	34,541	2,273
ERS estimate	6.4	1.9	34,541	2,215
Singapore	7.3	2.1	19,398	1,422
United Kingdom	9.1	3.8	24,260	2,214
Canada	9.6	3.4	27,761	2,679
Austria	10.1	3.3	25,908	2,617
Ireland	10.1	5.4	20,093	2,037

(Continues)

TABLE 1-1 Percent of Consumer Expenditures Spent on Food, Alcoholic Beverages, and Tobacco Consumed at Home, by Selected Countries, 2012[1]

	SHARE OF CONSUMER EXPENDITURES			
	FOOD[2]	ALCOHOLIC BEVERAGES AND TOBACCO	CONSUMER EXPENDITURES[3]	EXPENDITURE ON FOOD[2]
COUNTRY/TERRITORY	*PERCENT*		*U.S. DOLLARS PER PERSON*	
Australia	10.2	3.6	37,492	3,814
Germany	10.9	3.0	22,762	2,481
Switzerland	11.0	3.5	44,899	4,943
Denmark	11.1	3.8	27,306	3,036
Netherlands	11.6	3.3	20,625	2,388
Finland	12.0	4.7	24,927	3,001
Qatar	12.1	0.3	11,199	1,361
Sweden	12.2	3.7	26,146	3,193
South Korea	12.2	2.2	12,002	1,468
Norway	13.2	4.3	37,146	4,885
France	13.2	3.3	22,945	3,037
Czech Republic	13.3	9.3	9,643	1,279
Hong Kong, China	13.4	1.0	24,060	3,224
Taiwan	13.5	2.1	12,247	1,657
Japan	13.8	2.5	27,761	3,818
Belgium	13.8	3.8	22,208	3,075
Bahrain	13.9	0.4	10,200	1,422
Spain	14.0	3.0	17,713	2,483
Italy	14.2	2.8	20,362	2,892
United Arab Emirates	14.3	0.2	21,206	3,024
New Zealand	14.6	3.0	22,448	3,284
Slovenia	15.3	5.6	13,858	2,125
Brazil	15.9	1.4	7,063	1,123
Israel	15.9	2.6	17,491	2,783
Hungary	16.2	7.5	6,972	1,127

(*Continues*)

TABLE 1-1 Percent of Consumer Expenditures Spent on Food, Alcoholic Beverages, and Tobacco Consumed at Home, by Selected Countries, 2012[1]

	SHARE OF CONSUMER EXPENDITURES			
	FOOD[2]	ALCOHOLIC BEVERAGES AND TOBACCO	CONSUMER EXPENDITURES[3]	EXPENDITURE ON FOOD[2]
COUNTRY/TERRITORY	*PERCENT*		*U.S. DOLLARS PER PERSON*	
Chile	16.2	3.0	9,566	1,546
Greece	16.5	4.4	16,652	2,740
Portugal	16.5	3.0	13,473	2,225
Slovakia	16.8	4.9	9,556	1,603
Uruguay	18.3	3.4	10,272	1,878
Colombia	18.4	3.3	4,744	872
Kuwait	18.6	0.5	7,284	1,352
Venezuela	18.6	3.6	7,421	1,378
Latvia	18.8	6.8	8,612	1,619
South Africa	19.4	6.1	4,524	877
Malaysia	19.5	2.5	5,557	1,084
Poland	19.6	6.5	7,773	1,521
Estonia	19.6	8.6	8,923	1,753
Argentina	20.9	4.5	6,595	1,381
Bulgaria	21.2	4.0	4,718	999
Ecuador	21.9	0.9	3,526	771
Turkey	22.2	4.6	7,705	1,708
Costa Rica	23.3	0.4	6,754	1,577
Turkmenistan	23.5	2.0	2,503	589
Dominican Republic	24.5	4.2	5,192	1,272
Mexico	24.9	2.2	6,518	1,625
India	25.2	3.0	871	220
Iran	25.5	0.4	2,744	699
Lithuania	25.7	8.5	9,067	2,331
Saudi Arabia	25.8	0.6	6,220	1,607
China	26.9	3.6	2,149	577

(Continues)

TABLE 1-1 Percent of Consumer Expenditures Spent on Food, Alcoholic Beverages, and Tobacco Consumed at Home, by Selected Countries, 2012[1]

	SHARE OF CONSUMER EXPENDITURES			
	FOOD[2]	ALCOHOLIC BEVERAGES AND TOBACCO	CONSUMER EXPENDITURES[3]	EXPENDITURE ON FOOD[2]
COUNTRY/TERRITORY	*PERCENT*		*U.S. DOLLARS PER PERSON*	
Romania	28.6	3.4	4,827	1,382
Bolivia	28.7	1.9	1,567	450
Uzbekistan	31.0	2.5	908	281
Croatia	31.4	3.7	9,078	2,847
Bosnia-Herzegovina	31.4	6.3	4,057	1,275
Russia	31.6	7.8	6,709	2,120
Thailand	32.0	4.7	3,177	1,016
Jordan	32.2	4.5	3,743	1,205
Indonesia	33.4	5.4	1,964	655
Macedonia	34.4	3.5	3,626	1,247
Kazakhstan	35.1	2.6	5,483	1,925
Tunisia	35.5	1.0	2,660	943
Vietnam	35.9	2.8	962	345
Belarus	36.1	8.1	3,091	1,115
Peru	36.5	6.1	4,126	1,507
Ukraine	37.0	6.7	2,779	1,028
Guatemala	37.9	1.5	2,878	1,091
Nigeria	39.5	2.5	966	381
Georgia	40.4	5.1	2,663	1,076
Morocco	40.5	1.3	1,921	777
Azerbaijan	42.7	2.0	2,862	1,222
Egypt	42.7	2.2	2,410	1,030
Philippines	42.8	1.2	1,925	823
Algeria	43.7	2.0	1,749	764

(Continues)

TABLE 1-1 Percent of Consumer Expenditures Spent on Food, Alcoholic Beverages, and Tobacco Consumed at Home, by Selected Countries, 2012[1]

COUNTRY/TERRITORY	SHARE OF CONSUMER EXPENDITURES: FOOD[2]	ALCOHOLIC BEVERAGES AND TOBACCO	CONSUMER EXPENDITURES[3]	EXPENDITURE ON FOOD[2]
	PERCENT		U.S. DOLLARS PER PERSON	
Kenya	44.8	2.8	782	350
Cameroon	45.9	2.2	921	423
Pakistan	47.7	1.0	871	415

Source: ERS; USDA calculations based on annual household expenditure data from Euromonitor International, available at http://www.euromonitor.com/

NA = Not available.
[1]Data were computed based on Euromonitor International data extracted July 2013.
[2]Includes nonalcoholic beverages.
[3]Consumer expenditures include personal expenditures on goods and services. Consumption expenditures in the domestic market are equal to consumer expenditures by resident households plus direct purchases in the domestic market by nonresident households minus direct purchases abroad by resident households.

NOTE: Two sets of food-spending figures are shown for the United States. The first is from the ERS Food Expenditure series and is based on a comprehensive measure of the total value of all U.S. food expenditures. The second set is based on Euromonitor International Inc. data, which reports spending on food and nonalcoholic beverages and Consumption Expenditures for 84 countries, including the United States. The ERS estimate is lower partly because it excludes pet food, ice, and prepared feed, which are included in the food-spending data published by Euromonitor International.

Consumption **trends** change over time, and this influences what the food industry does in terms of production and advertising. According to loss-adjusted food availability data, Americans are consuming more calories per day than they did 40 years ago. In 1970, Americans consumed an estimated 2,109 calories per person per day; by 2010, they were consuming an estimated 2,569 calories (after adjusting for plate waste, spoilage, and other food losses) (see Figure 1-4). Of this 460-calorie increase, grains (mainly refined grains) accounted for 180 calories; added fats and oils, 225 calories; added sugar and sweeteners, 21 calories; dairy fats, 19 calories; fruits and vegetables, 12 calories; and meats, eggs, and nuts, 16 calories. Only dairy products declined (13 calories) during the time period. According to government recommendations from MyPlate, American diets fall short in the consumption of fruits, vegetables, and dairy (Figure 1-5).

However, demand for individual foods is more responsive to prices as consumers choose from alternative food commodities. Rising incomes mean consumers spend more on more expensive foods as they demand both more convenience and higher quality. Short-period changes in consumption reflect mostly changes in supply rather than changes in consumer

Flour and cereal products provided more calories per day for the average American than any other food group in 2010

Fruit and vegetables and dairy products provided smaller shares of calories per day for the average American

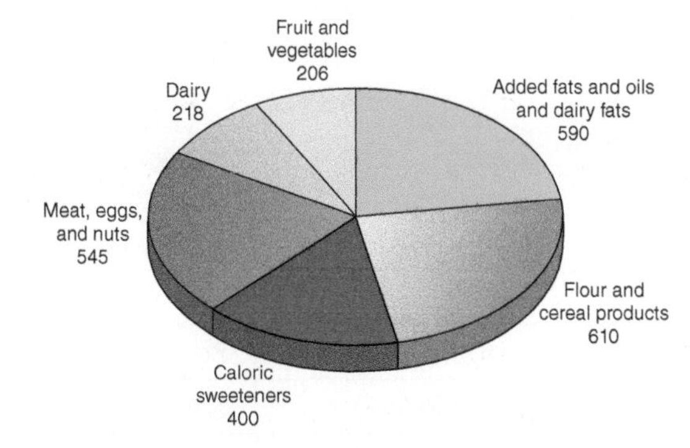

Notes: Added fats and oils and added sugar and sweeteners are added to foods during processing or preparation. They do not include naturally occurring fats and sugars in food (e.g., fats in meat or sugars in fruits).

Food availability data serve as proxies for food consumption.

Source: Calculated by USDA's Economic Research Service based on data from various sources (see Loss-Adjusted Food Availability Documentation). Data as of February 2014

FIGURE 1-4 In 2010 Americans consumed, on average, 2,569 calories per day. This is a considerable increase from 1970 when the average American consumed only 2,109 calories per day.

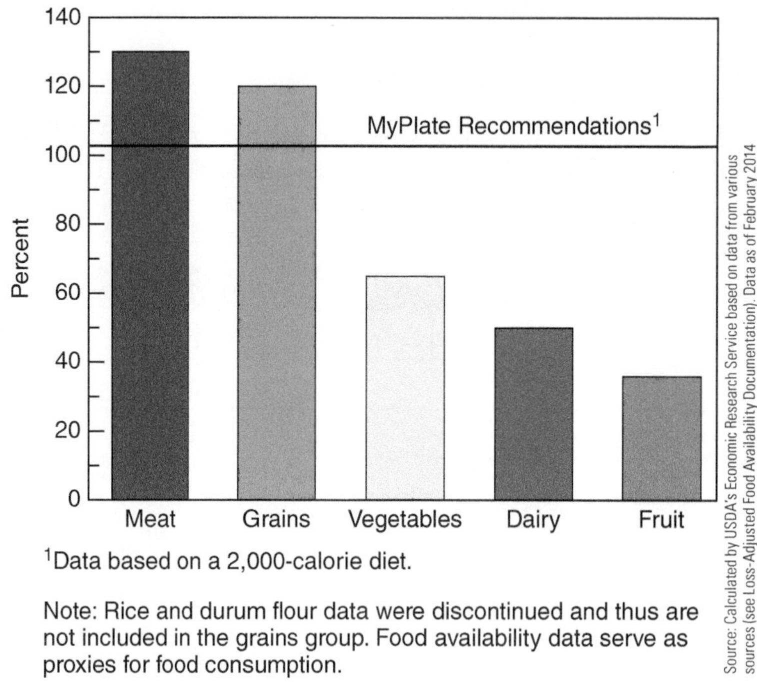

FIGURE 1-5 American diets are out of balance with dietary recommendations. In 2012, Americans consumed more than the recommended share of meat and grains in their diets but less than the recommended share of fruit, dairy, and vegetables.

tastes. Demographic factors, such as changes in household size and the population's age distribution, also can bring about changes in consumption.

Away-from-home meals and snacks now capture almost half (45%) of the U.S. food dollar. This is up from 34% in 1970. Fast food accounts for the largest and fastest rising share of sales in the food industry. Sales in fast-food industries now outpace sales in full-service restaurants. The top five U.S. fast-food chains and their sales are shown in Table 1-2.

The number of fast-food restaurant outlets in the United States has risen steadily since 1970. People want quick and convenient meals. They do not want to spend a lot of time preparing meals, traveling to

TABLE 1-2 Top 5 Restaurant Chains: U.S. Sales

			$ MILLION		
RANK	CHAIN	SEGMENT	2011	2010	2009
1	McDonald's	Sandwich	34,172	32,395	31,033
2	Subway	Sandwich	11,434.0	10,633	9,999
3	Starbucks Coffee	Beverage-snack	8,490	7,955	7,415
4	Burger King	Sandwich	8,131	8,433	8,799
5	Wendy's	Sandwich	8,108	7,943	8,023

Source: Adapted from Nation's Restaurant News (NRN). http://nrn.com/us-top-100/top-100-chains-us-sales.

NOTES: The rankings are based on U.S. system-wide food service sales, including company-store and franchised-unit sales, for the latest full fiscal years ended closest to December of the years listed.

- Tied results are given the same rank.
- The year 2011 reflects data for chain or company fiscal years ended closest to December 2011.
- The year 2010 reflects data for chain or company fiscal years ended closest to December 2010.
- The year 2009 reflects data for chain or company fiscal years ended closest to December 2009.
- Data are reported by a chain or parent company or estimated by Nation's Restaurant News.

pick up meals, or waiting for meals in restaurants. This trend has increased as more women have become part of the workforce. In more and more families, both parents work and no one stays at home. In addition, consumers more often combine meals with time engaged in activities such as shopping, working, and traveling. For example, McDonald's, Burger King, Taco Bell, and others are now located in outlets such as Wal-Mart stores and many gas stations.

Perhaps the current food service industry strategy was best stated in McDonald's 1994 annual report:

> *McDonald's wants to have a site wherever people live, work, shop, play, or gather. Our Convenience Strategy is to monitor the changing lifestyles of consumers and intercept them at every turn. As we expand our customer convenience, we gain market share.*

MATH CONNECTION!

Research the number of calories found in each item of your favorite fast-food restaurant meal. How do the total calories for that meal fit with a 2,000-calorie diet?

The food industry is big and it employs large numbers of people in a variety of occupations because everyone eats (Table 1-3), and they eat more prepared products at home and many meals away from home. Advertising (media) also plays an important role in influencing food trends (Figure 1-6).

TABLE 1-3 U.S Per Capita Food Expenditures

		U.S. PER CAPITA FOOD EXPENDITURES					
		CURRENT PRICES			1988 PRICES		
	U.S. RESIDENT POPULATION, JULY 1	AT HOME	AWAY FROM HOME	TOTAL	AT HOME	AWAY FROM HOME	TOTAL
YEAR	*MILLIONS*	*DOLLARS*					
1953	167.306	278	91	369	1,068	516	1,584
1960	179.979	306	109	415	1,132	522	1,654
1965	193.526	318	135	454	1,108	581	1,689
1970	203.984	387	194	581	1,130	630	1,760
1975	215.465	567	316	883	1,069	706	1,775
1980	227.225	828	529	1,357	1,092	773	1,865
1985	237.924	1,009	710	1,718	1,128	798	1,926
1990	249.464	1,301	982	2,283	1,147	897	2,044
1995	262.803	1,408	1,170	2,578	1,104	957	2,061
2000	282.172	1,571	1,396	2,966	1,091	1,003	2,093
2005	295.753	1,853	1,730	3,583	1,138	1,085	2,223
2010	309.326	2,065	1,979	4,043	1,116	1,054	2,170
2011	311.588	2,171	2,058	4,229	1,113	1,081	2,194
2012	313.914	2,215	2,167	4,382	1,114	1,109	2,223

Source: USDA, Economic Research Service (ERS)

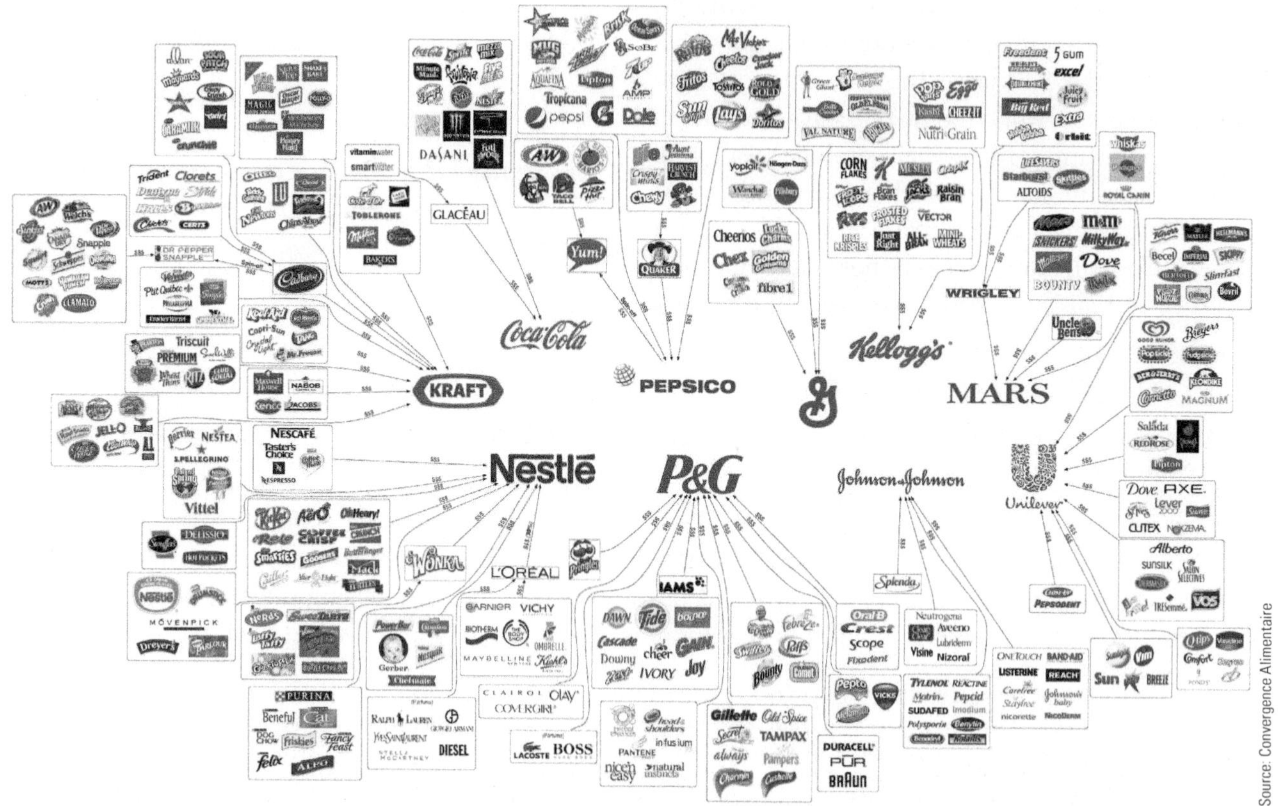

FIGURE 1-6 Illusion of choice. Most products we buy are controlled by just a few companies.

ALLIED INDUSTRIES

Many companies do not sell food directly but are still deeply involved in the food industry. These **allied industries** produce nonfood items that are necessary for marketing food. The packaging industry is a good example. Specific examples include cans, food colorings and flavorings, paper products, and plastic products (see Figure 1-7). Chemical manufacturers represent another group of allied industries. They supply the acidulants, preservatives, enzymes, stabilizers, and other chemicals used in foods.

SCIENCE CONNECTION!

Research food additives or preservatives commonly found in foods.

New food products and safe foods require new food-processing methods and systems. Food machinery and equipment manufacturers are more examples of allied industries. They develop pasteurizers, evaporators, microwave ovens, infrared cookers, freeze-drying systems, liquid nitrogen freezers, instrumentation, and computer controls (see Chapter 7).

Finally, keeping the food supply safe and healthy and consumers informed requires monitoring and regulatory agencies such as the Food and Drug Administration (FDA), attorneys, consumer

FIGURE 1-7 Plant physiologist prepares to make wheat-starch biodegradable containers.

action and information agencies, and other regulatory agencies.

If recent trends in the U.S. food industry continue, food production may be increasingly dominated by firms exercising control over most and even all stages of food production. Vertical coordination seems to be the way of the future, including how products are acquired or traded in the markets. Food industry firms form three basic types of vertical coordination:

1. **Open production.** A firm purchases a commodity from a producer at a market price determined at the time of purchase.
2. **Contract production.** A firm commits to purchase a commodity from a producer at a price formula established before the purchase is to be made.
3. **Vertical integration.** A single firm controls the flow of a commodity across two or more stages of food production.

The food industry has traditionally operated in an open production system. However, more discriminating consumers, plus new technological developments that allow the differentiation of farm products, are helping to lower open production and increase both contract production and vertical integration. Also fueling this trend are changing demographics and the increasing value of people's time, both of which have contributed to consumer preferences for a wider variety of safe, nutritious, and convenient food products.

Providing food products with specific characteristics preferred by discriminating consumers will likely involve producing more detailed raw commodities such as a frying chicken of a specific weight and size or a corn kernel with a specific protein content. This effort to carefully tailor raw commodities with processing in mind is already underway in some food industries and has been accompanied by changes in vertical coordination.

INTERNATIONAL ACTIVITIES

Food is an international commodity, with products being traded and shipped around the world. Most grocery stores now carry food items from other countries. Specialty and gourmet stores stock many international foods such as cheeses from Europe, beef from Australia, strawberries from Mexico, and apples from Argentina. In addition, gourmet and specialty foods can be purchased online. International food clubs and online buying are available to consumers worldwide.

Many U.S. companies also have established subsidiaries in other countries, and fast-food companies such as McDonald's and Pizza Hut continue to open outlets all around the world. Major food companies such as Kraft-General Foods, CPC International, H.J. Heinz, Borden, Campbell Soup, Nabisco Brands, Coca-Cola, PepsiCo, Beatrice Companies, Ralston Purina, and General Mills all have extensive overseas operations. Table 1-4 lists the top 50 international food-processing firms, their headquarters, and their annual sales.

The processed-food sector is a major participant in the global economy. The United States accounts for about one-fourth of the industrialized world's total production of processed foods. Six of the largest 10, and 21 of the largest 50 food-processing firms in the world are headquartered in the United States. The U.S. processed-foods market has become truly global in scope through a

TABLE 1-4 Top 50 of the World's Largest Food-Processing Firms

THIS YEAR	LAST YEAR	COMPANY	2013 FOOD SALES	2012 FOOD SALES	2013 TOTAL COMPANY SALES	2013 NET INCOME* (−LOSS)
1	1	Pepsico Inc.	37,806	37,618	66,415	6,740
2	2	Tyson Foods Inc. (9/28/13)	32,999	31,614	34,374	778
3	3	Nestle (U.S. & Canada)	27,300	27,200	103536C	11,000C
4	4	JBS USA	22,140	20,979E	41,000C	429
5	11	Coca-Cola Co.	21,600	21656R	46,854	8,626
6	5	Anheuser-Busch InBev	16,023	16,028	43,195	16,518
7	6	Kraft Foods Inc.	14,346	14,358R	18,218	2,715
8	8	Smithfield Foods Inc.	12,531	11753A	14,000	NA-Private
9	7	General Mills Inc. (5/25/14)	12,524	12,574	17,910	1,861
10	12	ConAgra Foods Inc. (5/25/14)	11,511	9,360R	17,703	315
11	10	Mars Inc.	11000E	11,000	33,000E	NA-Private
12	14	Kellogg Co.	9,716	9,539	14,792	1,808
13	9	Dean Foods Co.	9,016	11,462	9,016	819
14	15	Hormel Foods Corp.	8,752	8,231	8,752	530
15	13	Cargill Inc. (5/31/13)	8,500	8,500E	136,700	2,310
16	16	MillerCoors LLC	7,801	7,761	7,801	1,271
17	21	Saputo Inc. (3/31/14)	C7789	C6063	C9,233	C534

(Continues)

TABLE 1-4 Top 50 of the World's Largest Food-Processing Firms

THIS YEAR	LAST YEAR	COMPANY	2013 FOOD SALES	2012 FOOD SALES	2013 TOTAL COMPANY SALES	2013 NET INCOME* (–LOSS)
18	17	Pilgrim's Pride	7,500	7,249	8,411	550
19	20	Hershey Co.	7,146	6,644	7,146	820
20	19	Mondelez International	6,991	6,903	35,299	2,332
21	18	Unilever North America	6,876	7111E	68551 CC	7245 CC
22	22	Bimbo Bakeries USA	6,101	6,062C	13,464C	365C
23	23	Dr. Pepper Snapple Group	5,997	5,995	5,997	624
24	24	J.M. Smucker Co. (4/30/14)	5,611	5,898	5,611	565
25	29	Campbell Soup Co. (7/28/13)	4,910	4,110	8,052	449
26	38	Constellation Brands (2/28/14)	4,868	2,796	4,868	1,943
27	27	H.J. Heinz Co.	4,530	4,570		NA-Private
28	25	Maple Leaf Foods	4,406	4,552R	4,406	496
29	26	Land O'Lakes Inc.(2)	4,250	4,200E	14,236	306
30	31	Perdue Farms (3/30/14)	4,140	3,860E	6,729	NA-Private
31	32	Brown-Forman Corp.	3,946	3,784	3,946	659
32	30	Hillshire Brands (6/29/13)	3,920	3,958	3,920	184
33	37	Flowers Foods Inc.	3,751	3,046	3,751	231
34	86	Dairy Farmers of America (2)	3,700	3,500R	12,800	61
35	33	Agropur Cooperative	3,630	3,640	3,630	54
36	66	Lactalis American Group Inc.	3,500	3,230	3,500E	NA
37	35	E&J Gallo Winery	3,400	3,400E	3,600E	NA-Private
38	59	Parmalat Canada	3,161	2,848R		178

(Continues)

TABLE 1-4 Top 50 of the World's Largest Food-Processing Firms

THIS YEAR	LAST YEAR	COMPANY	2013 FOOD SALES	2012 FOOD SALES	2013 TOTAL COMPANY SALES	2013 NET INCOME* (−LOSS)
39	36	Chiquita Brands Intl.	3,057	3,078	3,057	(−16)
40	28	Dole Food Co. Inc.	2,800	4,247	23,800E	NA-Private
41	61	Prairie Farms Dairy Inc. (9/30/13)	2,800	2,700	2,800	NA
42	42	Sanderson Farms	2,683	2,386	2,683	131
43	39	Rich Products Corp.	2,661	2500E	3,300	NA-Private
44	48	Molson Coors Co. (Canada only)	2,575	2,675R	4,206	565
45	72	Beam Inc.	2,558	2,466		
46	52	WhiteWave Foods	2,542	2,289	2,542	NA-Private
47	44	Great Lakes Cheese Co.	2,500	2,250E	2,500E	NA-Private
48	39	McCain Foods (6/30/13)	C2,500E	C2,500E	C$6,000	NA-Private
49	41	Pinnacle Foods	2,464	2,478	2,462	89
50	68	Dannon Co. Inc.	2,305	1,800R	2,305	NA

Source: Food Processing, http://www.foodprocessing.com

combination of imports and exports of foods and food ingredients, foreign production by U.S. food firms, host production by foreign food firms, and other international commercial strategies. Easily recognized U.S. food brands are so well received internationally that many consumers in other countries accept them as leading local brands. In terms of international trade, the processed-foods sector surpasses agricultural commodities by a considerable margin.

World trade imports are also represented by products not grown in the United States such as coffee, tea, cocoa, and spices. The worldwide demand for cereal grains and soybeans has also increased, so the United States is the largest exporter of these foods.

Aside from the worldwide demand for food and food products, recent trends to decrease trade **tariffs** has stimulated international activities in the food industry, as have improvements in transportation and communication. Products now move around the world by air freight in hours or days. The World Wide Web has enabled communications to take place around the world in a matter of seconds.

A nation's infrastructure policies affect the ability of its firms to pursue global marketing strategies. For processed foods, particularly important linkages exist between the communications and transportation sectors. Technical innovations in both communications and transportation make the production and distribution of processed foods more efficient, improve managerial control and responsiveness, and help identify and fulfill new commercial opportunities. In the United States, policies that have reduced direct government control of these sectors and fostered the evolution of competitive communications and transportation industries are tied directly to international commercial gains in processed foods.

NEW FRUIT COMING TO A SUPERMARKET NEAR YOU

All across Southeast Asia, people eat durian or stinkfruit (*Durio zibethinus*) every day. It is so popular that Thais call it the King of Fruits. Though the flesh of the durian is sweet and mild, its aroma is so pungent and strong that many Westerners gag when trying to eat it. Many describe the smell as a blend of decayed onion, turpentine, garlic, Limburger cheese, and resin. Eating it on commercial flights has been banned by several Asian airlines.

This spiky-skinned, brownish green fruit can grow as big as cantaloupe. It can be found in many Asian import stores and will soon be more available in the United States and Europe.

For more information about the durian, visit these Web sites:

- http://www.smithsonianmag.com/science-nature/why-does-the-durian-fruit-smell-so-terrible-149205532/?no-ist
- http://www.nutrition-and-you.com/durian-fruit.html
- http://www.lifehack.org/articles/lifestyle/10-benefits-and-uses-durian-fruit-that-will-surprise-you.html

RESPONSIVENESS TO CHANGE

The total amount of food consumed by each individual (per capita food consumption) changes little from year to year. The kinds of foods consumed, however, change continually, contributing to competition and driving frequent changes in the food industry change. More than 10,000 new food products are introduced each year.

The kinds of foods people eat change for many reasons, including demographic shifts, supplies of ingredients, the availability and costs of energy, politics, and scientific advances in nutrition. In addition, health, food safety, and changes in lifestyle also influence what foods people choose to consume.

Attitudes toward foods also change consumption patterns. This means that the industry also must respond. For example, since 1990 fresh and frozen fruit consumption increased about 24% over the, whereas red meat consumption declined about 16%. The increase of smoothie and juicing trends as well as vegetarian diets can also influence these changes. Per capita consumption of poultry, carbonated soft drinks, and cheese increased markedly. During this same period of time, consumer attitudes about fat, cholesterol, and fiber also changed.

Changes in governmental regulation of food additives, food composition standards, and labeling also require the food industry to continually change. Finally, technical innovations such as ingredient modifications, new processing methods, new packaging methods, and cooking advances also force the industry to respond.

INTERRELATED OPERATIONS

Food production relies on a highly advanced and organized industry. Decisions to make and market a product are not random. The industry is a systematic and rhythmic process. Throughout production, manufacturing, processing, distribution, and marketing, costs and ingredient availabilities are carefully monitored and controlled.

Further, because the industry is high volume and low markup, small losses anywhere along the chain can mean large losses to the food producer.

Any trend toward contract production and vertical integration, as opposed to open production, implies that firms at one stage of production exert more control over the quality of output at other stages. For example, pasta processors who prefer a specific type of wheat for a specific type of pasta gain control over planting decisions and seed selection that were previously made by farmers who sold their wheat on the spot market. Farmers are compensated for relinquishing control through bonuses for quality and by reduced uncertainty.

Recent changes in vertical coordination have been accompanied by increased concentrations in the food sector. These developments have raised two primary policy concerns: market power in the processing sector and environmental protection.

SUMMARY

The food industry is divided into production, manufacturing and processing, distribution, and marketing. The industry is highly responsive to change and interrelated with others. Not only do consumers drive the food industry but also to some extent the food industry drives consumers, influencing changes in food consumption, food types, and meals purchased. Food is a global commodity because of changes in export and import laws, transportation, and processing and communication.

REVIEW QUESTIONS

Success in any career requires knowledge. Test your knowledge of this chapter by answering these questions and solving these problems.

1. What percentage of the U.S food dollar is captured by away-from-home meals?
2. Why have the international activities of food industries increased?
3. Name all seven product lines along which the food industry is divided.
4. List the four artificial divisions of the food industry.
5. Compare the consumption of cheese and red meat since 1990. Has consumption of each increased or declined?
6. List four influences on people that determine the kind of food they eat.
7. Approximately how many new food products are introduced each year?
8. Explain how consumers vote in the marketplace.
9. Define an allied industry.
10. Compare the spending on food in the United States with that spent in Spain and in Greece.

STUDENT ACTIVITIES

1. Make a list of foreign foods sold in a supermarket. Bring this list to class for discussion.
2. Pick one of the food companies listed in Table 1-4 and visit its Web site. Create a visual presentation using available consumer information.
3. Search the Internet or some other resource and look for trade tariffs that affect food. Each student will add one tariff description to a class poster.
4. Conduct a contest to see who can bring the most creative and novel food packaging to class. Students should vote and rank the quality of the packages that are brought in.

ADDITIONAL RESOURCES

- International Food Safety & Quality Network: http://www.ifsqn.com/
- Ten Food Trends Unveiled at I.F.T. 2014: http://www.foodbusinessnews.net/articles/news_home/Consumer_Trends/2014/06/Ten_food_trends_unveiled_at_IF.aspx?ID=%7BC8BEEAF6-9CD1-401B-9A19-972D66B243C4%7D&cck=1
- IFT: http://www.ift.org/knowledge-center/learn-about-food-science/what-is-food-science.aspx
- What the World Eats: http://www.nobelpeacecenter.org/en/exhibitions/hungry-planet/

Internet sites represent a vast resource of information. Although those provided in this chapter were vetted by industry experts, you may wish to further explore the topics discussed in this chapter using a search engine such as Google. Keywords or phrases may include the following: *food industry, food technology, imported food, exported food, food tariffs, food expenditures, international food companies, Food and Drug Administration*, and the names of particular food companies or brands. In addition, Table B-7 provides a listing of some useful Internet sites that can be used as a starting point.

REFERENCES

Potter, N. N., and J. H. Hotchkiss. 1995. *Food science*, 5th ed. New York: Chapman & Hall.

Vaclavik, V. A., and E. W. Christina. 1999. *Essentials of food science*. Gaithersburg, MD: Aspen Publishers.

Vieira, E. R. 1996. *Elementary food science*, 4th ed. New York: Chapman & Hall.

CHAPTER 2

Food Systems and Sustainability

OBJECTIVES

After reading this chapter, you should be able to:

- Describe food systems
- List five major segments of a food system
- Discuss the concept of sustainability
- Identify three allied industries of a food system
- List three effects of urbanization on a food system
- List 13 standards for sustainability
- Discuss why changes in the food system need to be science-based
- Identify three trends in global food demand that effect food systems
- Name two social or political aspects of a food system
- Identify three renewable energy sources that could be used in a food system

NATIONAL AFNR STANDARD

FPP.04

Explain the scope of the food industry and the historical and current development of food product and processing.

KEY TERMS

anaerobic
aquaponics
beneficials
bio-intensive
biosecurity
entomophagy
food insecure
food system
genomics
market niches
nutraceuticals
probiotics
risk
risk assessment
science
sustainable
urbanization

OBJECTIVES (continued)

- **Explain the role agriculture plays in water use and water consumption**
- **Discuss the role of income level on the food system**
- **Describe the importance of soil conservation to a successful food system**
- **Describe integrated pest management represents a sustainable method of pest control**
- **Identify how a food system contributes to strong communities**
- **List three types of water pollution that can adversely affect food systems**

What constitutes a food system and what sustainable means can both vary. In fact, both can even be divisive. In simple terms, a **food system** includes all processes and infrastructures involved in feeding a population. *Sustainable* refers to anything that can be maintained at a certain rate or level. When applied to food and food production, these definitions can have a wide range of interpretations and can become quite complex emotionally and politically. This chapter will attempt to avoid the emotional and political aspects of sustainability and avoid similar descriptions of alternative food systems.

Regardless of the definitions, emotions, and politics, the challenge is to feed a growing global population of nearly 7.5 billion people.

FOOD-SYSTEM DEFINITIONS

The food industry and food science involve more than grocery stores and restaurants. Food systems can be divided into five major segments: production, manufacture, distribution, marketing, and consumption. Perhaps a sixth segment of waste and disposal should also be added (Figure 2-1).

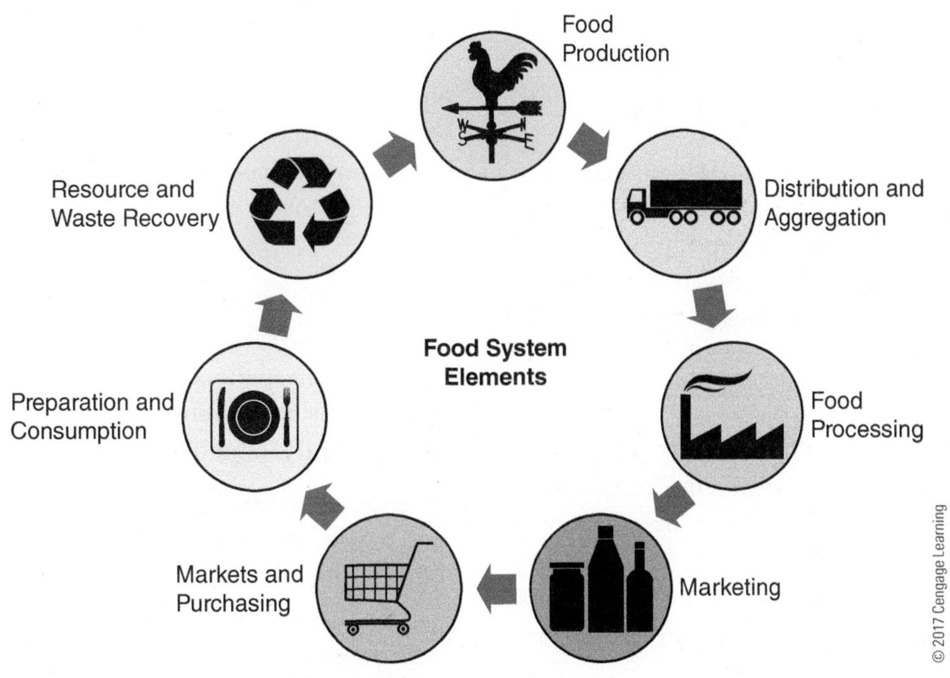

FIGURE 2-1 Components of a typical food system.

Production includes such industries as farming, ranching, orchard management, fishing, and aquaculture. Technologies involved in the production of raw materials include the selection of plant and animal varieties, cultivation, growth, harvest, slaughter, and the storage and handling of raw materials. Manufacturing converts raw agricultural products to more refined or finished products. Manufacturing requires many unit operations and processes that are at the core of food technology. Distribution deals with those aspects conducive to products sales, including product formation, weighing and bulk packaging, transportation, storage requirements, and storage stability. Marketing is the selling of foods and involves wholesale, retail, institutions, and restaurants. Once food is sold, it is consumed and provides carbohydrates, protein, fats, vitamins, and minerals. Each segment of the food system deals with some aspect of waste and disposal.

Because these five or six divisions overlap one another, a food system must be able to plan for and synchronize of its divisions to be successful.

Allied Industries

Many companies do not sell food directly but still are deeply involved in the food industry. As noted in Chapter 1, these are called *allied industries*. Allied industries produce nonfood items that are required for food marketing. The packaging industry is a good illustration. Specific examples include cans, food colorings and flavorings, paper products, and plastic products. Chemical manufacturers represent another group of allied industries. They supply acidulates, preservatives, enzymes, stabilizers, and other chemicals used in foods.

New food products and safe foods require new food-processing methods and systems. This is where food machinery and equipment manufacturers play an important role. They develop pasteurizers, evaporators, microwave ovens, infrared cookers, freeze-drying systems, liquid nitrogen freezers, instrumentation, and computer controls.

Finally, keeping the food supply safe and healthy and keeping consumers informed requires monitoring and regulatory agencies such as the Food and Drug Administration (FDA), lawyers, consumer action and information agencies, and other regulatory agencies.

International Activities

Food is a global commodity. Foods are traded and shipped around the world. Today's grocery store sells food from all over the world. These foods might include cheeses from Europe, beef from Australia, strawberries from Mexico, and apples from Argentina. Also, many U.S. companies have established subsidiaries in other countries. Fast-food companies such as McDonald's and Pizza Hut continue opening outlets all over the world. Major food companies such as Kraft General Foods, CPC International, H.J. Heinz, Borden, Campbell Soup, Nabisco Brands, Coca-Cola, PepsiCo, Beatrice Companies, Ralston Purina, and General Mills all have extensive overseas operations.

World trade imports are also represented by products not grown in the United States such as, coffee, tea, cocoa, and spices. Worldwide the demand for cereal grains and soybeans has increased, and the United States has become the largest exporter of these foods. Global food retail sales are about $4 trillion annually, with supermarkets and hypermarkets (combined grocery markets and department stores) accounting for the largest share of sales.

Aside from the worldwide demand for food and food products, recent moves by various countries and international trade organization to decrease tariffs have stimulated international activities throughout the food industry.

Improvements in transportation and communication have also helped increase the international activities of food industries. Products move around the world by air freight in hours or days and communications take place around the world in seconds.

MATH CONNECTION!

Kraft General Foods, CPC International, H.J. Heinz, Borden, Campbell Soup, Nabisco Brands, Coca-Cola, PepsiCo, Beatrice Companies, Ralston Purina, and General Mills are major food companies. Choose one company to research. What is the dollar amount of foods this company sells annually? In how many countries does your company sell its products? What economic impact does this company have in the United States? How many employees work for the company? Do your findings surprise you?

Interrelated Operations

Food production relies on a highly advanced and organized industry. The industry is a systematic and rhythmic process. Throughout production, the costs of manufacturing, distribution, and marketing as well as availability are carefully monitored and controlled. Furthermore, because the industry is high volume and low-markup, small losses anywhere along the chain can mean larger losses to the food producer.

FOOD-SYSTEM TRENDS

National and international trends drive changes in a food system. While the total food consumed by each individual (per capita food consumption) has changed little over the years, the kind of foods consumed continually change, contributing to competition and frequent changes in the food system.

According the data from the Economic Research Service (ERS) of the U.S. Department of Agriculture (USDA) (http://www.ers.usda.gov/), consumer food expenditures in recent years how shown a shift toward the consumption of higher value food products by every income group. As income grows, consumers in lower income countries shift their food purchases away from carbohydrate-rich staple foods toward more expensive sources of calories such as meat and dairy products (Figure 2-2). An examination of food expenditures and food sales data indicates that middle-income countries such as China and Mexico appear to be following trends in high-income countries as measured across several dimensions of food-system growth and change. These include trends in important food-expenditure categories, such as cereals and meat, as well as in

Source: USDA, Food and Nutrition Service (FNS), Supplemental Nutrition Assistance Program (SNAP).

FIGURE 2-2 As income grows, consumers shift their food purchases away from carbohydrate-rich staple foods toward more expensive sources of calories such as meat.

indicators of food-system modernization, such as supermarket and fast-food sales.

The changes in food-consumption patterns are largely driven by income growth and demographic factors, particularly lifestyle changes brought about by urbanization, away-from-home employment of women, and increased levels of information. Although income growth, which affects the purchasing power of consumers, is one of the most important factors contributing to demand changes, urbanization has been equally important in changing the types of foods consumed. Urban areas generally bring higher incomes, more women in the workforce, more education, and wider arrays of food products available for consumption.

While the economy goes in cycles, urbanization has so far been a one-way process. As typically occurs in developing countries, the rural share of the population eventually becomes so low that urbanization is no longer an important factor in food demand. Among developing countries with large shares of rural population and rapid rates of urbanization, urbanization is likely to significantly alter consumers' diets with greater consumption of meats, fruit, vegetables, and processed foods.

Urbanization and income growth are also associated with more household conveniences that enable consumers to purchase and store perishable food products. Since 2005, the percentage of households possessing refrigerators increased significantly in most developing countries. Similarly, the percentage of households owning microwave ovens is rising across countries, promoting sales of so-called ready meals.

ERS Analysis of Global Food Demand. ERS has estimated the income and price elasticities of demand for broad consumption and food categories across 144 countries. These data on food spending and other consumption categories and their respective budget shares reveal a lot about food demand trends across countries and improve the ability of forecasters to predict potential shifts in food product demands across different food groups.

The results of ERS research confirm many of the findings established by earlier studies:

- Low-income countries spend a greater portion of their budget on necessities such as food, whereas richer countries spend a greater proportion of their income on luxuries such as recreation.
- Low-value staples such as cereals account for a larger share of the food budget in poorer countries, whereas high-value food items make up a larger share of the food budget in richer countries.
- Overall, low-income countries are more responsive to changes in income and food prices and therefore make larger adjustments in their food-consumption patterns when incomes and prices change.
- Adjustments to price and income changes are not uniform across all food categories. Staple food consumption changes the least, whereas consumption of higher-value food items such as meats changes the most.

Retail Trends

Analysis of retail sales data reveals further trends in food-consumption patterns across countries. Packaged food products account for large shares of total food expenditures among consumers in high-income countries because of demand for convenience. The United States, the European Union, and Japan account for more than half of total global sales of packaged products. In developing countries, intermediate products such as vegetable oils, dry pasta, and other dried products account for the bulk of retail sales. But market trends indicate strong growth in sales of packaged food products among developing countries. This growth involves three-fourths of the world's consumers and is partly the result of rapidly growing incomes.

Trends in the soft drink and beverage sector often indicate consumers' ability to purchase higher-value foods, and foreign investment in the beverage sector often functions as a bellwether for the health of local food industries. Analysis of soft-drink retail sales data indicates a rapidly expanding sector with large sales growth in Eastern Europe and Asia. Markets in developed countries, however, are sluggish, particularly for carbonated drinks. Carbonated drinks face strong competition from fruit juices and various health and ethnic drinks. In many developing countries such as India, where growing affluence has prompted the demand for clean drinking water, increased demand for bottled water has further boosted total soft-drink sales.

Demand for unique or alternatively produced foods has increased the consumption of organic and natural foods in many developed countries and among smaller but wealthier segments of some developing countries (Figure 2-3). In many countries, particularly in Western Europe, this has meant

Source: USDA, Food and Nutrition Service (FNS), Supplemental Nutrition Assistance Program (SNAP).

FIGURE 2-3 Farmers' markets typically meet local demands for organic or naturally produced foods.

increased sales of private retail brands because retailers can set and enforce their own product quality standards. In developing countries, expansion of supermarket chains has also introduced private retailer brands, mainly as cheaper alternatives to major manufacturer brands.

Changes in government regulation of food additives, food composition standards, and labeling also require the food industry to be responsive to change. Finally, technical innovations such as ingredient modifications, new processing methods, new packaging methods, and cooking advances also create change in a food system.

DEFINING SUSTAINABILITY

Nowadays, every industry seems to promote "sustainability" in one way or another. Simply defined, **sustainable** means using a harvesting method or resource so that the resource is not depleted or permanently damaged.

Nevertheless, sustainability means different things to different people. For example, some people believe that any food production system that depends on nonrenewable resources such as oil is not sustainable. Other people argue that this is not an important aspect of sustainable food production because they believe that alternative energy sources will be found as oil becomes scarcer. Another definition of sustainability is meeting the needs of the present without compromising the ability of future generations to meet their own needs. That definition is quite complex and can have a wide range of interpretations.

In 1990, the U.S. Congress defined *sustainable* in its Food, Agriculture, and Trade Act. According to the act, sustainable food-production systems have seven key features:

1. Will meet human needs for food now and far into the future
2. Will integrate plant and animal production
3. Will rely as much as possible on natural processes and cycles
4. Are designed specifically to fit the biological, social, and economic conditions of specific places (i.e., they are site specific)
5. Will provide a livable income for farm families
6. Will protect natural resources
7. Will enhance the quality of life for farmers and for society as a whole

A literature search can reveal many more ideas about sustainability. Each idea contains elements of the other, as shown in Figure 2-4, and is used to describe sustainable agriculture.

Many scientists and leaders see sustainable food production as being environmentally, economically, and socially sustainable. The environmental and economic aspects of sustainability have been around for decades, even centuries, in agriculture. These can be measured and identified. The social aspect of sustainability is a newer concept that is gaining increasing focus in many

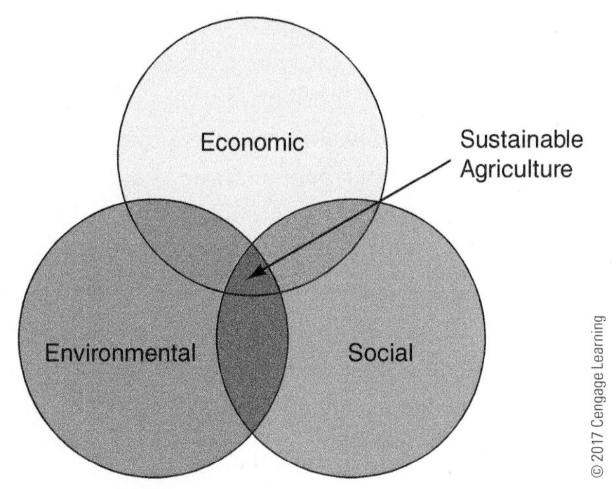

FIGURE 2-4 Typical diagram used to illustrate sustainable agriculture.

developed countries in the world, including the United States.

Many times, agricultural producers and others in a food system can agree on most of the economic and environmental aspects of sustainability, but they struggle more on the social aspects because many of these are not based on science and economic research but on concepts, opinions, business strategies, and personal preferences.

Many industries set and rely on standards to ensure consistency and uniformity. Standards are used to evaluate agricultural products by describing the ideal. To evaluate the sustainability of an operation or make changes in an operation or a system, some standards are needed. These set the ideal for comparison and allow progress to be tracked.

STANDARDS OF SUSTAINABLE FOOD PRODUCTION

Often people just seem to know that *sustainable* is good without really understanding what is involved. To help prevent misunderstandings about sustainable food systems, the concept is presented here within the framework of 13 standards that have been derived from a variety of sources.

A sustainable system of food production will meet the following 13 standards at some level:

1. Base direction and change on science
2. Follow market principles
3. Increase profitability and reduces risk
4. Satisfy human need for fiber and safe, nutritious food
5. Conserve and seek energy resources
6. Create and conserve healthy soil
7. Conserve and protect water resources
8. Recycle or manage waste products
9. Select animals and crops appropriate for environment and available resources
10. Manage pests with minimal environmental impact
11. Encourage strong communities
12. Use appropriate technology
13. Promote social and environmental responsibility

A sustainable food system focuses on and meets as many of these standards as possible. How well these standards are met determines sustainability. All 13 standards will be described and related to food production.

Standard 1: Base Direction and Changes Based on Science

The food system relies on **science**. Scientific research uses the scientific process or method (see Figure 2-5). The process has led to the creation of an organized body of knowledge that has been derived from the answers to practical and experimental approaches to specific problems. Science deals only with rational propositions—hypotheses—that can be verified or disproved by observation or experiment. Through scientific research, improvements continue to be made in production, manufacturing, distribution, marketing, consumption, and waste disposal, as well as in the allied industries.

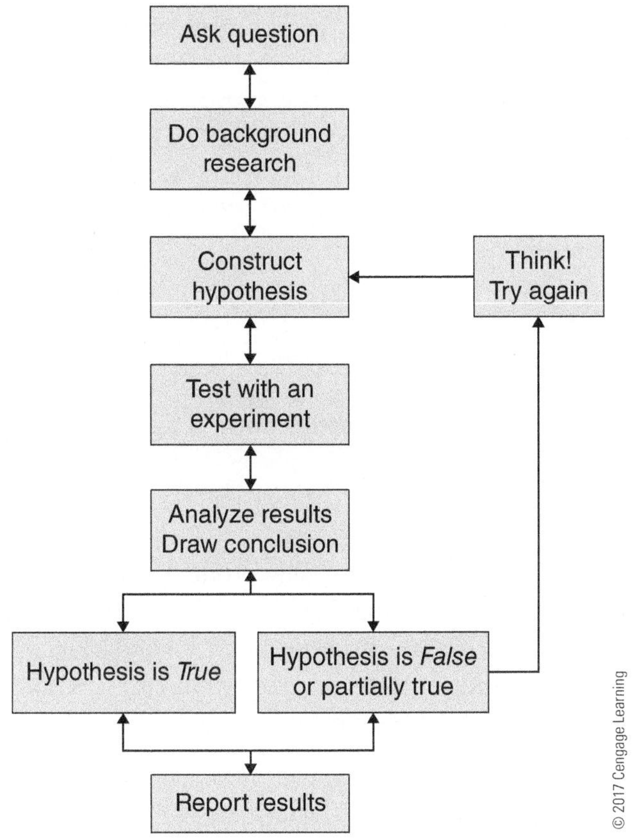

FIGURE 2-5 The scientific method.

Scientists working in the food system can never be absolutely certain that an experiment has eliminated all of the variables that might influence its results. To deal with such variables, scientists concentrate on experimental design and statistics. This increases the chance that an experiment will be repeatable and yield the same results, a critical requirement of the scientific method.

Why Science? Science is necessarily objective and must be approached without emotion or prejudice. This will help the method retain its basic nature and help it deal with contemporary problems.

Ongoing Research and the Use of Science. To improve food production and to address sustainability, ongoing scientific research addresses many issues. These include:

- Genetic improvement
- Animal health management
- Reproduction and early development
- Plant and animal growth, development, and nutrition
- Production systems
- Sustainable and environmental compatibilities
- The quality, safety, and variety of food products for consumers

Genetic improvement involves conserving, characterizing, and using genetic resources; selective breeding for economically important traits; and using genomic resources, specific breeding aids, bioinformatics, genetic engineering, and statistical analysis tools.

Animal health management involves identifying pathogens and diagnosing diseases, using vaccines and medicines, applying principles of immunology and disease resistance, understanding disease mechanisms, exploring epidemiology, and researching microbial and genomic principles).

Reproduction and early development involves controlling reproduction, gender, and fertility; ensuring gamete and zygote quality; developing methods of gamete and embryo storage; using cryopreservation (freezing); and studying early life-stage development and survival,

Plant and animal growth involves the use of nutrition and regulating feed intake; regulating tissue growth and development; developing sustainable sources of nutrients; managing nutrient use and evaluating feed; studying the interactions among gene regulation, nutrition, and the environment; managing interactions that affect reproduction; the effective use of **probiotics**; enhancing immune systems; and developing and using **nutraceuticals**.

Production systems (**biosecurity** involve production intensity; integrated production systems; predator control, live animal handling, transportation, and inventory; and the culture of new species.

Sustainable and environmental compatibilities involve food-system components such as water use and reuse, effluent management, waste disposal, social factors, and environmental issues,

Quality, safety, and variety of food products for consumers (food quality, interaction of genetics and nutrition, predicting product quality or defects, off-flavor, food borne diseases, harvesting, new uses for byproducts and processing)

Many universities, government agencies, and private and public organizations conduct scientific research to aid the food system. For details on specific research projects of the USDA's Agricultural Research Service, for example, reports can be on the ARS Web site: http://www.ars.usda.gov/research/programs.htm.

SCIENCE CONNECTION!

Google one of the preceding bulleted items. What type of research do you find? Read a few articles and highlight a few aspects from the article that you find especially interesting.

Standard 2: Follow Market Principles

A sustainable food system follows market principles: a market, marketing, a marketing plan, and profits from sales (Figure 2-6). Nothing happens until someone sells something. Obviously nothing can be sold if a market for the goods or services does not exist. Marketing is the process of planning and executing the conception, pricing, promotion, and distribution of ideas, goods, and services to create exchanges that satisfy individual and organizational objectives.

The first step in marketing is a plan—that is, a written statement that can guide the marketing process. Writing requires considerable thought, time, energy, and information. A new product plan is prepared before every new product is produced. Marketing plans are usually dynamic and can change as they are implemented. A good plan increases profits or the chance of profits from components of the food system.

FIGURE 2-6 The marketing mix.

Market Niches. Many of those who pursue certain components within a sustainable food system develop marketing niches—that is, specialization in a specific and limited market sector. These individuals develop marketing niches to take advantage of specific microclimates, regional demands, and their own special knowledge or skills. Marketing specialized products or services can be more difficult than those for traditional products and can also be more risky. But **market niches** can be more profitable.

Standard 3: Increase Profitability and Reduce Risks

To succeed, those engaged in the food system seek to increase profitability by reducing risk. Because of weather, consumer preferences, price fluctuations, and many other factors. Food production is a risky venture. Profitability often has high risks associated with it. To ensure sustainability, increased opportunities for profitability and reduced risks need to be evident. The ability to minimize risk is a major factor in a successful sustainable venture. Profits are the motivating force for anyone to go into business; they reward individuals for their willingness to take the risk of operating a sustainable food-system enterprise. Without risk, nothing sustainable can be developed; without the chance to create a profitable venture, no one will take risks.

Risk is manageable, and uncertainty must be accepted. Those within any area of the sustainable food system must learn to manage risk and accept uncertainty, which requires excellent business skills and the ability to respond to three guiding rules of risk management:

1. Do not risk more than you can afford to lose.
2. Do not risk a lot for a little.
3. Understand the likelihood and severity of possible losses.

To effectively manage risks, the business owner must know something about the likelihood and severity of loss from each possible risk and then decide on how best to manage it. This is called **risk assessment**. Options to manage risk include a wide range of techniques and approaches that can generally be grouped into two distinct categories: insurance and noninsurance options.

Noninsurance Risks. Noninsurance risk-management options include production risks and marketing risks. The management of production risks can include the following:

- Changing husbandry practices
- Building redundancy (backup) into the operation
- Improving management
- Employing stringent biosecurity measures
- Minimizing the chance of disease introduction and cross contamination
- Knowing appropriate treatments

Management of marketing risks can include the following:

- Diversification to include other products and categories of production
- Creation of unique identity (brand)
- Development of a niche market or value-added products

Although any number of options may exist to manage each potential risk, each option has associated costs and benefits that must be weighed against the risk.

Standard 4: Satisfy Human Need for Fiber and Safe, Nutritious Foods

Providing food and shelter are the primary purposes of a food system. Food and shelter (fiber) are basic human needs to satisfy. When sufficient fiber and food are not supplied, people rush to blame various issues: the population, the economy, politics, or the environment.

Sustainable agriculture contributes to providing sufficient fiber (for protection from the elements) and safe and nutritious food to people. This contribution can be local, regional, national, or international. Although adequate and continuous food and fiber production are essential to sustainability, transportation and distribution systems are key elements to the success of sustainability.

The United Nations Food and Agriculture Organization (FAO) reports on food insecurity worldwide (http://www.fao.org/hunger/en/). In its 2015 report, the FAO states that global hunger has continued to gradually decline to an estimated 795 million undernourished people, a reduction of 167 million hungry people over the preceding 10 years. This decline has been most pronounced in developing countries despite significant population growth. According to the FAO report, a key factor in reducing undernourishment has been economic growth, but only when it is inclusive—providing opportunities for the poor, who have meager assets and skills, to improve their livelihoods. Enhancing the productivity of family farmers and strengthening social-protection mechanisms are key factors in promoting inclusive growth, along with "well-functioning markets and governance in which all voices may be heard."

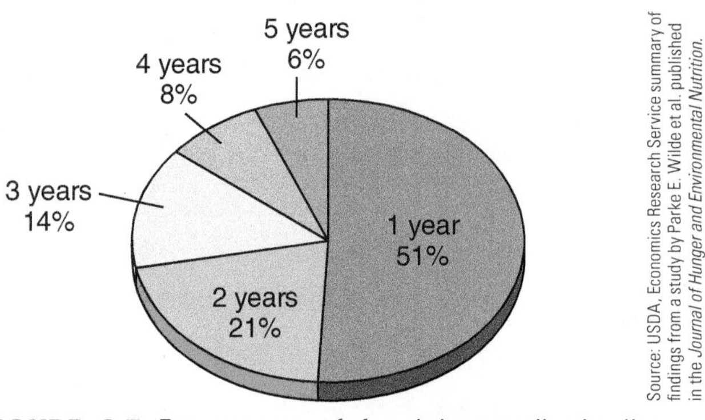

FIGURE 2-7 Frequency of food insecurity in the United States.

In the United States, the USDA's Economic Research Service collects data on food security and insecurity (http://www.ers.usda.gov/topics/food-nutrition-assistance/food-security-in-the-us.aspx), *Food security* means that households had access at all times to enough food so that all household members could maintain active and healthy lives. According to the ERS, this is about 86% of all households in the United States. **Food insecure** means that at some time during a year these households were uncertain whether they could acquire enough food to meet the needs of all members because they did not have enough money or had no other resources for food (Figure 2-7). This is about 14% of all U.S. households. The characteristics of food insecure households in the United States include:

- All households with children
- Households with children headed by single women
- Households with children headed by single men
- Black and non-Hispanic households
- Hispanic households
- Low-income households with incomes below the poverty threshold

Urbanization. Around the world, people are migrating from the country into cities. This is called **urbanization**. A continuous and massive amount of food is required to feed this growing urban population. Worldwide, the number of cities with populations of 10 or more million will increase to 26. Several of these cities are in North America. Every city this size requires 6,000 tons of food shipped in each day or about 2.2 million tons per year. Even cities with populations of less than 10 million require thousands of tons of food each day.

A sustainable food system will need to increase its ability to meet the human need for safe, abundant, and nutritious foods. Creating a sustainable food system is more complex than just producing and distributing food. Planning for a sustainable food system must include ways of ensuring a continuous, healthy, and adequate supply of safe food for all despite economic conditions, urbanization, and disruption from conflicts such as war or terrorism. This could require looking at local food systems to supply food and keeping national agriculture alive and healthy so as to not become dependent on food production in other countries where production and trade policies can change with little notice or little explanation.

SCIENCE CONNECTION!

Have you ever wondered how you could make a difference? What is something you, your agriculture department, or FFA chapter could do to help fight food insecurity? How can you make it a reality?

Standard 5: Conserve and Seek Energy Resources

A sustainable food system conserves energy resources and seeks new sources, recognizing that the energy resources of coal, petroleum, and natural gas are finite. Conservation can mean using less or using an alternative and renewable source.

Conservation of energy by using less energy uses often uses technology to increase the efficiency of some energy consuming practice associated with food production; for example, pumping, transportation, storage and harvesting. Obviously another way to conserve is to cutback on any operations that use energy. Conservation by increasing efficiency or reducing operations continues, and organizations such as the U.S. Department of Energy's Office of Energy Efficiency and Renewable Energy (http://www.eere.energy.gov/) provide information and resources to increase efficiency.

Energy Flow and Alternative Energy. Energy flow is enhanced through the increased capture of solar energy and strategies to effectively use and store it. Various parts of the food system are considering renewable and alternative energy resources, specifically wind, solar, bioenergy (biogas, butanol, and ethanol), hydrogen (fuel cells), and nuclear. Most of these alternative renewable energy sources require more development and research to become usable in the future for sustainable aquaculture and agriculture.

A sustainable food system maintains a focus on reducing energy flow and developing and using renewable and new energy sources. To succeed, this focus will likely need to be maintained by government programs and individual initiatives. The combination of new technology, demand, research, and economics will lead to the incorporation of renewable energy sources in the food system.

Current related research includes the following:

- Algal biofuels technology development
- Hydrogen production from algal systems
- Geothermal energy use
- Conservation and efficient use of energy
- Solar energy development
- Biomass energy use (Figure 2-8)

If all goes well, the future will see the development of clean energy technologies that strengthen the economy, protect the environment, and reduce dependence on foreign oil.

FIGURE 2-8 Besides its extraction for sugar, sugarcane produces a larger biomass that could be used as a renewable energy source.

Standard 6: Create and Conserve Healthy Soil

Healthy soils are crucial for ensuring the continued growth of natural and managed vegetation, providing feed, fiber, fuel, medicinal products, and other ecosystem services such as climate regulation and oxygen production. Soils and vegetation have a shared relationship. Fertile soil encourages plant growth by providing plants with nutrients; it also acts as a water-holding tank and serves as the material in which plants anchor their roots. In return, vegetation, tree cover, and forests prevent soil degradation and desertification by stabilizing the soil, maintaining water and nutrient cycling, and reducing water and wind erosion (Figure 2-9).

As global economic growth and demographic shifts increase the demand for vegetation, animal feed, and vegetation by-products such as wood, soils are put under tremendous pressure. The risk of degradation increases greatly. Managing vegetation sustainably—whether in forests, pastures, or grasslands—will boost its benefits, including timber, animal feed, and food, in a way meets society's needs while conserving and maintaining the soil for the benefit of current and future generations.

Obviously, a sustainable food system needs to create and conserve healthy soil. Erosion includes that by wind and water from rain and poor drainage,

FIGURE 2-9 Conservation tillage practices are used to prevent wind erosion of soil.

and it is accelerated by such human activities as forest removal, agriculture, grazing, construction, and mining. Whenever vegetation is removed and the ground is exposed to rainfall, soil erosion by water and wind may increase. On sloping land, erosion rates increase. Throughout the tropics, it is one of the most serious environmental and socioeconomic problems.

Conservation methods include the general categories of tillage, vegetation, mulching, and cropping patterns. Well-planned and managed vegetation cover can effectively control soil movement. Vegetation protects soil against erosion by reducing water movement and building soil structure. Mulching covers the soil with materials that reduce soil moisture evaporation and inhibit weed growth. Mulching slows rainfall infiltration and protects the soil from direct impact from rain. Changes to a production system help reduce soil movement through intercropping, alley farming, terracing, and the use of grass strips.

Soil is more than ground-up rock fragments. Minuscule creatures are the life force of a healthy soil. These organisms help purify water and air, detoxify pollutants, recycle crop nutrients, decompose plant residues, promote soil structure, and prevent disease outbreaks. Management of the land directly influences how well these organisms perform their important functions.

The food web does not just pertain to above-ground collections of carnivores and herbivores. In fact, above-ground herbivores are not responsible for most of the plant matter consumption in the world. Decomposers in the soil consume the most. Many organisms function in the soil food web such as arthropods, bacteria (actinomycetes and photosynthetic cyanobacteria), fungi, protozoa, nematodes, and earthworms. Some chemicals and production and tillage methods are detrimental to organisms in the soil food web.

Critical information for conservation and soil-health planning includes soil depth, soil type, drainage characteristics, slope of the land, chemical usage, and tillage practices. The practices selected to control erosion and improve soil health should be based on a combination of factors such as the type of soil, the crops being grown, irrigation practices and cultivation practices. To be sustainable, the practices must be within the means of the producer and be perceived to be of high economic and social value or they will not be implemented or maintained.

Standard 7: Conserves and Protects Water Resources

A sustainable food system conserves and protects water resources. Freshwater from streams, rivers, and underground wells has no substitute and is the most fundamental of the finite resources required for sustainability. It is expensive to transport. But freshwater sources are dwindling and becoming

FIGURE 2-10 A large field of lettuce requires water for a successful crop.

contaminated around the world. Chronic or acute water shortage is increasingly common in many countries with fast-growing populations, becoming a potential source of conflict.

Agriculture and food production are major users of groundwater and surface water in the United States, accounting for approximately 80% of the nation's water use and more than 90% in many Western states (Figure 2-10). Efficient irrigation systems and water-management practices can help maintain farm profitability in an era of increasingly limited and more costly water supplies. Improved water-management practices can also reduce the impacts of irrigation on offsite water quantities and quality as well as conserve water for growing nonagricultural demands.

Water Quality. Producing food and fiber involves many activities and practices that can affect the quality of water resources under and near the field (Table 2-1). For example, tilling the soil and leaving it without plant cover for extended periods of time can accelerate soil erosion. Chemical fertilizer and pesticide residues can wash into streams or leach through the soil into groundwater. Irrigation can move salt and other dissolved minerals to surface water. Livestock operations produce large amounts of waste that can threaten human health as well as contribute to excess nutrient problems in streams, rivers, lakes, and estuaries if not properly disposed. When pollutants degrade water quality, they impose costs on water users in the form of degraded ecosystems, reduced recreational opportunities, reduced commercial fishing catches and shellfish bed closings, increased water-treatment costs, threats to human health, and damage to reservoirs and water-conveyance systems. These costs provide the impetus for policies to reduce water pollution.

TABLE 2-1 Sources of Water Pollution and Their Impact on Both Fish and the Environment

TYPE OF WATER POLLUTION	EFFECT ON FISH AND THE ENVIRONMENT
Nutrients from yard waste, sewage, food processing, agriculture, livestock, and industry	Fish kills, low oxygen due to algal blooms, and reduced diversity of animals and plants
Heat from cities, power plants, industrial discharges, and the sun	Coldwater fish cannot survive; more disease occurs because of heat stress and lower oxygen levels
Toxic chemicals from landfills, agricultural and urban runoff, and industrial discharges	Diseases and death from chemicals being passed through the food chain
Disease organisms from raw or partially treated sewage and runoff from feedlots	Human health hazards and environmental contamination
Sediment from agricultural fields, logged areas, feedlots, degraded stream banks, and road construction	Filling of streams and ponds with sediments and reduction of plant and animal life

When water quality is a factor in the production of a market good, the benefits of changes in quality can be shown from changes in variables associated with the production of the good. If improvements in water quality reduce production costs and thus market prices to consumers, both producers and consumers benefit. An example could be a reduction in the price of vegetables because of lower salinity in irrigation water.

Food Processing. Besides the use of water in agricultural food production, water is extensively used in the food-processing industries for such functions as processing and preparation; ingredient mixing, brine formation, transporting, washing, rinsing, scalding, and chilling; blanching, pasteurizing, steam production, and cooling; and cleaning, sanitation, and disinfection. To conserve water and to maintain water quality, food processors:

- Develop unit operations that use less water or optimize water use
- Recycle or reuse water (following reconditioning)
- Find alternative sources of potable or clean water

As the world population and demand for food increases, the need for water conservation and preserving water quality will be increasingly important for a sustainable food system.

MATH CONNECTION!

Do you ever stop and think about how much water you use on a daily basis? Are there things you could do to conserve or reduce water use? Research your own water use and come up with four ways you could better conserve water.

Standard 8: Recycle and Reduce Waste Products

A sustainable food system recycles and reduces waste. Agricultural residues or wastes might better be regarded as "resources out of place" rather than simple waste. This change in perspective permits the evaluation of waste from a positive standpoint. With appropriate techniques, wastes can be recycled to produce an important source of energy, a natural fertilizer, or even a feed. The choice of recycling method will depend on the type of waste available and its intended end use. The mineral cycle is fostered by the cycling and recycling of wastes on a farm. There are many historical examples of waste products becoming valuable by-products. For example, slaughterhouse and processing plant wastes that were once dumped became valuable feeds. Whey, a by-product of cheese production, was once a waste product farmers had to deal with. New extraction and processing techniques have now made whey a valuable product.

Five popular recycling methods for agricultural wastes include: **anaerobic digestion**, refeeding, land application, composting (Figure 2-11), and incineration.

Anaerobic Decomposition. Biogas is produced when organic matter degrades in the absence of oxygen. This process is called *anaerobic*

FIGURE 2-11 Composting is successfully used with manure and other waste products to create material that can be reapplied to soils to increase their fertility.

decomposition or *anaerobic digestion.* Animal wastes in particular can be used to generate biogas, a mixture of methane and other gases that form from decomposing organic matter. Like other gas fuels, biogas can be used for cooking, lighting, and running small engines. Many other countries have spent considerable time and money in developing biogas production.

Standard 9: Select Animals and Crops Appropriate for an Environment and Available Resources

Producers in sustainable food systems look for animals and crops that are appropriate for the environment and available resources. This is especially true when considering some form of **aquaponics**—the combination of raising aquatic animals and producing some plants. Aquatic animal effluent accumulates in water as a by-product of keeping them in a closed system or tank. The effluent-rich water becomes high in plant nutrients that are toxic to aquatic animals. In a hydroponic system, plants are able to use the nutrient-rich water. The plants take up the nutrients, reducing or even eliminating the water's toxicity for aquatic animals.

Climate, resources, producer abilities, demand, distribution, and markets should determine the type of animal or plant grown in a sustainable food system. A wide selection of species and varieties of animals and plants exist from which producers can select.

With current demands and population growth, serious consideration is now being given to **entomophagy**. Entomophagy is the practice of eating insects—including arachnids (tarantulas) and myriapods (centipedes).

A SUSTAINABLE FOOD SOURCE OF THE FUTURE?

Will you have a bug in your future meal? Entomophagy is the practice of eating insects and certain arthropods such as arachnids (spiders) and myriapods (centipedes). Insects have served as a food source for people for thousands of years around the planet. Today insect eating is rare in the developed world, but insects remain a popular food in many developing regions of Central and South America, Africa, and Asia. For example, many inhabitants in Thailand, Vietnam, Cambodia, China, Africa, Mexico, Colombia, and New Guinea eat insects for nutritional value as well as for their taste.

According to the United Nations FAO, trends toward 2050 predict a steady population increase to 9 billion people, forcing increased food and feed output from available agricultural ecosystems, resulting in an even greater pressure on the environment. Scarcities are foreseen in agricultural land, water, forests, fisheries, and biodiversity resources, as well as nutrients and nonrenewable energy. Edible insects contain high-quality protein, vitamins, and amino acids for humans. Insects have a high food-conversion rate. For example, crickets need six times less feed than cattle, four times less than sheep, and twice less than pigs and broiler chickens to produce the same amount of protein. They also emit less in greenhouse gases and ammonia than conventional livestock. Insects also can be grown on organic waste. Insects are thus a potential source of the production of protein for direct human consumption or indirect recomposed foods (with extracted insect proteins) and as a protein source for feedstock mixtures.

An estimated 1,462 species of insects have been described as edible, including some arthropods. In all likelihood, there are hundreds if not thousands more that simply have not been sampled or discovered yet.

To some people, entomophagy may seem repulsive, but remember that many people in developed countries already eat a wide variety of arthropods, including crustaceans such as shrimp, prawns, crabs, lobsters, and crayfish.

Standard 10: Manage Pests with Minimal Environmental Impact

In a sustainable food system, managing pests to protect the food supply is a must. Sustainable practices manage pests with minimal environmental impact. Pest management is a challenge to all producers, but especially for those dedicated to sustainable, low-input practices. A wide array of techniques and control agents can effectively reduce or eliminate pest damage without sacrificing soil, water, or beneficial organisms. The new technologies of global positioning satellites and global information systems also provide ways of efficiently and effectively applying pesticides with minimal risk to the environment.

Integrated Pest Management. Much of the philosophy of pest management in sustainable agriculture is expressed as some variation of integrated pest management (IPM). Like many of the practices in sustainable agriculture, IPM suggests different things to different people. Simply put, IPM promotes minimized pesticide use, enhanced environmental stewardship, and sustainable systems.

Some individuals feel that traditional IPM still focuses too much on the use of pesticides. Perhaps a better term for IPM in a sustainable system is **bio-intensive** IPM. Bio-intensive IPM is a systems approach based on an understanding of pest ecology. Those who use bio-intensive IPM begin by accurately diagnosing the nature and source of pest problems. They then rely on a range of preventive tactics and biological controls to keep pests within acceptable limits. Pesticides are used carefully as a last resort when other tactics fail.

Planning is key to using bio-intensive IPM. Planning occurs before production because many pest strategies require steps or inputs, such as beneficial organism habitat management, that must be considered well in advance.

Biological control is the use of living organisms such as parasites, predators, and pathogens to maintain pest populations below economically damaging levels. Biological controls can be natural or applied. Natural biological control is generally characteristic of biodiverse systems. It results when naturally occurring enemies maintain pests at lower levels than would occur without them.

Mechanical or physical controls use some physical component of the environment such as temperature, humidity, and light that are detrimental to pests.

Correctly identifying pests is a crucial step in the effectiveness of any IPM program. Misidentification can be harmful and cost time and money. Monitoring allows the grower to gather information about the crop, pests, and natural enemies and even requires a producer to systematically check for pests and **beneficials** at regular intervals and at critical times.

About one-third of the worldwide food supply is lost to pests, so pest control is essential to a sustainable food system.

SCIENCE CONNECTION!

What do you know about integrated pest management? Research and find examples of IMP as used with crops. Are any IPM methods being used in your area?

Standard 11: Encourage Strong Communities

Sustainable food systems encourage strong communities, providing opportunities for young people, access to health care, living wage jobs, and access to quality education. This means that a sustainable food system must create and implement projects and industry that generate new wealth and jobs.

Lack of jobs (opportunities) and low-wage jobs are serious concerns. In general, jobs are viewed as a means to make communities vital and self-reliant and allow individuals and families to be self-reliant within the context of community.

Sustainable food systems as a part of sustainable community development help find solutions by increasing the capacity of individuals and communities to work together to respond to constant changes, which are common.

Standard 12: Use Appropriate Technology

The success or failure of a food system or any of its components often depends on selecting the right technology. Many technological advances have contributed to the success of the U.S. food system. Many of these technologies are not successful in developing countries and cannot help with small-scale food production. Good planning and judgment are needed to select those technologies that will benefit a food system—from the equipment used to the methods of pest control to the types of seed used. Advances in technology provide many options.

Among the factors to consider when selecting technology for a sustainable food system are the current technology being used, local and national infrastructure, local and national regulations, finances, local cultural customs, local institutions, the local environment.

Standard 13: Promote Social and Environmental Responsibility

Standard 13 is a capstone standard. In other words, a sustainable food system that addresses the first 12 standards will automatically address standard 13. Standard 13 has also helped many colleges and universities launch sustainable agriculture programs in response to emerging concerns about natural resources, the environment, economics, and the social dimensions of agriculture and the food system.

New technologies spawn economic, social, and political changes while creating shifting relationships among individuals. This often makes it difficult for people to commit to being socially and environmentally responsible. All of this is compounded by the shrinking of the world through modern communications and transportation.

Change is difficult. A diversity of strategies and approaches are necessary to create a more sustainable food system. Strategies and approaches will range from specific and concentrated efforts to alter specific policies or practices to the longer-term tasks of reforming key institutions, rethinking economic priorities, and challenging widely held social values.

Conversion of agricultural land to urban use is a particular concern because rapid growth and escalating land values threaten existing food-production operations and water supplies. Comprehensive new policies to protect sustainable enterprises and regulate development are often needed.

Policies and programs are needed to address labor, working toward socially just and safe employment that provides adequate wages, good working conditions, health benefits, and chances for economic stability.

Food and agriculture represent the union of human rights with community and environmental issues. The agriculture sector is by far the largest employer in the world, employing almost one-half of the world's workers. Every person relies on agricultural products for survival. Food and agriculture also constitute one of the most dangerous sectors in the world. For food and agriculture companies, the global marketplace creates increasingly complex supply chains along with ever-more challenging demands from the world's stakeholder communities.

Food and agriculture companies wanting to do business in a global economy are faced with an increasing assortment of international regulations governing the growing, production, and selling of agricultural products. Likewise, the demands of nongovernmental organizations (NGOs), shareholders, customers, governments, media, and other stakeholders have led to a dramatic growth in vendor codes of conduct and other external performance standards. In response, some food and agriculture companies are adopting international standards and certification schemes that address numerous environmental and social concerns. They are engaging with environmental, human rights, and development organizations to implement, monitor, and certify compliance with these largely voluntary codes of conduct. Furthermore, these companies are increasing the transparency of their operations, policies, and practices to broaden the trust of the public and encourage fair and ethical business practices.

SUSTAINABLE STANDARDS SCORE CARD

Considering all of its components, the difficulty in defining *sustainability* should be obvious. The scorecard in Figure 2-12 can be used to rate the sustainability level of each of the 13 standards, providing at least a starting point.

To approach sustainability, each standard must receive a score of at least 3. A total score of 53 to 65 indicates a high level of sustainability. A score of less than 53 suggests the system needs more work to become sustainable. Of course, a score of less than 3 for any standard should indicate that an operation is not meeting all 13 standards and is not truly sustainable.

STANDARD	SCORE (1 TO 5)
1. Base direction and change on science	
2. Follow market principles	
3. Increase profitability and reduces risk	
4. Satisfy human need for fiber and safe, nutritious food	
5. Conserve and seek energy resources	
6. Create and conserve healthy soil	
7. Conserve and protect water resources	
8. Recycle or manage waste products	
9. Select animals and crops appropriate for environment and available resources	
10. Manage pests with minimal environmental impact	
11. Encourage strong rural communities	
12. Use appropriate technology	
13. Promote social and environmental responsibility	
TOTAL	

FIGURE 2-12 Scorecard for rating the sustainability level of all 13 standards.

SUMMARY

Concerns about growing populations, increased food scarcity, and the environment have led researchers, farmers, not-for-profit organizations, governments, and industry representatives to work together to help find sustainable solutions to meet the world's growing demand for food, fuel, and water. A food system needs to be as sustainable as possible. This means getting more from natural resources while having a lighter environmental impact that is also economical.

Reliance on food production systems that are not environmentally sound could lead to the destruction of the natural resources needed to produce food. If food production systems are not economically viable and risks are not managed, then producers will go broke and their operations will cease to exist. If the systems are not socially sound, they will not meet the needs of people.

One key goal in sustainable food-production systems is to protect and conserve the natural resources on which food production depends. Water is a critical resource for all agricultural production and food processing. It must be available in sufficient quantity, at acceptable levels of quality, and at a cost that makes its useful for food production and processing.

The 13 standards provide guidelines for developing sustainable operations and a means for evaluating sustainable operations.

REVIEW QUESTIONS

Success in any career requires knowledge. Test your knowledge of this chapter by answering these questions or solving these problems.

1. List 13 standards for sustainability.
2. List five major segments of a food system.
3. What are three effects of urbanization on a food system?
4. Name three allied industries in a food system.
5. How does a food system contribute to strong communities?
6. List and describe three possible renewable energy sources that could be used in a food system.
7. Name three trends in global food demand.
8. Discuss the concept of sustainability.
9. Why is soil conservation so important to a food system?
10. Describe how integrated pest management contributes to a sustainable food system.
11. How can water pollution adversely affect food systems?
12. Explain why changes in a food system need to be science-based.

STUDENT ACTIVITIES

1. Using the Internet, research find four current projects where science is being used to improve food production. Describe your findings during a class discussion.
2. Create a digital presentation on water use in the United States from the United States Geological Service at: http://water.usgs.gov/
3. Visit the Web sites of large organizations involved in the food system and design a public service announcement that describes what they are doing to be sustainable.

4. Visit a business that represents a component of the food system (actual or online) and score its sustainability based on the score card at the end of this chapter.
5. Survey the class members to determine their food preferences—fast foods, convenience foods, cultural foods, home cooking, fresh fruits and vegetables, meats, etc. Create a chart showing the percentages of class preferences.
6. Depending on location visit a consumer supported agriculture (CSA) operation, community gardens, farmers' market, u-pick operation, or an organic farm or garden. Report on their role in the local food system.

ADDITIONAL RESOURCES

- University of California–Santa Cruz: Center for Agroecology & Sustainable Food Systems: http://casfs.ucsc.edu/
- Principles of a Healthy, Sustainable Food System: https://www.planning.org/nationalcenters/health/foodprinciples.htm
- Genetic Literacy Project: The Case for GMOs and Sustainability: https://www.geneticliteracyproject.org/2015/06/01/the-case-for-gmos-and-sustainability/
- National Sustainable Agriculture Coalition: http://sustainableagriculture.net/
- Syngenta: The Good Growth Plan: http://www.syngenta.com/global/corporate/en/goodgrowthplan/home/Pages/homepage.aspx
- USDA: Sustainable Agriculture Information Access Tools: http://afsic.nal.usda.gov/sustainable-agriculture-information-access-tools-1

Internet sites represent a vast resource of information. Although those provided in this chapter were vetted by industry experts, you may wish to further explore the topics discussed in this chapter using a search engine such as Google. Keywords or phrases may include the following: *sustainable agriculture, sustainable food system, sustainable practices, food production technology, aquaponics, bio-intensive, food market niches, integrated pest management, biomass, biofuels global food demand*, and *urbanization*. In addition, Table B-7 provides a listing of some useful Internet sites that can be used as a starting point.

REFERENCES

Burton, L. DeVere. 2009. *Environmental science: Fundamentals and applications*. Clifton Park, NY: Delmar, Cengage Learning.

McWilliams, James E. 2009. *Just food: Where locavores get it wrong and how we can truly eat responsibly*. New York: Little, Brown & Co.

National Research Council. 1989. *Alternative agriculture*. Washington, DC: National Academy Press.

National Research Council of the National Academies. 2010. *Toward sustainable agricultural systems in the 21st century*. Washington, DC: The National Academies Press.

Ruttan, Vernon W. (Ed.). 1994. *Agriculture environment and heath: Sustainable development in the 21st century.* Minneapolis: University of Minnesota Press.

Tarrant, John (Ed.). 1991. *Farming and food.* New York: Oxford University Press.

Zimmer, Gary F. 2000. *The biological farmer: A complete guide to the sustainable and profitable biological system of farming.* Austin: Acres U.S.A.

CHAPTER 3

Chemistry of Foods

OBJECTIVES

After reading this chapter, you should be able to:

- Name four carbohydrates and describe their chemical makeup
- Classify carbohydrates
- Compare the sweetness of various sugars
- Name three uses of carbohydrates in foods
- Describe the chemical makeup of proteins
- Discuss the use of proteins in foods
- List six functions of protein in the body
- Name three functions of protein in food
- Classify lipids
- Discuss the use of lipids or fats in foods
- Identify the difference between saturated and unsaturated fats
- List the fat- and water-soluble vitamins
- Name 10 minerals important to nutrition
- Describe two functions of water in the body
- Identify biotin, choline, and phytochemicals

NATIONAL AFNR STANDARD

FPP.02

Apply principles of nutrition, biology, microbiology, chemistry and human behavior to the development of food products.

KEY TERMS

amino acids
biotin
birefringence
caramelization
carbohydrates
cellulose
choline
crystallization
disaccharides
fatty acids
gelatinization
glucose
gum
homeostasis
hydrolysis
insoluble fiber
inversion
kilocalories
lipid
macrominerals
Maillard reaction

KEY TERMS (continued)

microminerals	peptide bond	rancidity
monosaccharides	phospholipid	saturated
oil	photosynthesis	soluble fiber
oligosaccharides	phytochemical	starch
osmotic pressure	polymer	triglycerides
oxidation	polysaccharide	unsaturated

Nutrition is the process by which the foods people eat provide the nutrients they need to grow and stay healthy. Nutrients are naturally occurring chemical substances found in food. There are six categories of nutrients: proteins, lipids, carbohydrates, vitamins, minerals, and water. The consumption of these nutrients can be beneficial or harmful to our bodies, depending on the amounts consumed. The chemistry of these nutrients influences the characteristics of our food.

Proteins, fats, and carbohydrates provide the energy as measured by **kilocalories** (kcal) that our bodies need to function. Food science uses the metric system—grams, milligrams, and micrograms—to measure the amounts of nutrients in foods. Each gram of protein and carbohydrate has 4 kilocalories; each gram of fat has 9 kilocalories.

SCIENCE CONNECTION!

Track the amount of carbohydrates you eat in a day. Categorize them as either simple carbohydrates (monosaccharides and disaccharides) or complex carbohydrates (polysaccharides). Of the two, which do you eat more of during the day?

CARBOHYDRATES

Carbohydrates are the main energy score for animals and humans. The carbohydrates in our diet come from plant foods. They cannot be produced by animals, only by green plants through the process of **photosynthesis**, which is the conversion of carbon dioxide and water to sugars. Along with starches and cellulose, sugars are among the better known varieties of carbohydrates.[1]

As plants convert carbon dioxide and water into carbohydrates and oxygen, carbohydrates are produced in units of sugar called **glucose**. After glucose is produced, the plant may convert glucose molecules into other sugars as well as starches or fibers. Simple carbohydrates include different forms of sugar: **monosaccharides** and **disaccharides**. More complex carbohydrates—**polysaccharides**—include starches and dietary fiber. Simple carbohydrates contain one or two sugars. Complex carbohydrates contain long chains of many sugars that the body has to break down into simple sugars. Glucose, fructose, and galactose are examples of monosaccharides. Lactose, maltose, and sucrose are examples of disaccharides.[2]

Carbohydrates are called *carbohydrates* because they are essentially hydrates of carbon. Specifically, they are composed of carbon and water and have a composition of $C_n(H_2O)_n$. The major nutritional role of carbohydrates is to provide energy; digestible carbohydrates provide 4 kilocalories per gram. No single carbohydrate is essential, but carbohydrates do participate in many required functions in the body.

Function in Food

Carbohydrates perform these functions in food:

- Flavor enhancing and sweetening due to caramelization
- Water binding
- Contributing to texture (starch, gluten)
- Hygroscopic nature (water absorption)
- Providing sources of yeast food
- Regulating gelation of pectin
- Dispersing molecules of protein or starch
- Acting to subdivide shortening for creaming control crystallization
- Preventing spoilage
- Delaying coagulation protein
- Giving structure due to crystals
- Affecting osmosis
- Affecting color of fruits
- Affecting texture (viscosity, structure)
- Contributing flavor other than sweetness

Depending on the food, carbohydrates can play many roles (Table 3-1). For example, in a lollipop, the sugars glucose (with or without glucose and fructose) will control crystallization, and sucrose will provide structure. All three sugars act as flavor enhancers and sweeteners. In a more complex system such as a pineapple upside-down cake, carbohydrates play many roles, consisting of all categories of carbohydrates—monosaccharides, disaccharides, and polysaccharides.

Monosaccharide

Monosaccharides with six carbon atoms are called *hexoses*; those with five carbons are called *pentoses* (Table 3-2). Glucose (sometimes called *dextrose*),

TABLE 3-1 Carbohydrate Applications in Food

INDUSTRY/PRODUCT	PROPERTIES AND BENEFITS
Baking and Snack Foods	
Cream-type fillings	Film forming, smooth texture
Fruit leather	Film forming, crystallization inhibitor, good humectant, sweetness moderator
Glazes and frostings	Crystallization inhibitor, film forming, improved adherence, sweetness control
Beverages	
Flavored drinks	Low sweetness, improved body
Infant formulas	Complete solubility, low sweetness, easy digestibility, bland flavor
Sports and special diets	Low osmolality, complete solubility, bland flavor, easy digestibility
Binders	
Cereals and snacks	Low hygroscopicity, good adhesion
Nut and snack coatings	Flavor carrier, film forming

(Continues)

TABLE 3-1 Carbohydrate Applications in Food

INDUSTRY/PRODUCT	PROPERTIES AND BENEFITS
Carriers	
Artificial sweeteners	Low hygroscopicity, neutral flavor, quick dispersion
Gums and hydrocolloids	Good solubility, standardizes viscosity
Confectionery	
Candies	Good solubility, humectant properties, inhibits bloom
Pan coatings	Binding, film forming, drying agent
Dairy	
Coffee whiteners	Fat dispersant, improved mouthfeel
Imitation cheeses	Processing aid, contributes texture
Dry Mixes	
Powdered drinks	Bulking agent, rapid dispersiblity, nonsweet carrier
Soup and sauce mixes	Low hydroscopicity, good solubility, provides body, protects flavor
Spice blends	Low hydgroscopicity, bulking agent, nonsweet carrier
Fat Reduction	
Baking	Film forming, humectant
Frozen desserts	Minimal freezing point depression, lactose and ice crystal inhibitor, provides body, improves melt
Meats	
Processed meats	Low sweetness, nonmeat solids, moisture retention, browning control
Spices and seasonings	Carrier, moisture management
Pharmaceutical	
Tableting	Good binding, low hygroscopicity, directly compressible
Sauces and Salad Dressings	
Cheese and white sauce	Low sweetness, smooth mouthfeel, does not mask flavors
Salad dressings	Provides body and cling, decreased gum "stringiness"
Tomato sauces	Provides body, brilliant sheen, intensifies color
Spray-Drying Aid	
Cheese and fats	Encapsulates fats and flavors, protects proteins, improves flowability
Flavors	Encapsulates, low hygroscopicity, bland flavor
Fruit juices and syrups	Low hygroscopicity, high solubility

TABLE 3-2 Carbohydrates and Characteristics

NAME/ CLASSIFICATION	END PRODUCTS (HYDROLYSIS)	SOURCE, FUNCTION, OR CHARACTERISTICS
Glucose	Glucose	Fruits, honey, corn syrup
Fructose	Fructose	Fruits, honey, corn syrup
Galactose	Galactose	Does not occur in free form in foods
Mannose	Mannose	Does not occur in free form in foods
Ribose	Ribose	Derived from pentoses of fruits and nucleic acids of meat products and seafood; does not occur in free forms in foods; an aldose
Xylose	Xylose	An aldose
Arabinose	Arabinose	An aldose
Sucrose	Glucose fructose	Beet and cane sugars, molasses, maple syrup; comes in many crystal sizes and grades
Lactose	Glucose galactose	Milk and milk products
Maltose	Glucose	Malt products; low concentrations in plants and processed foods
Starch	Glucose	Branches (amylopectin) contribute viscosity; linear (amylose) contributes gelling when gelatinized; granule is important to viscosity and gel formation
Dextrins	Glucose	Usually considered to be hydrolysis products of incompletely broken down starch fractions
Glycogen	Glucose	Meat products and seafood
Cellulose	Glucose	Comprises skeletal structure of plant cell; indigestible stable cell structural framework of stalks and leaves of vegetables, fruits, and coverings of seeds.
Hemicellulose	Glucose	Makes up some plant skeletal structure; amorphous heterogeneous substance; pentose and uronic acid predominate
Pectic substances	Galactose	Cell cementing compound; fruits and vegetables; pectin will form gel with appropriate concentration, amount of sugar, and pH. Amorphous substances in the matrix of plant skeletal structure; contains minor amounts of neutral monomers such as arabinose, amylose, galactose, mannose
Malin	Fructose	Matrix

(*Continues*)

TABLE 3-2 Carbohydrates and Characteristics

NAME/ CLASSIFICATION	END PRODUCTS (HYDROLYSIS)	SOURCE, FUNCTION, OR CHARACTERISTICS
Galactogens	Galactose	Monomers such as arabinose, xylose, mannose, raffinose
Mannosans	Mannose	Polymers of mannose found in plants
Raffinose	Glucose fructose galactose	Cottonseed meal, sugar beets, and molasses
Pentosans	Pentoses	Found with cellulose in woody plants

fructose, and galactose are three common hexoses. Ribose and deoxyribose are two common pentoses. Figure 3-1 shows some of the structure of some of the monosaccharides.

Disaccharides

Two monosaccharides may be linked together to form a disaccharide. Sucrose is the most common disaccharide and composed of one molecule each of glucose and fructose. Sucrose is commonly referred to as *sugar*. Lactose is the major sugar in milk and is made up of one molecule of glucose and one of galactose. Maltose is a disaccharide made from two molecules of glucose (Table 3-2). This linkage is formed by the removal of water (dehydration), and it can be broken by adding water back (**hydrolysis**).

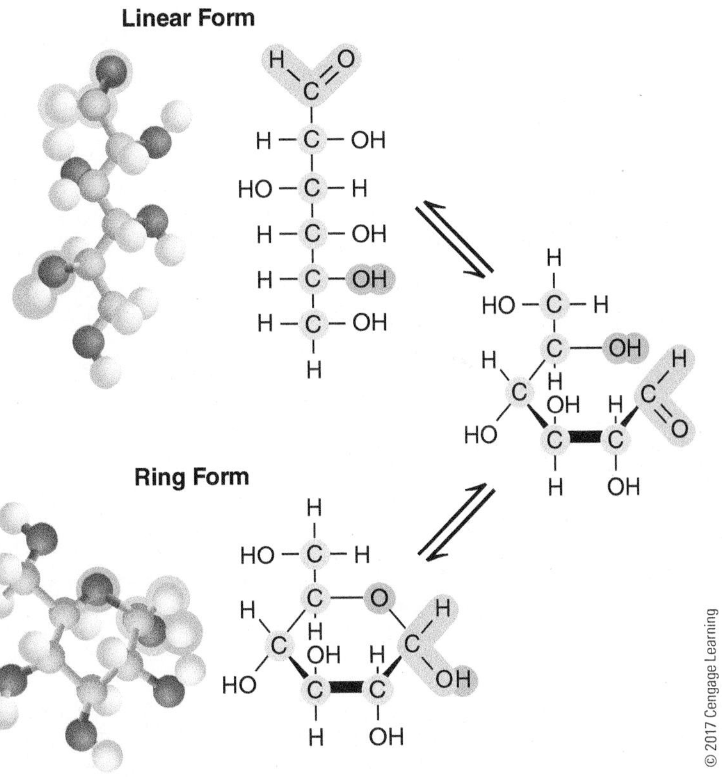

Linear Form

Ring Form

FIGURE 3-1 Representation of monosaccharides.

SCIENCE CONNECTION!

Draw the structural formulas for different monosaccharides, disaccharides, and polysaccharides.

TABLE 3-3 Relative Sweetness of Certain Sugars

SUGAR	RELATIVE SWEETNESS
Fructose	174
Invert Sugar	126
Sucrose	100
Glucose	74
Maltose	32
Galactose	32
Lactose	16

Not all sugars have the same sweetness. A cola-type soft drink has between 10% and 12% sugars. Depending on the formulation, the sugar might be all sucrose or a blend of sucrose, glucose, and fructose. Milk, on the other hand, contains a little less than half this much sugar (lactose) and is not sweet. Table 3-3 compares the sweetness of some common sugars.

Sugars in Food

Color, texture, and flavor are all sensory characteristics that sugar supplies to most foods. The study of sugars can be approached from their chemical structure, properties, characteristics, and variety or source. Of the three sensory characteristics, sugars generally play a major role as a sweetener or in developing texture. As a contributor to color, sugar participates in two phenomena: the **Maillard reaction**, a browning reaction between an amino group and a reducing group of a carbohydrate, and **caramelization**.

Honey, sorghum, molasses, maple syrup, and selected fruit juices and pulps serve as sweetener substitutes for cane sugar and sugar beet sugar. Sugar-based sweeteners are those developed from corn starch. Processing the cane and beet sugars in the United States produces a granulated sugar, a brown sugar, and liquid sugars. The sugar's source and type affect sweetness and their interactive functioning.

Acid will hydrolyze and invert a disaccharide sugar into its component monosaccharides. Any product with an acid compound can hydrolyze sucrose into fructose and glucose. Glucose and fructose are more soluble and more hygroscopic than sucrose, and they enhance browning.

Inversion of sugars refers to the hydrolysis of sucrose into fructose and glucose to form these sugars, which are sometimes referred to as *invert sugars.* This inversion is thought to take place because of the presence of either an enzyme or an acid.

SCIENCE CONNECTION!

Conduct an enzymatic browning lab using apple slices or bananas. Test different liquids and rate their effects on the browning of the fruit; for example plain water, lemon juice, different dilutions of vinegar or a soda.

MATH CONNECTION!

Compare sweetness between different products—maple syrup, corn syrup, honey, cola—and rate the levels from 1 to 10. After rating, compare nutritional labels and what type of sugar is listed on each ingredient list.

Caramelization. Caramelization is the application of heat to the point that sugars will dehydrate, break down, and polymerize (Figure 3-2).

Although a relatively complex reaction, caramelization can be as simple as making peanut brittle. Researchers attribute the caramelization reaction to a range of browning reactions and flavor development; making peanut brittle is an example. Once the melting point has been reached, sugars will caramelize and turn brown. Each sugar has its own caramelization temperature.

Crystallization. Crystallization of sugar can be a problem in a variety of products. For example, if too much milk solid is added to a frozen dessert, a gritty texture results because of lactose crystals. Crystallization of lactose is not desirable for the manufacture of powdered milk. When a lactose solution (milk) is rapidly dried, it does not have time to crystallize and instead forms a type of glass. Lactose glass exists in milk powders and causes clumping. The clumping is desirable because it results in a milk powder that dissolves instantly in water. Crystallization of sugar is a major process used in the candy-manufacturing industry (see Chapter 23).

Candies can be divided into two groups: crystalline and noncrystalline. Crystalline candies include fudge, fondant, and other candies that have crystals as important structural components. Although divinity is a crystalline candy, it is a special case because its crystals are dispersed in a foam. Noncrystalline candies include caramels, brittles, taffies, marshmallows, and **gum** drops. Marshmallows and gum drops are also special classes of candies because they contain a gelling substance.

SCIENCE CONNECTION!

Find a recipe online for making homemade marshmallows. What ingredient do you think makes the marshmallows gel-like? Look on YouTube to find a video on how to make marshmallows. Describe the process.

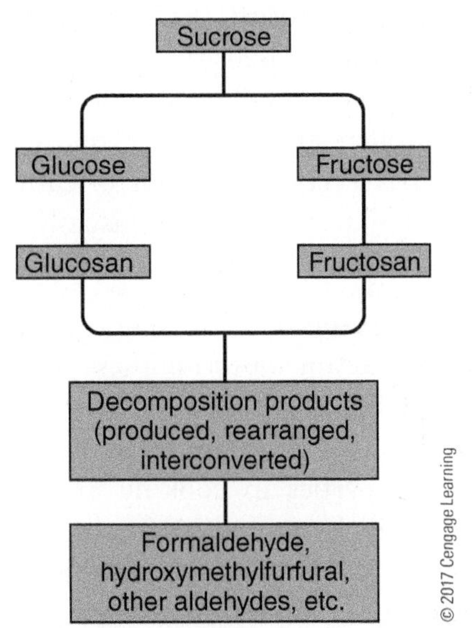

FIGURE 3-2 Representation of caramelization of sugars.

Crystallization is a complex process with many interrelated factors. The nature of the crystallizing substance is important for crystallization. The rate of crystallization is the speed at which nuclei grow into crystals. This rate depends on the concentration of the solute in the solution: A more concentrated (more supersaturated) syrup will crystallize more rapidly than a less concentrated syrup. At a higher temperature, the rate of crystallization is slow; the rate becomes more rapid at a lower temperature. Agitation distributes the crystal forming nuclei and hastens crystallization. Impurities in a solution usually delay crystallization and in some cases—caramels, for example—it may prevent crystal formation. Fats and proteins also decrease the number and size of crystals.

SCIENCE CONNECTION!

Using glass jars and string, set up a lab to form sugar or salt crystals in jars. Create a salt solution and a sugar solution and see how long it takes for crystals to form.

Polysaccharides

Combinations of more than two sugars are often referred to as **oligosaccharides**. If they are extremely large, they are called *polysaccharides*. Raffinose and stachyose are two oligosaccharides of interest because they are hard to digest. Raffinose contains one molecule each of glucose, fructose, and galactose. Stachyose is similar to raffinose but contains two molecules of galactose.

Polysaccharides may be added to foods for a variety of reasons. Nutritionally, they are generally added to increase dietary fiber content. Functionally, polysaccharides are added to thicken a substance, form gels, bind water, and stabilize proteins. Starch is the most common polysaccharide added to food products. For some uses, starch may be chemically modified to improve stability or alter its functional properties. Cellulose and cellulose derivatives are also added to many food products. The term *gum* is used to describe some of the naturally occurring polysaccharides added to food.

Naturally occurring polysaccharides added to foods include agar, gum tragacanth, algin, locust bean (carob) gum, carrageenan, starch, cellulose, pectin, guar gum, xanthan gum, and gum arabic.

Starch. Starch is a polysaccharide made up of glucose units linked together to form long chains. The number of glucose molecules joined in a single starch molecule varies from 500 to several hundred thousand, depending on the type of starch.

Starch is the storage form of energy for plants. Glycogen is the storage form of energy for animals. The plant directs the starch molecules to the amyloplasts, where they are deposited to form granules. In plants and in an extracted concentrate, starch exists as granules and varies in diameter from 2 to 130 microns (one micron is one millionth of a meter—unimaginably small). The size and shape of the granule is characteristic of the plant from which it came and serves as a way to identify the source of a particular starch. The structure of the grain granule is crystalline, with the starch molecules orienting in such a way as to form radially oriented crystals (Figure 3-3).

Two types of starch molecules exist: amylose and amylopectin. Amylose averages 20% to 30% of the total amount of starch in most native starches. Some starches—waxy cornstarch, for example—only contains amylopectin. Others may only contain amylose.

Amylose molecules contribute to gel formation because their linear chains can orient parallel to each other, becoming close enough to bond. Because of the ease with which they can slip past each other in the cooked paste, they do not contribute significantly to viscosity. The branched amylopectin molecules give viscosity to the cooked paste, partially because of the role it serves in maintaining the swollen granule. Their side chains and bulky shapes keep them from orienting closely enough to bond, so they do not usually contribute to gel formation.

Different plants have different relative amounts of amylose and amylopectin. These different proportions of the two types of starch within the starch grains of a plant give each starch its characteristic properties in cooking and gel formation. Starch in its processed commercial form is composed of starch grains or granules with most of the moisture removed. It is insoluble in water. When put in cold water, the grains may absorb a

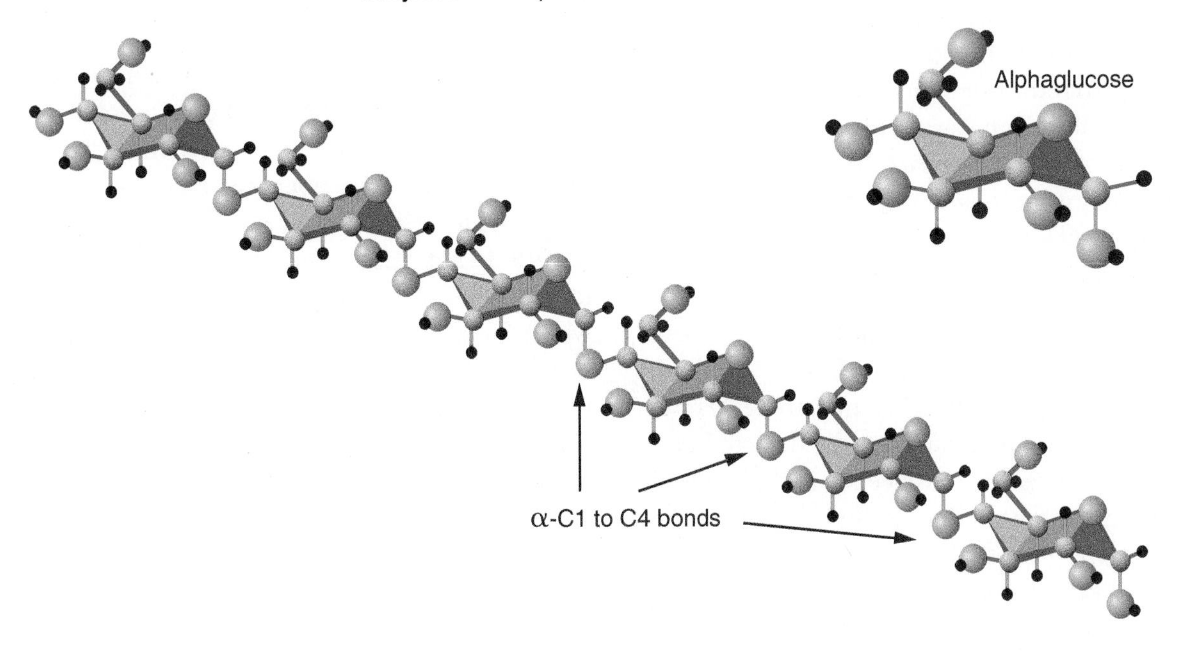

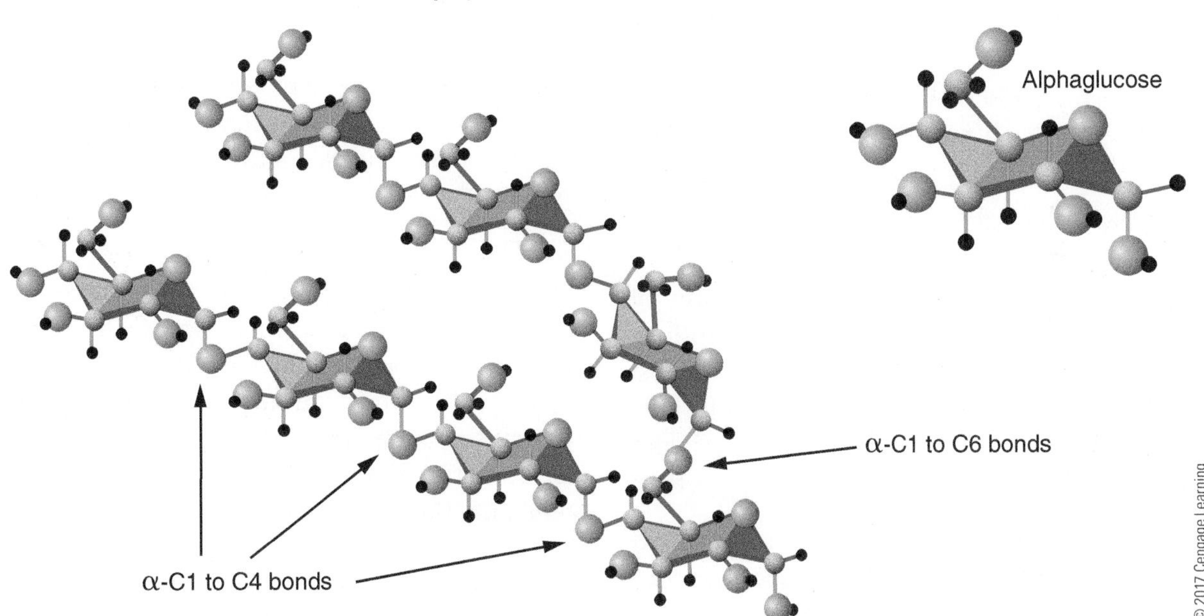

FIGURE 3-3 Straight chain of the starch amylose and the branched chain of the starch amylopectin.

small amount of the liquid. Up to 140 °F to 158 °F (60 °C to 70 °C) the swelling is reversible, the degree of reversibility depending on the particular starch. With higher temperatures, an irreversible swelling called **gelatinization** begins.

Starch begins to gelatinize between 140 °F and 158 °F (60 °C and 70 °C); the exact temperature depends on the specific starch. For example, different starches exhibit different granular densities; this affects how easily these granules can absorb water.

Because loss of **birefringence**—an optical property—occurs at the time of initial rapid gelatinization (swelling of the granule), loss of birefringence is a good indicator of the initial gelatinization temperature of a given starch. The largest granules, which are usually less compact, begin to swell first. Once optimal gelatinization of the grains has occurred, unnecessary agitation may fragment swollen starch grains and cause the paste to thin.

Gelatinization range refers to the temperature range over which all granules are fully swollen. This range is different for different starches. However, this gelatinization can often be detected and observed by increased translucency and increased viscosity. This is the result of water being absorbed away from the liquid phase and into the starch granule. If a typical starch paste is allowed to stand undisturbed, intermolecular bonds begin to form, causing the formation of a semirigid structure or gel. This gel is a structure of amylose molecules bonded to one another and only slightly to the branches of amylopectin molecules within the swollen granule. This phenomenon is sometimes called *retrogradation*.

Cellulose. Cellulose is the most common polysaccharide and the major component of plant cell walls. Cellulose is a **polymer** (long chain) of glucose molecules linked together by one to four linkages. Because it cannot by digested by humans, cellulose is a major component of dietary fiber. Pectin is a polymer of galacturonic acid and is also not digested. In plants, pectin "cements" cells together.

Complex carbohydrates that cannot be digested are generally called *fiber*. In the past, fiber was considered to be a non-nutritive substance. The Food and Drug Administration (FDA) recognizes label claims that a diet high in fiber may offer protection against some forms of cancer, especially of the large intestine. Dietary fiber is found in fruits, vegetables, whole grains, and legumes. Dietary fiber may be best known for helping to prevent or relieve constipation. However, dietary fiber can also help maintain a healthy weight as well as lower the risk of heart disease. Fiber is not absorbed by the body like proteins and carbohydrates but passes through the body.

Fiber is divided into two categories: soluble and insoluble. **Soluble fiber** can dissolve in water and can help lower blood-level cholesterol and glucose levels. Soluble fibers can be found in apples, citrus fruits, oats, beans, and peas. **Insoluble fibers** do not dissolve in water and can help move food particles through the digestive system. Whole wheat flour, nuts, green beans, and potatoes are all sources of insoluble fiber. Many plant-based foods can contain both soluble and insoluble fibers. Naturally occurring fibers are generally better for the body than processed foods with added fiber.[3]

PROTEINS

Proteins are polymers of **amino acids**, which are sometimes called the *building blocks of protein*. Dietary protein is supplied from both plant and animal sources. Proteins are needed to build and repair body tissues and as part of our bodies' metabolic functions. Figure 3-4 shows different proteins.

The shape and thus the function of a protein is determined by the sequence of its amino acids. Proteins must be broken down (hydrolyzed) to amino acids before they can be used. Once absorbed in the body, amino acids are used to make proteins and are converted to energy or stored as fat. About 20% of the human body is made of protein. Functions of proteins include:

- Enzymes (e.g., trypsin and pepsin)
- Storage (e.g., ovalbumin and ferritin)
- Transport (e.g., hemoglobin and lipoproteins)
- Contractile contraction (e.g., actin and myosin)
- Protective (e.g., antibodies and thrombin)
- Hormones (e.g., insulin and growth hormone)
- Structural (e.g., keratin, collagen, and elastin)
- Membranes

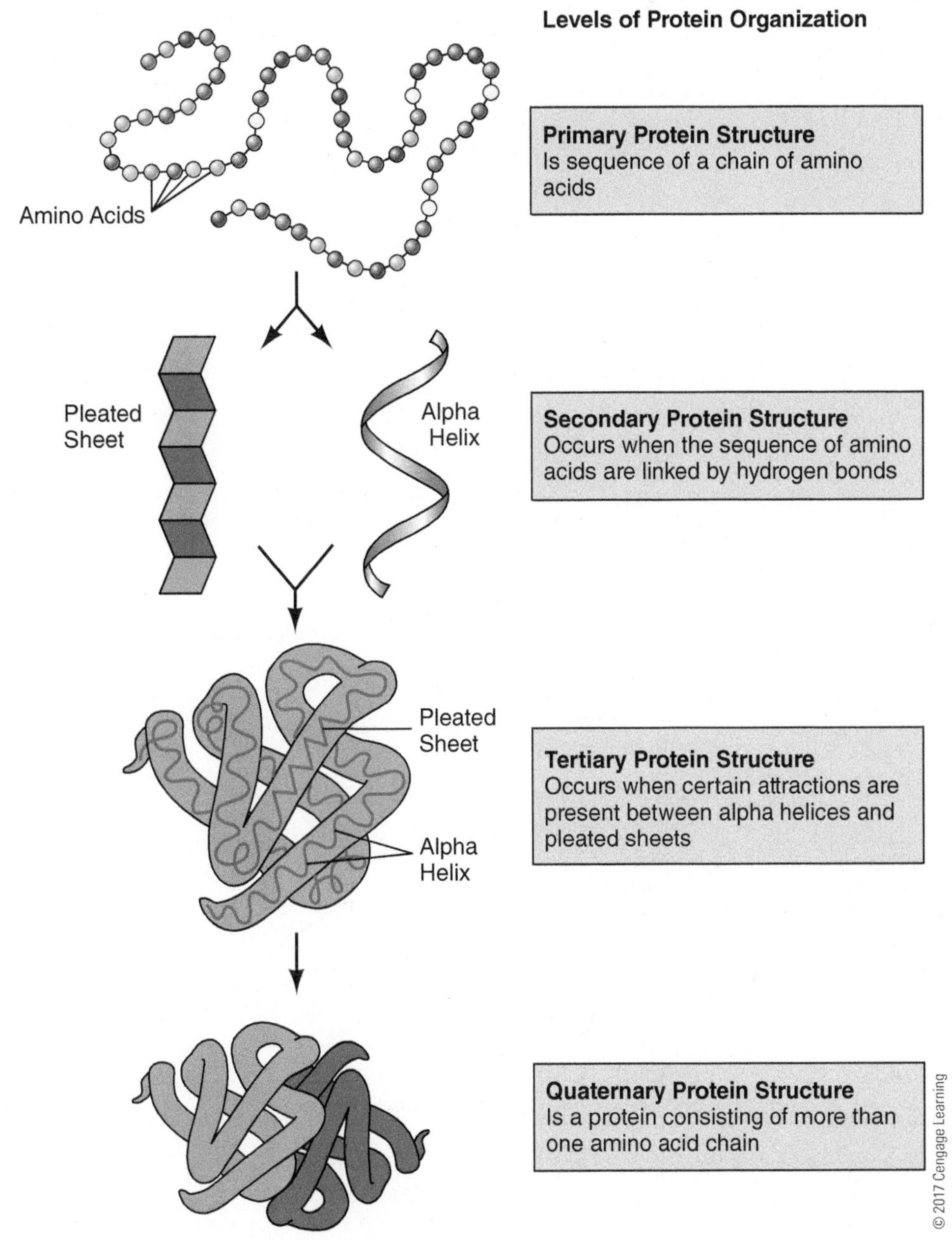

FIGURE 3-4 Four elements of protein structure.

Amino acids contain an amino group ($-NH_2$) and an acid group ($-COOH$). A total of 20 amino acids are found in proteins. Some of the 20 amino acids are shown in Figure 3-5.

Amino acids join by forming **peptide bonds**. A peptide bond is formed by the condensation of the amino group ($-NH_2$) of one amino acid with the acid group ($-COOH$) of another amino acid; loss of water molecules (H_2O) is the result. Condensation reactions involve the removal of water and the formation of a bond. The reversal of this process is *hydrolysis*, which involves the addition of water. Peptide bonds are not easily broken. Cooking would not normally result in the breaking of peptide bonds to yield amino acids from proteins.

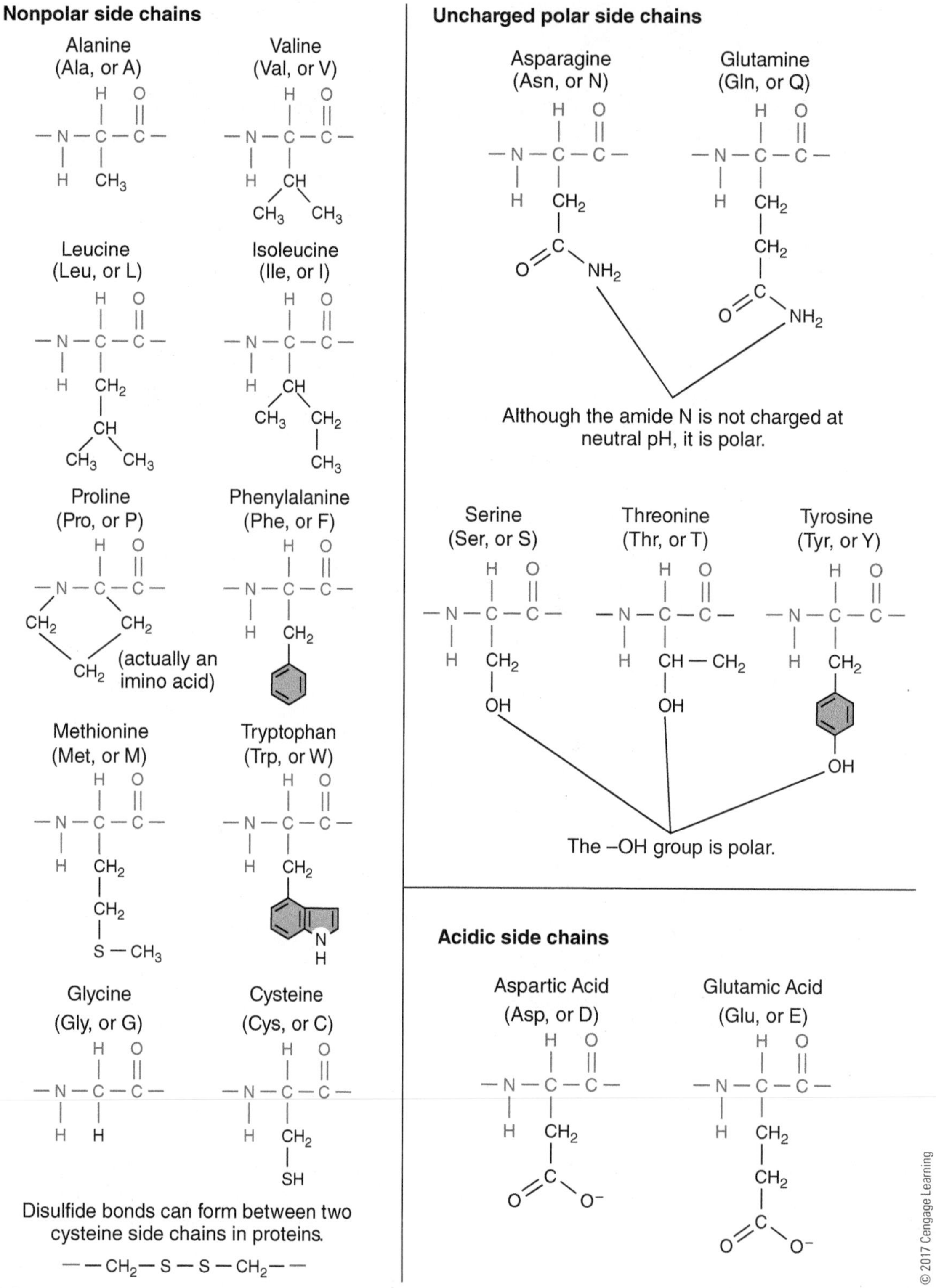

Nonpolar side chains

Alanine (Ala, or A)

Valine (Val, or V)

Leucine (Leu, or L)

Isoleucine (Ile, or I)

Proline (Pro, or P) (actually an imino acid)

Phenylalanine (Phe, or F)

Methionine (Met, or M)

Tryptophan (Trp, or W)

Glycine (Gly, or G)

Cysteine (Cys, or C)

Disulfide bonds can form between two cysteine side chains in proteins.

—CH_2—S—S—CH_2—

Uncharged polar side chains

Asparagine (Asn, or N)

Glutamine (Gln, or Q)

Although the amide N is not charged at neutral pH, it is polar.

Serine (Ser, or S)

Threonine (Thr, or T)

Tyrosine (Tyr, or Y)

The –OH group is polar.

Acidic side chains

Aspartic Acid (Asp, or D)

Glutamic Acid (Glu, or E)

FIGURE 3-5 Representation and classification of essential amino acids.

Amino acids form proteins because of the reaction between the amino group of one amino acid and the carboxyl group of another. The conformation of a protein molecule in the native state is determined by its primary structure, secondary structure, and tertiary structure.

Primary Structure. The primary structure of a protein molecule is the order of amino acids in a chain. This is the combination of amino acids in a proper sequence as determined by their peptide bonds. No other force or bond is implied by this structural-level designation.

Secondary Structure. The secondary structure of a protein molecule is the shape of the amino acid chains. They form a pleated or helix structure. The alpha helix is stabilized by hydrogen bonding between carboxyl and amide groups of the peptide bonds, which generally appear in a regular sequence along the amino acid chain.

Tertiary Structure. The tertiary structure of a protein molecule refers to the three-dimential structure of an amino acid chain. This is the folding of the coiled chain or chains. Covalent and hydrogen bonds and van der Waals forces may be involved in the structural organization of protein molecules.

SCIENCE CONNECTION!

Working in groups, research different amino acids assigned to you by your instructor. Share your findings with the class.

Functions of Proteins in Foods

Proteins are *amphophilic*—that is, they have polar (*hydrophilic*) and nonpolar (*hydrophobic*) side chains in one molecule, forming interfacial films similar to small molecular synthetic emulsifiers.

Proteins are fundamental food components, both functionally and nutritionally. A basic understanding of protein structure and characteristics is critical to understanding how these function in foods. The actual characteristics of proteins will be influenced if they can be whipped, beaten, or heated, with or without additional ingredients, to form a type of food. The characteristics of the proteins will influence how they will behave in colloidal systems and how they will contribute to the color, texture, and flavor of foods.

Color. The role of protein in color of foods is not clear-cut. In most instances, a protein can either play a role through its interaction or as part of a complex molecule. One of the biggest roles is seen in the Maillard reaction. The reactions between proteins and sugars is the most common of these browning reactions that occur in baked products and many other foods. Selected color pigments such as chlorophyll are bound in the chloroplasts in a protein–**lipid** matrix.

Texture. Proteins often contribute texture to foods. For example, custards are protein gels in which the strength of the gel is influenced by ovalbumin denaturation. Another clear example is in the production of yogurt. In this case, the texture of yogurt is influenced by the gelation of casein, a common milk protein.

Flavor. The contribution of proteins to flavor is not as clear-cut. Some proteins and amino acids may add flavor. Specifically, amino acids can contribute bitterness, sweetness, and other flavors.

LIPIDS

Lipids include fats and **oils** from both plants and animals. Cholesterol, however, is a fat found only in animal products. Lipids are of special interest because they are linked to the development of heart disease, the leading cause of death among Americans.

Lipids are substances in foods that are soluble in organic solvents. This category includes

triglycerides, fatty acids, phospholipids, some pigments, some vitamins, and cholesterol. Figure 3-6 illustrates the structure of some of the lipids found in food.

Role of Fats in Food

In food, fats provide a source of essential fatty acids, add caloric density (energy), act as carriers for flavors, carry fat-soluble vitamins, contribute

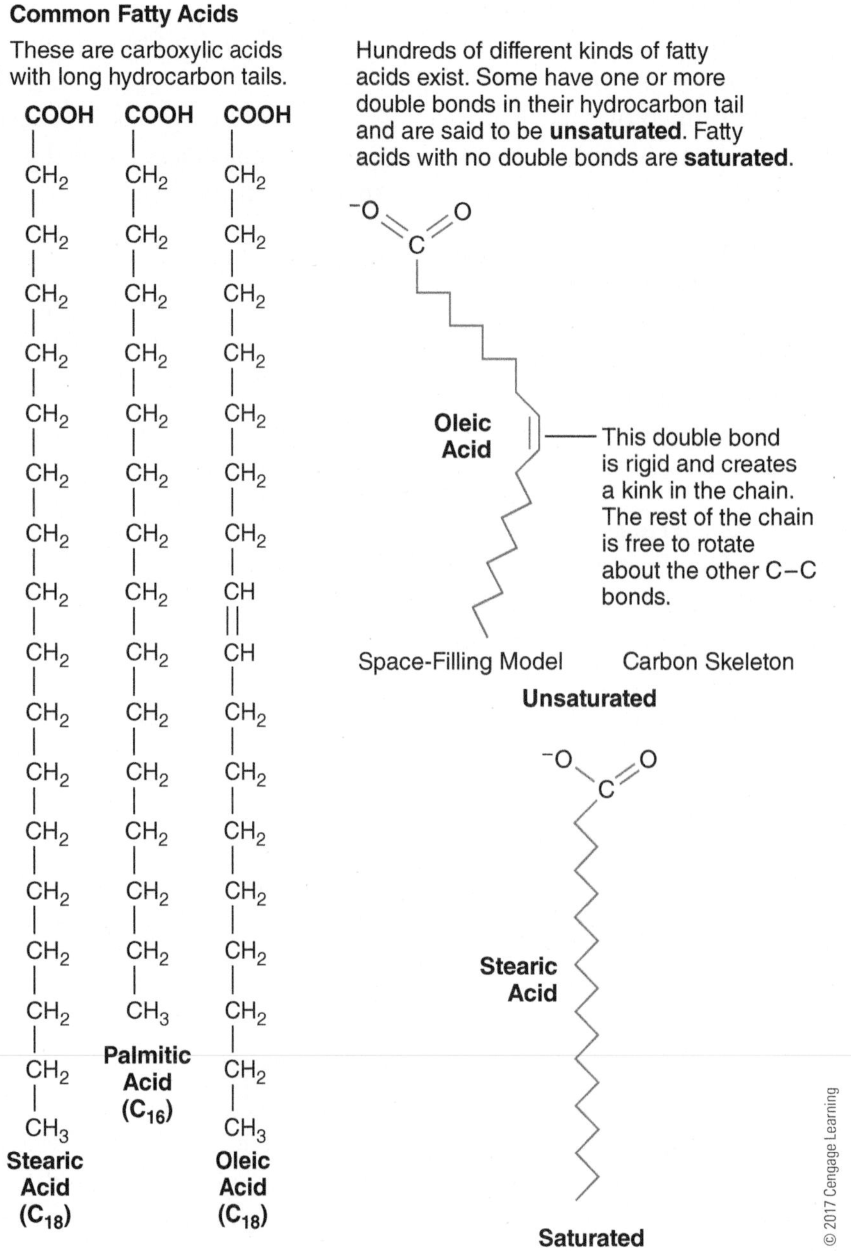

Common Fatty Acids

These are carboxylic acids with long hydrocarbon tails.

Stearic Acid (C_{18})	Palmitic Acid (C_{16})	Oleic Acid (C_{18})
COOH	COOH	COOH
CH_2	CH_2	CH_2
CH_2	CH_2	CH_2
CH_2	CH_2	CH_2
CH_2	CH_2	CH_2
CH_2	CH_2	CH_2
CH_2	CH_2	CH_2
CH_2	CH_2	CH_2
CH_2	CH_2	CH
CH_2	CH_2	‖ CH
CH_2	CH_2	CH_2
CH_2	CH_2	CH_2
CH_2	CH_2	CH_2
CH_2	CH_2	CH_2
CH_2	CH_2	CH_2
CH_2	CH_3	CH_2
CH_2		CH_2
CH_3		CH_3

Hundreds of different kinds of fatty acids exist. Some have one or more double bonds in their hydrocarbon tail and are said to be **unsaturated**. Fatty acids with no double bonds are **saturated**.

^{-}O C O

Oleic Acid

This double bond is rigid and creates a kink in the chain. The rest of the chain is free to rotate about the other C–C bonds.

Space-Filling Model   Carbon Skeleton

Unsaturated

^{-}O C O

Stearic Acid

Saturated

FIGURE 3-6 Representation of fatty acids.

TABLE 3-4 Fat Content of Some Foods

FOOD FAT	PERCENT FAT CONTENT
Oils and shortening	100
Butter and margarine	80
Most nuts	60
Peanut butter, bacon	50
Cheese, beef roasts	30–35
Franks	25–30
Lean pork, ice cream	12–14
Milk, shellfish	2–4

to texture and mouthfeel, become precursors of flavor, and provide a heat-transfer medium (especially through frying). See Table 3-4.

Fatty Acids

Naturally occurring fatty acids have even numbers of carbons. Short-chain fatty acids are important in odors. Longer-chain fatty acids are not volatile and do not contribute much to flavor, although their reaction products are especially important to the flavor of foods. Fatty acids may be **saturated** or **unsaturated** (Table 3-5).

TABLE 3-5 Characteristics of Fatty Acids

FATTY ACID	NUMBER OF CARBON ATOMS	MELTING POINT (°C)
Saturated Fatty Acids		
Butyric	4	−7.9
Caproic	4	−3.9
Caprylic	8	16.3
Capric	10	31.3
Lauric	12	44.0
Myristic	14	54.4
Palmitic	16	62.8
Stearic	18	69.6
Arachidic	20	75.4
Behenic	22	80.0
Lignoceric	24	84.2
Unsaturated Fatty Acids		
Palmitoleic	16	−0.5 to 0.5
Oleic	18	13
Linoleic	18	−5 to −12
Linolenic	18	−14.5
Arachidonic	20	−49.5

Double Bonds

Fatty acid molecules that are unsaturated contain what are known as *double bonds.* A fatty acid that contains one double bond is called *monounsaturated.* Fatty acids that contain two or more double bonds are called *polyunsaturated.* Unsaturated fatty acids can exist in two forms—*cis* and *trans*—depending on the arrangement of the portions of the fatty acid molecules around the double bonds. Naturally occurring fatty acids are in the cis conformation. The double bonds in lipid molecules are highly reactive with oxygen. The products of lipid **oxidation** have undesirable flavors; lipid oxidation leads to what is termed **rancidity**.

Some food additives function to inhibit the oxidation of food lipids. These molecules are called *antioxidants.* The antioxidants most commonly added to foods are:

- Butylated hydroxytoluene (BHT)
- Butylated hydroxy anisole (BHA)
- Vitamin C
- Vitamin E

Triglycerides

Food fats comprising three molecules of fatty acids connected to a molecule of glycerol are known as *triglycerides*. The vast majority of foods we consume contain fat in this form. Triglycerides can be broken apart by enzymes called *lipases*. The products of lipolysis often have soapy flavors. The food industry uses these products as emulsifiers. A triglyceride molecule that has had one fatty acid removed is called a *diglyceride*, and one that has had two fatty acids removed is called a *monoglyceride*. Both mono and diglycerides are used as emulsifiers.

Phospholipids

Some fatty acids are connected to glycerol molecules that contain a molecule of phosphorus. These special lipids are known as *phospholipids*. They play important roles in the body but are not essential nutrients because the body can synthesize them in adequate quantities. Probably the best known of the phospholipids is lecithin.

Cholesterol

Cholesterol is a compound produced by the body that has been strongly linked to heart disease. The reason cholesterol is not considered an essential nutrient is because the body can produce all the cholesterol it needs. The average American consumes from 400 to 800 mg/day of cholesterol and synthesizes from 1,000 to 2,000 mg/day. The more cholesterol that is consumed, the less the body produces and vice versa. This explains why it is so difficult to decrease serum cholesterol by dietary means alone. If you consume less cholesterol, your body produces more to keep the supply constant.

Cholesterol is used by the body for:

- Bile salts
- Membrane structure
- Myelin synthesis
- Vitamin D synthesis
- Steroid hormone synthesis

Some people have a genetic problem with the system that regulates cholesterol synthesis, and they produce excessive amounts. These people generally have greatly elevated serum cholesterol levels. This is of concern because high serum cholesterol is a risk factor for coronary heart disease.

VITAMINS

Vitamins are chemical compounds in our food that we need in tiny amounts (in milligrams and micrograms) to regulate chemical reactions in our bodies. They are divided into fat-soluble and water-soluble vitamins. Table B-8, in Appendix B, lists the vitamin content of common foods.

Fat-Soluble Vitamins

Fat-soluble vitamins include vitamins A, D, E, and K.

Vitamin A. Vitamin A occurs in a preformed state and as a precursor. Three active forms are retinol, retinal, and retinoic acid.

In food, most preformed vitamin A is found in the form of retinol. All three forms of vitamin A can be formed from plant pigments known as *carotenes*. The most common form is beta-carotene. Vitamin A is susceptible to oxidation but is relatively heat stable.

Vitamin D. The active form of vitamin D is cholecalciferol (vitamin D_3). It can be produced from cholesterol by the action of ultraviolet light. It can also be formed from a protovitamin. It is stored in the liver and functions in the absorption of the minerals calcium and phosphorus. Vitamin D also acts directly on bone and affects the reabsorption of calcium and phosphorus by the kidney.

Vitamin E. Vitamin E or alpha tocopherol is widely available in a normal diet. It functions to detoxify oxidizing radicals that arise in metabolism, to stabilize cell membranes, to regulate oxidation reactions, and to protect vitamin A and polyunsaturated fatty acids from oxidation.

Vitamin K. Dietary and intestinal bacterial sources contribute to the supply of vitamin K. It functions in normal blood clotting. Storage in the body is minimal.

Water-Soluble Vitamins

The water-soluble vitamins include the B vitamins and vitamin C. The B vitamins include thiamin, riboflavin, niacin, vitamin B_6, pantothenic acid, folic acid, **biotin**, and cobalamin (vitamin B_{12}).

Thiamin. Thiamin functions in carbohydrate metabolism. It makes ribose to form ribonucleic acid (RNA), and it maintains normal appetite and normal muscle tone in the digestive tract.

Riboflavin. Riboflavin functions as part of a coenzyme involved in oxidation–reduction reactions in energy production.

Niacin. Niacin functions as a component of two coenzymes involved in oxidation–reduction reactions that release energy from food.

Vitamin B_6. The functions of vitamin B_6 include the metabolism of amino acids and the conversion of glycogen to glucose.

Pantothenic Acid. Pantothenic acid is a part of coenzyme A, which is involved in synthesis and the breakdown of fats, carbohydrates, and proteins. It is also part of the enzyme known as *fatty acid synthetase.*

Folic Acid. The coenzyme form of folic acid is tetrahydrofolic acid. It functions in the transfer of formyl and hydroxymethyl groups. Folic acid is required for the synthesis of purines and pyrimidines and for the efficient use of the amino acid histidine.

Biotin. Biotin functions in fatty acid synthesis.

Cobalamin. Cobalamin or vitamin B_{12} is required for nucleic acid synthesis, amino acid synthesis, blood cell formation, neural function, and growth. Cobalamin is found only in animal products.

Vitamin C. Ascorbic acid or vitamin C functions in wound healing, collagen synthesis, and iron absorption, and it acts as an antioxidant. Vitamin C is necessary for the conversion of proline to hydroxyproline and lysine to hydroxylysine. It is involved in iron absorption and the conversion of amino acids to neurotransmitters. It is the least stable of all vitamins. It oxidizes readily in light or air, when heated, and in alkaline solutions. Degradation is enhanced by the presence of iron and copper.

MINERALS

Minerals, which are also needed only in small amounts, have many different functions. Some minerals assist in the body's chemical reactions, and others help form body structures. Minerals are important for energy transfer and as an integral part of vitamins, hormones, and amino acids. Depending on the amount in the body, minerals in the diet are classified as **macrominerals** or **microminerals** (also sometimes called *trace minerals*). Table B-8, in Appendix B, provides the mineral content of some common foods. The seven macrominerals are:

1. Calcium (Ca)
2. Phosphorus (P)
3. Potassium (K)
4. Sodium (Na)
5. Chloride (Cl)
6. Magnesium (Mg)
7. Sulfur (S)

Calcium

Calcium is involved in **homeostasis**, the functions that maintain life, including blood clotting and muscle contraction. Calcium also makes up 35% of the bone structure.

Phosphorus

This mineral makes up 14% to 17% of the human skeleton. Phosphorus is required for many energy-transfer reactions and for the synthesis of some lipids and proteins.

Potassium

Potassium maintains the acid–base balance and **osmotic pressure** inside the cells.

Sodium

Sodium maintains the acid–base balance outside the cells and regulates the osmosis of body fluids. Sodium is also involved in nerve and muscle function.

Chloride

In the diet, chloride normally accompanies sodium as NaCl or salt. This is an important extracellular anion (negative charge) involved in acid–base balance and osmotic regulation. Chloride is an essential component of bile, hydrochloric acid, and gastric secretions.

Magnesium

More than half of the magnesium found in the body exists in the skeleton. Magnesium is an activator of many enzymes.

Sulfur

Sulfur is a component of many biochemicals in the body, including amino acids, biotin, thiamin, insulin, and chondroitin sulfate.

Microminerals important in nutrition include:

- Chromium (Cr)
- Cobalt (Co)
- Copper (Cu)
- Fluorine (F)
- Iodine (I)
- Iron (Fe)
- Manganese (Mn)
- Molybdenum (Mo)
- Nickel (Ni)
- Selenium (Se)
- Silicon (Si)
- Tin (Sn)
- Vanadium (V)
- Zinc (Zn)

Chromium

In 1959, chromium was shown to be the factor responsible for improved glucose tolerance in rats. Now evidence supports the contention that chromium is essential for humans. It is involved in glucose tolerance, the stimulation of fatty acid synthesis, insulin metabolism, and protein digestion.

Cobalt

Cobalt is a part of vitamin B_{12}. Microflora in the cecum and colon use dietary cobalt to make vitamin B_{12}.

Copper

Copper is essential for several copper-dependent enzymes.

Fluorine

Fluorine is involved in bone and teeth development.

Iodine

Iodine is essential for the production of the thyroid hormones. These hormones regulate basal metabolism.

Iron

In the body, about 60% of the iron is in the red blood cells and 20% is in the muscles.

Manganese

Manganese is necessary for carbohydrate and fat metabolism and for the synthesis of cartilage.

Molybdenum

Molybdenum is part of the enzyme xanthine oxidase.

Nickel

Nickel is associated with the protein nickeloplasmin. In 1970, it was shown to be essential for chickens.

Selenium

Selenium is essential for detoxification of certain peroxides that are toxic to cell membranes. Selenium is closely connected with vitamin E; they work together to scavenge free radicals.

Silicon

In 1972, silicon was shown to be essential in young chickens. It may be important for bone development but is required in minute amounts by humans.

Tin

Tin may be required by humans. It has been shown to have a growth-promoting effect in rats. No requirement is known for humans.

Vanadium

Vanadium was shown to be essential in rats in 1971. In chickens, it increased growth rate and increased hematocrit. Probably a small amount is required for humans.

Zinc

Zinc is a component of many enzymes. Although not specifically discussed, cadmium (Cd), boron (B), and aluminum (Al) are also considered microminerals.

STUDYING THE LARGEST KNOWN PROTEIN

The largest known single-chain protein is found in muscle cells and is referred to by two names: titin and connectin. This huge molecule helps maintain resting tension in muscle tissue and takes part in the contraction of muscle fibers.

A molecule of titin can be nearly 1 micron long (0.000001 meter or 0.00004 inch); which is bigger than some cells. Each molecule consists of about 30,000 amino acids (the basic building blocks of proteins).

Scientists have recently used "optical tweezers" to study titin by carefully stretching individual molecules. Optical tweezers move and trap incredibly tiny objects using light. They can hold objects as small as single cells (or even viruses) by shining a focused laser beam onto them. With optical tweezers, it is possible to trap and move living cells (or even internal parts of cells) without damaging them. The technique has even been used to insert new genes into cells.

Using optical tweezers, scientists found that a molecule of titin is something like a series of springs connected by looser chains, allowing it to stretch and return to its original shape easily.

To find out about how titin's structure was revealed, visit these Web sites:

- http://jcs.biologists.org/content/127/4/858.full.pdf
- https://en.wikipedia.org/wiki/Titin

WATER

From 50% to 60% of human body weight consists of water. In the body, water performs these important functions:

- Carries nutrients and wastes
- Maintains structure of molecules
- Participates in chemical reactions
- Acts as a solvent for nutrients
- Lubricates and cushions joints, the spinal cord, and a fetus (during pregnancy)
- Helps regulate body temperature
- Maintains blood volume

Dehydration occurs when water output exceeds intake. Signs of dehydration include dry skin, dry mucous membranes, rapid heartbeat, low blood pressure, and weakness. Humans require 7 to 11 cups (56 to 88 fluid ounces) per day. This is why we are told to drink eight glasses of water per day. The so-called 8 × 8 rule (eight 8-ounce glasses of water) is easy to remember.

Water sources for the body include:

- Water (100%)
- Fruits and vegetables (90% to 99%)
- Fruit juices (80% to 89%)
- Pasta, legumes, beef, and dairy (10% to 60%)
- Crackers and cereals (1% to 9%)

The exact water content of specific foods can be determined by using Table B-8 in Appendix B.

BIOTIN

Biotin is also known as vitamin H and coenzyme R. It is found primarily in the liver, kidney, and muscle. Biotin functions as an essential cofactor for four carboxylases that catalyze the incorporation of cellular bicarbonate into the carbon backbone of organic compounds. Biotin is routinely provided to individuals who are receiving total intravenous feeding, and it is incorporated into almost all nutritionally complete dietary supplements and infant formulas. In larger doses, biotin is also used to treat inborn errors of metabolism.

Biotin is widely distributed in food stuffs, but the amounts are small relative to other vitamins. Biotin deficiency is rare in the absence of total intravenous feedings without added biotin or the chronic ingestion of raw egg white.

CHOLINE

Choline, a dietary component of many foods, is part of several major phospholipids (including phosphatidylcholine, which is also called *lecithin*) that are critical for normal membrane structure and function. The major precursor of betaine, choline is used by the kidney to maintain water balance and by the liver as a source of methyl groups for methionine formation. Choline is also used to produce the important neurotransmitter acetylcholine. In the body, choline is mainly found in phospholipids such as lecithin (phosphatidylcholine) and sphingomyelin.

A choline deficiency in healthy humans is difficult to demonstrate. Choline and choline esters can be found in significant amounts in many foods consumed by humans. Some of the choline is added during processing (especially in the preparation of infant formula).

PHYTOCHEMICALS

Plants manufacture chemicals known as **phytochemicals** that have multiple functions. Some attract insects to encourage fertilization; others provide defenses against predators such as viruses and animals. Phytochemicals exhibit diverse physiologic and pharmacologic effects. Active derivatives extracted from leaves, stems, roots, flowers, and fruits of plants may be classified into three main categories:

1. Toxic with no discernible therapeutic uses (compounds such as pyrrolizidine alkaloids, nicotine, and hydrazine derivatives)

2. Toxic but useful for treatment of disease when used in controlled amounts or for defined clinical conditions (compounds such as morphine, digitalis, and vinca alkaloids)
3. Chemopreventative activity (compounds useful against diseases such as atherosclerosis, cancer, and diverticular disease)

Most active chemopreventative phytochemicals are high-molecular-weight fibers such as celluloids, pectins, lignins, and low-molecular-weight compounds such as carotenoids, dithiolthiones, flavanoids, indole carbinols, isothiocyanates, mono- and triterpenoids, and thioallyl derivatives.

The majority of phytochemicals that have chemopreventative activity have no clearly defined role as essential nutrients except for the vitamins (ascorbate, tocopherols). Although phytochemical deficiencies have not been identified, their low concentrations in the diet have been associated with increased risks for cancer, cardiovascular diseases, and diabetes. A variety of data suggest that the best way to obtain chemopreventative phytochemicals is to include increased quantities of fruits and vegetables in the diet. All plants are sources of high-molecular-weight fibers. Specific low-molecular-weight phytochemicals with chemopreventative activity are contained within a variety of plants.

Plants such as cabbage and broccoli are excellent sources of indoles, dithiolthiones, isothiocyanates, and chlorophyllins. Legumes (soybeans, peanuts, beans, and peas) contain flavanoids, isoflavanoids, and other polyphenols that act as antioxidants and estrogenic agonists and antagonists. Citrus fruits and licorice root contain mono- and triterpenes that act as antioxidants and cholesterol synthesis inhibitors, and they stifle the growth of rapidly dividing cells. Thioallyl derivatives are found in garlic, leeks, and onion, and they prevent thrombi formation, decrease cholesterol synthesis, and prevent DNA damage.

SUMMARY

Carbohydrates, proteins, and lipids contribute to nutrition and other functions in foods. For example, carbohydrates enhance flavor, contribute to texture, prevent spoilage, and influence color. The function of carbohydrates in foods to some extent depends on their type—monosaccharides, disaccharides, or polysaccharides. Starch is a polysaccharide whose characteristics depend on the type of plant producing the starch. Cellulose is a nondigestible polysaccharide that contributes to the characteristics of food and demonstrates some health benefits. Proteins in food can act as emulsifiers and also influence the color, flavor, and texture of food. Lipids contribute to the texture, flavor, and heat transfer of foods. Lipids also carry flavors and fat-soluble vitamins. Food provides the vitamins and minerals necessary for normal growth and health. Although not a nutrient, water is necessary as a solvent for all nutrients. Biotin, choline, and phytochemicals are nutrients that seem to have some health benefits but do not have clearly defined requirements.

REVIEW QUESTIONS

Success in any career requires knowledge. Test your knowledge of this chapter by answering the following questions or solving the following problems.

1. What is the chemical composition of a carbohydrate?
2. List the three functions of proteins in food.
3. What is the difference between a monosaccharide and a disaccharide?

4. Name five functions that carbohydrates play in foods.
5. Explain two functions of water in the body.
6. Identify how the following are classed: triglycerides, fatty acids, phospholipids, some pigments, some vitamins, and cholesterol.
7. Which fatty acid molecules contain double bonds?
8. If a fatty acid contains only one double bond, what is it called? If it contains two or more?
9. List the fat- and water-soluble vitamins.
10. What major phospholipid is critical for normal membrane structure and function, is used by the kidney to maintain water balance, and is used to produce the important neurotransmitter acetylcholine?
11. Name 10 minerals important in nutrition.

STUDENT ACTIVITIES

1. Develop a report or presentation on one of the beneficial phytochemicals.
2. Taste samples of food that are pure or almost pure protein, starch, and lipid. Describe the flavors and sensation.
3. Using food-composition tables (see Table B-8), analyze the approximate amounts of energy, fat, protein, vitamins, and minerals in the food items from your most recent meal.
4. Calculate how much you could reduce your monthly fat intake by switching from whole milk (3.25% fat) to 1% fat milk.
5. With teacher supervision, use a match to burn a potato chip. Explain why it burns so readily.

ADDITIONAL RESOURCES

- Amino Acids Reference Chart: http://www.sigmaaldrich.com/life-science/metabolomics/learning-center/amino-acid-reference-chart.html
- The Carb Count Game: http://www.aboutkidshealth.ca/en/justforkids/health/diabetes/pages/thecarbgame.aspx
- Biochem Gems: http://www.spongelab.com/game_pages/biochem_gems.cfm
- Vitamins and Minerals: http://www.cnn.com/FOOD/resources/food.for.thought/vitamins.minerals/

Internet sites represent a vast resource of information. Although those provided in this chapter were vetted by industry experts, you may wish to further explore the topics discussed in this chapter using a search engine such as Google. Keywords or phrases may include the following: *carbohydrates, fats, proteins, crystallization, gelatinization, caramelization, cellulose, fiber, vitamins, minerals, triglycerides, monosaccharides, disaccharides*. In addition, Table B-7 provides a listing of useful Internet sites that can be used as a starting point.

REFERENCES

Brody, J. E. 1981. *Jane Brody's nutrition book.* New York: Bantam Books.

Corriher, S. O. 1997. *Cookwise: The hows and whys of successful cooking.* New York: William Morrow & Co.

Ensminger, A. H., M. E. Ensminger, J. E. Konlande, and J. R. Robson. 1994. *Foods and nutrition encyclopedia* (2 vols.). Boca Raton, FL: CRC Press.

Potter, N. N., and J. H. Hotchkiss. 1995. *Food science,* 5th ed. New York: Chapman & Hall.

Vaclavik, V. A., and E. W. Christina. 1999. *Essentials of food science,* 4th ed. New York: Springer.

Vieira, E. R. 1996. *Elementary food science,* 4th ed. New York: Chapman & Hall.

ENDNOTES

1. Science Clarified: http://www.scienceclarified.com/everyday/Real-Life-Physics-Vol-3-Biology-Vol-1/Carbohydrates-How-it-works.html#ixzz3gwlJiQwR. Last Accessed July 25, 2015.
2. WebMD http://www.webmd.com/food-recipes/carbohydrates. Last Accessed June 15, 2015.
3. Mayo Clinic http://www.mayoclinic.org/healthy-lifestyle/nutrition-and-healthy-eating/in-depth/fiber/art-20043983?pg=1. Last Accessed June 15, 2015.

CHAPTER 4

Nutrition and Digestion

OBJECTIVES

After reading this chapter, you should be able to:

- **Identify nutritional needs using RDA or DRI**
- **Discuss the functions of energy, carbohydrates, fats, and proteins in the body**
- **Provide the caloric content of proteins, carbohydrates, fats, and alcohol**
- **List the essential amino acids**
- **Name two protein-deficiency diseases**
- **Describe protein quality**
- **Name an essential fatty acid**
- **List the water- and fat-soluble vitamins and their functions**
- **List six minerals required by the body**
- **Describe the process of digestion**
- **Identify the organs involved in digestion**
- **Discuss the relationship of diet to health**

KEY TERMS

absorption
bioavailability
coenzyme
digestion
Dietary Reference Intakes (DRI)
enzyme
essential amino acid
fiber
limiting amino acid
protein quality
Recommended Dietary Allowance (RDA)
stability
vegan

NATIONAL AFNR STANDARD

FPP.02

Apply principles of nutrition, biology, microbiology, chemistry and human behavior to the development of food products.

People require energy and must obtain it from food and certain essential nutrients. These nutrients are essential because the body cannot make them. Essential nutrients include vitamins, minerals, certain amino acids, and certain fatty acids. Foods also contain other components such as fiber that are important for health. Although each nutritional component serves a specific function in the body, all of them together are required for overall health. The digestive system breaks down food into nutrients for absorption.

NUTRIENT NEEDS

In the United States, the nutritional needs of the public are estimated and expressed in the **Recommended Dietary Allowances (RDAs)**. These were initially established during World War II to determine what levels of nutrients were required to ensure the nutrition of the population would be safeguarded in a time of possible shortages. The RDAs are established by the Food and Nutrition Board of the National Research Council (NRC), whose members come from the National Academy of Sciences (NAS), the National Academy of Engineering, and the Institute of Medicine.

The first RDAs were published in 1943 by the National Nutrition Program, a forerunner of the Food and Nutrition Board. Initially, the RDAs were intended as guides for planning and procuring food supplies for national defense. Now RDAs are considered to be the desired goals for the average daily amounts of nutrients that population groups should consume over specified periods of time.

In the judgment of the Food and Nutrition Board using available scientific knowledge, the RDAs are the daily intake levels of essential nutrients considered to meet the known nutrition needs of practically every healthy person. The NAS and NRC recognize that diets are more than combinations of nutrients and should satisfy social and psychological needs as well. As the needs for nutrients have been clearly defined, the RDA has been revised several times. We are currently working under the 10th edition published in 1989. The requirement for a nutrient is the minimum intake that will maintain normal functions and health. In practice, estimates of nutrient requirements are determined by several techniques, including the following:

- Collection of data on nutrient intake from apparently normal, healthy people
- Determinations of the amount of nutrient required to prevent disease states (generally epidemiological data)
- Biochemical assessments of tissue saturation or adequacy of molecular function

The Institute of Medicine has developed and published the **Dietary Reference Intakes (DRIs)**. The DRIs represent the most current scientific knowledge about the nutrient needs of a healthy population. Individual DRIs can be calculated, and this is an overall system that individuals can follow. Figure 4-1 represents the DRI.

WATER

Water is essential to life. Approximately 65% of an adult body is made up of water. Water is in our cells, blood, bones, teeth, and skin. The lack of water can cause death more quickly than the lack of any other nutrient. Every chemical reaction in the body requires water. Water also regulates body temperature, transports nutrients and wastes, and dissolves nutrients. The human body typically loses a quart of water on a daily basis, and this must be replaced on a continual basis.

An adult should drink eight glasses of water each day. However, individuals who are more active than typical adults need to consume extra water to compensate for any losses that occurs as a result of additional activities.

Energy

The carbohydrates, fats, and proteins in food supply energy, which is measured in *calories*. Carbohydrates

Dietary Reference Intakes (DRIs): Estimated Average Requirements
Food and Nutrition Board, Institute of Medicine, National Academies

Life Stage Group	Calcium (mg/d)	CHO (g/d)	Protein (g/kg/d)	Vit A (μg/d)[a]	Vit C (mg/d)	Vit D (μg/d)	Vit E (mg/d)[b]	Thiamin (mg/d)	Ribo-flavin (mg/d)	Niacin (mg/d)[c]	Vit B_6 (mg/d)	Folate (μg/d)[d]	Vit B_{12} (μg/d)	Copper (μg/d)	Iodine (μg/d)	Iron (mg/d)	Magnes-ium (mg/d)	Molyb-denum (μg/d)	Phos-phorus (mg/d)	Sele-nium (μg/d)	Zinc (mg/d)
Infants																					
0 to 6 mo																					
6 to 12 mo			1.0													6.9					2.5
Children																					
1–3 y	500	100	0.87	210	13	10	5	0.4	0.4	5	0.4	120	0.7	260	65	3.0	65	13	380	17	2.5
4–8 y	800	100	0.76	275	22	10	6	0.5	0.5	6	0.5	160	1.0	340	65	4.1	110	17	405	23	4.0
Males																					
9–13 y	1,100	100	0.76	445	39	10	9	0.7	0.8	9	0.8	250	1.5	540	73	5.9	200	26	1,055	35	7.0
14–18 y	1,100	100	0.73	630	63	10	12	1.0	1.1	12	1.1	330	2.0	685	95	7.7	340	33	1,055	45	8.5
19–30 y	800	100	0.66	625	75	10	12	1.0	1.1	12	1.1	320	2.0	700	95	6	330	34	580	45	9.4
31–50 y	800	100	0.66	625	75	10	12	1.0	1.1	12	1.1	320	2.0	700	95	6	350	34	580	45	9.4
51–70 y	800	100	0.66	625	75	10	12	1.0	1.1	12	1.4	320	2.0	700	95	6	350	34	580	45	9.4
> 70 y	1,000	100	0.66	625	75	10	12	1.0	1.1	12	1.4	320	2.0	700	95	6	350	34	580	45	9.4
Females																					
9–13 y	1,100	100	0.76	420	39	10	9	0.7	0.8	9	0.8	250	1.5	540	73	5.7	200	26	1,055	35	7.0
14–18 y	1,100	100	0.71	485	56	10	12	0.9	0.9	11	1.0	330	2.0	685	95	7.9	300	33	1,055	45	7.3
19–30 y	800	100	0.66	500	60	10	12	0.9	0.9	11	1.1	320	2.0	700	95	8.1	255	34	580	45	6.8
31–50 y	800	100	0.66	500	60	10	12	0.9	0.9	11	1.1	320	2.0	700	95	8.1	265	34	580	45	6.8
51–70 y	1,000	100	0.66	500	60	10	12	0.9	0.9	11	1.3	320	2.0	700	95	5	265	34	580	45	6.8
> 70 y	1,000	100	0.66	500	60	10	12	0.9	0.9	11	1.3	320	2.0	700	95	5	265	34	580	45	6.8
Pregnancy																					
14–18 y	1,000	135	0.88	530	66	10	12	1.2	1.2	14	1.6	520	2.2	785	160	23	335	40	1,055	49	10.5
19–30 y	800	135	0.88	550	70	10	12	1.2	1.2	14	1.6	520	2.2	800	160	22	290	40	580	49	9.5
31–50 y	800	135	0.88	550	70	10	12	1.2	1.2	14	1.6	520	2.2	800	160	22	300	40	580	49	9.5
Lactation																					
14–18 y	1,000	160	1.05	885	96	10	16	1.2	1.3	13	1.7	450	2.4	985	209	7	300	35	1,055	59	10.9
19–30 y	800	160	1.05	900	100	10	16	1.2	1.3	13	1.7	450	2.4	1,000	209	6.5	255	36	580	59	10.4
31–50 y	800	160	1.05	900	100	10	16	1.2	1.3	13	1.7	450	2.4	1,000	209	6.5	265	36	580	59	10.4

NOTE: An Estimated Average Requirement (EAR) is the average daily nutrient intake level estimated to meet the requirements of half of the healthy individuals in a group. EARs have not been established for vitamin K, pantothenic acid, biotin, choline, chromium, fluoride, manganese, or other nutrients not yet evaluated via the DRI process.

[a] As retinol activity equivalents (RAEs). 1 RAE = 1 μg retinol, 12 μg β-carotene, 24 μg α-carotene, or 24 μg β-cryptoxanthin. The RAE for dietary provitamin A carotenoids is two-fold greater than retinol equivalents (RE), whereas the RAE for preformed vitamin A is the same as RE.

[b] As α-tocopherol. α-Tocopherol includes *RRR*-α-tocopherol, the only form of α-tocopherol that occurs naturally in foods, and the 2*R*-stereoisomeric forms of α-tocopherol (*RRR*-, *RSR*-, *RRS*-, and *RSS*-α-tocopherol) that occur in fortified foods and supplements. It does not include the 2*S*-stereoisomeric forms of α-tocopherol (*SRR*-, *SSR*-, *SRS*-, and *SSS*-α-tocopherol), also found in fortified foods and supplements.

[c] As niacin equivalents (NE). 1 mg of niacin = 60 mg of tryptophan.

[d] As dietary folate equivalents (DFE). 1 DFE = 1 μg food folate = 0.6 μg of folic acid from fortified food or asa supplement consumed with food = 0.5 μg of a supplement taken on an empty stomach.

SOURCES: *Dietary Reference Intakes for Calcium, Phosphorous, Magnesium, Vitamin D, and Fluoride* (1997); *Dietary Reference Intakes for Thiamin, Riboflavin, Niacin, Vitamin E_6, Folate, Vitamin B_{12}, Pantothenic Acid, Biotin, and Choline* (1998); *Dietary Reference Intakes for Vitamin C, Vitamin E, Selenium, and Carotenoids* (2000); *Dietary Reference Intakes for Vitamin A, Vitamin K, Arsenic, Boron, Chromium, Copper, Iodine, Iron, Manganese, Molybdenum, Nickel, Silicon, Vanadium, and Zinc* (2001); *Dietary Reference Intakes for Energy, Carbohydrate, Fiber, Fat, Fatty Acids, Cholesterol, Protein, and Amino Acids* (2002/2005); and *Dietary Reference Intakes for Calcium and Vitamin D* (2011). These reports may be accessed via www.nap.edu.

FIGURE 4-1 Recommended Dietary Allowances (RDA) and Dietary Reference Intakes (DRIs).

Adapted from complete document, which is available at http://iom.nationalacademies.org

and proteins provide about 4 calories per gram. Fat contributes more than twice as much—about 9 calories per gram. Foods that are high in fat are also high in calories. However, many low fat or nonfat foods can also be high in calories.

Individual calorie needs vary by age and level of activity (Figure 4-2). Many older adults need less food than younger and more active individuals, in part because of decreased activity. People who are trying to lose weight and eating little food may need to select more nutrient-dense foods to meet their nutrient needs in a satisfying diet.

Carbohydrates

Sugars are carbohydrates. Dietary carbohydrates include complex carbohydrates such as starch and **fiber**. During **digestion**, all carbohydrates except fiber are broken down into sugars. Sugars and starches occur naturally in many foods that also supply other nutrients. Examples of these foods include milk, fruits, some vegetables, breads, cereals, and grains. Americans eat sugars in many forms, and most people like their taste.

Some sugars are used as natural preservatives, thickeners, and baking aids in foods; they are often added during processing and preparation or when the foods are eaten. The body cannot tell the difference between naturally occurring and added sugars because they are chemically identical.

Fiber

Fiber is found only in plant-derived foods such as whole-grain breads and cereals, beans and peas, and other vegetables and fruits. Individuals should choose a variety of foods daily because the types of fiber in food vary. Eating a variety of fiber-containing plant foods is important for proper bowel function. Fiber can also reduce symptoms of chronic constipation, diverticular disease, and hemorrhoids, and it may lower the risk for heart disease and some cancers. Some of the health benefits associated with a high-fiber diet may come

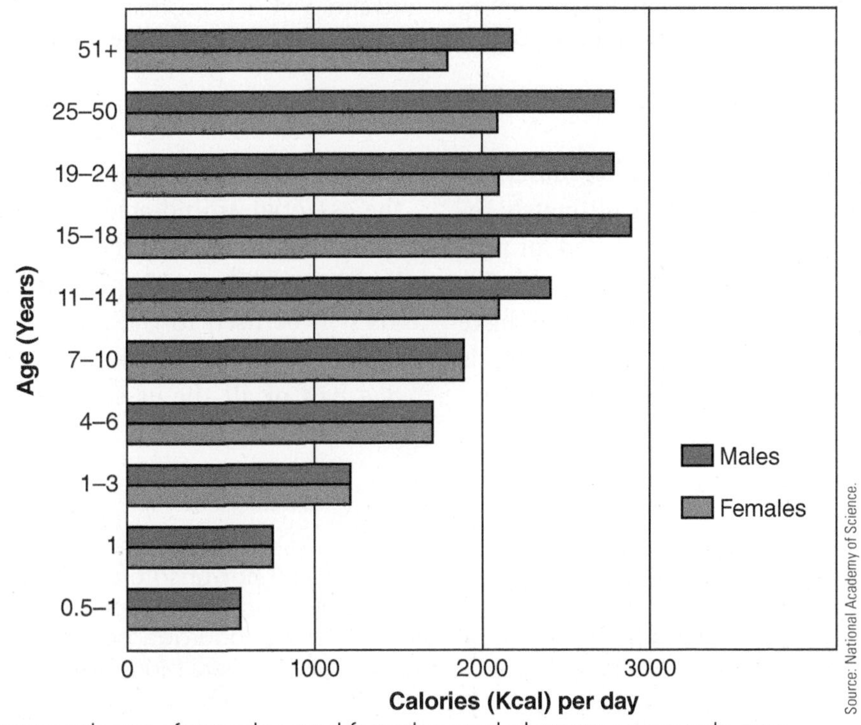

FIGURE 4-2 Energy needs vary for males and females and change as people age.

from other components present in these foods, not just from fiber itself. For this reason, fiber is best obtained from foods rather than supplements.

Protein

Depending on their ages and gender, humans require different levels of protein in their diet.

Requirement for Protein. Humans need specific amino acids that the body cannot synthesize in the amounts needed. These are known as **essential amino acids**. They include:

- Phenylalanine
- Tryptophan
- Histidine
- Valine
- Leucine
- Isoleucine
- Lysine
- Methionine
- Threonine
- Arginine

Additional Needs for Nitrogen in Protein. The nitrogen in protein is also used for the synthesis of purines, pyrimidines, porphyrin in nucleic acids, adenosine triphosphate, hemoglobin, and cytochromes.

Protein Deficiencies. Protein deficiencies in the diet of people from developing countries often leads to dietary diseases such as kwashiorkor and marasmus.

- Kwashiorkor results from a protein deficient diet that contains sufficient calories. Common symptoms include a bloated belly and extreme apathy.
- Marasmus results from a deficiency of both protein and calories. Symptoms include extremely low body weight and muscle wasting.

SCIENCE CONNECTION!

Research different protein deficiencies. Find images and share what happens to people's bodies when they are deficient in particular proteins or are malnourished. What protein deficiencies are common around the world?

Protein Quality. Protein quality describes the nutritive value of a protein. It is ultimately related to providing the amino acids needed for protein synthesis. The body cannot make part of a protein until all amino acids are available to make it. If all but one amino acid is present, no synthesis can occur. The amino acid that is present in the lowest quantity is called the **limiting amino acid**.

For example, assume that a protein provides all the essential amino acids in optimal proportions except one. If 90% of the required amount of the limiting amino acid is present, then all amino acids will be used to the point that 90% of the required protein is synthesized. If only 50%, then one half of all the essential amino acids will be used for energy.

GOOD FATS

Fish contain high concentrations of a unique type of fat, omega-3 polyunsaturated fatty acids (PUFAs), specifically docosahexaenoic acid (DHA) and eicospentanoic acid (EPA). Because of the health benefits associated with omega-3 fatty acids, health professionals encourage people to eat foods that contain high concentrations. Results of studies suggest that EPA lowers blood fats (triglycerides) and decreases the chance of blood clots.

(Continues)

GOOD FATS (continued)

Foods rich in EPA include flaxseed, legumes, green leafy vegetables, fish, and other seafood. Other studies are being conducted on the anti-inflammatory effects of omega-3 fatty acids. Omega-3 fatty acids could prove beneficial for the prevention of rheumatoid arthritis and systemic lupus erythematosus. Omega-3 fatty acids also demonstrate the ability to prevent fatal heart arrhythmia.

For more information on omega-3 fatty acids, DHA, or EPA, search the Web or visit these Web sites:

- http://www.webmd.com/healthy-aging/omega-3-fatty-acids-fact-sheet
- http://flaxcouncil.ca/
- http://www.hsph.harvard.edu/nutritionsource/omega-3-fats/
- https://www.hsph.harvard.edu/nutritionsource/omega-3-fats/
- http://lpi.oregonstate.edu/mic/other-nutrients/essential-fatty-acids
- https://www.urmc.rochester.edu/encyclopedia/content.aspx?ContentTypeID=1&ContentID=3054

Measurement of Protein Quality. The Food and Agriculture Organization (FAO) of the United Nations has released a report recommending a new and advanced method for assessing the quality of dietary proteins. The report, *Dietary Protein Quality Evaluation in Human Nutrition*, recommends that the Digestible Indispensable Amino Acid Score (DIAAS) replace the Protein Digestibility Corrected Amino Acid Score (PDCAAS) as the preferred method of measuring protein quality.

The report recommends that more data be developed to support full implementation. In the meanwhile, protein quality should be calculated using DIAAS values derived from fecal crude protein digestibility data. Under the current PDCAAS method, values are "truncated" to a maximum score of 1.00, even if scores derived are higher.

Protein is vital to supporting the health and well-being of all animal populations, including humans. However, not all proteins are alike because they vary according to their origin (animal or vegetable), their individual amino acid composition and their level of amino acid bioactivity. *High-quality proteins* are those that are readily digestible and contain essential amino acids in quantities that correspond to human dietary requirements.

Using the DIAAS method, researchers are now able to differentiate protein sources by their ability to supply amino acids for use by the body. For example, the DIAAS method was able to demonstrate the higher bioavailability of dairy proteins when compared to plant-based protein sources. Data in the FAO report showed whole milk powder to have a DIAAS score of 1.22, higher than the DIAAS score of 0.64 for peas and 0.40 for wheat.

DIAAS determines amino acid digestibility, at the end of the small intestine, providing a more accurate measure of the amounts of amino acids absorbed by the body and the contributions of proteins to human amino acid and nitrogen requirements. PDCAAS is based on an estimate of crude protein digestibility determined over the total digestive tract, and values determined using this method generally overestimate the amount of amino acids absorbed. Some food products may claim high protein content, but because the small intestine does not absorb all amino acids the same, they are not providing the same contribution to a human's nutritional requirements.[1]

Complementary Relationships. By combining a protein that is deficient in a given amino acid with a protein that has an excess of that amino acid, the protein quality can be increased. Complementary groups include the following:

- Grains + milk products
- Grains + legumes
- Seeds + legumes

Factors Affecting Protein Use. Four factors affect protein use in the body:

1. Ratios of essential amino acids
2. Amount of protein in the diet
3. Physiological state of the subject
4. Digestibility

If a protein is of poor quality, can a person eat more of it to satisfy needs? This can only be done to a certain extent. Adults must obtain 4% of their calories from high-quality protein, whereas an infant requires that 8% of its calories be from high-quality protein. For example, corn would be an adequate source of protein for an adult but not for a growing child. A child fed corn as the sole source of protein would develop kwashiorkor. No matter how much food it was fed, it would not be able to meet its protein requirements.

Lipids

In food, lipids can be categorized by their physical state at room temperature. Lipids that are solid at room temperatures are referred to as *fats*. If lipids are liquid at room temperature, then they are referred to as *oils*. Lipids provide a source of essential fatty acids. Solid fats contain more saturated fats and trans fats than oils. Oils contain more monounsaturated and polyunsaturated fats. Fats are denser than oils and require more energy to liquefy. Lipids add caloric density (energy), act as carriers for flavors, carry fat-soluble vitamins, and contribute to texture and flavor.[2]

Essential Fatty Acids. The body can produce most of the fatty acids that it requires. It cannot make some fatty acids that contain double bonds. Given an eighteen carbon fatty acid containing two double bonds called linoleic acid, humans can synthesize all the other fatty acids they require. Thus, linoleic acid is considered as an essential nutrient. The requirements for this essential fatty acid are not well established, and most adults have large amounts of it stored in their adipose tissue. It is estimated that adults should consume about 1% and infants 2% of their calories from linoleic acid.

Vitamins

Table 4-1 lists the fat- and water-soluble vitamins and their functions.

TABLE 4-1 Functions of Some Vitamins

VITAMINS	FUNCTIONS
Fat-Soluble Vitamins	
Vitamin A	Growth and development of bone and epithelial cells; vision
Vitamin D	Absorption of dietary calcium and phosphorus
Vitamin E	Antioxidant in tissues
Vitamin K	Aids in blood clotting
Water-Soluble Vitamins	
Thiamin	**Coenzyme** (a compound necessary for the functioning of an enzyme) in energy metabolism
Riboflavin	Coenzyme in many enzyme systems
Niacin	Coenzyme for cell respiration; release of energy from fat, carbohydrates, and proteins
Vitamin C	Metabolism of amino acids, fats, lipids, folic acid, and cholesterol control; collagen formation
Vitamin B_{12}	Coenzyme for red blood cell maintenance and nerve tissue; carbohydrate, fat, and protein metabolism

TABLE 4-2 Functions of Some Minerals

MINERAL (REQUIREMENT)	SOME FUNCTIONS
Calcium	Bone mineral; blood clotting; nerve, muscle, and gland function
Phosphorus	Bone mineral; part of many proteins involved in metabolism
Iron	Part of hemoglobin and some enzymes; oxygen transport
Copper	Iron absorption, hemoglobin synthesis, skin pigments, collagen metabolism
Magnesium	Bone mineral, enzyme activator, energy metabolism
Sodium chloride, potassium, chloride	Tissue fluid pressure and acid-base balance, passage of nutrients and water into cells, nerve and muscle function
Zinc	Activator of many enzymes
Iodine	Thyroid function
Manganese	Synthesis of bone and cartilage components, cholesterol metabolism
Selenium	Removal of peroxides from tissues, enzyme activation

Minerals

Table 4-2 lists some of the macrominerals and microminerals and their functions.

MYPLATE

Since 1916, the U.S. Department of Agriculture (USDA) has provided nutritional guidelines to help individuals regulate their food intake. In 2011, the USDA released the MyPlate program as the new nutritional guide to replace the food guide pyramids (see Figure 4-3). This visual approach focuses on the five main food groups per meal and is not intended to be a daily guide. A key message of the new MyPlate message is to make half your plate fruits and vegetables.

Food Groups

- **Fruits.** Any fruit or 100% fruit juice counts as part of the fruit group. Fruits may be fresh, canned, frozen, or dried. Fruits can be whole, cut up, or pureed for consumption (see Table 4-3).
- **Vegetables.** Any vegetable or 100% vegetable juice counts as part of the vegetable group.

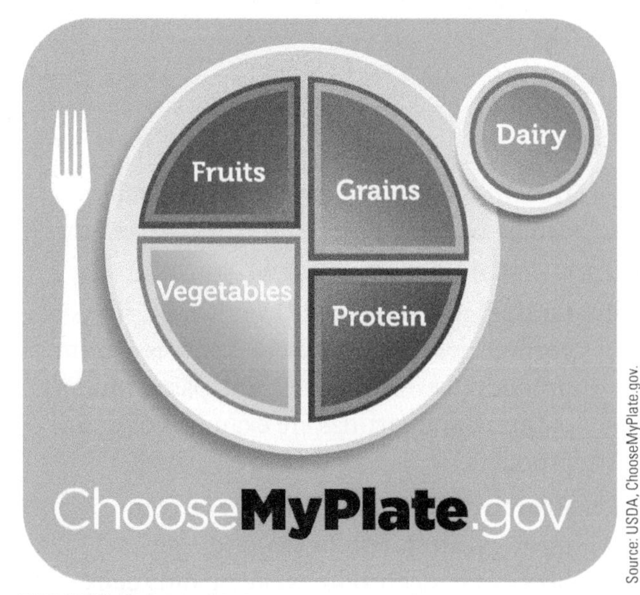

FIGURE 4-3 My Plate

TABLE 4-3 How Much Fruit Is Needed Daily?

The amount of fruit you need to eat depends on age, sex, and level of physical activity. Recommended daily amounts are shown in the chart.

DAILY RECOMMENDATION		
Children	2–3 years old	1 cup
	4–8 years old	1 to 1½ cups
Girls	9–13 years old	1½ cups
	14–18 years old	1½ cups
Boys	9–13 years old	1½ cups
	14–18 years old	2 cups
Women	19–30 years old	2 cups
	31–50 years old	1½ cups
	51+ years old	1½ cups
Men	19–30 years old	2 cups
	31–50 years old	2 cups
	51+ years old	2 cups

Source: USDA, ChooseMyPlate.gov

These amounts are appropriate for individuals who get less than 30 minutes per day of moderate physical activity, beyond normal daily activities. Those who are more physically active may be able to consume more while staying within calorie needs.

Vegetables may be raw, cooked, fresh, frozen, canned, or dehydrated. Vegetables can be whole, cut up, mashed, or pureed for consumption (see Tables 4a and 4b).

- **Grains.** Any food made from wheat, rice, oats, cornmeal, barley, or a cereal grain is a grain product (see Table 4-5). Bread, pasta, muffins, oatmeal, and tortillas are examples of grain products.
- **Protein Foods.** Any food made from meat, poultry, fish or seafood, eggs, processed soy products, nuts. and seeds are part of the protein food group. Beans and peas also provide protein even though they are considered vegetables (see Table 4-6).
- **Dairy.** All fluid milk products are considered part of the dairy food group. Many foods made from milk are also part of this group, as are foods made from milk that retain their calcium content—milk, yogurt, and cheeses, for example (see Table 4-7).

TABLE 4-4A How Many Vegetables Are Needed Daily?

The amount of Vegetables you need to eat depends on your age, sex, and level of physical activity. Recommended total daily amounts are shown in the first chart. Recommended weekly amounts from each Vegetable subgroup are shown in the second chart.

DAILY RECOMMENDATION		
Children	2–3 years old	1 cup
	4–8 years old	1½ cups
Girls	9–13 years old	2 cups
	14–18 years old	2½ cups
Boys	9–13 years old	2½ cups
	14–18 years old	3 cups
Women	19–30 years old	2½ cups
	31–50 years old	2½ cups
	51+ years old	2 cups
Men	19–30 years old	3 cups
	31–50 years old	3 cups
	51+ years old	2½ cups

 Source: USDA, ChooseMyPlate.gov

These amounts are appropriate for individuals who get less than 30 minutes per day of moderate physical activity, beyond normal daily activities. Those who are more physically active may be able to consume more while staying within calorie needs.

TABLE 4-4B How Many Vegetables Are Needed Weekly?

		DARK GREEN VEGETABLES	RED AND ORANGE VEGETABLES	BEAN AND PEAS	STARCHY VEGETABLES	OTHER VEGETABLES
		AMOUNT PER WEEK				
Children	2–3 yrs old	½ cup	2½ cups	½ cup	2 cups	1½ cups
	4–8 yrs old	1 cup	3 cups	½ cup	3½ cups	2½ cups
Girls	9–13 yrs old	1½ cups	4 cups	1 cup	4 cups	3½ cups
	14–18 yrs old	1½ cups	5½ cups	1½ cups	5 cups	4 cups
Boys	9–13 yrs old	1½ cups	5½ cups	1½ cups	5 cups	4 cups
	14–18 yrs old	2 cups	6 cups	2 cups	6 cups	5 cups
Women	19–30 yrs old	1½ cups	5½ cups	1½ cups	5 cups	4 cups
	31–50 yrs old	1½ cups	5½ cups	1½ cups	5 cups	4 cups
	51+ yrs old	1½ cups	4 cups	1 cup	4 cups	3½ cups
Men	19–30 yrs old	2 cups	6 cups	2 cups	6 cups	5 cups
	31–50 yrs old	2 cups	6 cups	2 cups	6 cups	5 cups
	51+ yrs old	1½ cups	5½ cups	1½ cups	5 cups	4 cups

Vegetable subgroup recommendations are given as amounts to eat weekly. It is not necessary to eat vegetables from each subgroup daily. However, over a week, try to consume the amounts listed from each subgroup as a way to reach your daily intake recommendation.

TABLE 4-5 How Many Grain Foods Are Needed Daily?

The amount of grain you need to eat depends on your age, sex, and level of physical activity. Recommended daily amounts are listed in the chart. Most Americans consume enough grains, but few are whole grains. **At least half of all the grains eaten should be whole grains.**

		DAILY RECOMMENDATION	DAILY MINIMUM AMOUNT OF WHOLE GRAINS
Children	2–3 years old	3 ounce equivalents	1½ ounce equivalents
	4–8 years old	5 ounce equivalents	2½ ounce equivalents
Girls	9–13 years old	5 ounce equivalents	3 ounce equivalents
	14–18 years old	6 ounce equivalents	3 ounce equivalents
Boys	9–13 years old	6 ounce equivalents	3 ounce equivalents
	14–18 years old	8 ounce equivalents	4 ounce equivalents
Women	19–30 years old	6 ounce equivalents	3 ounce equivalents
	31–50 years old	6 ounce equivalents	3 ounce equivalents
	51+ years old	5 ounce equivalents	3 ounce equivalents
Men	19–30 years old	8 ounce equivalents	4 ounce equivalents
	31–50 years old	7 ounce equivalents	3½ ounce equivalents
	51+ years old	6 ounce equivalents	3 ounce equivalents

Source: USDA, ChooseMyPlate.gov

These amounts are appropriate for individuals who get less than 30 minutes per day of moderate physical activity, beyond normal daily activities. Those who are more physically active may be able to consume more while staying within calorie needs.

TABLE 4-6 How Much Food from the Protein Foods Group is Needed Daily?

The amount of food from the protein foods group you need to eat depends on age, sex, and level of physical activity. Most Americans eat enough food from this group, but need to make leaner and more varied selections of these foods.

Recommended daily amounts are shown in the chart.

DAILY RECOMMENDATION		
Children	2–3 years old	2 ounce equivalents
	4–8 years old	4 ounce equivalents
Girls	9–13 years old	5 ounce equivalents
	14–18 years old	5 ounce equivalents
Boys	9–13 years old	5 ounce equivalents
	14–18 years old	6½ ounce equivalents
Women	19–30 years old	5½ ounce equivalents
	31–50 years old	5 ounce equivalents
	51+ years old	5 ounce equivalents
Men	19–30 years old	6½ ounce equivalents
	31–50 years old	6 ounce equivalents
	51+ years old	5½ ounce equivalents

These amounts are appropriate for individuals who get less than 30 minutes per day of moderate physical activity, beyond normal daily activities. Those who are more physically active may be able to consume more while staying within calorie needs.

Source: USDA, ChooseMyPlate.gov

TABLE 4-7 How Much Food from the Daily Group is Needed Daily?

The amount of food from the Daily Group you need to eat depends on age. Recommended daily amounts are shown in the chart below.

DAILY RECOMMENDATION					
Children	2–3 years old	2 cup	**Women**	19–30 years old	3 cups
	4–8 years old	2½ cups		31–50 years old	3 cups
Girls	9–13 years old	3 cups		51+ years old	3 cups
	14–18 years old	3 cups	**Men**	19–30 years old	3 cups
Boys	9–13 years old	3 cups		31–50 years old	3 cups
	14–18 years old	3 cups		51+ years old	3 cups

Source: USDA, ChooseMyPlate.gov

A BRIEF HISTORY OF USDA FOOD GUIDES

1916 to 1930s: "Food for Young Children" and "How to Select Food"

- Established guidance based on food groups and household measures
- Focus was on "protective foods"

1940s: A Guide to Good Eating (Basic Seven)

- Foundation diet for nutrient adequacy
- Included daily number of servings needed from each of seven food groups
- Lacked specific serving sizes
- Considered complex

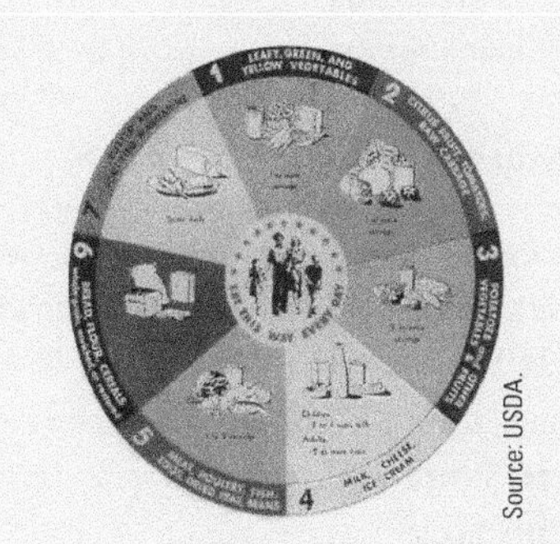

Source: USDA.

1956 to 1970s: Food for Fitness, A Daily Food Guide (Basic Four)

- Foundation diet approach—goals for nutrient adequacy
- Specified amounts from four food groups
- Did not include guidance on appropriate fats, sugars, and calorie intake

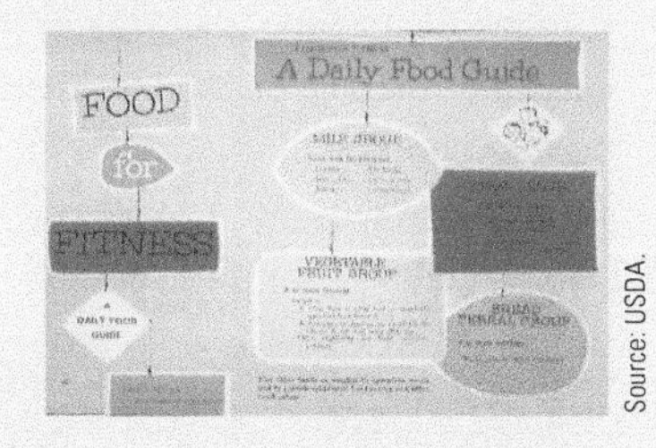

Source: USDA.

1979: Hassle-Free Daily Food Guide

- Developed after the 1977 Dietary Goals for the United States were released
- Based on the Basic Four, but also included a fifth group to highlight the need to moderate intake of fats, sweets, and alcohol

Source: USDA.

1984: Food Wheel: A Pattern for Daily Food Choices

- Total diet approach—Included goals for both nutrient adequacy and moderation
- Five food groups and amounts formed the basis for the Food Guide Pyramid
- Daily amounts of food provided at three calorie levels
- First illustrated for a Red Cross nutrition course as a food wheel

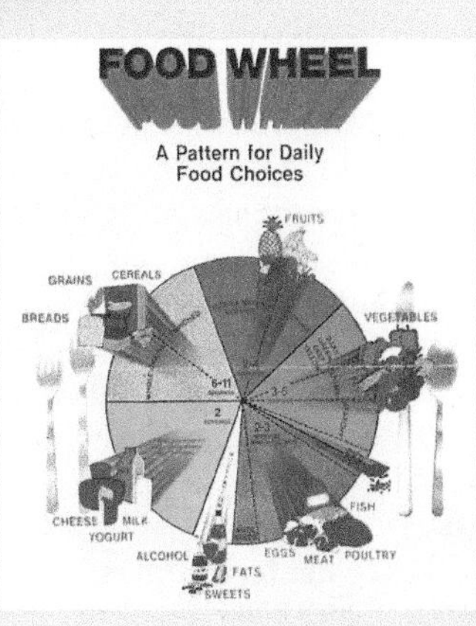

Source: USDA.

(Continues)

A BRIEF HISTORY OF USDA FOOD GUIDES (Continued)

1992: Food Guide Pyramid

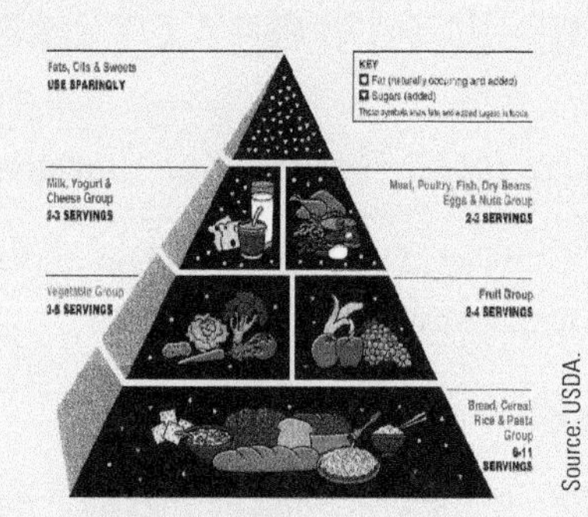

Source: USDA.

- Total diet approach—goals for both nutrient adequacy and moderation
- Developed using consumer research, to bring awareness to the new food patterns
- Illustration focused on concepts of variety, moderation, and proportion
- Included visualization of added fats and sugars throughout five food groups and in the tip
- Included range for daily amounts of food across three calorie levels

2005: MyPyramid Food Guidance System

Source: USDA.

- Introduced along with updating of Food Guide Pyramid food patterns for the *2005 Dietary Guidelines for Americans,* including daily amounts of food at 12 calorie levels
- Continued "pyramid" concept, based on consumer research, but simplified illustration. Detailed information provided on website "MyPyramid.gov"
- Added a band for oils and the concept of physical activity
- Illustration could be used to describe concepts of variety, moderation, and proportion

2011: MyPlate

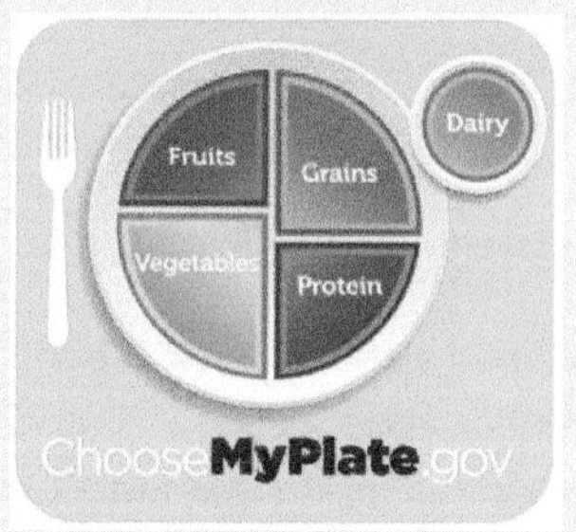

Source: USDA.

- Introduced along with updating of USDA food patterns for the *2010 Dietary Guidelines for Americans*
- Different shape to help grab consumers' attention with a new visual cue
- Icon that serves as a reminder for healthy eating, not intended to provide specific messages
- Visual is linked to food and is a familiar mealtime symbol in consumers' minds, as identified through testing
- "My" continues the personalization approach from MyPyramid
- In January 2016, the Secretary of Health and Human Services and Secretary of Agriculture released the 2015–2020 Dietary Guidelines for Americans. MyPlate continues to be based on this new release of the guidelines.

For more information:

- Welsh S, Davis C, Shaw A. A brief history of food guides in the United States. *Nutrition Today* November/December 1992:6-11.
- Welsh S, Davis C, Shaw A. Development of the Food Guide Pyramid. *Nutrition Today* November/December 1992:12-23.
- Haven J, Burns A, Britten P, Davis C. Developing the Consumer Interface for the MyPyramid Food Guidance System. *Journal of Nutrition Education and Behavior* 2006, 38: S124–S135.
- * Snap Shot of the 2015–2020 Dietary Guidelines for Americans: http://www.choosemyplate.gov/snapshot-2015–2020-dietary-guidelines-americans

Center for Nutrition Policy and Promotion

June 2011
Updated January 2016

Source: USDA, Agricultural Research Service (ARS), photo by Peggy Greb.

DIGESTIVE PROCESSES

The processing of food takes place in four stages:

1. Ingestion
2. Digestion
3. Absorption
4. Elimination

Ingestion

The act of eating—that is, chewing and swallowing—is the first of four main stages of food processing.

Digestion

Digestion breaks down food into molecules small enough to be absorbed. Polymers are broken into monomers that are easier to absorb and can be used to synthesize new polymers required by the organism.

Absorption

Cells that line the digestive tract take up the nutrients. Nutrients are transported to the cells, where they are incorporated and converted to energy that may be used immediately or stored until needed.

Elimination

In the last stage of food processing—elimination—undigested wastes pass out of the digestive tract.

Components of the Human Digestive System

Figure 4-4 illustrates the human digestive system. The following structures are considered parts of the digestive system:

- Mouth
- Tongue
- Pharynx
- Salivary glands
- Esophagus
- Stomach
- Liver
- Gall bladder
- Pancreas
- Small intestine
- Large intestine
- Rectum
- Anus

Mouth. Food enters the mouth and is reduced in size by teeth and tongue. Salivary glands secrete saliva, which lubricates and buffers. Saliva contains antimicrobial substances and amylase, which begins to digest starch.

Swallowing. Food passes to the pharynx, which contains both the trachea and the esophagus. The epiglottis prevents food from entering the trachea. Food passes through the esophagus into the stomach.

Stomach. The stomach stores and digests food. It contains pits that lead to gastric glands with three types of cells:

1. Mucous cells produce mucus to lubricate and protect the stomach lining,
2. Parietal cells secrete hydrochloric acid.
3. Chief cells that secrete pepsinogen.

Small Intestine. The small intestine receives food from the stomach. Bile enters from the liver via the gall bladder. **Enzymes** in the small intestine come from the pancreas. The small intestine is the site of most digestion and the site of most absorption.

Pancreas. The pancreas secretes many digestive enzymes into the small intestine. It also produces and secretes the hormones insulin and glucagon.

Liver. The liver produces bile but no digestive enzymes. Bile contains bile salt, which emulsifies fats. Bile is made from cholesterol and is stored in the gall bladder.

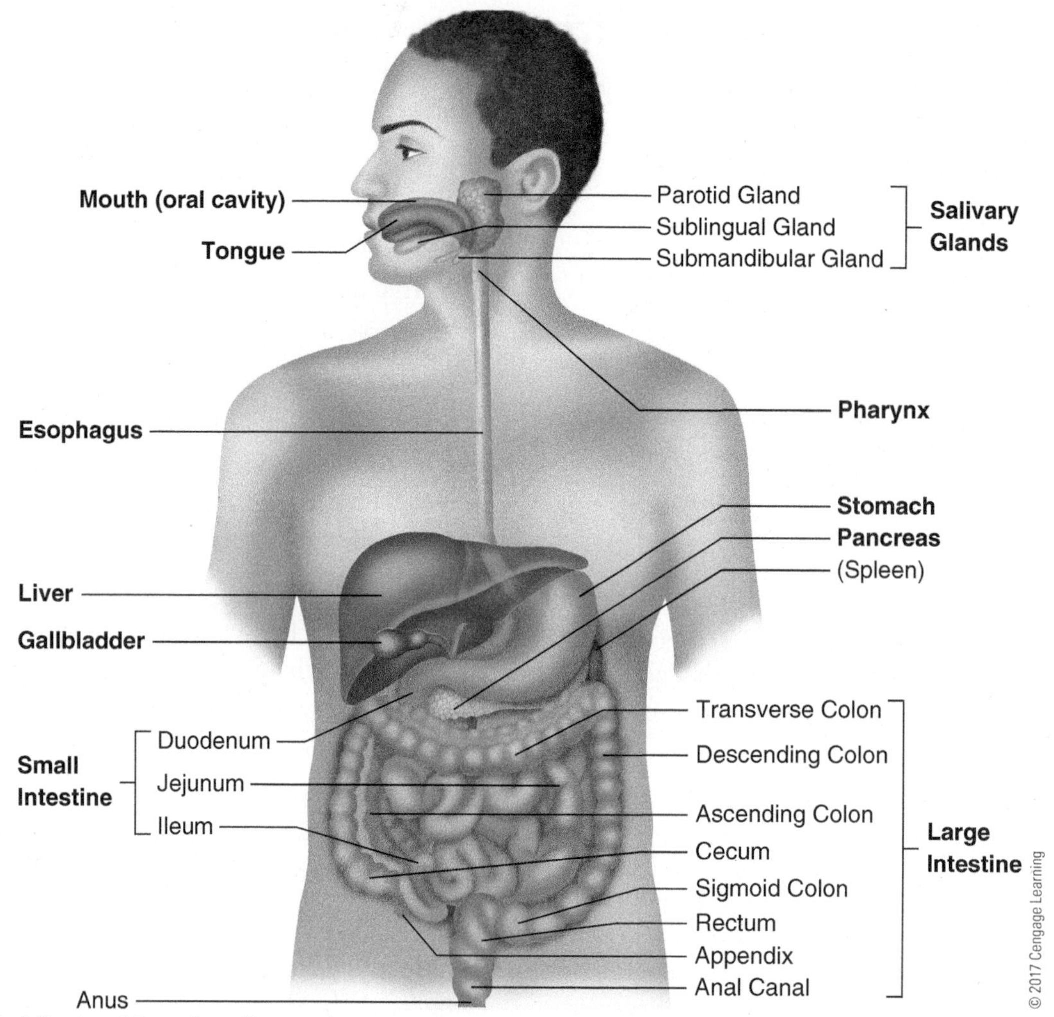

FIGURE 4-4 Parts of the digestive system.

Duodenum. The duodenum is the first 10 inches (25 centimeters) of small intestine. It receives enzymes from the pancreas and neutralizes acid from the stomach. The duodenum is where most chemical changes occur in food.

- **Enzymatic action.** Pancreatic amylases break starch down into maltose, a disaccharide. Disaccharides are converted to monosaccharides. Maltose is hydrolyzed by maltase to give glucose + glucose. Sucrose is hydrolyzed by sucrase to give glucose + fructose. Lactose is hydrolyzed by lactase to give glucose + galactose.

  Not all simple saccharides can be easily digested. Lactose intolerance is a common problem, one caused by a deficiency of lactase, the enzyme that catalyzes the hydrolysis of lactose to glucose and galactose. Lactose cannot be absorbed unless it is broken down first. If insufficient lactase is present, then some of the lactose will travel to the large intestine and cause problems. It will increase the flow of water into the intestine because of increased osmolality of the intestine's contents. This results in osmotic diarrhea.

  Emulsified fat is broken down by lipase into fatty acids and glycerol. Nucleic acid is converted by nucleases into nucleotides, which are converted to nitrogenous bases, sugars, and phosphates.

Proteins or polypeptides are converted to smaller peptides by trypsin and chymotrypsin, and small polypeptides are converted into amino acids by enzymes such as aminopeptidases, carboxypeptidases, and dipeptidases.

- **Absorption**. After enzymatic action in the duodenum, food is absorbed in the remainder of the small intestine. This portion of the small intestine has an enormous surface area. It contains villi and microvilli and is rich in capillaries and lymph vessels.

Large Intestine. Material not digested or absorbed passes into the large intestine. Most of the water in food (90%) is absorbed into the blood from the large intestine. Some vitamins are produced by bacteria in the large intestine and are absorbed. The residue is stored in the rectum and then eliminated through the anus.

VEGETARIAN DIETS

Some people eat vegetarian diets for reasons of culture, belief, or health. Most vegetarians eat milk products and eggs; as a group, these lacto-ovo-vegetarians enjoy excellent health. Vegetarians do not typically consume meat or meat products, although some consume fish or poultry. Vegetarian diets are consistent with the Dietary Guidelines for Americans can meet Recommended Dietary Allowances for nutrients. Individuals can get enough protein from a vegetarian diet as long as the variety and amounts of foods consumed are adequate. Meat, fish, and poultry are major contributors of iron, zinc, and B vitamins in most American diets, and thus vegetarians should pay special attention to getting enough of these nutrients.

Vegans follow a strict vegetarian diet. **Vegans** only eat food of plant origin. Vegans do not use animal products. Because animal products are the only food sources of vitamin B_{12}, vegans must supplement their diets with a source of this vitamin. In addition, vegan diets, particularly those of children, require care to ensure adequate intake of vitamin D and calcium, which most Americans obtain from milk products.

BIOAVAILABILITY OF NUTRIENTS

Although chemical analysis of a food can determine the presence of a nutrient, this factor alone can be misleading. A nutrient may be in a food, but whether it is available in a form that can be used by the metabolic processes of the body is another question. If the nutrient is in a form that can be used, it said to be *bioavailable.* Factors determining the **bioavailability** of a nutrient include digestibility, absorption, nutrient-to-nutrient interactions, binding to other substances, processing, and cooking procedures. Also, age, gender, health, nutritional status, drugs, and food combinations influence the bioavailability of carbohydrates, proteins, fats, vitamins, and minerals.

STABILITY OF NUTRIENTS

The nutritive value of food starts with the genetics of the plants or animals that will be consumed. Fertilization, weather, and maturity at harvest also influence the composition of the plant or animal being. Storage before processing affects nutrient levels, and then every processing step affects the nutrient levels in a food. Finally, preparation in the home or at a restaurant often further reduces a food's final nutritive value even before digestion begins.

A primary goal of food science is to preserve the nutrients through all phases of food harvesting, processing, storage, and preparation. To do this, food scientists need to know what nutrient **stability** is under varying conditions of pH, air, light, heat, and cold. Nutrient losses are small in most modern food-processing operations, but when nutrient losses are unavoidably high, U.S. law allows enrichment.

DIET AND CHRONIC DISEASE

Food choices also can help reduce the risk of chronic diseases such as heart disease, certain cancers, diabetes, stroke, and osteoporosis, which are leading causes of death and disability for Americans. Good diets can reduce major risk factors for chronic diseases such as obesity, high blood pressure, and high blood cholesterol.

Healthful diets contain the amounts of essential nutrients and calories needed to prevent nutritional deficiencies and excesses. Healthful diets also provide the right balance of carbohydrates, fats, and proteins to reduce the risks of chronic diseases. See Chapter 29, Health and Nutrition, for more complete information on this subject.

SUMMARY

The Food Nutrition Board of the National Research Council establishes Recommended Dietary Allowances and Dietary References Intakes. These guidelines provide daily nutrient levels for maintaining normal functions and health. The RDA lists recommendations for energy; protein; vitamins A, E, K, C, and the B vitamins; and the minerals iron, zinc, iodine, and selenium. These recommendations vary according to age, sex, pregnancy, and lactation. Protein requirements for humans should consider the essential amino acids. The MyPlate dietary guidelines provided by the USDA is easy for the average person to follow to ensure adequate nutrition.

The digestive process includes ingestion, digestion, absorption, and elimination. Nutrients in the diet are progressively broken down into smaller components by mechanical, chemical, and enzymatic means. Small molecules resulting from digestion are absorbed to supply the body with energy, protein, vitamins, and minerals.

REVIEW QUESTIONS

Success in any career requires knowledge. Test your knowledge of this chapter by answering these questions or solving these problems.

1. Name six minerals required by the body.
2. Identify the protein requirement for a 19-year-old male and a 19-year-old female.
3. Describe the function of protein in the diet.
4. How many calories are in 1 gram of protein? One gram of carbohydrate? One gram of fat? One gram of alcohol?
5. What acid is an essential fatty acid?
6. Identify the digestive organ that receives enzymes from the pancreas.
7. During digestion, enzymes such as aminopeptidases, carboxypeptidases, and dipeptidases convert polypeptides into what amino acid?
8. What nutritional deficiency causes kwashiorkor and marasmus?
9. List five essential amino acids.
10. What factor determines protein quality?

STUDENT ACTIVITIES

1. Use a log to track your diet for five days. Analyze how closely your diet conforms to the MyPlate dietary guidelines. Estimate your average daily consumption of calories, fats, and proteins by using Table B-8. Report your findings to the class.
2. Using Figure 4-1, look up the RDA for all the nutrients for your age and sex. Report your findings by creating an nutrient information sheet for someone your age and sex.
3. Develop a short verbal and visual presentation on one of these topics: bioavailability, fiber, nutrient stability in foods.
4. Collect labels from food products and write a report on the use of the RDAs on the label.
5. Develop a mnemonic that will help you remember the essential amino acids.
6. If possible, obtain a digestive tract of a pig and use this to discuss and describe the digestive process in humans. Dissect the tract and trace the passage of food. Record your steps and findings.
7. Plan a meal that meets the dietary guidelines of MyPlate or RDA. Take a digital picture of the meal and then describe how each item on the menu is digested and in what form it is absorbed.

ADDITIONAL RESOURCES

- Nutrition.gov: http://www.nutrition.gov/
- USDA ChooseMyPlate.gov: http://www.choosemyplate.gov/downloads/NutritionFactsLabel.pdf
- Digestion Animation: http://kitses.com/animation/swfs/digestion.swf
- Nutrient Recommendations: Dietary Reference Intakes (DRI): http://ods.od.nih.gov/Health_Information/Dietary_Reference_Intakes.aspx
- 2015–2020 Dietary Guidelines for Americans: http://health.gov/dietaryguidelines/2015/

Internet sites represent a vast resource of information. Although those provided in this chapter were vetted by industry experts, you may wish to further explore the topics discussed in this chapter using a search engine such as Google. Keywords or phrases may include the following: *food pyramid, Recommended Dietary Allowances, Dietary Reference Intakes, protein deficiencies, protein quality, amino acids, , biological value, human digestive system, vegetarian diets, vitamins, minerals, fiber.* In addition, Table B-7 provides a listing of some useful Internet sites that can be used as a starting point.

REFERENCES

Brody, J. E. 1981. *Jane Brody's nutrition book.* New York: Bantam.

Drummond, K. E. 1994. *Nutrition for the food service professional*, 2nd ed. New York: Van Nostrand Reinhold.

Ensminger, A. H., M. E. Ensminger, J. E. Konlande, and J. R. Robson. 1994. *Foods and nutrition encyclopedia* (2 vols.). Boca Raton, FL: CRC Press.

Gardner, J. E. (Ed.). 1982. Eat better, live better. *Reader's Digest.* Pleasantville, NY: Reader's Digest Association.

U.S. Department of Agriculture (n.d.). Dietary reference intakes. The National Library: http://fnic.nal.usda.gov/dietary-guidance/dietary-reference-intakes. Accessed June 9, 2015.

U.S. Department of Agriculture (n.d.). Choose MyPlate: http://www.choosemyplate.gov/food-groups/. Accessed June 15, 2015.

Potter, N. N., and J. H. Hotchkiss. 1995. *Food science*, 5th ed. New York: Chapman & Hall.

Vaclavik, V. A., and E. W. Christina. 1999. *Essentials of food science*, 4th ed. New York: Springer.

Vieira, E. R. 1996. *Elementary food science*, 4th ed. New York: Chapman & Hall.

ENDNOTES

1. Institute of Food Technologies, March 2013, http://www.ift.org/Food-Technology/Daily-News/2013/March/07/FAO-proposes-new-protein-quality-measurement.aspx.
2. http://www.choosemyplate.gov/food-groups/oils.html. Last accessed June 15, 2015.

CHAPTER 5

Food Composition

OBJECTIVES

After reading this chapter, you should be able to:

- Find foods in a food composition table and describe their nutritional value
- List three methods of determining the composition of foods
- Describe the method for determining the caloric content of foods
- Explain the difference between Calorie and calorie
- Identify common abbreviations and terms used in a food-composition table
- Discuss the use of food-composition tables
- List four factors that affect the nutrient content of foods

KEY TERMS

bomb calorimeter
Calorie
chromatography
energy
ether extract
proximate analysis
spectrophotometry

NATIONAL AFNR STANDARD

FPP.03

Select and process food products for storage, distribution, and consumption.

Food-composition tables are used to evaluate the nutritional value of food supplies, develop food-distribution programs, plan and evaluate food-consumption surveys, provide nutritional counseling, and estimate the nutritional content of individual diets.

DETERMINING THE COMPOSITION OF FOODS

The nutrient content of foods is influenced by variety, season, geographical differences, stage of harvesting, handling, commercial processing, packaging, storage, display, home preparation, cooking, and serving. The composition of foods is determined by a variety of scientifically sound, standardized methods. The first system of approximating the value of a food or feed for nutritional purposes was developed at the Weende Experiment Station in Germany more than 100 years ago. This system separates a food into nutritive fractions through a series of chemical determinations. These determinations reflect a food's nutritive value. The different fractions included water or dry matter, crude protein, **ether extract** or fat, crude fiber, nitrogen-free extract (sugars and starches), and ash or total mineral. This system became known as **proximate analysis**. Newer methods of determining the composition of foods have replaced or supplemented the old proximate analysis and allowed determination of more specific nutrients in foods. Among these newer methods are **spectrophotometry** and two types of **chromatography**—liquid and gas (Figure 5-1). These new methods allow the determination of fatty acids, cholesterol, amino acids, specific minerals, and vitamins.

Soruce: USDA, Agricultural Research Service (ARS), photo by Peggy Greb.

FIGURE 5-1 A geneticist and Iowa State University food science professor examine a vial of oil extracted from Tripsacum-introgressed corn before loading it into an autosampler for gas chromatography analysis.

ENERGY IN FOOD

Energy in food is measured in terms of calories. The **Calorie** is a metric unit of heat measurement. The small calorie (cal) is the amount of heat required to raise the temperature of 1 gram of water from 14.5 °C to 15.5 °C. The definition now generally accepted in the United States, and the standard in thermochemistry, is 1 cal equals 4.1840 joules (J).

A large calorie, or kilocalorie (Cal), usually referred to as a *calorie* and sometimes as a *kilogram calorie*, equals 1,000 cal. This unit is used to express the amount of energy that a food provides when consumed.

Calorimeters measure the heat developed during the combustion of food. **Bomb calorimeters** have been used to determine the calorie content of foods. Basically, a bomb calorimeter consists of an enclosure in which the reaction takes place, surrounded by a liquid, such as water, that absorbs the heat of the reaction and thus increases in temperature. Measurement of this temperature rise for a known weight of food permits the total amount of heat generated to be calculated.

The design of a typical bomb calorimeter is shown in the Figure 5-2. The food to be analyzed is placed inside a steel reaction vessel called a bomb. The steel bomb is placed inside a bucket filled with water, which is kept at a constant temperature relative to the entire calorimeter by use of a heater and a stirrer. The temperature of the water is monitored with a thermometer fitted with a magnifying eyepiece, which allows accurate readings to be taken.

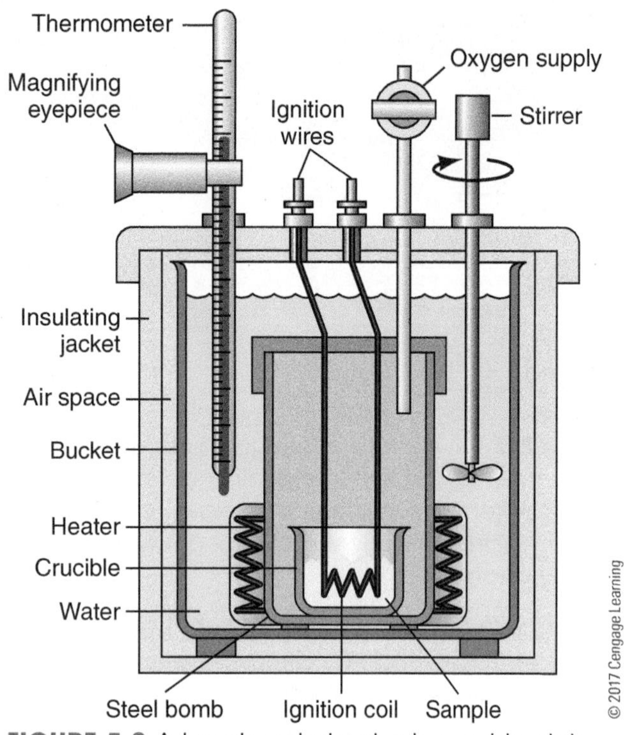

FIGURE 5-2 A bomb calorimeter is used to determine the energy content of food.

Inserting an air space between the bucket and an exterior insulating jacket minimizes heat losses. Slots at the top of the steel bomb allow ignition wires and an oxygen supply to enter the vessel. When an electric current passes through the ignition coil, a combustion reaction occurs. The heat released from the sample is largely absorbed by the water, which results in an increase in temperature.

FOOD-COMPOSITION TABLES

The Food Composition Table, Appendix Table B-8, is from the U.S. Department of Agriculture's Agricultural Research Service. Other food-composition tables can be found at the USDA Web site: http://fnic.nal.usda.gov/food-composition.

Table 5-1 explains the abbreviations used in Table B-8.

MUSCLE CRAMPS AND DIET

Almost everyone has experienced a muscle cramp. The muscle becomes contracted and rigid, and this is usually quite painful. The contracted muscle gets locked into a self-sustaining knot, which can last for a few minutes, hours, or even days. The informal name for this involuntary muscle cramp is "charley horse." The term *charley horse* is most often used in connection with athletes, especially baseball players.

Many muscle cramps are associated with exercise. These cramps are often the result of a depletion or imbalance of salts in the muscle tissue, especially calcium, sodium, and potassium, which are lost through perspiration (sweat). A buildup of lactic acid, one of the by-products of heavy exercise, also can contribute to cramping. Drinking so-called electrolyte drinks that restore the salt balance can often relieve cramps. Also, eating foods high in sodium and potassium can relieve cramps.

Check out the foods in the Food Composition Table, Table B-8, and suggest those that would be good for muscle cramps.

To learn more about muscle cramps and diet, search the Web or start by visiting these sites:

- http://www.mayoclinic.org/diseases-conditions/muscle-cramp/basics/causes/con-20014594
- http://www.humankinetics.com/excerpts/excerpts/learn-the-connection-between-diet-and-muscle-cramping
- http://www.nlm.nih.gov/medlineplus/musclecramps.html

TABLE 5-1 Explanation of Abbreviations Used in the Food Composition Table (Table B-8)

DESCRIPTION	ABBREVIATION
Food item number	No.
Description of food and measure	Description of Food
Percentage of water	Water (%)
Food energy in kilocalories	Energy (kcal)
Protein in grams	Prot (g)
Fat in grams	Fat (g)
Saturated fatty acid in grams	Sat (g)
Monounsaturated fatty acid in grams	Mono (g)
Polyunsaturated fatty acid in grams	Poly (g)
Cholesterol in milligrams	Chols (mg)
Carbohydrate in grams	Carb (g)
Calcium in milligrams	Ca (mg)
Phosphorus in milligrams	P (mg)
Iron in milligrams	Fe (mg)
Potassium in milligrams	K (mg)
Sodium in milligrams	Na (mg)
Vitamin A in International Units	Vit A (IU)
Vitamin A in retinol equivalents (REs)	Vit A (RE)
Thiamin in milligrams	Thmn (mg)
Riboflavin in milligrams	Ribof (mg)
Niacin in milligrams	Niacin (mg)
Ascorbic acid in milligrams	Vit C (mg)

SUMMARY

Food-composition tables are used to evaluate diets and food supplies. Methods such as spectrophotometry, liquid chromatography, and gas chromatography determine the composition of foods. The bomb calorimeter measures the caloric content of foods. Many food-composition tables are available, but the USDA's table maintains and updates data on food composition.

REVIEW QUESTIONS

Success in any career requires knowledge. Test your knowledge of this chapter by answering these questions or solving these problems.

Note: The answers to some of these questions is found by using Table B-8.

1. How many Calories and grams of protein are in 3 oz. of Froot Loops® cereal?
2. How many grams of fat are in one slice of cheese pizza?
3. Describe item number 4270.
4. List three methods for determining the composition of foods.
5. What term describes the amount of heat required to raise the temperature of 1 gram by 1 °C?
6. Describe two uses of a food-composition table.
7. Name four factors that affect the nutrient content of foods.
8. Explain the relationship between Calorie, Kcal, calorie, and cal.
9. Identify the following abbreviations: oz., mg, IU, RE, mono, sat, poly, carb, and chols.
10. In terms of energy and protein, what is the difference between a slice of white bread and a slice of whole wheat bread?

STUDENT ACTIVITIES

1. Check out the USDA's online Nutrient Database for Standard Reference. Containing nearly 6,000 foods and more than 70 components, the site is the nation's primary source of food-composition data:

   http://www.nal.usda.gov/fnic/pubs/foodcomp.pdf

   Use this Web site or Table B-8 to answer the following questions.

   a. Energy (calories) is reported in what two units? Search the home page to find the difference between these units.
   b. What is another name for fat?
   c. What is another name for vitamin C?
   d. What minerals are listed in the database?
   e. What are the three major classes of fatty acids?
   f. How much sodium is in a teaspoon of salt and in a large double cheeseburger with everything?
   g. Deep yellow and dark green leafy vegetables are among the best sources of vitamin A. List three of these.
   h. Compare the fat contents in several popular snack foods. List them from most fat to least fat.
   i. Find three fruits low in fat.
   j. Compare the vitamin C content in five different beverages. Use 1-cup portions.

k. Which of the following is highest in cholesterol—2 Tbsp. chunky peanut butter, 1 cup orange juice, a batter-fried chicken drumstick, 3 cups rice, or ½ cup salsa?

l. Provide product descriptions and dietary fiber values for NDB number 09200 (1 large) and NDB number 16005 (1 cup).

m. Which has more calcium—a cup of 1% low-fat milk or ½ cup 1% low-fat cottage cheese? Record values and item descriptions.

2. Develop a report that describes other ways you could use the USDA's Nutrient Database to manage nutrient intake.

3. Create a chart for tracking the foods you eat for five days.

ADDITIONAL RESOURCES

- What Is a Calorie? http://www.superkidsnutrition.com/nw_whatisacalorie/
- USDA—Food Composition: http://fnic.nal.usda.gov/food-composition
- How Does a Spectrophotometer Work? http://www.lsteam.org/projects/videos/how-does-spectrophotometer-work
- Journal of Food Composition and Analysis: http://www.journals.elsevier.com/journal-of-food-composition-and-analysis/

Internet sites represent a vast resource of information. Although those provided in this chapter were vetted by industry experts, you may wish to further explore the topics discussed in this chapter using a search engine such as Google. Keywords or phrases may include the following: *food composition, proximate analysis, calorimetry, nutritional value, dietary intake*. In addition, Table B-7 lists useful Internet sites that can be used as a starting point.

REFERENCES

Brody, J. E. 1981. *Jane Brody's nutrition book*. New York: Bantam Books.

Cremer, M. L. 1998. *Quality food in quantity. Management and science*. Berkeley, CA: McCutchan.

Drummond, K. E. 1994. *Nutrition for the food service professional*, 2nd ed. New York: Van Nostrand Reinhold.

Ensminger, A. H., M. E. Ensminger, J. E. Konlande, and J. R. Robson. 1994. *Foods and nutrition encyclopedia* (2 vols.). Boca Raton, FL: CRC Press.

Gardner, J. E., Ed. 1982. *Reader's digest. Eat better, live better.* Pleasantville, NY: Reader's Digest Association.

Horn, J., J. Fletcher, and A. Gooch. 1997. *Cooking A to Z: The complete culinary reference source*. Glen Ellen, CA: Cole Publishing Group.

Wagner, S. (Ed.). 1999. *The recipe encyclopedia: The complete illustrated guide to cooking*. San Diego: Thunder Bay Press.

CHAPTER 6

Quality Factors in Foods

OBJECTIVES

After reading this chapter, you should be able to:

- Describe the influence of color on food quality
- Identify the instrument that can measure food texture
- Discuss the influence of color, texture, size, and shape on consumer acceptance
- Describe how water changes the texture of food products
- Identify six words used to describe food flavor
- Describe sensory methods humans use to determine food flavor
- Discuss three factors that can affect food flavor
- Explain three means for maintaining or assessing quality in foods
- Describe the role the U.S. Department of Agriculture plays in food quality

KEY TERMS

astringency

chroma

Current Good Manufacturing Practices (CGMPs)

Hazard Analysis and Critical Control Point (HACCP)

hue

Maillard reaction

phenolic compounds

pigment

rheology

sensory evaluation

standards

texture

Total Quality Management (TQM)

value

volatile

NATIONAL AFNR STANDARD

FPP.02

Apply principles of nutrition, biology, microbiology, chemistry, and human behavior to the development of food products.

What determines an individual's food choices? Price, availability, religion, and personal preference can all play a role in a consumer's choices. Most consumers, however, want high-quality, nutritious food at the best value. The quality of a food product involves maintenance or improvement of a product's key attributes. These can include a food's color, flavor, and texture as well as the food's safety, healthfulness, shelf life, and convenience of purchasing and preparing. To maintain food quality, it is important to control or manage microbiological spoilage, enzymatic degradation, and chemical degradation. These components of quality depend on the composition of the food, processing methods, packaging, and storage. Internationally, different levels of quality are considered. What might be considered a delicacy in one country may not be considered a food source at all in another.

DOES YOUR CULTURE MAKES A DIFFERENCE?

Have you ever stopped to think about the foods you eat? Even though the United States is considered a melting pot of cultural diversity, cultural and regional patterns influence what most Americans eat. Religious beliefs can also play a role. For example, Hindu followers typically do not consume beef and devout Jews and Muslims do not consume pork. The cultural background may influence the fruits and vegetables that are consumed and probably influences the seasoning and spices used for different dishes. A Hispanic influence may include mango, papaya, and guava fruits, whereas a French influence may mean more breads and diverse types of cheeses. The cultural background may also dictate the typical time and size of meals. In some cultures, lunch is the largest meal of the day, while in other cultures the evening meal is the largest meal of the day.

Even though some of the same products are consumed in all regions of the United States, they may be called different names. For example, is it a soda, a coke, or a pop? Depending on where you live, you may refer to soft drinks by different names. Is it a casserole dish or a hot dish? It probably depends if you live in the southern United States or the Great Lakes region. Some common yet distinct foods by U.S. regions include the following:

Far West—fish tacos, avocado pie, California rolls, saimin

Great Lakes—lutefisk, cheese curds, kringle, chicken booyah

Deep South—sweet tea, hush puppies, Cuban sandwich, boiled peanuts

Mid-Atlantic—Buffalo chicken wings, reuben sandwich, Whoopie pie, Philadelphia cheese steak

Midwest—buffalo burgers, potato candy, chili, St. Louis toasted ravioli

New England—lobster roll, clam chowder, Boston cream pie, johnnycakes

Pacific Northwest—apple candy, smelt, sourdough pancakes, huckleberry pie

Southeast—hot brown sandwich, shrimp and grits, biscuits and gravy, Brunswick stew

Southwest—chicken enchilada soup, tamales, beef tongue, steak fajitas

Are any of your favorite foods listed? Research the origin and history of some of your favorite foods and see where they came from.

To see more food items by region:

- http://whatscookingamerica.net/
- http://whatscookingamerica.net/AmericanRegionalFoods/RegionalAmericanIndex.htm

MATH CONNECTION!

Compare rare foods from different countries. Compare their prices. How do they compare to food prices in the United States?

SENSORY EVALUATION

Humans have five senses: taste, touch, smell, vision, and hearing. All of these are important when eating. **Sensory evaluation** is the process of human analysis of the look, taste, smell, sound, and feel of food. We have certain expectations about how certain foods in our diet should be. Imagine eating a potato chip. Before eating it, do you inspect the chip to see if there are burned edges? You expect the potato chip to be salty rather than sweet. You also expect a crunch when eating the chip; if it does not crush, then it might be stale or old. Typically, you also expect an oily texture from the chip unless it has been prepared a different way.

SCIENCE CONNECTION!

Compare a name brand potato chip with a generic chip. Use all all five senses. How do the chips differ and how are they alike in look, taste, smell, sound, and touch?

Appearance Factors

A food's *appearance* refers to its shape, size, condition, and color. Although the appearance of foods is subjective to consumers, it remains important. Consumers have certain expectations about how a food should look. If it does not appear as a consumer thinks it should, it is often skipped over and not purchased. Two separate categories of appearance are (1) color attributes and (2) geometric attributes (size and shape).

Color. Color can influence a consumer's perception of his or her other senses. Consumers expect meat to be red, apple juice to be light brown and clear, orange juice to be orange, egg yolks to be bright yellow-orange, and so on. If colors do not match the perception of the consumer, they may not purchase that product. Food companies have often tried varying food color in the marketplace. Have you ever seen green ketchup or clear cola? Typically, these alternate color choices are short lived.

Food-color measurements provide an objective index of food quality. Color can be an indication of ripeness or spoilage. The results of all cooking processes are judged by color. Unexpected changes in colors can also indicate problems with processing or packaging.

Browns and blackish colors can be either enzymatic or nonenzymatic reactions. The major nonenzymatic reaction is the **Maillard reaction**, which is the dominant browning reaction in meat. The Maillard reaction—also called the *browning reaction*—occurs when the denatured proteins on the surface of the meat combine with sugars that are present. This combination creates the meaty flavor and changes in color. The reaction occurs between 300 °F and 500 °F.[1] Other reactions include blackening in potatoes and browning in orange juice. The enzymatic browning found widespread in fruits and selected vegetables is the result of enzymatic catalyzed oxidation of **phenolic compounds**. Most consumers consider bruised fruit and vegetables to be lower in quality and will not select them for purchase.

Naturally occurring **pigments** also play a role in food color. Water-soluble pigments may be categorized as anthocyanins and anthoxthanins.

Lesser known water-soluble pigments include the leucoanthocyanins. Fat-soluble plant pigments are primarily categorized into the chlorophyll and carotenoid pigments. These green and orange-yellow pigments considerably affect color. Myoglobins contribute to the color of meat.

Measuring Color. To maintain quality, the colors of food products must be measured and standardized. If a food is transparent, like a juice or a colored extract, colorimeters or spectrophotometers can be used to measure the color. Colors of nontransparent liquids and solid foods can be measured by comparing their reflected color to defined (standardized) color tiles or chips. For a further measurement of color, reflected light from a food can be divided into three components: **value**, **hue**, and **chroma**. The color of a food can be precisely defined with numbers for these three components with tristimulus colorimetry. Instruments such as the Hunterlab spectrophotometers and other color difference meters measure the value, hue, and chroma of foods for comparisons.

Size and Shape. Depending on the product, consumers also expect foods to have certain sizes and shapes (Figure 6-1). For example, consumers have some idea of what an ideal french fry should look like—or an apple, or a cookie, or a pickle. Size and shape are easily measured. Fruits and vegetables are graded based on their size and shape, and this is done by the openings they will pass through during grading. Computerized electronic equipment now determines the size and shape of most foods.

FIGURE 6-1 Consumers expect foods to have specific shapes.

SCIENCE CONNECTION!

Compare fruit gum drops to spice gum drops. When looking at their color, what flavor do you expect the gum drops to be? Record your hypothesis and findings!

Textural Factors

Consumers expect gum to be chewy, crackers to be crisp, steak to be tender, cookies to be soft, and breakfast cereal to be crunchy. The **texture** of food refers to those qualities felt with the fingers, tongue, or teeth. Textures in food vary widely, but any departure from what the consumer expects is a quality defect.

Texture is a mechanical behavior of foods measured by sensory (physiological or psychological) or physical (**rheology**) means. Rheology is the study of the science of deformation of matter. The four main reasons for studying rheology include:

1. Insight into structure
2. Information used in raw material and process control in industry
3. Applications to machine design
4. Relation to consumer acceptance

Regardless of the reason for studying texture, classification and understanding are difficult because of the enormous range of materials. Moreover, food materials behave differently under different conditions (Figure 6-2).

Texture testing in foods is based on the actions of stress and strain. Many of the methods use compression, shearing, shear pressure, cutting, or tensile strength to determine texture. For example, the compressimeter was used to determine the compressibility of cakes and other spongelike products. Historically, the penetrometer, has been used to

Source: USDA, Agricultural Research Service (ARS), photo by Keith Weller

FIGURE 6-2 Texture is important to consumers, and scientists have found ways to accurately measure the textures of different foods.

measure gel strength. The Warner–Bratzler shear apparatus has been the standard method of evaluating meat tenderness. The Instron company has developed and adapted many historical texture-measuring instruments. One, for example, measures elasticity. A viscometer developed by Brookfield Engineering measures viscosity in terms of Brookfield units. Other instruments used to measure texture include succulometers and tenderometers.

Changes in texture are often the result of water status. Fresh fruits and vegetables become soggy as cells break down and lose water. On the other hand, when dried fruits absorb water, their texture also changes. Breads and cakes lose water as they become stale. If crackers, cookies, and pretzels take up water, they usually become soft and undesirable.

Various methods are used to control the texture of processed foods. Lipids (fats) are softeners and lubricants used in cakes. Starch and gums are used as thickeners. Protein can also be a thickener or it can form a rigid structure if it coagulates in baked bread, for example. Depending on its concentration in a product, sugar can add body to soft drinks or chewiness to other products; in greater concentrations, it can thicken and add chewiness or brittleness.

Flavor Factors

Food flavor includes taste sensations perceived by the tongue—sweet, salty, sour, and bitter—and smells perceived by the nose. Often the terms *flavor* and *smell* (or *aroma*) are used interchangeably. Food flavor and aroma are difficult to measure and difficult to get people to agree on. One part of food science called *sensory science* is dedicated to finding ways to help humans accurately describe the flavors and other sensory properties of their food.

Flavor, like color and texture, is a quality factor. It influences the decision to purchase and consume a food product. Food flavor is a combination of taste and smell, and it is highly subjective and difficult to measure. People not only differ in their ability to detect tastes and odors but also in their preferences for flavors.

Besides the tastes of sweet, salty, sour, and bitter (see Figure 6-3), an endless number of compounds give food characteristic aromas such as:

- Fruity
- **Astringency**
- Sulfur
- Hot

Sweetness may result from sugars such as arabinose, fructose, galactose, glucose, riboses, xylose, and other sweeteners. Organic acids may be perceived on the bottom of the tongue. Some of these common acids are citric, isocitric, malic, oxalic, tartaric, and succinic acids. The fruity flavors are often esters, alcohols, ethers, or ketones. Many of these are **volatile** and are associated with acids.

Phenolic compounds are closely related to plants' sensory and nutritional qualities. They are found in many fruits, including apples, apricots, peaches, pears, bananas, and grapes—and vegetables such

SCIENCE CONNECTION!

Compare and contrast a regular food item to its reduced fat version. Do the textures feel the same? Can you tell which item is regular and which is reduced fat? Record your hypothesis and findings!

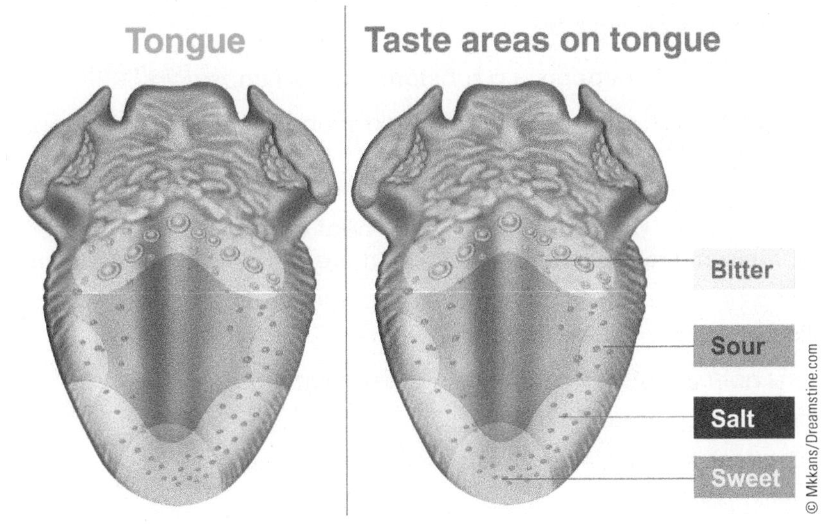

FIGURE 6-3 Taste areas on our tongues help perceive flavor.

as avocado, eggplant, and potatoes. These compounds contribute to color, astringency, bitterness, and aroma. Most phenolic compounds are found around the vascular tissues in plants, but they have the potential to react with other components in the plant as damage to the structure occurs during handling and processing. Loss of nutrients and changes in color and flavor occur in foods because phenolic compounds react with polyphenol oxidase, an enzyme that catalyzes oxidation.

SCIENCE CONNECTION!

As part of an FFA Food Science Career Development Event, evaluate aromas for the following:

Cinnamon	Oregano	Peppermint
Basil	Lemon	Clove
Vanilla	Lime	Nutmeg
Onion	Orange	Molasses
Garlic	Smoke (liquid)	Wintergreen
Ginger	Cherry	Banana
Coconut	Pine	Lilac
Licorice (anise)	Butter	Raspberry
Chocolate	Menthol	Strawberry
Maple	Grape	

Bring in different samples of aromas to compare in class.

TASTE OF HOT CHILI PEPPERS

The active ingredient in hot chili peppers is a substance called *capsaicin.*

Capsaicin is so potent that even a minute amount has a strong effect. Why does capsaicin taste so hot? Receptor molecules in the membranes of certain pain sensor nerves respond strongly to capsaicin molecules, evoking the same pain as heat, acids, and various chemical or physical stimuli, including injuries.

Capsaicin's heat is created by direct activation of the heat and pain sensory system. Once capsaicin triggers a receptor, the same receptor becomes even more sensitive to heat, making warm soup taste even hotter.

For more information about capsaicin, search the Web or visit these Web sites:

- http://acapulcos.net/capsaicin-why-do-hot-peppers-burn/
- http://www.chm.bris.ac.uk/molm/chilli/
- http://www.chipotlechiles.com/hot-chili-pepper-capsaicin.htm/

The sense of taste is a powerful predictor of food selection. Of the four main tastes, humans most prefer sweet-tasting foods. This preference may be an evolutionary holdover from our ancestors, who found that sweetness indicated a food provided energy.

Judgment of flavor is often influenced by color and texture. Flavors such as cherry, raspberry, and strawberry are associated with the color red. Beef flavor is brown. Actually, the flavor essences are colorless. As for texture, people expect potato chips to be crunchy and gelatin to be soft and cool. Part of a food science technologist's job is to test the textures of food products (see Figure 6-4). Food textures are also tested with different preservation methods and cooking methods. For example, frozen foods are warmed in a microwave to make sure they are ready to eat and have the appropriate texture.

Source: USDA, Agricultural Research Service (ARS), photo by Peggy Greb

FIGURE 6-4 Food technologists discuss fiber orientation of a chicken breast sample. Texture measurements are recorded as a texture analyzer blade shears each meat strip and the results are compared to sensory texture data.

Depending on the food, flavor can also be influenced by:

- Bacteria
- Yeasts
- Molds
- Enzymes
- Heat and cold
- Moisture and dryness
- Light
- Time
- Additives

Finally, depending on the product, the influence these factors have on a food flavor can be positive or negative—which sometimes differs depending on the person doing the tasting.

Taste Panels. For acceptable consumer quality, the best method of measuring taste is to have people sample the products. Taste panels can be groups of professionals or groups of customers. Have you ever seen a commercial for a blind taste test? A good example is the Pepsi–Coca Cola challenge, which frequently determines whether consumers can tell the difference between these popular soft drinks. Typically, those taking part in taste panels are in separate booths so they cannot influence each other. Food samples are coded with letters and numbers, and tasters are given an evaluation form to complete as they taste and evaluate a product.

Proteins

Heat denaturation changes foods' solubility and texture, and light oxidation of protein causes off flavors. In addition, enzymatic degradation of protein can cause changes in body and texture and also bitter flavors. Freezing can alter protein conformation and solubility in some cases.

SCIENCE CONNECTION!

As part of the FFA Food Science Career Development Event conduct Triangle Tests:

To create a Triangle Test, you will need two of the same samples and one different sample. This can be done with name brand and generic products, regular and low calorie or reduced fat products or different brands of the same product. Conduct a Triangle Test on another student in class or on someone at home. (Were they able to pick out the different product?)

NOTE

Be aware of food allergens and consult with your teacher before conducting the test.

ADDITIONAL QUALITY FACTORS

Additional quality factors include shelf life, safety, healthfulness, and convenience. To extend a product's storage or shelf life generally involves heat treatment, irradiation, refrigeration, freezing, or reduction of water activity. Water activity can be changed either by adding water-binding agents such as sugars or by drying. In many cases, compromises are made to achieve a desired shelf life or convenience. Such processes, although they can improve shelf life, almost always have other effects on food components. The factors that influence changes in ingredients in a food include proteins, lipids, carbohydrates, vitamins, chemicals, and microbiological characteristics.

Lipids

Enzymatic hydrolysis of lipids can cause such off flavors as soapy or goaty, depending on the type of oil. This process also makes frying oils unsuitable for use, and it can change an oil's functionality and crystallization properties. The oxidation of unsaturated fatty acids also causes off flavors.

Carbohydrates

High-heat treatments cause interactions between reducing sugars and amino groups to give the Maillard browning reaction and changes in flavor. Hydrolysis of starch and gums can also change the texture of foods. Some starches also can be degraded by enzymes or by acidic conditions.

MATH CONNECTION!

Research a fruit or vegetable that can be purchased fresh, frozen, and canned. What is the storage or shelf life of the product in these different forms? Compare and contrast the pros and cons of extending the shelf life of a fruit or vegetable.

Vitamins

Depending on the vitamin, losses can occur when a food is heated or exposed to light or oxygen.

Chemicals and Microbiological Characteristics

Ensuring the safety of all foods involves carefully controlling all processes from the farm to the consumer. Safety includes controlling both chemical and microbiological characteristics of a product. Most processing emphasizes microbial control, and it often has as the objective of eliminating organisms or preventing their growth.

Processes used to prevent growth include:

- Irradiation
- Refrigeration
- Freezing
- Drying
- Control water activity (adding salt, sugars, polyols, and so forth)

Processes used to minimize organisms include:

- Pasteurization
- Sterilization (canning)
- Cleaning and sanitizing
- Membrane processing

Additional methods of processing that seek to control undesirable microflora are the deliberate addition of microorganisms and the use of fermentation.

Safety from a chemical viewpoint generally relates to keeping undesirable chemicals such as pesticides, insecticides, and antibiotics out of the food supply. Making sure that food products are free from extraneous matter (metal, glass, wood, etc.) is another facet of food safety.

QUALITY STANDARDS

Today's consumers want food products that are convenient to use and still have all the qualities of a fresh product. Quality **standards** help ensure such food quality (see Figure 6-5). Types of standards include research standards, trade standards, and government standards. Research standards are set up by a company to help ensure the quality of its products in a competitive market. Trade standards are established by members of an industry. These are voluntary and ensure at least minimum acceptable quality. As for government standards, some are mandatory and others optional. Grade standards established by the government provide a common language for producers, dealers, and consumers for buying and selling food products.

Quality Standards: USDA and the AMS

In cooperation with companies throughout the food industry, the Agricultural Marketing Service (AMS) of the U.S. Department of Agriculture (USDA) develops and maintains official U.S. quality standards and grades for hundreds of agricultural products (see Figure 6-6). These standards are based on descriptions of the value, utility, and entire range of quality for each product in the following categories:

- Nuts and specialty crops
- Dairy
- Poultry and eggs (including rabbits)
- Fresh fruits and vegetables (including fresh fruits and vegetables for processing)
- Processed fruits and vegetables (including juices and sugar products)
- Livestock (including wool and mohair)

The AMS Web site is http://www.ams.usda.gov/grades-standards.

Official Listing of Approved USDA Process Verified Programs

Company	Claims Verified	Program Scope	Verification Information
AgriTrax, LLC 39244 Kiowa Bennett Road Kiowa, CO 80117 Contact: RaeMarie Gordon-Knowles Phone: (303) 621-7747 Email: raemarie@agritrax.com	• Age Verification • Source Verification • Never Ever 3 (NE3) • Non-Hormone Treated Verified Beef Cattle (NHTC) with approval number PV5174LLA	**Livestock:** *Cattle (Beef)* **Location(s):** *Producers* **Service(s):** *Program Compliant Tags*	PVP Certificate No.: PV5174LLA Effective Date: June 23, 2015 Renewal Date: March 23, 2016
Amick Farms, LLC 2079 Batesburg Highway Batesburg, SC 29006 Contact: Christian Ouzts Phone: (803) 532-1400 Email: couzts@amickfarms.com	• Birds are never given antibiotics.	**Poultry:** Broilers **Location(s):** Hatchery, Grow-out Farms, Feed Mill and Harvest/Processing facility Harvest/Processing Facility: Amick Farms Batesburg, SC, Est. P-7987	PVP Certificate No.: PV5139ACA Effective Date May 19, 2015 Renewal Date: November 19, 2015
AngusSource® – American Angus Association 3201 Frederick Avenue St. Joseph, MO 64506 Contact: Ginette Kurtz Phone: (816) 383-5100 Email: gkurtz@angus.org	AngusSource o Angus sired cattle have a minimum of 50% Angus Genetics o Source Verified to Ranch of Origin o Group Age Verified Gateway o Second tier program to verify Group Age & Source to Ranch of Origin	**Livestock:** *Cattle (Beef)* **Location(s):** *Producers & Feedyards* **Service(s):** *Program Compliant Tags*	PVP Certificate No. PV4205AJA Rev 01 Effective Date: October 18, 2005 Renewal Date: July 31, 2015 Extension Date: September 15, 2015
Archer Daniels Midland **(ADM)** Agricultural Services 4666 Fairies Parkway Decatur, IL 62521 Contact: Brett Madison, Management Representative Phone: (815) 942-3700 Email: Brett.Madison@adm.com	Claims Verified	**Locations:** Refer to specific points **Services:** Supplies specialty grains in accordance with customers' contract requirements and internal product specifications	PVP Certificate No.: PV5083TEA Effective Date: March 30, 2008 Renewal Date: March 30, 2016

Archer Daniels Midland **(ADM)** Agricultural Services - Claims Verified

1. Sale by Decatur Specialty Grains Group of Ocean Vessels from our Gulf Operations (Ama, Destrehan, St. Elmo and Reserve, LA export elevators) of specialty grains that are certified to be segregated and delivered in an identity preserved manner to barge or rail by a signed certificate from an approved Rail or Barge Terminal Supplier.

Page 1 of 31
Last Revised August 5, 2015

Adapted: USDA, Agricultural Marketing Service (AMS)

FIGURE 6-5 Standards ensure the production of consistent products. For more information, visit: www.ams.usda.gov/services/auditing/process-verified-programs.

TABLE II
ALLOWANCES FOR DEFECTS IN RAISINS WITH SEEDS
EXCEPT LAYER OR CLUSTER

Defects	U.S. Grade A	U.S. Grade B	U.S. Grade C
	Maximum count (per 32 ounces)		
Pieces of stem	1	2	3
	Maximum count (per 16 ounces)		
Capstems in other than uncapstemmed types	10	15	20
Seeds in seeded types	12	15	20
Loose capstems in uncapstemmed types	20	20	20
	Maximum (percent by weight)		
Sugar	5	10	15
Discolored, damaged, or moldy raisins	5	7	9
Provided these limits are not exceeded:			
Damaged	3	4	5
Moldy	2	3	4
Substandard Development and Undeveloped	2	5	8
	Appearance or edibility of product:		
Slightly discolored or damaged by fermentation or any other defect not described above	May not be affected.	May not be more than slightly affected.	May not be more than materially affected.
Grit, sand, or silt	None of any consequence may be present that affects the appearance or edibility of the product.		Not more than a trace may be present that affects the appearance or edibility of the product.

Source: USDA, Agricultural Marketing Service (AMS)

FIGURE 6-6 The USDA's Agricultural Marketing Service provides guidelines for grade standards.

Grading and Certification

Quality grading is based on the standards developed for each product (see Figure 6-7). Grading services are often operated cooperatively with state departments of agriculture but typically require the payment of user fees. Quality grades provide a common language among buyers and sellers, which in turn ensures consistent quality for consumers.

Certification services facilitate the ordering and purchasing of products used by large-volume buyers. These services ensure that these buyers will be purchasing products that meet the terms of a contract with respect to quality, processing, size, packaging, and delivery. They are typical for the following commodities:

- Fresh fruits, vegetables, and specialty crops
- Processed fruits and vegetables
- Milk and other dairy products
- Livestock and meat
- Poultry
- Eggs
- Cotton
- Tobacco

Table I—Classification of Flavor		
Identification of Flavor Characteristics	**U.S. Extra Grade**	**U.S. Standard Grade**
Cooked	Definite	Definite
Feed	Slight	Definite
Bitter	—	Slight
Oxidized	—	Slight
Scorched	—	Slight
Stale	—	Slight
Storage	—	Slight

Table II—Classification of Physical Appearance

Identification of Physical Appearance Characteristics	U.S. Extra Grade	U.S. Standard Grade
Dry product:		
Unnatural color	None	Slight
Lumps	Slight pressure	Moderate pressure
Visible dark particles	Practically free	Reasonably free
Reconstituted product:		
Grainy	Free	Reasonably free

Table III—Classification According to Laboratory Analysis

Laboratory Tests	U.S. Extra Grade	U.S. Standard Grade
Bacterial estimate, SPC/gram	50,000	100,000
Coliform estimate/gram	10	10
Milkfat content, percent	Not less than 26.0, but less than 40.0	Not less than 26.0, but less than 40.0
Moisture content, percent[1]	4.5	5.0
Scorched particle content, mg:		
Spray process	15.0	22.5
Roller process	22.5	32.5
Solubility index, ml:		
Spray process	1.0	1.5
Roller process	15.0	15.0

Source: USDA

FIGURE 6-7 Tables provided by the USDA indicate differences between extra grade and standard grade.

Mission

The Agricultural Marketing Service also facilitates the strategic marketing of agricultural products in domestic and international markets by grading, inspecting, and certifying product quality in accordance with official USDA standards or contract specifications.

U.S. grade standards are quality driven and provide a foundation for uniform grading of agricultural commodities nationwide. Uniform standards provide identification, measurement, and control of quality characteristics that are important to marketing. In addition, they provide a common language for marketing, a means of establishing the value or basis for prices, and a gauge of consumer acceptance.

USDA grade standards also form the basis for quality certification services that buyers and sellers of agricultural products use in domestic and international contracting. AMS provides the following fee services on request:

- Quality standards for more than 200 agricultural commodities to help buyers and sellers trade on agreed-on quality levels
- Grading, inspection, quality assurance, and acceptance services to certify the grade or quality of products for buyers and sellers
- Inspection of facilities involved in the processing of agricultural commodities
- Assessment and registration of product and service quality-management systems to established internationally recognized standards for some commodities

A variety of quality-management services for some commodities are based on the International Organization for Standardization's audit-based quality-assurance standards. These services provide additional and alternative approaches to verifying compliance with voluntary standards and contractual requirements.

Food Quality Assurance

The AMS food-quality assurance staff manages the federal food product description system as well as associated quality-assurance policies and procedures for food procured by federal agencies. Staff members work with user agencies, research and development groups, and industry on food-specification issues. This work has led to the development of commercial item descriptions (CIDs) and quality-assurance procedures that will better serve government user needs.

Commercial Item Descriptions

In cooperation with industry, AMS develops and maintains commercial item descriptions (CIDs) for hundreds of food items. A CID is a simplified product description that concisely describes key product characteristics of an available and acceptable commercial product. These CIDs are based on attributes that describe each product's odor, flavor, color, texture, analytical requirements, and so on (see Figure 6-8). The product areas include:

- Meat, poultry, fish, and shellfish
- Dairy foods and eggs
- Fruit, juices, nectars, and vegetables
- Bakery and cereal products
- Confectionery, nuts, and sugar
- Jams, jellies, nectars, and preserves
- Bouillions and soups
- Dietary foods and food specialty preparations
- Fats and oils
- Condiments and related products
- Coffee, tea, and cocoa
- Beverages, nonalcoholic
- Composite food packages

QUALITY CONTROL

Regardless of government, research, or trade standards, most food-manufacturing plants have some type of internal, formal, quality-control or quality-assurance department. These departments perform a wide variety of functions to ensure that a consistent, quality product is produced. Quality control may perform inspection duties and laboratory tests, oversee sanitation and microbiological aspects, and guide research and development.

COMMERCIAL ITEM DESCRIPTION

CHICKEN NUGGETS, FINGERS, STRIPS, FRITTERS, AND PATTIES, FULLY COOKED, INDIVIDUALLY FROZEN

The U.S. Department of Agriculture (USDA) has authorized the use of this Commercial Item Description.

1. SCOPE

1.1 This Commercial Item Description (CID) covers individually frozen, fully cooked, solid muscle, chunked and formed or ground/chopped and formed, breaded or unbreaded, seasoned or unseasoned, chicken nuggets, fingers, strips, fritters, and patties (chicken products) packed in commercially acceptable containers, suitable for use by Federal, State, local governments, and other interested parties.

2. CLASSIFICATION

2.1 The frozen, fully cooked chicken products shall conform to the classifications in the following list and shall comply with USDA, Food Safety and Inspection Service (FSIS), Meat and Poultry Inspection Regulations, (9 CFR Part 381) and applicable State regulations. When applicable, the frozen, fully cooked chicken products shall comply with the USDA, Food and Nutrition Service (FNS), Child Nutrition Programs, National School Lunch Program

Source: USDA

FIGURE 6-8 Numerous food items purchased by the government have commercial item descriptions to ensure uniform products.

For example, the processing plant that manufactures hot dogs may check each production line on an hourly basis. They may pull packages to test the weight of the products, how well packages are sealed, and the quality of the product. **Total Quality Management (TQM)** and **Hazard Analysis and Critical Control Point (HACCP)** are two programs for controlling quality and safety.

TQM seeks continuous overall improvement in product quality by making small changes in ingredients, manufacturing, handling, or storage. All workers at a plant are involved in and responsible for a product's quality improvements.

HACCP, on the other hand, is a preventative management food-safety system. Potential biological, chemical, and physical hazards are analyzed throughout processing. A step-by-step analysis of processes for manufacturing, storing, and distributing a food product is conducted, and then tight controls are established at potential problem points. Control measures are put in place before problems occur. HACCP is covered more thoroughly in Chapter 26, "Food Safety."

Current Good Manufacturing Practices (CGMPs) are guidelines that a company uses to evaluate the design and construction of food-processing plants and equipment. These standards require that all stainless steel and plastics used during processing must meet food-grade specifications. Agencies such as the USDA and the Food and Drug Administration (FDA) will help food companies select appropriate equipment.

The CGMPs also require that standards for hygiene and food contact procedures must be met. These procedures include employees wearing white uniforms, hairnets, disposable gloves, face masks, and other protective gear. Standards for cleaning and sanitizing practices in food-processing plants and equipment are also outlined in the CGMPs.

The CGMPS also address treating water to make it drinkable, filtering air, and treating

HOW TO READ A MARKET REPORT

The following terms and definitions are frequently used in Fruit and Vegetable Market News reports:

QUALITY includes size, color, shape, texture, cleanness, freedom from defects, and other more permanent physical properties of a product which can affect its market value.

The following terms, when used in conjunction with "quality," are interpreted as meaning:

FINE: Better than good. Superior in appearance, color, and other quality factors.

GOOD: In general, stock which has a high degree of merchantability with a small percentage of defects. This term includes U.S. No. 1 stock, generally 85 percent U.S. No. 1 or better quality on some commodities, such as tomatoes.

FAIR: Having a higher percentage of defects than "good." From a quality standpoint, having roughly 75 percent U.S. No. 1 quality with some leeway in either direction.

ORDINARY: Having a heavy percentage of defects as compared to "good". Roughly 50 to 65 percent U.S. NO. 1 quality.

POOR: Having a heavy percentage of defects, with a low degree of salability, except to "low priced" trade. More than 50 percent grade defects.

Adapted: USDA, Agricultural Marketing Service (AMS)

FIGURE 6-9 Knowledge of government standards is necessary to read market reports. For additional information, visit www.marketnews.usda.gov.

food-processing wastes. The management of unavoidable pests in food-processing plants and warehouses is also done to ensure that CGMPs are used.

In effect, the CGMPs cover every aspect of food processing (see Figure 6-9). The FDA and the USDA use these guidelines when inspecting a plant to ensure it is in compliance with the regulations set forth in the federal Food, Drug, and Cosmetic Act. The FDA provides copies of the CGMP regulations.[2]

SUMMARY

Consumers expect certain qualities in their food, including color, flavor, texture, and size. When any of these are missing or different than expected, the food is rejected. Food science determines and uses methods to measure food-quality factors. These methods ensure a consistent and reliable product. Some evaluation methods use chemical and mechanical techniques. Others are completely human, such as taste panels. The USDA's AMS establishes quality and grading standards. Also, in cooperation with industry, the AMS develops and maintains commercial item descriptions for hundreds of items. Within the food industry, methods such as HACCP, TQM, and CGMP monitor quality.

REVIEW QUESTIONS

Success in any career requires knowledge. Test your knowledge of this chapter by answering or solving the following questions and problems.

1. List the three components of reflected light used to define colors.
2. Name one instrument used to measure texture and describe how it functions.
3. Discuss what humans can taste and what they smell and how this forms food flavor.
4. What do the following acronyms stand for? AMS, HACCP, TQM, CGMP, CID.
5. What are CIDs? Why are they maintained in the food industry?
6. List six factors that can influence the flavor of food.
7. Explain what usually causes changes in the texture of food products.
8. Describe common qualities that consumers expect of their food.
9. What is rheology?
10. How do fats and lipids affect the texture of food?

STUDENT ACTIVITIES

1. Cut an apple or a potato and record how long it takes for browning to occur on the cut surface. Experiment with different environmental factors such as light and temperature and compare their times.
2. Make a list of foods you eat and describe their color. Discuss what would happen to your consumption if a food's color changed.
3. Leave a slice of bread on a plate for a couple of days. Record the textural changes. Explain what activities are taking place in the bread as the texture changes.
4. Conduct a taste test. This can involve taste alone or taste and food appearance. For example, find out how red color affects food choice, or compare the taste of a name brand product with a generic product. A taste test could also be designed around a food's preferred texture. Make a hypothesis, record the data from the test, and report your findings to the class.
5. Many charts are available that visually explain government grading standards. Obtain one of these charts and display it.
6. Remove potato chips from their packaging and place them in a plastic bag exposed to light. Explain the changes after few days.
7. Visit the USDA's AMS Web site on the Internet and describe the quality standards for one of the product groups or find a CID for a food item. Create a computer-generated poster that explains the quality standards.

ADDITIONAL RESOURCES

- Food Quality and Safety: http://www.foodqualityandsafety.com/
- Food Quality and Safety in the World Health Program: http://foodqualityandsafety. wfp.org/
- Why Food Browns: http://www.scienceofcooking.com/maillard_reaction.htm
- The Impact of Texture: http://www.starchefs.com/features/textures/html/ index.shtml

Internet sites represent a vast resource of information. Those provided in this chapter were vetted by industry experts, but you may also wish to further explore the topics discussed in this chapter using a search engine such as Google. Keywords or phrases may include the following: *HACCP, Maillard reaction, rheology, phenolic compounds, taste panels, pasteurization, quality food standards, quality grading, food quality assurance, commercial items descriptions, quality control.* In addition, Table B-7 provides a listing of useful Internet sites that can be used as starting points.

REFERENCES

Corriher, S. O. 1997. *Cookwise: The hows and whys of successful cooking.* New York: William Morrow & Co.

Cremer, M. L. 1998. *Quality food in quantity. Management and science.* Berkeley, CA: McCutchan Publishing.

Drummond, K. E. 1994. *Nutrition for the food service professional,* 2nd ed. New York: Van Nostrand Reinhold.

Gardner, J. E., Ed. 1982. *Reader's digest. Eat better, live better.* Pleasantville, NY: Reader's Digest Association.

McGee, H. 2004. *On food and cooking. The science and lore of the kitchen.* New York: Scribner.

Vaclavik, V. A., and E. W. Christina. 2013. *Essentials of food science,* 4th ed. New York: Springer.

ENDNOTES

1. The Science of Cooking; https://www.exploratorium.edu/cooking/meat/INT-what-makes-flavor.html. Last accessed June 18, 2015.
2. U.S. Food and Drug Administration; http://www.fda.gov/Food/GuidanceRegulation/default.htm. Last accessed June 18, 2015.

CHAPTER 7

Unit Operations in Food Processing

OBJECTIVES

After reading this chapter, you should be able to:

- Describe materials handling in the food industry
- Name three methods of reducing the size of a food product
- Name three methods for separating food products
- Identify two general types of pumps
- Describe four factors that affect mixing
- Describe five factors that influence heat transfer
- Identify five unit processes that include heat transfer
- Understand the uses of food drying methods
- List two examples of a formed food
- Describe the purposes of concentration
- Identify reasons for packaging food products
- Discuss why some unit operations overlap

NATIONAL AFNR STANDARD

FPP.01

Develop and implement procedures to ensure safety, sanitation and quality in food product and processing facilities.

KEY TERMS

agglomeration
aggregation
centrifuge
concentration
conduction
convection
disinfecting
extrusion
food irradiation
food processing
freeze-drying
gelation
gravity flow
heat transfer
high hydrostatic pressure (HHP)
impeller
laminar
microfiltration
micron
ohmic heating

KEY TERMS (continued)

permeate
reciprocating
retentate
reverse osmosis (RO)
sanitizing
solutes
specific heat
standard operating procedures (SOPs)
supercritical fluids
supercritical fluid (SCF) extraction
ultrafiltration (UF)
viscosity

Food processing is the procedure of taking raw materials and preparing them so that they become foods for human consumption. Most food processing comprises a series of physical or chemical processes that can be broken down into many basic operations. These unit operations can stand alone but depend on logical physical principles. Unit operations include materials handling, cleaning, separating, size reduction, fluid flow, mixing, heat transfer, concentration, drying, forming, packaging, and controlling.

MATERIALS HANDLING

Materials handling includes the range of operations from harvesting on the farm or ranch, refrigerated trucking of perishable produce, and transporting live animals to conveying a product such as flour from a railcar or truck to a bakery storage bin (see Figure 7-1). During all of these operations, sanitary and safety conditions must be maintained, and bacterial growth must be minimized. In addition, quality must be maintained while minimizing product loss. Also, all transfers and deliveries of materials must be on time and that time kept to a minimum for efficiency and quality.

Materials handling involves trucks and trailers, harvesting equipment, railcars, a variety of conveyors, forklifts, storage bins, pneumatic (air) lift systems, and so on.

© Jovan Jaric/Dreamstime.com

FIGURE 7-1 Dumping wheat grain, a materials handling operation.

CLEANING

The way foods are grown or produced in open environments on the farm or ranch often requires that the products be cleaned before they are used. Cleaning ranges from removing dirt to removing bacteria from liquid foods. Brushes, high-velocity air, steam, water, vacuum, magnets, microfiltration, and mechanical separation are all used to clean foods. Cleaning methods are prescribed by the type of surface a food product has (see Figure 7-2).

Washing is the physical removal of soil, dust, organic matter, and pathogens from a surface, including the hands of people who work with food products. The basic steps of washing are wet, scrub, rinse, and then dry. Soap, cleaners, and detergents are used in the scrub process to lift contaminants

© Richard Thornton/Shutterstock.com

FIGURE 7-2 Potatoes arriving at a packing plant on trucks are transferred to conveyers to be washed. Cleaning is just another step involved in processing food.

from product surfaces so they can be rinsed away. Washing does not necessarily kill bacteria and microbes, but it will remove many of these as scrubbing and rinsing remove soil or dust particles to which they are attached. **Sanitizing** and **disinfecting** are words that are often used interchangeably, but there are slight differences. Sanitizers are used to reduce germs from surfaces but do not entirely remove them; enough germs are removed that surfaces are considered safe. Disinfectants are chemical products that destroy or inactivate germs and prevent them from growing. They have no effect on dirt, soil, or dust. Disinfectants and sanitizers are regulated by the U.S. Environmental Protection Agency.

Food-processing equipment also requires frequent, thorough, and special cleaning to maintain the quality of the product. The floors and walls of processing facilities also must be cleaned. Most farms will set their own **standard operating procedures (SOPs)** for scheduled cleaning and sanitizing. SOPs are put into place to reduce personal risk, reduce the potential spread of foodborne illnesses, keep facilities clean, and maintain records for farm traceability. Each farm should post SOPs at all times and train workers on the required steps and procedures. Workers should know the appropriate steps to take in case an employee is absent so that the processes can still be followed properly.

SEPARATING

Separating items can be achieved on the basis of density or size and shape. Separations based on density differences include the separation of cream from milk, the recovery of solids from suspensions, and the removal of bacteria from fluids.

Cream Separator

Milk can be separated into skim milk and cream based on the density difference between fat and nonfat solids of milk. A cream separator obtains the cream from milk using a disc-type **centrifuge** in which the fluid is separated into low- and high-density fluid streams, which permits the separate collection of cream and skim milk.

Clarification

Sediment and microorganisms can be removed centrifugally in a clarifier, which is generally a disc-type centrifuge that applies forces of 5,000 to 10,000 times gravity and forces the denser material to the outside. By periodically opening the bowl, the solids can be continuously removed from the remainder of the fluids. This same principle has been used to recover yeast cells from spent fermentation broths and to continuously concentrate baker's cheese from whey.

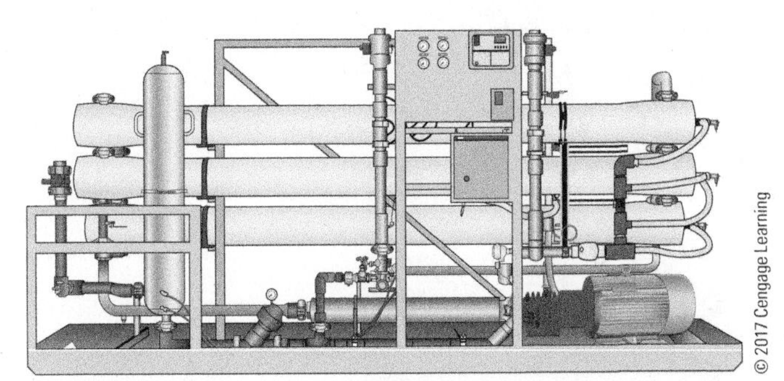

FIGURE 7-3 Reverse osmosis unit, a separation process.

Membrane Processes

Reverse osmosis (RO), **ultrafiltration (UF)**, and **microfiltration (MF)** are processes that use membranes with varying pore sizes to separate particles on the basis of size and shape (see Figure 7-3). Reverse osmosis uses membranes with the smallest pores to separate water from other **solutes** or dissolved substances. Ultrafiltration uses membranes with larger pores to retain proteins, lipids, and colloidal salts while allowing smaller molecules to pass through to the permeate phase. Microfiltration, with pores less than 0.1 **micron**, is used to separate fat from proteins and to reduce microorganisms from fluid food systems. High-pressure pumps are required for RO, and low-pressure pumps for UF and MF.

SCIENCE CONNECTION!

Research different processes used in growing, handling, or processing foods. What food safety concerns are present in the process?

SIZE REDUCTION

Size reduction can be through the use of high-shear forces, graters, cutters, or slicers (see Figure 7-4). Emulsions with minute fat globule droplets are frequently made with a homogenizer, which is a high-shear positive pump that forces fluid through a tiny opening at extremely high pressure to form an emulsion or reduce its size. The positive pump uses a **reciprocating** or rotating cavity between two lobes or gears between a stationary cavity and a rotor. The fluid forms the seal between the rotating parts.

Typical equipment for size reduction in meat products and their component parts include:

- Grinder
- Bacon slicer
- Sausage stuffer
- Vertical chopper

In addition, ball mills grind products into fine particles.

Sometimes this process is better thought of as size adjustment either through size reductions using such methods as slicing, dicing, cutting, or grinding, or as size enlargement by **aggregation**, **agglomeration**, or **gelation**.

PUMPING (FLUID FLOW)

The transporting of fluids is achieved either by **gravity flow** or through the use of pumps (see Figure 7-5). In gravity flow, the flow is **laminar**, where the flow is transferred from the fluid to the wall between adjacent layers of adjacent molecules, which do not mix. Often, though, fluids are transported from one unit operation or process to another by pumps that create turbulent flow where

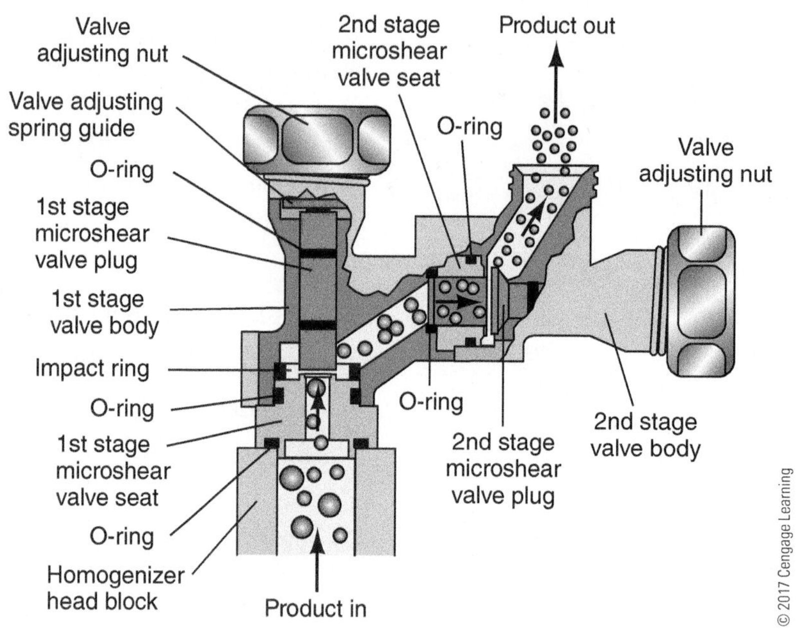

FIGURE 7-4 Homogenization, a size-reduction operation.

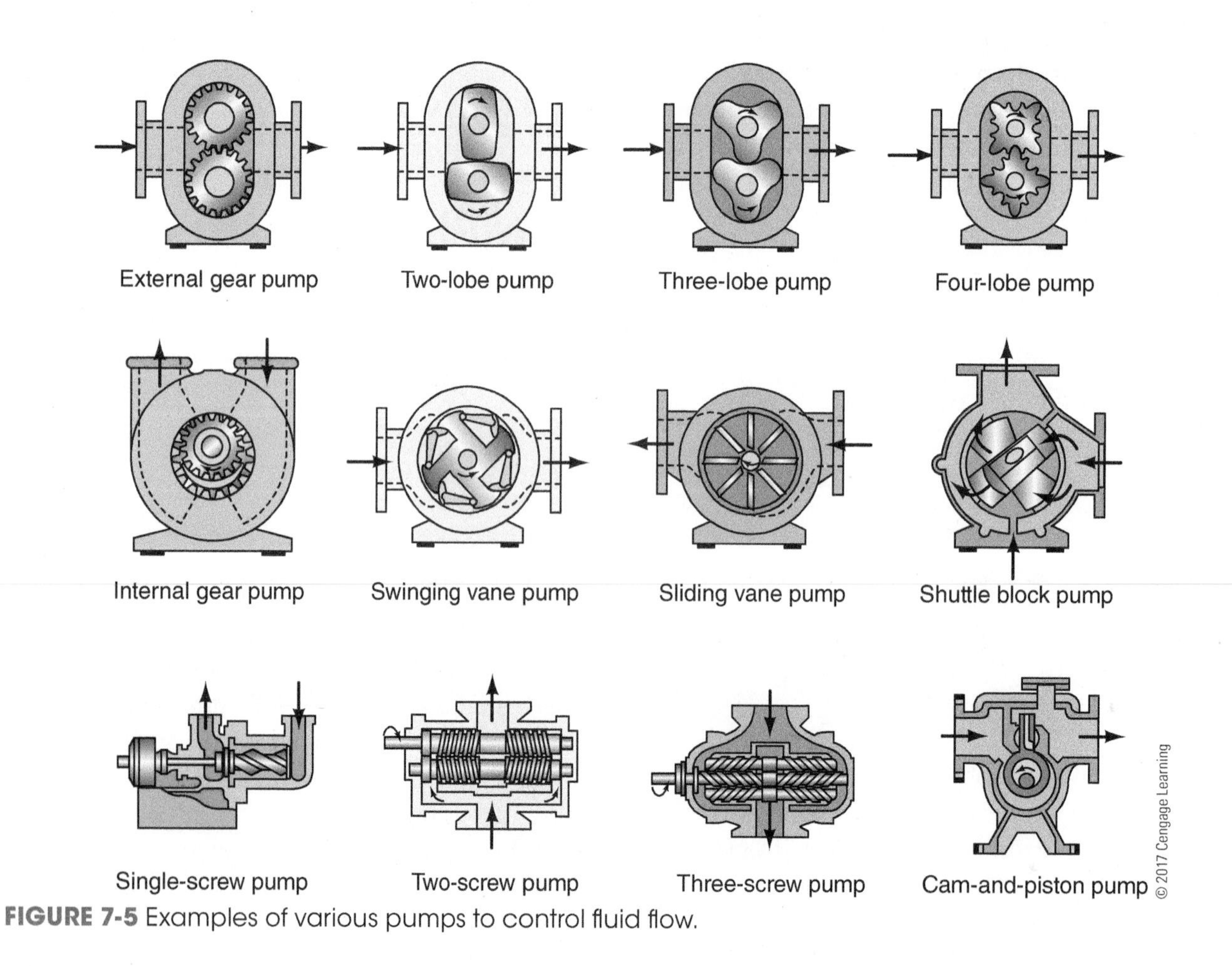

FIGURE 7-5 Examples of various pumps to control fluid flow.

adjacent particles mix. Two different types of pumps are commonly used for different purposes:

1. A centrifugal pump uses a rotating **impeller** to create a centrifugal force within the pump cavity. The flow is controlled by the choice of impeller diameter and the rotary speed of the pump drive. The capacity of a centrifugal pump depends on the impeller's speed and length and the pump's inlet and outlet diameters.
2. A positive pump consists of a reciprocating or rotating cavity between two lobes or gears and a rotor. Fluid enters by gravity or a difference in pressure, and the fluid forms the seals between the rotating parts. The rotating movement of the rotor produces the pressure that causes the fluid to flow.

MIXING

An agitation (mixing) device may be placed in a tank for several reasons. The two major reasons to mix are to provide **heat transfer** and to incorporate ingredients. Different mixer configurations are used to achieve different goals. The efficiency of mixing will depend on the following:

- Design of the impeller
- Diameter of the impeller
- Speed
- Baffles

HEAT EXCHANGING

Heat is either transferred into a product (heating) or removed from it (cooling). Heating is used to destroy microorganisms, produce a healthful food, prolong shelf life through the destruction of certain enzymes, and promote a product with acceptable taste, odor, and appearance.

Five factors influence the heat transfer into or out of a product:

1. Heat-exchanger design (see Figure 7-6)
2. Heat-transfer properties of the product, such as:
   - **Specific heat** (the amount of heat required to change the temperature of a unit mass of product a specific temperature without changing the state of the material),

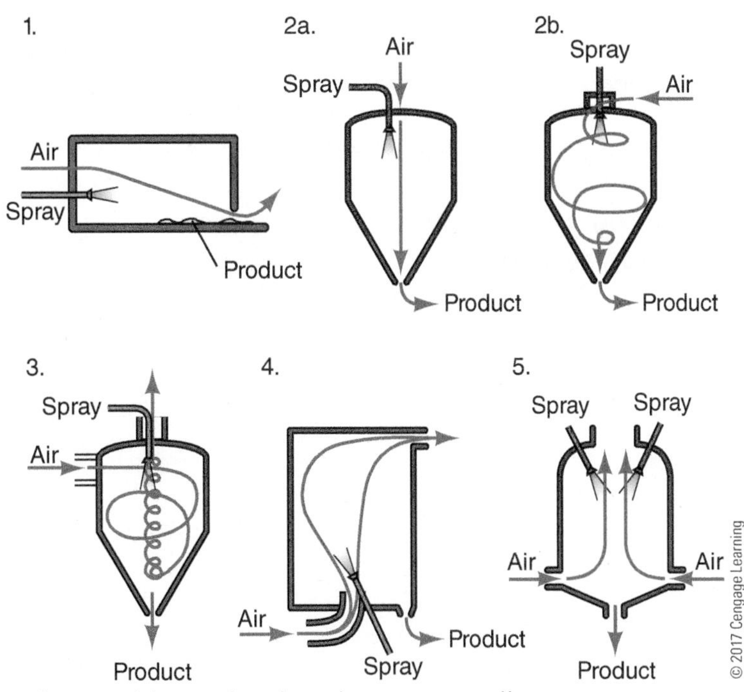

FIGURE 7-6 Examples of spray driers, a heat-exchange operation.

- Thermal conductivity (the rate by which heat is transferred through a material), and
- Latent heat (the heat required to change the state of a material)

3. Density (weight per unit volume)
4. Method of heat transfer such as:
   - **Conduction** (transfer from molecule to molecule through the material),
   - Radiation (transfer from electromagnetic radiation of a body because of the vibration of its molecules), or
   - **Convection** (transfer through movement of mass)
5. **Viscosity** (related to the amount of force required to move the fluid product)

A variety of heat exchanges are used in the food industry, including:

- Plate heat exchanges
- Tubular heat exchanges
- Swept surface heat exchangers

Plate heat exchanges pass fluid over a plate the same time as a heating or cooling medium is passed up or down the other side of the plate. The thin film makes for rapid heat transfer and is the most efficient method of heating fluids of low viscosity.

HISTORICAL FREEZE-DRYING

The Incas of Peru, who stored their vegetables near the peaks of high mountains, first used the process of freeze-drying. The vegetables froze solid but over time the frozen water sublimated into the thin mountain air (converted directly to vapor without passing through the liquid state), leaving behind the perfectly preserved, desiccated vegetables.

Modern freeze-drying started during World War II to preserve blood plasma for use on the frontlines. Freeze-drying grew in popularity in the United States as the space program grew; some of became known as astronaut food.

Today, freeze-drying is done using flash-freezing and vacuum dehydration. Freeze-drying preserves almost all the nutrients of foods, as well as the important flavor elements. Freeze-dried foods reconstitute to their original state when placed in water. They are a shelf staple at room temperatures and do not require cold storage. In addition, the weight of a freeze-dried food is reduced by 70% to 90% without a change in its volume. This allows products to be lightweight and easy to handle. Freeze-drying can now be done at home with a vacuum chamber.

To learn more about freeze-drying, search the Web or visit these sites:

- http://www.freeze-dry.com/biotech.html
- http://www.eufic.org/article/en/food-technology/food-processing/artid/Freeze-drying-value-quality-products
- http://www.lyotechnology.com/fd-milestones.html
- http://www.freezedry.com/r_process.htm

Tubular heat exchanges in general are composed of a tube within a tube; a product and a heating or cooling medium flow in opposite (countercurrent) directions. This low-cost method of heating or cooling is applied to fluids of higher viscosities that generally pass through a plate heat exchanger.

Swept surface heat exchanges have blades that scrape the surface of the heat exchanger and bring new product continuously to the heating or cooling surface. They are used for fluids of extremely high viscosity. An ice cream freezer is an example of a swept surface heat exchanger.

Common unit processes that include heat transfer as a unit operation include:

- Pasteurization (heat)
- Sterilization (heat)
- Drying (heat)
- Evaporation (heat)
- Refrigeration (cold)
- Freezing (cold)

CONCENTRATION

Concentration can be achieved through both evaporation and reverse osmosis (RO). Evaporation generally involves heating the fluid in a vessel under a vacuum to cause water to change state from liquid to vapor; water is then recovered by passing the vapor through a condenser. In some products, evaporation causes the loss of flavor volatiles. When this happens, a low-temperature unit is added to recover the flavor volatiles so that they can be added back to the product.

To reduce operating costs, multiple-effect evaporators are used, which have two or more evaporators placed in a series to continuously concentrate a fluid product. This increases the efficiency of the evaporation process.

In RO, fluid passes through a semipermeable membrane with tiny pores that permit only the transfer of water. Most systems consist of a membrane cast on a solid porous backing—usually in the form of a tube. High pressure is applied to force the water (called **permeate**) through the membrane. The concentrated fluid (called **retentate**) is retained in the tubing. The rate of water removal decreases as the fluid is concentrated, until it is no longer economically feasible to remove more water.

Concentration is often used as a preliminary step to drying to reduce the amount of water that needs to be removed during drying, thus reducing drying costs. Evaporation can achieve higher solids economically than can RO. RO is preferred over evaporation for heat-sensitive fluids.

Contact equilibrium processes or mass transfer may or may not require a change in state. Generally, a type of molecule is transferred to or from a product. Processes that use mass transfer include distillation, gas absorption, crystallization, membrane processes, drying, and evaporation (see Figure 7-7).

FIGURE 7-7 Drying extracted potato starch.

DRYING

Three common methods of drying are (1) sun or tray drying, (2) spray drying, and (3) **freeze-drying**. Sun or tray drying is least expensive, followed by spray drying and freeze-drying. The drying method of choice is generally based on a product's characteristics.

Products that have already been solid lend themselves to sun or tray drying, including fruits and vegetables. The products may be dried by exposure to sunlight or placed in trays and dried in a current of warm or hot air. This method is used to make raisins from grapes.

Products that are highly heat sensitive are freeze-dried. Commercially, instant coffee is the most widely freeze-dried product. Fruits, vegetables, meats, and grains all can be freeze-dried. This is expensive process and the most costly food-preservation method, however. In freeze-drying, the moisture is removed without a phase change (sublimation). Many emergency foods and meals ready-to-eat (MREs) foods are freeze-dried.[1]

The most common drying method is spray drying, which is applied to fluid products. The bulk density (weight per unit volume) is controlled to a large extent by the solids that are sent to the dryer. Several different designs of spray nozzles are used to atomize the fluid into the heated air. These generally are either centrifugal nozzles or high-pressure spray nozzles. The type of nozzles will vary with the product being dried.

For highly hygroscopic products (that is, they take on water from the air), the dried product may be partially rewetted and then redried. This produces agglomerated products that are easily dispersed into solution. Spray dried powders with a surfactant are also a method for improving dispersion.

An older method is roller drying. The product was allowed to flow over a hot, rotating drum; the dried product was then scraped off. This was a low-cost method of drying, but it created considerable heat damage to the product.

MATH CONNECTION!

Conduct an experiment with different slices of different fruits. How long does it take to dry the fruit in the sun? Record your findings. (Unless you have food-grade materials, this is not an edible lab.) Compare the color, feel, and size of the fruit slices. For more information, see http://nchfp.uga.edu/how/dry/sun.html.

FORMING

Often foods need to be formed into specific shapes—for example, hamburger patties, chocolates, jellies, tablets, snack foods, breakfast cereals, butter and margarine bars, cheeses, variety breads, and sausages. Processes used to form foods include compacting, pressure **extrusion**, molding, the use of powdering and binding agents, heat and pressure, and extrusion cooking. In extrusion, products are forced through a die to create a formed shape. For example, you can extrude pasta dough to form many shapes from spaghetti to macaroni.

PACKAGING

Packaging is used for a variety of purposes, including shipping, dispensing, and improving the usefulness of the product. In addition, packaging helps protect food from microbial contamination, dirt, insects, light, moisture, drying, flavor changes, and physical alterations. The principal role of packaging is to ensure the food product is safe. Packaging also provides important information to the consumer about ingredients, nutritional facts, and allergy concerns. Packaging is marketed to target audiences, and food companies spend a significant amount of money to ensure quality and attractiveness to consumers.

Foods are packaged in metal cans, glass bottles, plastic bottles, paper, cardboard, and plastic and metallic films. Many foods are packaged in a combination of materials. In the past, rigid containers of glass and metal were commonly used. Now, more products are being packaged in flexible and formable containers, such as retortable pouches used for fruit juices and chewy bars.

Machines that automatically package food products operate at high speeds, and the complete process is stepwise and automated from the forming of the container, filling of the container, to the sealing, labeling, and stacking.

FIGURE 7-8 Controls help combine unit operations into a complex process.

SCIENCE CONNECTION!

Design a new package for your favorite food product! What is it made of? What makes it better than the existing packaging? How will this new packaging help market your product?

CONTROLLING

Producing a food often requires that the unit operations are combined into a complex processing operation. To ensure the quality of a food product, food processors need to measure and control these operations. The tools used in controlling and measuring include valves, thermometers, scales, thermostats, and other instruments that measure and control pressure, temperature, fluid flow, acidity, weight, viscosity, humidity, time, and specific gravity.

In modern processing plants, the instrumentation and controls are automatic and computer controlled. An operator oversees and controls the processing from a remote console (see Figure 7-8).

OVERLAPPING OPERATIONS

Some food-processing operations may use a single-unit operation, but most food processing includes a combination of unit operations to achieve the total process. For example, the manufacture of a dried coffee creamer from a combination of fluid and dry ingredients includes the following unit operations in sequence:

- Mixing
- Fluid flow
- Size reduction (homogenization)
- Heat transfer (heating)
- Fluid flow
- Heat transfer (cooling)
- Mass transfer (conversion of water to vapor during drying)
- Pasteurization of milk (including unit operations of fluid flow and heat transfer, both heating and cooling)

Other examples of unit processes and associated unit operations are described below:

- Freeze-drying involves heat transfer and mass transfer.
- Extrusion requires fluid flow, heat transfer, mass transfer, and size reduction in the case of cereals and snack foods.

- Ice cream manufacturing comprises two processes: (1) mix-making, which uses mixing, fluid flow, heat transfer, and size reduction; and (2) freezing, which involves fluid flow, heat transfer, and mass transfer of air into the ice cream.

CONSERVING ENERGY

Food processing is energy intensive, so the energy represents a significant share of the cost of the final product. Food processors always look for new ways to optimize the use of energy. For example, heat that is used or removed is captured and used elsewhere in the process. Food processes such as dehydration, concentration, freezing, and sterilization are reevaluated for times and temperature adjustments. Processors continuously monitor the energy required for all the unit processes and look for more efficient ways. Also, such simple measures as improved temperature control, insulation, controlling ventilation rates, reducing lighting, and checking sensors save energy.

OTHER PROCESSES

Major goals of food scientists and food-processing engineers are to develop new methods that improve quality, increase efficiency, or both. New processes are constantly being tried in the unit operations. Newer processes include ohmic heating, irradiation, supercritical fluid extraction, and high hydrostatic pressure.

In **ohmic heating** (see Figure 7-9), an advanced thermal-processing method, food material—which serves as an electrical resistor—is heated by passing electricity through it. Electrical energy is dissipated as heat, which results in rapid and uniform heating. This process destroys microorganisms and maintains food quality. Ohmic heating can be used for heating liquid foods containing large particulates, such as soups, stews, and fruit slices in syrups and sauces, and heat-sensitive liquids. The technology is useful for treating foods with considerable protein, which tends to denature and coagulate when thermally processed. For example, liquid egg can be

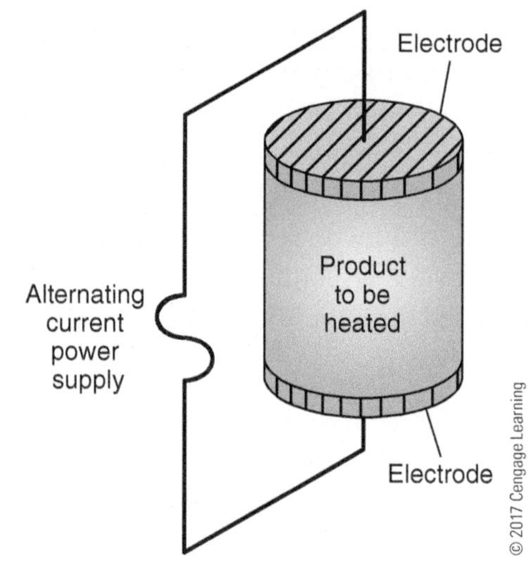

FIGURE 7-9 Diagram of the principle of ohmic heating. An alternating electrical current is passed through a food in a conducting fluid.

ohmically heated in a fraction of a second without the egg coagulating. Juices can be treated to inactivate enzymes without affecting flavor. Other potential applications of ohmic heating include blanching, thawing, online detection of starch gelatinization, fermentation, peeling, dehydration, and extraction.[2]

Food irradiation is a technology for controlling spoilage and eliminating food-borne pathogens (see Figure 7-10). It is similar to conventional pasteurization and is often called *cold pasteurization* or *irradiation pasteurization.* Like pasteurization, irradiation kills bacteria and other pathogens. The difference between the two methods is the source of the energy used to destroy the microbes—pasteurization using heat and irradiation using the energy of ionizing radiation.[3]

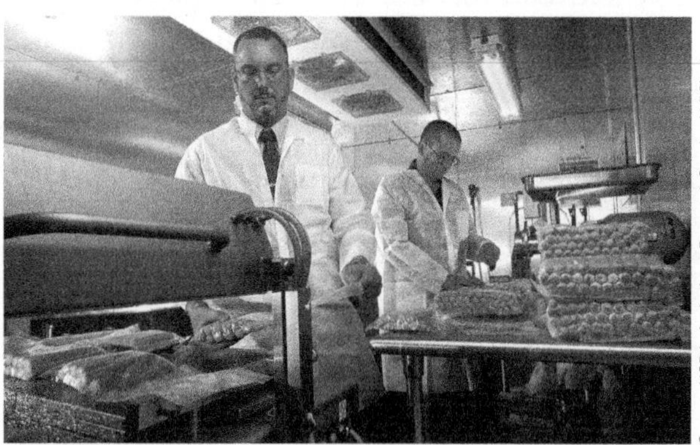

Source: USDA, Agriculture Research Service (ARS), photo by Stephen Ausmus

FIGURE 7-10 Scientists vacuum-seal hot dogs to get them ready for irradiation.

Source: USDA, Agricultural Research Service, photo by Keith Weller

FIGURE 7-11 To remove oil from corn hull fiber, this chemist pours a sample into a supercritical fluid extractor.

Supercritical fluid (SCF) extraction uses a gas such as carbon dioxide at high pressure to extract or separate the food components (see Figure 7-11). **Supercritical fluids** are increasingly replacing organic solvents used in industrial purification and recrystallization operations because of regulatory and environmental pressures on hydrocarbon and ozone-depleting emissions. SCF-based processes have helped eliminate the use of hexane and methylene chloride as solvents. Coffee and tea are decaffeinated using SCF extraction, and most major brewers in the United States and Europe use flavors that are extracted from hops with supercritical fluids.[4]

High hydrostatic pressure (HHP) is one of the newest food-preservation technologies and is used to inactivate microorganisms (see Figure 7-12). This technology subjects foods, which can be liquid or solid, packaged or unpackaged, to pressures between 40,000 and 80,000 pounds per square inch for five minutes or less. The high pressure does not destroy the food, but it does inactivate microorganisms living on both the surface and in the interior of the food. The technology is increasingly popular in the fish and shellfish industries because it accomplishes the near elimination of pathogenic or spoilage microorganisms at room temperature. This process only uses a small amount of energy as compared to heating to 212 °F in traditional methods.

HHP offers several advantages such as reduced process times; reduced physical and chemical changes; retention of freshness, flavor, texture, appearance, and color; and elimination of vitamin C loss. For example, it can be used to accomplish nonthermal treatment of raw or fresh oysters. Oysters are a natural fit, as are products where disease outbreaks have been associated with minimal processing of the product. The expense of this technology is the largest barrier to its integration.[5]

Source: USDA, Agriculture Research Service (ARS), photo by Peggy Greb

FIGURE 7-12 A food technologist prepares ingredients for high-pressure homogenization, a process that can increase the fiber content in fluid foods.

Food scientists will continue to develop new processes to control food pathogens while improving the shelf life and quality of foods. In an increasingly health-conscious world, an uncontaminated well-preserved food source is the ultimate goal. The expense of developing and implementing new technologies will continue to be a concern because of the food costs it passes to consumers.

SUMMARY

Unit operations make up the basics of food processing. These include materials handling, cleaning, separating, size reduction, fluid flow, mixing, heat transfer, concentration, drying, forming, packaging, and controlling. Most food processing involves a combination or an overlap of these unit operations. Where unit operations overlap or are combined, complex controls ensure the proper function of each operation. Many of the unit processes discussed in this chapter are discussed in more detail in the following chapters.

REVIEW QUESTIONS

Success in any career requires knowledge. Test your knowledge of this chapter by answering these questions or solving these problems.

1. Explain how ice cream is an example of overlapping operations.
2. Why are foods packaged?
3. Describe the meaning of specific heat.
4. Name the three methods for separating foods.
5. What are the two types of fluid flow pumps?
6. How does the method of plate heat function?
7. List the four factors that affect the mixing of food products.
8. Why is it important to handle food materials carefully?
9. Explain the three most common methods of drying foods.
10. Identify three membrane processes for separating food products.

STUDENT ACTIVITIES

1. Bring in seeds (beans or wheat) directly from a combine. Clean the product and separate the broken seeds. Report on the processes needed to convert the seed into various products.
2. Develop a demonstration of conduction, radiation, and convection and present it to the class.
3. Build a simple food drier and dry some fruits. Record the changes in shape, texture, and taste.
4. Bring a collection of food packages to class and discuss the purposes of the packaging. Explain which information you believe is most important.
5. Identify a food product you eat. List the unit processes required to produce that food.
6. Pick a food product that has been formed. Describe to the class how is it formed.

ADDITIONAL RESOURCES

- Food Science: http://bulletins.psu.edu/undergrad/courses/F/FD%20SC/430/200910FA
- Engineering Aspects of Food Processing: http://www.eolss.net/sample-chapters/c06/e6-34-09-10.pdf
- Food-Processing Equipment: http://www.angrau.ac.in/media/10829/fden223fpequii.pdf
- Sanitary Design and Construction of Food Processing and Handling Facilities: http://ucfoodsafety.ucdavis.edu/files/26503.pdf

Internet sites represent a vast resource of information. Those provided in this chapter were vetted by industry experts, but you may also wish to further explore the topics discussed in this chapter using a search engine such as Google. Keywords or phrases may include the following: *food processing, unit operations, freeze-dried, specific heat, ultrafiltration, food-processing equipment, pasteurization, food sterilization, food evaporation, refrigeration of food, freezing food, food forming, food packaging, ohmic heating of food, supercritical fluid extraction, heat exchanges (plate, tubular, swept surface).* In addition, Table B-7 provides a listing of some useful Internet sites that can be used as a starting point.

REFERENCES

Cremer, M. L. 1998. *Quality food in quantity. Management and science.* Berkeley, CA: McCutchan Publishing.

Potter, N. N., and J. H. Hotchkiss. 1995. *Food science,* 5th ed. New York: Chapman & Hall.

Vaclavik, V. A., and E. W. Christina. 2013. *Essentials of food science*, 4th ed. New York: Springer.

Vieira, E. R. 1996. *Elementary food science,* 4th ed. New York: Chapman & Hall.

ENDNOTES

1. National Center for Home Preservation: http://nchfp.uga.edu/how/dry/sun.html. Last accessed June 26, 2015.
2. The Ohio State University: http://ohioline.osu.edu/fse-fact/0004.html. Last accessed June 27, 2015.
3. US Environmental Protection Agency: http://www.epa.gov/radiation/sources/food_irrad.html. Last accessed June 27, 2015.
4. University of Illinois at Chicago: http://tigger.uic.edu/~mansoori/SCF.and.SFE.by.TRL.at.UIC.pdf. Last accessed June 27, 2015.
5. Virginia Cooperative Extension Service: http://seafood.oregonstate.edu/.pdf%20Links/High-Hydrostatic-Pressure-has-Potential.pdf. Last accessed June 27, 2015.

CHAPTER 8

Food Deterioration

OBJECTIVES

After reading this chapter, you should be able to:

- **List three general categories of food deterioration**
- **Discuss shelf life and dating**
- **Name six factors that cause food deterioration**
- **Understand six preservation techniques that prevent deterioration**
- **Describe normal changes in food products following harvest or slaughter**
- **Identify four food enzymes and describe their function**

NATIONAL AFNR STANDARD

FPP.01

Develop and implement procedures to ensure safety, sanitation, and quality in food product and processing facilities.

KEY TERMS

denature
deterioration
emulsion
facultative
food-borne disease
mesophilic
mycotoxins
obligative
organoleptic
osmosis
pyschrophilic
radiation
shelf life
thermophilic
water activity (A_w)

Deterioration is the breakdown of food's composition. This includes changes in organoleptic quality, nutritional value, food safety, aesthetic appeal, color, texture, and flavor. To some degree, all foods deteriorate after harvest. The role of food science is to minimize negative changes as much as possible.

TABLE 8-1 Useful Life at 70 °F

FOOD	DAYS
Meat	1 to 2
Fish	1 to 2
Poultry	1 to 2
Dried, smoked meat	360+
Fruits	1 to 7
Dried fruits	360+
Leafy vegetables	1 to 2
Root crops	7 to 20
Dried seeds	360+

TYPES OF FOOD DETERIORATION

The three general categories of food **deterioration** are (1) physical, (2) chemical, and (3) biological.

Among the many factors that cause food deterioration are light, cold, heat, oxygen, moisture, dryness, types of **radiation**, enzymes, microorganisms, time, industrial contaminants, and macroorganisms (insects, mice, and so on).

SCIENCE CONNECTION!

Conduct a research experiment on several fresh fruits or vegetables. Write your hypothesis. What time frame do you anticipate for the foods to deteriorate?

SHELF LIFE AND DATING OF FOODS

All foods have a time limit on their usefulness. This time limit depends on the type of food, its storage conditions, and other factors. If food is held at about 70 °F (21°C), its useful life varies by type, as shown in Table 8-1.

Shelf life is the time required for a food product to deteriorate to an unacceptable level of quality. How quickly that happens depends on the food item (Table 8-1), the processing method, packaging, and storage conditions. Food manufacturers put code dates on their products (see Figure 8-1). "Pack date" is the date of manufacture. The date of display is called the "display date," and the "sell-by date" is the last day a product should be sold. Some foods have a "best used by date," or the last date of maximum quality. The "expiration date" indicates when the food is no longer acceptable.

LEARN THE LINGO OF EXPIRATION DATES

The dates listed on food packaging can cause confusion. Few food items *must* be labeled—in fact, only infant formula and a few other foods are required by federal law to be labeled. The actual term *expiration date* refers to the last date a food should be eaten or used. *Last* means *last*—proceed at your own risk.

(continues)

LEARN THE LINGO OF EXPIRATION DATES (continued)

Other, more commonly spotted terms are:

- **Sell-by date.** The labeling *sell by* tells a store how long to display a product for sale. You should buy the product before the date expires. This is basically a guide for the retailer, so the store knows when to pull the item. This is not mandatory, so reach in back and get the freshest. The issue is quality of the item (freshness, taste, and consistency) rather than whether it is on the verge of spoiling.
- **Best if used by (or before) date.** This refers strictly to quality, not safety. This date is recommended for best flavor or quality. It is not a purchase or safety date. Sour cream, for instance, is already sour, but it can still have a fresh taste when freshly sour.
- **Guaranteed fresh date.** This usually refers to bakery items. They will still be edible after the date but will not be at peak freshness.
- **Use-by date.** This is the last date recommended for use of a product while at peak quality. The date has been determined by the manufacturer of the product.
- **Pack date.** In general, this date is found on canned or packaged goods, but it can be tricky, especially if it is in code. It can be month-day-year (usually, MMDDYY), or the manufacturer might revert to a variation on the Julian calendar. January would then be 001–0031 and December 334–365.

Suggested rules for food use at home include the following:

- **Milk.** Usually fine until a week after the sell-by date.
- **Eggs.** Good for 3 to 5 weeks after purchase (assuming they were bought before the sell-by date).
- **Poultry and seafood.** Cook or freeze this within 1 or 2 days.
- **Beef and pork.** Cook or freeze within 3 to 5 days.
- **Canned goods.** Highly acidic foods such as tomato sauce can keep 18 months or longer. Low-acid foods like canned green beans are probably risk-free for up to 5 years.

Source: http://www.webmd.com/a-to-z-guides/features/do-food-expiration-dates-matter?page=2.

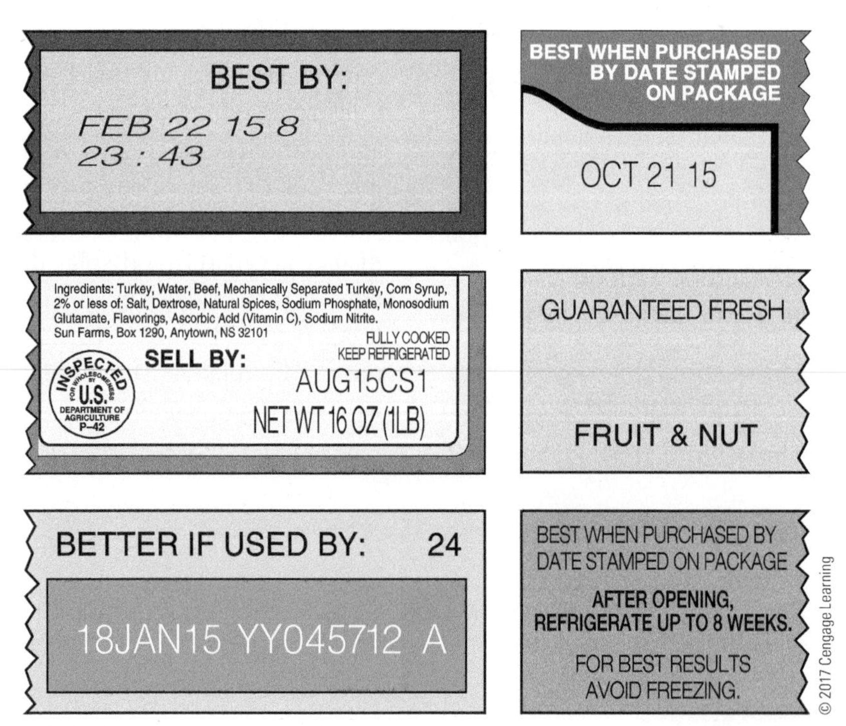

FIGURE 8-1 Dates on labels help ensure quality.

CAUSES OF FOOD DETERIORATION

Specific causes of food deterioration include the following:

- Microorganisms such as bacteria, yeast, and molds
- Activity of food enzymes
- Infestations by insects, parasites, and rodents
- Inappropriate temperatures during processing and storage
- Gain or loss of moisture
- Reaction with oxygen
- Light
- Physical stress or abuse
- Time

Deterioration can be caused by these items individually or in any combination.

Bacteria are single-celled organisms occurring in three shapes: round (cocci), rod (bacilli), and spiral (spirilla and vibrios). Some produce spores, and these spores may be resistant to heat, chemicals, and other adverse conditions.

Yeasts are the largest of the microorganisms but are still single cells, and some produce spores. Molds are larger than bacteria. They are often filamentous, and they all produce spores.

In foods, these microorganisms attack basically every type of food component: sugars, starches, cellulose, fats, and proteins. Depending on the food and the microorganism, the action on food could be to produce acids, making the food sour, or to produce alcohol. Some microorganisms produce gas, making a food foamy; still others produce unwanted pigments or toxins.

FUNGI AS A LIFE FORM

Although life forms in the fungus kingdom (molds, mushrooms, and yeasts) may seem like plants, they are actually more closely related to animals. In the history of life, fungi and primitive animals branched apart after plants had evolved.

Like some kinds of animal cells, walls made out of chitin surround fungal cells. Chitin is a complex molecule made of sugar and nitrogen. Plants use cellulose for their cell walls.

Fungi get their energy by breaking down organic molecules that they absorb directly from their environment. They emit substances that chop complex organic molecules into smaller ones that are easier to absorb. Fungi are vital to the global ecosystem because they are so good at breaking down anything organic.

To learn more about fungi, search the Web or visit this site:

- http://www.kew.org/science-conservation/plants-fungi/fungi/about
- http://www.apsnet.org/edcenter/intropp/pathogengroups/pages/introfungi.aspx

Bacteria, Yeast, and Mold

Thousands of species of microorganisms exist, and a few hundred are associated with foods. Not all are bad because some are desirable in food preservation. Microorganisms are found in soil, water, and air as well as on animal skins, on plant surfaces, and in digestive tracts. Most are usually not found in healthy tissue.

Environmental conditions that affect microbial growth include temperature and oxygen. Microbes that prefer cold temperatures are said to be **pyschrophilic. Mesophilic** microorganisms prefer normal temperatures; **thermophilic** microorganisms prefer hot temperatures. Bacteria or molds that require atmospheric oxygen are said to be *aerobic*, and those yeast and bacteria that do not require atmospheric oxygen are called *anaerobic*. **Facultative** microorganisms are both aerobic and anaerobic; **obligative** microorganisms can be either.

Food-Borne Disease

Food-borne disease is caused when microorganisms is present in food and cause an infection in the human who consumes the food. Food-borne illness is commonly referred to as *food poisoning*. Each year, one in six Americans are sickened by consuming contaminated foods or beverages. Infections can be caused by *Clostridium perfringen*, *Salmonella sp.*, *Escherichia coli* (*E. coli* 0157), and several others (see Chapter 26). Food intoxication occurs when a consumed food contains a chemical toxic to humans. *Staphylococcus aureus* and *Clostridium botulinum* produce toxins. Molds in foods produce **mycotoxins** such as aflatoxin. These toxins are not destroyed by heat.

Insects

Insect damage can be minor, but wounds to plant and animal tissues allow for additional damage by microorganisms. Insect damage and infestation can be so complete as to render a food inedible. Pesticides control insects, as do an inert atmosphere and cold storage.

Food Enzymes

All foods from living tissues have enzymes. Most of these enzymes will survive harvest or slaughter. In fact, at harvest or slaughter time, enzymes that control digestion and respiration proceed uncontrolled and cause tissue damage. Some postharvest enzymatic reactions are actually desirable, however. Good examples are the ripening of tomatoes and the aging or tenderizing of beef. Enzyme action can be controlled by heat, chemicals, and radiation. Food enzymes are also used in food processing (Table 8-2).

TABLE 8-2 Important Food-Processing Enzymes

ENZYME	SOURCE	IMPORTANCE	ACTION
Ascorbic acid oxidase	Citrus fruit, vegetables (squash, cabbage, cucumbers)	Browning and off flavor in fruits; destruction vitamin C activity	Oxidizes ascorbic acid to dehydro form, destroying the browning prevention ability.
Beta-amylase	Grains, wheat, barley, sweet potatoes	Produces maltose and glucose to support yeast growth or to sweeten or both	Beta-amylase in flour with fungal glucoamylase produces mixtures of fermentable sugars such as glucose and maltose.
Bromelain	Pineapple	Meat tenderizing; chill-proofing beef	Acts on collagen to hydrolyze peptides, amides, and esters from the nonreducing end.
Catalase	Meat, liver, blood, molds, bacteria, milk	Removes excess H_2O_2 from treated milk in cheese making; cheese making	Removes residual H_2O_2 (hydrogen peroxide) from treated foods; converts H_2O_2 to H_2O and oxygen. Used when pasteurization is not feasible.
Cathepsins	Animal tissue, liver, muscle	Aging of meat or game when hung	Autolytic reaction during aging of meat (on protein substrate)

(Continues)

TABLE 8-2 Important Food-Processing Enzymes

ENZYME	SOURCE	IMPORTANCE	ACTION
Cellulase	Digestive juice of invertebrates, snails	Hydrolysis cell wall breaks glucosidic link in cellulose; clarifies fruit juices; removes graininess from pears	Converts native cellulose and soluble, low-molecular-weight products such as maltose and glucose.
Chlorophyllase	Spinach; other selected vegetables, greens	Chlorophyll is made water soluble; loses color	Makes chlorophyll water soluble.
Collagenase	Clostridia bacterial	Tenderizing of meat—collagen	Hydrolyzes native collagen and denatured collagen such as gelatin and hide powder.
Elastase	Mammalian organs (hog, pancreas), microorganisms	Breaks down meat muscle fiber	Hydrolyzes elastin, hemoglobin, fibrin, albumin casein, and soy protein.
Ficin	Fig trees	Meat tenderizer	Hydrolyzes muscle fiber proteins and both types of connective tissue.
Glucoxidase	Molds, fungi	Removes glucose and oxygen in foods; impaired by residual reducing sugars such as egg white	Will catalyze the oxidation of glucose to gluconic acid with the formation of H_2O_2 (hydrogen peroxide) in the presence of moisture and atmospheric oxygen.
Lactase	Plants; almonds, peaches; calf intestine	Cheese making; removing lactose from products for allergy-sensitive persons	Splits lactose and other galactosides yielding glucose and galactose.
Lysozyme	Egg whites	Protects eggs from microbacterial action.	Dissolves cells of certain bacteria.
Maltase	Yeast and other microorganisms, grains, gastrointestinal tract	Baking—lean dough with little added sugar	Hydrolyzes glucose in starch.
Papain	Papaya melons	Chill-proofing beer; meat tenderizing	Acts on muscle fiber proteins and on elastin and to a lesser extent on collagen.
Pectinase	Citrus fruit, fungus *Aspergillus niger*	Production of clear jellies and juices; greater yields (reduces viscosity)	Hydrolyzes pectins or pectic acids; lowers viscosity, eliminates the protective colloid action of pectin.
Pectin methyl esterase	Citrus, tomatoes, apples	Commercial pectinase to remove pectin from fruit juice used in diabetic low-sugar jellies	Demethylation of pectins results in methanol and pectate (polygalacturonic acid).

(Continues)

TABLE 8-2 Important Food-Processing Enzymes

ENZYME	SOURCE	IMPORTANCE	ACTION
Peroxidase	Liver, milk, horseradish	Presence or absence can be used to judge adequacy of pasteurization.	Splits H_2O_2 to H_2O and O_2 when an oxygen acceptor is present.
Polyphenol oxidase, phenolase, catecholase	Plants, mushrooms	Tea fermentation; production of colored end products and flavors associated with browning reaction in fruits and vegetables	Catalyzes reactions in which molecular oxygen is the H acceptor and phenols act as H donors; oxidizes phenoliccompounds to quinines.
Phosphatase	Baby food, milk, brewing	Detects effectiveness of pasteurization as denatured slightly higher temperature than tubercle bacillus.	Liberates inorganic phosphate from phenyl phosphate.
Rennin	Salt extract of calves' fourth stomach	Coagulation of milk in cheese preparation; gel desserts	Hydrolyzes peptides. Acts on the surface of the kappa casein molecule.
Sucrase	Yeast	Slow crystallization in candy; hydrolyzes sucrose; artificial honey; development of creamy centers in candy	Splits sucrose to the invert sugars.
Zymase	Yeast	Fermentation yeast, maltozymase	Zymase is the heat-labile fraction of the enzymes system responsible for alcoholic fermentation. Fermentation converts glucose into CO_2 and ethanol.

Heat and Cold

Normal harvest temperatures range from 50 °F to 100 °F. The higher the temperature, the faster that biochemical reactions occur. In fact, the rate of chemical reactions doubles with each 10-degree Fahrenheit rise in temperature. On the other hand, subfreezing temperatures damage tissues. Cold temperatures may also cause discoloration, change textures, break **emulsions**, and **denature** proteins. Chilling can injure the tissues of fruits.

Oxygen

Chemical oxidation reactions can destroy vitamins (especially A and C), alter food colors, cause off flavors, and promote the growth of molds.

POSTHARVEST BIOCHEMICAL CHANGES

The high water content of fruits and vegetables promotes bacterial, yeast, and mold growth. If fruits and vegetables become partially dehydrated or wilted because of bruising or rough handling, their economic value sharply decreases (Figure 8-2). To avoid this, processors must carefully handle and transport these commodities, and processing techniques must minimize natural enzymatic deterioration.

When processors realized that freshly picked produce needed to be immediately cooled, mobile processing units that supercool produce immediately after harvest were developed. Jet streams wash products and begin to remove internal heat, and supercoolers remove the remaining heat.

Much of today's harvesting is accomplished by mechanical methods. Because fruits and vegetables are delicate items, harvesting is usually conducted before maturity is reached. Some harvesting is still conducted by hand and is highly labor intensive. Ripening chambers containing ethylene gas are an important part of the processing of certain fresh fruits and vegetables. Other controlled atmospheric conditions such as temperature, humidity, oxygen levels, and light are used to regulate shelf life.

FIGURE 8-2 Environmentally controlled spud cellars maintain the quality of potatoes during storage.

POSTSLAUGHTER BIOCHEMICAL CHANGES

Rigor mortis is an essential process in the conversion of live muscle to meat. After death, the biochemistry of muscle tissue changes. The muscle will use up energy from glycogen—a complex carbohydrate found in animal tissue. Because the blood is no longer flowing to remove the by-products of metabolism, lactic acid builds up in the muscle. This reduces the pH, causing a complex series of reactions that results in the contraction of muscle fibers. This contraction makes muscles hard or stiff, thus the name *rigor mortis*.

After more time, muscle fibers begin to relax. The relaxation of muscle postrigor is sometimes called the *resolution* of rigor. This process is greatly influenced by temperature, being faster at higher temperatures. In processing meats, the time and temperature during rigor are carefully controlled to maximize tenderness. This part of the process is sometimes called *aging*.

PRINCIPLES OF FOOD PRESERVATION

Food preservation involves the use of heat, cold, drying (**water activity**, or $\mathbf{A_w}$), acid (pH), sugar, salt, smoke, atmosphere, chemicals, radiation, and mechanical methods (Table 8-3).

Heat

Most bacteria are killed at 180 °F to 200 °F, but spores are not. To ensure sterility requires wet heat at 250 °F for 15 minutes. High-acid foods require less heat. (See Chapter 9 for a complete discussion.)

Cold

Microbial growth slows at temperatures below 50 °F, but some psychrophiles will continue slow growth. Foods frozen at less than 14 °F usually do

TABLE 8-3 Summary of Processing Methods to Control Factors Affecting Food Safety, Quality, and Convenience

	METHOD AND EFFECT				
TO BE CONTROLLED	HEAT	COLD	CHEMICALS	WATER ACTIVITY (A_w)	MECHANICAL
Microorganisms	Prevents growth	Reduces growth rate	Preservatives retard growth	Do not grow below A_w of 0.6	Reduces numbers
Enzymes	Destroyed by heat activity	Decreases reaction rate	Modify activity	Alters rate of enzyme activity	Increases enzyme-substrate complex formation
Chemical reactions	Increases chemical rate; browning oxidation	Reduces reaction rate	May inhibit or activate	Can alter rate of reaction, especially oxidation	Not applicable
Physical structure	Increases effects	Decreases effects	May modify structure	High A_w may cause caking	Can destroy structures

not have any free water, so these foods also benefit from low water activity to help protect against microbial growth. Freezing kills some but not all microorganisms (see Tables 8-4 and 8-5). (See Chapter 10 for more information.)

are required to destroy *C. botulinum* spores. The pH of pears is about 3.8, so only 5 minutes are necessary to destroy *C. botulinum* at 226 °F. Acid may occur naturally in foods, be produced by fermentation, or be added artificially.

MATH CONNECTION!

Using Tables 8-4 and 8-5, compare some of your favorite foods. How long can they be stored in the refrigerator or freezer? Compare different foods. Do you write dates on food packaging?

Drying

Drying reduces the water activity (A_w) in a food. Because microorganisms contain about 80% moisture, drying or dehydrating the food also dehydrates the microorganism. Changing the amount of water in a food also alters the rate of enzyme activity and other chemical reactions. (See Chapter 11 for a more complete discussion.)

Acid

As foods become more acidic (lower pH), the heat required for sterilization is reduced. For example, the pH of corn is about 6.5. At 226 °F, 15 minutes

Sugar and Salt

Sugar, salt, and smoke are chemical means of controlling food deterioration. The addition of sugar or salt to a food item increases the affinity of the food for water. This removes the water from the microorganism through **osmosis**.

Smoke

Smoke contains formaldehyde and other preservatives. The heat involved with adding the smoke helps reduce microbial populations and dries the food somewhat. Chapter 14 provides more details on the use of chemicals.

TABLE 8-4 Approximate Storage Life of Frozen Foods

PRODUCT	0 °F	10 °F	20 °F
Orange juice (heated)	27	10	4 months
Peaches	12	<2	6 days
Strawberries	12	2.4	10 days
Cauliflower	12	2.4	10 days
Green beans	11–12	3	1 month
Peas	11–12	3	1 month
Spinach	6–7	3	21 days
Chicken, raw	27	15½	<8 months
Fried chicken	<3	<1	18 days
Turkey pies	>30	9½	2½ months
Beef, raw	13–14	5	<2 months
Pork, raw	10	<4	<1½ months
Lean fish, raw	3	<2½	<1½ months
Fat fish, raw	2	1½	24 days

Source: USDA

Atmosphere

Changing the storage atmosphere reduces food deterioration. The growth of aerobic bacteria is slowed by removing the oxygen; providing oxygen limits the growth of anaerobic organisms. Adding carbon dioxide or nitrogen also slows deterioration. (See Chapter 20 for more details.)

Chemicals

Chemical additives such as sodium benzoate, sorbic acid, sodium or calcium propionate, and sulfur dioxide retard the growth of microorganisms, modify enzyme activity, inhibit chemical reactions, and modify the structure of foods. (See Chapter 14 for a discussion on the use of chemicals.)

Radiation

Radiation includes X-rays, microwave, ultraviolet light, and gamma rays. Radiation can destroy the microorganisms and inactivate enzymes. Chapter 12 provides complete details on irradiation.

TABLE 8-5 Maximum Cold Storage Time for Meat

MEAT PRODUCT	REFRIGERATOR (38–40 °F)	FREEZER (AT 0 °F OR LOWER)
Beef (fresh)	2 to 4 days	6 to 12 months
Veal (fresh)	2 to 4 days	6 to 9 months
Pork (fresh)	2 to 4 days	3 to 6 months
Lamb (fresh)	2 to 4 days	6 to 9 months
Poultry (fresh)	2 to 3 days	3 to 6 months
Ground beef, veal, lamb	1 to 2 days	3 to 4 months
Ground pork	1 to 2 days	1 to 3 months
Variety meats	1 to 2 days	3 to 4 months
Luncheon meats	7 days	Not recommended
Sausage, fresh pork	7 days	2 months
Sausage, smoked	3 to 7 days	1 month
Frankfurters	4 to 5 days	1 month
Bacon	5 to 7 days	1 month
Smoked ham, whole	1 week	2 months
Smoked ham, slices	3 to 4 days	1 month
Beef, corned	7 days	2 weeks
Leftover cooked meat	4 to 5 days	2 to 3 months
FROZEN COMBINATION FOODS		
Meat pies (cooked)		3 months
Swiss steak (cooked)		3 months
Stews (cooked)		3 to 4 months
Prepared meat dinners		2 to 6 months

SUMMARY

All foods undergo deterioration—physical, chemical, and biological. There are many ways to control this deterioration—from proper handling in the initial stages of harvesting to correct food-preservation techniques. Some deterioration produces toxins that are not destroyed by heat. Some of these toxins can cause infections in humans. All foods from living tissues have enzymes. Some postharvest enzymes are desirable and are controlled by heat, chemicals, and radiation. Food processors realize the importance of controlling deterioration through such means as heat, cold, drying, acids, sugar, atmosphere, chemicals, and radiation.

Because product dates are not a guide to the safe use of products, how long can consumers store certain foods and still use them when they are at peak quality? Follow these tips:

- Purchase the product before the sell-by date expires.
- Follow handling recommendations on product.
- Keep beef in its package until using.

It is safe to freeze beef in its original packaging. If anything will be frozen longer than 2 months, the packages should be overwrapped with airtight heavy-duty foil, plastic wrap, or freezer paper or placed inside a plastic bag.

For poultry and fish, pick them up just before checking out at the store's register. Put in a disposable plastic bag (if available) to contain leakage that could cross-contaminate cooked foods or produce. Make the grocery store your last stop before going home.

REVIEW QUESTIONS

Success in any career requires knowledge. Test your knowledge of this chapter by answering these questions or solving these problems.

1. Name the two environmental conditions that affect microbial growth on food.
2. Name the three general categories of food deterioration.
3. Which bacteria can be desirable in food preservation?
4. Why do foods have a shelf life?
5. Which factors affect the growth of aerobic and anaerobic microorganisms?
6. Explain four preservation techniques that prevent food deterioration.
7. Why are some fruits and vegetables washed immediately after being picked?
8. Name four food enzymes and describe their function.

STUDENT ACTIVITIES

1. Make a list of the date codes on five different foods. List the sell-by, best used by, and expiration dates. If possible, bring the labels and discuss these in class.

2. Leave a food such as meat, bread, fruit, and so on at room temperature and describe the changes in food quality over the course of 2 weeks. Create a chart to categorize the changes in each food and the causes.
3. Develop a report or presentation on why the occurrence of *E. coli* 0157 in food is such a concern. Site examples of the impact *E. coli* can have on consumers.
4. What is the chemical makeup of enzymes? How many enzymes exist?
5. Describe the **organoleptic** (sensory) properties of six common foods. As a term, *organoleptic* describes the major quality factors of appearance, texture, and flavor as perceived by the senses. Find an enzyme or a food product produced by an enzyme described in Table 8-2. Develop a visual presentation that gives information on this enzyme or food product.
6. Obtain foods that are processed with several methods—for example, dried, canned, fresh, and frozen. Describe what method extends their shelf life. Refer to Table 8-3.

ADDITIONAL RESOURCES

- Packaging and Shelf-Life Testing: http://fic.oregonstate.edu/packaging-shelf-life-testing
- Food Storage and Shelf-Life Guidelines: http://site.foodshare.org/site/DocServer/Food_Storage_and_Shelf_Life_Guidelines.pdf?docID=5822
- Refrigeration and Food Safety: http://www.fsis.usda.gov/shared/PDF/Refrigeration_and_Food_Safety.pdf
- Turkeys and Antibiotics Use: http://www.foodsafetynews.com/
- FoodSafety.gov: http://www.foodsafety.gov/

Internet sites represent a vast resource of information. Although those provided in this chapter were vetted by industry experts, you may also wish to further explore the topics discussed in this chapter using a search engine such as Google. Keywords or phrases may include the following: *shelf life of foods, food deterioration, food dating, food-borne disease, food enzymes, chemical oxidation, mycotoxins, organoleptic, pyschrophilic, thermophilic*. In addition, Table B-7 provides a list of useful Internet sites that can be used as a starting point.

REFERENCES

Cremer, M. L. 1998. *Quality food in quantity: Management and science.* Berkeley, CA: McCutchan Publishing.

Ensminger, A. H., M. E. Ensminger, J. E. Konlande, and J. R. Robson. 1994. *Foods and nutrition encyclopedia* (2 vols.). Boca Raton, FL: CRC Press.

Potter, N. N., and J. H. Hotchkiss. 1995. *Food science,* 5th ed. New York: Chapman & Hall.

Vieira, E. R. 1996. *Elementary food science,* 4th ed. New York: Chapman & Hall.

SECTION *Two*

Preservation

CHAPTER 9

Heat

OBJECTIVES

After reading this chapter, you should be able to:

- Identify the four degrees of preservation achieved by heating
- Describe how specific heat treatments are selected
- Identify how the heat resistance of microorganisms is determined
- Discuss methods of heating foods before or after packaging
- Describe one type of retort
- Compare conduction and convection heating
- Explain time–temperature combinations
- List three factors that influence how foods heat
- Describe a thermal death curve
- Compare the acidity of various foods to the heat treatment required

KEY TERMS

aseptic packaging
batch
blanching
commercial sterility
D value
hot fill
hot pack
hydrostatic retort
logarithmic
pasteurization
retort
spores
sterilization
still retort
Z value

NATIONAL AFNR STANDARD

FPP.01

Develop and implement procedures to ensure safety, sanitation, and quality in food products and food-processing facilities.

Heating or cooking foods kills some microorganisms, destroys most enzymes, and improves shelf life. The goal of any food-preservation method is to extend the shelf life of food. It cannot preserve a food indefinitely. Heating methods create various degrees of preservation, depending on the product. Packaging is also important to prevent recontamination of foods.

HEAT

According to the laws of physics, heat is the transfer of energy from one part of a substance to another by virtue of a difference in temperature. Heat is energy in transit; it always flows from a substance at higher temperature to a substance at lower temperature, raising the temperature of the latter substance and lowering that of the former, provided the volume of the bodies remains constant. The sensation of warmth or coldness of a substance on contact is determined by the property known as *temperature*. Adding heat to a substance, however, not only raises its temperature, causing it to impart a more acute sensation of warmth, but also expands or contracts a substance and alters its electrical resistance properties. Heat also increases the pressure exerted by gases. Temperature depends on the average kinetic energy (motion) of the molecules of a substance.

DEGREES OF PRESERVATION

Depending on the product and the use of the product, heat can be used to create varying degrees of preservation, including **sterilization, commercial sterility, pasteurization**, and **blanching**. Table 9-1 categorizes and compares these heat treatments as mild and severe.

Sterilization

Sterilization refers to the complete destruction of microorganisms. At least 250 °F (121 °C) for 15 minutes is needed to destroy all **spores**. In practice, food would require much longer times because there is a lag between when a product is exposed to heat and when it reaches the desired temperature. The time it takes to sterilize food depends on the amount of food. Often complete sterilization takes several hours and destroys the quality of a food product, so sterilization is limited to foods that are in pureed or liquid form.

Commercial Sterility

In commercial sterility, all pathogenic and toxin-forming organisms have been destroyed, as well as other organisms capable of growth and spoilage under normal handling and storage conditions.

TABLE 9-1 Mild and Severe Heat Treatments

COMPARISONS	MILD	SEVERE
Aims	Kill pathogens; reduce bacterial count	Kill all bacteria; food will be commercially sterile
Advantages	Minimal damage to flavor, texture, nutritional quality	Long shelf life; no other preservation method is necessary
Disadvantages	Short shelf life; another preservation method must be used such as refrigeration or freezing	Food is overcooked; major changes in texture, flavor, nutritional quality
Examples	Pasteurization; blanching	Canning

Such products may contain viable spores, but they will not grow under normal conditions.

Pasteurization

Pasteurization is a comparatively low-energy thermal process with two main objectives: (1) destroy all pathogenic microorganisms that might grow in a specific product, and (2) extend shelf life by decreasing the number of spoilage organisms present. The product is not sterile and will remain subject to spoiling. Milk and eggs are pasteurized to kill microbes.

Blanching

Blanching is a mild heat treatment generally applied to fruits and vegetables to inactivate enzymes that might decrease product quality. It may also destroy some microorganisms and thus lead to increased product shelf life. The primary objective, however, is enzyme inactivation. Blanching also loosens the skins of some fruits and vegetables such as peaches and tomatoes, making them easier to peel.

SELECTING HEAT TREATMENTS

A heat treatment sufficient to destroy all microorganisms and enzymes is detrimental to the food qualities of color, flavor, texture, nutrition, and consistency. To pick the right heat treatment for a specific food, two factors must first be determined:

1. The right time and temperature combination required to inactivate the most resistant microbe
2. The heat-penetration characteristics of the food and the container

Heat-penetration characteristics of a food vary with its consistency and particle size; the heat-penetration characteristics of the container vary with its size, shape, and constituent material.

HEAT RESISTANCE OF MICROORGANISMS

The most resistant microbe in canned foods is *Clostridium botulinum* (botulism), so food processors must use a time–temperature combination that is adequate for killing this deadly microbe.

Heat kills any bacteria **logarithmically.** For example, if 90% are killed in the first minute at a certain temperature, then 90% of those remaining alive will die during the second minute, and 90% of those remaining alive will die during the third minute, and so on. (*Note:* Spores are more heat resistant than bacterial cells.)

In foods where the type of contamination is unknown, a margin of safety is applied. Especially in foods that are low acid, processors assume *C. botulinum* to be present and treat the food accordingly.

SCIENCE CONNECTION!

Find three food items at home that you think used heat processing. What method was used?

HEAT TRANSFER

Food processors cannot assume that a product heated in an environment at 250°F (121°C) for 10 minutes has been exposed to that temperature for that time. Product at the edge of the container will reach temperature much sooner than will the product at the center of the container.

Many factors influence the time required for the coldest portion of the product to reach the required temperature. The size and shape of the container are important as are the physical properties of the product within the container (Figure 9-1). To overcome some of the problems, different methods of heating are used. These include

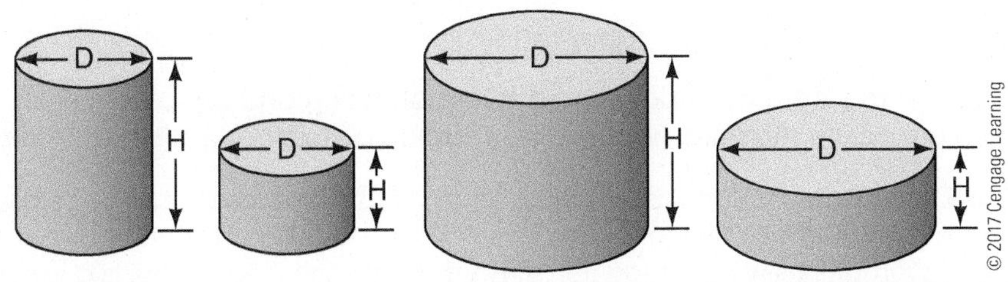

FIGURE 9-1 The size and shape of a container (can), expressed by diameter (D) and height (H), determine how the product will heat.

conduction, convection, radiation, and combinations of these three.

Conduction is the transfer of heat from one particle to another by contact. Food particles in a can do not move. *Convection* heating means that the movement inside a can distributes the heat (Figure 9-2). *Radiation* heating occurs when the energy transfers through a medium that itself is not heated.

Convection-Conduction

Conduction heating is thermal transfer resulting from collisions of hot food particles with cooler ones. Convection heating involves the circulation of warm molecules and results in more effective thermal transfer. Foods that are heated by convection require less time to reach temperature than do foods that are heated by conduction.

Heating starts out rapidly by convection. A change in the texture of the food occurs that may cause further heating by conduction. This often involves the gelatinization of starch in products such as pork and beans. The longer that heating is by convection, the shorter the total process time.

Conduction-Convection

Heating starts as conduction, then a change in food makes the heating convective. An example would be heating foods containing large pieces of meat. At first the heating occurs as conduction. As juices are released by the meat, heating becomes more convective.

Radiation

Radiation is the transfer of energy in the form of electromagnetic waves. It is the fastest method of heat transfer. Radiation transfers heat directly from

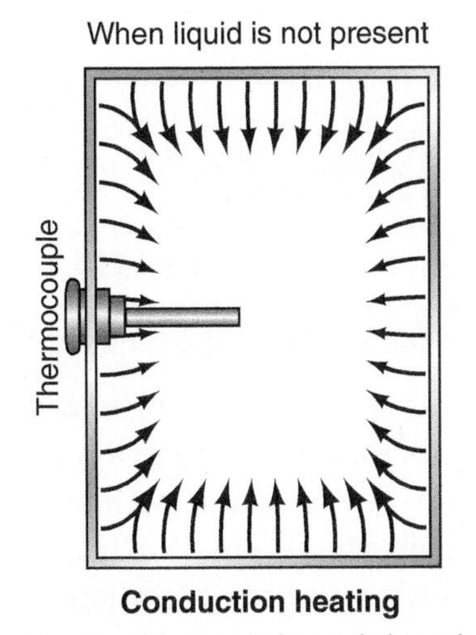

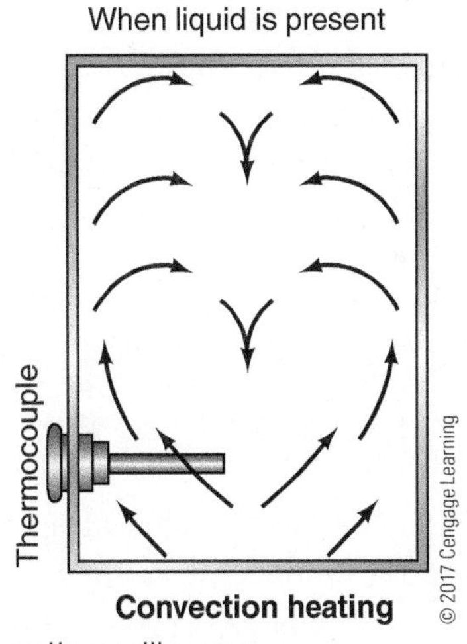

FIGURE 9-2 The contents of a container determine how heating will occur.

FOODS FOR THE SPACE PROGRAM

Foods going into space require special packaging and processing and are categorized as rehydratable, thermostabilized, intermediate moisture, natural form, irradiated, frozen, fresh, and refrigerated.

Rehydratable Food. In rehydratable foods, water is removed to make them easier to store. This process of dehydration is also known as freeze-drying. Water is replaced in the foods before they are eaten. Rehydratable items include beverages as well as food items. Hot cereal such as oatmeal is a rehydratable food.

Thermostabilized Food. Thermostabilized foods are heat processed so that they can be stored at room temperature. Most of the fruits and fish (tuna) are thermostabilized in cans. The cans open with easy-open pull tabs similar to fruit cups that can be purchased in the local grocery store. Puddings are packaged in plastic cups.

Intermediate Moisture Food. Intermediate moisture foods have some but not all water removed from the product, leaving enough in to maintain the soft texture. These foods can be eaten without any preparation.

These foods include dried peaches, pears, apricots, and beef jerky.

Natural Form Food. Natural form foods are ready to eat and packaged in flexible pouches. Examples include nuts, granola bars, and cookies.

Irradiated Food. Beefsteak and smoked turkey are the only irradiated food products currently being used. These products are cooked and packaged in flexible foil pouches and sterilized by ionizing radiation so that they can be kept at room temperature.

Frozen Food. Frozen foods have been quickly frozen to prevent the buildup of large ice crystals. This process maintains a food's original texture and helps it taste fresh. Examples include quiches, casseroles, and chicken potpie.

Fresh Food. Fresh foods are neither processed nor artificially preserved. Examples include apples and bananas.

Refrigerated Food. Refrigerated foods require cold or cool temperatures to prevent spoilage. Examples include cream cheese and sour cream.

For more information on space foods, visit the NASA Web site at http://www.nasa.gov/audience/forstudents/postsecondary/features/F_Food_for_Space_Flight.html

a radiant heat source such as a broiler plate to the food being heated. Surfaces between the heat source and the food being heated reduce the amount of energy transmitted by radiation. Because electromagnetic waves fan out as they travel, the food farthest from the source takes longer to heat.

PROTECTIVE EFFECTS OF FOOD CONSTITUENTS

Some food constituents protect bacterial spores. Sugar protects bacterial spores in canned fruit. Starch, protein, fats, and oils also protect bacterial spores. This makes some foods difficult to sterilize—for example, meat products, fish packed in oil, and ice cream.

DIFFERENT TEMPERATURE–TIME COMBINATIONS

Food processors must determine the time–temperature combination required to kill the most heat-resistant pathogen or spoilage organism in the product of interest. Processors must know the heat-penetration characteristics of the food and its container.

Organisms

The most heat resistant of all pathogens is *Clostridium botulinum.* Some nonpathogenic spore formers are even more heat resistant—for example, putrefactive anaerobe 3679 and *Bacillus stearothermophilus* 1518. If these organisms are destroyed, the pathogens will also be killed.

Thermal Death Curves

Like microbial growth, microbial death is generally described by a logarithmic equation. Food processors are concerned with two values: the **D value** and the **Z value** (Figure 9-3). These numbers relate the death of microorganisms to the time and temperature of heating.

D Value

The D value is the time in minutes at a specified temperature that reduces the number of microorganisms by one log cycle (by a factor of 10). The units of time (minutes or seconds) and a temperature must be specified—for example, D121 = 3.5 minutes on the Celsius scale (D250 on the Fahrenheit scale).

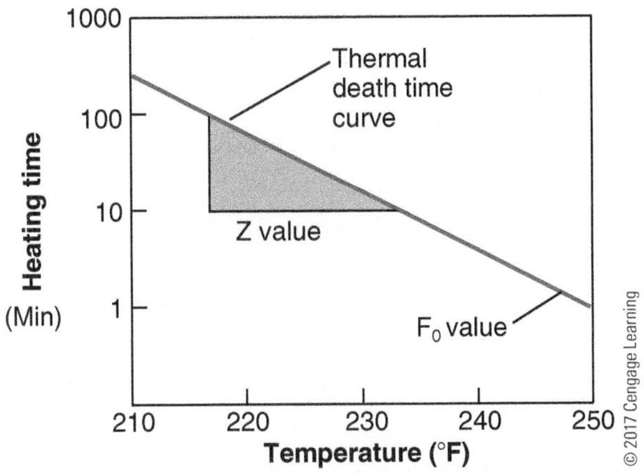

FIGURE 9-3 Graphing heating time and temperature provides a thermal death curve.

Z Value

The Z value is the temperature required to decrease the time necessary to obtain a one-log reduction in cell numbers to 1/10th of the original value. For example, if 100 minutes at 220 °F (104 °C) killed 90% of the organisms and 10 minutes at 238 °F (114 °C) also killed 90% of the organisms, then the Z value would be equal to 18 °F (10 °C).

Time-Temperature Combinations

Many combinations of heating times and temperatures will give the same lethality. For *C. botulinum*, the following are equivalent:

0.78 min at 261 °F (127 °C)	10 min at 241 °F (116 °C)
1.45 min at 255 °F (124 °C)	36 min at 230 °F (110 °C)
2.78 min at 250 °F (121 °C)	150 min at 219 °F (104 °C)
5.27 min at 244 °F (118 °C)	330 min at 212 °F (100 °C)

Low-Acid Foods

The exact number of spores in a food is not known. In the United States, the legal processing required for low-acid foods is exposure to a temperature for a period equal to 12 D values for *Clostridium botulinum.* This is sufficient to decrease any population of *Clostridium botulinum* through 12 log cycles, thus providing an adequate safety margin. The acid level of a food affects its processing times and temperatures as shown in Table 9-2.

HEATING BEFORE OR AFTER PACKAGING

Heating before packaging is the simplest and oldest form of heat preservation. Heating after packaging is less damaging to the food but requires aseptic (germ-free) packaging.

TABLE 9-2 Processing Requirements and pH

HEAT AND PROCESSING REQUIREMENTS	ACIDITY	pH	FOOD
High-temperature processing: 240–250 °F (116–121 °C)	Low acid	7.0	Hominy, ripe olives, crabmeat, eggs, oysters, milk, corn, duck, chickens, codfish, beef, sardines
		6.0	Corned beef, lima beans, peas, carrots, beets, asparagus, potatoes
		5.0	Figs, tomato soup
	Medium acid	4.5	Ravioli, pimentos
Boiling water processing 212 °F (100 °C)	Acid	—	Potato salad, tomatoes, pears, apricots, peaches, oranges
		3.7	sauerkraut, pineapple, strawberry, grapefruit
	High acid	3.0	Pickles, relish, cranberry juice, lemon juice, lime juice

The **batch**, continuous, and **hot-pack** and **hot-fill** methods are examples of heating food before packaging. In the batch method, the product is heated in a steam-jacketed kettle to a specific temperature and then rapidly cooled. For example, milk is heated to 145 °F for 30 minutes. For continuous pasteurization, the food is heated to a high temperature for a short time. For example, milk is heated to 161 °F for 15 seconds. In the hot-pack and hot-fill method, food processors fill unsterilized containers with sterilized food that is still hot enough to render the package commercially sterile. Retorts are large pressure canners used by commercial food processors. In **retorts**, the product is heated after being placed in a container. Several types of retorts process food this way.

Still Retort

Still retorts are the simplest commercial canners and are similar to home pressure cookers for canning at home. In the **still retort** process, the product is placed in a container and then heated in a steam atmosphere without agitation (Figure 9-4). If the temperature is above 250 °F (121 °C), considerable burning of the product on the side of the can will occur. Heating times in still retorts are often in the range of 30 to 45 minutes.

Agitating Retort

In some retorts, the product is agitated during cooking. This allows for the use of high temperatures during processing. The agitation will allow for convection heating and reduce the time necessary to reach final temperature. Cooking times are only 10% to 20% of those in still retorts.

Hydrostatic Retort

In a **hydrostatic retort**, the cans flow continuously (Figure 9-4). The retort uses hydrostatic heat to control pressure. As an agitating system, the retort moves food through a U-shaped tube filled with water and steam. This retort has a conveyor belt and keeps food moving through the system of increased and then reduced temperature and pressure.

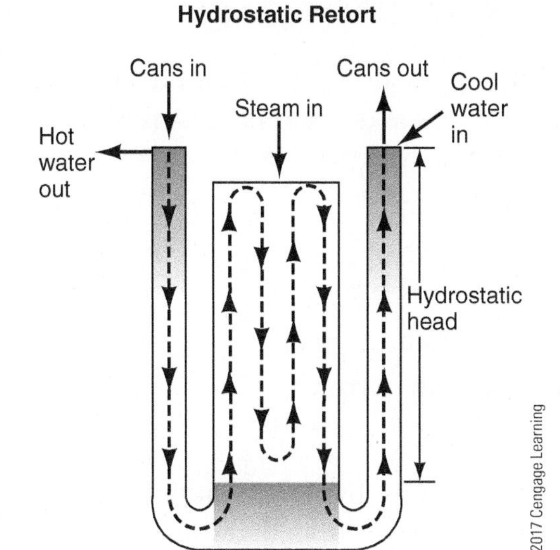

FIGURE 9-4 Retorts are used to heat cans.

Aseptic Packaging

For **aseptic packaging**, the food is sterilized outside of the container (Figure 9-5) and then placed into a sterile container and sealed under aseptic conditions. Paper, plastic, and aluminum packaging materials are most commonly used, and this method is most suitable for liquid-based food products. Silicon has also been used in layers to replace the aluminum to allow for microwave packaging.

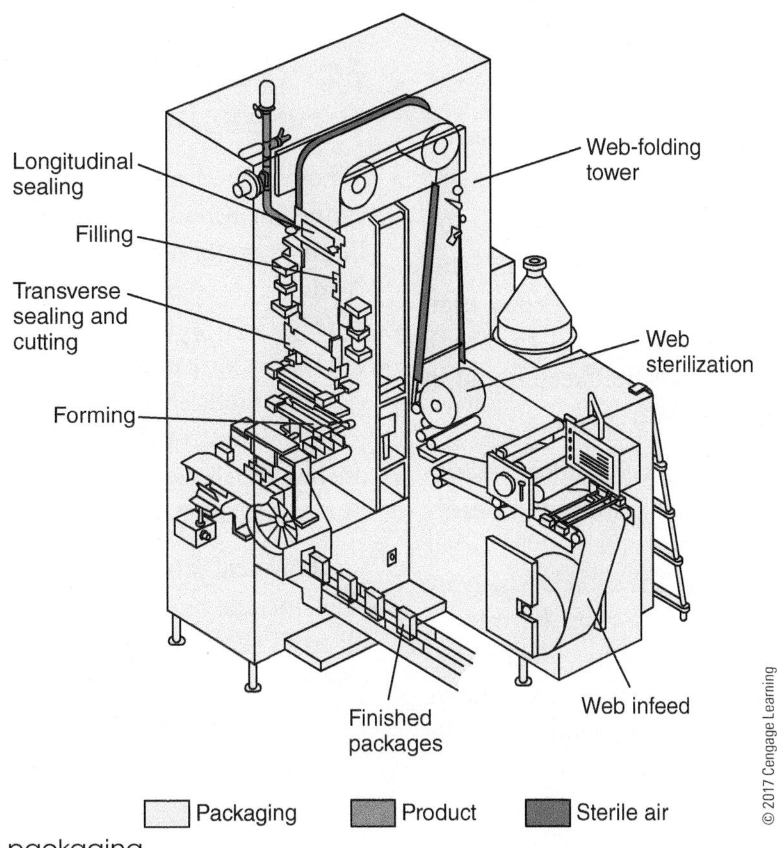

FIGURE 9-5 Aseptic packaging.

HOME CANNING

Many different foods can be canned at home. The high percentage of water in most fresh foods makes them highly perishable. They spoil or lose their quality for several reasons:

- Growth of undesirable microorganisms such as bacteria, molds, and yeasts
- Activity of food enzymes
- Reactions with oxygen
- Moisture loss

Microorganisms live and multiply quickly on the surfaces of fresh food and on the inside of bruised, insect-damaged, and diseased food. Oxygen and enzymes are present throughout fresh food tissues. Proper home canning practices include:

- Carefully selecting and washing fresh food
- Peeling some fresh foods
- Hot packing many foods
- Adding acids (lemon juice or vinegar) to some foods
- Using acceptable jars and self-sealing lids
- Processing jars in a boiling-water or pressure canner for the correct period of time

Together, these practices remove oxygen; destroy enzymes; prevent the growth of undesirable bacteria, yeasts, and molds; and help form a high vacuum in jars (similar to commercial canning). Good vacuums form tight seals that keep liquid in and air and microorganisms out.

Growth of the bacterium *Clostridium botulinum* in canned food can cause botulism, a deadly form of food poisoning. These bacteria exist either as spores or as vegetative cells. The spores, which are comparable to plant seeds, can survive harmlessly in soil and water for many years. When ideal conditions exist for growth, the spores produce vegetative cells that multiply rapidly and may produce a deadly toxin within 3 to 4 days of growth in an environment consisting of:

- A moist, low-acid food
- A temperature between 40 °F and 120 °F
- Less than 2% oxygen

Botulinum spores are on most fresh food surfaces. Because they grow only in the absence of air, they are harmless on fresh foods.

Most bacteria, yeasts, and molds are difficult to remove from food surfaces. Washing fresh food reduces their numbers only slightly. Peeling root crops, underground stem crops, and tomatoes reduces their numbers greatly. Blanching also helps, but the vital controls are the method of canning and making sure that the recommended research-based process times are used. Correct processing times ensure destruction of the largest expected number of heat-resistant microorganisms in home-canned foods. Properly sterilized canned food will be free of spoilage if lids seal and jars are stored below 95 °F. Storing jars at 50 °F to 70 °F helps retain food quality.

Food Acidity and Processing Methods

Whether food should be processed in a pressure canner or a boiling-water canner to control botulinum bacteria depends on the acidity in the food. Acidity may be natural, as in most fruits, or added, as in pickled food. Low-acid canned foods contain too little acidity to prevent the growth of these bacteria. Acid foods contain enough acidity to block their growth or destroy them more rapidly when the foods are heated. The acidity level in foods can be increased by adding lemon juice, citric acid, or vinegar.

Low-acid foods have pH values higher than 4.6. They include red meats, seafood, poultry, milk, and all fresh vegetables except for most tomatoes. Most mixtures of low-acid and acid foods also have pH values above 4.6 unless their recipes include enough lemon juice, citric acid, or vinegar to make them acid foods. Acid foods have a pH of 4.6 or lower. They include fruits, pickles, sauerkraut, jams, jellies, marmalades, and fruit butters.

Although tomatoes usually are considered an acid food, some are now known to have pH values slightly above 4.6. Figs also have pH values slightly above 4.6. If they are to be canned as acid foods, these products must be acidified to a pH of 4.6 or lower with lemon juice or citric acid. Properly acidified tomatoes and figs are acid foods and can be safely processed in a boiling-water canner.

Botulinum spores are extremely hard to destroy at boiling-water temperatures. The higher the canner temperature, however, the more easily the spores are destroyed. All low-acid foods should be sterilized at temperatures of 240 °F to 250 °F. These temperatures are possible with pressure canners operated at 10 to 15 pounds per square inch of pressure as measured by gauge (psig). At temperatures of 240 °F to 250 °F, the time needed to destroy bacteria in low-acid canned food ranges from 20 to 100 minutes. The exact time depends on the kind of food being canned, the way it is packed into jars, and the size of jars. The time needed to safely process low-acid foods in a boiling-water canner ranges from 7 to 11 hours; the time needed to process acid foods in boiling water varies from 5 to 85 minutes.

Process Adjustments at High Altitudes

Using the process time for canning food at sea level may result in spoilage at altitudes of 1,000 feet or more. Water boils at lower temperatures as altitude increases. Lower boiling temperatures are less effective for killing bacteria. Increasing the process time or canner pressure compensates for lower boiling temperatures.

For more information about home canning, check out the online "USDA Complete Guide to Home Canning, 2009" at http://nchfp.uga.edu/publications/publications_usda.html.

SCIENCE CONNECTION!

Research and find recipes for home canning. Compare canning fruits and vegetables. Is the process difference? What is similar and what is different? Compare making jam and pickles. What is similar and what is different?

SUMMARY

Heat produces varying degrees of preservation, depending on the product and the use of the product. Heat treatments can be selected on the basis of time and temperature combination to inactivate the most resistant microbe and the heat-penetration characteristics of both food and container. The most heat-resistant microbe in canned foods is *Clostridium botulinum*. Thermal death curves describe the time-temperature relationships for preserving food.

Depending on the type of food and its acidity, the heat treatment can be mild or severe. Convection, conduction, and radiation transfer heat to the processed foods. Commercially, foods are heated before or after packaging. Different types of retorts efficiently process commercial foods. Aseptic packaging is used when the food is sterilized outside the container. Home canning follows the same principles as commercial heat-preservation methods.

REVIEW QUESTIONS

Success in any career requires knowledge. Test your knowledge of this chapter by answering these questions or solving these problems.

1. What is the most heat-resistant microbe in canned foods?
2. Describe the two main objectives of pasteurization.
3. Name four types of preservatives achieved by heating.
4. Explain the thermal death curve.
5. Heating after packaging requires what type of packaging?
6. What type of heating results in thermal transfer because of collisions between hot food particles and cooler ones?
7. What is the difference between a still retort and an agitating retort?
8. Identify the two factors to pick the right heat-treatment severity for a specific food.
9. Define conduction heating.
10. Describe how radiation functions in the food industry.

STUDENT ACTIVITIES

1. Using an electronic spreadsheet, develop a chart that shows the logarithmic growth of bacteria.
2. Develop a computer-generated visual presentation on botulism.
3. Create a sketch that describes a spore.
4. Research a food product and contact the processing company such as Del Monte® or Campbell's® and request information on the use of heat in some of the company's products.
5. Culture some bacteria in petri dishes on agar or obtain some prepared microscope slides of bacteria and observe the different types through a microscope.
6. Choose a food that is easy to process and demonstrate home canning. Create a digital video of the process to share with the class.

ADDITIONAL RESOURCES

- Food Preservation FAQs: http://www.extension.umn.edu/food/food-safety/preserving/general/food-preservation-faqs/
- Heat Treatment: https://meathaccp.wisc.edu/validation/heat_treatment.html
- Ohmic Heating of Foods: http://ohioline.osu.edu/fse-fact/0004.html
- Food Packaging Technology: http://gardoonism.com/files/60/root/books/Food%20Packaging%20Technology.pdf

Internet sites represent a vast resource of information. Although those provided in this chapter were vetted by industry experts, you may also wish to further explore the topics discussed in this chapter using a search engine such as Google. Keywords or phrases may include the following: *aseptic packaging, hydrostatic retort, agitating retort, still retort, food preservation, conduction/convection heating, radiation transfer of heat, low-acid foods, home canning.* In addition, Table B-7 provides a listing of some useful Internet sites that can be used as a starting point.

REFERENCES

Asimov, I. 1988. *Understanding physics.* New York: Hippocrene Books.

Potter, N. N., and J. H. Hotchkiss. 1995. *Food science,* 5th ed. New York: Chapman & Hall.

Vaclavik, V. A., and E. W. Christina. 2014. *Essentials of food science,* 4th ed. New York: Springer.

Vieira, E. R. 1999. *Elementary food science,* 4th ed. New York: Chapman & Hall.

CHAPTER 10

Cold

OBJECTIVES

After reading this chapter, you should be able to:

- **Compare cooling, refrigeration, and freezing**
- **Identify four requirements for refrigeration**
- **Correlate storage temperature to length of storage**
- **Understand requirements for refrigeration and freezing**
- **List three methods of freezing**
- **Describe changes in food quality that may occur during refrigeration or freezing**
- **Compare home freezing to commercial freezing**

KEY TERMS

cool storage
cryogenic
freezer burn
hydrocooling
hypobaric
immersion freezing
refrigeration
vacuum cooling

NATIONAL AFNR STANDARD

FPP.01

Develop and implement procedures to ensure safety, sanitation, and quality in food product and processing facilities.

Cold (cool) storage, refrigeration, and frozen storage are methods of food preservation and processing that differ primarily in temperature and time. Various methods are used to take food to temperatures necessary for cold storage or frozen storage.

FIGURE 10-1 Commercial freezer unit.

REFRIGERATION VERSUS FREEZING

Cool storage is considered any temperature from 68 °F to 28 °F (16 °C to −2 °C). Refrigerator temperatures range from 40 °F to 45 °F (4.5 °C to 7 °C); frozen storage temperatures range from 32 °F to 0 °F (0 °C to −18 °C) (see Figure 10-1).

Microbes grow more rapidly at temperatures above 50 °F (10 °C). Some growth occurs at subfreezing temperatures as long as water is available. Little growth occurs below 15 °F (−9.5 °C).

Historically, the first mechanical ammonia **refrigeration** system was invented in 1875. In the 1920s, Clarence Birdseye started frozen food packaging.

REFRIGERATION AND COOL STORAGE

Cool storage is the gentlest of all food-preservation methods. It affects taste, texture, nutritive value, and color the least, but it is only a short-time preservation method.

CLARENCE BIRDSEYE

Clarence Birdseye was born in Brooklyn, New York, on December 9, 1886. He attended Amherst College, majoring in biology but did not graduate. Instead, Clarence pursued a career as a field naturalist for the U.S. government. The job took him far north, near the Arctic, where he made a discovery that changed the history of the food industry. He noticed that freshly caught fish, when placed onto the Arctic ice and exposed to the icy wind and frigid temperatures, froze solid almost immediately. Clarence also found that when thawed, the fish still had all of its fresh characteristics and was fine to eat.

In September 1922, Clarence organized his own company, Birdseye Seafoods, Inc., where he began processing chilled fish fillets at a plant near the Fulton Fish Market in New York City. In 1924, he developed the process of packing dressed fish or other food in cartons, then freezing the contents between two flat, refrigerated surfaces under pressure.

On July 3, 1924, he organized the General Seafood Corporation, with the financial help of Wetmore Hodges, Basset Jones, I. L. Rice, William Gamage, and J. J. Barry. This was the beginning of the wholesale frozen foods industry. The retail frozen foods business began March 6, 1930, in Springfield, Massachusetts, when the Springfield Experiment Test Market produced and packaged 26 different vegetables, fruits, fish, and meats.

For more information, visit the Birdseye Web site at https://www.birdseye.com/birds-eye-view/history

Refrigeration is also a gentle method of food preservation. It has minimal effects on the taste, texture, and nutritional value of foods. Refrigeration has a limited contribution toward preserving food, however. For most foods, refrigeration extends the shelf life by a few days or week. In many cases, refrigeration is not the sole means of preserving the food. Refrigeration slows down but does not stop the spoilage of food. A good example is milk. This liquid is first heat processed (pasteurized) and then refrigerated. The heat treatment destroys the pathogenic microorganisms. reduces the total microbial load, and inactivates the enzymes in the milk. Refrigeration will keep the spoilage reactions (microbial or enzymatic) to a minimum. Refrigeration does not kill microorganisms or inactivate enzymes, but it does slow down their deteriorative effects.

Refrigeration temperature is a key factor in predicting the length of storage time (refer to Tables 8-4 and 8-5). For example, meat will last 6 to 10 days at 32 °F (0 °C), one day at 72 °F (22 °C), and less than one day at 100 °F (38 °C). Household refrigerators usually run at 40.5 °F to 44.6 °F (4.7 °C to 7 °C). Commercial refrigerators operate at slightly lower temperatures. A group of microorganisms called *psychrophilics* will grow at refrigeration temperatures.

When cooling is used as a preserving technique, the food must be maintained at the proper cold temperature during manufacturing, transport, display, and home storage. Signs of spoilage in refrigerated foods vary with the food product. In fruits and vegetables, loss of firmness or crispness takes place. Red meat will change in color; fish will get softer and show noticeable drippage.

Refrigerated food will last longer if it is cleaned and properly packaged before refrigeration and, of course, maintained at the proper temperature with minimum exposure to surrounding temperatures.

The timelines of refrigeration and food processing require that heat be rapidly removed from foods at harvest or slaughter. To do this, such methods as moving air, **hydrocooling** (water), **vacuum cooling**, and liquid nitrogen are used.

Because refrigeration is an energy-demanding process that must be maintained throughout the life of the product, refrigerated foods tend to cost more than nonrefrigerated foods. Yet today's consumer is buying more refrigerated foods because of their fresh quality. When attempting to preserve the high quality of refrigerated foods, the consumer needs to keep cold food at temperatures between 33 °F and 45 °F.

Requirements of Refrigerated Storage

Refrigerated storage requires low temperatures, air circulation, humidity control (80% to 95%), and modified gas atmosphere. Insulation, frequency of door opening, quantity of hot product added daily, and the respiration rate of the food product can affect the requirements.

In controlled atmosphere storage, standard cold-storage temperatures and humidity are maintained, but the oxygen levels are reduced and the carbon dioxide levels are elevated to reduce respiration in the food product. Some controlled atmosphere storage units are **hypobaric**, meaning the pressure is also reduced. This reduces the availability of oxygen.

Changes in Food During Refrigerated Storage

During refrigerated storage, foods can experience chill injury, flavor (odor) absorption, and loss of firmness, color, flavor, and sugar. Refrigeration is not good for every food. For example, bananas turn black when they are chilled. Packaging is also important for refrigeration. Proper packaging can reduce odor absorption. This is one reason why many home refrigerators have different compartments to separate foods for better storage.

FREEZING AND FROZEN STORAGE

Freezing can be thought of as a continuation of refrigeration. The freezing point for pure water is 32 °F (0 °C). For food, however, the freezing point is below that. Chemistry and physics teach that the presence of solutes in water will lower its freezing point. The addition of one mole (molecular weight in grams) of any nonionic substance (solute) to one liter of water lowers the freezing point by 1.885 degrees C (Figure 10-2). Water contained in foods has many solutes, although these are mainly salts and sugars.

Freezing is similar to refrigeration because it will not destroy microorganisms or inactivate enzymes but only slow their deteriorative effect. In some cases, it will completely inhibit the activity of microorganisms, even though they will still be alive. Enzymes, on the other hand, will maintain a certain level of activity during freezing. For this reason, food processors blanch vegetables prior to freezing them. Blanching is a mild heat treatment designed to inactivate enzymes.

Freezing has been a key technology in bringing convenience foods to homes and restaurants. It causes minimal changes in the quality of food in terms of size, shape, texture, color, flavor, and microbial load. This assumes, of course, that the freezing process is carried out properly and that a food can be frozen.

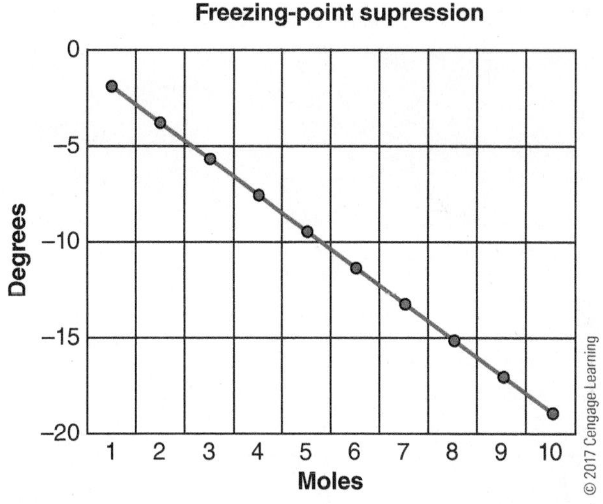

FIGURE 10-2 Each mole of a solute in water suppresses the freezing point by 1.885 °C.

Food processors freeze foods to an internal temperature of 0 °F (−18 °C). The food must be maintained at this temperature or slightly lower during transport and storage. Many fruits and vegetables will retain good quality at the above temperature for as long as 12 months and even longer. The expected frozen storage will vary with temperature. For example, frozen orange juice will last 27 months at 0 °F (−18 °C), 10 months at 10 °F (−12 °C), and only 4 months at 20 °F (−6.7 °C). Table 10-1 provides a list of products and their frozen storage times.

SCIENCE CONNECTION!

Conduct a refrigeration experiment. Place food samples in plastic bags. Place one of the samples in a freezer and one sample in a refrigerator; leave one sample at room temperature. Observe the food daily for 5 days. Write down your observations of the changes that occur. (*Caution:* Do not eat the food product!) What changes occur?

TABLE 10-1 Suggested Maximum Home-Storage Periods for Commercially Frozen Foods

FOOD	APPROXIMATE MONTHS STORAGE (0 °F)
Fruits and Vegetables	
Fruits:	
Cherries	12
Peaches	12
Raspberries	12
Strawberries	12
Fruit Juice concentrates:	
Apple	12
Grape	12
Orange	12
Vegetables:	
Beans	8
Corn	8
Peas	8
Spinach	8
Baked Goods	
Bread and yeast rolls:	
White bread	3
Plain rolls	3
Cakes:	
Angel or chiffon	2
Chocolate layer	4
Fruit	12
Pound or yellow	6
Danish pastry	3
Doughnuts:	
Cake type	3
Yeast raised	3
Pies, Fruit (unbaked)	8
Meat	
Beef:	
Ground beef	4
Roast	12
Steaks	12

FOOD	APPROXIMATE MONTHS STORAGE (0 °F)
Lamb:	
Steaks	9
Roast	9
Pork, cured	2
Pork, fresh:	
Roasts	8
Sausage	2
Veal:	
Cutlets, chops	9
Roasts	9
Cooked meat:	
Meat dinners	3
Meat pie	3
Swiss steak	3
Poultry	
Chicken:	
Cut-up	6
Livers	3
Whole	12
Duck, whole	6
Goose, whole	6
Turkey:	
Cut-up	6
Whole	12
Cooked chicken and turkey:	
Chicken or turkey: (sliced meat and gravy)	6
Chicken or turkey pes	6
Fried chicken	3
Fried chicken dinners	3

FOOD	APPROXIMATE MONTHS STORAGE (0 °F)
Fish and Shellfish	
Fish Fillets:	
"Lean" fish: Cod, flounder, haddock, halibut	6
"Fatty" fish:	
Mullet, ocean perch, sea trout, striped bass	3
Shellfish:	
Clams, shucked	3
Crabmeat	
King or Dungeness	2
Oyster, shucked	1
Shrimp (unbreaded)	12
Cooked fish and shellfish:	
Fish with cheese sauce	3
Fish with lemon butter sauce	3
Fried fish dinner, fried fish sticks, scallops, or shrimp	3
Shrimp creole	3
Tuna pie	3
Frozen desserts	
Ice cream	1

Source: Bulletin 989, "So Easy to Preserve," produced by the Cooperative Extension Service, the University of Georgia, College of Agricultural and Environmental Sciences. Sixth Edition published in 2014.

Chemical Changes During Freezing

Enzymes in fruits and vegetables are slowed down but not destroyed during freezing. If not inactivated, these enzymes can cause color and flavor changes as well as loss of nutrients.

Enzymes in Vegetables. Blanching—the exposure of a vegetable to boiling water or steam for a brief period of time—inactivates enzymes in vegetables. The vegetable must then be rapidly cooled in ice water to prevent further cooking. Contrary to statements in some publications on home freezing, blanching is essential for top-quality frozen vegetables. Blanching also helps destroy microorganisms on the surface of the vegetables. It makes vegetables such as broccoli and spinach more compact, so they do not take up as much room in the freezer. Overblanching results in a cooked product and a loss of flavor, color, and nutrients. Underblanching stimulates enzyme activity and is worse than no blanching at all.

Enzymes in Fruits. Enzymes in fruits can cause browning and loss of vitamin C. Because fruits are usually served raw and people like their uncooked texture, they are not usually blanched. Instead, the use of chemical compounds controls enzymes in frozen fruits. The most common control chemical is ascorbic acid (vitamin C). Ascorbic acid may be used in its pure form or in commercial mixtures of ascorbic acid and other compounds.

Some directions for freezing fruits also include temporary measures to control browning. Such temporary measures include placing the fruit in citric acid or lemon juice solutions or in a sugar syrup. These methods do not prevent browning as effectively as treatment with ascorbic acid.

Rancidity in Foods. Another type of chemical change that can take place in frozen products is the development of rancid off flavors. This can occur when fat, such as that found in meat, is exposed to air over a period of time. Using a wrapping material that does not permit air to reach the product can control rancidity. Removing as much air as possible from the freezer container reduces the amount of air in contact with the product.

Textural Changes During Freezing

Freezing actually consists of freezing the water contained in the food. When the water freezes, it expands and the resulting ice crystals cause the cell walls to rupture. Consequently, the texture of the product will be much softer when the product thaws.

These textural changes are more noticeable in fruits and vegetables that have a higher water content and those that are usually eaten raw. For example, when a frozen tomato is thawed, it turns into mush and liquid. This explains why celery, lettuce, and tomatoes are not usually frozen. For this reason, frozen fruits, usually consumed raw, are best served before they have completely thawed. In the partially thawed state, the effect of freezing on the fruit tissue is less noticeable.

Textural changes caused by freezing are not as apparent in products that are cooked before eating because cooking also softens cell walls. These changes are less noticeable in high-starch vegetables such as peas, corn, and lima beans.

SCIENCE CONNECTION!

Conduct an oxidation rancidity lab using potato chips! Place samples of chips in glass jars. Cover one jar with aluminum foil and leave one uncovered. What changes do you notice? Set up the lab for 14 days. It is safe to use all your senses. What happened? Why? Record your hypothesis and findings!

Rate of Freezing. Freezing products as quickly as possible can control the extent of cell wall rupture. In rapid freezing, a large number of small ice crystals are formed. These small ice crystals cause less cell wall rupture than slow freezing, which produces only a few large ice crystals. This is why some home freezer manuals recommend that the temperature of the freezer be set at the coldest setting several hours before foods will be placed in the freezer.

Changes Caused by Fluctuating Temperatures. Storing frozen foods at temperatures higher than 0 °F increases the rate at which deterioration can take place and can shorten the shelf life of frozen foods. For example, the same loss of quality in frozen beans stored at 0 °F for 1 year will occur in 3 months at 10 °F, in 3 weeks at 20 °F, and in 5 days at 30 °F.

Fluctuating temperatures can cause the ice in the foods to thaw slightly and then refreeze. Each time this happens, the smaller ice crystals form larger ones, further damaging cells and creating a mushier product. Fluctuating temperatures can also cause water to migrate from the product. This defect may also be seen in commercially frozen foods that have been handled improperly.

Moisture Loss. Moisture loss—or ice crystals evaporating from the surface area of a product—produces **freezer burn**, a grainy, brownish spot where the tissues become dry and tough. Freezer burn is caused by ice evaporating out of a frozen food product. This surface freeze-dried area typically develops off flavors, but it will not cause illness. Once the food loses moisture, it cannot be regained. Packaging in heavyweight, moisture-resistant wrap will prevent freezer burn. Airtight packaging can also limit freezer burn. Avoiding thawing and refreezing can also reduce the chance of freezer burn.

Microbial Growth in the Freezer

The freezing process does not destroy microorganisms that may be present on fruits and vegetables. Although blanching destroys some of them and there is a gradual decline in the total number of microorganisms during freezer storage, sufficient numbers remain to multiply and cause spoilage of the product when it thaws. For this reason, it is necessary to carefully inspect any frozen products that have accidentally thawed by the freezer losing power or the freezer door being left open.

Clostridium botulinum, the microorganism that causes the greatest problem in canning low-acid foods, does not grow and produce toxin at 0 °F. Therefore, freezing provides a safe and easy alternative to pressure canning low-acid foods.

Freezing Methods

Freezing methods that freeze the product in air include the still-air sharp freezer, the blast freezer, and the fluidized-bed freezer (Figure 10-3). Single-plate, double-plate, pressure-plate, and slush freezer all are methods of freezing where the food or food packages directly contact a surface that is cooled by a refrigerant. In **immersion freezing**, intimate contact occurs between the food or package and the refrigerant. Types of immersion freezing include a heat-exchange fluid, compressed gas, or refrigerant spray. Extremely rapid freezing occurs with the use of liquid nitrogen, a **cryogenic** liquid.

Packaging

Packaging for frozen foods protects against dehydration, light, and air. It needs to be strong, flexible, and liquid tight.

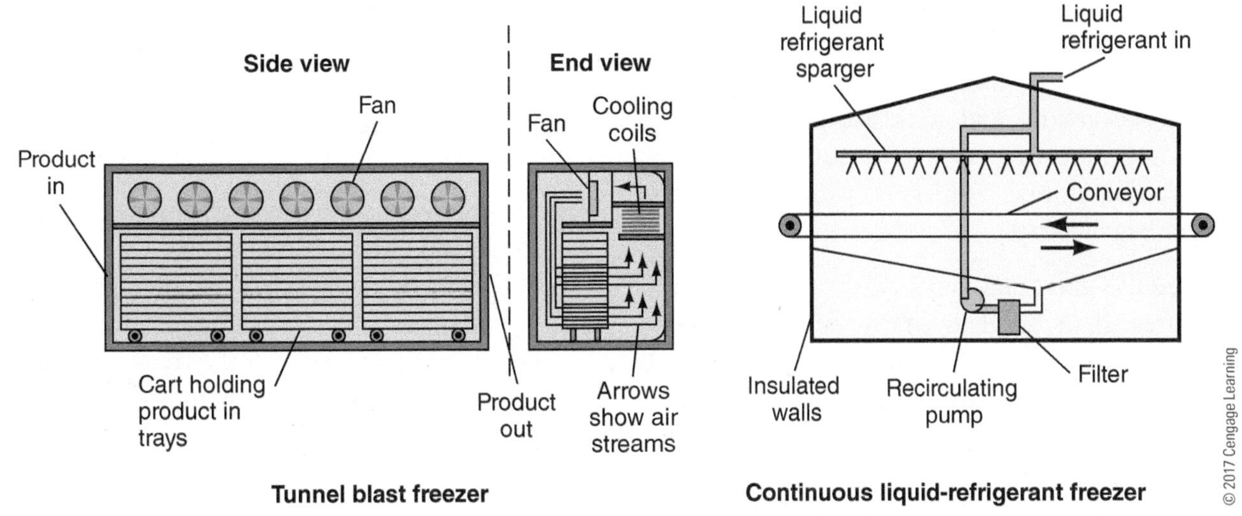

FIGURE 10-3 Tunnel blast freezers and continuous liquid-refrigerant freezers preserve large quantities of product.

NEW DEVELOPMENTS

The success of freezing technology has expanded the field for food processors. They can prepare complete meals and freeze them until the consumer is ready to thaw and heat them. Many of these meals are sold in their serving dishes. Meat and pasta dishes as well as seafood and vegetables are all available as frozen meals. Other popular frozen foods are pot pies, pizza, desserts, breads, and potatoes. Many frozen meal options boast healthy choices and are advertised as "lean," "healthy," or "smart." No other form of food preservation currently offers the convenience that frozen foods do.

HOME FREEZING

Freezing is one of the easiest, most convenient, and least time-consuming methods to preserve food. Freezing does not sterilize foods; the extreme cold simply retards the growth of mcroorganisms and slows down chemical changes that affect quality or cause food to spoil. By following well-established procedures, high-quality nutritious frozen food can be enjoyed year round.

Foods for the home freezer must have proper packaging materials to protect their flavor, color, moisture content, and nutritive value from the dry climate of the freezer. The selection of containers depends on the type of food to be frozen, personal preference, and types that are readily available. Fruits and vegetables should not be frozen in containers with a capacity of more than one-half gallon. Foods in larger containers freeze too slowly to result in a satisfactory product. In general, packaging materials must have these characteristics:

- Moisture-vapor resistant
- Durable and leak-proof
- Not become brittle and crack at low temperatures
- Resistant to oil, grease, or water
- Protect foods from absorption of off flavors or odors
- Easy to seal
- Easy to mark

Two types of packaging materials are most commonly used for home preservation: rigid containers and flexible bags or wrappings.

Rigid Containers

Rigid containers made of plastic or glass are suitable for all packs and are especially good for liquid packs. Straight sides on rigid containers make the frozen food much easier to get out. Rigid containers are often reusable and can make stacking food in a freezer easier. Cardboard cartons for cottage cheese, ice cream, and milk are not sufficiently resistant o moisture vapor to be suitable for long-term freezer storage—unless they are lined with a freezer bag or wrap.

Regular glass jars break easily at freezer temperatures. If using glass jars, wide-mouth, dual-purpose jars made for freezing and canning must be used. These jars have been tempered to withstand extremes in temperatures. The wide mouth allows easy removal of partially thawed foods. If standard canning jars (those with narrow mouths) are used for freezing, extra head space allows for expansion of foods during freezing, although expanding liquid could cause the jars to break at the neck. Some foods will need to be thawed completely before removal from the jar is possible. Covers for rigid containers should fit tightly. If they do not, freezer tape can seal and reinforce the covers. This tape is especially designed to stick at freezing temperatures.

Flexible Bags or Wrappings

Bags and sheets of materials resistant to moisture vapor and heavy-duty aluminum foil are suitable for dry-packed vegetables and fruits, meats, fish, or poultry. Bags can also be used for liquid packs.

Protective cardboard cartons can be used to protect bags and sheets against tearing and to make stacking easier.

Laminated papers made of various combinations of paper, metal foil, glassine, cellophane, and rubber latex are also suitable for dry-packed vegetables and fruits, meats, fish, and poultry. Laminated papers also can be used as protective overwraps.

Freezer Pointers

The following freezer pointers will help any individual enjoy the convenience of home freezing of foods:

- Freeze foods at 0 °F or lower. To facilitate more rapid freezing, set the temperature control at −10 °F or lower 24 hours in advance.
- Freeze foods as soon as they are packaged and sealed.
- Do not overload the freezer with unfrozen food. Add only the amount that will freeze within 24 hours, which is usually 2 to 3 pounds of food per cubic foot of storage space. Overloading slows down the freezing rate, and foods that freeze too slowly may lose quality.
- Place packages in contact with refrigerated surfaces in the coldest part of the freezer.
- Leave a little space between packages so that air can circulate freely. When the food is frozen, store the packages close together.

Foods to Freeze for Quality

Freezing cannot improve the flavor or texture of any food, but it can preserve most of the quality of the fresh product when properly done. Only the best-quality fruits and vegetables at their peak of maturity should be frozen.

Fruit should be firm yet ripe—firm for texture and ripe for flavor. Vegetables should be young, tender, unwilted, and garden fresh. If fruits or vegetables must be held before freezing, they should be stored in the refrigerator to prevent deterioration. Some varieties of fruits or vegetables are more suitable for freezing than others.

Foods found in Table 10-2 do not freeze well and are best preserved by another method or left out of mixed dishes that are to be frozen.

Effect of Freezing on Spices and Seasonings

Freezing spices and seasonings can create some undesirable effects, including the following:

- Pepper, cloves, garlic, green pepper, imitation vanilla, and some herbs tend to become strong and bitter.

TABLE 10-2 Foods Showing Reduced Quality after Freezing

FOODS	USUAL USE	CONDITION AFTER THAWING
Cabbage,[1] celery, cress, cucumbers,[1] endive, lettuce, parsley, radishes	As raw salad	Limp, waterlogged, quickly develops oxidized color, aroma, and flavor
Irish potatoes, baked or boiled	In soups, salads, sauces with butter	Soft, crumbly, waterlogged, mealy
Cooked macaroni, spaghetti, or rice	When frozen alone for later use	Mushy, tastes warmed over
Egg whites, cooked	In salads, creamed foods, sandwiches, sauces, gravy, or desserts	Soft, tough, rubbery, spongy
Meringue	In desserts	Toughens
Icings made from egg whites	Cakes, cookies	Frothy, weepy
Cream or custard fillings	Pies, baked goods	Separates, watery, lumpy
Milk sauces	For casseroles or gravies	May curdle or separate
Sour cream	As topping, in salads	Separates, watery
Cheese or crumb toppings	On casseroles	Soggy
Mayonnaise or salad dressing	On sandwiches (not in salads)	Separates
Gelatin	In salads or desserts	Weeps
Fruit jelly	Sandwiches	May soak bread
Fried foods	All except french-fried potatoes and onion rings	Lose crispness, become soggy

[1]Cucumbers and cabbage can be frozen as marinated products such as "freezer slaw" or "freezer pickles." These do not have the same texture as regular slaw or pickles.

Source: Bulletin 989, "So Easy to Preserve," produced by the Cooperative Extension Service, the University of Georgia, College of Agricultural and Environmental Sciences. Sixth Edition, 2014.

- Onion and paprika change flavor during freezing.
- Celery seasonings become stronger.
- Curry may develop a musty off flavor.
- Salt loses flavor and has the tendency to increase the rancidity of any item containing fat.

When using seasonings and spices, foods can be seasoned lightly before freezing, and additional seasonings can be added when the food is reheated or served.

Freezer Management

A full freezer is most energy-efficient, and refilling your freezer several times a year is most cost-efficient. If the freezer is filled and emptied only once each year, the energy cost per package is especially high. The cost for each pound of stored food can be reduced by filling and emptying a freezer two, three, and even more times each year.

Posting a frozen foods inventory near the freezer and keeping it up to date by listing the foods and dates of freezing keeps foods from being forgotten. This inventory should show the exact amounts and kinds of foods in the freezer at all times. Also, foods in a freezer should be organized into food groups for ease in locating. Packages in the freezer the longest need to be the first ones used.

Storage temperature must be maintained at 0 °F or lower. At higher temperatures, foods lose quality much faster. A freezer thermometer in the freezer allows frequent checking of the temperature.

For more details on home freezing, search the Web or visit the following Web sites:

How Do I Freeze? http://nchfp.uga.edu/how/freeze.html

Food Freezing Guide: https://www.ag.ndsu.edu/burleighcountyextension/pdfs/fcs/fcs-publications/fn-403-food-freezing-guide.

SUMMARY

Freezing and refrigeration preserve foods not because they destroy microbes but because they slow or stop microbial growth. Refrigeration temperatures range from 40 °F to 45 °F; frozen storage temperatures range from 32 °F to 0 °F. Refrigerated storage also requires air circulation, humidity control, and a modified gas atmosphere. During refrigerated storage, foods can experience the absorption of flavor and a loss of firmness, color, flavor, and sugar.

Freezing technology has been key to the development of convenience foods. Foods properly frozen and stored experience minimal changes in food quality. Frozen storage varies with temperature and the type of food being stored. General freezing methods include still air, blast freezer, and fluidized-bed freezer. Single-plate, double-plate, and slush freezers freeze foods or packages that directly contact a cold surface. Immersion freezing is the direct contact of the food or package with a refrigerant such as liquid nitrogen. To maintain quality, frozen foods must be packaged in airtight and liquid-tight strong and flexible containers.

Home freezing of foods follows the same general principles as commercial freezing.

REVIEW QUESTIONS

Success in any career requires knowledge. Test your knowledge of this chapter by answering these questions or solving these problems.

1. Describe the three methods of freezing.
2. List the four requirements of refrigerated storage.
3. Identify four changes in food during refrigeration.
4. What are the key factors in food freezing?
5. Describe the temperature difference between cooling, refrigeration, and freezing.
6. Why do food processors blanch vegetables before freezing them?
7. Name the two types of containers commonly used for home freezing use.
8. What food characteristics can freezing preserve?
9. Explain why a freezer should not be overloaded with unfrozen food.
10. List the three things that frozen food packaging protects against.

STUDENT ACTIVITIES

1. Freeze one of the following and report what you observe when the food thaws: basil, artificial vanilla, egg custard, fresh apples or peaches, lettuce. Use your senses of taste, smell, and sight to describe what you observe. Compare the items to their unfrozen counterparts.
2. Develop a report or presentation on what happens as water freezes.
3. Under instructor supervision, freeze some food products using liquid nitrogen or dry ice and compare the speed of freezing. Also, compare the speed of freezing to that in a normal freezer.
4. Develop an informational report on the use of refrigeration and freezing in other countries, include visual examples in the report.
5. Keep some of the following foods at refrigerator temperatures for extended times and describe changes in food quality: hamburger, celery, apple, tomato, potato, strawberries, and cheese.

ADDITIONAL RESOURCES

- Cold Storage Times: http://www.extension.umn.edu/food/food-safety/preserving/storage/cold-storage-times/
- Microbiology of Food Preservation: https://www.tamu.edu/faculty/acastillo/Handouts/Preservation.pdf
- Home Food Preservation Links and Resources: http://cag.uconn.edu/nutsci/nutsci/foodsafety/ConsumersHome_Cooks_Landing_Page/Preserve_food_safely_at_home_/Home_Food_Preservation.php
- Food Preservation: http://dbs.extension.iastate.edu/answers/longform.cfm?PID=3&CID=16

Internet sites represent a vast resource of information. Although those provided in this chapter were vetted by industry experts, you may wish to further explore the topics discussed in this chapter using a search engine such as Google. Keywords or phrases may include the following: *air cooling, cool storage, hydrocooling, hypobaric, immersion freezing, vacuum cooling, frozen food storage, freezing point, freezing methods.* In addition, Table B-7 provides a listing of some useful Internet sites that can be used as a starting point.

REFERENCES

National Council for Agricultural Education. 1993. *Food science, safety, and nutrition.* Madison, WI: National FFA Foundation.

Potter, N. N., and J. H. Hotchkiss. 1995. *Food science*, 5th ed. New York: Chapman & Hall.

Vaclavik, V. A., and E. W. Christina. 2014. *Essentials of food science*, 4th ed. New York: Springer.

Vieira, E. R. 1996. *Elementary food science*, 4th ed. New York: Chapman & Hall.

CHAPTER 11

Drying and Dehydration

OBJECTIVES

After reading this chapter, you should be able to:

- Discuss several reasons for dehydrating foods
- Describe changes that occur during dehydrating
- List three factors that affect dehydration
- Identify chemical changes that can occur in food during drying
- Explain two problems that can occur during drying
- Identify reasons for food concentration
- Describe three drying methods
- Describe the methods of food concentration
- Compare home drying of foods to commercial drying

KEY TERMS

atmospheric pressure
bars
caramelization
case hardening
dehydration
enzymatic browning
rehydrated
sublimation
surface area
vacuum

NATIONAL AFNR STANDARD

FPP.03

Select and process food products for storage, distribution, and consumption.

Drying and dehydration both remove water from foods. Dehydration is one of the oldest and easiest preservation methods. There are several ways to dehydrate foods. Dehydration occurs under natural conditions in the field and during cooking. Dehydrated and dried foods are lighter, take up less space, and cost less to ship. They often make great foods on the go and for camping or hiking because they do not require refrigeration.

DEHYDRATION

Water is removed from foods under natural conditions in the field in the case of grains, raisins, and seeds. Water is also removed during cooking and as part of a controlled **dehydration** process. Dehydration is the almost complete removal of water. It is commonly used to produce dried milk, coffee, soups, and corn flakes (Figure 11-1). Dehydration also results in:

- Decreased weight
- Increased amount of product per container
- Decreased shipping costs

The purpose of drying is to remove enough moisture to prevent microbial growth. Figure 11-2 shows several types of commercial driers. In some cases, such as sun drying, the rate may be too slow and organisms may cause spoilage before the product can be thoroughly dried. In these cases, salt or smoke may be added to the product before it is dried.

The lower limit of moisture content by the sun-drying method is approximately 15%. Sun drying can have problems with contamination, and it only works in areas of low humidity. Dried foods are not sterile. Many spores survive in dry food, and drying never completely removes all water.

FIGURE 11-1 Wide varieties of dried and dehydrated soups are available at many grocery stores.

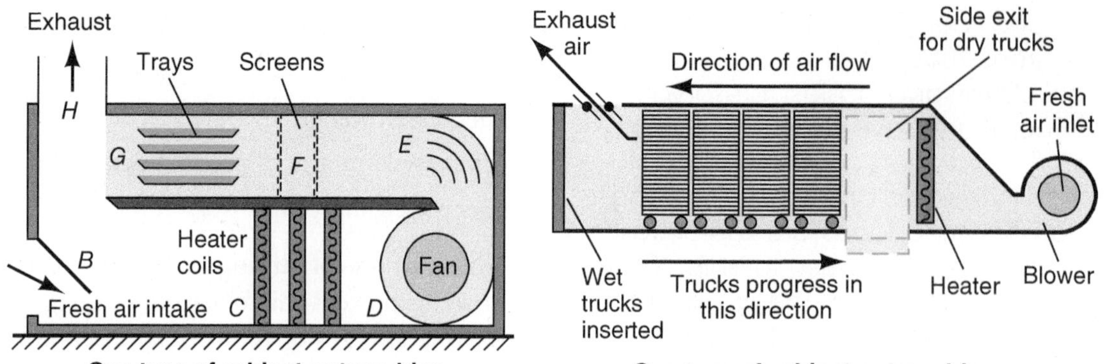

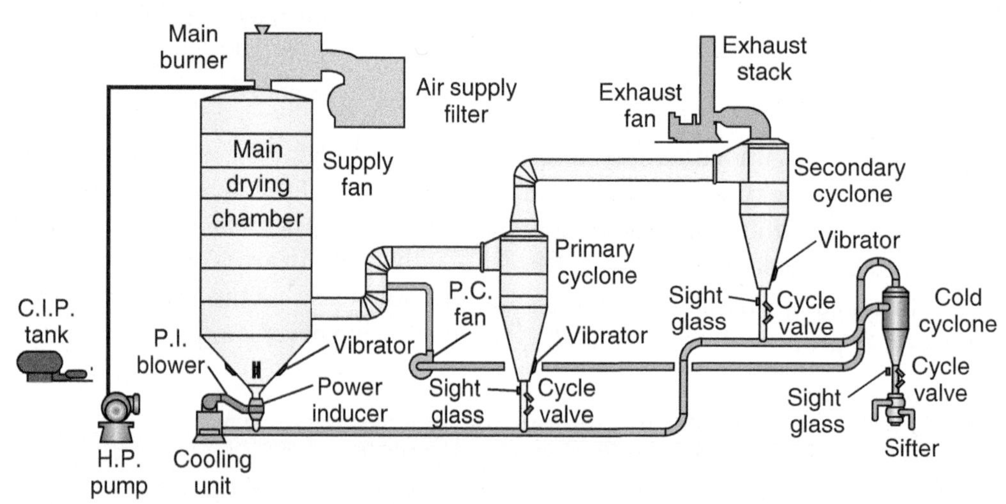

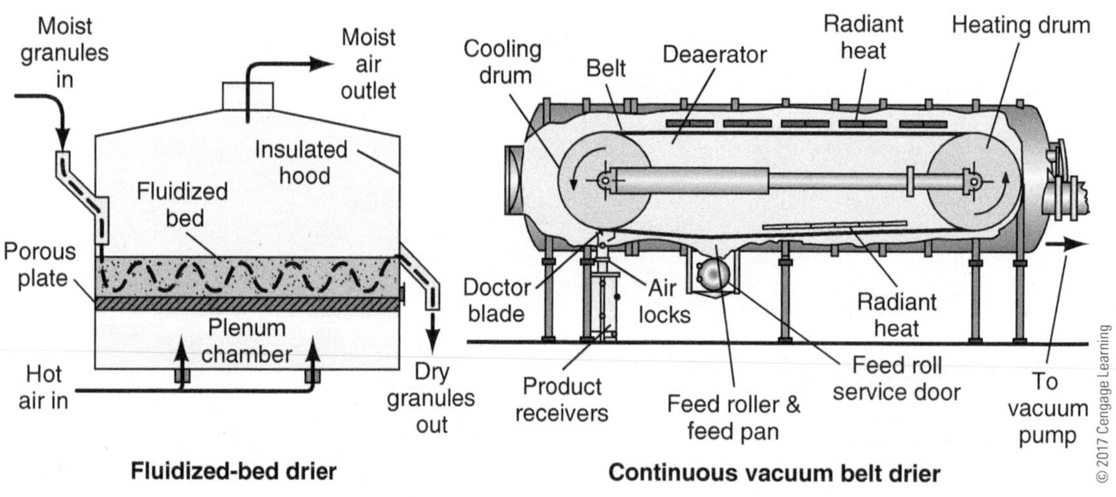

FIGURE 11-2 Diagrams of typical commercial dehydrators.

Four factors affect heat and liquid transfer in food products during drying:

1. Surface area
2. Temperature
3. Humidity
4. Atmospheric pressure

The greater its **surface area**, the faster a product dries. The greater the temperature differences between a product and the drying medium, then the greater the rate of drying. The higher the humidity, the slower the drying. **Atmospheric pressure** controls the temperature at which water boils: The lower the pressure, the lower the temperature required to remove water. To overcome this, a vacuum can be used. So, dehydration can be enhanced by increasing the surface area of the food product, increasing the temperature and air velocity, and reducing the humidity and pressure (vacuum).

Solute Concentration

Foods high in sugar or other solutes dry more slowly. In addition, as drying proceeds, the concentration of solutes becomes greater in the water that remains. This is another reason the drying rate slows.

Binding of Water

As a product dries, its free water is removed. This water is the easiest to remove, and it evaporates first. Some water is held by absorption to

SCIENCE CONNECTION!

Conduct a dehydration experiment. This could be done with fruit, vegetables, or meat. Does the size of the slices affect the amount of time it takes to dehydrate the food? Record your hypothesis and findings!

MATH CONNECTION!

Research the average humidity and atmospheric pressure across different regions in the United States. What effects can these differences have on the foods we eat?

Drying Curve

When foods are dehydrated, they lose water at a changing rate. Water is lost rapidly at first because it is being lost from the surface of the food. As the food develops a dried layer on the outside, the remaining moisture is confined to the center. The outer dry layer creates an insulation barrier that prevents rapid heat transfer into the food. Also, the moisture in the center has farther to travel. Eventually, the food piece reaches its normal equilibrium in relative humidity. It is picking up moisture from the drying atmosphere as fast as it loses moisture. So, when the moisture content of a food is graphed against the time of drying (see Figure 11-3), a curve with a typical shape develops. The graph shows moisture being lost rapidly at first and then being lost quite slowly toward the end of the drying period.

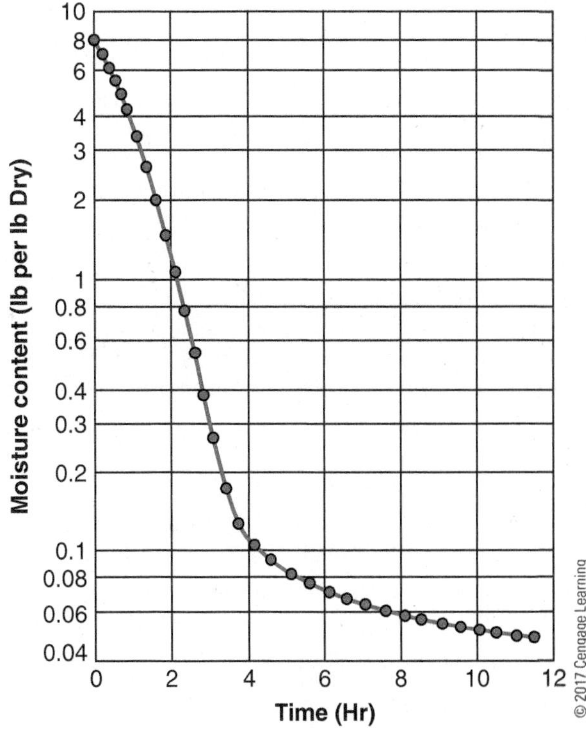

FIGURE 11-3 A typical drying curve shows a rapid water loss at first that slows as the product dries.

food solids. Water that is in colloidal gels such as starch, pectin, and other gums is more difficult to remove. The most difficult water to remove is that chemically bound in the form of hydrates such as glucose monohydrate or inorganic salt hydrates.

Chemical Changes

Several chemical changes can occur during drying, including:

- Caramelization
- Enzymatic browning
- Nonenzymatic browning
- Loss of ease of rehydration
- Loss of flavor

The extent of these chemical changes depends on the foods and the type of drying method used.

Caramelization of sugars occurs if the temperature is too high. This is when sugars on the surface of the food start to brown during cooking. **Enzymatic browning** is caused by enzymes and can be prevented by inactivating the enzymes before drying. This is usually done by pasteurizing or blanching the food first. Nonenzymatic browning or Maillard browning is controlled by dehydrating the food rapidly through the moisture ranges that are optimal for the Maillard browning. Physical changes, denatured proteins, and loss of sugars and salts during dehydration make the reabsorption of water by the dried product less than the original. For this reason, **rehydrated** dried products have an altered texture. Preventing a loss of flavor during dehydration is almost impossible. Food processors develop ways to trap the vapors carrying the flavor and then adding them back after dehydration. Some foods such as meats and eggs will harden as they are dried or cooked. On the other hand, some foods such as onions will soften while dried or cooked.

DRINKS FOR THE SPACE PROGRAM

At the beginning of the space program in the early 1960s, astronauts wanted to be able to drink orange juice and milk in space. Neither pure orange juice or whole milk can be dehydrated. When rehydrated, orange drink crystals just make orange "rocks" in water. A freeze-dried orange juice is available, but it is difficult to rehydrate. Whole dried milk does not dissolve properly. It floats around in lumps and has a disagreeable taste. Therefore, nonfat dry milk must be used in space packaging. To solve the orange juice problem, a new product had to be developed.

During the 1960s, General Foods developed a synthetic orange-flavored juice called Tang for the space program. It can be used in place of orange juice. Sales of Tang products declined by 2007 and Kraft Foods revitalized Tang through different flavors, packaging, and advertisement. Today, more than 30 flavors of Tang are sold and consumed around the world.

For more information on space food or Tang, search the Web or visit these Web sites:

- Kraft: http://www.kraftfoodservice.com/ProductsandBrands/ColdBeverages/BulkPackets.aspx
- NASA—Food for Space Flight: http://www.nasa.gov/audience/forstudents/postsecondary/features/F_Food_for_Space_Flight.html
- General Foods Corporate Timeline: http://www.kraftfoodsgroup.com/SiteCollectionDocuments/pdf/CorporateTimeline_GeneralFoods_FINAL.pdf
- Space Food Hall of Fame http://education.ssc.nasa.gov/fft_halloffame.asp

Drying Methods

Common drying methods are:

- Air convection
- Drum
- Vacuum
- Freeze

Air Convection. A typical air convection drier has an insulated enclosure, a way of circulating air through the enclosure, and a way to heat this air. The food is supported within the enclosure, and the movement of air is controlled by fans, blowers, and baffles. Dried product is collected by specially designed devices. Types of air driers include kiln, tunnel, continuous conveyor belt, air lift, fluidized bed, spray, cabinet, tray or pan, and belt trough. Figure 11-4 is an example of a milk products spray dryer. These driers can dehydrate food pieces, purees, liquids, and granules.

Drum. Drum or roller driers are used to dry liquid foods, purees, pastes, and mashes. These food products are continuously applied in a thin layer onto the surface of a revolving heated drum. As the drum rotates, the thin layer of food dries. The speed of the drum is regulated so that when the food reaches a point where a scraper is located, it will be dry. This type of drier can have one or two drums. Milk, potato mash, tomato paste, and animal feeds are typically dried on drums. The high surface temperature of the drum restricts the foods that can be dried this way.

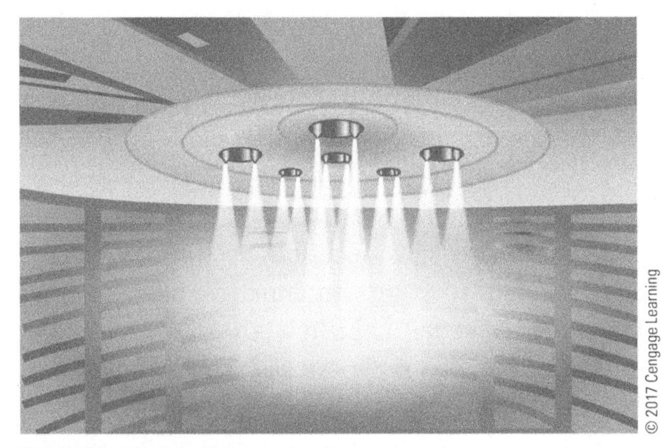

FIGURE 11-4 Milk products spray dryer.

Vacuum. Vacuum drying produces the highest quality of product, but it is also the most costly. Essential elements of vacuum-drying systems include a vacuum chamber, a heat source, a device for maintaining a vacuum, and a component to collect (condense) water vapor as it leaves the food. Shelves or other supports suspend the food in the vacuum chamber. Vacuum shelf driers (batch driers) and the continuous vacuum belt drier are two main types of vacuum driers. Fruit juices, instant tea, milk, and delicate liquid foods are dried in vacuum driers.

Freeze-Drying. Freeze-drying is used to dehydrate sensitive, high-value foods such as coffee, juices, strawberries, whole shrimp, diced chicken, mushrooms, steaks, and chops. Freeze-drying protects the delicate flavors, colors, texture, and appearance of foods. To dry foods with this method, they are first frozen so that the frozen food does not shrink or distort while giving up its moisture. The principle of freeze-drying is that under conditions of low vapor pressure (a vacuum), water evaporates from ice without the ice melting. In other words, through a process called **sublimation**, water goes from a solid to a gas without passing through its liquid phase.

FOOD CONCENTRATION

Evaporation concentrates food (Figure 11-5) by removing approximately one-third to two-thirds of the water present. It has some preservative effects, but mainly it reduces weight and volume. Depending on the method, during concentration food may take on a cooked flavor, darken somewhat, change in nutritional value, and ensure some microbial destruction. Methods of concentration include:

- Solar (sun)
- Open kettles
- Flash evaporators
- Thin film evaporators

FIGURE 11-5 Concentrated products produced by evaporation.

- Vacuum evaporators
- Freeze concentration
- Ultrafiltration and reverse osmosis

Reduced Weight and Volume

Reducing volume and weight saves money. For example, most of the liquid foods that are to be dehydrated are first concentrated because the early stages of water removal by some form of evaporation are more economical. Reducing the volume and weight saves money by removing water that would otherwise have to be contained and shipped. For example, a soup producer needing tomato solids has little need for the original tomato pulp, which is only about 6% solids. By reducing the water content, the solids can be increased to 32%. Food concentration saves container costs, transportation costs, warehousing costs, and handling costs. Foods that are commonly concentrated include evaporated and sweetened condensed milks, fruit and vegetable juices, sugar syrups, jams and jellies, tomato paste, and other types of purees, buttermilk, whey, and yeast. Some food by-products are concentrated and used as animal feed. The chart in Figure 11-6 indicates the method used according to the size of material.

Solar Evaporation

Solar evaporation is the oldest method of food concentration. It is slow and only used today to concentrate salt solutions in human-made lagoons. From the Great Salt Lake in Utah, companies draw highly saline waters from the lake's remote areas into especially shallow solar evaporation ponds to produce salt, potassium sulfate, and magnesium.

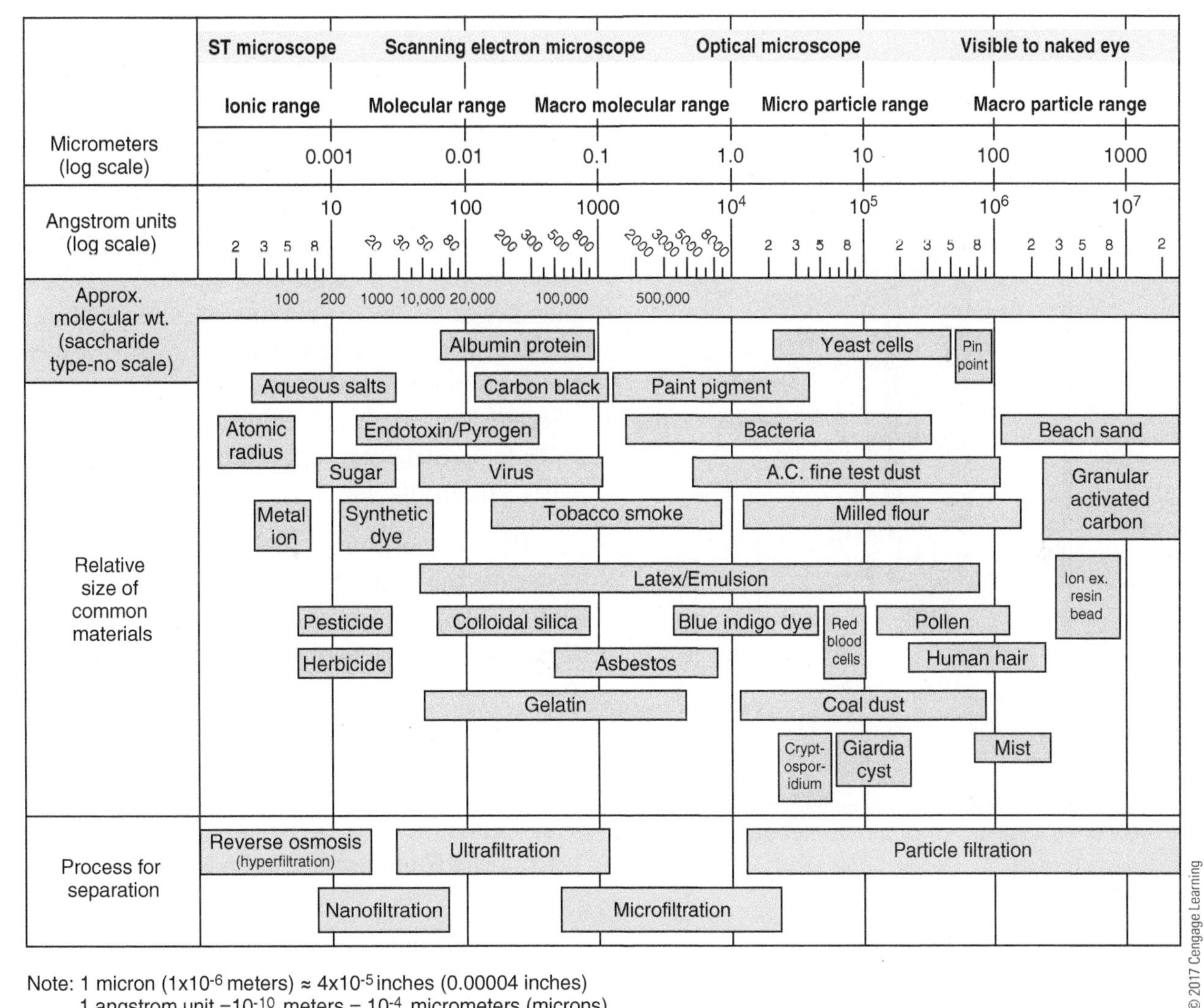

FIGURE 11-6 Chart showing the relative size of materials and the method for filtration separation. Note that protein requires ultrafiltration and sugar (lactose) requires reverse osmosis.

Open Kettles

Open, heated kettles are used for jellies, jams, and some soups. High temperatures and long concentration times damage many foods. Kettles are also still used to make maple syrup; the high heat produces the desirable color and typical flavor of maple syrup.

Flash Evaporators

Flash evaporators subdivide the food and bring it into direct contact with the steam. The concentrated food is drawn off at the bottom of the evaporator, and the steam and water vapor are removed through a separate outlet.

Thin-Film Evaporators

In thin-film evaporators (Figure 11-7), the food is pumped onto a rotating cylinder and spread in a thin layer. Steam quickly removes the water from the thin layer, and the concentrated food is then wiped from the cylinder wall. The concentrated food and water vapor are continuously removed to an external separator. Here the food product is taken and the water vapor condensed.

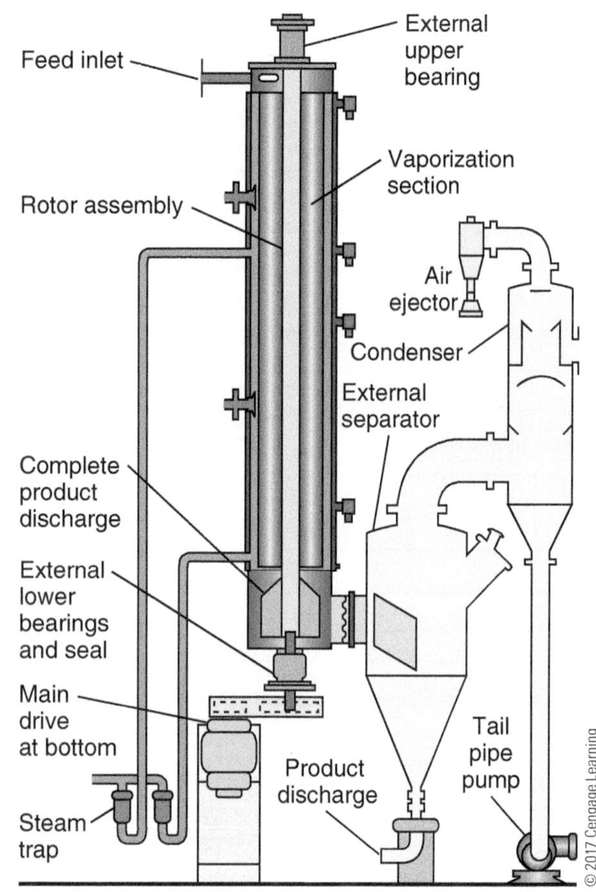

FIGURE 11-7 Diagram of a thin-film evaporator.

Vacuum Evaporators

Low-temperature vacuum evaporators are used for heat-sensitive foods. With a vacuum, lower temperatures can be used to remove water from the food. Often vacuum chambers are in a series, and the food product becomes more concentrated as it moves through the chambers.

Freeze Concentration

All of the components of a food do not freeze at the same time. First to freeze is the water. It forms ice crystals in a mixture. This allows the initial ice crystals to be removed before the entire mixture freezes. To do this, the partially frozen mixture is centrifuged, then the frozen slush is put through a fine-mesh screen. Frozen water crystals are held back by the screen and discarded. Freeze concentration is used commercially in orange juice production.

Ultrafiltration and Reverse Osmosis

Ultrafiltration (Figure 11-8) is a membrane-filtration process operating from 2 to 10 **bars** (international unit of pressure that equals 29.531 in. of mercury at 32 °F) of pressure, allowing molecules the size of salts and sugars to pass through the membrane pores while molecules the size of proteins are rejected. Typical ultrafiltration applications include milk for protein standardization, cheeses, yogurt, whey, buttermilk, eggs, gelatin, and fruit juice.

The reverse-osmosis membrane-filtration process uses the tightest membranes and operates at 10 to 100 bars pressure, allowing only water to pass through the pores of the membrane. It is used to concentrate whey and reduce transportation costs of milk by removing the water. Reverse osmosis can also be used to recover rinsing water by concentrating it from tanks, pipes, and so on to help recover milk solids; it is also used to concentrate eggs, blood, gelatin, and fruit juices (Refer to Figure 11-6).

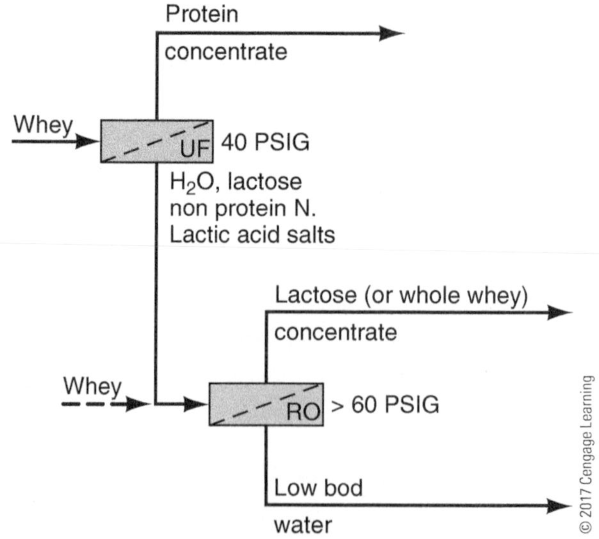

FIGURE 11-8 Schematic showing the use of ultrafiltration and reverse osmosis to separate the components of cheese whey.

Besides the reduction in water and the recovery of valuable food components, ultrafiltration and reverse osmosis discharge water that is low in organic matter. This decreases the potential for pollution from discharge water.

HOME DRYING

Most foods can be dried indoors using modern food dehydrators, countertop convection ovens, or conventional ovens. Microwave ovens are recommended only for drying herbs, because not enough airflow can be created in them. Some foods can be dried outdoors in the sun.

Foods can be dried in an oven or in a food dehydrator by using the right combination of warmth, low humidity, and air current. In drying, a warm temperature allows the moisture to evaporate. Air current speeds up drying by moving the surrounding moist air away from the food. Low humidity allows moisture to move from the food to the air.

Drying food is a slow process. It will take six or more hours in a dehydrator and eight or more hours in an oven. Drying time depends on type of food, thickness, and type of dryer. Turning up the oven temperature cannot speed up drying. This cooks the food on the outside before it dries on the inside. This is called **case hardening**. The food may appear dry on the outside but it may still be wet on the inside. Remaining moisture in the food will cause it to mold.

Food Dehydrators

A food dehydrator is a small electrical appliance for drying foods indoors. It has an electric element for heat and a fan and vents for air circulation. Dehydrators are efficiently designed to dry foods fast at 140 °F. Food dehydrators are available from department stores, online catalogs, small appliance sections of department stores, natural food stores, and seed and garden supply catalogs. Twelve square feet of drying space dries about one-half bushel of produce. The major disadvantage of a dehydrator is its limited capacity. Many dehydrators have shelves or racks and can adjust the number based on how much food is being dehydrated.

Instructions are available from county extension offices and various books offer plans for building homemade dehydrators. Building a dehydrator could save money, but the final unit will not be as efficient as a commercial dehydrator.

Oven Drying

Everyone who has an oven has a food dehydrator. By combining the factors of heat, low humidity, and air current, an oven can be used as a dehydrator. An oven is ideal for occasional drying of meat jerkies, fruit leathers, banana chips, or for preserving excess produce such as celery and mushrooms. Because the oven may also be needed for everyday cooking, it may not be satisfactory for preserving abundant garden produce.

Oven drying is slower than dehydrators because the oven does not have a built-in fan for air movement. Drying food in an oven takes twice as long as in a dehydrator.

Room Drying

The room-drying method takes place indoors in a well-ventilated attic, room, car, camper, or screened-in porch. Herbs, hot peppers, nuts in the shell, and partially sun-dried fruits are the most common air-dried items.

Herbs and peppers can be strung on a string or tied in bundles and suspended from overhead racks in the air until dry. Enclosing them in paper bags, with openings for air circulation, protects them from dust, loose insulation, and other pollutants. Nuts are spread on papers, a single layer thick. Partially sun-dried fruits should be left on their drying trays.

Sun Drying

The high sugar and acid content of fruits make them safe to dry outdoors when conditions are

FIGURE 11-9 Dried fruits, especially raisins and prunes are familiar products. Here, raisins are dried on a farm in California, where many of the world's raisins are produced.

favorable for drying (Figure 11-9). Vegetables (with the exception of vine-dried beans) and meats are not recommended for outdoor drying. Vegetables are low in sugar and acid, increasing the risk for food spoilage. Meats are high in protein, making them ideal for microbial growth when heat and humidity cannot be controlled.

Sun-dried raisins are the best known of all dried foods. California produces much of the world's supply of raisins.

Drying fruits outdoors is best done on a hot, dry, and breezy day. A minimum temperature of 85 °F is needed; higher temperatures are even better. Drying foods outdoors requires several days. Because the weather is uncontrollable, drying fruits outdoors can be risky. Rain in California while the grapes are drying destroys an entire production of raisins. High humidity in the South is a problem for drying fruits outdoors. Humidity below 60% is best. Often these ideal conditions are not available when the fruit ripens, and other alternatives to dry the food are needed.

Racks or screens placed on blocks allow for better air movement around the food. Because the ground may be moist, it is best to place the racks or screens on a concrete driveway or if possible over a sheet of aluminum or tin. The reflection of the sun on the metal increases the drying temperature.

Because birds and insects are attracted to dried fruits, two screens are best for drying food. One screen acts as a shelf and the other as a protective cover. Cheesecloth could also be used to cover the food. Solar dryers may need turning or tilting throughout the day to capture the direct, full sun. Food on the shelves needs to be stirred and turned several times a day.

More information about drying can be found at these Web sites:

Preserving Food—Drying: www.ext.nodak.edu/extnews/askext/foodss/4214.htm

How to Preserve Your Garden Herbs: http://www.hgtvgardens.com/herbs/out-to-dry-how-to-preserve-your-garden-herbs

Drying Herbs: http://nchfp.uga.edu/how/dry/herbs.html

SCIENCE CONNECTION!

What is your favorite dried or dehydrated fruit? How is it commercially produced? Can it be done at home? Research your favorite fruit and find out!

SUMMARY

Besides preserving a product, drying and dehydration decrease its weight and volume of and thereby decrease shipping costs. Drying is affected by surface area, temperature, humidity, and atmospheric pressure. As foods dry, they demonstrate a typical drying curve. Chemical changes that can occur during dehydration include caramelization, browning, loss of rehydration, and loss of flavor. Foods can be dried by air convection, drum, vacuum, or freeze-drying.

Food concentration removes one-third to two-thirds of the water and reduces weight and volume. Methods of concentration include solar, open-kettle, flash evaporators, thin-film evaporators, freeze concentration, and ultrafiltration or reverse osmosis.

Home drying of foods follows the same general principles as commercial processes. Small home dehydrators are available already assembled or can be built from kits. The kitchen oven (or microwave) can also be used to dry foods.

REVIEW QUESTIONS

Success in any career requires knowledge. Test your knowledge of this chapter by answering these questions or solving these problems.

1. List the three drying methods.
2. Describe the results of dehydrating food products.
3. Identify the pros and cons of vacuum drying food products.
4. What is ultrafiltration?
5. Explain the principle of freeze-drying.
6. The purpose of drying is to remove enough moisture to prevent the growth of what?
7. Define sublimation.
8. What types of foods are dried using a drum or roller driers?
9. Discuss the two problems with drying of a food product.
10. List three chemical changes that occur during drying.

STUDENT ACTIVITIES

1. Develop a presentation on the dehydration of foods in underdeveloped countries. Demonstrate the use of this method.
2. Identify a product that has been freeze-dried. If possible, bring a sample to the class and reconstitute it.
3. Research reverse osmosis or ultrafiltration and develop a report on its use in the food industry.
4. Using a microwave, a small food dehydrator, or a conventional oven at a low temperature, dehydrate a variety of fruits and vegetables. Describe any desirable or undesirable changes in the food, including taste. Then rehydrate some of the foods. Also calculate the water percentage of the foods by using the original wet weight and the dry weight. Record all findings and report to the instructor.

ADDITIONAL RESOURCES

- Drying Food: http://www.aces.uiuc.edu/vista/html_pubs/DRYING/dryfood.html
- Quality for Keeps—Drying Foods: http://extension.missouri.edu/p/GH1562
- Food Safety—Drying: http://www.extension.umn.edu/food/food-safety/preserving/drying/

Internet sites represent a vast resource of information. Although those provided in this chapter were vetted by industry experts, you may also wish to further explore the topics discussed in this chapter using a search engine such as Google. Keywords or phrases may include the following: *dehydration, food drying, freeze-drying, caramelization, enzymatic browning, nonenzymatic browning, air convection drying, food concentration, freeze concentration, solar drying, ultrafiltration, reverse osmosis, evaporation, sublimation.* In addition, Table B-7 provides a listing of some useful Internet sites that can be used as a starting point.

REFERENCES

Ensminger, A. H., M. E. Ensminger, J. E. Konlande, and J. R. Robson. 1994. *Foods and nutrition encyclopedia.* 2 Vols. Boca Raton, FL: CRC Press.

Potter, N. N., and J. H. Hotchkiss. 1995. *Food science,* 5th ed. New York: Chapman & Hall.

Vaclavik, V. A., and E. W. Christina. 2014. *Essentials of food science,* 4th ed. New York: Springer.

Vieira, E. R. 1999. *Elementary food science,* 4th ed. New York: Chapman & Hall.

CHAPTER 12

Radiant and Electrical Energy

OBJECTIVES

After reading this chapter, you should be able to:

- Describe ionizing radiation
- Identify two requirements for the irradiation process
- Discuss the four areas in which irradiation is most useful
- List the three specific ways the Food and Drug Administration has approved irradiation uses
- Explain in scientific detail how microwaves heat food
- Describe ohmic (electrical) heating and its major advantage
- Explain why salt and water content are important in microwave heating

NATIONAL AFNR STANDARD

FPP.03

Select and process food products for storage, distribution, and consumption.

KEY TERMS

electromagnetic energy
ionization
ions
irradiation
microwaves
ohmic (joule) heating
polar
radioisotopes

Ionizing radiation and microwaves are invisible energy waves moving through space. Food processors use these two forms of radiation. Electrical or ohmic heating of foods is a relatively new method for heating and preserving foods.

FOOD IRRADIATION

Radiation is broadly defined as energy moving through space in invisible waves (Figure 12-1). Radiant energy has differing wavelengths and degrees of power. Light, infrared heat, and **microwaves** are forms of radiant energy. So are the waves that bring radio and television broadcasts into our homes. Broiling and toasting use low-level radiant energy to cook food.

The radiation of interest in food preservation is ionizing radiation, also known as **irradiation**. These shorter wavelengths are capable of damaging microorganisms such as those that contaminate food or cause food spoilage and deterioration. Scientists have experimented with irradiation as a method of food preservation since 1950. They found it to be a controllable and highly predictable process.

As in the heat pasteurization of milk, the irradiation process greatly reduces but does not eliminate all bacteria. Irradiated poultry, for example, still requires refrigeration but would be safe longer than untreated poultry. Strawberries that have been irradiated will last 2 to 3 weeks in a refrigerator compared to only a few days for untreated berries. Irradiation complements, but does not replace, the need for proper food-handling practices by producers, processors, and consumers.

The Electromagnetic Spectrum

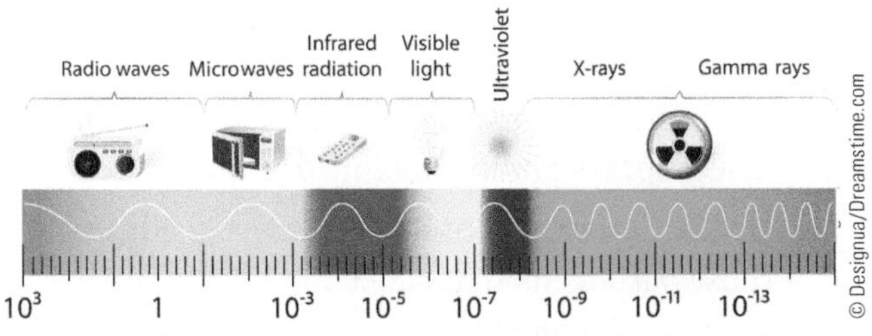

FIGURE 12-1 The electromagnetic spectrum showing wavelengths from radio through visible light to visible gamma rays.

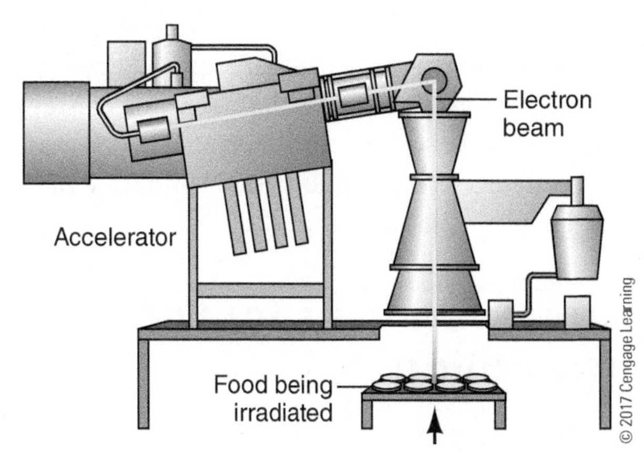

FIGURE 12-2 Irradiation equipment using an accelerator.

Two requirements for the irradiation process are a source of radiant energy and a way to confine that energy.

For food irradiation, the two sources of radiant energy are (1) **radioisotopes** (radioactive materials), and (2) machines that produce high-energy beams. Specially constructed containers or compartments are used to confine the beams so that people will not be exposed. Radioisotopes are used in medical research and therapy in many hospitals and universities. They require careful handling, tracking, and disposal. Machines that produce high-energy beams offer greater flexibility (Figure 12-2). For example, they can be turned on and off unlike the constant emission of gamma rays from radioisotopes.

Food Irradiation Process

Irradiation is known as a *cold* process. It does not significantly increase the temperature or change the physical or sensory characteristics of most foods. An irradiated apple, for example, will still be crisp and juicy. Fresh or frozen meat can be irradiated without cooking it.

During irradiation, the energy waves affect unwanted organisms but are not retained in the food. Food cooked in a microwave oven is similar to teeth or bones that have been X-rayed: They do not retain the energy waves. Bulk or packaged food passes through a radiation chamber on a conveyor belt. The food does not come in contact with radioactive materials but rather passes through a radiation beam similar to a large flashlight.

Approved Uses for Food Irradiation

Irradiation has been approved for many uses in more than 35 countries, but only a few applications are currently being used because of consumer concern and the cost of building facilities.

In the United States, the Food and Drug Administration (FDA) has approved irradiation for eliminating insects from wheat, potatoes, flour, spices, tea, fruits, and vegetables. Irradiation also can be used to control sprouting and ripening. Approval was given in 1985 to use irradiation on pork to control trichinosis. Using irradiation to control *Salmonella* and other harmful bacteria in chicken, turkey, and other fresh and frozen uncooked poultry was approved in May 1990. In December 1997, the FDA approved the use of irradiation to control pathogens (disease-causing microorganisms such as *E. coli* and *Salmonella* species) in fresh and frozen red meats such as beef, lamb, and pork. In 2000, the FDA approved irradiation for shell eggs and seeds for growing sprouts. In 2005, molluscan shellfish was approved for irradiation by the FDA to help control *Vibrio* species and other foodborne pathogens in fresh or frozen oysters, mussels, and clams.

Applications for Food Irradiation

Because the irradiation process works with both large and small quantities, it has a wide range of potential uses. For example, a single serving of poultry can be irradiated for use on a spaceflight. Or a large quantity of potatoes can be treated to reduce sprouting during warehouse storage.

Irradiation cannot be used with all foods. It causes undesirable flavor changes in dairy products, and it causes tissue softening in some fruits, especially peaches and nectarines.

Irradiation is most useful in four areas: (1) preservation; (2) sterilization; (3) control of sprouting, ripening, and insect damage; and (4) control of foodborne illness.

1. Irradiation can be used to destroy or inactivate organisms that cause spoilage and decomposition, thereby extending the shelf life of foods. It is an energy-efficient food-preservation method that has several advantages over traditional canning. The resulting products are closer to their fresh state in texture, flavor, and color. Using irradiation to preserve foods requires no additional liquid and does not cause the loss of natural juices. Both large and small containers can be used, and food can be irradiated after being packaged or frozen.
2. Foods that are sterilized by irradiation can be stored for years without refrigeration—just like canned (heat-sterilized) foods. Irradiation makes it possible to develop new shelf-stable products. Sterilized food is useful in hospitals for patients with severely impaired immune systems such as some with cancer or AIDS. These foods can be used by the military and for spaceflights.
3. In the role of the control of sprouting, ripening, and insect damage, irradiation offers an alternative to chemicals for use with potatoes, tropical and citrus fruits, grains, spices, and seasonings. However, because no residue is left in the food, irradiation does not protect against reinfestation the way that insect sprays and fumigants do.
4. Irradiation can be used to effectively eliminate those pathogens that cause foodborne illness such as *Salmonella*.

FIGURE 12-3 International symbol placed on irradiated food.

Nutritional Quality of Irradiated Foods

Scientists believe that irradiation produces no greater nutrient loss than what occurs in other processing methods such as canning.

Regulation of Food Irradiation

Since 1986, all irradiated products must carry an international symbol called a *radura*, which resembles a stylized flower (Figure 12-3). Accurate records are essential to regulation because there is no way to verify or detect if a product has been irradiated or how much radiation it has received. Irradiation labeling requirements only applies to foods sold in stores. Irradiation does not apply to restaurant foods.

MICROWAVE HEATING

Microwaves are another method for heating foods. Conventional methods heat foods by conducting heat from an external source to the inside. Microwaves generate heat inside a food using water friction. Conventional methods brown or crust foods on their surfaces, but microwaves do not.

A microwave oven generates microwaves from electricity. Microwaves are **electromagnetic energy**—that is, they have an electric and a magnetic component. (Originally, microwave ovens were called "radar" ranges.) Other forms of electromagnetic energy are radio waves, sunlight, and electricity.

Scientists refer to microwave energy as *radiation* because, similar to other radiation energy, it travels through space. Microwave radiation is often called *nonionizing radiation* to distinguish it from other forms of radiation, like X-rays, which are called *ionizing radiation*. In **ionization**, an atom or a molecule is broken into two electrically charged groups.

Microwaves themselves are generated by a magnetron tube that converts electrical energy at 60 cycles/second into an electromagnetic field with positive and negative charges that change direction millions of times per second (915+ million times per second). Essentially, it consists of **polar** and nonpolar molecules switching back and forth. This switching back and forth causes friction and makes heat. This friction is caused by the disruption of hydrogen bonds between neighboring water molecules, and the heat is really "molecular friction." To have microwave heating of foods, there must be a polar substance available. Water serves as that polar substance.

Other constituents in foods are also a factor in microwave heating. The positive and negative **ions** of table salt in foods will interact with the electrical field by migrating to the oppositely charged regions of the electrical field and also disrupting the hydrogen bonds with water to generate heat.

Many factors influence the speed of cooking using a microwave oven. The electrical and physical properties of foods themselves will determine microwave penetration depth, the dielectric constant, heat capacity, density, and the conventional heat transfer and heating rate. The food dielectric properties of interest in microwave heating are determined by the moisture and heating.

PAST, PRESENT, AND POTENTIAL OF IRRADIATION

Research on food irradiation dates back to the 1920s. The U.S. Army used the process on fruits, vegetables, dairy products, and meat during World War II, and NASA routinely sends irradiated food on U.S. spaceflights.

Irradiation to control microorganisms on beef, lamb, and pork is safe and eventually could mean that consumers will have less risk of becoming ill from contaminated meat. Irradiation of fruits and vegetables will also mean longer shelf life.

Research shows that specific doses of radiation can kill rapidly growing cells such as those of insect pests and spoilage bacteria. Irradiation kills foodborne pathogens such as *E. coli* 0157:H7 and *Salmonella*.

A recent survey conducted by the International Food Information Council (IFIC) in Dallas, New York, and Los Angeles found that consumers were willing to buy irradiated foods for themselves and their families, including children. Food irradiation is one food-safety tool whose time has come. The informed consumer will make choices based on irradiated products identified with clear labeling. The IFIC survey found that consumers appreciated the safety benefit of irradiation to eliminate harmful bacteria rather than extend shelf life.

In addition to the FDA and the U.S. Department of Agriculture (USDA), the list of irradiation endorsers includes the U.S. Public Health Service, the American Medical Association, and the National Food Processors Association. The World Health Organization (WHO) also sanctions irradiation, which is currently being used in some 40 countries. In 1992, WHO called irradiation a "perfectly sound food-preservation technology." In 1997, WHO again endorsed food irradiation, joined this time by the United Nations Food and Agriculture Organization (FAO).

For more information about the irradiation of food, search the Web or visit these Web sites:

- Radiation Protection: http://www2.epa.gov/radiation
- Essential Tips and Tricks for Eating a Healthy Diet: www.food-irradiation.com/
- Food Irradiation—What You Need to Know: http://www.fda.gov/Food/ResourcesForYou/Consumers/ucm261680.htm
- Food Irradiation: http://www.physics.isu.edu/radinf/food.htm

Like all forms of electromagnetic radiation, microwave energy travels in a wave pattern. The waves are reflected by metals but pass through air, glass, paper, and plastic to be absorbed by food. Most microwave containers are designed to transmit microwave energy without reflecting or absorbing it and thus are made of paper or plastic. The microwaves will travel through the container to the food.

When food is exposed to microwaves, it absorbs that energy and converts it to heat. Food composition (mainly water content) is a key factor that determines how fast it will heat in a microwave environment. The higher the water content of a food, the higher its loss factor and thus the faster it will heat. Solutes such as sugar and salt also influence the loss factor of foods exposed to microwave energy. For foods to heat in a microwave oven, the electromagnetic waves must penetrate the food. There are limits to the depth of their penetration. This makes container geometry a critical factor in heating foods by microwave energy.

Food composition only influences the loss factor but also the penetration depth. For example, when salt is added to water, it will change its microwave heating characteristics in two different directions. On one hand, it will increase the water's loss factor, causing the water to heat faster. On the other hand, salt will decrease the penetration depth of microwaves into the salt and water solution, decreasing the heating rate. If salt is added to water to enhance its loss factor and, at the same time, the container geometry changes to minimize the drop in penetration depth, the heating rate will be greatly increased.

SCIENCE CONNECTION!

Have you ever wondered how microwave popcorn packets were created? Design an experiment to test different materials for popcorn packaging. Research different package ideas that could be used.

Food-Processing Applications

Food scientists are dedicated to developing microwavable foods that can heat quickly and evenly while remaining high quality. Heating foods evenly in a microwave oven is difficult at best, particularly with solid foods of different compositions. A frozen dinner is an example of a solid food with varying composition. The use of plastic lining helps heat foods evenly. It is important to always read food packaging to see the proper cooking directions for each food item.

Understanding how food composition influences microwave heating is a skill needed by those who develop microwavable foods. Developing new microwavable meals and snacks is a growing segment of the food industry. Aside from using microwave energy to heat meals at home, researchers have made many attempts to use microwave energy to heat food in processing plants. Some of these uses include baking, concentrating, cooking, curing, drying, enzyme inactivation (blanching), finish drying, freeze-drying, heating, pasteurizing, precooking, puffing and foaming, solvent removal, sterilizing, tempering, and thawing. The best use of microwaves depends on product quality and cost.

OHMIC (ELECTRICAL) HEATING

Ohmic heating, also referred to as **Joule heating**, is the heating of a food product using an alternating current flowing between two electrodes (refer to Figure 7-9). It is one of the newest methods of heating foods. Ohmic heating's major advantage is that it simultaneously heats solid pieces and liquids in a food. Also, the ohmic method works well as a continuous heating system for foods with particles such as beef stew. It is also suitable for aseptic packaging methods. Ohmic heating can be used to heat liquid foods containing large particulates such as soups, stews, and fruit slices in syrups and sauces, as well as heat-sensitive liquids. The technology is useful for treating proteinaceous foods, which tend to denature and coagulate when thermally processed. For example, liquid egg can be ohmically heated in a fraction of a second without coagulating it. Juices can be treated to inactivate enzymes without affecting flavor. Other potential applications of ohmic heating include blanching, thawing, online detection of starch gelatinization, fermentation, peeling, dehydration, and extraction.

Many processing plants currently produce sliced, diced, and whole fruit within sauces in various countries, including Italy, Greece, France, Mexico, and Japan. In the United States, ohmic heating has been used to produce a low-acid particulate product in a can as well as pasteurized liquid egg. For continuous flow processing with aseptic

SCIENCE CONNECTION!

What is your favorite microwaved food? How is it prepared? What is the history of the food item? If you could change something about it, what would you change?

packaging, the approaches are currently in development in a project funded by the USDA's National Integrated Food Safety Initiative. The FDA is responsible for evaluating and monitoring the safety of ohmically processed foods, unless the product contains a specified minimum amount of meat and poultry. In those cases, evaluation and monitoring fall under the USDA purview.[1]

SUMMARY

Ionizing radiation (irradiation) and microwaves are both used to preserve food. Although irradiation has been in use since 1950, the FDA must approve its uses on each additional food. The latest approval for irradiation occurred in2005 for molluscan shellfish. Irradiation serves four uses in the food industry: preservation; sterilization; control of sprouting, ripening, and insect damage; and control of foodborne illness. Though irradiation demonstrates little effect on nutrient content, the greatest challenge will be consumer acceptance. Microwaves are also used to heat foods by generating heat inside the food due to water friction. Scientists refer to microwaves as nonionizing radiation.

Microwave development has led to a whole new group of convenience foods. Microwave energy has been used in food processing to bake, concentrate, cook, cure, dry, heat, puff, foam, temper, and thaw. The best use depends on product quality and cost.

Ohmic heating heats foods between electrodes using alternating current. It is one of the newest methods of heating and is useful because solid and liquid in a food heat at the same time.

REVIEW QUESTIONS

Success in any career requires knowledge. Test your knowledge of this chapter by answering these questions or solving these problems.

1. Describe ohmic heating.
2. Name the two requirements for irradiation.
3. Define radiation.
4. Explain ionizing radiation.
5. List the four ways in which irradiation is most useful.
6. Describe how microwaves heat food.
7. Explain what changes when salt is added to water during the microwaving process.
8. List three specific ways irradiation has been approved for use by the FDA.
9. Food composition influences microwave heating of food in what two ways?
10. Irradiation cannot be used on what two specific products?

STUDENT ACTIVITIES

1. Research the use of irradiation on food. Develop a report or presentation defending the advantages or disadvantages of its use.
2. Collect news articles from newspapers, TV, magazines, or the Internet that describe an outbreak of a foodborne disease. Summarize the articles and use them for a classroom discussion on how irradiation could have prevented the outbreak.
3. Develop a visual presentation on the use of microwaves in U.S. households. Include how microwaves were introduced and the growth in their sales.
4. Create a cookbook that gives recipes for microwave meals.
5. Using a microwave, demonstrate heating and cooking small quantities of different types of food, including liquids and items of differing shapes. Also demonstrate some of the unique features of microwaves. Record your findings.
6. Contact an appliance repair shop and ask if they have any microwaves that cannot be repaired. Disassemble these old microwaves and identify the parts. Present the parts on a display board.

ADDITIONAL RESOURCES

- Radiation Protection: http://www2.epa.gov/radiation
- Food Irradiation—What You Need to Know: http://www.fda.gov/Food/ResourcesForYou/Consumers/ucm261680.htm
- Ohmic Heating of Foods: http://ohioline.osu.edu/fse-fact/0004.html
- Probing Question—How Do Microwaves Cook Food? http://news.psu.edu/story/141277/2005/11/28/research/probing-question-how-do-microwaves-cook-food

Please note that internet sites represent a vast resource of information. Those provided in this chapter were vetted by industry experts but you may also wish to explore the topics discussed in this chapter further using a search engine, such as Google. Keywords or phrases may include the following: *food irradiation, microwaves, ohmic heating, ionizing food.* In addition, Table B-7 provides a listing of some useful Internet sites that can be used as a starting point.

REFERENCES

Asimov, I. 1988. *Understanding physics.* New York: Hippocrene Books.

Center for Food Safety and Applied Nutrition. 2007. *FDA Food safety A to Z reference guide.*

FDA. 2000. *Food irradiation: A safe measure.* Publication No. 00-2329. Washington, DC: USDA.

Potter, N. N., and J. H. Hotchkiss. 1995. *Food science,* 5th ed. New York: Chapman & Hall.

ENDNOTE

1. The Ohio State University Extension Fact Sheets: http://ohioline.osu.edu/fse-fact/0004.html. Last accessed July 7, 2015.

CHAPTER 13

Fermentation, Microorganisms, and Biotechnology

OBJECTIVES

After reading this chapter, you should be able to:

- Discuss the use of fermentation in food preservation
- Provide the general reactions for fermentation
- Understand three methods for controlling fermentation
- List six foods produced by fermentation
- Identify the uses of acetic acid bacteria and lactic acid bacteria
- Describe fermentation use in bread making
- Identify four uses of acetic acid
- Describe the use of microorganisms as food
- Discuss a role of biotechnology in the food industry

NATIONAL AFNR STANDARD

FPP.01

Develop and implement procedures to ensure safety, sanitation, and quality in food product and processing facilities.

KEY TERMS

anaerobic
biosensors
brewing
brine
coagulation
fermentation
hops
inoculation
kimchi
leavening
malt
mashing
microsensors
pathogenic
single-celled protein (SCP)
starter culture
syneresis
yeast

Fermentation allows the growth of microorganisms in order to produce a stable product. Products commonly produced, at least in part, by fermentation include: beer, pickles, olives, some meat products, bread, cheese, coffee, cocoa, soy sauce, sauerkraut, kimchi and wine. Micro-organisms can become food, and biotechnology changes the way microorganisms are used.

FERMENTATIONS

Fermentation is the oldest form of food preservation. The mechanism of fermentation is the breakdown of carbohydrate materials by bacteria and **yeasts** under **anaerobic** conditions—that is, without atmospheric oxygen. Fermentation produces acids and alcohols along with some aldehydes, ketones, and flavorings. Products produced by fermentation help preserve foods against microbial degradation. Fermentation by lactic acid bacteria produces the following:

- Pickles
- Olives
- Some meat products (sausage and salami)
- Sour cream
- Cottage cheese
- Cheddar cheese
- Coffee

Acetic acid bacteria produces cooking wine and cider. Lactic acid bacteria with propionic acid bacteria produces Swiss cheese. Molds produce blue cheeses. Yeasts are involved in the production of beer, wine, rum, whiskey, and bread. Yeasts with acetic acid bacteria act on cacao beans (chocolate).

Fermentation follows these general reactions:

Glucose → pyruvic acid → acetaldehyde + carbon dioxide → alcohol (ethanol)

A **starter culture** is a concentrated number of the organisms desired to start the fermentation. This will ensure that the proper organisms are growing right from the start.

Benefits

The main benefit of fermentation is preserving a product. For example, the acid produced by the process can prevent spoilage by some microorganisms. Fermentation also can add flavor to wine. It may remove or alter existing flavors as in soya. And it can alter the chemical characteristics of a food by converting sugar to ethanol, ethanol to acetic acid, and sugar to lactic acid.

Fermented food can be more nutritious than their unfermented original forms. Fermentation microorganisms produce vitamins and growth factors in the food. They also may liberate nutrients locked in plant cells and structures by indigestible materials. Finally, fermentation can enzymatically split polymers such as cellulose into simpler sugars that are digestible by humans.

Control

To encourage the growth of certain microorganisms, fermentation can be controlled by:

- pH
- Salt content
- Temperature

Acid or a low pH has an inhibitory effect. Acid may be added to the food or produced by the microorganisms that cause fermentation. It must be added or formed quickly to prevent spoilage or the growth of undesirable microorganisms.

Microorganisms exhibit differing tolerances for salt. Lactic acid–producing microorganisms are used to ferment olives, pickles, sauerkraut, and some meats, although they must be are tolerant of salt concentrations from 10% to 18%. Many of the microorganisms that cause spoilage cannot tolerate salt concentrations above 2.5%, and they are especially intolerant of salt and acid combinations. This gives lactic acid-producing microorganisms an advantage. Salt in a solution is called **brine**. Salt also serves another purpose when fermenting vegetables. It draws water and sugar out of the vegetables. The sugar entering the brine is available to the microorganisms for continued fermentation.

Depending on the temperature, different microorganisms dominate fermentation. Temperature can affect the final acid concentration and the time it takes to reach different acidities. For example, in sauerkraut fermentation, three different types of microorganisms are used, and the actions of each depends on the temperature, so the temperature starts low and is gradually increased to allow each microorganism to grow. Controlling fermentation is an attempt to favor desired organisms. Fermentation is stopped by pasteurizing and cooling.

USES OF FERMENTATION

Fermentation produces dairy products, breads, pickles, processed meats, vinegar, wine, and beer (Figure 13-1). The following paragraphs describe a few specific uses of fermentation in foods.

Fermented Dairy Products

Cheese making relies on the fermentation of lactose (milk sugar) by lactic acid bacteria. This bacteria produces lactic acid, which lowers the pH and in turn assists **coagulation** (clot formation), promotes **syneresis** (expulsion of water from clots), helps prevent spoilage and **pathogenic** bacteria from growing, and contributes to cheese texture, flavor, and keeping quality. Lactic acid bacteria also produce growth factors that encourage the growth of nonstarter organisms and provides enzymes that act on fats (lipases) and proteins (proteases), which is necessary for flavor development during curing.

FIGURE 13-1 Products produced by fermentation.

Yogurt is a semisolid fermented milk product that originated centuries ago in what is now Bulgaria. Its popularity has grown, and is now consumed in most parts of the world. Although the consistency, flavor, and aroma may vary from region to region, the basic ingredients of yogurt are milk and a starter culture. The fermentation products of lactic acid, acetaldehyde, acetic acid, and diacetyl contribute to the flavor.

Cultured buttermilk was originally the fermented by-product of butter manufacture, but today it is more common to produce cultured buttermilks from skim or whole milk. The fermentation is allowed to proceed for 16 to 20 hours, to an acidity of 0.9% lactic acid. This product is frequently used as an ingredient in the baking industry, in addition to being packaged for sale in the retail trade.

Acidophilus milk is a traditional milk fermented with *Lactobacillus acidophilus*, which is now thought to have therapeutic benefits for the human gastrointestinal tract.

Sour cream usually has a fat content between 12% and 30%, depending on the required properties. The starter is similar to that used for cultured buttermilk. **Inoculation** and fermentation conditions are also similar to those for cultured buttermilk, but the fermentation is stopped at an acidity of 0.6%.

Other fermented dairy products include such products as kefir and kumiss. Many of these have developed in regional areas and, depending on the starter organisms used, have various flavors, textures, and components from the fermentation process.

Dairy products are covered more completely in Chapter 16.

Bread Making

Bread is leavened with yeast. Baker's yeast is composed of the living cells of *Saccharomyces cerevisiae*, a unicellular microorganism. Yeast's **leavening** function is done by fermenting carbohydrates such as sugars—glucose, fructose, maltose, and sucrose—to produce carbon dioxide and ethanol. (Yeast cannot metabolize lactose, the predominant sugar in milk.) It is the carbon dioxide that produces the leavening effect. Yeast also produces many other chemical substances that flavor baked products and change a dough's physical properties.

Although most breads and rolls are leavened by yeast, some breadlike products such as corn bread and certain kinds of muffins are leavened by chemicals, usually baking soda or baking powder (Figure 13-2). Chapter 20 contains more information about bread making.

FIGURE 13-2 When yeast ferments sugars, bread rises.

Pickling

Pickles are made by covering a fruit with a sweetened vinegar solution that contains a flavoring spice or spices such as cloves. Some vegetable pickles are produced by fermentation. The fruit or vegetable is placed in a covered crock and allowed to ferment in a brine solution for a period of time ranging from a few days to several weeks. Because of the acidity of the pickling solution, heat processing under pressure to kill microorganisms is not required commercially or in the home.

SCIENCE CONNECTION!

Research fermentation labs online and see if you can find an experiment you can create at home. Perhaps making root beer, sauerkraut, or bread can be activities you can pursue. What are the necessary steps for the experiment? How does the fermentation take place?

THE ART OF PICKLING

Pickling is a global culinary art. Anyone going on an international food-tasting tour would find pickled foods just about everywhere: kosher cucumber pickles in New York City, chutneys in India, kimchi in Korea, miso pickles in Japan, salted duck eggs in China, pickled herring in Scandinavia, corned beef in Ireland, salsas in Mexico, pickled pigs feet in the southern United States, and much more.

What makes a pickle a pickle? On a most general level, pickles are foods soaked in solutions that help prevent spoilage.

There are two basic categories of pickles. The first type includes pickles preserved in vinegar, a strong acid in which few bacteria can survive. Most of the bottled kosher cucumber pickles available in the supermarket are preserved in vinegar.

(Continues)

THE ART OF PICKLING (continued)

The other category includes pickles soaked in a salt brine to encourage **fermentation**—the growth of beneficial bacteria that make a food less vulnerable to spoilage-causing bacteria. Common examples of fermented pickles include **kimchi** and many cucumber dill pickles.

Pickling is not only an international food-preservation technique but also an ancient one. For thousands of years, our ancestors have explored ways to pickle foods, following an instinct to secure surplus food supplies for long winters, famine, and other times of need. Historians know, for instance, that more than 2,000 years ago, workers building the Great Wall of China ate sauerkraut, a kind of fermented cabbage.

But pickling foods does much more than simply preserve them. It can also change their taste and texture in a profusion of interesting—and yummy—ways. It is no surprise that cultures across the globe enjoy such an assortment of pickled foods. In fact, food experts say, the evolution of diverse pickled foods in different cultures has contributed to unique cultural food preferences such as spicy sour tastes in Southeast Asia and acidic flavors in Eastern Europe.

Source: Modified from "What Is Pickling? https://www.exploratorium.edu/cooking/pickles/pickling.html

Vegetables, fruits, meats, eggs, and even nuts can be pickled. Some well-known pickled foods include sauerkraut (cabbage fermented in brine), dill or sweet pickles made from cucumbers, peach pickles, and pickled watermelon rind. Among the favorite regional mixtures of pickled vegetables are piccalilli, chowchow, and assorted other relishes (Figure 13-3).

Processed Meats

Some processed meats have microbial starter cultures added to achieve fermentation to enhance preservation and create a unique "tangy" flavor from the production of lactic acid. Fermentation inhibits the growth of spoilage and pathogenic microorganisms.

MATH CONNECTION!

What pickled foods are common in your region? Is there a cultural influence for that food to be popular? Visit your local grocery store or supermarket. How many pickled products can you find in the store? What do you think is the most unusual pickled food product you found? Where is it produced? What can you find out about the sales of this food product?

FIGURE 13-3 Some vegetable pickles are produced by fermentation.

Vinegar

Vinegar (from the French *vinaigre*, meaning "sour wine") is an acidic liquid obtained from the fermentation of alcohol and used either as a condiment or as a preservative. Vinegar usually has an acetic acid content of between 4% and 8%. Its flavor may be sharp, rich, or mellow, depending on the original product used. Vinegar is made by combining sugary materials (or materials produced by the breakdown of starches) with vinegar or acetic acid bacteria and air. The sugars or starches are converted to alcohol by yeasts, and the bacteria make enzymes that oxidize the alcohol to acetic acid.

Several varieties of vinegar are manufactured for commercial sale. Wine vinegars, produced in grape-growing regions, are used for salad dressings and relishes and may be reddish or white, depending on the wine used in the fermentation process. Tarragon vinegar has the distinctive flavor of the herb. **Malt** vinegar, popular in Great Britain, is known for its earthy quality. Apple cider vinegar is typically a pale amber color and used for dressings, food preservatives, and chutneys. White vinegar, also called *distilled vinegar*, is made from industrial alcohol; it is often used as a preservative or in mayonnaise because of its less distinctive flavor and clear, untinted appearance. Rice vinegar, which has an agreeably pungent quality, is often used in Asian cuisines for marinades and salad dressings.

Vinegar may be used as an ingredient of sweet-and-sour sauces for meat and vegetable dishes, as a minor ingredient in candies, or as an ingredient in baking as a part of the leavening process. Vinegar is also added to milk if sour milk is needed in a home recipe. Commercially and in the home, the most common use of vinegar is in making salad dressings.

Wine Making

When red wine is made, the grapes are crushed immediately after harvest and usually after the stems have been removed. Naturally occurring yeasts on the skins come into contact with the grape sugars, and fermentation begins naturally. Cultured yeasts, however, are sometimes added. During fermentation, the sugars are converted by the yeasts to ethyl alcohol and carbon dioxide. The alcohol extracts color from the skins. Glycerol and some of the esters, aldehydes, and acids that contribute to the character, bouquet (aroma), and taste of the wine are by-products of fermentation. In most cases, the wine is matured over years in 50-gallon oak casks. During this time, the wine is drawn off three or four times into fresh casks to avoid bacterial spoilage. Further aging naturally occurs after bottling.

Brewing

Brewing is a centuries-old technique that involves four steps: (1) **mashing**, (2) boiling, (3) fermentation, and (4) aging.

1. **Mashing.** This step infuses malt and crushed cereal grains with water at high enough temperatures to encourage the complete conversion of cereal starches into sugars (Figure 13-4).
2. **Boiling.** This step concentrates the resulting *wort* (liquid) and sees the addition of **hops**.
3. **Fermentation.** This step begins with the addition of yeast to the wort; the production of alcohol and carbon dioxide gas then occurs, the by-products of the action of yeast on sugar.
4. **Aging.** In this step, proteins that settle out of beer or are "digested" by enzymatic action. The aging process may last from 2 weeks to 24 weeks.

The uniform clarity of modern beers results from filtration systems that use such agents as cellulose and diatomaceous earth (ancient deposits of trillions of diatoms). Additives are frequently used to stabilize foam and to maintain freshness. With few exceptions, bottled and canned beer is pasteurized in the container to ensure that the yeast that may have passed through the filters is incapable of continued fermentation. Genuine draft beer is not

FIGURE 13-4 Copper kettles used for mashing during the fermentation process of making beer.

pasteurized and must be stored at low temperature. Chapter 24 provides more details on brewing and wine making.

In the future, technology will design more systems that call for continuous fermentations. Also, genetically engineered organisms will provide better control of the process.

MICROORGANISMS AS FOODS

Besides producing desirable changes in food, some microorganisms are used for animal feed and human food. These microorganisms are selected for their rapid growth, nutritional value, and other properties that make them good food or feed. Sometimes these are called **single-celled proteins (SCPs)**. Examples include brewer's yeast and baker's yeast. The practice of growing yeasts for food goes back many years.

GENETIC ENGINEERING AND BIOTECHNOLOGY

In the food industry, genetic engineering is improving the yields and efficiencies of traditional fermentation products and to convert unused or underused raw materials to useful products. Researchers now can isolate a known trait from any living species—plant, animal, or microbe—and incorporate it into another species. These traits are contained in genes, or segments of the DNA found in all living cells. The process of recombining genes bearing a chosen trait into the DNA of a new host is called *recombinant DNA technology* (Figure 13-5).

Although recombinant DNA has presented concerns to some people, its ability to transfer genes from a wide variety of species shows the power of genetic engineering. DNA is a universal molecule, so genes spliced into one organism from another can produce a specific and desired trait or quality. Because the chemical building blocks of DNA are

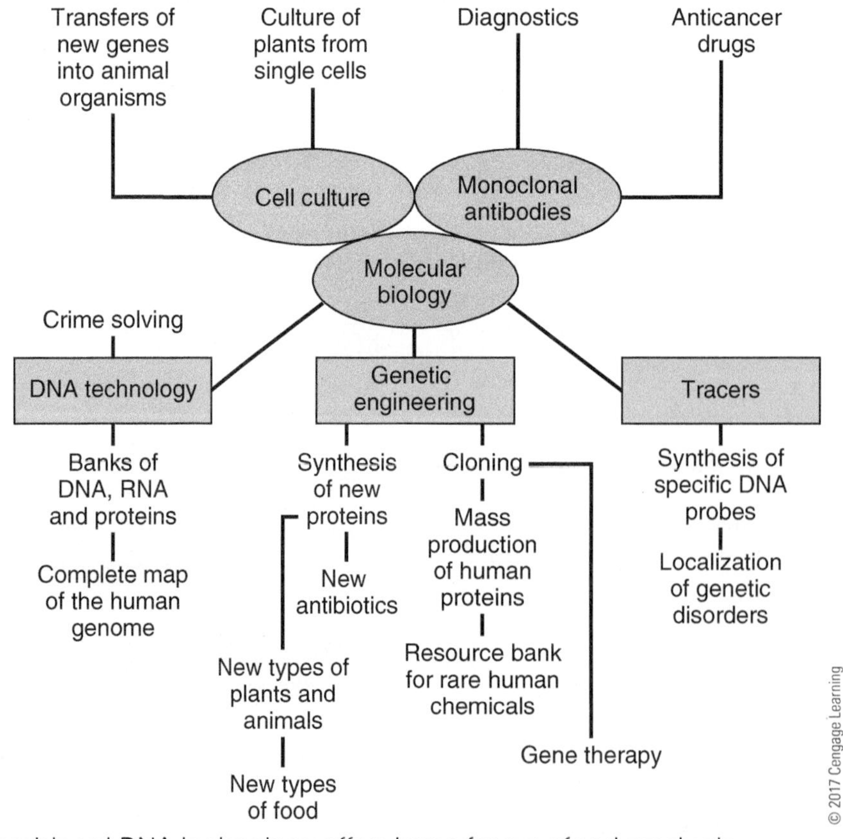

FIGURE 13-5 Recombinant DNA technology offers hope for new food products.

identical in all living things, a wide variety of desirable genes provides developers with a much larger selection of valuable traits.

Genetic engineering and biotechnology are also being used to produce new and improved enzymes, flavors, sweeteners, and other food ingredients.

Chymosin (Rennin)

Rennin is used in cheese manufacturing. It used to be extracted from the stomachs of calves, and demand far exceeded supply. Scientists inserted the gene for producing rennin into microorganisms. Rennin thus became the first genetically engineered product approved for use in food.

Bovine Somatotropin (BST)

Bovine (cattle) somatotropin (BST) or growth hormone is produced by genetically engineered microorganisms. When administered to dairy cows, it increases milk production by 10% to 25%, whereas feed intake increases by only some 6%, thus increasing the efficiency of the dairy cow. Many dairy marketing companies, however, have banned the use of BST by farms that produce milk.

Tomatoes

Flavr Savr tomatoes produced by Calgene contains an antisense gene for a specific enzyme—polygalacuronase. The engineered tomatoes ripen normally with full flavor but much more slowly. For example, the fruit can be harvested ripe in California and be shipped some 2,000 miles to Ohio and still be consumable for 2 weeks.

Cloning

In theory, every cell contains the genetic information to produce a complete new organism that is identical to the original. Plants can be cloned. It is also possible to produce many genetically identical engineered animals from somatic (body) cells. These animals and plants might provide food in the future.[1]

GMO Debate

Currently, the debate rages about the use of *genetically modified organisms*—more commonly known as GMOs—in our food supply. Though genetic engineering promises better and more plentiful products, genetically engineered foods does encounter some obstacles to widespread public acceptance. A few consumer advocacy groups and their supporters have voiced concern about the safety and environmental impacts of these new food products. Some urge an outright ban on any genetically engineered food. Others support mandatory labeling that discloses the use of genetic engineering. Still others advocate more stringent testing of these products before marketing.

From the standpoint of the Food and Drug Administration (FDA), the important thing for consumers to know about these new foods is that they will be every bit as safe as the foods now on store shelves. All foods, whether traditionally bred or genetically engineered, must meet the provisions of the federal Food, Drug, and Cosmetic Act. Whether genetically engineered foods succeed or fail ultimately depends on public acceptance. FDA scientists and others in the field blame some negative consumer reaction on the recombinant DNA technique's complexity. The technology is difficult to understand, so there is a fear of the unknown.[2]

GOLDEN RICE: THE CONTROVERSY OF GMOs

Given the task to end childhood blindness in Asian countries because of nutritionally incomplete diets that lack beta carotene and vitamin A, the idea of Golden Rice was born more than 30 years ago. Considerable controversy has followed this project over the years with ethical debates about GMOs and whether food corporations developing genetic foods have ulterior motives. Today, Golden Rice is not owned by any company and is being developed by a not-for-profit group called the International Rice Research Institute with the aim of providing a new source of vitamin A to people in the Philippines,

(Continues)

GOLDEN RICE: THE CONTROVERSY OF GMOs (continued)

where the majority of households get most of their calories from rice. The goal is to eventually get this saffron-colored grain distributed to millions more people who eat rice every day, roughly half the world's population.

Lack of vital nutrients causes blindness in a quarter-million to a half-million children each year. It affects millions of people in Asia and Africa and so weakens the immune system that some 2 million die each year of otherwise survivable diseases. This development brings its share of controversy, however, because some cultures and religions are opposed to the scientific development of foods. In addition, there are concerns about what could happen in the future if we consume mostly genetically engineered foods.

If Golden Rice is approved by various governments, then it will be available for farmers to plant and grow. First, however, the government regulatory agencies must officially determine that Golden Rice as safe for the food supply.

For more information on Golden Rice, see the following Web sites:

- Golden Rice Project: http://www.goldenrice.org/
- In a Grain of Golden Rice a World of Controversy over GMO Foods: http://www.npr.org/sections/thesalt/2013/03/07/173611461/in-a-grain-of-golden-rice-a-world-of-controversy-over-gmo-foods
- Golden Rice—A Lifesaver? http://www.nytimes.com/2013/08/25/sunday-review/golden-rice-life-saver.html?_r=0

The Future

Biotechnology is also used in some food-processing related areas, including processing aids, ingredients, rapid-detection systems, and **biosensors**. Enzymes acting as protein catalysts are used extensively in the food-processing industry to control texture, appearance, and nutritive value, as well as to generate desirable flavors and aromas. Because they are isolated from plants, animals, or microorganisms, their availability depends on the availability of the source material. Using genetically engineered microorganisms for the production of enzymes eliminates the need to rely on source materials while ensuring a continuous supply of enzymes.

New technologies allow researchers to target the genetics of plants, animals, and microorganisms and to manipulate them to our food-production advantage. Predictions for the future include the following:

- Environmentally hardy food-producing plants that are naturally resistant to pests and diseases and capable of growing under extreme conditions of temperature, moisture, and salinity
- An array of fresh fruits and vegetables with excellent flavor, appealing texture, and optimum nutritional content, that stay fresh for several weeks
- Custom-designed plants with defined structural and functional properties for specific food-processing applications
- Cultures of microorganisms that are programmed to express or shut off certain genes at specific times during fermentation in response to environmental triggers
- Strains engineered to serve as delivery systems for digestive enzymes for individuals who have reduced digestive capacity
- Cultures capable of implanting and surviving in the human gastrointestinal tract for delivery of antigens to stimulate the immune response or protect the gut from invasion by pathogenic organisms
- Microbially derived, high-value, "natural" food ingredients with unique functional properties
- **Microsensors** that accurately measure the physiological state of plants, temperature-abuse

indicators for refrigerated foods, and shelf-life monitors built into food packages

- Online sensors that monitor fermentation processes or determine the concentration of nutrients throughout processing
- Biotechnologically designed foods to supply nutritional needs, meat with reduced saturated fat, eggs with decreased levels of cholesterol, and milk with improved calcium bioavailability

But the future depends primarily on how the GMO controversy is resolved. Additional information on genetically modified foods is provided in Chapter 26.

SUMMARY

Fermentation is the breakdown of carbohydrate materials by bacteria and yeasts under anaerobic (without atmospheric oxygen) conditions. The main benefit of fermentation is that it preserves a product. For example, acid produced may prevent spoilage by some microorganisms. Fermentation may add flavor—for example, wine. It may remove or alter existing flavors—for example, soy. Or fermentation can alter the chemical characteristics of a food—as in sugar to ethanol, ethanol to acetic acid, and sugar to lactic acid. To encourage the growth of certain microorganisms, fermentation can be controlled by pH, salt content, and temperature. Controlling fermentation is an attempt to favor desired organisms. Fermentation is stopped by pasteurizing and cooling. Microorganisms such as single-celled proteins(SCPs) can become food. Biotechnology can bring great changes to food production, although not without controversy.

REVIEW QUESTIONS

Success in any career requires knowledge. Test your knowledge of this chapter by answering these questions or solving these problems.

1. Describe SCP.
2. Explain fermentation.
3. Show the general reactions that fermentation follows.
4. List six foods produced by fermentation.
5. Name three factors that control fermentation.
6. What is BST and how is it produced?
7. Describe the difference between acetic acid and lactic acid bacteria.
8. What action does yeast perform in bread making?
9. What was the first genetically engineered product approved for use in food?
10. Define recombinant DNA technology.

STUDENT ACTIVITIES

1. Make a list of 10 cheeses. Compare with other members of the class and see who has the least duplicated list.
2. Describe how yeast is prepared for bread making. Prepare a yeast for making bread.

3. Develop a chart that lists and describes the chemical products of fermentation.
4. Conduct an in-class taste test of some of the variety of foods produced by fermentation. Record your findings.
5. Demonstrate the use of fermentation on a product such as bread, cucumber pickles, or sauerkraut. Find instructions using the Internet.
6. Conduct a class debate about using genetic engineering and biotechnology (genetically modified organisms or GMOs) in foods.

ADDITIONAL RESOURCES

- Fermenting Veggies at Home: http://www.foodsafetynews.com/2014/03/fermenting-veggies-at-home-follow-food-safety-abcs/#.VlK2H3arQdU
- Canning Pickles and Fermented Foods: http://food.unl.edu/canning-pickles-fermented-foods

Internet sites represent a vast resource of information. Although those provided in this chapter were vetted by industry experts, but you may wish to further explore the topics discussed in this chapter using a search engine such as Google. Keywords or phrases may include the following: *fermentation, food preservation, lactic acid bacteria, acetic acid bacteria, coagulation, syneresis, biotechnology in foods, single-celled protein (SCP), wine making, beer making, bread making, genetic engineering of food*. In addition, Table B-7 provides a listing of useful Internet sites that can be used as a starting point.

REFERENCES

Brody, J. E. 1981. *Jane Brody's nutrition book*. New York: Bantam Books.

Mathewson, P. R. 1998. *Enzymes*. St. Paul, MN: Eagan Press.

McGee, H. 2004. *On food and cooking. The science and lore of the kitchen*. New York: Simon & Schuster.

Vieira, E. R. 1999. *Elementary food science*, 4th ed. New York: Chapman & Hall.

ENDNOTES

1. Cloning a Steak: http://www.agweb.com/livestock/beef/article/cloning-a-steak-naa-steve-cornett/.
2. See the FDA Web site on genetic engineering: http://www.fda.gov/AnimalVeterinary/DevelopmentApprovalProcess/GeneticEngineering/default.htm).

CHAPTER 14

Food Additives

OBJECTIVES

After reading this chapter, you should be able to:

- Describe three reasons for using food additives
- Discuss how food additives are monitored and controlled
- List five general categories of intentional food additives
- Identify the five specific uses of food additives and give examples
- Discuss the use of nutritional additives
- Identify five uses of food additives that would be considered an abuse
- Name two methods used to reduce fat intake
- Identify three color additives exempt from certification
- Understand food labels and identify the additives and provide a reason for their use

KEY TERMS

acidulants
antimicrobial agents
antioxidants
chelating agents
Delaney clause
emulsifier
Generally Recognized as Safe (GRAS)
lakes
sequestrants
stabilizer
surface active agents

NATIONAL AFNR STANDARD

FPP.02

Apply principles of nutrition, biology, microbiology, chemistry, and human behavior to the development of food products.

Food additives (chemicals) are used only to maintain or to improve the quality of food or to give it added qualities that consumers want. The Food and Drug Administration (FDA) monitors the use of additives and allows them to be used only if proven information has shown that the additive will accomplish the intended effect in the food. The amount used cannot be more than is needed to accomplish the intended effect.

REASONS FOR USE

Food additives are any substance used intentionally in food and that may reasonably be expected to, directly or indirectly, become a component of food or affect the characteristic of any food. This includes any substance intended for use in producing, manufacturing, packaging, processing, preparing, treating, transporting, or holding food. Food additives can be natural (such as sugar, salt, and other spices—see Figure 14-1A) or they may be chemical (such as food colorants—see Figure 14-1B). The uses of food additives are governed by the Food, Drug, and Cosmetic Act.

Intentional food additives include the following:

- Flavors
- Colors
- Vitamins

FIGURE 14-1A Natural Food Additives.

FIGURE 14-1B Chemical Food Additives.

- Minerals
- Amino acids
- Antioxidants
- Antimicrobial agents
- Acidulants
- Gums
- Sequestrants
- Surface active agents
- Sweeteners

The use of food additives is controlled by the **Delaney clause**, a 1958 amendment to the Food, Drug, and Cosmetic Act. This clause basically states that the food industry cannot add any substance to food if it induces cancer when ingested by humans or animals or if tests that evaluate the safety of food additives show it induces cancer in humans or animals.

Additives are used to achieve one or a combination of four purposes:

1. To maintain or improve nutritional value
2. To maintain freshness
3. To aid in processing or preparation
4. To make food more appealing

Without additives or preservatives, some foods would spoil quickly or taste bland. If some foods were not made storable, food would be wasted. In addition, foods that are not stored properly can cause illnesses. Some of the general categories of additives (chemicals) that benefit food include preservatives, nutritional additives, color modifiers, flavoring agents, texturing agents, and processing aids.

Additives also have many other functions, including hardening, drying, leavening, antifoaming, firming, crisping, antisticking, whipping, creaming, clarifying, and sterilizing (Tables 14-1 and 14-2).

TABLE 14-1 Additives Classes and Functions

ADDITIVE CLASS	FUNCTION
Curing and pickling agents	Impart a unique flavor or color to food (or both); often increase shelf life and stability
Dough conditioners	Modify starch and gluten to improve baking quality of yeast-leavened dough
Drying agents	Absorb moisture
Enzymes	Improve food processing and the quality of finished foods
Firming agents	Maintain the shape or crispness of fruits and vegetables
Flavor enhancers	Supplement, enhance, or modify the original flavor or aroma of a food without contributing their own flavors
Flour-treating agents	Improve the color or baking qualities of flour (or both)
Formulation aids	Promote or produce a desired physical state or texture
Fumigants	Control insects or pests
Leavening agents	Produce or stimulate CO_2 production in baked goods
Lubricants and release agents	Prevent sticking of food to contact surfaces
Nutritive sweeteners	Provide greater than 2% of the caloric value of sucrose per equivalent unit of sweetening capacity when used to sweeten food
Oxidizing and reducing agents	Produce a more stable product
pH control agents	Change or maintain active acidity or alkalinity
Processing aids	Enhance the appeal or utility of a food or food components
Propellants, aerating agents, and gases	Supply force needed to expel a product from a can or other container
Solvents and vehicles	Extract or dissolve another substance
Surface active agents	Modify surface properties of liquid food components
Surface-finishing agents	Increase palatability, preserve gloss, and inhibit discoloration
Synergists	Produce a total effect different or greater than the sum of the effects produced by the individual food ingredients
Texturizers	Affect the feel or appearance of foods

TABLE 14-2 Additives, Classes, Examples and Functions in Foods

ADDITIVES	EXAMPLE	FUNCTION IN FOOD
Anticaking free-flowing agents	Calcium silicate, magnesium carbonate	Used to keep food dry and to prevent caking as moisture is absorbed in foods. This often occurs in foods such as salt, powdered sugar, and baking powder.
Antimicrobial agents	Salt, sugar, sodium nitrate, sodium propionate, potassium sorbate	Prevent growth of microorganisms in foods such as breads, carbonated beverages, and margarine. Nitrites preserve color, enhance flavor, and prevent rancidity.
Antioxidants	BHT, BHA, ascorbic acid	Retard rancidity of unsaturated oils; prevent browning in fruits and vegetables that occurs during exposure to oxygen.
Buffers	Sodium bicarbonate, malic acid, citric acid, delactose whey	Cause leavening (rising) of starch products; impart tart taste in carbonated drinks and candies.
Coloring agents and adjuncts	Beta-carotene, Red No. 3, Yellow No. 6	Used to enhance or correct the colors of foods, making them more visually appealing; most criticized nutritive additives.
Flavoring agents	Natural extracts, essential oils, MSG, benzaldehyde	Some 2,000 agents are now added to many types of food to enhance flavor; largest group of additives. These substances modify the original flavor or aroma without contributing flavors of their own.
Emulsifiers	Lecithin, monocyglycerides	Used to evenly distribute fat- and water-soluble ingredients throughout food products; used in margarine, bakery products, and chocolate.
Humectants	Glycerine, propylene glycol, sorbitol	Retain moisture, keep foods soft; used in marshmallows, flaked coconut, cake icings, soft and chewy cookies.
Maturing and bleaching agents	Chlorine, benzoyl peroxide, acetone peroxide	Improve baking properties and whiten appearance of wheat flour and cheese.
Sequestrants	EDTA, citric acid	Added to bind with metals such as iron, calcium, and copper to prevent color, flavor, and appearance changes in food products such as wine and cider.
Stabilizers and thickeners	Gums, pectins, alginates, modified starch, carrageenan, dextrins, gelatin, guar, protein derivatives, cellulose	Used to maintain the texture and body of many food products such as ice cream, pudding, candy, and milk products.
Nonnutritive sweeteners	Aspartame, saccharin	Used to give sweet flavor to items such as beverages, cereals, and dietetic foods without adding the calories of ordinary sweeteners.

HOW MUCH IS ONE TRILLIONTH?

The analytical capabilities of the 1950s and 1960s could detect approximately 100 parts per billion (0.00001%) of a chemical in food, and any amount less than 100 ppb was then equal to zero. With vastly improved detection methods in analytical chemistry, however, it is now possible to detect amounts as low as 2 ppb (0.0000002%) and, in some cases, parts per trillion (ppt). These small concentrations are difficult to even imagine, but one analogy sets the amount that 1 part per million equals about 1/32 ounce (approximately 1 gram) in 1 ton of food.

One part per billion is about one drop in a 10,000-gallon tank or can, and 1 ppt is one grain of sugar in an Olympic-size swimming pool. This increase in the sensitivity of detection has produced evidence of trace contaminants in foods that were previously unsuspected. This presents a dilemma. Chasing an ever-receding zero level in foods as analytical methods continually improve could bring us eventually to ask this question: "Does the presence of one molecule of a carcinogen constitute grounds for removing a food from the marketplace?"

To find out more about the Delaney clause, search the Web or visit these Web sites:

- Delaney Clause: http://www2.ca.uky.edu/agripedia/glossary/delaney.htm
- Overview of Food Ingredients, Additives, and Colors: http://www.fda.gov/Food/IngredientsPackagingLabeling/FoodAdditivesIngredients/ucm094211.htm#foodadd
- The Nation Is Going Metric: http://www.fda.gov/iceci/inspections/inspectionguides/inspectiontechnicalguides/ucm093577.htm
- Nanotechnology: http://fsrio.nal.usda.gov/food-processing-and-technology/nanotechnology

Without additives to help process food, today's grocery stores would need a lot less room to sell the foods they could keep on the shelf for any length of time. Tested and approved food additives are a part of today's modern food technology.

Over the past few years, the food industry has worked to reduce the use of food additives with special emphasis on certain groups of additives such as artificial colors and preservatives (Figure 14-2).

TORTILLAS: ENRICHED FLOUR (FLOUR, NIACIN, REDUCED IRON, THIAMIN MONONITRATE, RIBOFLAVIN), WATER, HYDROGENATED SOYBEAN OIL WITH BHT ADDED TO PROTECT FLAVOR, GLYCERIN, SALT, DEXTROSE, CALCIUM PROPIONATE (PRESERVATIVE), WHEY, MONO- AND DIGLYCERIDES, SORBIC ACID (PRESERVATIVE), CITRIC ACID, L-CYSTEINE (DOUGH CONDITIONER).

SAUCE MIX: TOMATO POWDER, MALTODEXTRIN, SALT, CHILI PEPPER, FRUCTOSE, SUGAR, MODIFIED CORN STARCH, MONOSODIUM GLUTAMATE, ONION POWDER, PAPRIKA, GARLIC POWDER, SPICE, SILICON DIOXIDE, HYDROGENATED SOYBEAN OIL (WITH BHT ADDED TO PROTECT FLAVOR, MALIC ACID, AUTOLYZED YEAST EXTRACT, ARTIFICIAL COLOR, ASCORBIC ACID, NATURAL FLAVOR.

RICE PACKET: RICE

FIGURE 14-2 Food label with additives circled.

This effort has been driven by consumer desires. Food companies spend millions of dollars to find out what consumers want, and they adjust their products to attempt to meet these demands. The group of food additives that add the most value to foods are the flavoring agents, and these are generally used in minute amounts. Sweeteners, on the other hand, are the most heavily used additives, especially sucrose, high-fructose corn syrup, dextrose, and salt. The per capita consumption of the others is less than 1 pound per capita per year.

Over the years, chemicals have been added to food for unacceptable reasons, including:

- To disguise inferior products
- To deceive the consumer
- To provide otherwise desirable results that lower the nutritional value
- To replace good manufacturing practices
- To use in amounts greater than are necessary

Tables 14-1 and 14-2 provide complete descriptions of food additives, their use, and some examples.

PRESERVATIVES

One principal function of food additives is to maintain food freshness. Preservatives are substances added to food to slow down or prevent food spoilage. A good example of a commonly used preservative is salt (see Figure 14-3). Preservatives can also help maintain natural colors and flavors. Preservatives include **antioxidants**, sequestrants, and **antimicrobial agents** (see Table 14-3). Antioxidants help protect food from changes when exposed to oxygen. Oxidation can change a food's color, texture, flavor, and nutritional value. Common antioxidants are butylated hydroxy anisole (BHA), butylated hydroxytoluene (BHT), tert-butylhydroquinone (TBHQ), erythorbic acid, sodium erythorbate, tocopherols, and ascorbic acid. Some of the common antimicrobial agents include benzoic acid or sodium benzoate, calcium propionate, potassium, and sorbate or sorbic acid.

FIGURE 14-3 Salt piles on a salt farm.

SCIENCE CONNECTION!

Look at a few packages of food items at home from the refrigerator or pantry. What food additives do you see listed in the ingredients list? What do you think would happen to the food if it did not have the additives? How long would it last without the additives? Research a few of the additives and record your findings.

TABLE 14-3 Common Preservatives and Their Uses

PRESERVATIVE	USES
Salt	Retards bacterial growth on cheese and butter
Nitrites and nitrates of sodium and potassium	Add to flavor, maintain pink color in cured meats, and prevent botulism in canned foods
Sulfur dioxide and sulfites	Used as bleaches and antioxidants to prevent browning in alcoholic beverages, fruit juices, dried fruits and vegetables; prevent yeast growth
Benzoic acid and sodium benzoate	In oyster sauce, fish sauce, ketchup, alcoholic beverages, fruit juices, margarine, salads, jams, and pickled products
Propionic acid and propionates	In bread, chocolate products, cheese
Sorbic acid and sorbates	Prevent mold formation in cheese and flour confectioneries
BHA and BHT	Prevent oxidative spoilage of unsaturated fats and oils in potato chips, cheese balls, and so on
Ascorbic acid and ascorbates	In pork sausages
Sodium citrate	In cooked, cured meats and canned baby foods

Sequestrants are **chelating agents**. They are organic compounds that react with metallic ions to bind in a relatively inactive structure. **Sequestrants** prevent metals from catalyzing reactions of fat oxidation, pigment discoloration, flavor loss, and odor loss. Common sequestrants include ethylenediaminetetraacetate acid (EDTA), citric acid and its salts, and phosphoric acid and its salts.

FIGURE 14-4 Evaporated milk with vitamin D added.

NUTRITIONAL ADDITIVES

Vitamins and minerals are added to foods to make them more nutritious and sometimes to replace nutrients lost during processing. Examples include bread that has been enriched, milk with vitamin D added (Figure 14-4), and margarine with vitamins A and D added. Table 14-4 provides a list of some of the common nutritional additives.

MATH CONNECTION!

We all hear that we need daily vitamins, but what does that mean? Research the suggested daily intakes of vitamins for you. (Age and gender may play a role in amounts suggested). What foods do you eat that help meet these needs? Are there vitamins lacking in your diet? What can you do about it?

TABLE 14-4 Some Common Nutritional Additives

NAME OR TYPE	NUTRIENT AND USE
Alpha tocopherols	Vitamin E; antioxidant; nutrient; used in vegetable oil
Ascorbic acid	Vitamin C; antioxidant reacts with unwanted oxygen; stabilizing colors, flavors, oily foods; nutrient added to beverages, breakfast cereals, cured meats; prevents formation of nitrosamines
Beta-carotene	Vitamin A; coloring agent added to butter, margarine, shortening
Calcium pantothenate	Added for calcium
Ferrous gluconate, ferric orthophosphate, ferric sodium pyrophosphate, ferrous fumarate, ferrous lactate	Added to supply iron; some black color
Minerals	Added to improve nutritional value; zinc and iodine
Vitamins	Added to improve nutritional value; thiamin, riboflavin, niacin, pyridoxine, biotin, folate, and vitamins A and D

COLOR MODIFIERS

Colors include both natural and synthetic colorants. "FD&C" stands for *foods, drugs, and cosmetics*, and this means that this colorant can be used in all three of these products. If the color is marked "D&C" or "C colorants," it is used in household products such as shampoos.

The FDA is responsible for controlling all color additives used in the country. All color additives permitted for use in foods are classified either as *certifiable* or *exempt from certification* (Figure 14-5).

Color additives permitted for direct addition to human food in the United States includes these certified colors:

- FD&C Blue No. 1 (dye and lake)
- FD&C Blue No. 2 (dye and lake)
- FD&C Green No. 3 (dye and lake)
- FD&C Red No. 3 (dye)
- FD&C Red No. 40 (dye and lake)
- FD&C Yellow No. 5 (dye and lake)
- FD&C Yellow No. 6 (dye and lake)
- Orange B (restricted to specified use)
- Citrus Red No. 2 (restricted to specified use)

Colors exempt from certification include the following:

- Annatto extract
- Beta-carotene
- Beet powder
- Canthaxanthin
- Caramel color
- Carrot oil
- Cochineal extract (carmine)
- Cottonseed flour, toasted partially defatted, cooked
- Ferrous gluconate (restricted to specified use)
- Fruit juice
- Grape color extract or enocianina (restricted to a specified use)

Source: USDA, Agricultural Research Service (ARS)

FIGURE 14-5 Plant physiologist takes color reading from mangoes. Circled areas allow researchers to measure a fruit's color, hue, and intensity at the same spot before and after treatment. Scientists measure the color of food because it is important to how well consumers accept a food.

- Paprika and paprika oleoresin
- Riboflavin
- Saffron
- Titanium dioxide (restricted to specified use)
- Turmeric and turmeric oleoresin
- Vegetable juice

As previously indicated, certified color additives are available for use in food as either dyes or **lakes**. Dyes dissolve in water and are made as powders, granules (small hard pieces), liquids, and other special-purpose forms. They can be used in beverages, dry mixes, baked goods, confections (food made with sweet ingredients), dairy products, pet foods, and a variety of other products. Lakes are water-insoluble forms of dyes. They are more stable than dyes and ideal for coloring products containing fats, oils, and items lacking enough moisture to dissolve dyes. Typical uses include coated tablets, cake and donut mixes, and hard candies. Table 14-5 describes some certifiable colors and shows their uses in food.

Colors are used in food products for the following reasons:

- To offset color losses from exposure to light, air, extremes of temperature, moisture, and storage conditions
- To correct natural variations in color (off-colored foods are often incorrectly associated with poor quality—for example, some oranges are sprayed with Citrus Red No. 2 to correct the natural orangy-brown or the patches of green color of their peels; however, masking poor quality is an unacceptable use of colors
- To strengthen colors that occur naturally but at levels weaker than those usually associated with a given food
- To provide a color identity to foods that would otherwise be colorless (e.g., red colors provide a pleasant identity to strawberry ice cream, and lime sherbet is known by its bright green color)
- To provide a colorful appearance to certain "fun foods" (e.g., many candies and holiday treats are colored to create a festive appearance)
- To protect flavors and vitamins that may be affected by sunlight during storage
- To provide an appealing variety of healthy and nutritious foods that meet consumers' demands

TABLE 14-5 Color Additives Certifiable for Food Use

NAME/COMMON NAME	HUE	COMMON FOOD USES
FD&C Blue No. 1/Brilliant Blue FCF	Bright blue	Beverages, dairy products, dessert powders, jellies, confections, condiments, icings, syrups, extracts
FD&C Blue No. 2/Indigotine	Royal blue	Baked goods, cereals, snack foods, ice cream, confections, cherries
FD&C Green No. 3/Fast Green FCF	Sea green	Beverages, puddings, ice cream, sherbet, cherries, confections, baked goods, dairy products
FD&C Red No. 40/Allura Red AC	Orange-red	Gelatins, puddings, dairy products, confections, beverages, condiments
FD&C Red No. 3/Erythrosine	Cherry red	Cherries in fruit cocktail and in canned fruits for salads, confections, baked goods, dairy products, snack foods
FD&C Yellow No. 5/Tartrazine	Lemon yellow	Custards, beverages, ice cream, confections, preserves, cereals
FD&C Yellow No. 6/Sunset Yellow	Orange	Cereals, baked goods, snack foods, ice cream, beverages, dessert powders, confections

FLAVORING AGENTS

Flavoring agents are the largest group of food additives. They include both natural and synthetic flavors. Some agents such as spices and liquid derivatives of onion, garlic, cloves, and peppermint enhance flavor (see Figure 14-6). Synthetic flavorings that resemble natural flavors have been developed, and these have the advantage of being more stable than natural flavors (see Figure 14-7). Some of these are described in Table 14-6. Flavors cost the most to use but add the most value to products.

FIGURE 14-7 Vanilla extract, a popular baking ingredient, can be natural or synthetic, as shown here.

SCIENCE CONNECTION!

Vanilla extract is a popular ingredient in cakes and cookies. What is the difference between vanilla extract and imitation vanilla extract? Research and record your findings.

FIGURE 14-6 Chefs in restaurants add natural flavorings to enhance the taste of their dishes.

TABLE 14-6 Some Common Flavorings Used as Food Additives

FLAVOR	FOOD ADDITIVE
Camphor	Bornyl acetate
Cinnamon	Cinnam aldehyde
Ginger	Ginger oil
Grape	Methyl anthranilate
Lemon	Citral
Orange	Orange oil
Pear	Amyl butyrate
Peppermint	Menthol
Rum	Ethyl formate
Spearmint	Carvone
Spicy	Ethyl cinnamate
Vanilla	Ethyl vanillin
Wintergreen	Methyl salicylate

TEXTURING AGENTS

Have you ever heard the expression "Oil and water don't mix"? This is true unless you have an emulsifier. **Emulsifiers** are sometime called

surface active agents. These improve the uniformity of a food—the fineness of grain—the smoothness and body of foods such as bakery goods, ice creams, and confectionery products. Emulsifiers help disperse flavors evenly in foods and beverages. Mono- and diglyceride emulsifiers are derived from vegetable and animal fats. They help stabilize margarine. Lecithin is an emulsifier found in milk and egg yolks and is used in ice cream, chocolates, and mayonnaise. Polysorbate 60 and 80, as well as proteins, are also considered emulsifiers.

Stabilizers and thickeners add smoothness, color uniformity, and flavor uniformity to such foods as ice creams, chocolate milk, and artificially sweetened beverages. Stabilizers and thickeners are labeled as *pectin*, *vegetable gums*, and *gelatins*. Table 14-7 lists some common emulsifiers, stabilizers, thickeners, and their uses.

TABLE 14-7 Emulsifiers, Stabilizers, Thickeners, and Their Uses

TEXTURING AGENT	USE
Carboxymethyl-cellulose	Batter coating, frozen chips, and fish sticks
Xanthan gum	Seafood dressings, frozen pizza, and packet dessert toppings
Pectin	Jams, marmalades, and jellies
Dextrins	Icings, frozen desserts, confectioneries, whipped cream, cake mixtures, mayonnaise, and salad dressings
Lecithins	Milk chocolate and powdered milk
Sodium alginate	Ice cream, yogurt, sauces, and syrups
Mono- and diglycerides of stearic acid	Low cholesterol margarine, hot chocolate mix, dehydrated mashed potato, aerosol cream
Potassium dihydrogencitrate	Processed cheese; condensed and evaporated milk

ACIDULANTS

Acidulants make a food acid or sour. They are added to foods primarily to change the taste and to control microbial growth. These include citric acid, acetic acid, phosphoric acid, and hydrochloric acid, among others.

FAT REPLACERS

To help reduce consumers' fat intake, the food industry is attempting to modify fat itself. These approaches generally fall into two main categories: (1) decreasing fat content and (2) using fat replacers, substitutes, extenders, mimetics, or synthetic fat. (For more information refer to the section on "Fat Substitutes" in Chapter 22.)

Fat replacers include various carbohydrate-based, protein-based, and fat-based replacers for different food categories. Consumers constantly demand new and improved fat replacers. Some of these are:

- Olean (sucrose polyester)
- Olestra
- Amalean I and Amalean II (modified high-amylose corn)
- Cellulose and hemicelluloses
- Chitosan (fiber of crustaceans)
- Hydrocolloids[1]

IRRADIATION

The FDA considers irradiation of food to be an additive. Irradiated food is safe and will last longer. Irradiation to control microorganisms on beef, lamb, and pork is safe and eventually could mean that consumers will face less risk of becoming ill from contaminated meats.

Irradiation of fruits and vegetables means longer shelf life for those items. Irradiation passes through food without leaving any residue. The ionizing radiation kills bacteria and other pathogens in food, but the food never comes into contact with radioactive materials. The process does not make the food radioactive.

Research on food irradiation dates back to the 1920s. The U.S. Army used the process on fruits, vegetables, dairy products, and meat during World War II, and NASA routinely sends irradiated food on U.S. spaceflights to the International Space Station. Moreover, research shows that any changes in irradiated food are similar to the effects of canning, cooking, or freezing. Nutritionally, irradiated food is virtually identical to nonirradiated food.

Chapter 12 covers more details of irradiation.

HAZARDS

Additives remain a public concern. Contrary to popular perception, the majority of direct food additives are **Generally Recognized as Safe (GRAS)** substances. GRAS substances are ingredients that may be added to food without extensive prior testing; the category was established to prevent companies from having to prove the safety of substances already regarded as safe. The banning of cyclamates in 1969 produced a presidential directive to review the safety of the GRAS substances and led to the FDA cyclic review of all direct and indirect additives.

A review of GRAS substances revealed that about 90% present no significant hazard with normal human food uses. Most of those remaining to be tested have not been associated with hazards to humans. The other direct food additives have been approved, and their uses are regulated by the FDA. Indirect additives—such as those used in production, processing, and packaging and that might migrate to food— are numerous but normally occur in foods at trace levels, if at all. Many may occur at parts per billion or less. Examination of severity, incidence, and onset of effects indicates that additives are the lowest-ranking hazard.

Some chemicals are naturally a part of the food, but they are at such low levels they are harmless (Table 14-8).

TABLE 14-8 Common Foodborne Toxicants

FOOD TYPE	SUBSTANCE	EFFECT
Peaches, lima beans, apple seeds, apricot pits, certain grains, legumes	Cyanide	Neurological degenerative disease, death
Nutmeg, sassafras tea	Safrole	Liver cancer
Coffee, tea, red wine	Tannin	Liver cancer
Coffee, tea, colas	Caffeine	Birth defects
Green potato skins	Solanine	Nervous system disorders
Cabbage, mustard, brussel sprouts, cauliflower, turnips	Glucosinolates	Goiter

SUMMARY

Food additives are used to maintain or improve the quality of food. The Food and Drug Administration monitors the use of additives. According to the FDA, an additive is any substance that is intentionally or indirectly a component of food. Intentional food additives include the general categories of flavor, colors, vitamins, minerals, amino acids, antioxidants, antimicrobial agents, acidulants, sequestrants, gums (thickeners), sweeteners, and surface active agents. Some additional additives include fat replacers and irradiation. The majority of direct food additives fall into the category of Generally Recognized as Safe. Unacceptable uses of food additives include any uses to deceive, disguise, lower nutritional value, or avoid good manufacturing practices. For years, the use of food additives has been controlled by the Delaney clause of the federal Food, Drug, and Cosmetic Act.

REVIEW QUESTIONS

Success in any career requires knowledge. Test your knowledge of this chapter by answering these questions or solving these problems.

1. Name the two methods used to reduce fat intake.
2. List three reasons for using food additives.
3. Using Table 14-1, choose three additives and describe their functions.
4. Describe the use of BHT.
5. What is the function of nitrates and nitrites of sodium and potassium in the production of cured meats?
6. What organization is responsible for controlling all color additives used on foods in the United States?
7. Why are vitamins and minerals added to foods?
8. List three examples of common foods that contain food-borne toxicants.
9. List five categories of intentional food additives.
10. List three colors that are exempt from FDA certification.

STUDENT ACTIVITIES

1. Conduct a contest in which everyone collects five food labels. The objective is to use the labels to identify the most additives and their functions.
2. Find an example for each of the following uses of color additives: identification for otherwise colorless foods, fun foods, and correction of natural variations in foods. Present your examples to the class.
3. Identify at least two food products that use one of the flavors listed in Table 14-6. Conduct a taste test challenge with your classmates to determine if they can identify the correct flavors in the food products.
4. Visit the Web site for the FDA (www.fda.gov) and develop a research report on food additives based on information from the site.

ADDITIONAL RESOURCES

- Our Food—Packaging and Public Health: http://www.ncbi.nlm.nih.gov/pmc/articles/PMC3385451/
- What Are All Those Chemicals in My Food? https://www.extension.purdue.edu/extmedia/he/he-625.html
- GRAS Flavoring Substances 20: http://www.ift.org/~/media/Food%20Technology/pdf/2001/12/1201feat_gras20.pdf
- Fat Substitutes, Fat Mimetics, and Bulking Agents http://www.nutrientdataconf.org/PastConf/NDBC19/4-2_Gordon.pdf
- Food (FDA): http://www.fda.gov/Food/default.htm

Internet sites represent a vast resource of information. Although those provided in this chapter were vetted by industry experts, you may wish to further explore the topics discussed in this chapter using a search engine such as Google. Keywords or phrases may include the following: *food additives, antioxidants* (*BHA, BHT, TBHQ, erythorbic acid, sodium erythorbate, tocopherols, ascorbic acid*), *antimicrobial agents, acidulants, sequestrants, humectant, Food and Drug Administration, food coloring, food preservatives, nutritional additives, FD&C (food, drug, and cosmetics), certifiable color additives, flavoring agents, irradiation, GRAS (Generally Recognized As Safe)*. In addition, Table B-7 provides a listing of useful Internet sites that can be used as starting points.

REFERENCES

Alexander, R. J. 1998. *Sweeteners: Nutritive.* St. Paul, MN: Eagan Press.

Cremer, M. L. 1998. *Quality food in quantity. Management and science.* Berkeley, CA: McCutchan Publishing.

Ensminger, A. H., M. E. Ensminger, J. E. Konlande, and J. R. Robson. 1994. *Foods and nutrition encyclopedia* (2 vols.). Boca Raton, FL: CRC Press.

Francis, F. J. 1998. *Colorants.* St. Paul, MN: Eagan Press.

McGee, H. 2004. *On food and cooking. The science and lore of the kitchen.* New York: Simon & Schuster.

Stauffer, C. E. 1999. *Emulsifiers.* St. Paul, MN: Eagan Press.

Thomas, D. J., and W. A. Atwell. 1999. *Starches.* St. Paul, MN: Eagan Press.

ENDNOTE

1. Information to Keep You Current: http://www.ift.org/knowledge-center/read-ift-publications/science-reports/scientific-status-summaries/fat-replacers.aspx

CHAPTER 15

Packaging

OBJECTIVES

After reading this chapter, you should be able to:

- **Identify three types of food packaging**
- **Name and describe the use of four basic packaging materials**
- **List 10 features or requirements of packaging material**
- **Understand tests that measure the properties of packaging material**
- **Identify packages with special features**
- **Discuss how packaging addresses environmental concerns**
- **Identify and describe a packaging innovation**

NATIONAL AFNR STANDARD

FPP.04FPP.03

Select and process food products for storage, distribution, and consumption.

KEY TERMS

aseptically
copolymer
extratries
hermetically
ionomer
polyethylene terephithalate (PET)
primary
printability
recycled
secondary
tertiary

Food-packaging development started with humans' earliest beginnings. Early forms of packaging ranged from gourds to seashells to animal skins. Later came pottery, cloth, and wooden containers. These packages were created to facilitate transportation and trade. Food packaging today serves many purposes, including protecting food from contaminants as well as protecting food during transportation and storage.

Aside from protecting the food, the package serves as a vehicle through which the manufacturer can communicate with the consumer. Nutritional information and ingredients are always found on a food label, and many manufacturers offer recipes. The package is also used as a marketing tool designed to attract your attention at the store. This makes printability an important property of a package. Packaging is an industry in itself, and some universities offer degree programs in package engineering.

TYPES OF CONTAINERS

Food packaging can be divided into three general types:

1. Primary
2. Secondary
3. Tertiary

Primary containers come into direct contact with food. A **secondary** container is an outer box or wrap that holds several primary containers together. **Tertiary** containers group several secondary holders together into shipping units.

Think about your favorite breakfast cereal. Typically, the primary container is a plastic bag that holds the cereal. The secondary container is the box that holds the bag, and the tertiary container is the large shipping boxes that brought the cereal from the warehouse to the grocery store or supermarket.

Many containers used in the food industry are part of form–fill–seal packaging. Containers may be preformed at another site and then filled at the processing plant formed in the production line just ahead of the filling operation. This is called form–fill–seal, and it is one of the most efficient ways to package food.

To protect the food against exchanges of gases and vapors and contamination from bacteria, yeasts, molds, and dirt, containers are **hermetically** sealed. The most common hermetic containers are cans and glass bottles.

FOOD-PACKAGING MATERIALS AND FORMS

There are many packaging options to consider. Packaging materials are usually selected based on several variables, including the type of food being packaged, the function of the package, and the purpose the packaging is serving. The food industry uses four main packaging materials: metal, plant matter (paper and wood), glass, and plastic. The combination of different packing materials is often combined to create better packaging. The fruit drink box is an example where plastic, paper, and metal are combined in a laminate to give an ideal package. Another example is a peanut butter jar. The primary package contains the food and is made of glass (or plastic); the lid is made of metal lined with plastic; and the label is made of paper. All food packaging must meet food-grade container specifications. This means that the container will not transfer nonfood chemicals into the food and does not contain chemicals that could be hazardous to human health.

METALS

The three most popular metals used to hold food are tin, steel and aluminum. Cans (Figure 15-1) are formed at a food-processing factory or shipped

FIGURE 15-1 Various types and sizes of cans used by the food industry.

with their bottoms attached and lids separate so they can be seamed onto the cans later. The outside of the steel can is protected from rust by a thin layer of tin (0.025% by weight). The inside of the can is protected by a thin layer of tin or baked-on enamel. Steel is corrosive and can react with oxygen, so these thin layers allow the use of steel while being food safe. Tin-free steel and thermoplastic adhesive-bonded seams have become more common. These do away with the need for solder, which can leave traces of lead in food.

Factory equipment allows hermetically sealed sanitary steel cans to be manufactured and later sealed at the rate of 1,000 units per minute. Rigid aluminum, tin-plate, and tin-free containers also can be readily formed without side seams or bottom end seams by pressure extrusion. Aluminum is used as a packaging metal because of its light weight, low levels of corrosion (rust), recyclability, and ease of shipping. However, aluminum has less structural strength than steel cans. This limitation has been overcome by the injection of a small amount of liquid nitrogen into the can before it is closed. This gas provides for internal pressure that adds rigidity to the can.

Glass

Historically, glass has been used for food packaging for many years. Glass is still used today primarily for home canning. Glass provides a chemically inert and noncorrosive recyclable food packaging material which does not react to food. Its transparency is an advantage as well. The disadvantage is that glass breaks, and it is too heavy for some processing uses. Glass is heavier than other packaging materials and cost more to ship. Also, recycling is not easy, except in the case of home canning use.

Paper

Paper is lightweight, low in cost, and recyclable. However, paper must be combined with other substances to provide sealable packaging. As a primary container, it must be treated, coated, or laminated (see Figure 15-2). Paper from wood pulp and reprocessed waste paper is bleached and coated or impregnated with waxes, resins, lacquers, plastics, and laminations of aluminum to improve its water strength and gas impermeability, flexibility, tear resistance, burst strength, wet strength, grease resistance, sealability, appearance, and **printability** for labels and advertising. Papers treated for primary contact with food are reduced in their ability to be **recycled**. Paper that comes into contact with foods must meet Food and Drug Administration (FDA) standards for chemical purity, and paper used for milk cartons must come from sanitary virgin pulp. The major safety concern FDA is that a container must not allow punctures or tears that will allow the outside environment to contaminate a packaged food.

FIGURE 15-2 Various types of paper and plastic packaging used by the food industry.

Plastics

Plastics are lightweight and economical packaging choices. One of its major advantages is its versatility to form a variety of shapes that meet the diverse needs of food packaging. Popular plastics include cellophane, cellulose acetate, nylon, mylar, saran, and polyvinyl chloride. Polymers are made up of many molecules strung together to form chains. Think of decorative paper chains you might have made when you were younger. **Copolymer** plastics extend the range of useful food-packaging applications. **Ionomer** (ionic bonds) plastic materials are improved food-handling materials that function well even with greater amounts of oil and grease, are resistant to solvents, and have higher melting strengths. Newer plastic materials even contain cornstarch, which makes them more biodegradable.

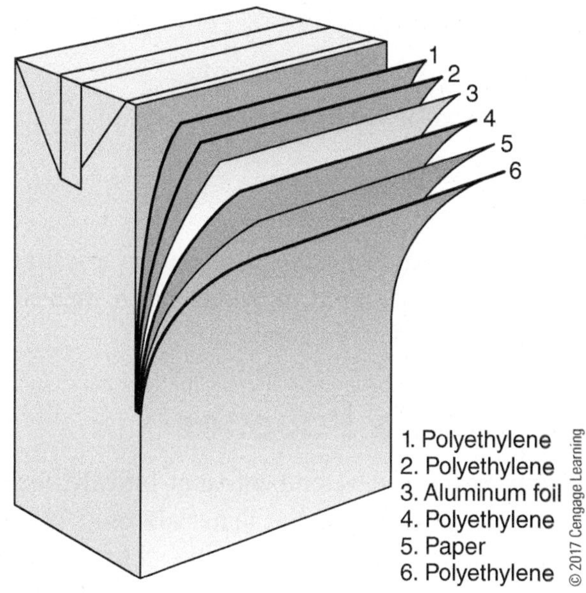

FIGURE 15-3 Diagram of a laminate cross-section.

SCIENCE CONNECTION!

Have you ever wondered how pots and pans can cook food but the handles do not usually burn your hand when you pick them up? Or how the handle of the spoon stays cool while you are stirring hot foods? How do you think that works? Research the construction of cookware. Record your hypothesis and your findings.

Laminates

Commercial laminates with as many as eight different layers can be custom designed for a specific product. In the case of prepackaged dry beverages, the laminate (from outside of package to inside) may have a special cellophane that is printable, polyethylene for a moisture barrier, treated paper for stiffness, a layer for bonding, and aluminum foil (prime gas barrier) inside a layer of polyethylene for an additional water vapor barrier (Figure 15-3).

Each basic packaging material has advantages and disadvantages. Metal is strong and a good overall barrier, but it is heavy and prone to corrosion. Paper is economical and has good printing properties, but it is not strong and absorbs water. Because paper absorbs water, it gains moisture from milk, for example, becomes weaker and ultimately fails. Glass is transparent, allowing the consumer to see the product, but it is breakable. Plastics are versatile but often expensive.

Overall, the requirements, functions, and considerations of food containers include:

1. Nontoxic and compatible with food
2. Sanitary protection
3. Moisture protection
4. Resistance to impact
5. Light protection
6. Gas and odor protection
7. Ease of opening and closing
8. Tamper resistant and tamper evident
9. Pouring features
10. Size, shape, and weight limitations

11. Reseal features
12. Ease of disposal
13. Appearance and printability
14. Transparency
15. Affordability

Plastics can also serve as secondary packages. The case in which many milk cartons are delivered to the supermarket is a good example.

Retortable Pouches

Twenty years of development went into the technology to ensure that the materials used in the production of retortable pouches (Figure 15-4) would protect the food and not contribute harmful **extratries** (chemical interactions with foods) to foods. The three layers consist of an outer layer of polyester film for strength, temperature resistance, and printability; a middle layer of aluminum film for barrier properties; and an inner layer of polypropylene film that provides for a heat seal. The advantages to retortable pouches include reduced processing time, reduced shipping cost, and lower storage space. Pouches also have good shelf appeal and a growing acceptance by consumers. However, pouches have more processing challenges compared to other packaging products. They do not sit, fill, and seal like cans do. In addition, the equipment to work with retortable pouches is more sophisticated and expensive than other packaging processing equipment.[1]

In today's food market, we find retortable pouches in use for everything from baby food to drink pouches, soup containers, and pasta dishes. Many ready-to-eat products are being packaged in retortable containers.

FIGURE 15-4 Retort packaging.

Edible Films

Sausage casings are an example of an edible film. By spraying gelatin, gum arabic, starch, monoglycerides, proteins, or other edible materials, a thin protective coating can be formed around food particles. For example, raisins in breakfast cereals are sprayed to prevent them from moistening the cereal in the box, and nuts are coated to protect them from oxidative rancidity. An edible wax film is used to coat the surface of vegetables to reduce moisture loss and provide increased resistance to the growth of molds. All edible films must be approved by the FDA for human consumption.

PACKAGE TESTING

Many tests measure the protective properties of packaging materials. Basically, the tests can be divided into chemical and mechanical. Chemical tests are used to determine if any of the packaging materials such as plastic migrate into a packaged food and to measure resistance to greases, acids, alkalis, and other solvents. Mechanical tests measure barrier properties, strength, heat-seal ability, and clarity.

FAKE FOODS

According to FDA investigations, some companies and individuals have made hundreds of thousands, even millions, of dollars off of fraudulent foods. These early cases helped lead to passage of the 1938 Federal Food, Drug, and Cosmetic Act, which specifically bars the economic adulteration of food.

1922: The Bureau of Chemistry in the U.S. Department of Agriculture—FDA precursor—investigated reports that various makers of Eskimo Pie coatings were illegally substituting coconut oil and cottonseed sterine for the required cocoa butter and milk fat. One report claimed that the imitation coating, unlike the real coating, did not snap off when bitten into but instead would break in such a way that it allowed melting ice cream to leak out.

1923: Two businesses dealing in adulterated olive oil were convicted and fined in federal court of the Southern District of New York. This was the second conviction for one of the companies, which was fined in 1922 for selling so-called pure olive oil adulterated with peanut oil.

1926: Chicago police arrested two Chicago men following a complaint from New York bakers that the two men were making and selling a butter substitute for genuine butter. Their substitute butter included melted low-grade butter and lard that was churned with buttermilk. The federal government seized 32 tubs of the product, and the case was turned over to the Internal Revenue Bureau of the U.S. Treasury Department.

1933: A San Francisco man was fined for adulterating dried egg yolks with artificial color and skim milk powder. He imported the dried egg yolks from China, unpacked them and added the adulterants, and then repacked them in containers closely matching those in which they were imported.

An Atlanta company was fined for selling supposed vanilla extract that contained insufficient vanilla but was artificially colored to make it look like real vanilla extract. It was sold to Army posts in Kansas, Nebraska, and Oklahoma.

1936: The new FDA recorded the seizure of 62 cases of sardines labeled as "packed in olive oil." An inspector learned that the sardines were actually packed in a mixture of one-third sesame and two-thirds olive oil.

1938: A complaint from an assistant to a quartermaster supply officer at Fort Sam Houston in Texas led to the seizure of more than 500 bottles of a product labeled "Vanilla Extract." It contained artificial color and little, if any, true vanilla. Fortunately, the product had not been paid for.

Under the supervision of a state inspector, 60 cases of a product labeled tomato catsup were poured down a sewer after the catsup was found to consist largely of water, starch, and cochineal—a red dye made from the dried bodies of female cochineal insects.

In recent years, FDA has sought and won convictions against companies and individuals engaged in making and selling bogus orange juice, apple juice, maple syrup, honey, cream, olive oil, and seafood.

1990s: An orange juice manufacturer defrauded consumers of more than $45 million during an estimated 20-year period. Another orange juice company netted $2 million in 2 years by substituting invert beet sugar for frozen orange juice concentrate. Still another orange juice manufacturer saw its earnings rise from zero in the company's second year of operation to $57 million in its fifth year before being convicted and sentenced for adulterating orange juice concentrate with liquid beet sugar.

Actual tests consist of subjecting a few samples of food-filled packages through the complete processing, shipping, warehousing, and merchandising sequence. This allows the packaging material to be subjected to every type of normally occurring abuse. These packages are recovered and analyzed to see how they withstood the vibrations, humidity, temperature, and handling. Similar tests are sometimes performed in simulations.

PACKAGES WITH SPECIAL FEATURES

With the new foods being produced, packages frequently have some type of added convenience feature. Often a package must withstand freezer temperatures as well as boiling or steam heat without bursting. The properties of polyester and nylon films allow this type of packaging.

Microwave oven packaging presents another challenge. This packaging not only must meet all the standard requirements for packaging but also be transparent to microwaves and be able to withstand high temperatures.

Squeezable plastic bottles are used for all types of packaging. These bottles have the high barrier properties of glass with less than one-fourth the weight, and they do not break.

Composite paper cartons can be sterilized and then **aseptically** filled with sterile liquid products. The process is called *aseptic packaging.* This type of packaging allows foods such as milk and fruit juices to be packaged in inexpensive flexible containers that require no refrigeration. From the outside of the container inward, this packaging material is made from the lamination of polyethylene, paper, polyethylene, aluminum foil, and polyethylene and then a coating of ionomer resin.

Supplying food to the military has always created special problems. The packaging must provide protection, and it must simplify preparation and consumption under adverse conditions.

ENVIRONMENTAL CONSIDERATIONS

Packaging waste can have profound environmental effects. Although recycling is a sound approach to pollution, problems often come with the collection, separation, and purification of consumers' discarded food packages. This mode of recycling is called *postconsumer recycling.* Though it offers a logistic challenge, recycling is gaining in popularity, and the packaging industry is cooperating in that effort. Aluminum cans are currently the most recycled container. Plastic recycling is increasing, yet most plastic is recycled during manufacturing of the containers—not as postconsumer recycling (Figure 15-5). For example, trimmings from plastic bottles are reground and reprocessed into new ones.

FIGURE 15-5 Crushed and baled cans at a recycling center. Environmental concerns require that many packaging materials must be recycled.

The plastics industry facilitates consumer recycling by identifying the type of plastic from which the container is made. A number from 1 to 7 is placed within the recycling logo on the container's bottom. For example, 1 refers to **polyethylene terephthalate (PET)**, the plastic used for large 2-liter soft-drink bottles.

Plastics have the advantage of being light. This helps conserve fuel during transport and also reduces the amount of package waste.

Environmental issues have gained importance because of regulatory requirements. Increasingly, it is not possible to sell a new packaging material without addressing every environmental issue. If someone feels that a package does not meet environmental standards, the brand name could suffer.

Environmental regulations also play an important part in beverage packaging. An important trend in beverage packaging is the use of nonreturnable bottles and cans. The PET bottle is also used more and more in both returnable and nonreturnable applications. With the increased use of plastic containers worldwide, recycling and return concepts become an absolute necessity.

Along with the desire for convenience, microwavable food has also become more popular. For these foods, some packaging companies have developed expanded polypropylene trays that are convenient to handle. These trays provide foods with extended shelf life, fresh product presentation, and environmental savings. The material can also be heated safely in the microwave, and the insulating feel of the foam material allows the consumer to comfortably handle the package after heating.

SCIENCE CONNECTION!

What packaging is used for your favorite foods? How could it be improved? Design an innovative package for a peanut butter and jelly sandwich as well as a pepperoni pizza. How are the packages similar? How would they be different? How long would the package last in a freezer setting? Record your ideas and comparisons.

INNOVATIONS IN PACKAGING

Packaging innovation is becoming more about convenience than cost. Consumers want convenience, and food companies are developing packages that provide it. Globalization and continued environmental pressures provide new challenges to the food-packaging industry.

Cost is no longer the main driver behind packaging innovations. Consumers want convenience, whether it is the elderly consumer who needs to read the label on a box of cookies or the working mother who does not have time to cook a fresh meal. More households are small, and the number of elderly consumers is increasing. This means that packaging not only should be easy to open but also easy to close again so that the food will keep longer.

Ready-to-eat meals have grown tremendously in popularity in recent years. This phenomenon has swept across the United States and is becoming more important in European markets. Food processors see a growing demand for convenience food—for example, kitchen-ready preparations of pasta products, oven-ready preparations of pizza, and related products.

Consumer attitudes toward packaging have also changed. Though consumers want convenience, they also demand higher levels of package security. A similar conflict arises with child-resistant packages, which must be too difficult for children to open but possible for infirm or elderly adults.

Processors recognize that foods must meet consumer quality standards and also appeal to the different palates around the world. For example, McDonald's recognized that many consumers in India look at cows differently than do people in Westernized countries. The company adjusted the menu to meet local needs. This is happening in other areas as well, such as Japan. Although the Japanese have developed a taste for beef products, they do not eat large portions, so the package size is adjusted. In China, which has millions of pigs, more sophisticated pork products are being produced.

MATH CONNECTION!

Take a trip to the local grocery store or supermarket. Document the type and number of retortable pouches you see. What type of food uses the pouches the most? How to you think this compares to 20 years ago? What predictions do you have about the future of food packaging?

SUMMARY

Using modern technology, manufacturers have created an overwhelming number of new packages containing a multitude of food products. A modern food package has many functions, but its main purpose is to physically protect a product during transport. The package also acts as a barrier against potential spoilage agents, which vary with the food product. Practically all foods should be protected from filth, microorganisms, moisture, and objectionable odors. Consumers rely on the package to offer that protection.

Aside from protecting food, a package serves as a vehicle for the manufacturer to communicate with consumers. Nutritional information, ingredients, and often recipes are found on food labels. The package is also used as a marketing tool to attract attention in the store. This makes printability an important property of any package. Globalization of the food industry and the consumers are driving the development of innovations in packaging.

REVIEW QUESTIONS

Success in any career requires knowledge. Test your knowledge of this chapter by answering these questions or solving these problems.

1. What two basic tests are used to measure the protective properties of packaging materials?
2. Name the three general types of food packaging.
3. Explain the three layers in a retortable package.
4. List 10 features of packaging materials.
5. Name the four basic packaging materials used by the food industry.
6. What is the main problem with used packaging materials?
7. Explain any safety concerns that can affect packaging decisions.
8. Packages with special features have what requirements?
9. List two examples of edible film packaging.
10. Describe the functions of primary, secondary, and tertiary packaging.

STUDENT ACTIVITIES

1. Form teams to collect and display 10 different types of food packaging. On the display, indicate whether the packaging is primary, secondary, or tertiary and what type of material was used.
2. Identify a raw fruit or vegetable and develop possible types of packaging for fresh and processed forms. Present your packaging ideas to the class.
3. Create a list of food products that use each of the following for primary packaging: metal, glass, plastic, wood, paper.
4. Search the Internet for information about the unique packaging challenges presented by the space program and the military. Report on your findings.
5. Conduct a contest to see who can bring to class an actual example of the most innovative packaging for a product. Innovation includes not only bright colors and design but also flexibility and convenience. All students must defend their product choices, and then the class will vote on the most innovative.
6. Let potato chips set outside their package for a few days. Describe the changes in the chips and relate this to the function of the packaging.

ADDITIONAL RESOURCES

- Food, Nutrition, and Packaging Sciences: http://www.clemson.edu/cafls/departments/fnps/
- Packaging Engineering: http://www.abe.ufl.edu/academics/undergraduate/packaging-engineering.shtml
- Environmental Impacts of Food Production and Consumption: http://www.ifr.ac.uk/waste/Reports/DEFRA-Environmental%20Impacts%20of%20Food%20Production%20%20Consumption.pdf
- Oregon State University Food Innovation Center: http://fic.oregonstate.edu/

Internet sites represent a vast resource of information. Although those provided in this chapter were vetted by industry experts, you may also wish to further explore the topics discussed in this chapter using a search engine such as Google. Keywords or phrases may include the following: *food packaging, recycling, paper packaging, extratries, postconsumer recycling, edible film packaging.* In addition, Table B-7 provides a listing of useful Internet sites that can be used as a starting point.

REFERENCES

Potter, N. N., and J. H. Hotchkiss. 1999. *Food science*, 5th ed. New York: Springer.

Vaclavik, V. A., and E. W. Christina. 2014. *Essentials of food science.* New York: Springer.

Vieira, E. R. 1999. *Elementary food science*, 4th ed. New York: Springer.

ENDNOTE

1. Allpax: http://www.retorts.com/white-papers/pouch-processing/. Last accessed July 8, 2015.

SECTION *Three*

Foods and Food Products

CHAPTER 16

Milk

OBJECTIVES

After reading this chapter, you should be able to:

- Define the term *milk*
- Describe quality control during the production of milk and milk products
- Explain pasteurization and homogenization
- Identify three methods of pasteurization
- Describe the composition of milk solids
- Discuss the separation of butterfat and its uses
- List four drinkable milk products
- Describe the process of making butter
- Name five concentrated or dried dairy products
- List the steps in cheese making
- Identify three bacteria used to produce dairy products
- Name five fermented dairy products
- List the steps in making ice cream
- Describe three USDA quality grade shields

NATIONAL AFNR STANDARD

FPP.01

Develop and implement procedures to ensure safety, sanitation, and quality in food products and processing facilities.

KEY TERMS

buttermilk
churning
coalesce
curd
high-temperature short time (HTST)
lipolysis
low-temperature longer time (LTLT)
mastitis
rennet
ripening
ropey
solids-not-fat
somatic cell count (SCC)
standard plate count
standardized
thermization
ultrahigh temperature (UHT)
ultrapasteurization
vacuum evaporation
whey

Milk is the first food of young mammals. It provides a high-quality protein, a source of energy, and vitamins and minerals. Worldwide, many mammalian species are used to produce milk and milk products. These include goats, sheep, horses, and yaks. The focus of this chapter is milk from dairy cows.

Most milk in the United States is produced by cows. The dairy industry produces milk as a fluid product and in a variety of manufactured products, including butter, cheeses, condensed milk, dry milk, and cultured milk.

FLUID MILK

Milk is composed primarily of water (87% to 89%). Like other foods, removing water from milk extends its shelf life. The term *total milk solids* describes the remaining 12% to 13% of milk. This includes the carbohydrates, lactose, fat, protein, and minerals of milk. The term milk **solids-not-fat** excludes the fat and includes the lactose, caseins, **whey**, proteins, and minerals (calcium, phosphorus, magnesium, potassium, sodium, chloride, and sulfur). Table B-8 provides the composition of milk and milk products.

Milk also contains water-soluble vitamins (thiamin and riboflavin) and mineral salts. It is described as a colloidal dispersion of the protein casein and the whey proteins. Finally, fluid milk is an emulsion with fat globules suspended in the water phase of milk.

Legal Description

The standard of identity under Title 21 of the Code of Federal Regulations [131.110(a)] describes milk as follows:

> *Milk is the lacteal secretion, practically free from colostrum, obtained by the complete milking of one or more healthy cows. Milk that is in the final package form for beverage use shall have been pasteurized or ultrapasteurized, and shall contain not less than 8¼% milk solids not fat and no less than 3¼% milk fat. Milk may have been adjusted by separating part of the milk fat therefrom, or by adding thereto cream, concentrated milk, dry whole milk, skim milk, concentrated skim milk or nonfat dry milk. Milk may be homogenized.*

Production Practices

Since 1970, milk production in the United States has risen by almost half, even though milk cow numbers have declined by about a fourth (from about 12 million in 1970 to roughly 9.3 million in 2015), and the dairies are becoming larger. In major production areas, dairies of 1,000 cows or more are common (Figure 16-1).

Milk fresh from a cow is virtually sterile. All postmilking handlings must maintain the milk's nutritional value and prevent deterioration caused by numerous physical and biological elements. In addition, equipment on the farm must be maintained to government and industry standards. Most cows are milked twice a day, although some farms milk three or four times per day. The milk is immediately cooled from the cow's body temperature to below 41 °F (5 °C) and then stored at the farm under refrigeration until picked up by insulated tanker trucks at least every other day (Figure 16-2). When milk is pumped into the tanker, a sample is collected for later laboratory analysis.

FIGURE 16-1 Modern dairies milk 1,000 cows or more in clean, convenient environments.

© Blondsteve/Dreamstine.com

FIGURE 16-2 Large milk tanker trucks haul milk from a dairy to a processing plant.

Grades and Classes of Milk

The price farmers receive for raw, unprocessed, and unpasteurized milk is largely determined by supply-and-demand forces that are influenced by federal and state dairy programs. Usually, the minimum farm milk price is established based on the value of the products made from it. Those products are categorized into four classes of milk:

- Class I milk is used for fluid, or beverage, milk products.
- Class II milk goes into "soft" manufactured products such as sour cream, cottage cheese, ice cream, and yogurt.
- Class III milk is used for making hard cheeses.
- Class IV milk is used to make butter and dry products such as *nonfat dry milk* (NFDM).

One factor of grading milk is its **somatic cell count (SCC)**. The SCC is used throughout the world as an indicator of milk quality. Poor quality milk has a high number of somatic cells and is an inferior product with reduced processing properties resulting in dairy products with reduced shelf life. High quality milk has an extremely low number of somatic cells, a longer shelf life, a better taste, and higher nutritional value. SCC can be used to gauge a dairy herd's udder infection status and also provides a good indication of the loss in milk production in a herd due to mastitis (infection in the udder). The bacteria in milk is determined by the **standard plate count (SPC)**, which is an estimate of the total number of viable aerobic bacteria present in raw milk. The overall count of bacteria in the milk is an indication of cleanliness, dirt in the milk, the temperature at which the milk has been stored, and the milk's age.

Factors Necessary to Produce Quality Milk

The objective of good dairy farming practice is the on-farm production of safe, quality milk from healthy animals under generally acceptable conditions. To achieve this, dairy producers need to apply *good agricultural practices* (GAPs) in the following areas.

- **Animal health.** Producers must manage a herd's resistance to disease, prevent the entry of disease onto the farm, establish effective herd health management, and use all chemicals and veterinary medicines as directed.
- **Milking hygiene.** Producers must ensure that milking routines do not injure the animals or introduce contaminants into the milk, that milking is carried out under hygienic conditions, and that milk is handled properly after milking.
- **Nutrition (feed and water).** Producers must meet the requirements for feed and nutrients that dairy animals depend on according to their physiological states, milk production levels, ages, sex, and body condition. Because dairy animals consume large amounts of water for milk production and pregnancy, access to water has a great effect on milk production.
- **Animal welfare.** Producers must follow dairy farming practices that aim to keep animals free from hunger, thirst, malnutrition, discomfort, pain, injury, stress, and disease.
- **Environment.** Producers must follow good dairy farming practices for the environment and implement environmentally sustainable

farming systems, have an appropriate waste-management system, and ensure that dairy farming practices do not have an adverse impact on the local environment.

Mastitis Detection and Control

Mastitis occurs when white blood cells (leukocytes) are released into a cow's mammary gland, usually in response to bacteria entering through the teat canal of the udder. Milk-secreting tissues in the gland are damaged by toxins released from the bacteria. This disease can be identified by abnormalities in the udder such as swelling, heat, redness, hardness, or pain. Other indications of mastitis may be abnormalities in milk such as a watery appearance, flakes, or clots. A common infection on dairy farms is subclinical mastitis, in which a cow has no visible signs of infection.

Mastitis may be classified according two different criteria—according to clinical symptoms or by the mode of transmission (contagious or environmental mastitis). Mastitis is most often transmitted by repetitive contact with the milking machine, and through contaminated hands or materials. Infection results in decreased casein, the major protein in milk; the disruption of casein synthesis contributes to lowered calcium in milk. Milk from cows with mastitis also has a higher somatic cell count. Generally speaking, the higher the somatic cell count, the lower the milk quality.

Detection. The most common mastitis causing bacteria is *Staphylococcus aureus*. Cattle affected by mastitis can be detected by examining the udder for inflammation and swelling or by observing the consistency of the milk, which often develops clots or changes color when a cow is infected.

One method of detection through milk consistency is the *California mastitis test*, which is designed to measure milk's somatic cell count as a means of detecting inflammation and infection of the udder.

Treatment. Treatment is possible with long-acting antibiotics, but milk from such cows is not marketable until drug residues have left the cow's system. Cows being treated may be marked with tape to alert dairy workers, and their milk is syphoned off and discarded.

Control. Some practices can help reduce the incidence of mastitis such as good nutrition, proper milking hygiene, and the culling of chronically infected cows. Making sure that cows have clean, dry bedding decreases the risk of infection and transmission. Staff should wear rubber gloves while milking, and machines should be cleaned regularly to decrease the incidence of transmission.

A good milking routine is critical to reducing mastitis cases. This usually consists of applying a teat dip or spray, such as iodine, and wiping teats dry before milking. The milking machine is then applied. After milking, the teats can be cleaned and sprayed again to ensure there will be no post-milking infections. The control of mastitis is not just a health practice but also an economic necessity because this disease costs the U.S. dairy industry about $1.7 billion to $2 billion each year.

Quality Control on the Farm

At the dairy farm, inspectors monitor herd health, farm water supplies, sanitation of milking equipment, milk temperature, holding times, and milk bacterial counts. Violations of health standards result in heavy penalties up to and including suspension from business. Inspections, whether at the farm or at the processing plant, and regulatory inspections occur on an ongoing basis. Inspectors have full authority to suspend plant operations in order to allow detailed examination of all equipment, facilities, and products. The dairy industry works hard to ensure that its members comply with or exceed all regulations.

Several tools and practices help ensure proper cleaning of milking equipment. Backflushers have been developed to sanitize the liners and claws between milkings. These units have several cycles. The first cycle is a water rinse, followed by an iodine rinse, a clear-water rinse, and an air-dry cycle. Research has demonstrated that backflushers reduce the number of bacteria on the liners between cows but do not reduce the number of bacteria on teats.

A cleaning and sanitizing regime is typically made up of the following chemical processes, some of which may be combined into a single step and some of which may contain several steps. The majority of the residual milk in the system should be removed before cleaning by allowing components to drain. The processes involved in a proper equipment-cleaning system are as follows: Water rinses remove residual and suspended materials and bring the equipment to a specific temperature; alkaline detergents remove organic soils such as milk fat and proteins; and an acid-rinse cycle removes mineral deposits from water and milk, When done properly, this disinfection routine will reduce microorganisms to acceptably low levels.

The following is an example of a milking-equipment sanitation checklist:

Milking Machine Inspection and Maintenance Checklist[1]

Before each milking:

Check

- ☐ Vacuum controller
- ☐ Milking vacuum
- ☐ Hoses and teat-cup liners for holes or tears
- ☐ Pulsators
- ☐ Air-admission holes in claw or tailpiece of liner

Weekly (or every 50 hours of operation):

Check

- ☐ And clean vacuum controller
- ☐ And clean pulsator filters
- ☐ Belts on vacuum pumps
- ☐ Oil reserve on vacuum pump
- ☐ If it is time to change liners (every 1,000–2,000 milkings, or as recommended)
- ☐ And clean moisture trap
- ☐ Automatic take off-equipment (especially vacuum shutoff)

Monthly (or every 250 hours):

Check

- ☐ And clean the pulsators
- ☐ And clean vacuum pulsation lines
- ☐ Vacuum pump(s)
- ☐ Belts for wear and tension
- ☐ And clean screens
- ☐ And change filters on vacuum pump tank
- ☐ And change oil if needed
- ☐ And change liners if it is time

Maintenance items:

- ☐ Check all gaskets, flappers, O-rings, and caps that come into contact with milk. Replace if worn.
- ☐ Clean all electric pulsator selectors and activators. Check all solenoids and coils. Clean all plungers and vacuum lines.
- ☐ Overhaul pneumatic pulsators. Repair milk supports, milker units, and so on.

Every year (or 2,500 hours of use):

- ☐ Change all solenoid coils, plungers, hoses, diaphragms, caps, gaskets, flappers, and rubber vacuum connectors.
- ☐ Check electric timer controls, switches, motors, and parts. Grease all bearings.
- ☐ Check milk pump seal, rubber spring, and clearances. Change gaskets.

Every two years (or 5,000 hours of use):

- ☐ Recondition the entire system, including all motors, pumps, selectors, timers, and starters.
- ☐ Replace all rubber coils, hoses, gaskets, O-rings, springs and plungers. Clean the entire pipeline system.

All finished dairy products are tested regularly by state inspectors to ensure compliance with each of the following: (1) *standards of identity*, which refers to such criteria as moisture, butterfat, and protein content; and (2) *purity*, which refers to such criteria as the presence or absence of pathogens and residues. The Food and Drug Administration (FDA) sets standards of identity for beverage milk products.

Milk Pricing Economics and Trends

The four classes (uses) from federal milk marketing orders include the following.

- Class I: Grade A milk used in all beverage milks.
- Class II: Grade A milk used in fluid cream products, yogurts, and perishable manufactured products (ice cream, cottage cheese, and others).
- Class III: Grade A milk used to produce cream cheese and hard manufactured cheeses.
- Class IV: Grade A milk used to produce butter and any milk in dried form.

The minimum prices for the different classes of raw milk are calculated by valuing cheese, dry whey, NFDM, and butter using weekly average wholesale market prices monitored by the U.S. Department of Agriculture (USDA). The Class IV price is based on NFDM and butter prices; the Class III price is based on cheese, dry whey, and butter prices; the Class II price is based on NFDM and butter prices plus an added differential equal to 70 cents per 100 pounds of raw milk; and the Class I price is based on a location differential currently ranging from $1.60 to $4.30 per 100 pounds (cwt) of raw milk. Every county in every state in the country has a location differential assigned by the USDA. The actual minimum price received by farmers is a blend of these prices weighted by the percentage of milk used in each class.

In a significant policy change contained in the 2014 federal farm bill, the USDA will no longer purchase dairy products at announced prices under the dairy price-support program. In addition, USDA's Agricultural Marketing Service announced it will no longer collect and publish the announced cooperative Class I price for 30 U.S. cities.

Marketing and payments made to farmers are usually handled by a cooperative. Many dairy farms belong to cooperatives through which they sell their milk. A cooperative is an agency owned and controlled by members to conduct their business. Each member farm fully retains its economic individuality and independence, and the board of directors is elected from among member farmers. Farmers organize cooperatives to perform functions that cannot be satisfactorily carried out alone by individual farmers. Cooperatives make it feasible for farmers to join together to gain greater market access.

Processing

When milk arrives at a milk plant, it is checked to make sure it meets the standards for temperature, total acidity, flavor, odor, tanker cleanliness, and absence of antibiotics. The butterfat and solids-not-fat contents of this raw milk are also analyzed. The amounts of butterfat (BF) and solids-not-fat (SNF) in the milk will vary according to time of year, breed of cow, and feed supply. BF and SNF contents and volume are used to determine the amount of money paid each producer.

Once the load passes these receiving tests, it is pumped into large refrigerated storage silos (nearly half-million pounds capacity) at the processing plant.

Pasteurizing

All raw milk must be processed within 72 hours of receipt at the plant. Milk is such a nutritious food that many naturally occurring bacteria are always present. The milk is pasteurized, which is a process of heating the raw milk to kill all pathogenic microorganisms that may be present. Pasteurization is not sterilization (sterilization eliminates all viable life forms, whereas pasteurization does not). After pasteurization, some harmless bacteria may survive the heating process. These bacteria will cause milk to go sour. Keeping milk refrigerated is the best way to slow the growth of these bacteria.

The batch method of pasteurization heats the milk to at least 145 °F and holds it at that temperature for at least 30 minutes. This method is called **low-temperature longer time (LTLT)** pasteurization. Because it can cause a "cooked" flavor, this process is not used by some milk plants for fluid milk products.

High-temperature short time (HTST) pasteurization heats the milk to at least 161 °F for at least 15 seconds. The milk is immediately cooled to below 40 °F and packaged into plastic jugs or plastic-coated cartons.

Heating milk to 280 °F or higher for 2 seconds followed by rapid cooling to 45 °F or lower is called **ultrapasteurization.**

Sterilization of milk occurs when it is heated from 280 °F to 302 °F for 2 to 6 seconds. This is called **ultrahigh temperature (UHT)** processing. With this treatment, the sterilized milk is aseptically packaged, and the milk does not require refrigeration until it is opened.

Identifying Diseases Transmitted to Consumers Through Milk

Knowledge of diseases that can be transmitted through milk is necessary to ensure that safe milk is produced and processed. The following are diseases that can be transmitted through milk: viral diseases such as polio, foot and mouth disease, infectious hepatitis, and tick-borne encephalitis; and bacterial diseases such as anthrax, botulism, brucellosis (undulant fever), cholera, and *Escherichia coli.*

Recognizing Off Flavors. A pleasantly sweet, refreshing milk flavor is the key to consumer acceptance. The following are off flavors and their causes.[2]

- Feed: aromatic or odorous flavors. Caused by cows eating or inhaling odors of strong feeds, sudden changes in roughage, or poor ventilation in building.
- Oxidized: cardboard or metallic flavors. Caused by exposure to rusty surfaces on milking equipment; copper, iron, or manganese in the water supply; excessive use of chlorine sanitizers; lighting; and exposure of milk to sunlight or fluorescent lighting.
- Spoiled or unclean: unpleasant aftertaste. Caused by spoilage, dirty cows because of poorly maintained feeding areas, improperly prepared cows for milking, and failure to dry teats before milking.
- Rancid: bitter, soapy flavors. Caused by excess agitation or foaming of raw milk, late lactation or low-production cows, low protein content of cow feed.
- Foreign flavors: Medicinal, chemical flavors. Caused by medications, chemicals, or insecticides improperly managed; excessive concentration or improper use; inadequate, strong-smelling sanitizers and disinfectants.

Butterfat

Butterfat content accounts for several different types of products. Whole milk, 2%, 1%, nonfat, and half-and-half are examples (see Figure 16-3). A machine called a *separator* separates the cream

FIGURE 16-3 Whole milk, 2%, 1%, and nonfat products are derived from butterfat content.

FIGURE 16-4 Cream is used for the production of such products as ice cream and butter.

and skim portions of the milk. Two streams are produced during the separation of whole milk: the fat-depleted stream, which produces the beverage milks just described; skim milk for evaporation and possibly for subsequent drying; and the fat-rich stream, the cream. This usually comes off the separator with fat contents in the range of 35% to 45%.

Cream is used for further processing in the dairy industry to make ice cream and butter (see Figure 16-4), or it can be sold to other food-processing industries. These industrial products normally have higher fat contents than creams for retail sale, normally in the range of 45% to 50% fat. A product known as *plastic* cream can be produced from certain types of milk separators. This product has a fat content approaching 80% fat, but it remains as an oil-in-water emulsion (the fat is still in the form of globules, and the skim milk is the continuous phase of the emulsion), unlike butter, which also has a fat content of 80% but which has been churned so that the fat occupies the continuous phase and the skim milk is dispersed throughout in the form of tiny droplets (a water-in-oil emulsion).

Homogenization

Milk is homogenized to prevent the cream portion (butter fat) from rising to the top of the package. Cream is lighter in weight than milk and would form a cream layer at the top of a carton of nonhomogenized milk. A homogenizer forces the milk under high pressure through a valve that breaks up the butterfat globules to such small sizes they cannot **coalesce** (stick together). Homogenization does not affect a product's nutrition or quality.

Beverage Milk

While the fat content of most raw milk is 4% or higher, the fat content in most beverage milks has been reduced to 3.4%. Lower-fat alternatives, such as 2% fat, 1% fat, or skim milk (less than 0.1% fat), are also available in most markets. These products are either produced by partially skimming the whole milk, or by completely skimming it and then adding an appropriate amount of cream back to achieve the desired final fat content.

Nutritional Qualities

Vitamins may be added to both full fat and reduced fat milks. Fat-soluble vitamins A and D are often added to milk (see Figure 16-5). Vitamin A is lost during fat separation and heating, and vitamin D is not present in milk. These are supplemented in the form of a water-soluble emulsion. Many states have milk standards that require the addition of milk solids. These solids represent the natural mineral (calcium and iron), protein (casein), and sugar (lactose) portion of nonfat dry milk.

Table B-8 provides the details on the nutritional qualities of beverage milk.

SCIENCE CONNECTION!

Can you tell the difference between whole milk, 2%, 1%, and skim milk? How do they differ? How are they alike? Set up a sensory analysis blind taste test. Can you correctly identify each type?

FIGURE 16-5 Vitamins A and D are often added to milk.

SCIENCE CONNECTION!

Why is vitamin D added to milk? What are some other food products that have vitamin D? Research and record your findings.

Quality Control During Processing

Quality-control personnel conduct many tests on the raw and pasteurized products to ensure optimal quality and nutrition. A sample is analyzed for the presence of microbiological organisms with a standard plate count and **ropey** milk test. The equipment used to analyze BF and SNF is calibrated on a regular basis to ensure a consistent, quality product that meets or exceeds government requirements.

All milk products have a sell-by date printed on their packaging. This is the last day the item should be offered for sale. However, most companies guarantee the quality and freshness of the product for at least 7 days past the printed date. Samples of each product packaged each day are saved to confirm that they maintain their freshness 7 days after the sell-by date.

Packaging

Once the milk has been separated, **standardized**, homogenized, and pasteurized, it is held below 40 °F in insulated storage tanks before being packaged into gallon, half-gallon, quart, pint, and half-pint containers. The packaging machines are maintained under strict sanitation specifications to prevent bacteria from being introduced into the pasteurized products. All equipment that comes into contact with product (raw or pasteurized) is washed daily. Sophisticated automatic clean-in-place systems guarantee consistent sanitation with a minimum of manual handling, thus reducing the risk of contamination.

Once packaged, the products are quickly conveyed to a cold-storage warehouse. They are stored there for a short time and shipped to the supermarket on refrigerated trailers. Once at the store, the milk is immediately placed into a cold-storage room or refrigerated display case.

MILK PRODUCTS AND BY-PRODUCTS

Milk products and by-products include butter, concentrated and dried products, cheese, whey products, yogurt and other fermented products, and ice cream.

Butter

Butter is made by **churning** pasteurized cream. Churning breaks the fat globule membrane so the emulsion breaks, fat coalesces, and water (in the form of **buttermilk**) escapes. Federal law requires that it contain at least 80% milk fat. Salt and coloring may be added. Nutritionally, butter is a fat (see Table B-8). Whipped butter is regular butter whipped with air for easier spreading and to increase the volume of butter per pound.

Today's commercial butter making is a product of the knowledge and experience gained over the years in such matters as hygiene, bacterial acidifying, and heat treatment, as well as the rapid technical development that has led to the advanced machinery now used.

The principal constituents of a normal salted butter are fat (80% to 82%), water (15.6% to 17.6%), salt (approximately 1.2%) as well as protein, calcium, and phosphorous (approximately 1.2%). Butter also contains the fat-soluble vitamins A, D, and E.

Butter should have a uniform color, be dense, and taste clean. The water content should be dispersed in fine droplets so that the butter looks dry. The consistency should be smooth so that the butter is easy to spread and melts readily on the tongue.

The butter-making process involves many stages. The continuous butter maker has become the most common type of equipment used.

From the storage tanks, the cream goes to pasteurization. This destroys enzymes and microorganisms that would impair the keeping quality of the butter.

In the **ripening** tank, the cream is subjected to a program of heat treatment designed to give the fat the required crystalline structure when it solidifies on cooling. The program is chosen to deal with factors such as the composition of the butterfat, which is expressed in terms of the iodine value, which is a measure of the butter's unsaturated fat content. The treatment can even be modified to obtain butter with good consistency despite a low iodine value—that is, when the unsaturated proportion of the fat is low.

As a rule, ripening takes 12 to 15 hours. From the ripening tank, the cream is pumped to the churn or continuous butter maker via a plate heat exchanger that brings it to the requisite temperature. In the churning process, the cream is violently agitated to break down the fat globules, causing the fat to coagulate into butter grains, while the fat content of the remaining liquid, the buttermilk, decreases.

Thus, the cream is split into two fractions: butter grains and buttermilk. In traditional churning, the machine stops when the grains have reached a certain size; at this point, the buttermilk is drained off. With the continuous butter maker, the draining of buttermilk is also continuous.

After draining, the butter is worked to a continuous fat phase containing a finely dispersed water phase. It used to be common practice to wash the butter after churning to remove any residual buttermilk and milk solids, but this is rarely done today. If the butter is to be salted, salt is spread over its surface in the case of batch production. In the continuous butter maker, a salt slurry is added to the butter.

After salting, the butter must be worked vigorously to ensure even distribution of the salt. The working of the butter also influences the characteristics by which the product is judged—aroma, taste, keeping quality, appearance, and color. The finished butter is discharged into the packaging unit, and from there to cold storage (Figure 16-6).

Continuous Butter Making. Methods of continuous butter making were introduced at the

FIGURE 16-6 Butter in cold storage.

end of the 19th century, but their application was limited. In the 1940s, the work was resumed and resulted in three different processes, all based on the traditional methods of churning, centrifuging, and concentration or emulsifying. One process based on conventional churning was the Fritz method, which is the one now used predominantly in Western Europe. In machines based on this method, butter is made in generally the same way as traditional methods. The butter is basically the same except for being denser as a result of uniform and fine water dispersion.

In today's market, butter is sold in many varieties. Salted and unsalted are the most common forms used for baking. Whipped butter is not common in baking, but is used mostly for spreading on toast or in finishing dishes. It has a lighter texture than regular butter. Cultured butter is made from fermented cream and has a tangy taste. It is usually used for creating croissants and flaky breads. Spreadable butters are usually a combination of cream and vegetable oil. Various flavorings such as olive oil, garlic. and herbs are also added to butter varieties and used for various sautéing and cooking applications.

ORGANIC DAIRY: PERCEPTION AND MARKETING

Organic dairying represents the organic food industry's largest and fastest-growing segment, showing a consistent increase in sales in recent years. In 2011, nearly 2.1 billion pounds of organic milk products were sold, a 14.5% increase from the previous year. This is rapid growth compared to the general food industry.

Figures indicate that more than 40% of U.S. homes now purchase some type of organic product. In the past, most organic products were sold through natural food stores, but recently more grocery stores offer them. Organic foods seemingly are becoming a mainstream idea, not a niche anymore. This says something about the changing U.S. consumer base.

With products from organic dairies pushing beyond specialty stores and moving into supermarkets, large organic dairy operations attribute their success and growth to seven major trends:

1. Mainstreaming of organic consumers, products, and retailers
2. Increased knowledge of the relationship between diet and health
3. Environmental awareness

(Continues)

ORGANIC DAIRY: PERCEPTION AND MARKETING (continued)

4. Search for alternatives to genetically modified organisms (GMOs)
5. Declining organic food-production costs
6. Development and distribution of organic standards
7. Capital investments in organic agriculture by the financial community

Organic dairy products—milk, blended yogurts, and cheese—require that cows in the dairy not be given antibiotics or hormones for a minimum of 1 year before they are milked. Organic dairies often forbid the use of pesticides and GMOs.

Some organic dairies also market eggs and calcium-enriched orange and grapefruit juices. These companies market their products by emphasizing organic and pushing brand identity.

For more information about organic dairy products, visit these Web sites:

- Horizon Organic: https://www.horizondairy.com/
- USDA National Organic Program: www.ams.usda.gov/nop/
- USDA Available Services: http://www.ams.usda.gov/AMSv1.0/getfile?dDocName=STELPRDC5100590

The Manufacturing Process. The cream is prepared in the same way as for conventional churning before being fed continuously from the ripening tanks to the butter maker.

The cream is first fed into a churning cylinder fitted with beaters that are driven by a variable speed motor. Rapid conversion takes place in the cylinder and, when finished, the butter grains and buttermilk pass on to a draining section. The first washing of the butter grains sometimes takes place en route—either with water or recirculated chilled buttermilk. The working of the butter commences in the draining section by means of a screw that also conveys the butter to the next stage. On leaving the working section, the butter passes through a conical channel to remove any remaining buttermilk. Immediately afterward, the butter may be given its second washing, this time by two rows of adjustable high-pressure nozzles.

The water pressure is so high that the ribbon of butter is broken down into grains, thus effectively removing any residual milk solids. Following this stage, salt may be added through a high-pressure injector.

The third section in the working cylinder is connected to a vacuum pump. Here it is possible to reduce the air content of the butter to the same level as conventionally churned butter.

In the final or mixing section, the butter passes a series of perforated disks and star wheels. There is also an injector for final adjustment of the water content. Once regulated, the water content of the butter deviates by ±0.1%, provided the characteristics of the cream remain the same.

The finished butter is discharged in a continuous ribbon from the end nozzle of the machine and then moves into the packaging unit.

Concentrated and Dried Dairy Products

Fluid milk contains approximately 88% water. Concentrated milk products are obtained through partial water removal (see Figure 16-7). Dried dairy products have even greater amounts of water removed (usually to less than 4%). The benefits of both these processes include increased shelf life, convenience, product flexibility, decreased transportation costs, and easier storage. Concentrated dairy products include:

- Evaporated skim or whole milk
- Sweetened condensed milk
- Condensed buttermilk
- Condensed whey

FIGURE 16-7 Condensed milk and evaporated milk are examples of concentrated products.

Dried dairy products include:

- Milk powder
- Whey powder
- Whey protein concentrates

The principles of evaporation and dehydration can be found in Chapter 11.

To extend the shelf life, evaporated milk can be packaged in cans and then sterilized in an autoclave. Continuous flow sterilization is followed by packaging under aseptic conditions. Though the sterilization process produces a light brown coloration, the product can be successfully stored for up to a year.

Sweetened Condensed Milk. Where evaporated milk uses sterilization to extend its shelf life, sweetened condensed milk has an extended shelf life because of the addition of sugar. Sucrose, in the form of crystals or solution, increases the osmotic pressure of the liquid. This prevents the growth of microorganisms. The only real heat treatment this product receives—185 °F to 194 °F (85 °C to 90 °C) for several seconds—is after the raw milk has been clarified and standardized. This treatment destroys osmiophilic and thermophilic microorganisms, inactivates lipases and proteases, decreases fat separation, and inhibits oxidative changes.

SCIENCE CONNECTION!

Why are there so many different dairy products? What are they all used for? Research some of these different dairy products and how they are used. What food products are they used in?

Evaporated Skim or Whole Milk. After raw milk is clarified and standardized, it is given a preheating treatment. Milk is then concentrated at low temperatures by **vacuum evaporation**. The vacuum lowers the boiling point to approximately 104 °F to 113 °F (40 °C to 45 °C). This results in little to no cooked flavor.

The milk is concentrated to 30% to 40% total solids. A second standardization is done at this time to ensure that the proper salt balance is present. The ability of milk to withstand intensive heat treatment depends to a great degree on its salt balance.

The product at this point is quite perishable. The fat is easily oxidized. The evaporated milk at this stage is often shipped by the tanker for use in other products.

The milk is evaporated in a manner similar to evaporated milk. Although sugar may be added before evaporation, addition after evaporation is recommended to avoid undesirable viscosity changes during storage. Enough sugar is added so that the final concentration of sugar is approximately 45%.

The sweetened evaporated milk is then cooled and lactose crystallization is induced. The milk is inoculated, or seeded, with powdered lactose crystals, then rapidly cooled while being agitated. The product is packaged in smaller containers such as cans for retail sales and bulk containers for industrial sales.

Condensed Buttermilk

As previously described, buttermilk is a by-product of the butter industry. It can be evaporated on its own or blended with skim milk and dried to produce skim milk powder. This blended product may oxidize readily because of the higher fat content. Condensed buttermilk is perishable and must be stored cool.

Condensed Whey. The process of cheese making creates a lot of whey that needs disposal. One way to treat cheese whey is to condense it by evaporation. The whey contains fat, lactose, lactoglobulin, lactalbumin, and water. The fat is generally removed by centrifugation and churned as whey cream or used in ice cream. Evaporation is the first step in producing whey powder.

Milk Powder. Milk used in the production of milk powder is first clarified, standardized, and then given a heat treatment. This heat treatment is usually more severe than that required for pasteurization. Besides destroying all the pathogenic and most of the spoilage microorganisms, it also inactivates the enzyme lipase, which could otherwise cause **lipolysis** during storage.

The milk is then evaporated before drying. Homogenization may be applied to decrease the free fat content. Spray drying is the most used method for producing milk powders. After drying, the powder must be packaged in containers able to provide protection from moisture, air, and light. Whole milk powder can then be stored for long periods (up to 6 months or so) at ambient temperatures. Skim milk powder processing is similar.

Instant milk powder is produced by partially rehydrating the dried milk powder particles, which causes them to become sticky and agglomerate. The water is then removed by drying, resulting in an increased amount of air incorporated between the powder particles.

Whey Powder. Converting whey into powder has led to its incorporation into many products (see Figure 16-8).

FIGURE 16-8 Many food products contain whey or whey products.

Whey powder is essentially produced by the same method as other milk powders. Reverse osmosis can be used to partially concentrate the whey before vacuum evaporation. Before the whey concentrate is spray dried, lactose crystallization is induced to decrease the hygroscopicity. This is accomplished by quick cooling in flash coolers after evaporation. Crystallization continues in agitated tanks for 4 to 24 hours.

A fluidized bed may be used to produce large agglomerated particles with free-flowing, nonhygroscopic, no-caking characteristics.

Whey Protein Concentrates. Whey disposal problems and high-quality animal protein shortages have increased worldwide interest in whey protein concentrates.

After clarification and pasteurization, the whey is cooled and held to stabilize the calcium phosphate complex. The whey is commonly processed using ultrafiltration, although reverse osmosis, microfiltration, and demineralization methods can be used. During ultrafiltration, the low-molecular-weight compounds such as lactose, minerals, vitamins, and nonprotein nitrogen are removed in the permeate while the proteins become concentrated in the retentate. After ultrafiltration, the retentate is pasteurized, maybe evaporated, then dried. Drying—usually spray drying—is done at lower temperatures than for milk so that protein denaturation can be largely avoided.

Cheese

Traditionally, cheese was made as a way to preserve milk's nutrients. Simply defined, cheese is the fresh or ripened product obtained after coagulation and whey separation of milk, cream, or partly skimmed milk, buttermilk, or a mixture of these products. It is essentially the product of selective concentration of milk (Figure 16-9). Thousands of varieties of cheeses have evolved that are characteristic of various regions of the world.

Some common cheese-making steps include:

1. Treatment of milk
2. Additives
3. Inoculation and milk ripening
4. Coagulation
5. Enzyme
6. Acid
7. Heat-acid
8. Curd treatment
9. Cheese ripening

Like most dairy products, cheese milk must first be clarified, separated, and standardized. An initial **thermization** treatment results in a reduction of high initial bacteria counts before storage. It must be followed by proper pasteurization. HTST pasteurization is often used. The addition of hydrogen peroxide is sometimes used as an alternative treatment for full pasteurization.

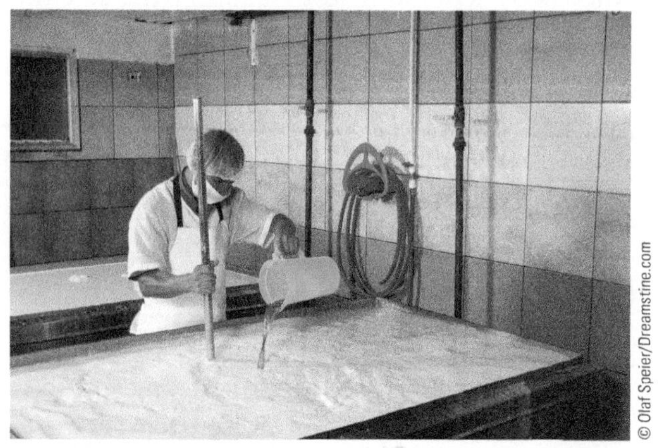

FIGURE 16-9 A cheese-production worker adds liquid rennet to milk to initiate the separation of curds from whey.

© Olaf Speier/Dreamstine.com

Homogenization is not usually done for most cheese milk. It disrupts the fat globules and increases the fat surface area where casein particles adsorb. This results in a soft, weak **curd** at **renneting** and increased hydrolytic rancidity.

Calcium chloride is added to replace calcium lost during pasteurization. The calcium assists in coagulation and reduces the amount of rennet required.

Because milk color varies from season to season, color may added to standardize the color of the cheese throughout the year. Annato, beta-carotene, and paprika are used.

Lipases, normally present in raw milk, are inactivated during pasteurization. The addition of lipases are common to ensure proper flavor development through fat hydrolysis.

The basis of cheese making relies on the fermentation of lactose by lactic acid bacteria (LAB). Lactic acid lowers the pH and helps coagulation, promotes syneresis, helps prevent spoilage and pathogenic bacteria from growing, and contributes to cheese texture, flavor, and keeping quality. LAB also produce growth factors that encourages the growth of nonstarter organisms and provides lipases and proteases necessary for flavor development during curing.

After inoculation with the starter culture, the milk is held for 45 minutes to 60 minutes at 77 °F to 86 °F (25 °C to 30 °C) to ensure that the bacteria are active, growing, and have developed acidity. This stage is called *ripening the milk* and is done before renneting.

Coagulation is essentially the formation of a gel by destabilizing the casein micelles, causing them to aggregate and form a network that partially immobilizes the water and traps the fat globules in the newly formed matrix. This may be accomplished with enzymes, acid treatment, or heat–acid treatment. Chymosin, or rennet, is most often used for enzyme coagulation. Lowering the pH of the milk results in aggregation. Acid curd is more fragile than rennet curd because of the loss of calcium. Acid coagulation can be achieved naturally with the starter culture or artificially with the addition of gluconodeltalactone. Acid-coagulated fresh cheeses may include cottage cheese, quark (fresh curd), and cream cheese.

Heat causes denaturation of the whey proteins (see Figure 16-10). The denatured proteins then interact with the caseins. With the addition of acid, the caseins precipitate with the whey proteins. In rennet coagulation, only 76% to 78% of the protein is recovered; in heat–acid coagulation, 90% of protein can be recovered. Examples of cheeses made by this method include paneer, ricotta, and queso blanco.

Source: USDA, photo by Bob Nichols

FIGURE 16-10 This milk curd of fresh raw milk has to be constantly stirred so the milk curd that has been cut with curd knives does not set until whey has been separated from the curd.

After the milk gel has been allowed to reach the desired firmness, it is carefully cut into small pieces with knives or wires. Cutting shortens the distance and increases the available area for whey to be released. The curd pieces immediately begin to shrink and expel the greenish liquid called *whey* in a process known as *syneresis*. This process is further driven by a cooking stage. The increase in temperature causes the protein matrix to shrink because of increased hydrophobic interactions; the temperature rise also increases the rate of fermentation of lactose to lactic acid. The increased acidity also contributes to shrinkage of the curd particles. The final moisture content depends on the time and temperature of the cook stage.

When the curds have reached the desired moisture and acidity, they are separated from the whey. The whey can then be removed from the top or drained by gravity. The curd–whey mixture may also be placed in molds for draining. Some cheese varieties, such as Colby, Gouda, and Brine Brick include a curd washing that increases the moisture content, reduces the lactose content and final acidity, decreases firmness, and increases the openness of the texture.

Curd handling from this point on is highly specific for each cheese variety. Salting may be achieved through brine as with Gouda, surface salt as with feta, or vat salt as with cheddar. To achieve the characteristics of cheddar, cheddaring (curd manipulation), milling (cutting into shreds), and pressing at high pressure are crucial.

Except for fresh cheese, the curd is ripened, or matured, at various temperatures and times until the characteristic flavor, body, and texture profile is achieved. During ripening, degradation of lactose, proteins, and fat is carried out by ripening agents. The ripening agents in cheese include: bacteria and enzymes of the milk, lactic culture, rennet, lipases, added molds or yeasts, and environmental contaminants.

FIGURE 16-11 A variety of aged cheeses on display at a grocery store. Aging can vary from weeks to years, according to the cheese variety.

FIGURE 16-12 A variety of yogurts. Flavor and aroma can vary between regions, but the basic ingredients and manufacturing processes are essentially consistent.

The microbiological content, the biochemical composition of the curd, and temperature and humidity all affect the final product. This final stage—aging—varies from weeks to years, according to the cheese variety (see Figure 16-11).

SCIENCE CONNECTION!

Cheese is creating by separating curds from whey and letting it ripen with bacteria or certain molds. How many different cheeses can you name? Research different types of cheese. Are they soft, semisoft, or hard cheeses?

Yogurt

Yogurt (also spelled *yoghurt*) is a semisolid fermented milk product that originated centuries ago in Bulgaria. Yogurt flavor and aroma vary from region to region, but the basic ingredients and manufacturing are essentially consistent (see Figure 16-12). Although the milk of various animals has been used for yogurt production in various parts of the world, most of the industrialized yogurt production uses cow's milk. Whole milk, partially skimmed milk, skim milk, or cream can be used.

The starter culture for most yogurt production in North America is a blend of *Streptococcus salivarius thermophilus* (ST) and *Lactobacillus delbrueckii bulgaricus* (LB). Although both cultures can grow independently, the rate of acid production is much higher when they are used together. ST grows faster and produces both acid and carbon dioxide. These microorganisms are ultimately responsible for the formation of typical yogurt flavor and texture. The yogurt mixture coagulates during fermentation because of the drop in pH. The ST is responsible for the initial pH drop of the yogurt mix to approximately 5.0, and the LB are responsible for a further decrease to pH 4.0. The fermentation products of lactic acid, acetaldehyde, acetic acid, and diacetyl contribute to flavor.

The milk is clarified and separated into cream and skim milk, then standardized to achieve the desired fat content. The various ingredients are then blended together in a mix tank equipped with a powder funnel and an agitation system. The mixture is then pasteurized. Once the homogenized mix has cooled to an optimal growth temperature, the yogurt starter culture is added.

conditions are also similar to those for cultured buttermilk, but the fermentation is stopped at an acidity of 0.6%.

SCIENCE CONNECTION!

What is the difference between regular yogurt and Greek yogurt? How many different yogurt products are on the market today?

Other Fermented Milk Beverages. Cultured buttermilk was originally the fermented by-product of butter manufacture, but today it is more common to produce cultured buttermilks from skim or whole milk. Milk is usually heated to 203 °F (95 °C) and cooled to 68 °F to 77 °F (20 °C to 25 °C) before the starter culture is added. Once starter is added, the fermentation proceeds from 16 hours to 20 hours to an acidity of 0.9% lactic acid. This product is frequently used as an ingredient in the baking industry, in addition to being packaged for sale in the retail trade.

Acidophilus Milk. Acidophilus milk is a traditional milk fermented with *Lactobacillus acidophilus* (LA), which is believed to have therapeutic benefits in the human gastrointestinal tract. Skim or whole milk may be used. The milk is heated to a high temperature (203 °F or 95 °C) for 1 hour to reduce the microbial load and favor the slow-growing LA culture. Milk is inoculated with LA and incubated at 99 °F (37 °C) until coagulated. Some acidophilus milk has an acidity as high as 1% lactic acid, but for therapeutic purposes 0.6% to 0.7% is more common. Another variation has been the introduction of a sweet acidophilus milk, one in which the LA culture has been added but there has been no incubation.

Sour Cream. Cultured cream usually has a fat content between 12% and 30%, depending on the required properties. The starter is similar to that used for cultured buttermilk. The cream after standardization is usually heated to 167 °F to 176 °F (75 °C to 80 °C) and is homogenized to improve the texture. Inoculation and fermentation

Other Fermented Products. Many other fermented dairy products are on the market, including kefir, koumiss, beverages based on *bulgaricus* or *bifidus* strains, and labneh. Many of these have developed in regional areas and, depending on the starter organisms used, have various flavors, textures, and components from the fermentation process.

Ice Cream

Ice cream is greater than 10% milk fat by legal definition and as high as 16% fat in some premium ice creams. It typically contains 9% to 12% milk solids-not-fat (MSNF). This component, also known as the *serum solids*, contains the proteins (caseins and whey proteins) and carbohydrates (lactose) found in milk. Ice cream also contains 12% to 16% sweeteners, which are usually a combination of sucrose and glucose-based corn syrup sweeteners, and 0.2% to 0.5% stabilizers and emulsifiers. Finally, by weight, 55% to 64% of ice cream, whether mix or frozen, is water that comes from the milk or other ingredients (Figure 16-13). However, when frozen, approximately one-half of the volume of ice cream is air. All ice cream is made from a basic white mix.

The basic steps in the manufacturing of ice cream generally include the following:

1. Blending of the mix ingredients
2. Pasteurization
3. Homogenization
4. Aging the mix

FIGURE 16-13 Ice cream contains more than 10% milk fat and sweeteners in the form of sugars and corn syrup.

5. Freezing
6. Packaging
7. Hardening

Butterfat in ice cream increases the richness of flavor, produces a characteristic smooth texture by lubricating the palate, gives body to the ice cream, aids in good melting properties, and aids in lubricating the freezer barrel during manufacturing. Nonfat mixes are extremely hard on freezing equipment.

MSNFs contain the lactose, caseins, whey proteins, minerals, and ash content of the product from which they were derived. They are important ingredients because they improve the texture of the ice cream, help give body and chew resistance to the finished product, allow higher incorporation of air, and provide an inexpensive source of total solids.

Consumers like a sweet ice cream. As a result, sweetening agents are added to ice cream mix at a rate of usually 12% to 16% by weight. Sweeteners improve the texture and palatability of the ice cream and enhance flavors, and they are usually the cheapest source of total solids. In addition, the sugars, including the lactose from the milk components, contribute to a depressed freezing point so that the ice cream has some unfrozen water associated with it at their typical low serving temperatures, 5 °F to 0 °F (−15 °C to −18 °C). Without this unfrozen water, the ice cream would be too hard to scoop.

The stabilizers in ice cream are a group of compounds, usually polysaccharides, that are responsible for adding viscosity to the unfrozen portion of the water, thus holding this water so that it cannot migrate within the product. This produces an ice cream that is firmer. Without the stabilizers, the ice cream would rapidly become coarse and icy because of the migration of this free water and the growth of existing ice crystals.

The emulsifiers are a group of compounds in ice cream that aid in developing the appropriate fat structure and air distribution necessary for the desired smooth eating and good meltdown characteristics. Because each molecule of an emulsifier contains a hydrophilic portion and a lipophilic portion, they reside at the interface between fat and water. They act to reduce the interfacial tension or the force that exists between the two phases of the emulsion. The emulsifiers actually promote a destabilization of the fat emulsion, which leads to a smooth, dry product with good meltdown properties.

The original ice cream emulsifier was egg yolk, which was used in most of the original recipes. Today, two emulsifiers predominate most ice cream formulations: monoglycerides and diglycerides, which are derived from the partial hydrolysis of fats or oils of animal or vegetable origin.

QUALITY PRODUCTS

The USDA establishes U.S. grade standards to describe different grades of quality in butter, cheese (cheddar, Colby, Monterey, and Swiss), and instant

FIGURE 16-14 USDA grade shields commonly found on butter and cheese.

nonfat dry milk. The FDA established the Grade A designation for fluid milk products, yogurt, and cottage cheese. Manufacturers use the grade standards to identify levels of quality, provide a basis for establishing prices at wholesale, and supply consumers with a choice of quality levels.

The USDA also provides inspection and grading services on request and for a fee to manufacturers, wholesalers, and other distributors. Only products that are officially graded may carry the USDA grade shield.

The U.S. Grade AA or Grade A shield (Figure 16-14) is most commonly found on butter and sometimes on cheddar cheese. U.S. Extra Grade is the grade name for instant nonfat dry milk of high quality. Processors who use the USDA's grading and inspection service may use the official grade name or shield on the package. The Quality Approved shield may be used on other dairy products (such as cottage cheese) or other cheeses for which no official U.S. grade standards exist, as long as the products have been inspected for quality under USDA's grading and inspection program.

Milk available in stores today is usually pasteurized and homogenized. Little raw milk is sold today. Federal, state, and local laws or regulations control the composition, processing, and handling of milk. Federal laws apply when packaged or bottled milk is shipped interstate. Raw milk is prohibited from being sold interstate.

The FDA's Pasteurized Milk Ordinance requires that all packaged or bottled milk shipped interstate be pasteurized to protect consumers. Milk can be labeled Grade A if it meets FDA or state standards under the ordinance.

The Grade A rating designates wholesomeness or safety rather than a level of quality. According to the standards recommended in the ordinance, Grade A pasteurized milk must come from healthy cows and be produced, pasteurized, and handled under strict sanitary controls that are enforced by state and local milk-sanitation officials.

Once the consumer purchases milk and milk products, proper storage conditions and time maintain quality and safety. Guidelines of storage times for maintaining the quality of some products in the home refrigerator include the following:

- Fresh milk—5 days
- Buttermilk—10 to 30 days
- Condensed or evaporated milk—opened 4 to 5 days
- Half-and-half, light cream, and heavy cream—10 days
- Sour cream—2 to 4 weeks

MILK SUBSTITUTES

One well-known substitute for a milk product is margarine. It is made from vegetable fat. Other substitutes include frozen desserts, coffee whiteners, whipped toppings, and imitation milk (soy milk or almond milk). These substitutes are made by combining nondairy fats or oils with certain classes of milk components.

MATH CONNECTION!

How do milk and milk substitute products differ in price? Besides price, what would be another advantage of a milk substitute?

REDUCED FAT PRODUCTS

To reduce calories, saturated fat, and cholesterol in the diet, a wide variety of low-fat, reduced fat, and no-fat products have appeared on the market. Where fat is replaced in a milk product, the replacement must perform the same function as the original fat. In other words, fat replacements must give the product the same texture (mouthfeel). Fat substitutes have been made of proteins processed into miniscule particles and from carbohydrates that bind large amounts of water. New fat substitutes being developed and marketed are fats that are nonabsorbable or nondigestible. These do not contribute to the calories or fat intake of the individual. Refer to the section on "Fat Substitutes" in Chapter 22, Fats and Oils for more information.

SCIENCE CONNECTION!

How do reduced fat products compare to regular products? Create a lab to compare reduced fat and regular cheese. Compare the taste. Do they have the same melting qualities? Would you consume the reduced fat product? Why or why not?

MILK QUALITY EVALUATION AND FLAVOR DEFECTS

The following are flavor and sensation descriptions for seven different off flavors in fluid milk:

1. *Bitter* or *bitter aftertaste* describes a piercing, throbbing taste toward the back of the throat.
2. *Cooked* describes an eggy, sulfurous, or custardy taste.
3. *Feed* describes a grassy, stalky, or hay taste.
4. *Flat* or *watered down* describes a thin mouthfeel and less dairy fattiness taste.
5. *Foreign* describes a bleach-like or swimming pool smell.
6. *Light oxidized* describes a cardboard, pasty taste, or mouth-drying sensation with a wet brown paper towel odor.
7. *Metal* describes an oxidized metallic or penny coin taste and a tingling sensation at the back of the tongue.

FFA MILK QUALITY AND PRODUCTS CAREER DEVELOPMENT EVENT

The National FFA Organization (FFA) Milk Quality and Products Career Development Event requires participants to demonstrate knowledge in dairy science, including factors of quality production, processing, distribution, promotion, and marketing of milk and dairy products. Milk grading, sensory identification, and equipment identification and sanitation, as well as dairy products economics and pricing practicums will test FFA members' dairy foods knowledge. Knowledge of processed dairy products is essential for success in the Milk Quality and Products CDE. For more information and contest rules, visit the National FFA Organization's Web site for milk quality and products (https://www.ffa.org/participate/cdes/milk-quality-and-products).

SUMMARY

Milk provides high-quality protein, energy, vitamins, and minerals to human nutrition. Besides fluid milk, the dairy industry produces a variety of milk products, including butter, cheeses, condensed and dried products, and cultured products. The USDA and the FDA maintain quality standards. To protect consumers against pathogenic microorganisms in milk, it must be pasteurized. Butterfat globules in homogenized milk are reduced in size to prevent coalescence. Butterfat is also separated from milk and added back to produce beverage milk with specific fat contents or used in the production of butter and creams. Butter is produced by churning butterfat, and a by-product is buttermilk.

Concentrated or dried dairy products such as evaporated milk, sweetened condensed milk, condensed whey, milk powder, or whey powder increase shelf life and convenience and decrease transportation costs. Traditionally, cheese developed as a way to preserve the nutrients in milk. Many varieties of cheese have evolved over the centuries. Cheese production basically involves coagulating the milk and separating the whey using enzymes, acid, or heat. Yogurt is a fermented dairy product, as are acidophilus milk, sour cream, and kefir. Fermented dairy products require a starter culture for fermentation. By legal definition, ice cream contains 10% or more butterfat. It also relies on sweeteners at a level of 12% to 16%.

The USDA has established grade standards for butter, cheese, and instant nonfat dry milk. The FDA established grade designations for fluid milk, yogurt, and cottage cheese. Only officially graded products may carry the grade shield.

To meet new consumer demands, the food industry developed milk and milk product substitutes such as coffee whiteners, whipped toppings, imitation milk, and reduced fat products.

REVIEW QUESTIONS

Success in any career requires knowledge. Test your knowledge of this chapter by answering these questions or solving these problems.

1. Describe the process of pasteurization in detail.
2. Name four beverage milk products.
3. What dairy product is processed by churning pasteurized cream?
4. Define milk.
5. List the steps in cheese making.
6. Why is milk homogenized?
7. Name the three reasons manufacturers use the grade standards.
8. List the six areas of inspection that occur at the dairy farm for quality control of milk production.
9. After water, what are the components of milk solids?
10. Name three dried milk products.
11. Which bacteria aids in the production of cheese?
12. List the steps in making ice cream.

STUDENT ACTIVITIES

1. Track all the milk and milk products you consume in a week. Using Table B-8, develop a report or presentation on the amount of nutrition they provided.
2. Develop a report or presentation to compare butter to margarine. Use Table B-8 as a resource.
3. How is advertising used to encourage consumers to drink more milk? Find examples of current milk or milk product advertising campaigns and post these in the classroom.
4. Make butter, cheese, yogurt, or ice cream as an in-class project. These products are easy to make, and instructions can be found in many resources.
5. Develop a taste test to identify various types of cheese.
6. Taste and describe various fermented dairy products.
7. Develop a report or presentation on the competition that dairy products face from nondairy substitutes.
8. Compare the taste of a dairy product to that of a substitute such as margarine, dairy creamer, or soy milk.
9. Develop a report or presentation on the uses of whey.

ADDITIONAL RESOURCES

- Dairy PC Cleaning and Sanitizing Guidelines: http://www.dairypc.org/catalog/guidelines/cleaning-sanitizing
- Detecting and Correcting Off Flavors in Milk: https://www.ag.ndsu.edu/pubs/ansci/dairy/as1083.pdf
- USDA Economic Research Service—Dairy: http://www.ers.usda.gov/topics/animal-products/dairy.aspx
- National FFA Organization Milk Quality and Products: https://www.ffa.org/participate/cdes/milk-quality-and-products

Internet sites represent a vast resource of information. Although those provided in this chapter were vetted by industry experts, you may wish to further explore the topics discussed in this chapter using a search engine such as Google. Keywords or phrases may include the following: *milk, buttermilk, ice cream, names of specific cheese types or brands, milk industry, milk production, specific brand names of milk and milk products, pasteurization, quality control of milk, whey protein concentrates, homogenization, cheese making, milk grading, milk substitutes.* In addition, Table B-7 provides a listing of useful Internet sites that can be used as a starting point.

REFERENCES

Bartlett, J. 1996. *The cook's dictionary and culinary reference.* Chicago: Contemporary Books.

Ensminger, A. H., M. E. Ensminger, J. E. Konlande, and J. R. Robson. 1994. *Foods and nutrition encyclopedia* (2 vols.). Boca Raton, FL: CRC Press.

Horn, J., J. Fletcher, and A. Gooch. 1997. *Cooking A to Z: The complete culinary reference source.* Glen Ellen, CA: Cole Publishing Group.

Medved, E. 1973. *The world of food.* Lexington, MA: Ginn & Co.

Shaw, I.C. 2013. *Food Safety: The science of keeping food safe.* West Sussex, UK: Wiley-Blackwell.

Vaclavik, V. A., and E. W. Christina. 1999. *Essentials of food science*, 4th ed. New York: Springer.

ENDNOTES

1. Source: Adapted from "The Modern Way to Effective Milking," published by the Milking Machine Manufacturers Council of the Farm and Industrial Equipment Institute, 410 N. Michigan Avenue, Chicago, IL 60611-4251.
2. Adapted from D. K. Bandler et al., *Milk Flavor Handbook*, Cornell, Rutgers, and Pennsylvania State University, 1976, with revisions by F. W. Bodyfelt, Extension Dairy Processing Specialist, Oregon State University, 1985.

CHAPTER 17

Meat

OBJECTIVES

After reading this chapter, you should be able to:

- Name the types of red meat
- Identify meat products: wholesale or primal cuts and retail cuts from cattle, sheep, and pigs
- Explain the difference between wholesale or primal and retail cuts
- Compare quality grade to yield grades
- Define dressing percentage
- Identify the government agencies involve in meat production
- Describe the general composition of meat and meat products
- List five factors affecting meat tenderness
- Describe the types of cooking methods that can be used on meats
- Discuss the production of meat substitutes

NATIONAL AFNR STANDARD

FPP.01

Develop and implement procedures to ensure safety, sanitation, and quality in food products and processing facilities.

KEY TERMS

aging
albumen
bromelin
by-products
cold shortening
curing
cutability
deboning
electrical stimulation
eviscerated
ficin
marbling
marinating
myoglobin
offal
palatability
papain
primal cuts
processed meats
quality grades

KEY TERMS (continued)

retail cuts
rigor mortis
smoking
textured protein
wholesale cuts
yield grade

In general, meat is the tissues of an animal body used for food. A wide definition includes cattle, sheep, pigs, poultry, fish, and shellfish. This chapter deals with red meats, including:

- Beef from cattle
- Veal from calves 3 months or younger
- Pork from swine, pigs, or hogs
- Mutton from mature sheep
- Lamb from young sheep
- Chevon or goat meat

The first meat packers in the United States were colonial New England farmers who packed meats in salt to preserve them. In the 19th century, the beef industry came into being, moving from large metropolitan areas to be near commercial feedlots in the central United States in such states as Texas, Oklahoma, Kansas, and Nebraska. The pork industry remains centrally located in the Midwest, principally in Iowa, Illinois, Minnesota, Michigan, and Nebraska, but it is making a move west and southeast. Rapid growth and vertical integration characterize the poultry industry.

MEAT AND MEAT PRODUCTS

Livestock slaughter occurs in federally inspected plants that do most all of the processing in the United States. A few large packers dominate the industry. Approximately 62% of all beef is consumed as beef cuts, 24% is ground for hamburger, and 14% is processed into meat products. In the case of pork, more than 65% is consumed in the form of processed meat such as ham, bacon, and sausage. In addition, the meatpacking industry produces such valuable **by-products** as cosmetics, glues, gelatins, tallow, variety meats, and meat and bonemeal.

Livestock marketing and prices are affected by weather, feed prices, federal import policies, and consumer demands.

Traditionally, meat packers sold carcasses as sides and quarters (hinds or fores) or as wholesale or **primal cuts** (large cuts such as entire rounds, loins, ribs, or chucks). Most beef today is sold as *boxed beef*. Boxed beef is prepared at a packing plant by removing more of the bone and fat from the carcass as it is cut into smaller portions, vacuum-packed to reduce spoilage and shrinkage, and placed into boxes that are easier to ship and handle than quarters (Figure 17-1). Boxed beef reduces shipping costs and labor costs, and increases the value of the fat and bone for the packer. Some large packers prepare consumer-ready meat cuts such as precut steaks and roasts in vacuum packages that are placed directly in supermarket meat cases.

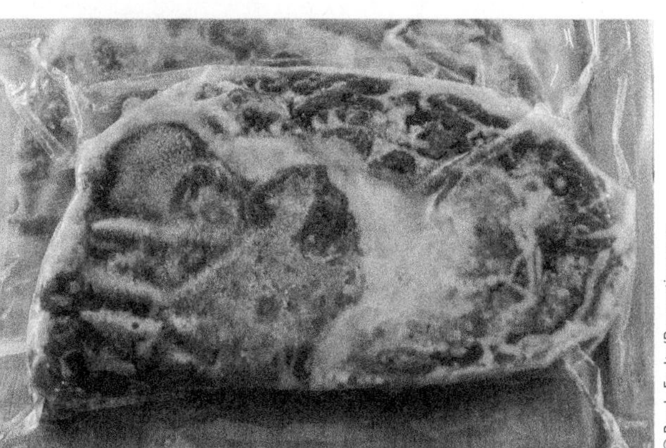

FIGURE 17-1 Frozen, vacuum-sealed beef rib-eye steaks.

Government Oversight

Inspection takes place at practically every step of the livestock procurement and meatpacking processes. Inspection is intended to ensure that harmful additives and ingredients are kept out of manufactured meat products, that sick and diseased animals are excluded from the market, that misleading labeling and packaging are eliminated, and that contaminated and unwholesome meats are stopped from reaching consumers.

Federal meat inspection was authorized by the federal Meat Inspection Act (1906) and is administered by the Food Safety and Inspection Service of the U.S. Department of Agriculture (USDA; see http://www.fsis.usda.gov/wps/portal/fsis/topics/inspection/mpi-directory). Meat that is to be used entirely within a given state may be inspected only by that state's department of agriculture. All meats entering interstate commerce must be federally inspected.

Grading

Unlike inspection, which is mandatory, meat grading is a service offered to packers on a voluntary basis by the Agricultural and Marketing Service (AMS) of the USDA. Grading is operated on a self-supporting basis and is funded from fees paid by the users. Grading establishes and maintains uniform trading standards and aids in the determination of the value of various cuts of meat.

Carcasses are given both a quality and a **yield grade** (Figures 17-2 and 17-3). **Quality grades** for beef carcasses are Prime, Choice, Good, Standard, Commercial, Utility, Cutter, and Canner and are assigned in terms of carcass characteristics associated with palatability. These grades are assigned on the basis of carcass **marbling** (fat flecks or streaks within the lean), color and texture of the lean, and maturity, which is determined according to the color, size, and texture of the cartilage and bones. Although grading was not originally intended to provide estimates of **palatability** (taste and tenderness) for consumers, it has become a consumer rating for beef.

Five yield grades are applied to all classes of beef, denoted by 1 through 5, with yield grade 1 representing the highest degree of **cutability**—that is,

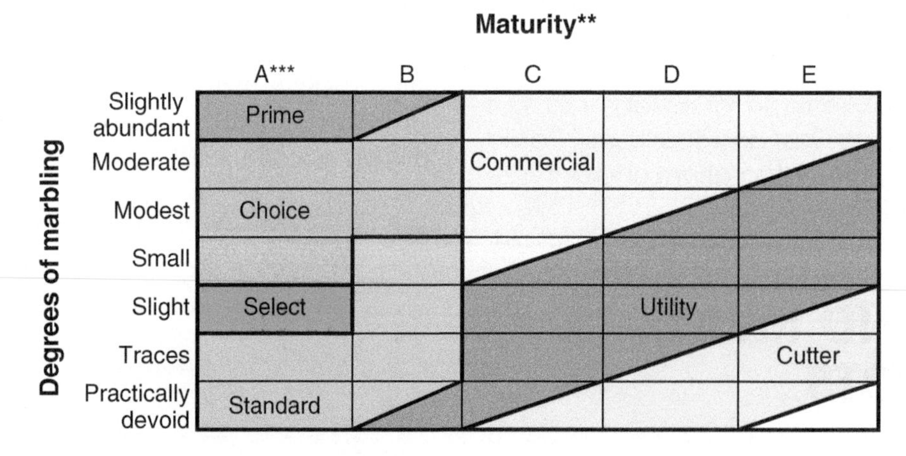

FIGURE 17-2 Graphical representation of the USDA quality grades for beef.

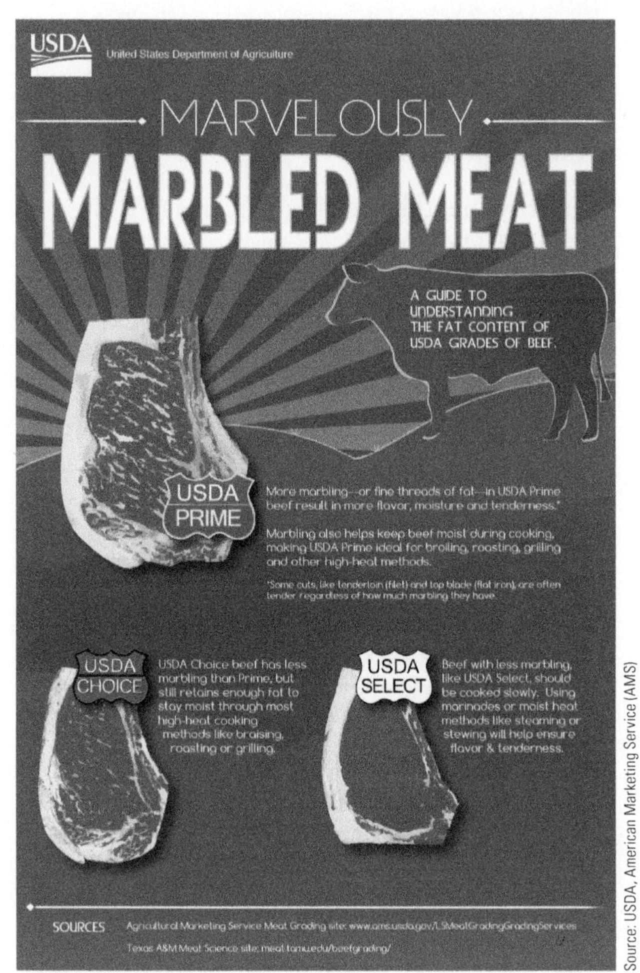

Source: USDA, American Marketing Service (AMS)

FIGURE 17-3 Degree of marbling is the primary difference between the quality grades of prime, choice, and select.

the percentage of closely trimmed, boneless, retail cuts from the round, loin, rib, and chuck, which are the four major beef wholesale cuts.

Carcasses below the choice grade have rarely been stamped with a grade because they were thought to be less palatable. The belief held by some consumers that leaner meats (those with a lower fat content) are more healthful has led to an increased demand for the select grade. Yield grades classify carcasses on the basis of the proportion of usable meat to bone and fat and are used in conjunction with quality grades to determine the monetary value of a carcass based on two general considerations: (1) quality, which includes characteristics of the lean and fat; and (2) the expected yield of the four lean cuts (ham, loin, picnic shoulder, and Boston butt).

The standards for grades of hog carcasses are based on (1) differences in yields of lean cuts and fat cuts and (2) differences in the quality of cuts.

Lamb falls into four classifications: Prime, Choice, Good, and Utility. Carcasses are evaluated by muscle tone, fat streaking, size of the flat rib bone, and firmness.

The USDA AMS Web site for quality and yield grade standards is http://www.ams.usda.gov/grades-standards.

Grading Formulas

The following are specifications for official United States standards of grades for beef, pork, and lamb.

Beef. The yield grade of a beef carcass is determined on the basis of the following formulas. Yield grade: 2.50 + (2.50 × adjusted fat thickness, inches) + (0.20 × percent kidney, pelvic, and heart fat) + (0.0038 × hot carcass weight, pounds − 0.32 × area ribeye, square inches).

Pork. The grade of a barrow or gilt carcass is determined on the basis of the following equation: Carcass grade = (4.0 × backfat thickness over the last rib, inches) (1.0 × muscling score). To apply this equation, muscling should be scored as follows: thin muscling = 1, average muscling = 2, and thick muscling = 3. Carcasses with thin muscling cannot be graded U.S. No. 1.

Lamb. The yield grade of lamb is determined on the basis of the adjusted fat thickness over the ribeye muscle between the 12th and 13th ribs. The adjusted fat-thickness range for each yield grade is as follows: yield grade 1—0.00 to 0.15 inch, yield grade 2—0.16 to 0.25 inch, yield grade 3—0.26 to 0.35 inch, yield grade 4—0.36 to 0.45 inch, and yield grade 5—0.46 inch and greater.

Value-Based Beef Pricing

Various methods are used to price beef cattle, but two of the most common methods are *showlist pricing* and *carcass merit pricing* (also called *grid pricing*). When pricing an entire show list of market-ready cattle at one price, profit on an individual pen of cattle can be calculated as:

$$\text{Profit (showlist)} = \text{Dressed price} \times \text{Dressed weight} - \text{Feeding costs} - \text{Feeder price} \times \text{Feeder weight} \quad (1)$$

Using this method each variable is the average for the pen. The dressed price is a function of the overall supply and demand forces determining the general market level, but it does not account for the carcass characteristics of the cattle.

If cattle are sold on a carcass merit, value-based pricing system, then profit on an individual pen of cattle can be calculated as:

$$\text{Profit (grid)} = \text{Grid price (carcass characteristics)} \times \text{Dressed weight} - \text{Feeding costs} - \text{Feeder price} \times \text{Feeder weight} \quad (2)$$

The grid or value-based price accounts for the carcass characteristics for that pen of cattle and gives value to those characteristics. The grid price is still a function of the general market level and is determined by the same supply-and-demand forces as the average dressed price.

Dressing Percentage

Dressing percentage is the percentage yield of chilled carcass in relation to the weight of the live animal.

MATH CONNECTION!

What effect do yield grades and quality grades have on the economic value of the animals sold? What returns do farmers and ranchers see based on these grading differences?

In beef cattle, the dressing percentage depends on the quality grade of the animal:

- Prime—62%
- Choice—60%
- Select—59%
- Standard—57%

For example, a 1,000-pound choice steer would produce a 600-pound carcass. (The head, feet, hide, internal organs, and some fat trim make up the rest of the 400 pounds.)

For swine, the dressing percentage is the percentage yield of chilled carcass:

- U.S. No. 1—70%
- U.S. No. 2—71%
- U.S. No. 3—72%
- U.S. No. 4—73%
- Utility—69%

Grades with poorer yields have higher percentages because fatter hogs produce heavier carcasses in relation to their live weight. In addition, hogs may be dressed in two ways: (1) *packer style* in which the head, kidneys, and leaf fat are removed; and *shipper style* in which the head, kidneys, and leaf fat are left on the carcass. These hogs dress from 4% to 8% higher because there is more weight.

The dressing percentage of lamb is the percentage yield of the chilled carcass:

- Prime—52%
- Choice—50%
- Select—47%
- Utility—45%

Usually lambs are sheared before slaughtering.

Slaughtering Practices

The Humane Slaughter Act (1960) requires that before being slaughtered animals must be rendered completely unconscious with a minimum of excitement and discomfort by mechanical, electrical, or chemical (carbon dioxide gas) methods.

After being bled, skinned, and **eviscerated** (internal organs removed), carcasses are chilled for 24 hours to 48 hours before being graded and processed. Meat items such as brains, kidneys, sweetbreads (calf thymus glands), the tail, and the tongue do not accompany a carcass and are considered by-products to be sold separately as specialty items. These parts, and all other items removed from the carcass, such as feet, hide, and intestines, are called **offal** and are an important source of income for meat packers.

Wholesale and Retail Cuts of Meat

After slaughter, carcasses are cut into wholesale cut or retail cuts for use by customers. **Wholesale cuts** are larger primal cuts of meat that are shipped to grocery stores and meat markets. **Retail cuts** are family-sized or single-serving cuts purchased at the market. These are described in Tables 17-1, 17-2 and 17-3.

UNINTENDED CONSEQUENCES

The Jungle is a novel written in 1906 by American journalist and novelist Upton Sinclair (1878–1968). Sinclair wrote the novel to show the lives of immigrants in Chicago and similar industrialized American cities. The book depicts working-class poverty, the absence of social programs, harsh and unpleasant living and working conditions, and a hopelessness among many workers. Sinclair had spent 7 weeks gathering information for his newspaper while working incognito in the meatpacking plants of the Chicago stockyards. The resulting book was published by Doubleday on February 26, 1906.

The public misunderstood the point of Sinclair's book. Many readers were most concerned with his exposure of health violations and unsanitary practices in the U.S. meatpacking industry during the early 20th century. Additional investigations and public pressure led to the passage of the Meat Inspection Act and the Pure Food and Drug Act of 1906 and the establishment of the Bureau of Chemistry, which was renamed the Food and Drug Administration in 1930.

The Jungle is still a good read and reveals the social conditions of the time and the changes that were brought to an industry that most of us depend on to produce a wholesome and safe product.

TABLE 17-1 Primal Wholesale and Retail Cuts of Beef

PRIMAL WHOLESALE	RETAIL
Chuck	Ground beef, stew meat, pot roast, blade roast, short ribs, boneless chuck roast
Rib	Rib-eye roast or steak
Short loin	Top loin steak, T-bone steak, porterhouse steak, filet mignon
Sirloin	Sirloin steaks or roast
Round	Eye of round, ground beef, cubed steak, rump roast, top round steak, round steak
Fore shank	Stew meat, shank crosscuts
Brisket	Brisket (corned beef if cured)
Short plate	Short ribs, ground beef, stew meat
Flank	Ground beef, flank steak

TABLE 17-2 Primal Wholesale and Retail Cuts of Pork

PRIMAL WHOLESALE	RETAIL
Jowl	Smoked jowl
Picnic shoulder	Fresh arm roast, sausage, neck bones, fresh or smoked picnic shoulder
Belly	Spare ribs, salt pork, bacon
Leg	Fresh leg or ham
Loin	Center loin roast, sirloin roast, tenderloin, back ribs, blade chops, loin chops, Canadian-style bacon
Boston butt	Blade roast, blade steak

TABLE 17-3 Primal Wholesale and Retail Cuts of Lamb

PRIMAL WHOLESALE	RETAIL
Shoulder	Neck slices, arm chops, blade chops, stew meat, shoulder roast
Rib	Rib roast, rib chops
Loin	Loin roast, loin chops, sirloin chops
Leg	Leg (many styles of preparation—French style, American style, boneless)
Breast	Riblets, breast
Shank	Fore and hind shanks

Knowledge of the various wholesale primal and retail cuts of meat is necessary in preparation for the Meats Evaluation and Technology FFA Career Development Event and 4H contests. While preparing for the contest, students can look at photographs and diagrams on recommended Web sites to practice identifying retail cuts as well as work with local butchers to practice identification of real meat cuts. Retail cuts of beef and pork are illustrated in Figures 17-4 and 17-5. For focused practice and preparation for meat identification, refer to the following Web site: http://aggiemeat.tamu.edu/4-hffa-retail-cut-id-test-2-2012/

SCIENCE CONNECTION!

Meat can be cooked in many ways: grilled, pan seared, oven roasted, fried, and baked are just a few. What determines the best way to cook each cut of meat? Research different cuts of meat and find recommended cooking methods. Why?

Beef Cuts AND RECOMMENDED COOKING METHODS

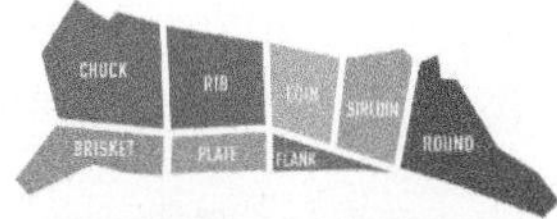

Courtesy of the Beef Checkoff Program

FIGURE 17-4 Retail cuts of beef.

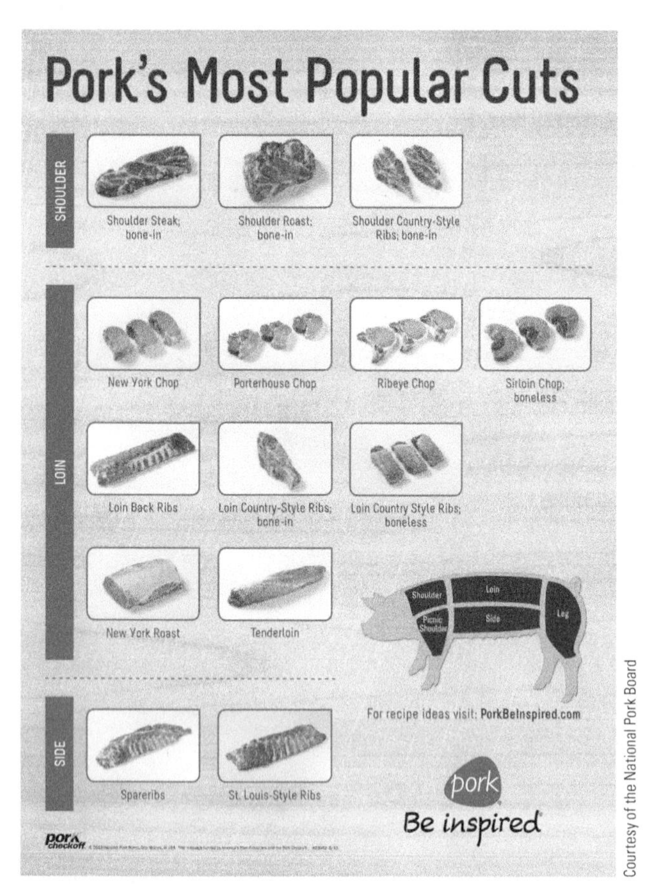

FIGURE 17-5 Retail cuts of pork.

source that contains, in favorable quantities, every essential amino acid. Proteins from plant sources are frequently deficient in one or more of these amino acids. Overall, protein use by the body is more efficient when complete proteins are consumed.

Meats, including processed meat products, are an excellent source of iron, which is more biologically available than iron from plant sources or that added through fortification. The iron in meat also improves the absorption and utilization of iron from other sources.

Fat contributes to a product's juiciness, tenderness, and flavor of meat and all **processed meats**. In addition, fat reduces formulation costs. The fat content of processed meat products is regulated by the USDA. For example, products such as hot dogs cannot contain more than 30% fat. Some specialty loaf items, however, may contain more than 30% fat, although many processed products contain considerably less fat. The meat industry is responding to consumer demands for leaner products by providing new reduced fat options.

MATH CONNECTION!

Why do different cuts of meat cost have different prices at the grocery store? What are your theories? Research your ideas and record your findings.

Structure and Composition of Meat

The word *meat* generally refers to the skeletal muscle from the carcasses of animals—beef and veal (cattle), pork (hogs), and lamb (sheep). The general composition of such meat is approximately 70% water, 21% protein, 8% fat, and 1% ash (mineral). For the specific composition of meats, refer to Table B-8.

Meat and processed meat products, and other foods of animal origin, provide a complete protein

Meat Products Formulations

Various formulations and methods are used to mix ingredients and develop meat products. Understanding these methods is necessary in any processing facility or food-service business that formulates its own products. One example of a meat products formulation method is the *least cost ingredient formulation*.

In formulating any meat product, different combinations of meat ingredients can be used to achieve a specific lean-to-fat percentage in a ground product. Aside from specialty value-added products, deciding ingredient proportions is done on the basis of cost. Usually, for complex combinations where more than two ingredients are being considered, a computer software is used. In simple situations, a method called *Pearson's square* can be used to achieve the correct proportions.

The following example uses Pearson's square calculation to determine the amount of each product needed to process regular, 25% fat ground beef. Subtract the desired fat percentage (middle of square) from the fat percentages of the meats on left of square across the diagonal (top left − middle = bottom right; bottom left − middle = top right). Repeat this for both meats. Make any negative numbers on the right side of the square positive. The answers on the right side of the square are the parts of each meat to include in the ground beef mixture After subtracting across the diagonal, sum the parts of the two meats to get the total. Then, divide each part by the sum of the parts to calculate the percentage of each meat in the in the ground beef mixture to get 25% fat.

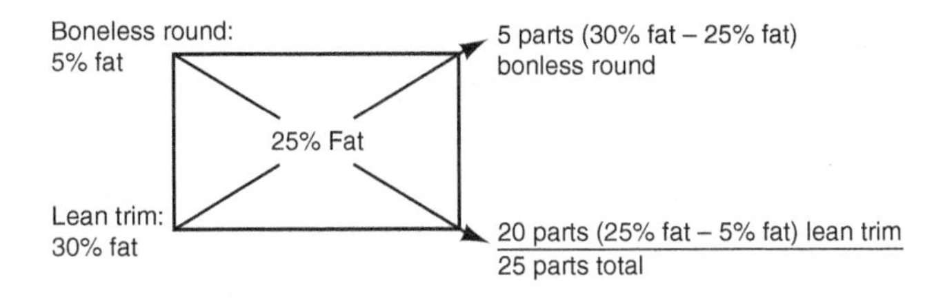

Pearson's Square calculations:

1. Subtract across the diagonal:

   25% − 30% = 5 parts boneless trim

   25% − 5% = 20 parts lean trim
2. Sum the parts:

   5 parts boneless round + 20 parts lean trim = 25 total parts
3. Divide each part by the total to calculate the percentage of each meat to include in the ground beef mixture:

   5/25 × 100 = 20% boneless round

   20/25 × 100 = 80% lean trim
4. To make 485 lbs of a ground beef that is 25% fat you will need:

   .2 × 485 = 97 lbs of the 5% fat boneless round

   .8 × 485 = 388 lbs of the 30% fat lean trim
5. To check 97 + 388 = 485

Once the total amount of each product needed is determined, the current market price of each meat product can be multiplied by the amount needed to determine the product's final total cost.

Chilling

Immediately after slaughter, many changes take place in muscle that converts that muscle to meat. One change is the contraction and stiffening of muscle known as **rigor mortis**. Muscle is extremely tender at the time of slaughter. However, as rigor mortis begins, it becomes progressively less tender and more rigid until rigor mortis is complete. In the case of beef, 6 to 12 hours are required for rigor mortis to be complete. Pork requires only 1 to 6 hours to complete rigor. The carcass is chilled immediately after slaughter to prevent spoilage.

If a carcass is chilled too rapidly, however, the result is **cold shortening** and subsequent toughness. Cold shortening occurs when the muscle is chilled to less than 60 °F (16 °C) before rigor mortis has been completed. If the carcass is frozen before rigor is complete, the result is "thaw rigor" and extremely tough meat.

Aging of Meat

After rigor mortis has passed, changes in the meat result in beef that becomes progressively more tender. Holding the beef in a cooler or refrigerator is commonly referred to as the **aging** period. The increase in tenderness is the result of natural enzymatic changes taking place in the muscle. The increase in tenderness continues only for 7 to 10 days after slaughter when the beef is held at approximately 35 °F (2 °C). Beef held at higher temperatures will tenderize more rapidly, but it also may spoil and develop off flavors. Lamb and pork are rarely aged.

A lack of tenderness usually is not encountered because of the animals' relatively young age at slaughter.

Tenderizing

Tenderness, juiciness, and flavor are components of meat palatability. Although juiciness and flavor normally do not vary a great deal, tenderness can vary considerably from one cut to the next. The principal causes of variation in the tenderness of beef, pork, lamb, and veal cuts are genetics, species and age, feeding, muscle type, suspension of the carcass, **electrical stimulation** (an electric current transmitted through the carcass of freshly slaughtered and eviscerated animals), chilling rate, aging, mechanical tenderizing, chemical tenderizing, freezing and thawing, cooking and carving.

Genetics. About 45% of the observed variation in the tenderness of cooked beef is determined by genetics or the parents of the animal from which the beef came. Although many other factors are involved in tenderness, genetics is one main reason there are such wide differences in tenderness among identical grades and cuts of meat.

Species and Age. Beef usually is the most variable in tenderness followed by lamb, pork, and veal. The tenderness variation from species to species is primarily the result of an animal's age at slaughter. Beef normally is processed at approximately 20 months of age, lamb at 8 months, pork at 5 months, and veal at approximately 2 months of age.

Within a given species such as beef, an animal's age at slaughter also influences the meat's tenderness. Beef normally is slaughtered between 9 and 30 months of age. Usually, the meat from these animals is fairly tender. However, meat becomes progressively less tender as the animal gets older. The decrease in tenderness with increasing age is due to the changing nature of collagen (gristle), the connective tissue protein found in meat. Pork and lamb from older animals normally are processed into sausage items, so age-related toughness is not usually a problem.

Feeding. What an animal is fed does not directly influence tenderness. In the case of beef, feeding can have an indirect effect on tenderness. Animals that are finished with grain tend to reach a given slaughter weight sooner than animals that are finished to the same slaughter weight on pasture. Thus, grain-fed animals usually are slightly more tender because they are slaughtered at a younger age.

Muscle to Muscle. Within any species, considerable variation exists in tenderness among muscles. For example, tenderloin is much more tender than the fore shank or heel of round in beef. This difference results from the amount of connective tissue in the various cuts. The tenderloin usually has a small amount of connective tissue compared with the fore shank or heel of round. How much connective tissue is present is a function of muscle use in the live animal (see Figure 17-6).

Muscle Identification

A beef carcass has more than 100 different muscles. These muscles have properties that affect the processing characteristics and how well consumers accept the resulting products. Today the majority of the cuts found in retail meat counters are boneless; this means that those who process carcasses understand the musculature structure of the beef animal. The *institutional meat purchaser's specifications* (IMPSs) dictate the expectations of meat processing, so knowing the muscular anatomy of the beef carcass is required for consistency in cutting procedures.

Recent research has profiled the physical and chemical characteristics of beef muscles to more fully realize their value. This information will aid processors in developing and preparing new products based on the inherent properties of each muscle. For a detailed description of muscles and their characteristics, check out Major Muscles of the Carcass (http://www.aps.uoguelph.ca/~swatland/ch4_1.htm).

Table 17-4 lists the common names for beef muscles and the resulting common name associated with specific cuts of meat.

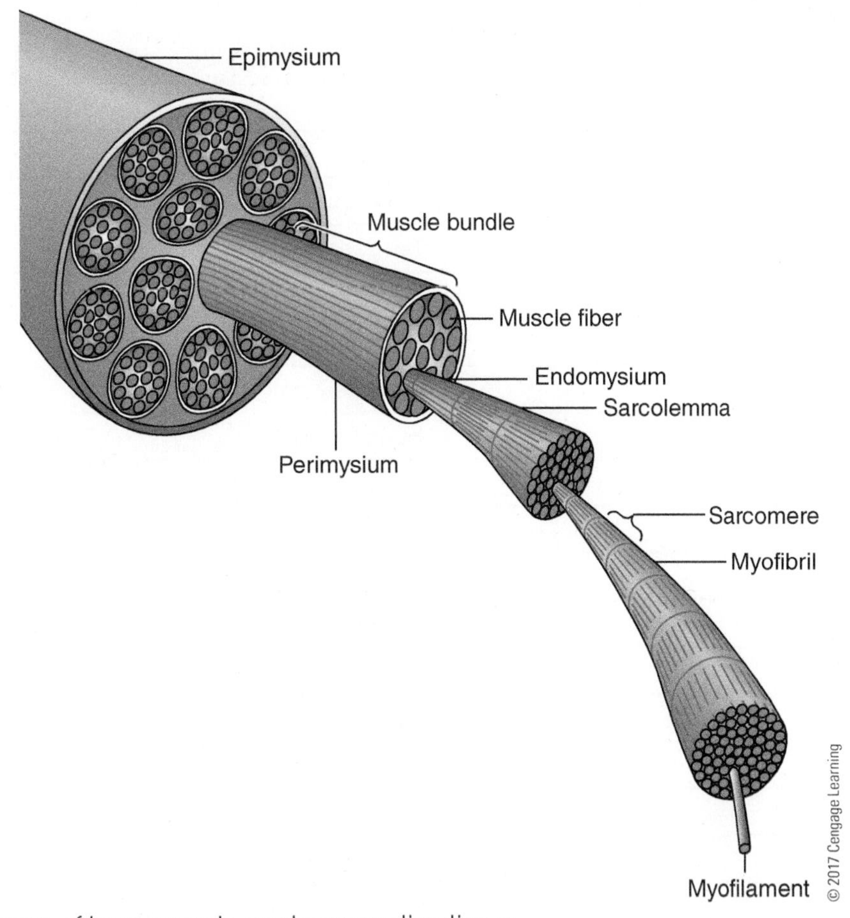

FIGURE 17-6 Diagram of lean muscle and connective tissue.

TABLE 17-4 Common Name for Beef Muscles

MUSCLE	COMMON NAME
Adductor	Top (inside) round
Biceps femoris	Bottom (outside) round
Cutaneous-omo brachialis	Shoulder rose
Deep pectoral	Brisket
Deltoideus	Outside chuck (chuck)
Gastrocnemius	Round heel
Gluteus medius	Top sirloin
Gracilis	Inside round cap
Infraspinatus	Top blade, flat iron, triangle
Longissimus dorsi	Rib eye, loin eye
Longissimus dorsi (chuck)	Chuck eye
Longissimus lumborum	Loin eye

(Continues)

TABLE 17-4 Common Name for Beef Muscles

MUSCLE	COMMON NAME
Longissimus thoracis	Rib eye
Multifidus dorsi	Sub eye
Obliquus internus abdominus	Sirloin butt flap
Psoas major	Tenderloin
Quadriceps femoris	Knuckle, sirloin tip
Rectus abdominus	Flank
Rectus femoris	Knuckle center
Rhomboideus	Hump meat
Semimembranosus	Top (inside) round
Semitendinosus	Eye of round
Serratus ventralis	Boneless short ribs, inside chuck, underblade
Spinalis dorsi	Rib cap
Superficial pectoral	Brisket
Supraspinatus	Mock tender; Chuck tender; Scotch tender
Tensor fascia latae	Tri-tip
Teres major	Shoulder tender, petite tender
Trapezius	Outside chuck
Triceps brachii	Clod heart, shoulder center, shoulder top, ranch cut
Vastus lateralis	Knuckle side

Source: *Ranking of Beef Muscles for Tenderness Fact Sheet*. Written by Chris Calkins and Gary Sullivan, University of Nebraska. Funded by The Beef Checkoff. Available at www.beefresearch.org

Suspending the Carcass. Stretching muscles during chilling of the carcass affects their tenderness; the amount depends on the muscles' anatomical location in the carcass. Although most carcasses are hung from the hind leg, a new method of hanging the carcass from the pelvic or hip bone changes the tension applied to some muscles.

Electrical Stimulation. Electrical stimulation of the hot carcass immediately after slaughter increases muscle tenderness. Approximately 1 minute of high-voltage electrical current through the carcass improves the tenderness of many of the resulting cuts.

Chilling Rate. If the carcass is chilled too rapidly, the result is cold shortening and subsequent toughness. Cold shortening occurs when the muscle is chilled to less than 60 °F (16 °C) before rigor mortis has passed.

Quality Grade. An animal's age also plays a major role in tenderness when it comes to quality grading in beef. The quality grades of USDA Prime, Choice, Select, Standard, Utility, and Commercial are applied to the carcasses of young animals.

Mechanical. Grinding is commonly used to increase meat tenderness, especially with beef (Figure 17-7). The popularity of hamburger and ground beef is the result of their more uniform texture and tenderness compared to steaks and roasts. Cubing is another way to mechanically tenderize meat. The small blades of a cuber simply sever connective tissue in boneless retail cuts so that the connective tissue is broken into smaller pieces.

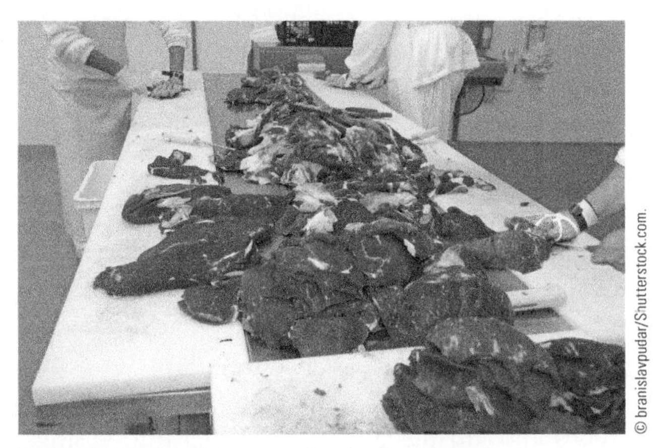
© branislavpudar/Shutterstock.com.

FIGURE 17-7 Mixing and grinding beef.

Chemical. At certain concentrations, salt increases meat tenderness. The presence of salt is one reason why cured meats such as ham are more tender than uncured meats. Salt exerts its influence on tenderness by softening collagen, the protein of connective tissues, into a more tender form.

Vegetable enzymes such as **papain** (from papaya), **bromelin** (from pineapple), and **ficin** (from figs) tenderize meat both commercially and in the home kitchen (refer to Table 8-2). These tenderizers dissolve or degrade the connective tissues collagen and elastin. The limitation of vegetable enzymes is that their tenderizing action is often restricted to the meat's surface.

Marinating. Consumers can improve tenderness and add taste variety to the meat component of meals by **marinating**. The basic ingredients of a marinade include salt (or soy sauce), acid (vinegar, lemon, Italian salad dressing, or soy sauce), and enzymes (papain, bromelin, ficin, or fresh ginger root). Some marinade recipes call for the addition of an alcohol (wine or brandy) for enhanced flavor. The addition of several tablespoonfuls of olive oil will seal the surfaces from the air and thus result in the meat staying fresher and brighter in color for a longer period of time.

The tenderizing action of marinades occurs through the softening of collagen by the salt, increased water uptake, and hydrolysis and breakage of the cross-links of the connective tissue by the acids and alcohols.

Freezing. Freezing rate plays a small role in tenderness. When meat is frozen rapidly, small ice crystals form; when meat is frozen slowly, large ice crystals form. Large ice crystals may disrupt components of the muscle fibers in meat and increase tenderness slightly, but they also increase the loss of juices when the meat thaws. This means the meat is less juicy at cooking and usually less tender.

Thawing. Thawing meat slowly in the refrigerator generally results in greater tenderness compared with cooking from the frozen state. Slow thawing minimizes the toughening effect from cold shortening (when present) and reduces the amount of moisture loss. Thawing in a microwave should be done on a lower power setting or by manually alternating cooking and standing times.

Cooking. As cooking progresses, the contractile proteins in meat become less tender, and the major connective tissue protein (collagen) becomes more tender.

Carving. Muscles, muscle bundles, and muscle fibers are all surrounded by connective tissue. When cuts are made from carcasses and wholesale cuts, the normal procedure is to cut at right angles to the length of the muscle. This procedure severs the maximum amount of connective tissue and distributes the bone more evenly among all cuts in that area. Likewise, consumers should carve cooked meat at right angles to the length of the muscle fibers or "against the grain" to achieve maximum tenderness. Cutting with the grain results in stringy pieces that are less tender.

Curing

Curing meat was used as one form of preservation. Now it is used more to enhance flavor and color (Figure 17-8). Curing agents include salt, sodium nitrate, sodium nitrite, sugar, and spices. Salt is added to preserve the meat and add flavor. Sodium nitrate and sodium nitrite fix the red color of

© Andrei Stancu/Dreamstime.com

FIGURE 17-8 A variety of cured meats in a butcher shop.

meat, act as a preservative, and prevent botulism. Sugar also provides color stability and flavor, and spices obviously produce desired flavors.

Color

The primary color pigment of meat is a protein called **myoglobin**. Its function is to store oxygen in the muscle tissue. When oxygen is present, meat is a bright red color. When oxygen is absent the meat is a purplish color. Myoglobin is denatured by cooking and by prolonged exposure to air, making the meat brown and less attractive.

Smoking

Like many other meat-processing practices, **smoking** has been practiced since the beginning of recorded history. The highly smoked products of the past have largely given way to milder smoking methods that have reduced but not eliminated the effectiveness of smoke as an inhibitor of microbial growth. As a microbial growth inhibitor, smoke is most effective when combined with other preservation techniques. Smoke also protects fat from rancidity, contributes to characteristic colors, and creates unique flavors in processed meats.

Liquid smoke, used widely in industry, not only avoids many of the questionable compounds found in wood smoke but also eliminates virtually all the emissions associated with burning wood or sawdust.

Meat Specialties

Dry sausages may or may not be characterized by a bacterial fermentation. For fermented items, the intentional encouragement of lactic acid bacterial growth produces a typical tangy flavor and preserves the meat. The ingredients are mixed with spices and curing materials, stuffed into casings, and put through a carefully controlled, long, continuous air-drying process. Dry sausages include salami or pepperoni.

Semidry sausages are usually heated in the smokehouse to fully cook the product and partially dry it. Semidry sausages are semisoft sausages with good keeping qualities because of their lactic acid fermentation. *Summer sausage* (another name for *cervelat*) is the general classification for mildly seasoned, smoked, semidry sausages such as mortadella and Lebanon bologna.

Most consumers purchase their meat and poultry from retail stores. Other sources include retailing at farms, auction markets, direct sales, and cooperatives and sometimes even door-to-door vendors. Consumers must always know something about a dealer or company before making a decision they might later regret.

Freezing

When properly wrapped, fresh meat cuts can be frozen and held in storage at 0 °F (−18 °C) or less for months if the cut is a fatty meat like pork. Beef can be held for years. Storage time for pork and fatty meats is limited because fat gradually oxidizes at freezer temperatures, producing off flavors. Once frozen, meat should not be thawed and refrozen. Few cured meats or sausages are frozen because the salt in their formulations increases the rate of the development of rancid flavors. The flavor of spices used in sausage also may change during frozen storage.

Storage

Storage times for frozen meat and refrigerator storage are given in Tables 10-1 and 8-5.

Cooking

For cuts that have less connective tissue—such as steaks and chops from the rib and loin—the recommended cooking method is dry heat, including pan frying, broiling, roasting, grilling, and barbecuing (Figure 17-9). Dry heat raises temperatures rapidly, and the flavor of a meat will develop before the contractile proteins become significantly less tender. For cuts with more connective tissue—such as those from the fore shank, heel of round, and chuck—the recommended cooking method is long and slow at low temperatures using moist heat such as braising. The application of moist heat for a long time at low temperatures [275 °F to 325 °F (135 °C to 163°C)] results in the conversion of tough collagen into tender gelatin and makes this type of cut more tender than dry heat cooking. Degree of doneness also significantly affects tenderness.

As lean meat heats, the contractile proteins (muscle proper) toughen and lose moisture, decreasing tenderness. Tender cuts of meat cooked to a rare degree of doneness [140 °F (60 °C)] are more tender than when cooked to medium [155 °F (68 °C)], and medium is more tender than well done [170 °F (77 °C])]. Degree of doneness is especially important in the case of beef.

Pork is cooked to an internal temperature of approximately 160 °F or 170 °F (71 °C to 77 °C) for desirable flavor. Although this temperature range corresponds to well done in beef, pork still may be slightly pink. Because *Trichinella spiralis* (trichinosis) is destroyed at 137 °F (58 °C), reaching an internal temperature of 160 °F to 170 °F (71 °C to 77 °C) will result in pork that is definitely safe to eat. More cooking will result in dehydration, loss of juices, and unnecessary toughening. Lamb is usually cooked to well done [approximately 160 °F to 170 °F (71 °C to 77 °C) internal temperature] because the flavor is more desirable compared with lower temperatures.

Precooked meat products such as beef roasts are increasingly being prepared for institutional use, especially in the fast-food industry. The products generally are precooked to a rare state and then chilled and vacuum-packaged. Precooked meat products offer the advantages of closely predictable yields and rapid warming for service.

Cooking time and temperature of roasts is carefully controlled to maintain microbial safety. For example, an internal temperature of 145 °F (63 °C) must be reached or longer times at lower temperatures are required to control potentially pathogenic microorganisms.

Precooked products should be handled carefully to avoid recontamination and incubation of pathogenic and spoilage microorganisms. These problems are most serious with repeated warming and chilling. Canning subjects meat products to sufficient heating to control pathogenic and spoilage microorganisms, and sealing prevents recontamination and conditions favorable for the growth of microorganisms.

FFA MEATS EVALUATION AND TECHNOLOGY CAREER DEVELOPMENT EVENT

Those who participate in an FFA Meats Evaluation and Technology Career Development Event (CDE) must be able to evaluate and place various classes of carcass. Four carcasses are typically compared. To accurately place a class, begin by evaluating each carcass individually. An initial evaluation of marbling and maturity must be done to separate the carcasses into categories based on their overall quality grade. Once the category is determined, a carcass can be evaluated on its lean yield or cutability, which is a measure of trimness and muscling.

Examine all sides of a carcass to see differences in trimness and muscle shape between carcasses.

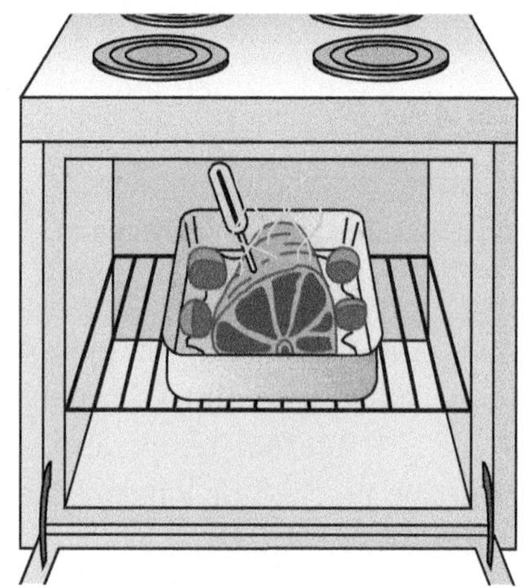

Roasting, suitable for large tender cuts of beef, veal, pork, and lamb.

Broiling, suitable for tender beef steaks, lamb chops, pork chops, sliced ham, bacon, and ground beef or lamb.

Panfrying, suitable for comparatively thin pieces of tender meats.

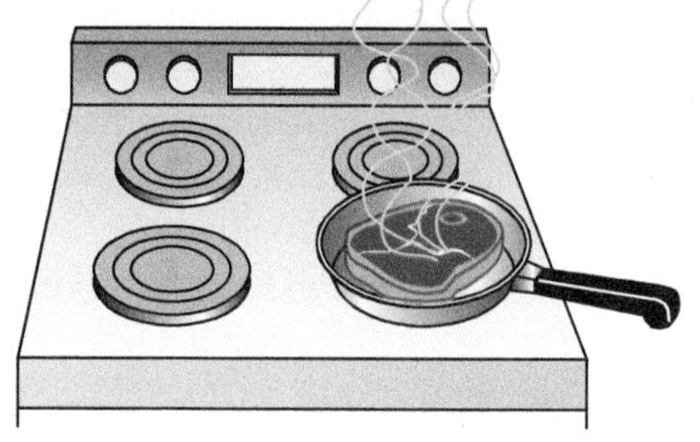

Panbroiling, suitable for tender cuts when cut 1 in. or less thick.

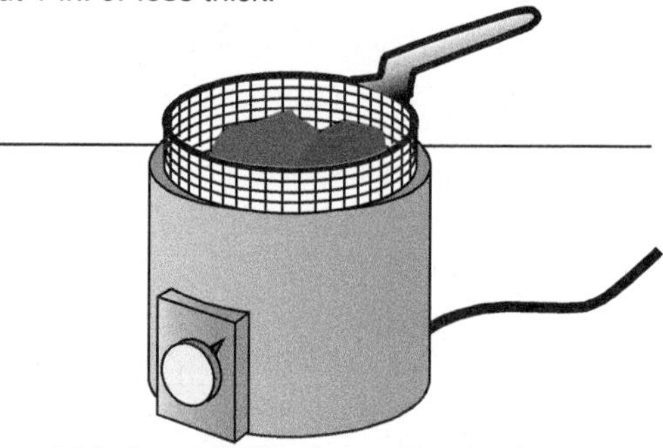

Deep-fat frying, suitable for cooking brains, sweetbreads, liver, croquettes, and leftover meat.

Braising, suitable for less tender cuts of meat.

Cooking in liquid, suitable for large less tender cuts and stews.

FIGURE 17-9 Different methods are used to cook meat, depending on the cut.

Pairs or groups of exhibits within the same quality grade are sorted by trimness and muscling, with higher cutability carcasses sorting up.

After placing a group of beef carcasses in an FFA CDE, participants will be required to defend their placings with specific reasons or verbal descriptions of the factors that influenced the order of placings. The American Meat Science Association provides a Web site with many meat judging resources: http://www.meatscience.org/students/meat-judging-program/meat-judging-resources.

MEAT SUBSTITUTES

Modern techniques for manufacturing protein products from plant materials are largely the result of two notable advances in processing: (1) a method invented in the 1950s for spinning vegetable proteins into solution and (2) extruding the solution through spinnerets and using coagulation to form bundles of fibers. Flavor and color compounds that simulate meat can be added to these fibers. A binder such as egg **albumen** or vegetable gum and such other additives as fats, emulsifiers, and nutrients are also added. The fibers can then be processed into a variety of shapes and textures. In the extrusion method, the vegetable protein is combined with flavorings, color, and other ingredients and then formed into a plastic mass in a cooker extruder. Under pressure, this mass is forced through a die to form beef-like strips or other shapes that are characteristic of meats.

Textured protein products are usually at least 50% protein and contain the eight essential amino acids and the vitamins and minerals found in meats. Although soybean protein is most commonly used, other plant proteins—wheat gluten, yeast protein, and most other edible proteins—can be used singly or in combination. The use of textured protein products will probably increase in the future as populations grow and conventional sources of protein become scarcer or more expensive.

SUMMARY

The word *meat* generally refers to the skeletal muscle from carcasses of animals—beef and veal (cattle), pork (hogs), and lamb (sheep). Inspection takes place at practically every step of the livestock procurement and meatpacking processes. Grading establishes and maintains uniform trading standards and aids in determining the value of various cuts of meat. Carcasses are given both a quality grade and a yield grade. Meat and processed meat products, and other foods of animal origin, provide a complete protein source that contains all essential amino acids in favorable quantities. The most causes of variation in the tenderness of beef, pork, lamb, and veal include an animal's genetics, its species and age, its feeding, its muscle type, how the carcass was suspended, whether electrical stimulation was used, the chilling rate, aging, mechanical tenderizing, chemical tenderizing, freezing and thawing, and cooking and carving.

REVIEW QUESTIONS

Success in any career requires knowledge. Test your knowledge of this chapter by answering these questions or solving these problems.

1. The general composition of meat includes what percentage water, protein, fat, and ash (mineral)?
2. Who authorizes meat inspection?
3. Name the types of red meat.

4. List three general meat by-products.
5. List the five factors that affect meat tenderness.
6. Identify wholesale or primal cuts of meat from cattle, sheep, and pigs.
7. Compare quality grades to yield grades.
8. Calculate the dressing percentage of a steer that weighed 1,200 lbs at slaughter and produced a 700-lb carcass.
9. Identify the roles of government agencies involved in meat production.
10. Describe the types of cooking methods for five different retail cuts of meat.

STUDENT ACTIVITIES

1. Visit the Web site of the USDA Agricultural and Marketing Service and compare the quality grades to the yield grades of beef and pork. Report on your findings.
2. Leave a small piece of meat sitting at room temperature for a couple of days. Describe and explain the changes.
3. Make a list of processed meats. Compare your list to other class members' lists to see who has the least duplicated list.
4. Read and research magazine articles and magazine and television advertisements and draw some conclusions about the modern consumers of meat. What do they want in a product? Report on your findings.
5. Display and discuss the colored wall charts of the cuts of beef and pork. These are available from the National Live Stock and Meat Board.
6. Cook a portion of one retail cut of meat by the suggested method and the other portion by a method that is not recommended. Report on the results.
7. Practice calculating yield and quality grades with different values.
8. Watch the YouTube video featuring Temple Grandin giving a tour of a beef plant: https://youtu.be/VMqYYXswono. Report on what you learned.

ADDITIONAL RESOURCES

- USDA Available Services: http://www.ams.usda.gov/AMSv1.0/getfile?dDocName=STELPRDC5065728&acct=grddairy
- Profitability of Ground Meat: http://www.canadabeef.ca/pdf/profitability.pdf
- Meat Identification Pictures: http://aggiemeat.tamu.edu/meat-identification-pictures/
- Meat Identification Tutorial: http://shop.ffa.org/item/MID05-0000/MEAT-IDENTIFICATION-TUTORIAL/
- Meat Judging Resources: http://www.meatscience.org/students/meat-judging-program/meat-judging-resources

Internet sites represent a vast resource of information. Although those provided in this chapter were vetted by industry experts, but may wish to further explore the topics discussed in this chapter using a search engine such as Google. Keywords or phrases may include the following: *meat processing, names of specific cuts, specific meat product names, federal meat inspection, meat grading, processed meats, beef processing, meat cooking, meat substitutes, USDA Quality Grade Standards.* In addition, Table B-7 provides a listing of useful Internet sites that can be used as starting points.

REFERENCES

Aberle, E. D., J. C. Forrest, D. E. Gerrard, and E.W. Mills. 2012. *Principles of meat science*, 5th ed. Dubuque, IA: Kendall Hunt Publishing.

Corriher, S. O. 1997. *Cookwise: The hows and whys of successful cooking.* New York: William Morrow & Co.

Cremer, M. L. 1998. *Quality food in quantity. Management and science.* Berkeley, CA: McCutchan Publishing.

Ensminger, A. H., M. E. Ensminger, J. E. Konlande, and J. R. Robson. 1994. *Foods and nutrition encyclopedia* (2 vols.). Boca Raton, FL: CRC Press.

Savell, J., and G. W. Smith. 2009. *Meat science lab manual*, 8th ed. Lake Charles, LA: American Press.

CHAPTER 18

Poultry and Eggs

OBJECTIVES

After reading this chapter, you should be able to:

- **Outline the production of poultry**
- **List the steps in the processing of poultry**
- **Recognize factors that affect the quality of poultry meat**
- **Identify meat products from poultry**
- **Name factors affecting the tenderness and the flavor of poultry meat**
- **Discuss the general composition of poultry meat and products**
- **Describe the purpose of a Contract Acceptance Certification**
- **Describe egg production**
- **Identify factors affecting egg quality, both exterior and interior**

KEY TERMS

albumen
antemortem
broiler
blood spot
brooding
chalaza
deboning
emulsify
integrated
Julian date
mechanically separated
postmortem
poults
toms
vitelline membrane

NATIONAL AFNR STANDARD

FPP.01

Develop and implement procedures to ensure safety, sanitation, and quality in food products and processing facilities.

OBJECTIVES (continued)

- **List the cooking functions of eggs**
- **Discuss egg grading**
- **Define these egg sizes: jumbo, extra large, large, medium, small, and peewee**

Scientific research has led to genetic improvement of animals, improved management practices, better feeding, and disease control—and the poultry industry has benefited to become a food-science success story. In the mid-1990s, per capita consumption of **broilers** surpassed the per capita consumption of both beef and pork. Now per capita consumption continues to be far ahead of beef and pork consumption. In fact, total per capita consumption of poultry (chickens and turkeys) more or less equals the total per capita consumption of red meat (beef and pork).

FIGURE 18-1 Broilers in well-ventilated housing where thousands are raised.

POULTRY PRODUCTION

Meat chicken production—that is, broiler production—is dominated by large **integrated** companies that typically control hatching egg production and the hatching, growing, processing, and marketing of the birds. They often mill their own feed and render the offal and feathers to produce feed ingredients. Any of these steps may be controlled by contract. The company owns all functions except live production. With a production contract, the farmer may provide the growing facility, equipment, litter, brooder, fuel, electricity, and labor. The company usually provides the chicks, feed, medications, bird loading and hauling, and some grow-out supervision. Contract payments are based on a set amount per pound of chicken marketed.

Growing houses are buildings that are often 40 feet to 50 feet wide and 400 feet to 500 feet long (see Figure 18-1). Modern facilities control the air entering the sides of the building, and exhaust fans blow air over the birds during hot weather and overhead fogger lines cool chickens in hot weather. Space allowances range from 0.7 square foot to 1.0 square foot per bird, depending on the season, the house type, and the age of the bird when marketed.

A poultry house should be located away from other farm structures, and the ground should allow good water drainage. The housing and management of layer hens can be carried out using one of two methods: caged-layer production or floor production. Either method will help keep hens in production throughout the year if proper environmental and nutritional needs are provided.

The caged-layer production method consists of placing hens in wire cages; feed and water are provided to each cage. The birds are housed at a capacity of two to three hens in each cage, which measures approximately 12″ × 16″ × 18″. The cages are organized in rows that rest on leg supports or they are suspended from the ceiling, creating space so the floors of the cages are 2 feet to 3 feet above the ground. The cages are designed so the eggs will roll out of the cage to a holding area. This method of housing is used primarily with egg-type layers used for egg production.

The floor-production method is designed for either egg-type or broiler-type birds kept for fertile or infertile eggs. In commercial flocks, this method is used when fertile eggs for hatching are needed. The birds are maintained in the house on a litter-covered floor, giving the term *floor production*.

Feed is moved to the birds by mechanical conveyers that drop it into attached pans. Water is supplied by bird-activated nipples attached to water pipes running the length of the building. Three diets are typically used: starter, grower, and finisher.

Young chicks require a room temperature of about 87 °F (31 °C) during the first week of life. This is decreased approximately 5 degrees per week until an ambient temperature is reached. Diseases are controlled by vaccination, medicated feed to control coccidiosis control, exclusion of animals that transmit disease (vectors), and sanitation. One current consumer-driven movement is the removal of medications (antibiotics) from feed.

Typical broiler production costs assume that large operations are more efficient. Individual grower operations raise thousands of broilers at a time.

Turkey production is also integrated. Most meat-production units brood and grow from 50,000 to 75,000 birds three and one-half times per year. Many larger facilities have single brooding complexes that have the capacity to brood 50,000 to 100,000 **poults** (young turkeys), and these serve two separate grow-out facilities with the same capacity. In this scheme, the facility broods seven times each year and furnishes the poults needed to fill both growing facilities three and one-half times per year.

Poults are **brooded** (cared for) with an average density of 1.0 square foot per bird. Toms (males) are placed in grow-out facilities at a density of 3.0 to 4.0 square feet per bird. Hens (females) receive about 2.5 square feet in their grow-out facilities.

Turkeys are no longer produced seasonally. Further processing and the structure of the turkey industry has made turkey production a year-round activity. Almost all birds are produced on a contract basis. The producer or grower furnishes the land, facilities, and labor and is paid based on the weight, grade, and feed conversion of the birds delivered to a processing plant.

Processing

Meat chickens can be marketed as broilers, roasters, or game hens (Figure 18-2), although most are broilers. Typical eviscerated weights are shown in Table 18-1. Modern commercial meat strains reach an average live weight of 4.0 pounds to 5.0 pounds at 6 weeks and 6 pounds to 10 pounds at 8 to 12 weeks of age. The chickens are slaughtered at an appropriate age to get the eviscerated weight desired by the customer.

FIGURE 18-2 Processing chicken parts for wholesale and retail markets.

TABLE 18-1 Weights for Different Types of Meat Chickens

TYPE	LIVE WEIGHT (LBS.)	EVISCERATED WEIGHT (LBS.)
Cornish game hen	1.2 to 3.1	0.75 to 2.0
Broiler, fryer	4.0 to 6.3	2.8 to 4.4
Roaster	7.4 to 10.0	5.0 to 7.0

THE HORMONE MYTH

Since the 1950s, the poultry industry has been incredibly successful. Broilers and turkeys grow fast and are brought to market more quickly than ever before. This makes people skeptical, leading to speculation about the ways meat poultry are raised. Most popular in this speculation is that growers use hormones to achieve rapid growth. More and more consumers ask why hormones are used in broiler production. The myth of hormone use in chickens and turkeys persists, probably because of a lack of information and understanding on the part of the general consumer. The truth is that hormones are not used in the production of chickens and turkeys. This myth continues in part because consumers are taking greater interest in how their food is produced.

So if hormones are not used, then why do chickens and turkeys grow faster than ever? The extremely rapid growth of broilers is easy to explain. For the past several decades, birds have reached a specified market weight one day earlier each year. The rapid growth of turkeys and chickens is the result of improved genetics, the use of high-quality feed (proteins, vitamins, minerals, etc.), and the provision of healthy environments that are conducive to rapid growth. So the short and the correct answer is that the rapid growth of modern chickens and turkeys represents a logical consequence of slow but consistent improvements in genetics, nutrition, management, and disease control—not hormones!

The commercial turkey is a white hybrid that is so large that it can no longer efficiently breed by natural means. As a result, artificial insemination is now routinely used. Turkey hens are marketed at 12 to 14 weeks of age. At this age, hens typically weigh from 14 pounds to 20 pounds. **Toms** (males) are often marketed between 16 weeks and 19 weeks of age and will weigh from 35 pounds to 42 pounds. Market age is determined by the product being produced. Approximately 70% of all turkeys grown are further processed. For this market, the industry prefers to grow toms, because they are larger. Hens are also further processed even though the unit cost is higher. Some 16% of all turkeys are processed for the whole body market.

Approximately 14% of all turkeys produced are processed as parts. In the past, parts such as wings and drums were often sold at greatly reduced prices. Today, these parts are used extensively in further processing and often end up as part of another processed product such as ground meat.

Processing Steps

The slaughter and processing of broilers and turkeys is an assembly-line operation conducted under sanitary conditions. Inspecting, classifying, and grading are parts of the processing operation. Although the processing procedures may vary from plant to plant and between broilers and turkeys, the steps usually include the following:

1. Spot inspection of each lot of birds before slaughter (**antemortem** inspection)
2. Suspension and shackling of each bird by its legs
3. Stunning with electrical shock
4. Bleeding
5. Scalding to loosen the feathers
6. Machine removal of feathers
7. Removal of pinfeathers
8. Removal of internal organs (evisceration)
9. Chilling (in ice water)
10. **Postmortem** (after death) inspection
11. Grading
12. Packaging

Meat Properties

Meat from chickens and turkeys provides a high-quality protein that is low in fat. The protein is an excellent source of essential amino acids. Poultry meat is also a good source of phosphorus, iron, copper, zinc, and vitamins B12 and B6. Dark meat is higher in fat than white meat, and fat is also high in the skin. A 3-ounce piece of roasted chicken or turkey breast provides about 140 kcal of energy, and 3 ounces of dark turkey meat provides about 160 kcal of energy. Table B-8 provides the composition of many poultry products.

Whether a poultry product meets a consumer's expectations depends on the conditions surrounding various stages in the bird's development from fertilized egg through production and processing to consumption. Appearance, texture, and flavor are primary concerns in the food industry and to the consumer.

Appearance (Color). The color of cooked or raw poultry meat is important because consumers associate it with product freshness. Poultry is unique because it is sold with and without skin. In addition, poultry has muscles that show dramatic extremes of color—white breast meat and dark thigh and leg meat.

Poultry meat color is affected by factors such as bird age, sex, strain, diet, intramuscular fat, meat moisture content, preslaughter conditions, and various processing variables. The color of meat depends on the presence of the muscle pigments myoglobin and hemoglobin. Discoloration of poultry can be related to the amount of these pigments that are present in the meat, the chemical state of the pigments, or the way in which light reflects off the meat. The discoloration can occur in an entire muscle, or it can be limited to a specific area.

When an entire muscle is discolored, it is frequently the breast muscle. This occurs because breast muscle accounts for a large portion of the live weight. It is more sensitive to factors that contribute to discoloration, and its already light color makes small changes in color more noticeable. Extreme environmental temperatures or stress from live handling before processing can cause broiler and turkey breast meat to be discolored. The extent of the discoloration is related to each bird's individual response to the conditions.

Another major cause of poultry meat discoloration is bruising. Approximately 29% of all carcasses processed in the United States are downgraded (show reduced quality), and the majority of these defects are from bruises. The poultry industry generally tries to identify where this happens (field or plant), how it happens, and when the injuries occur, but this is often difficult to determine. The color of the bruise, the amount of "blood" present, and the extent of the "blood clot" formation in the affected area are good indicators of the age of the injury and may give clues as to its origin.

A bruise will vary in appearance from a fresh, bloody red color with no clotting minutes after the injury to a normal flesh color 120 hours later. The amount of blood present and the extent of clot formation are useful in determining whether an injury occurred during catching or transportation or during processing. Injuries that occur in the field are usually magnified by processing-plant equipment or handling conditions in the plant.

Texture (Tenderness). After consumers buy a poultry product, they relate the quality of that product to its texture and flavor when they are eating it. Whether poultry meat is tender depends

SCIENCE CONNECTION!

Have you ever wondered why poultry has light meat and dark meat? Why? Which parts of their bodies do poultry use to move? Is movement related to the color of the meat?

FIGURE 18-3 Whole chickens being inspected in a processing plant.

on the rate and extent of the chemical and physical changes occurring in the muscle as it becomes meat (Figure 18-3). When an animal dies, blood stops circulating and muscles receive no new supplies of oxygen or nutrients. Without oxygen and nutrients, muscles run out of energy, and they contract and become stiff. As noted in Chapter 17, this stiffening is called *rigor mortis*. Eventually, muscles become soft again, which means that they are tender when cooked.

Anything that interferes with the formation of rigor mortis or the softening process that follows it will affect meat tenderness. For example, birds that struggle before or during slaughter cause their muscles to run out of energy more quickly, and rigor mortis thus beings much sooner than normal. The texture of these muscles tends to be tough because energy was reduced in the live bird. Exposure to environmental stress (hot or cold temperatures) before slaughter creates a similar situation. High preslaughter stunning temperatures, high scalding temperatures, longer scalding times, and machine picking can also cause poultry meat to be tough.

The tenderness of boneless or portioned cuts of poultry is influenced by the amount of time between death (postmortem) and **deboning**. Muscles that are deboned during early postmortem still have energy available for contraction. When these muscles are removed from the carcass, they contract and become tough. To avoid this toughening, meat can be aged for 6 hours to 24 hours before deboning. This is costly for the processor, however.

The poultry industry started using postslaughter electrical stimulation immediately after death to hasten rigor development of carcasses and reduce aging time before deboning. When electricity is applied to the dead bird, the treatment acts like a nerve impulse and causes a muscle to contract, use up energy, and enter rigor mortis at a faster rate. Meat can be boned within 2 hours postmortem instead of the 4 hours to 6 hours required with normal aging.

Flavor. Flavor is another quality attribute that consumers use to determine the acceptability of poultry meat. Both taste and odor contribute to the flavor of poultry. Generally, distinguishing between the two during consumption is difficult. When poultry is cooked, flavor develops from sugar and amino acid interactions, lipid and thermal (heat) oxidation, and thiamin degradation. These chemical changes are not unique to poultry, but the lipids and fats in poultry are unique and combine with odor to account for the characteristic poultry flavor.

Few factors during production and processing actually affect poultry meat flavor. The age of the bird at slaughter (both young and mature birds) most affects meat flavor. Minor effects on meat flavor are related to bird strain, diet, environmental conditions (litter, ventilation, etc.), scalding temperatures, chilling, product packaging, and storage. Overall, these effects are too small for consumers to notice.

The most important aspect of poultry meat is its eating quality, a function of the combined effects of appearance, texture, and flavor. Live production affects poultry meat quality by determining the state of the animal at slaughter. Poultry processing affects meat quality by establishing the chemistry of the muscle constituents and their interactions within the muscle structure. The producer,

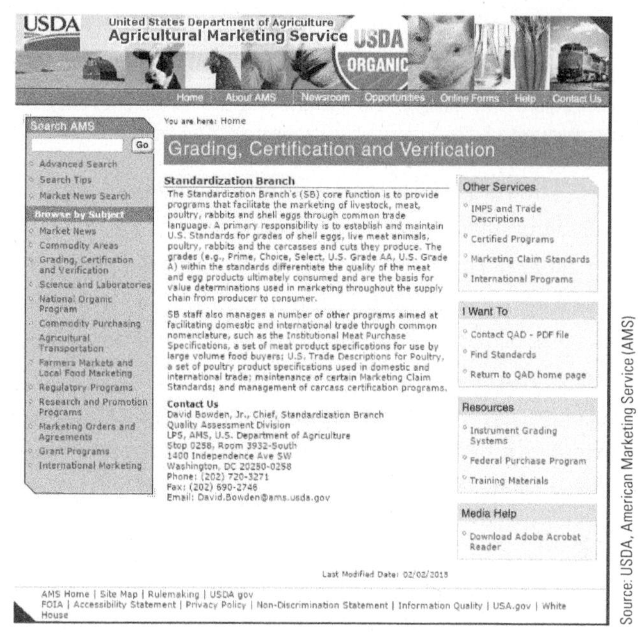

Source: USDA, American Marketing Service (AMS)

FIGURE 18-4 The Web site for the USDA's AMS shows the quality standards for poultry products.

processor, retailer, and consumer all have specific expectations for the quality attributes of poultry.

Grading

Chickens, turkeys, ducks, geese, guineas, and pigeons are all eligible for grading and certification services provided by the USDA Agricultural Marketing Service (AMS) poultry program's grading branch (Figure 18-4). These services are provided in accordance with federal poultry-grading regulations.

Chickens and turkeys are often sold as value-added products. Poultry parts and an increasing number of skinless and boneless products are meeting consumer demand for convenient, lower-fat, portion-controlled items. This shift away from whole carcass birds creates special challenges for buyers and sellers, whether they are poultry producers or processors, wholesalers, food manufacturers, food-service operators, food retailers, or consumers. All these traders rely on USDA's poultry-grading services to ensure that the requirements for quality, weight, condition, and other factors are met.

Grading and USDA Quality Grade Standards. Grading provides a standardized means of describing the marketability of a particular food product. For poultry to be eligible for an official USDA grade designation, each carcass or part must be individually graded by a plant grader and then a sample must be certified by a USDA grader. Officially graded poultry that passes this examination and evaluation process is eligible for the grade shield and may be identified as USDA Grade A, B, or C (Figure 18-5). Poultry standards are frequently reviewed, revised, and updated as needed to keep pace with changes in processing and merchandising.

For poultry, the USDA has developed quality grade standards for whole carcasses and parts, as well as boneless and skinless parts and products. Depending on the product, the standards define and measure quality in terms of meat yield, fat covering, and freedom from defects such as cuts and tears in the skin, broken bones, and discolorations on the meat and skin. The intensity, aggregate area, location, and number of defects encountered for each quality factor are determined. The final quality rating (A, B, or C) is based on the factor with the lowest rating.

Quality standards for poultry can be found on the Web site for the AMS.

MATH CONNECTION!

What effect does quality grading have on the price of poultry? Are all grades of poultry sold at grocery stores or are some grades sold in other ways?

FIGURE 18-5 Various inspection and grade shields found on poultry products.

Evaluation of Ready-to-Cook Carcasses of Chickens and Turkeys. All poultry products for human consumption in the United States must be processed, handled, packaged, and labeled in accordance with federal law. The Poultry Products Inspection Act of 1957 and the Wholesome Poultry Products Act of 1968 are the government regulations that spell out the requirements by which poultry products must be evaluated.

The quality grade of poultry is determined by evaluating the following factors: conformation, fleshing, fat covering, defeathering, exposed flesh, discolorations, disjointed and broken bones, and freezing defects. Grade A is generally the only grade sold at retail. Lower grades are used mostly in processed poultry products where the meat is cut up, chopped, or ground into other products. See Figures 18-6 and 18-7.

Contract Acceptance Certification. The contract acceptance service ensures the integrity and quality of poultry and further processed poultry products bought by quantity food buyers such as food manufacturers, food-service operators, and food retailers. USDA specialists help institutional buyers develop and prepare explicit poultry specifications tailored to their requirements. USDA

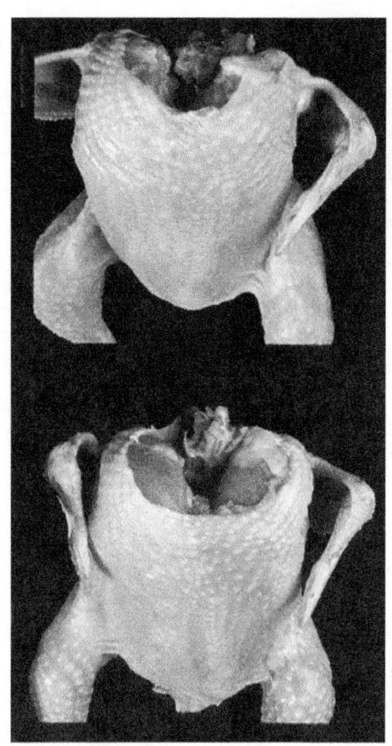

FIGURE 18-6 Grade A chicken (top). The carcass shown in the bottom photo is Grade B because excess neck skin was removed during processing.

Source: USDA

graders then provide certification that purchases comply with these specifications. Specific items that may be part of a product specification include:

- Kind and class (species and age of the poultry)
- Type (frozen, chilled) and style (cut-up parts, whole muscle)
- Formula, processing, and fabrication
- Laboratory analysis
- Net weight
- Labeling and marking, packing and packaging
- Storage and transportation

Products meeting specified requirements are eligible for the contract compliance identification mark. This official grading certificate accompanies each shipment to the receiving agency.

Products

Over the years, the per capita consumption of poultry has increased, primarily because of the increased availability of poultry and the large variety of products being made from poultry meat (Figure 18-8). Often these products are similar to traditional products from red meats—for example, frankfurters (hot dogs), hams, sausages, bologna, salami, pastrami, ham, and other lunchmeats. Many new products use **mechanically separated** poultry meat, which is ground to a fine emulsion for curing, seasoning, smoking, and processing.

Forming products requires a change in particle size and often includes the addition of ingredients to add flavor. Forming products also requires the use of a mold to obtain desired shapes. Formed products include hot dogs, chicken nuggets, and sausage.

One unusual phenomenon of poultry consumption is the chicken wing. Until the last few

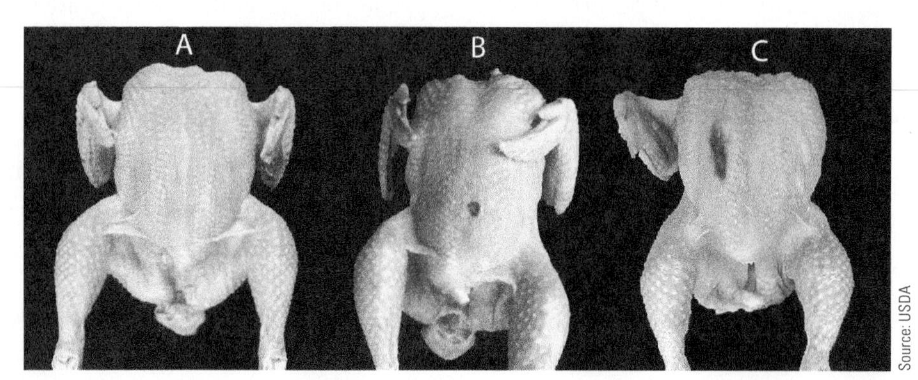

Source: USDA

FIGURE 18-7 Chicken carcasses—Grades A, B, and C The carcass in the middle has exposed flesh greater than ¼ inch. Grade C, on the right, has an entire wing missing and discoloration that is greater than 2 inches and more than moderate in color.

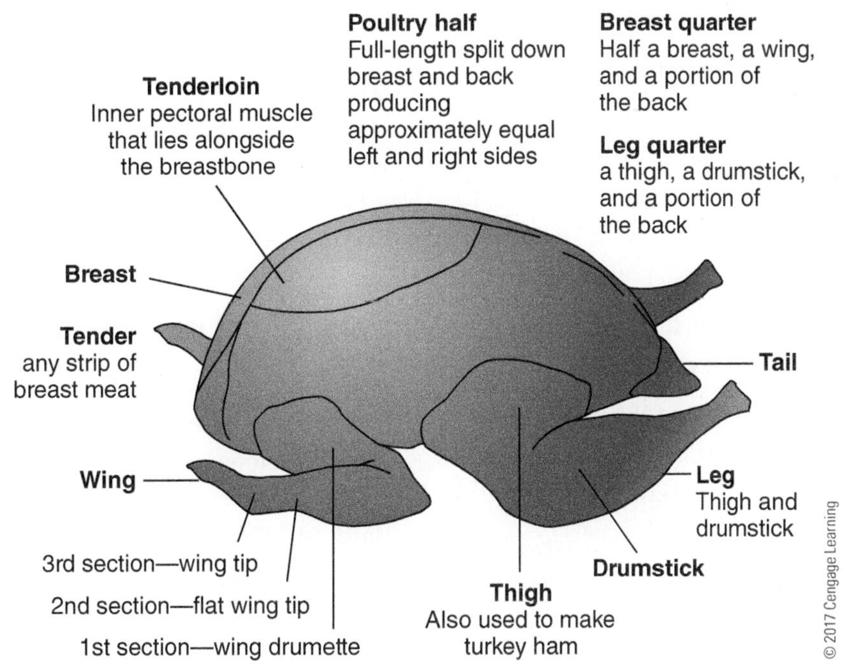

FIGURE 18-8 Products obtained from a poultry carcass.

decades of the 20th century, wings were not considered an important part of the chicken when it came to sales. Now chicken wings are on a variety of fast-food menus as well as stores. In fact, wings have become America's party food. They seem to be a staple at parties ranging from birthdays to Super Bowl get-togethers and are big sellers for many restaurants throughout the entire fall and winter football seasons. According to the industry, sales go up some 50% during football season.

Label Requirements. Labels for poultry must meet certain requirements and show the following:

- A sell-by date, which is preferred by stores and used for quality assurance
- A plant code that can be used for recalls in case of a bacterial outbreak
- The grade of the chicken (usually Grade A, which means plump, free of bruises, and without any broken bones)
- A warning to keep the product refrigerated below 40 °F to reduce or inhibit microbial growth
- Directions to follow all label instructions
- Nutrition facts such as serving size, fat, sugar, sodium, and carbohydrate content

Factors Governing the Evaluation of Precooked, Further Processed Poultry Products. Each type of precooked and further processed poultry product must meet specifications and be evaluated on industry standards. An example is the meat portion of a patty or nugget that contains chicken breast meat and natural proportion skin that is chopped or ground and formed by a machine.

Six Quality Characteristics of Chicken Patties or Nuggets

- Batter or breading texture
- Meat color
- Meat texture
- Batter or breading color
- Shape, size, completeness
- No foreign materials

Grades of Chicken Patties or Nuggets

- Minor defect: ¼″ to ¾″ of batter or breading is missing
- Major defect: ¾″ to 1″ of batter or breading texture is missing

  ½″ to 1″ of meat is missing

  Very light batter or breading color
- Critical defect: More than 1″ of batter or breading texture is missing

  More than ½″ of meat is red or pink (undercooked)

  More than 1″ of meat is missing

  Black or burned area on batter or breading

  Broken meat patty or nugget

  Foreign material located in or on meat

Identification of Poultry Carcass Parts. The FFA Poultry Evaluation CDE also requires participants to identify the carcass parts of various birds. Knowledge of the location, shape, and uses of each part is important information and a good skill to develop. Many restaurants save money on meats by processing their own whole carcasses into parts, making it necessary that the kitchen staff knows these chicken parts as well.

EGGS

Maximum production of top-quality eggs starts with a closely controlled breeding program that emphasizes favorable genetic factors. The White Leghorn type of hen dominates today's egg industry. This breed reaches maturity early, uses its feed efficiently, has a relatively small body size, adapts well to different climates, and produces a relatively large number of white-shelled eggs, the color preferred by most consumers. In the major egg-producing states, flocks of 100,000 laying hens are not unusual, and some flocks number more than 1 million. Each of the 283 million laying birds in the United States produces from 250 to 300 eggs a year.

In today's egg-laying facilities control temperature, humidity, light, and air circulation. The buildings are well insulated, windowless (to aid light control), and force ventilated. Most new construction favors the cage system because of its sanitation and efficiency. Because care and feeding of hens, maintenance, sanitation, and egg gathering all take time and money, automation is used whenever possible.

Caged Layers in California

Beginning January 2015, all eggs sold in California were required to come from chickens that live in more spacious quarters—almost twice as spacious, in fact, as the cages that have been the industry standard. In 2008, California voters passed Proposition 2, which changed cage standards for laying hens.

Processing

The moment an egg is laid, physical and chemical changes begin to reduce its freshness. In most production facilities, automated gathering belts collect and refrigerate eggs frequently. Gathered eggs are moved into refrigerated holding rooms where temperatures are maintained between 40 °F and 45 °F (5 °C and 7 °C). Humidity is relatively high to minimize moisture loss.

Carton Dates. Egg cartons from USDA-inspected plants must display a **Julian date** (a number 1 through 365) indicating the date the eggs were packed. Although not required, they may also carry an expiration date beyond which the eggs should not be sold. In USDA-inspected plants, this date cannot exceed 30 days after the pack date. Plants not under USDA inspection are governed by the laws of their states. Fresh shell eggs can be stored in their cartons in the refrigerator for 4 weeks to 5 weeks beyond the Julian date with insignificant quality loss.

Formation and Structure

The structure and characteristics of an egg include its color, shell, white, yolk, air cell, chalaza, germinal disc, and membranes (see Figure 18-9).

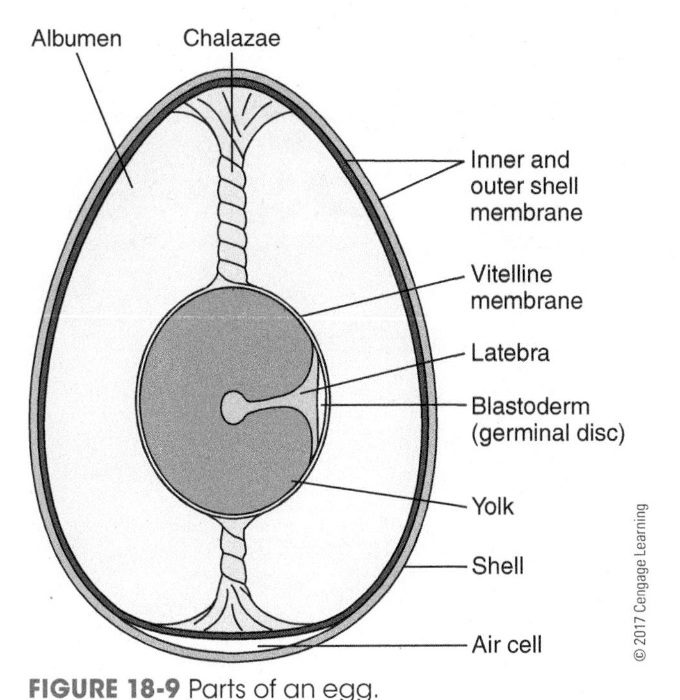

FIGURE 18-9 Parts of an egg.

Color. Egg shell and yolk color may vary, but color has nothing to do with egg quality, flavor, nutritive value, cooking characteristics, or shell thickness. The color comes from pigments in the outer layer of the shell and may range in various breeds from white to deep brown. The breed of hen determines the color of the shell.

Shell. The egg's outer covering accounts for some 9% to 12% of its total weight, depending on egg size (Figure 18-9). The shell is the egg's first line of defense against bacterial contamination. It is largely composed of calcium carbonate (94%) with small amounts of magnesium carbonate, calcium phosphate, and other organic matter, including protein.

White. Egg **albumen** or egg white in raw eggs is opalescent and does not appear white until it is beaten or cooked. A yellow or greenish cast in a raw white may indicate the presence of riboflavin. Cloudiness of the raw white is due to the presence of carbon dioxide that has not had time to escape through the shell and thus indicates an extremely fresh egg.

Yolk. Yolk color depends on a hen's diet. Artificial color additives are not permitted. Gold or lemon-colored yolks are preferred by most buyers in the United States. Yolk pigments are relatively stable and are not lost or changed in cooking.

Air Cell. The air cell (Figure 18-9) is the empty space between the white and shell at the large end of the egg. When an egg is first laid, it is warm. As it cools, the contents contract and the inner shell membrane separates from the outer shell membrane to form the air cell.

Chalaza. Chalaza are ropey strands of egg white that anchor the yolk in place in the center of the thick white. They are neither imperfections nor beginning embryos. The more prominent the chalazae, the fresher the egg.

Germinal Disc. The germinal disc is the channel leading to the center of the yolk. When the egg is fertilized, sperm enter by way of the germinal disc and travel to the center and a chick embryo starts to form.

Membranes. Just inside the shell are two shell membranes—inner and outer. After the egg is laid and begins to cool, an air cell forms between these two layers at the large end of the egg. The **vitelline membrane** is the covering of the yolk. Its strength protects the yolk from breaking. The vitelline membrane is weakest at the germinal disc and tends to become more fragile as the egg ages.

Composition

The yolk, or yellow portion, makes up approximately 33% of the liquid weight of the egg. It contains all the fat in the egg and a little less than half its protein. With the exception of riboflavin and niacin, the yolk contains a higher proportion of the egg's vitamins than the white.

The yolk also contains more phosphorus, manganese, iron, iodine, copper, and calcium than the white, and it contains all the zinc. The yolk of a large egg contains some 59 calories; the whole egg, including the albumen contains approximately

75 calories (kcal) of energy. Table B-8 lists the composition of eggs and egg products.

Also known as *egg white*, albumen contains more than half the egg's total protein, niacin, riboflavin, chlorine, magnesium, potassium, sodium, and sulfur. Protein from the yolk and the albumen provides humans with a high-quality protein containing every essential amino acid.

Albumen is more opalescent than truly white. The cloudy appearance comes from carbon dioxide. As the egg ages, carbon dioxide escapes, so the albumen of older eggs is more transparent than that of fresher eggs.

Cooking Functions. Although eggs are widely known as breakfast entrees, they also perform many other functions for the knowledgeable cook. Their cooking properties are so varied that they have been called "the cement that holds the castle of cuisine together."

Eggs can bind ingredients together—for example, meat loaves and croquettes. They can also leaven such baked high-rise dishes such as soufflés and sponge cakes. Their thickening talent is seen in custards and sauces. They **emulsify** (allow water and oil to mix) mayonnaise, salad dressings, and Hollandaise sauce and are frequently used to coat or glaze breads and cookies. They clarify soups and coffee. In boiled candies and frostings, they retard crystallization. As a finishing touch, they can be hard cooked and used as a garnish.

Grading

Classification is determined by interior and exterior quality and designated by the letters AA, A, and B. In many egg-packing plants, the USDA provides a grading service for shell eggs. Its official grade shield certifies that the eggs have been graded under federal supervision according to USDA standards and regulations (Figure 18-10). The grading service is not mandatory. Other eggs are packed under state regulations, which must meet or exceed federal standards.

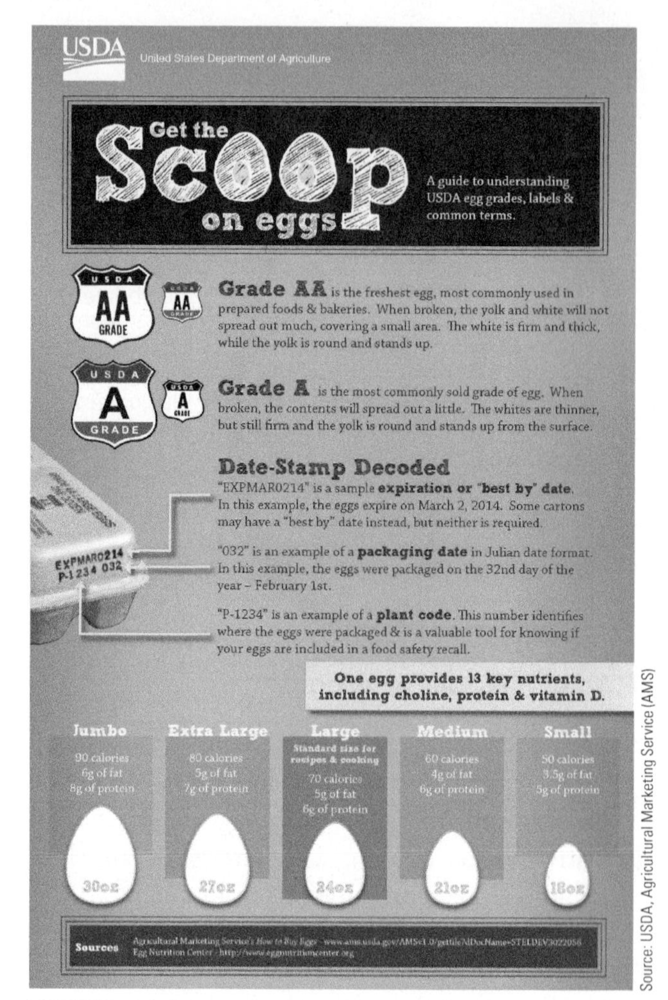

FIGURE 18-10 As explained in this informational poster, the official USDA grade shield certifies that eggs have been graded under federal supervision according to USDA standards and regulations.

Source: USDA, Agricultural Marketing Service (AMS)

In the grading process, eggs are examined for both interior and exterior quality and are sorted according to weight (size). Grade quality and size are not related to one another. In descending order of quality, grades are AA, A, and B. No difference in nutritive value exists between the different grades.

Because production and marketing methods have become incredibly efficient, eggs move so rapidly from laying house to market that consumers find little difference in quality between grades AA and A. Although Grade B eggs are just as wholesome to eat, they rate lower in appearance when broken out. Almost no Grade B's find their way to the retail supermarket. Some go to institutional egg users such as bakeries and food service operations, but most go to egg breakers for use in egg products.

Grade AA. When cracked onto a surface, a Grade AA egg will stand up tall. The yolk is firm, and the area covered by the white is small. A large proportion of thick white to thin white exists. The shell approximates the usual shape for an egg. It is generally clean and unbroken. Ridges or rough spots that do not affect the shell's strength are permitted.

Grade A. When cracked onto a surface, a Grade A egg covers a relatively small area. The yolk is round and upstanding. The thick white is large in proportion to the thin white and stands fairly well around the yolk. The shell approximates the usual shape for an egg. It is generally clean and unbroken. Ridges/rough spots that do not affect the shell strength are permitted.

Both Grades AA and A are ideal for any use but are especially desirable for poaching, frying, and cooking in their shells.

TABLE 18-2 Weight Classes for Shell Eggs

SIZE	WEIGHT PER DOZEN (OZ.)
Jumbo	30
Extra large	27
Large	24
Medium	21
Small	18
Peewee	15

MATH CONNECTION!

Do egg sizes and quality grading affect price? What other factors affect the selling price of eggs? Record your findings.

Grade B. When cracked onto a surface, a Grade B egg spreads out more. The yolk is flattened and there is about as much (or more) thin white as thick white. The shell has an abnormal shape; some slight stained areas are permitted. It is unbroken, and pronounced ridges or thin spots are permitted.

Size. Several factors influence the size of an egg. The major factor is the age of the hen. As the hen ages, her eggs increase in size. The breed of hen from which the egg comes is a second factor. Weight of the bird is another. Environmental factors that lower egg weights are heat, stress, overcrowding, and poor nutrition. All these variables are of great importance to the egg producer. A slight shift in egg weight influences size classification. Size is one of the factors considered when eggs are priced.

Egg sizes are jumbo, extra large, large, medium, small, and peewee. Medium, large, and extra large are the sizes most commonly available. Sizes are classified according to minimum net weight expressed in ounces per dozen, as shown in Table 18-2.

FFA POULTRY EVALUATION CAREER DEVELOPMENT EVENT

The FFA Poultry Evaluation Career Development Event (CDE) not only requires participants to evaluate poultry products but also have the ability to evaluate and place various classes of live poultry animals. In live chicken evaluation, a chicken is judged for its production purposes, whether it is an egg-laying breed or a meat breed. For success when judging live chickens, knowing the common breeds and the ability to handle them properly is necessary. This includes

FIGURE 18-11 Parts of a male chicken.

knowing the names of the various parts of the chicken and their quality characteristics (see Figure 18-11).

When judging egg-laying birds for production, evaluate their current and past productions and their rate of production. This can be determined by evaluating the appearance and condition of certain parts of the body. When judging for egg production, the birds must be handled individually. The past egg-laying production is determined by examining the amount of yellow pigment that is left in the hen's body and the time of her molt. A chicken that has laid a large amount of eggs will have less yellow and mostly bleached color in all parts of her body. The bird's current production is determined by the condition of the comb, wattles, eyes, pubic bones, abdomen, and vent. If the hen is currently in production, the wattles will be large, soft, and bright red; the eyes will be bright and prominent (not sunken in); the pubic bones will be well spread; the abdomen will be full and pliable; and the vent will be moist and bleached.

Blood Spots. Blood spots are also called *meat spots*. These are occasionally found on an egg yolk. These tiny spots do not indicate a fertilized egg. Rather, they are caused by the rupture of a blood vessel on the yolk surface during formation of the egg or by a similar accident in the wall of the oviduct. Less than 1% of all eggs produced have blood spots.

Quality Factors of Eggs

Several factors separate eggs from each other besides size. There are quality characteristics that are judged before packaging in cartons to determine the quality. An exterior evaluation of the egg can be done visually, but most modern processing facilities

TABLE 18-3 USDA Grade Standard Chart

QUALITY FACTOR	AA QUALITY	A QUALITY	B QUALITY	INEDIBLE
Air cell	1/8″ or less in depth	3/16″ or less in depth	More than 3/16″	Does not apply
White	Clear, firm	Clean, may be reasonably firm	Clean, may be weak and watery	Does not apply
Yolk	Outline slightly defined	Outline may be fairly well defined	Outline clearly visible	Does not apply
Spots (blood or meat)	None	None	Blood or meat spots aggregating not more than 1/8″ in diameter	Blood or meat spots aggregating more than 1/8″ in diameter

Source: USDA

use high-tech scanning and laser equipment for detection. The following are quality indicators:

- Soundness
- Stains
- Adhering dirt or foreign material
- Egg shape
- Shell texture
- Shell thickness
- No defects

Table 18-3 illustrates the quality factors for each grade of egg.

Evaluation of Egg Interior and Exterior Quality

The interior quality of an egg is determined by a process called *candling.* The eggs are examined by being held up to a high-intensity light. Candling can identify the size of the air cell and other interior quality factors. Figures 18-12, 18-13, and 18-14 show the various grades of eggs based on the area that a cracked eggs covers.

FFA Meats Evaluation and Technology Career Development Event

The FFA Meats Evaluation and Technology Career Development Event (CDE) requires knowledge of carcass evaluation and grading, retail meat cut identification, as well as meat-product formulation and pricing. Explore the FFA Web site for more information about the contest requirements and participant preparation at https://www.ffa.org/participate/cdes/meats-evaluation-and-technology.

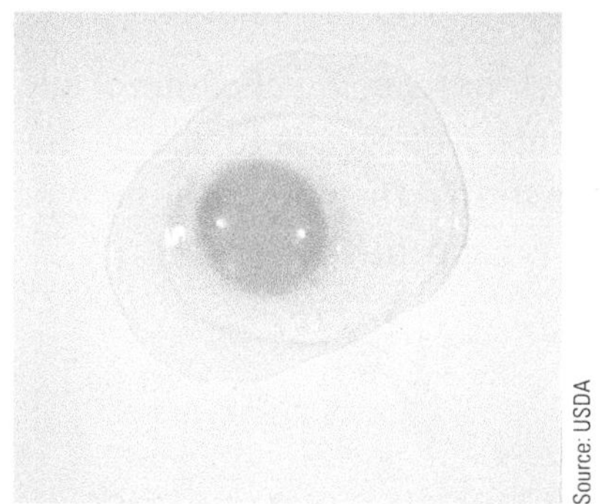
Source: USDA

FIGURE 18-12 Grade AA egg, top view. The egg content covers a small area.

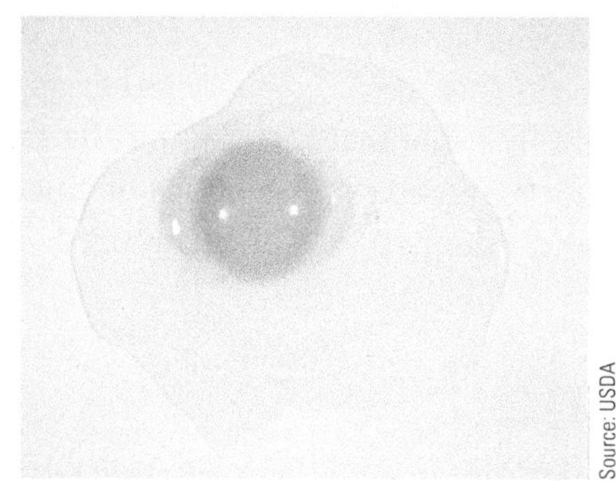
Source: USDA

FIGURE 18-13 Grade A egg, top view. The egg content covers a moderate area.

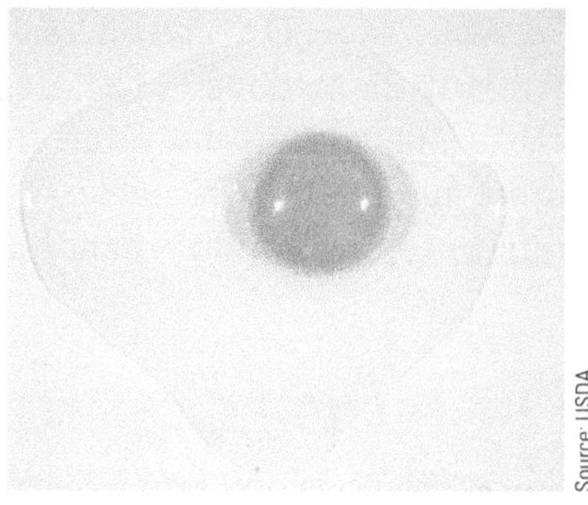
Source: USDA

FIGURE 18-14 Grade B egg, top view. The egg content covers a large area.

The Poultry Evaluation CD, tests FFA members' knowledge of live bird evaluation and anatomy identification, carcass evaluation and grading, carcass parts identification, and processed poultry product evaluation. For more information and contest rules, explore the National FFA Poultry Evaluation Web page (https://www.ffa.org/participate/cdes/poultry).

Storage

Eggs can be stored at 30 °F (−1 °C) for as long as 6 months in the shell. They can be frozen out of the shell for extended storage. The large quantities of eggs required by the food-manufacturing industry are preserved by freezing. Eggs can be frozen whole (minus the shell), separately as white and yolk, or in varying combinations.

After removal from the shell, eggs also can be dried (dehydrated) as whole eggs, whites, or yolks. The methods of dehydrating eggs include spray drying, tray drying, foam drying, and freeze-drying.

Salmonella. The inside of the egg was once considered almost sterile. Recently, however, a bacterial organism, *Salmonella enteritidis,* has been found inside some eggs. Only a miniscule number of eggs might actually contain *Salmonella enteritidis*. Even in areas where outbreaks of salmonellosis have occurred, tested flocks show an average of only about two or three infected eggs out of each 10,000 produced.

Still, the FDA considers these foods "potentially hazardous." The designation is not cause for alarm, however. It simply means these foods are perishable and should receive refrigeration, sanitary handling, and adequate cooking. Lack of attention can make any food a hazardous food.

Fertile Eggs

Eggs that can be incubated develop into chicks. Fertile eggs are not more nutritious than nonfertile eggs, do not keep as well as nonfertile eggs, and are more expensive to produce.

Organic Eggs

Organic eggs are from hens fed rations having ingredients that were grown without pesticides, fungicides, herbicides, or commercial fertilizers, according to the USDA Organic Certification Standards. Rations for commercial laying hens never contain hormones. Because of higher production costs and lower volume per farm, organic eggs are more expensive than eggs from hens fed conventional feed. Whether or not feed is organic does not affect any egg's nutrient content.

Egg Substitutes

With all the attention paid to cholesterol in the last decades of the 20th century, the level of cholesterol (about 240 mg) in eggs caused consumers to reduce their consumption of eggs. Food manufacturers have taken different approaches to reducing cholesterol in eggs—from physically separating cholesterol from the yolk to formulating yolks from other products and combining these with the albumen. These products are sold as egg substitutes.

Another approach to reducing cholesterol and changing the fat content of eggs is to change the genetics of chickens so that they produce the type of egg desired. This is happening.

SCIENCE CONNECTION!

How are eggs and egg substitutes similar? How are they different? What is the advantage and disadvantage of both eggs and egg substitutes? Record your research findings.

SUMMARY

Poultry includes meat from chickens and turkeys. Meat from chickens and turkeys provides a high-quality protein that is low in fat. The protein is an excellent source of essential amino acids. Appearance, texture, and flavor of poultry meat are a primary concern in the food industry and to the consumer. Poultry meat color is affected by factors such as bird age, sex, strain, diet, intramuscular fat, meat, moisture content, preslaughter conditions, and processing variables. Chickens, turkeys, ducks, geese, guineas, and pigeons are all eligible for grading and certification services provided by the USDA Agricultural Marketing Service Poultry Program's Grading Branch.

Although eggs are widely known as breakfast entrees, they also perform in many other ways for the knowledgeable cook. Eggs are an excellent source of amino acids. The structure and characteristics of an egg include its color, shell, white, yolk, air cell, chalaza, germinal disc, and membranes. Classification is determined by interior and exterior quality and designated by the letters AA, A, and B. In many egg-packing plants, the USDA provides a grading service for shell eggs. Food manufacturers have taken different approaches to reducing the cholesterol in eggs—from physically separating cholesterol from the yolk to formulating yolks from other products and combining these with the albumen. Producers are also trying to reduce the cholesterol and change the fat content of eggs by changing the genetics of chickens.

REVIEW QUESTIONS

Success in any career requires knowledge. Test your knowledge of this chapter by answering these questions or solving these problems.

1. What is another word for egg white?
2. A Julian date of 250 represents what month and day?
3. Name three cooking functions of eggs.
4. Identify four meat products from poultry.
5. One dozen eggs weighs 21 ounces. What is the size of these eggs?
6. What are the parts of an integrated meat chicken production company?
7. Explain the difference between Grades AA, A, and B eggs.
8. Discuss factors affecting the tenderness and the flavor of poultry meat.
9. Describe the purpose of a Contract Acceptance Certification.
10. How does the composition of poultry meat compare to beef and pork?

STUDENT ACTIVITIES

1. Visit a grocery store and make a list of all poultry products. Separate list items in terms of canned, frozen, and processed. Report your findings.
2. Visit restaurants and find out how poultry is featured in the menu. Develop a report or presentation on your findings.

3. Conduct a taste test of some of the wide variety of processed poultry meats such as hams, hot dogs, lunchmeats, sausages, and salami. Compare the taste of these to the same products traditionally prepared from red meats.
4. Compare the nutrients in one egg to those found in a fast-food hamburger, a candy bar, or another food you eat often. Report your findings. Consult Table B-8.
5. Obtain a dozen organic eggs and a dozen regular commercial eggs. Compare cost per dozen, packaging, and the appearance of the eggs. Cook some of each type and compare their tastes.
6. Watch the YouTube video featuring Temple Grandin giving a tour of a turkey farm and processing plant: https://youtu.be/852zxDEAR-Q. Report on what you learned.
7. Compare the cost of purchasing a whole chicken to that of purchasing the individual parts.
8. Evaluate and compare the nutritional value of 1 pound of beef steak and 1 pound of turkey breast meat with the skin on: protein, fat, and calories. Use Table B-8.

ADDITIONAL RESOURCES

- Judging Live Birds: http://www.caes.uga.edu/extension/dawson/4h/documents/JudgingLiveBirds.pdf
- FFA Poultry Judging: https://www.youtube.com/watch?v=v_G-6ENNNd8
- American Egg Board: http://www.aeb.org/
- U.S. Poultry & Egg Association: http://www.uspoultry.org/
- National Chicken Council: http://www.nationalchickencouncil.org/
- The Incredible Edible Egg: http://www.incredibleegg.org/

Internet sites represent a vast resource of information. Although those provided in this chapter were vetted by industry experts, you may wish to further explore the topics discussed in this chapter using a search engine such as Google. Keywords or phrases may include the following: *poultry production, egg production, poultry processing, nutritional value of poultry, nutritional value of eggs, turkey production, organic eggs, poultry inspection*, and *poultry products*. In addition, Table B-7 provides a listing of useful Internet sites that can be used as a starting point.

REFERENCES

Bell, D. D., and W. D. Weaver. 2002. *Commercial chicken meat and egg production*, 5th ed. New York: Springer

Rose, S. P. 1996. *Principles of poultry science*. Oxfordshire, UK: CABI.

Scanes, C. G., G. Brant, and M. E. Ensminger. 2003. *Poultry science*, 4th ed. Upper Saddle River, NJ: Prentice Hall.

United States Department of Agriculture, Agricultural Marketing Service. 1998. *Poultry-grading manual.* Agricultural handbook number 31. Washington, DC: Author.

United States Department of Agriculture, Agricultural Marketing Service. 2000. *Egg-grading manual.* Agricultural handbook number 75. Washington, DC: Author.

U.S. Poultry and Egg Association. (n.d.) Poultry and egg production curriculum: http://www.uspoultry.org/educationprograms/PandEP_Curriculum/menu.html

CHAPTER 19

Fish and Shellfish

OBJECTIVES

After reading this chapter, you should be able to:

- Identify three varieties of fish and three varieties of shellfish used for food
- Describe the aquaculture industry
- Recognize the methods used in processing fish and shellfish products
- Discuss the composition of fish and shellfish
- Identify three spoilage issues associated with fish
- Describe two processes that ensure quality
- List four factors that affect the grading of fish
- Identify four fish products and by-products
- Describe two methods for preserving fish
- Explain the methods of inspection during processing

KEY TERMS

aquaculture
breaded
crustaceans
fish protein concentrate (FPC)
freezer burn
glazing
Hazard Analysis and Critical Control Point (HACCP)
mollusks
roe
shucked
surimi

NATIONAL AFNR STANDARD

FPP.01

Develop and implement procedures to ensure safety, sanitation, and quality in food product and processing facilities.

Fish and shellfish provide a source of high-quality protein in the human diet. Because of the demand and popularity of fish and shellfish, many are commercially cultured and processed. Processed fish and shellfish are checked for quality and graded. A variety of fish and shellfish are commercially raised across the United States.

FISH AND SHELLFISH: SALTWATER AND FRESHWATER

Fish (finfish) are classified into saltwater and freshwater varieties. Their flavor depends on the water in which they lived. Fish are also classified on the basis of their fat content—lean being less than 2% fat and fat being more than 5% fat. Common species of edible fish include catfish, trout, cod, halibut, haddock, pollock, salmon, tuna, mackerel, herring, shad, tilapia, and eel. Fish are vertebrates (Figure 19-1).

Shellfish include **mollusks** and **crustaceans.** Mollusks are soft-bodied and partially or wholly enclosed in a hard shell composed of minerals. Examples are oysters, clams, abalone, scallops, and mussels. Crustaceans are covered in a crustlike shell and have segmented bodies (similar to insects). Common crustaceans used for food include lobsters, crabs, shrimp, prawns, and crayfish.

© Halabala2010/Dreamstine.com

FIGURE 19-1 Catfish ready for processing.

FISHING VERSUS CULTURE

Firms that produce, process, and distribute fish and shellfish are located throughout the United States, which is ranked as the third-largest consumer of fish and shellfish behind China and Japan. This supply is provided by commercial fishermen, **aquaculture** producers, and imports.

Aquaculture

Many popular fish and shellfish products in the United States are harvested to their full biological capacity in U.S. waters (Figure 19-2). To help meet the demand for some of these products, several

USDA, Agricultural Research Service (ARS), photo by Stephen Ausmus

FIGURE 19-2 This basket containing 2,000 pounds of catfish will be hauled to a fish-processing plant.

varieties of fish and shellfish are grown in both freshwater and marine aquaculture facilities around the country. These facilities cultivate approximately 30 different species of fish and shellfish and grow a variety of aquatic plants. Some of these products include the following.

Catfish farming is concentrated in Mississippi, Arkansas, Alabama, and Louisiana.

Rainbow trout are grown throughout the United States, with significant production in Idaho.

Oyster and clam production is found along the mid-Atlantic coast, the Gulf of Mexico, and Washington state.

Shrimp and prawns are harvested in facilities in the southern United States, Hawaii, Southeast Asia, South America, and Central America.

Salmon are cultivated in ocean pens in Washington and Maine. Salmon also are partially cultivated in hatcheries on the east and west coasts and are released into the wild.

Other products include baitfish, crayfish (crawfish), hybrid striped bass, tilapia, yellow perch, wall-eye, bass, sturgeon, and shrimp. Alligators are also commercially grown and are included in this category.

COMPOSITION, FLAVOR, AND TEXTURE

On average, Americans eat about 15 pounds of fish and shellfish each year. Scientific reports and government guides commonly cite fish and shellfish as low in fat, easily digestible, and good sources of protein, important minerals, and vitamins.

Fish and shellfish contain high-quality protein with every essential amino acid, just like red meat and poultry. It is also low in fat—and most of that fat is unsaturated. Because many diets now specify unsaturated fat, rather than saturated fat, fish and shellfish make excellent main dishes. Some fish are relatively high in fat such as salmon, mackerel, and catfish. However, the fat is primarily unsaturated.

The cholesterol content of most fish is similar to red meat and poultry—about 20 milligrams per ounce. Some shellfish contain more cholesterol than red meat. Because the fat is mainly polyunsaturated, shellfish may be allowed for some fat- and cholesterol-restricted diets.

Fish is also a good source of B vitamins—B_6, B_{12}, biotin, and niacin. Vitamins D and A are found mainly in fish liver oils, but some high-fat fish are good sources of vitamin A. Fish is a good source of several minerals—especially iodine, phosphorus, potassium, and zinc. Canned fish with edible bones, such as salmon or sardines, are also good sources of calcium. Oysters are a good source of iron and copper. Saltwater fish and shellfish are also excellent sources of iodine. Table B-8 provides the composition of a variety of fish and shellfish products.

Fish, shellfish, and products possess unique flavors and texture. Table 19-1 compares the flavors and textures of a variety of fish, shellfish, and related products.

SPOILAGE

Fresh fish held at 61 °F (16 °C) remains good for only a day—and sometimes even less. At 32 °F (0 °C) finfish may remain good for 14 to 28 days, depending on the species. Some species spoil even faster. All fish spoil more quickly than other meats because bacteria on the skin and in the digestive tract attack all of a fish's tissues once it dies, and these bacteria are often well adapted to cold temperatures. Fish struggle when they are caught, and they convert all their glycogen to lactic acid before death. The fat of fish contains phospholipids with a compound known as trimethylamine. Bacteria and enzymes from the fish split the trimethylamine from the phospholipids to produce the characteristic fishy odor.

Shellfish such as mussels, clams, and oysters that are purchased live in their shells should be put in a shallow pan without water, covered with

TABLE 19-1 Fish and Shellfish Classified by Flavor and Texture

TEXTURE	FLAVOR		
	MILD	MODERATE	FULL
Delicate	Cod Crabmeat Flounder Haddock Pollock Scallops Skate Sole	Black cod Buffalo Butterfish Lake perch Lingcod Whitefish Whiting	Bluefish Mussel Oysters
Moderate	Crawfish Lobster Rockfish Sheepshead Shrimp Walleye pike	Canned tuna Conch Mullet Ocean perch Shad Smelt Surimi products Trout	Canned salmon Canned sardines Mackerel Smoked fish
Firm	Grouper Halibut Monkfish Catfish Sea bass Snapper Squid Tautog Tilefish	Amberjack Catfish Drum Mahimahi Octopus Pompano Shark Sturgeon	Clams Marlin Salmon Swordfish Tuna

Source: U.S. Department of Commerce, Seafood Inspection Program.

moistened paper towels, and refrigerated. Mussels and clams should be consumed within 2 to 3 days and oysters within 7 to 8 days. Shucked shellfish can be placed in sealed containers and frozen. Live lobster and crabs should be cooked the day they are purchased. Proper handling of fish and shellfish is important and cross-contamination should be avoided to lower the risk of food-borne illness.

When purchasing fish, a few guidelines will help the consumer identify fresh and good quality fish. (Figure 19-3). The skin of the fish should be shiny, almost metallic, with color that has not faded. As the fish decomposes, its skin markings and colors become less distinctive. Scales of the fish should be brightly colored and tightly attached to the skin. The gills should be red and free from slime. As fish ages, the gills change color, fading gradually to a light pink, then becoming gray and eventually brownish and greenish.

© tishomir/Shutterstock.com

FIGURE 19-3 Fish and shellfish are common even in small grocery outlets.

If the head is still on the fish, the eyes should be bright, clear, transparent, and full—sometimes even protruding. As the fish decomposes, its eyes become cloudy and may turn pink and shrink. The flesh of whole or dressed fish should be firm, elastic, and not separated from the bones. As the fish decomposes, the flesh becomes soft, slimy, and slips away from the bones. Fish fillets should have a freshly cut appearance and color that resembles freshly dressed fish. Last, but not least, the odor should be fresh or mild—not fishy.

PROCESSING

Most domestic and imported fish and shellfish are processed before they reach consumers. The National Marine Fisheries Service (NMFS) estimates that approximately 1,500 plants manufacture fish and shellfish in the United States. Most are small businesses, and many are family owned. A few concentrate on a single species such as tuna, salmon, and menhaden. Most process several different species to take advantage of the different fisheries in their regions.

Although most of the fish produced by commercial fishermen is processed into seafood products, some fish are processed into products such as animal feeds, fish oil, and a wide variety of other products.

Aquaculture Processing

Processing fish through several steps turns it into a salable product (Figure 19-4). The following steps are typical for catfish processing, but the steps are similar for trout, salmon, and other finfish. Live fish are:

1. Received and weighed at the processing plant
2. Held alive until needed
3. Stunned
4. Deheaded
5. Eviscerated
6. Skinned
7. Chilled
8. Graded for size
9. Frozen or packed in ice
10. Packaged
11. Warehoused for safe keeping
12. Iced
13. Shipped as final product

The finished product may vary, depending on the type of fish and how the fish is being prepared and cooked. Typical finished products include whole fish (no processing), dressed fish (scaled and eviscerated), filleted fish (scaled, eviscerated, and cut), fish steaks, and sticks and cakes (pieces of fish usually preformed in specific shapes).

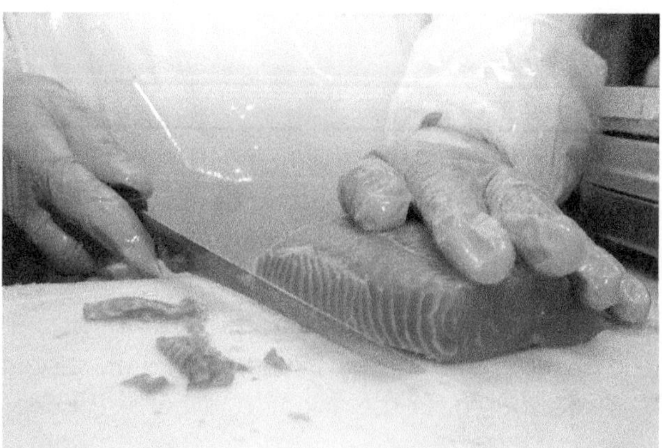

© Marco Jimenez/Dreamstine.com

FIGURE 19-4 Salmon filleting.

Inspection

Unlike the red meat and poultry processing industries, fish processing does not fall under the regulations of the U.S. Department of Agriculture. Before beginning operation, fish processors must contact local county health officials to comply with county health regulations and to obtain health permits. Fish-processing operations also must adhere to standards set forth by the Good Manufacturing Practice Code of Federal Regulations, Title 21, Part 110, and are subject to announced and unannounced inspections by the Food and Drug Administration (FDA).

HACCP. Traditionally, industry and regulators have depended on spot checks of manufacturing conditions and random sampling of final products to ensure safe food. This system tends to be reactive rather than preventative and can be less efficient than the system known as **Hazard Analysis and Critical Control Point**, or **HACCP** (pronounced "hass-sip"). HACCP is a management system in which food safety is addressed through the analysis and control of biological, chemical, and physical hazards from raw material production, procurement, and handling to manufacturing, distribution, and consumption of the finished product.[1] Many of its principles already are in place in the FDA-regulated low-acid canned food industry. In December 1995, the FDA issued a final rule establishing HACCP for the seafood industry. Those regulations took effect December 18, 1997.

Quality

As in other industries, the aquaculture industry considers quality its number-one priority. Without a quality product, product sales would quickly decrease.

To maintain a quality product and promote consumer confidence, the major commercial fish processors contracted voluntarily with National Marine Fisheries Service to have their plants inspected. NMFS is an agency service of the National Oceanic and Atmospheric Administration, an agency within the U.S. Department of Commerce (USDC). Federal inspectors with the NMFS perform unbiased, official inspections of plants, procedures, and products for firms that pay for these services. The inspectors issue certificates that indicate the quality and condition of a facility's products.

The NMFS voluntary inspection program provides for the inspection of products and facilities and the grading of products.

Inspection is the examination of fish (seafood) products by a USDC inspector or a cross-licensed state or USDA inspector. They determine whether the product is safe, clean, wholesome, and properly labeled. The equipment, facility, and food-handling personnel must also meet established sanitation and hygienic standards.

HACCP EQUALS TWO PLUS FIVE PLUS SEVEN

HACCP stands for *Hazard Analysis Critical Control Point*; it is pronounced "hass-sip." Its concept supports the goals of producers, manufacturers, food-service operators, retailers, and distributors to supply safe food—one that will not cause illness or injury to a consumer when used as intended. HACCP is a systematic, two-part process, and it has five preliminary steps and seven principles.

The Two-Part Process

1. Assess hazards, taking into consideration factors that contribute to most outbreaks, and use risk-evaluation techniques to identify and prioritize hazards.
2. Identify control measures that focus on prevention, establishing controls to reduce, prevent, or eliminate safety risks.

(Continues)

HACCP EQUALS TWO PLUS FIVE PLUS SEVEN (continued)

The Five Primary Steps

1. Assemble the HACCP team.
2. Describe the food and its distribution.
3. Identify intended use and consumers.
4. Develop the flow diagram.
5. Verify the flow diagram.

The Seven Principles

1. Conduct a hazard analysis.
2. Determine the critical control points.
3. Establish critical limits.
4. Establish monitoring procedures.
5. Establish corrective actions.
6. Establish verification procedures.
7. Establish recordkeeping and documentation procedures.

HACCP is as simple as 2, 5, 7.

To learn more about HACCP, search the Web or visit these Web sites:

- Hazard Analysis Critical Control Point (HACCP): http://www.fda.gov/Food/GuidanceRegulation/HACCP/
- Food Service HACCP: http://fsrio.nal.usda.gov/haccp/food-service-haccp
- International HACCP Alliance: http://www.haccpalliance.org/sub/index.html

SCIENCE CONNECTION!

Research a fish or shellfish commodity. Design a HACCP plan for that commodity. What is important? How do you minimize food-borne illness risks associated with that product?

Grading

After inspection, grading determines the quality level. Only products that have an established grade standard can be graded. Industry uses the grade standards to buy and sell products. Consumers rely on grading as a guide to purchasing products of high quality. Graded products can bear a U.S. grade mark that shows their quality level. The U.S. Grade A mark indicates that the product is of high quality—uniform in size, practically free of blemishes and defects, in excellent condition, and with good flavor and odor.

A grading scheme used by trout processors is a good example of how grading works to provide a Grade A mark. In determining the grade of processed fish, each fish is scored for the following factors.

- Appearance: The overall general appearance of the fish, including consistency of flesh, odor, eyes, gills, and skin.
- Discoloration: This refers to any color not characteristic of the species.
- Surface defects: These include the presence of fins; ragged, torn, or loose fins; bruises; and damaged portions of fish muscle.
- Cutting and trimming defects: These include body cavity cuts, improper washing, improper deheading, and evisceration defects.

- Improper boning: For boned styles (fillet) only, this refers to the presence of an unspecified bone or piece of bone.

After inspecting each fish, the number of defects are totaled. Grade A is given when the maximum number of minor defects is three or less and there are no major defects. Grade B is given to fish with as many as five minor defects and one major defect. Grade A fish must also possess good flavor and odor for the species, and Grade B must possess reasonably good flavor and odor for the species.

Products

Fresh or frozen fish can be marketed as whole or round, dressed or pan-dressed, fillets, drawn fish, steaks, sticks, or nuggets (Figure 19-5). Whole fish or round fish are just as they come out of the water. Drawn fish have only the entrails removed. Dressed fish are scaled and eviscerated, and they usually have the head, tail, and fins removed. Steaks are crosscut sections of the larger sizes of dressed fish. Fillets are sides of fish cut lengthwise away from the backbone. Sticks are uniform pieces of fish cut lengthwise or crosswise from fillets or steaks. Nuggets are like fillets only smaller. Some fish is manufactured into products such as **breaded**, formed, and imitation products. Some fish is cured, and some is canned.

PRESERVATION

Fish are preserved by drying, salting, curing, or smoking, but not all people like fish preserved in these ways. Some progress is being made on the use of irradiation. Refrigeration, freezing, and canning remain the best methods for preserving the quality of fish.

Large boxes of fish are frozen at temperatures of −22 °F (−30 °C) or below. When individual fish are

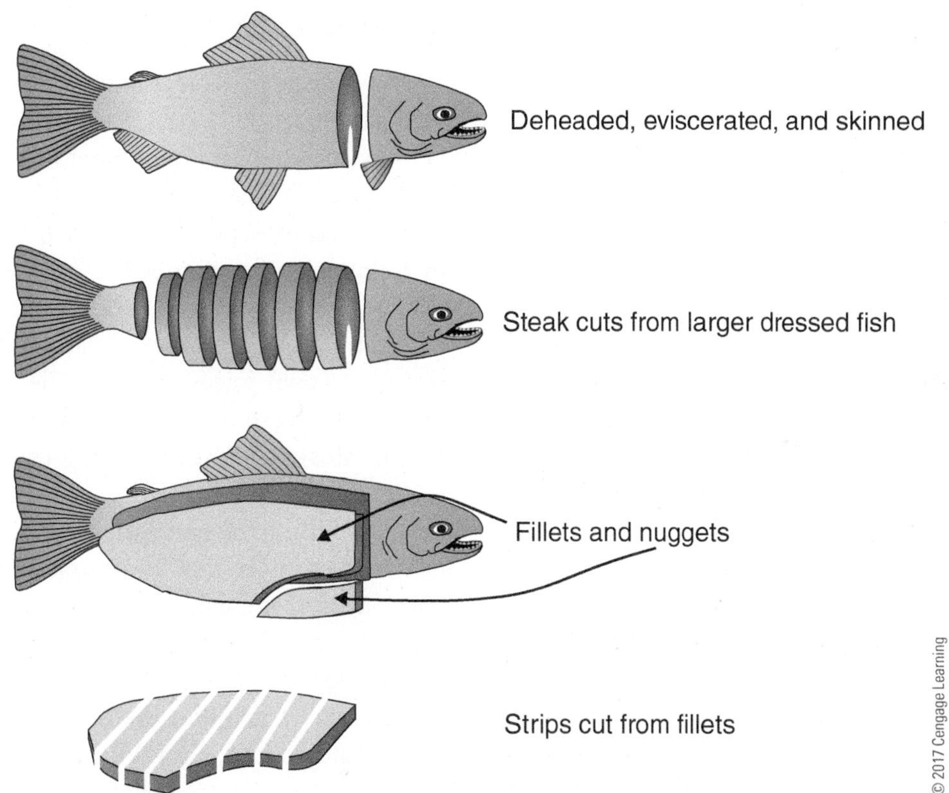

FIGURE 19-5 Catfish cuts are available fresh, packed, on ice, or frozen.

frozen, they are sometimes glazed with layers of ice to protect the surface of the fish from oxidation and from **freezer burn** (drying out). **Glazing** is done by dipping the fish in cold water and then freezing a layer before dipping the fish again. Shrimp are also glazed. Even with glazing, frozen fish require packaging in materials that are airtight and moisture-tight. Prebreaded, precooked fish sticks and individual portions are also frozen. High-quality, low-fat fish can be held frozen at −6 °F (−21 °C) for as long as 2 years.

Fish with higher fat content such as salmon, tuna, and sardines are often canned. Additional fish oil, vegetable oil, or water is often added to the can before it is sealed. Canned fish products have a shelf life of several years. A typical canning operation involves the following steps:

1. Thaw the partially frozen fish received from a fishing vessel.
2. Eviscerate, clean, and sort.
3. Precook.
4. Cool and separate the meat (usually done by hand).
5. Compact meat into shapes to fit cans.
6. Add salt, oil, or water to cans.
7. Vacuum seal cans and sterilize in a retort.

SHELLFISH

Shellfish may be marketed in the shell, **shucked** (removed from the shell), headless (shrimp), and as cooked meat. Shrimp is also sold as peeled, cleaned, and breaded. Shrimp are designated jumbo, large, medium, and small based on the number per pound. Oysters receive similar designations.

FISH BY-PRODUCTS

Parts of the fish—typically the intestines, heads, and gills—as well as less favored fish are not used for human food. These are often ground up, dried, and converted to fish meal for use as fish meal, which is used both as animal feed and as fertilizer.

Fish protein concentrate (FPC) or fish flour is produced from the dehydrated and defatted fish. It is an excellent source of high-quality protein, which can be used to supplement the breads and cereal products of people in many parts of the world.

Roe

Roe is the mass of eggs and sacs of connective tissue enclosing the thousands of eggs. Some people eat the roe of such fish as the shad. Fresh roe is usually cooked by parboiling. Caviar is sturgeon roe preserved in brine.

STORAGE

Fish and shellfish must never sit unrefrigerated for long. If necessary, they should be transported in an ice chest. Seafood with bruises or punctures will spoil more rapidly.

As soon as possible, finfish needs to be refrigerated as close to 32 °F (0 °C) as possible. Fish can be held twice as long at that temperature as it can be at 37 °F (2.8 °C). Fish and shellfish should be cooked within 2 days of purchase. If not, it can be stored by following a few guidelines. Before storing fresh fish, the package should be removed, and the fish rinsed under cold water and patted dry. When fish sets in its own juices, the flesh deteriorates more rapidly. To prevent this, cleaned finfish, whole, fillets, or steaks are placed onto a cake rack so that the fish do not overlap. This rack is placed in a shallow pan. Filling the pan with crushed ice allows the fish to keep more that 24 hours. Ice leaches color and flavor from fish and should not come into contact with the fish. The covered pan is placed in a refrigerator and then drained and iced again as necessary. Each day, the fish, pan, and rack are rinsed and re-iced. Fish with a fishy or ammonia smell after being rinsed should be discarded.

Fish that will not be used within a day or so should be immediately frozen. After rinsing the fish under cold water and patting it dry, the fish should first be wrapped tightly in plastic wrap, squeezing all the air out. It should then be wrapped tightly in aluminum foil and frozen. It should be used within 2 weeks to enjoy its peak quality.

When frozen fish is thawed, it is always thawed in the refrigerator. Thawing at temperatures higher than 40 °F (4.4 °C) causes excessive drip loss and adversely affects taste, texture, aroma, and appearance.

Live oysters, clams, and mussels are stored in the refrigerator at a temperature of about 35 °F (1.7 °C). They should be kept damp but not be placed on ice. Freshwater or an airtight container will kill them.

Freshly shucked oysters, scallops, and clams are stored in their own containers and stored in a refrigerator around 32 °F (0 °C). Surrounding the containers with ice gives the best results.

Live lobster and crab are stored in the refrigerator in moist packaging (seaweed or damp paper strips) but not in airtight containers, water, or salted water. Lobsters should generally remain alive for 24 hours.

Just before opening or cooking scallops, mussels, clams, or oysters in the shell, they should be scrubbed under cold water to clean them. Soaking them in water with flour or cornmeal to encourage the creatures to eat to clean out the grit only shortens their life.

Frozen fish and seafood should be stored at 0 °F (−18 °C) or below. For fish purchased frozen, it is best when used within 2 months.

NEW PRODUCTS

Machines similar to deboning machines are being used to obtain minced fish flesh from filleting wastes and underutilized fish species. The minced fish flesh is washed to remove solubles, including pigments (color) and flavors. This leaves an odorless, flavorless, high-protein product called **surimi**. This can be combined with other flavors and colors. Then it can be extruded in shapes resembling other products such as crabmeat and lobster (Figure 19-6).

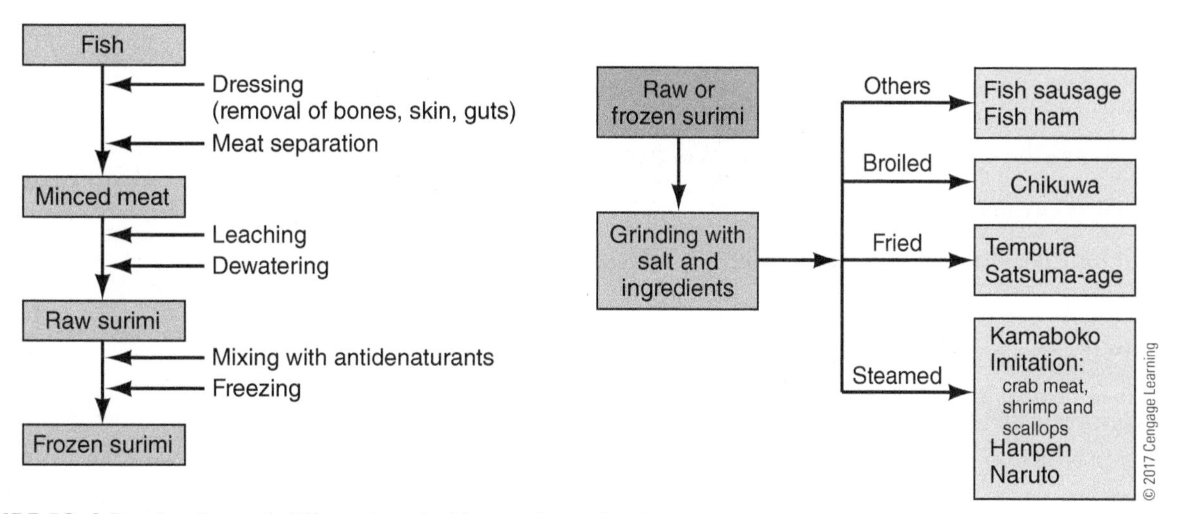

FIGURE 19-6 Production of different surimi-based products.

SUMMARY

Fish includes saltwater and freshwater finfish such as catfish, trout, halibut, salmon, tuna, herring, and eel. Shellfish include mollusks and crustaceans such as clams, oysters, lobsters, crabs, shrimp, and crayfish. Fish and shellfish are provided by commercial fishing and aquaculture producers. Fish and shellfish provide a high-quality protein and are also good sources of B vitamins, calcium, phosphorus, iodine, and potassium.

Fish spoil easily, so they require strict processing and preservation procedures to maintain quality. Fish processing operations adhere to standards in the Good Manufacturing Practice Code and processors use the HACCP method to monitor quality. After inspection, grading determines the quality. Grade A indicates a product of high quality that is uniform in size, free of blemishes and defects, in excellent condition, with good flavor and odor.

Fish and shellfish are marketed fresh or frozen. Fish can be marketed as whole, dressed, pan-dressed, filleted, steaks, sticks, or nuggets. Fish protein concentrate or fish flour are fish by-products used for humans. Roe is fish eggs used for human food also. Surimi represents a new manufactured product made from pieces of fish.

REVIEW QUESTIONS

Success in any career requires knowledge. Test your knowledge of this chapter by answering these questions or solving these problems.

1. Fish are classified based on what factor?
2. Describe the difference between fish and shellfish.
3. Besides protein, what substances are fish a good source of?
4. Discuss three indications of spoilage in fish.
5. List the steps for catfish processing.
6. What does HACCP stand for?
7. Name the four grading factors for fish.
8. What are the head, gills, and intestines of fish used for?
9. What is caviar?
10. List four methods of preserving fish.

STUDENT ACTIVITIES

1. Track your diet for 1 week and report on how much fish and what type of fish you eat in a week. Include any product containing any fish ingredients. Create a list of products and how much was consumed.
2. Compare the protein and energy content of a type of fish and a shellfish to that of steak and chicken breast. Use Table B-8. Create a comparison chart.
3. Name some common lobsters and crabs frequently seen on restaurant menus. Make a list that compares the various methods in which these products are prepared.

4. Create a classroom display of color pictures of several types of edible saltwater and freshwater fish and types of mollusks and crustaceans.
5. Create a sample (taste) test of some unique types of fish and shellfish. Use Table 19-1 for ideas. When conducting the taste test, inform the instructor so that student allergies will be taken into consideration.

ADDITIONAL RESOURCES

- Food and Agriculture Organization of the United Nations—Fisheries and Aquaculture Department: http://www.fao.org/fishery/countrysector/naso_usa/en
- Meaty Fish: http://www.fooduniversity.com/foodu/seafood_c/resources/composition/meaty_fish.htm
- Comparative Study of Three Different Bangladeshi Smoke-Dried Lean Fishes: http://pubs.sciepub.com/ajfst/2/6/7/
- Shellfish Processing: http://www.freshfromflorida.com/Divisions-Offices/Aquaculture/Agriculture-Industry/Shellfish/Shellfish-Processing
- What Is Aquaculture? http://www.nmfs.noaa.gov/aquaculture/what_is_aquaculture.html

Internet sites represent a vast resource of information. Although those provided in this chapter were vetted by industry experts, you may wish to further explore the topics discussed in this chapter using a search engine such as Google. Keywords or phrases may include the following: *specific types of fish and shellfish, saltwater fish, freshwater fish, aquaculture, HACCP, fish processing, fish grading, fish by-products, fish roe, FPC (fish protein concentrate).* In addition, Table B-7 provides a listing of useful Internet sites that can be used as a starting point.

REFERENCES

Bartlett, J. 1996. *The cook's dictionary and culinary reference.* Chicago: Contemporary Books.

Brody, J. E. 1981. *Jane Brody's nutrition book.* New York: Bantam Books.

Cremer, M. L. 1998. *Quality food in quantity. Management and science.* Berkeley, CA: McCutchan Publishing.

Ensminger, A. H., M. E. Ensminger, J. E. Konlande, and J. R. Robson. 1994. *Foods and nutrition encyclopedia* (2 vols.). Boca Raton, FL: CRC Press.

Gardner, J. E. (Ed.). 1982. *Reader's digest. Eat better, live better.* Pleasantville, NY: Reader's Digest Association.

Horn, J., J. Fletcher, and A. Gooch. 1996. *Cooking A to Z: The complete culinary reference source.* Glen Ellen, CA: Cole Publishing Group.

Parker, R. 2012. *Aquaculture science*, 3rd ed. Clifton Park, NY: Cengage.

ENDNOTE

1. FDA, HACCP: http://www.fda.gov/Food/GuidanceRegulation/HACCP/ucm2006764.htm. Last accessed July 8, 2015

CHAPTER 20

Cereal Grains, Legumes, and Oilseeds

OBJECTIVES

After reading this chapter, you should be able to:

- Diagram the general structure of a grain
- Name three cereal grains
- Describe the general composition of grains, legumes, and oilseeds
- Identify three properties of starch
- List four factors that must be controlled when cooking starch
- Describe the process of milling grain to flour
- Identify five types of wheat flour and describe their differences
- Explain the classes of wheat and grades of flour
- Identify the type of flours other than wheat flour
- List the steps in corn refining
- Name four products derived from corn
- Explain the processes that take place during baking

NATIONAL AFNR STANDARD

FPP.03

Select and process food products for storage, distribution, and consumption

KEY TERMS

amylopectin
amylose
bioproducts
bleached
bran
dextrinization
endosperm
extraction
germ
insoluble
middlings
milling
mill starch
refining
retrogradation
shorts
soluble
starch
steeping
straight grade
waxy
weeping

OBJECTIVES (continued)

- **List four oilseeds and indicate the use of their products**
- **Define legumes and discuss their general use in the food industry**
- **Name four general categories of products from soybean extraction**
- **Identify five food products of soybean extraction**

Cereal grains and legumes supply energy (starch) and protein to billions of people. As foods, they often are consumed as seeds, but more often they are consumed in some processed form such as flour, syrup, or vegetable protein extract. Corn refining and soybean extraction provide a wide variety of products for food and technical uses.

© 2017 Cengage Learning

FIGURE 20-1 Because of the appeal of various types of grains, manufacturers have sought to meet consumer demand by expanding cereal markets to use many different types of grains and seeds.

CEREAL GRAINS

Many types of grains and seeds are used throughout the world. For example, consumers can purchase bean flour, peanut flour, sunflower flour, buckwheat flour, soy flour, and many others. Cereal markets have expanded the range of uses of these sources because of their consumer appeal and because manufacturers respond to demands for these products with improved efficiencies and productivity as well as advertising, marketing, lowering prices, and cutting costs (see Figure 20-1). The focus of this chapter is on one of the most important grains—wheat—with some discussion of the characteristics of others such as corn, rice, oats, and rye.

General Structure and Composition

All whole grains have similar structures—outer bran coats, a **germ**, and a starchy endosperm portion (see Figure 20-2). Cereal products vary in composition, depending on the parts of the grain used.

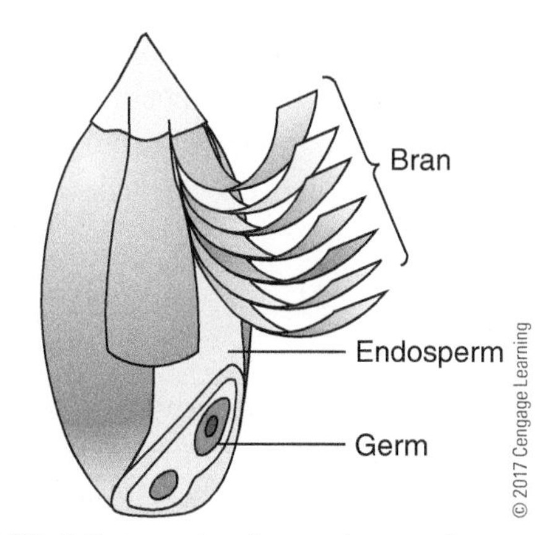

FIGURE 20-2 The parts of a grain seed.

The outer layers of a kernel—the **bran**—makes up approximately 5% of the kernel. The bran is chiefly cellulose but contains much of the mineral and some of the vitamins in the kernel.

As milled, it may also contain some germ and a small amount of the aleurone layer.

The aleurone layer beneath the bran layer is rich in proteins, phosphorus, and thiamine. It also contains **starch**. Approximately 8% of the kernel is aleurone layer.

The **endosperm** is the large central portion of the kernel and contains most of the starch. It also contains most of the protein of the kernel but minimal fiber or minerals and only a trace of fat. The endosperm constitutes some 82% of the kernel.

The germ is the small structure at the end of the kernel. It is rich in fat, protein, and minerals and contains most of the kernel's riboflavin content.

The chaffy coat that covers the kernel during the grain's growth is eliminated during harvesting or during milling.

Cereals are processed grains that are generally 75% to 80% carbohydrates. Fiber is also an important attribute of cereal, especially bran cereals, which may contain 10 grams to 26 grams of fiber per cup. Cereals contain both **soluble** and **insoluble** fiber. Insoluble fiber is good for the human digestive tract and helps reduce the risk of certain cancers. Soluble fiber, the type that lowers blood cholesterol, originates in the endosperm of grain and is found in oats, legumes, fruits, and vegetables.

STARCH

Starch is a storage form of carbohydrate deposited as granules or aggregates of granules in the cells of plants. Sizes and shapes of granules differ in starches from various sources, but all are microscopic in size (see Figure 20-3).

The parts of plants that serve most prominently in the storage of starch are seeds, such as cereal grains and legumes, and roots and tubers such as parsnips, potatoes, and sweet potatoes. Some starches are derived from the cassava root

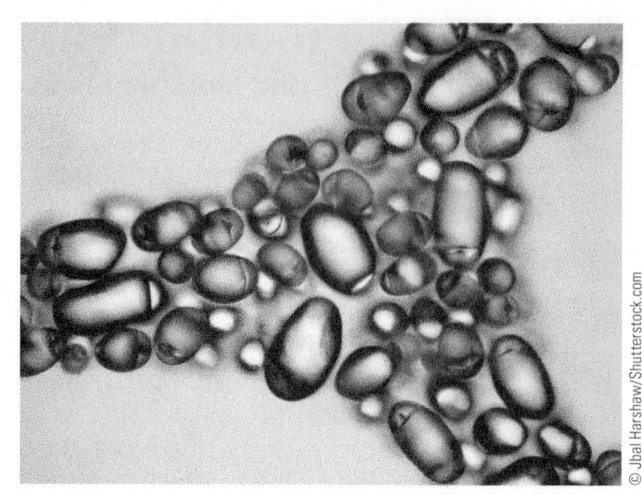

FIGURE 20-3 Highly magnified starch granules from a potato.

(marketed as tapioca) and from the pith of a tropical palm (marketed as sago). Starch may be hydrolyzed to form glucose, but several intermediate products, such as dextrin and maltose, are formed first (see Chapter 3).

Starch granules are made up of many starch molecules arranged in an organized manner. These molecules are of two types, called fractions, of starch: **amylose** and **amylopectin** (refer to Figure 3-3). The amylose is a polysaccharide of glucose. It contributes gelling characteristics to cooked and cooled starch mixtures. Amylopectin is a highly branched polysaccharide of glucose that provides thickened properties but does not usually contribute to gel formation. Most starches are mixtures of the two fractions. Corn, rice, and wheat starches contain 16% to 24% amylose with 76% to 84% being amylopectin. Tapioca and potato starches are lower in amylose content than corn, rice, and wheat. Certain strains of corn, rice, grain sorghum, and barley have been developed that are practically devoid of amylose. These are called **waxy** varieties because of the waxy appearance of the cut grain.

Properties of Starch

Starch granules are insoluble in cold water: A nonviscous suspension is formed in which the granules gradually settle to the bottom. When cooked, a colloidal dispersion results in a starch paste. Some pastes form gels, and some are nongelling. Some are opaque, and some are clear, semiclear, or cloudy in appearance and soft or cohesive in texture. In general, the pastes made with cereal starches, like corn and wheat, are cloudy in appearance, whereas those from root starches, like potato and tapioca, are clearer. Cooked and cooled mixtures of starches containing somewhat larger proportions of amylose, like ordinary cornstarch, tend to become rigid on standing or to gel. Tapioca and potato starch, containing a little less amylose, have fewer tendencies to gel.

Tapioca and potato starch pastes are cohesive and tend to be stringier, partly because of the chain length of the amylose molecules. The "skin" on the surface of cooked starches and cereals mainly results from amylose reverting to an insoluble state. Waxy varieties of starch, like waxy cornstarch, form thickened viscous pastes that do not gel on cooling. The stringy characteristics of some of these waxy starch pastes may be eliminated if the starch is modified to produce cross bonding between branches of the amylopectin molecules. High-amylose starches have also been produced. These offer possibilities for the development of edible protective coatings for individual pieces of food such as dried fruits, nuts, beans, and candies.

Effect of Dry Heat. When dry heat is applied to starch or starch-containing foods, the color changes to brown, the flavor changes, and the starch becomes more soluble and has reduced thickening power. This process is called **dextrinization**. Brown gravy is usually relatively thin in consistency if the flour is browned in the process of making the gravy. Dry-heat dextrins are formed in the crust of baked flour mixtures, on toast, on fried starchy or starch-coated foods, and on various ready-to-eat cereals.

Effect of Moist Heat. When starches are heated with water, the granules swell and the dispersion increases in viscosity until it reaches a peak thickness. The dispersion also increases in translucency. The term *gelatinization* is used to describe these changes. The changes appear to be gradual over a temperature range during gelatinization. The granules are of varying sizes and do not swell at the same rates. The gelatinization temperature ranges also vary from one kind of starch to another.

Potato starch begins to gelatinize at a lower temperature than does cornstarch. Gelatinization is usually complete by 190 °F to 194 °F (88 °C to 90 °C). After maximum swelling has occurred, the granule ruptures. Continued heating under controlled conditions after gelatinization results in decreased thickness. Boiling or cooking starchy sauces and puddings in the home for longer periods of time usually does not produce thinner mixtures because the loss of moisture by evaporation is usually not controlled. The loss of moisture results in increased concentration of the starch and causes increased thickness, which offsets the decreasing thickness.

Gel Formation. The presence of amylose encourages the formation of a gel in cooked and cooled starch mixtures. Waxy varieties of starch without amylose do not form gels. The amylose molecules become more soluble as the granules are disrupted and swell during gelatinization. On cooling, they tend toward **retrogradation**, or gel formation, as the starch mixture stands. Cornstarch puddings become thicker and more gelled when stored overnight in the refrigerator. Overall, the thickness of a starch paste on heating is not directly related to the strength of gel formed on cooling.

Factors Requiring Control

To obtain uniformity in the cooking of starches, five conditions must be standardized and controlled:

1. Temperature of heating
2. Time of heating
3. Intensity of stirring
4. pH of the mixture
5. Addition of other ingredients

Temperature and Time of Heating. Gelatinization temperatures vary for different starches.

The larger granules start to swell first and at lower temperatures than the smaller sizes, which explains why there is no exact temperature of gelatinization. It is a change that occurs over a range of temperatures. More concentrated mixtures show higher viscosity at lower temperatures than do less concentrated mixtures because of the larger number of granules that swell. Under controlled conditions, starch pastes that are heated rapidly are slightly thicker than similar pastes that are heated slowly.

Intensity of Stirring. To obtain a uniform consistency, stirring while cooking starch mixtures is desirable in the early stages, but it also has value in accelerating gelatinization. However, if stirring is too intense or continues too long, it accelerates the breakdown or rupturing of the starch and decreases viscosity.

pH of the Mixture. A low pH or acidity causes some fragmentation of starch granules, affecting swelling and thus decreasing the viscosity of starch pastes. Mixtures with both low and high pH values—pH 2.5 and pH 10.0—gelatinize more rapidly than those with intermediate pH values—pH 4.0 to pH 7.0—and they also break down more rapidly with a decrease in viscosity. In lemon pie filling, for example, cooking the starch paste with lemon juice might be expected to have some effect in decreasing the thickness of the pudding. This degree of acidity does not seem to have a major effect. However, a better natural flavor is obtained when lemon juice is added at the end of the cooking period, reducing the loss of volatile flavor constituents.

Addition of Other Ingredients. Various other ingredients are commonly used with starch in food preparation. Some ingredients have a distinct effect on gelatinization and the breakdown of starch and on the gel strength of the cooled mixture. Sugar (disaccharides or monosaccharides) is one ingredient used in many starchy mixtures. If used in a relatively large amount, it interferes with the complete gelatinization of the starch and decreases the thickness of the pastes. It competes with the starch for water. If not enough water is available for the starch granules, they cannot swell sufficiently.

High concentrations of disaccharides (like sucrose) are more effective in inhibiting gelatinization than are equal concentrations of monosaccharides (like glucose). At a concentration of 20% or more, all sugars and syrups cause a noticeable decrease in the gel strength of starch pastes.

Egg, fat, salt, and dry milk solids also hinder the gelatinization of starch granules.

Handling of Cooked Starch

Cooked mixtures that are not to be used immediately should be protected against bacterial contamination because they are a good medium for bacterial growth.

Acid mixtures should be cooled quickly to minimize the hydrolyzing effect of acid on the starch. The manner of cooling a starch paste affects its properties. Placing the vessel used for cooking in a refrigerator or over cold water or an ice bath and stirring gently cools the mixture in a short time.

Weeping

Starch granules absorb water and swell during cooking. During the early stages of swelling, the water is so loosely held that undercooked starch releases the water easily while cooling and during storage. If the undercooked mixture is refrigerated, **weeping** is accelerated. At the same time, the viscosity of the paste may decrease, sometimes suddenly. This is called *collapse* of the gel.

MILLING OF GRAINS

Primitive people used stones, wood, and other natural objects to grind their cereals. Eventually, they converted to water-driven mills with large mill stones. Modern **milling** replaces these mill stones with rollers (Figure 20-4).

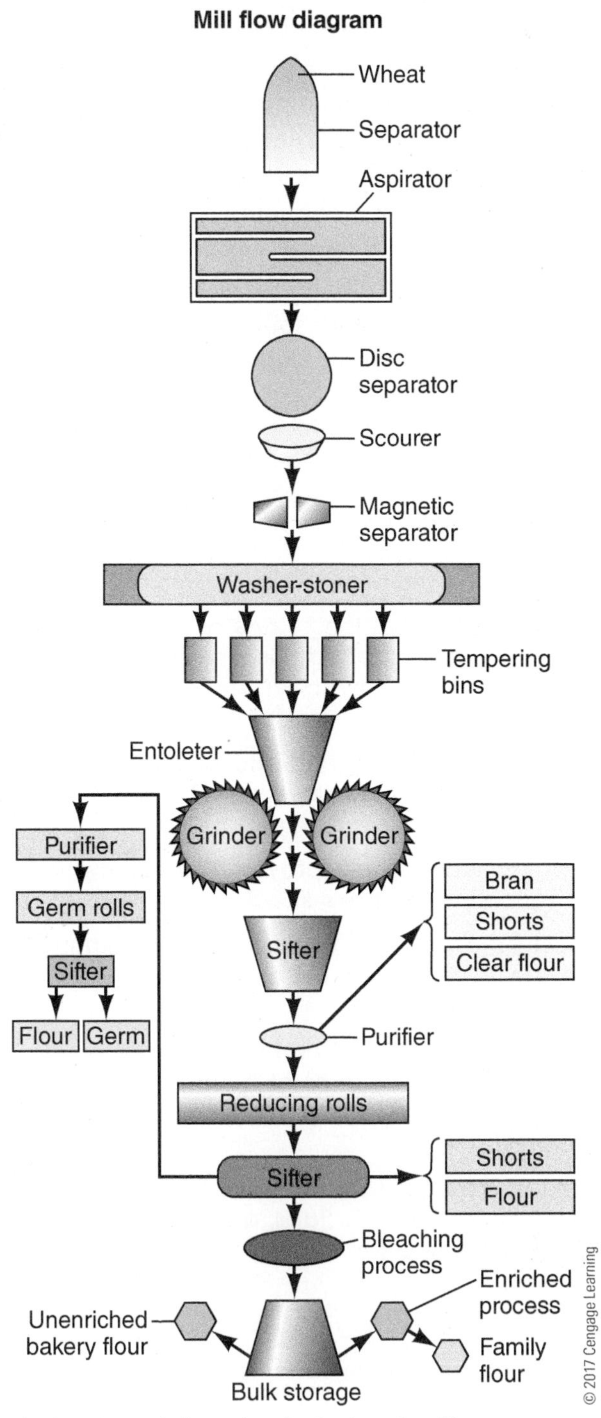

FIGURE 20-4 A flowchart of wheat milling.

Although not all grain is milled, most of the wheat used in the United States has been milled to some extent. The aim of milling is to separate the bran covering, germ, and endosperm to the extent desired. Generally, the endosperm is then pulverized. If milling is adequate and correct, then 100 pounds of wheat will yield 72% to 75% straight flour. The rest of the wheat kernel, the bran, the germ, and **shorts** used to be for cattle feed only. Now nutritionists, health personnel, and a variety of food faddists have expanded the use of these waste products into a "viable health food." A sizable market exists for bran and germ. The bran is indigestible but is used in high-fiber diets.

The inner portion of the kernel, which is granulated with difficulty, is known as the **middlings**. After separation from the bran, it is fed through a series of smooth rollers that further reduce the size of the particles and produce finer flour. About six to eight streams of flour are obtained from the rolling and sifting of purified middlings.

From the many streams of flour resulting from the modern roller process, various grades and types of flours are made. The streams vary in their content of bran, germ, and gluten.

A final stage in the production of white flour is often bleaching and maturing. Freshly milled, unbleached flour is yellowish in color chiefly because of the presence of carotenoid pigments (carotin). When used for baking bread, it produces a small and fairly coarse-textured loaf. If the flour is stored for several months, the color becomes lighter and the baking qualities improve. Food scientists found that the addition of certain chemical substances to the freshly milled flour will produce similar effects in a much shorter period of time. The Food and Drug Administration (FDA) permits the use of nitrogen trichloride and nitrogen tetroxide, chlorine dioxide, benzyl peroxide, acetone peroxides, and azodicarbonamide to bleach and mature flour. The flour must then be labeled **bleached**. Most of these substances have a maturing effect that improves baking qualities besides bleaching the flour.

Because the endosperm represents approximately 84% of the total kernel, theoretically about that much white flour should be obtained by milling, but in actual practice only 72% to 75% is separated as white flour. The kind and composition of flour depends on these characteristics:

1. Class of wheat used
2. Conditions under which the wheat is grown
3. Degree of fractionation

Classes of Wheat

Wheats are classed as hard, soft, and durum (a special class of hard wheat). Durum wheat is used almost exclusively for producing semolina—granular flour of high gluten content used in the manufacture of macaroni or pasta products. The geographical areas producing most of the hard spring wheats are the north central part of the United States and western Canada. Hard winter wheats are grown mainly in the south central and middle central states. Soft winter wheat is grown east of the Mississippi River and in the Pacific Northwest. Because climatic and soil conditions affect the composition of wheat, wide variations may occur within the classes.

BREADS WITH SYMBOLIC MEANING

Pretzels and croissants are two popular breads whose origins have been forgotten.

Pretzels are shaped like knots. Traditionally, pretzels are made out of long strips of dough folded over into a loose, trefoil knot before being baked. They have been shaped this way since the seventh century. Thought to bring good luck and prosperity, pretzels have been called the world's oldest snack food.

Medieval monks invented pretzels to carry deep, religious meanings. The folded strips of dough resemble the folded arms of someone who is praying in the usual manner in those days, while the three holes represent the Christian Holy Trinity.

In medieval times, pretzels were given to children as rewards for learning their prayers. Today, they have lost their religious meaning, but they are still among the world's most popular snacks.

The delicate, flaky croissant or crescent roll is a baked pastry that is curved with pointed tips. Although its popular name in English speaking countries is French, the roll itself is of Austrian origin, commemorating a Turkish shape.

According to the most popular story, in 1683 the Ottoman Turks invaded Vienna by trying to tunnel under the city's walls. The Turks were successfully repelled, thanks to the vigilance of the only people who were awake during the night time raid: the bakers. In celebration of the victory, the bakers created the croissant, shaping it like the crescent found on the Ottoman flag.

Because there are several different stories of this event, the true details may be different. But all sources agree that the croissant's shape is the crescent found on Islamic flags, and that it was created in celebration of the Austrian victory over the Ottomans. The Ottoman siege of Vienna lasted 60 days.

For more about croissants and pretzels, as well as recipes, visit the following Web sites:

- Interesting History of Croissants: http://labadiane-hanoi.com/spice-conners/interesting-history-croissants/
- The Pretzel—A Twisted History: http://www.history.com/news/hungry-history/the-pretzel-a-twisted-history
- Is the Croissant Really French? http://www.smithsonianmag.com/travel/croissant-really-french-180955130/?no-ist
- The Formidable Story Behind the French Croissant: http://www.frenchmoments.eu/the-formidable-story-behind-the-french-croissant/
- The Pretzel Museum: http://www.ushistory.org/tour/pretzel-museum.htm

Grades of Flour

Millers grade white flours on the basis of the four streams used to make them. **Straight grade** theoretically should contain all flour streams from milling, but actually 2% to 3% of the poorest streams is withheld. The poorest streams are those containing most of the outer layers of the endosperm and fine bran particles. Little flour on the market is straight grade. Patent flours come from the more refined portion of the endosperm and may be made from any class of wheat. They are divided into the following, in order of quality:

- First patent
- Second patent
- First clear
- Second clear
- Red dog

Most patent flours on the market include about 85% of the straight flour. Clear grade is made from streams withheld in the making of patent flours.

Types of White Flour

Wheat flour contains a high percentage of starch and 10% to 14% protein, most of which is gluten. Gluten is important for binding ingredients together and for its elastic or stretching property. Within limitations, various types of flour may be used interchangeably by altering the proportions of the other ingredients of the mixture. Four common types of white flour include bread, all-purpose, pastry, and cake. White flours produced from wheats grown in some areas are of such poor baking quality that improvers (a gum such as xanthan or guar) are added.

Bread flour has a slightly higher percentage of gluten but it is much stronger and more elastic than in other types of flour. Most flours are blended; a strong bread flour is made chiefly from hard wheat. Such flour is valuable chiefly for yeast breads but may be used in quick breads. It is considered stronger than all-purpose flour because it has more protein in it.

FIGURE 20-5 All-purpose flour is just one of the many types of flours that are available to the consumer.

All-purpose flour (Figure 20-5) has a less strong and elastic gluten than bread flour. It may be a blend of hard and soft wheat flour or may be made entirely from hard or soft winter wheats. Although designated all-purpose flour, it is more popular for some quick breads than for other uses.

Pastry flour or cake flour is made from soft winter wheat and contains a weaker quality of gluten and a slightly lower percentage of gluten than is found in bread and all-purpose flours. It is softer than the other flours. Its chief use is for cakes and pastries, although it is also useful for all quick breads. Pastry or cake flour provides a less chewy and more tender texture.

Whole-grain and stone-ground flour is milled to leave some of the bran intact. Both typically retain more oil and are more flavorful than other flours.

Cake flours are specially prepared to reduce the gluten content to approximately 7%. They are best made from soft wheat and are finely ground. They are usually highly bleached with chlorine. High starch content and the weak quality of their gluten make cake flours best and most widely used for cakes.

According to a federal ruling, the terms *whole wheat* and *graham* are synonymous terms and refer to products made from the whole-wheat kernel with nothing added or removed. Graham flour was named for Sylvester Graham, an American food-reform advocate.

Enriched Flour

Enriched flour is white flour to which specified B vitamins and iron have been added. Optional ingredients are calcium and vitamin D. The enrichment of bakers' white bread and rolls was made compulsory by the federal government in 1941 as a war measure to improve the nutritional status of the people. After the war, enrichment became voluntary, although many states have passed laws requiring that all white flour sold within their boundaries be enriched. If enrichment is practiced, it must conform to FDA standards.

Gluten

The proteins of wheat are so important to the usefulness of flour in baked products that they have been studied for many years. A variety of proteins have been extracted from wheat. Some 85% of the proteins of white flour are relatively insoluble. These insoluble proteins separate into two fractions called *gliadin* and *glutenin*. When flour is moistened with water and thoroughly mixed or kneaded, these insoluble proteins combine to form gluten. Gluten may be extracted from dough by a thorough washing with water to remove the starch. The extracted gluten has elastic and cohesive properties. When gliadin and glutenin are separated from the gluten, the gliadin is a syrupy substance that may bind the mass together and the glutenin exhibits toughness and rubberiness that probably contributes strength.

WHAT'S ALL THE FUSS ABOUT GLUTEN?

A gluten-free diet is a diet that excludes the protein gluten. Gluten is found in grains such as wheat, barley, rye, and a cross between wheat and rye called *triticale*.

A gluten-free diet is primarily used to treat celiac disease. Gluten causes inflammation in the small intestines of people with celiac disease. Eating a gluten-free diet helps celiac sufferers control their signs and symptoms and prevent complications.

Initially, following a gluten-free diet may be frustrating, but with time, patience, and creativity, people find there are many foods they already eat that are gluten free, and they typically find enjoyable substitutes for gluten-containing foods.

People who do not have celiac disease also may have symptoms when they eat gluten. This is called *nonceliac gluten sensitivity*.

People with nonceliac gluten sensitivity may benefit from a gluten-free diet. But people with celiac disease must be gluten free to prevent symptoms and disease-related complications.

Anyone who is considering a gluten-free diet should first consult a dietitian who can answer questions and offer advice about how to avoid gluten while still eating a healthy, balanced diet.

Foods Allowed

Many healthy and delicious foods are naturally gluten free:

- Beans, seeds, and nuts in their natural, unprocessed forms
- Fresh eggs
- Fresh meats, fish and poultry (as long as they are not breaded, batter coated, or marinated)
- Fruits and vegetables
- Most dairy products

(Continues)

WHAT'S ALL THE FUSS ABOUT GLUTEN? (continued)

It is critical to make sure these foods are not processed or mixed with gluten-containing grains, additives or preservatives. Many grains and starches can be part of a gluten-free diet, such as:

- Amaranth
- Arrowroot
- Buckwheat
- Corn and cornmeal
- Flax
- Gluten-free flours (rice, soy, corn, potato, bean)
- Hominy (corn)
- Millet
- Quinoa
- Rice
- Sorghum
- Soy
- Tapioca
- Teff

People with celiac disease who eat a gluten-free diet experience fewer symptoms and complications from the disease. They must continue to eat a strictly gluten-free diet for the remainder of their lives. In some severe cases, a gluten-free diet alone cannot stop the symptoms and complications of celiac disease, and additional treatment is needed.

For more information on a gluten free diet, check out the Mayo Clinic's Nutrition and Healthy Eating at http://www.mayoclinic.org/healthy-lifestyle/nutrition-and-healthy-eating/in-depth/gluten-free-diet/art-20048530?pg=1

Other Flours

Cornmeal, a granular product made from either white or yellow corn, is commonly used in several types of quick breads (Figure 20-6). Its chief protein, zein, has none of the properties of wheat gluten. To avoid a crumbly product, however, cornmeal must be combined with some white flour to bind it. Corn flour has the same properties as cornmeal except that it is finer. It is used chiefly in commercial pancake mixes.

Barley flour is rarely used in baked products requiring a gluten structure because it contains no gluten-forming proteins. Barley flour can be used, however, in extruded cereals, cakes, cake donuts, cookies, and crackers. The oligosacchrides and pentosans (types of carbohydrates) are useful in these flours. However, it is actually used as a cereal itself, although it is not currently as popular as in previous generations.

FIGURE 20-6 Wheat, rice, corn, and other seeds are used in a variety of breads and pastas.

Although oat flour is not a common flour, it does have some use in extruded cereal products, cakes, cookies, and crackers. The oat flakes have been used in cookies and oat breads. There are different size of flakes themselves. In addition, some products use coarsely ground groats and dehulled oats. The milling and cleaning of oats is similar to that of wheat.

Rice flour has been used for many products as a substitute flour for those who have an allergy to wheat flour. It cannot be used in products that require gluten, however. Rice flour is basically rice starch. Although other flours are available, they have little use, either because they lack gluten or some similar constituent or because they are little known except in certain areas. Potato flour is used in some countries and, like rice flour, is chiefly starch.

Buckwheat is not a seed of the grass family. It is the seed of an herbaceous plant, but because it contains a glutenous substance and is made into flour, it is commonly considered with grain products. Fine buckwheat flour includes little of the thick fiber coating and in that respect is similar to refined white flour. It is prized for its distinctive flavor and is commonly used in the making of pancakes.

Rice

Rice is frequently categorized into short, medium, and long grain. American long-grain white rice has large, fluffy grains with a woody flavor. Glutinous white rice is a short-grained, sticky, and chewy rice used to make balls or sushi. Long-grain brown (Texmati) rice is a beige, nutty-flavored, fluffy, light rice. Rice bran is ground bran and a soluble fiber. Rice flour is ground from brown or white rice and it is used as a thickener and dusting powder. Rizcous is a cracked rice that may be used interchangeably with bulgur. Short-grain brown rice is unpolished rice with a layer of bran. Brown rice requires more water and a longer time to cook providing a nutty, sweet, dense, and chewy food.

CORN REFINING

Corn **refining** is the leading example of modern value-added agriculture. In 2014, more than 900 million tons of corn were used to produce a broad array of food, industrial, and feed products for the world market. Corn refiners use shelled corn, which has been stripped from the cob during harvesting. Refiners separate the corn into its components—starch, oil, protein, and fiber—and convert them into higher-value products (see Figure 20-7).

Inspection and Cleaning

Refinery staff inspect arriving corn shipments and clean them twice to remove cobs, dust, chaff, and foreign materials before the first processing step, **steeping**, begins.

SCIENCE CONNECTION!

What type of flour is in your kitchen at home? How does it feel? Research different types of flour discussed in this chapter. How do you think they differ in feel, weight, and taste? Why do you think there are so many options?

Steeping

Each stainless steel steep tank holds some 3,000 bushels of corn for 30 hours to 40 hours of soaking in 50 °F (10 °C) water. During steeping, the kernels absorb water, increasing their moisture levels from 15% to 45% and more than doubling in size. The addition of 0.1% sulfur dioxide to the water prevents excessive bacterial growth in the warm environment. As the corn swells and softens, the mild

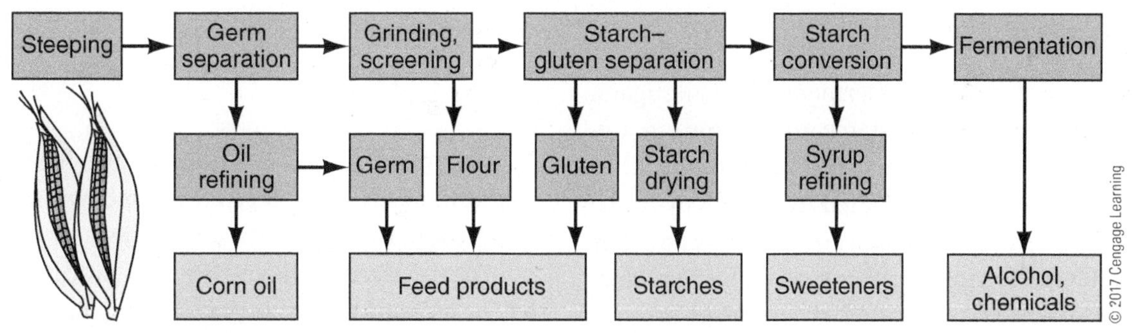

FIGURE 20-7 Flowchart of corn refining.

acidity of the steepwater begins to loosen the gluten bonds within the corn and release the starch. After steeping, the corn is coarsely ground to break the germ loose from other components. Steepwater is condensed to capture nutrients in the water for use in animal feeds and as a nutrient for later fermentation processes. The ground corn flows to the germ separators in a water slurry.

Germ Separation

Cyclone separators spin the low-density corn germ out of the slurry. The germs, containing about 85% of the corn's oil, are pumped onto screens and washed repeatedly to remove any starch left in the mixture. A combination of mechanical and solvent processes extracts the oil from the germ. The oil is then refined and filtered into finished corn oil. The germ residue is saved as another useful component of animal feeds.

Fine Grinding and Screening

The corn and water slurry leaves the germ separator for a second and more thorough grinding in an impact or attrition-impact mill to release the starch and gluten from the fiber in the kernel. The suspension of starch, gluten, and fiber flows over fixed concave screens that catch fiber but allow starch and gluten to pass through. The fiber is collected, slurried, and screened again to reclaim any residual starch or protein before being piped to the feed house as a major ingredient of animal feeds. The starch–gluten suspension, called **mill starch**, is piped to the starch separators.

Starch Separation

Gluten has a low density compared to starch. By passing mill starch through a centrifuge, the gluten is readily spun out for use in animal feeds. The starch, with just 1% to 2% protein remaining, is diluted, washed 8 to 14 times, rediluted, and washed again in hydroclones to remove the last trace of protein and produce high-quality starch, typically more than 99.5% pure.

Some of the starch is dried and marketed as unmodified cornstarch, some is modified into specialty starches, but most is converted into corn syrups and dextrose.

Syrup Conversion

Suspended in water, starch is liquefied in the presence of acid or enzymes (or both) that convert the starch to a low-dextrose solution. Treatment with another enzyme continues the conversion process. Throughout the process, refiners can halt acid or enzyme actions at key points to produce the right mixture of sugars such as dextrose and maltose for syrups to meet different needs. In some syrups, the conversion of starch to sugars is halted at an early stage to produce low- to medium-sweetness syrups. In others, the conversion is allowed to proceed until the syrup is nearly all dextrose. The syrup is refined in filters, centrifuges, and ion-exchange columns, and excess water is evaporated. Syrups are sold directly, crystallized into pure dextrose, or processed further to create high-fructose corn syrup.

Fermentation

Dextrose is one of the most fermentable of all sugars. Following conversion of starch to dextrose, many corn refiners pipe dextrose to fermentation facilities where the dextrose is converted to alcohol by traditional yeast fermentation or to amino acids and other **bioproducts** through either yeast or bacterial fermentation. After fermentation, the resulting broth is distilled to recover alcohol or concentrated through membrane separation to produce other bioproducts. Carbon dioxide from fermentation is recaptured for sale, and nutrients remaining after fermentation are used as components of animal feed ingredients.

Bioproducts

The term *bioproducts* designates a wide variety of corn-refining products made from natural, renewable raw materials that replace products made from nonrenewable resources or that are produced by chemical synthesis. The most recognized bioproduct is ethanol—a motor fuel additive fermented from corn.

MATH CONNECTION!

Are by-products a lucrative business? Is it possible to earn more money from a by-product than from the original product? Research some of the by-products discussed in this chapter to see the value added to selling by-products.

Fermentation of corn-derived dextrose has created an entirely new group of bioproducts: organic acids, amino acids, vitamins, and food gums.

Citric and lactic acid from corn can be found in hundreds of food and industrial products. They provide tartness to foods and confections, help control pH, and are themselves feedstocks for further products.

Amino acids from corn provide a vital link in animal nutrition systems. Most grain feeds do not have the amount of lysine required by swine and poultry for optimal nutrition. Economical corn-based lysine is now available worldwide to help supplement animal feeds. Threonine and tryptophan for feed supplements also come from corn.

Vitamins C and E—human nutritional supplements—are now derived from corn, replacing old production systems that relied on chemical synthesis. Even well-known food additives such as monosodium glutamate and xanthan gum are now produced by fermenting a dextrose feedstock.

Corn refiners now have fully commercial products to help deal with the plastic disposal problem and are developing an increasing array of degradable plastic products. Extrusion, the same process used to make snack foods, can alter the physical structure of cornstarch to make totally biodegradable packaging peanuts such as Eco-foam™. Other biodegradable plastics such as Eco-Pla™ are being made by modifying lactic acid.

BREAKFAST CEREALS

Breakfast foods made from cereal grains vary widely in composition, depending on the kind of grain, the part of the grain used, the method of milling, and the method of preparation. Many breakfast foods of the ready-to-serve type are a mixture of several cereals. Breakfast foods may be raw, partially cooked, or completely cooked. Some have added sweeteners, such as sugar, syrup, molasses, or honey. The carbohydrate may be changed by the use of malt or may be browned (dextrinized) by dry heat. Some cereals are reinforced with vitamins (especially the B vitamins) and with iron, calcium, and other minerals.

FIGURE 20-8 There are many ready-to-eat cereals on the market.

Raw cereals cooked at home may be in various forms: whole grains, cracked or crushed grains, granular products made from either whole grain or the endosperm section of the kernel, and rolled or flaked whole grains. The finely cut flaked grains cook in a shorter period of time. Disodium phosphate is sometimes added for quick cooking. It changes the pH of the cereal and causes it to swell faster and cook in a shorter time.

There are many ready-to-eat cereals on the market (Figure 20-8). These include granulated, flaked, shredded, and puffed cereals in many shapes and sizes with various seasonings, sugar, malt, vitamins, and minerals often added. The average American eats more than 160 bowls of cereal in a year. The *Encyclopedia of Business* estimates this is approximately 10 pounds of cereal per person per year, amounting to more than 2.7 billion packages sold in grocery stores annually.[1]

Table B-8 provides the composition of many cereals, some by brand name.

PRINCIPLES OF BAKING

The basic foundation of baked products is usually flour and liquid. Fat, sugar, salt, eggs, leavening agents, and flavorings are other common ingredients that may or may not be used in a recipe, depending on the product desired. Each ingredient has its own role and function in baked products. Roles will vary somewhat from one type of batter or dough to another. Liquid in a product may be milk, water, orange juice, or any other. Although these liquids in the baked product may function differently, generally liquid serves as a solvent for salt, sugar, and other solutes; assists in dispersing all colloids and suspensions; assists in developing gluten; and contributes to both the leavening and gelatinization phenomena during baking.

In addition to sweetening baked products, sugar facilitates air incorporation by shortening, inhibits the development of gluten and the gelatinization of starch, and elevates the temperature at which egg and flour proteins heat denature. Eggs contribute to the structure of a baked product by helping to heat denature proteins, producing steam for leavening, or producing moisture for starch gelatinization. Egg yolk is also a rich source of emulsifying agents and thus facilitates the incorporation of air, inhibits starch gelatinization, and contributes to flavor. The leavening source used in a baked product may serve to produce gas by physical, chemical, or biological methods.

Which leavening is selected usually depends on the balance and kind of ingredients in the formula and the manipulation methods used. Salt and a wide variety of flavorings are used to obtain the type and variety of product wanted. In addition to being used as a flavoring, salt controls yeast metabolism in yeast bread.

The roles and functions of the ingredients selected may be maximized or minimized, depending on the method of manipulation chosen. The number of different approaches used in preparing baked products is almost infinite; however, several basic methods are designated. Anyone who prepares baked products should know the procedures for foods made using the biscuit, pastry, muffin, conventional cake, and straight dough methods of mixing. Each method and its many variations is selected considering both the ingredients used and the characteristics desired in the end products.

Many factors affect the quality of flour mixtures, including:

- The nature of ingredients in a formula, their correct proportion, and their exact measurements
- Proper methods and environment for combining ingredients
- Correct heating temperature, time, and method
- Process melting of fat
- Increased fluidity of batter or dough
- Dissolved ingredients
- Chemical, physical, or biological leavening
- Denaturation of protein
- Gelatinization of starch
- Steam leavening
- Maillard reaction

Different flour mixtures bring about delicate balances between the firming of structure, leavening action, and the development of optimal flavors and colors. The particular heating process, as with the selection and proportion of ingredients, is specific for each formula.

Mixing is an important factor in producing any baked product. The mixing blades themselves make a difference. These influence viscosity, degree of dispersion, air incorporation, and other quality characteristics.

General objectives in mixing batters and doughs include the following:

- Uniform distribution of ingredients
- Minimum loss of the leavening agent
- Optimum blending to produce characteristic textures
- Optimum development of gluten for various products

Many different mixing methods and beating utensils exist. Each method serves to prepare a product of particular quality characteristics can be adapted to particular ingredients and conditions.

Flour millers and bakers use a mixograph that actually records the changes as the wheat flour and water mass gradually becomes coherent, loses its wet and rough sticky feel, and becomes a smooth homogenous mass.

SCIENCE CONNECTION!

San Francisco is famous for its sourdough bread with its tangy sour taste. What makes sourdough bread different? What provides its unique taste? How is it different from other breads?

Pour batters are mechanically leavened. In cream puffs, eggs are an important constituent, serving to contribute leavening with water, emulsifying the high percentage of fat, as well as structure. Both egg protein and wheat starch are the primary structural components. Popovers also have equal parts of flour and liquid that formulate a pour batter. Again, egg and starch are the primary structural components.

Pastry is another mechanical steam-leavened product. In mixing a pastry product, the incorporation of fat is critical. Finely mixed fat and flour will make a more mealy pastry. A coarsely mixed flour and fat formula will likely make a more flaky pastry.

LEGUMES

Legumes and grains provide protein and energy to much of the world's population. Legumes are plants with a double-seamed pod containing a single row of seeds. Edible legumes are found almost everywhere in the world. Whether people eat only the seeds or the pods and seeds together depends on the legume. Some common legumes and their processing, preparation, and uses are described in Table 20-1. Dried legumes that are longer than they are round are called *beans*. Round legumes are called *peas*. Lentils are round disks.

TABLE 20-1 Common Legumes

POPULAR NAME(S)/ *SCIENTIFIC NAME*	PROCESSING	PREPARATION	USES
Adzuki bean *Phaseolus angularis* © f2.8/Shutterstock.com	Picked when mature; may be left whole or pounded into a meal	Whole beans boiled, then mashed	Vegetable dish; bean flour for cakes and dessert
Alfalfa (Lucerne) *Medicago sativa* © wjarek/Shutterstock.com	Mature seeds sprouted; flour made from the dried leaves; protein concentrate from the juice of fresh leaves	Sprouts cooked or raw	Sprouts used in salads, soups, sandwiches; flour added in small amounts to some cereal products; protein concentrate in livestock feed
Bean (common, French, kidney, navy, pea, pinto, snap, stringless, green) *Phaseolus vulgaris* © quingquing/Shutterstock.com	String and mature beans sold fresh, canned, frozen; immature as pods and seeds	Boil, bake, fry	Vegetable dishes, casseroles, soups, stews
Broad bean (fava bean) *Vicia faba* © paulista/ Shutterstock.com	Picked when mostly mature then shelled, dried, canned, or frozen	Boiled, cooked, or steamed in other foods; immature pods cooked whole or sliced	Vegetable dishes, soups, stews, casseroles
Chickpea (garbanzo bean) *Cicer arietinum* © Anna Shepulova/ Shutterstock.com	Dehulling then drying	Boil, fry, roast	Snacks, vegetable dishes, soups, salads, stews

(Continues)

TABLE 20-1 Common Legumes

POPULAR NAME(S)/ *SCIENTIFIC NAME*	PROCESSING	PREPARATION	USES
Cowpea (black-eyed pea) *Vigna sinensis, V. unguiculata* © Ozgur Coskun/ Shutterstock.com	Mature beans dried, canned, frozen; picked as immature pods	Boil or bake	Vegetable dishes with or without pork
Field pea *Pisum arvense* © Antonsov85/Shutterstock.com	Shelling, drying; whole plant plowed under as green manure or forage peas	Boil or bake; do not soften like garden peas	Vegetable dishes, livestock feed
Garden pea *Pisum sativum* © Malivan_Iuliia/ Shutterstock.com	Picked immature; sold fresh, frozen, cooked, or canned; mature seeds dried whole or after splitting	Boil or pressure cook	Vegetable dishes, casseroles, soups, stews
Lentil *Lens esculenta* © Zigzag Mountain Art/ Shutterstock.com	Picked mature, can be left whole or dehulled, usually dried; sometimes ground into flour	Boil or stew	Vegetable dishes, soups, stews; flour mixed with cereals
Lima bean (butter bean) *Phaseolus lunatus* © Alvaro German Vilela/ Shutterstock.com	Picked immature or mature; sold fresh, cooked, canned, frozen, dried, or as flour	Boil or bake	Vegetable dishes, casseroles, soups, stews

(Continues)

TABLE 20-1 Common Legumes

POPULAR NAME(S)/ *SCIENTIFIC NAME*	PROCESSING	PREPARATION	USES
Mung bean (golden gram, green gram) *Phaseolus aureus* © Evlakhov Valeriy/ Shutterstock.com	Picked mature	Boil or sprout for eating raw or cooked	Cooked as vegetable dish; sprouts in salad, soups, sandwiches
Peanut (groundnut) *Archis hypogaea* © Elena M. Tarasova/ Shutterstock.com	Vines harvested, nuts shelled and blanched; oil may be expressed or extracted then cake ground into defatted flour	Boil or roast whole nuts with or without shell	Snack food; flour added to cereal mixtures
Soybean *Glycine max* © Only background/ Shutterstock.com	Picked mature or immature; immature seeds sold fresh or cooked or canned; oil expressed or extracted; press cake ground into defatted flour; mature seeds roasted or salted	See Figure 20-8.	Refer to Table 20-2.

Nutritional Composition

Table B-8 lists the nutrient composition of many legumes. In general, the seeds of food legumes are good sources of carbohydrates, fats, proteins, minerals, and vitamins. Mixtures of legumes and grains have a protein quality that comes close to that of animal proteins.

Legume Products

Products made from legumes include fermented foods, flours, imitation meats, infant formulas, oils, and sprouts (see Figure 20-9).

Fermented Foods. Soy sauce is produced by fermentation. Soybean products tempeh (mold-ripened

SCIENCE CONNECTION!

Are any legumes in Table 20-1 grown in your area? Research the legumes listed in the table. Do you currently consume any of the legumes listed? How could you add more legumes to your diet?

FIGURE 20-9 Some of the products made from legumes include fermented foods, infant formulas and oils.

soybean cake that is fried in deep fat) and tofu (a cheeselike item made from coagulated soybean milk) are now being produced and sold in many U.S. supermarkets.

Flours. Soybean flour is used in the United States to make soybean milk and low-gluten baked goods. Similar legume flours have been made from locally grown beans and peas in many countries of the world.

Imitation Meat. Textured vegetable protein (TVP) products are meatlike in taste and texture. Most TVP items are made from soy protein extracted from soy flour and spun into fibrous strands. Some of the more popular items are made from mixtures of soy proteins, wheat protein, egg albumin, and various additives.

Infant Formulas. Many infant formulas are made of soy protein concentrates.

Oil. Seeds of legumes such as soybeans and peanuts are extracted to yield oil that is used for cooking. The by-product of this process—a protein meal—is often fed to livestock.

Sprouts. Legume seeds allowed to germinate are called *sprouts*. They are sold in the fresh produce section of many supermarkets. The most popular types are made from alfalfa seeds, mung beans, and soybeans. Some bean sprouts are also canned by manufacturers of Chinese foods.

SOYBEANS

As a soybean matures in its pod, it ripens into a hard, dry bean. Most soybeans are yellow, but some varieties are brown and black. Whole soybeans can be cooked and used in sauces, stews, and soups. They are an excellent source of protein. Whole soybeans that have been soaked can be roasted for snacks and can be purchased in natural food stores and some supermarkets. But soybeans are more than this.

The soybean is a versatile agricultural product. After preparation and **extraction**, it has edible and technical uses (Table 20-2). Although they can be eaten whole after being boiled or roasted, most soybeans are transformed into a great variety of foods. In addition, many foods already found in a kitchen cupboard contain soy foods such as soy oil (often called vegetable oil), lecithin, soy protein concentrates, textured soy protein, and many more. Figure 20-10 illustrates the preparation and extraction of soybeans to produce a wide variety of useful products.

Green Vegetable Soybeans

These large soybeans are harvested when the beans are still green and sweet tasting and can be served as a snack or a main vegetable dish, after boiling in slightly salted water for 15 to 20 minutes. They are high in protein and fiber and contain no cholesterol.

Hydrolyzed Vegetable Protein

Hydrolyzed vegetable protein (HVP) is a protein obtained from any vegetable, including soybeans. The protein is broken down into amino acids by a chemical process called acid hydrolysis. HVP is a flavor enhancer that can be used in soups, broths, sauces, gravies, flavoring and spice blends, canned and frozen vegetables, meats, and poultry.

TABLE 20-2 Soybean Products and Uses

	OIL PRODUCTS			SOYBEAN PROTEIN PRODUCTS	
WHOLE SOYBEAN PRODUCTS	**GLYCEROL, STEROLS, FATTY ACIDS**	**REFINED SOY OIL**	**SOYBEAN LECITHIN**	**SOY FLOUR CONCENTRATES AND ISOLATES**	**SOYBEAN MEAL**
Edible Uses	Oleochemistry Soy diesel Solvents	Edible Uses	Edible Uses	Edible Uses	Feed Uses
Seed Stock feed Soy sprouts Baked soybeans Full-fat soy flour Bread Candy Doughnut mix Frozen desserts Instant milk drinks Pancake flour Pan grease extender Pie crust Sweet goods		Coffee creamers Cooking oils Filled milks Margarine Mayonnaise Medicinals Pharmaceuticals Salad dressings Salad oils Sandwich spreads Shortenings	Emulsifying agent Bakery products Candy, chocolate coatings Pharmaceuticals Nutritional uses Dietary Medical	Alimentary pastes Baby food Bakery ingredients Candy products Cereals Diet food products Food drinks Hypoallergenic milk Meat products Noodles Prepared mixes Sausage casings Yeast Beer and ale	Aquaculture Bee foods Calf milk replacement Fish food Fox and mink feed Livestock feeds Poultry feeds Protein concentrates Pet foods
Roasted Soybeans		Technical Uses	Technical Uses	Technical Uses	Hulls
Candies, confections Cookie ingredient, topping Crackers Dietary items Soy nut butter Soy coffee		Anticorrosion agents Antistatic agents Caulking compounds Composite building material Concrete-release agents Core oils Crayons Dust-control agent Electrical insulation Epoxies Fungicides Hydraulic fluids Linoleum backing Lubricants Metal casting, metal working Oiled fabrics Paints Pesticides Plasticizers Printing inks Protective coatings Putty Soap, shampoos, detergents Vinyl plastics Waterproof cement	Antifoam agents Alcohol Yeast Antispattering agents Margarine Dispersing agents Paint Ink Insecticides Magnetic tape Paper Rubber Stabilizing agent Shortening Wetting agents Calf milk replacements Cosmetics Paint pigments	Adhesives Antibiotics Asphalt Emulsions Composite building material Fermentation aids Nutrients Fibers Films for packaging Firefighting foams Inks Leather substitutes Paper coatings Particle boards Plastics Polyesters Pharmaceuticals Pesticides Fungicides Textiles Water-based paints	Dairy feed
Soybean Derivatives					
Miso Soy milk Tempeh Tofu					

Source: American Soybean Association

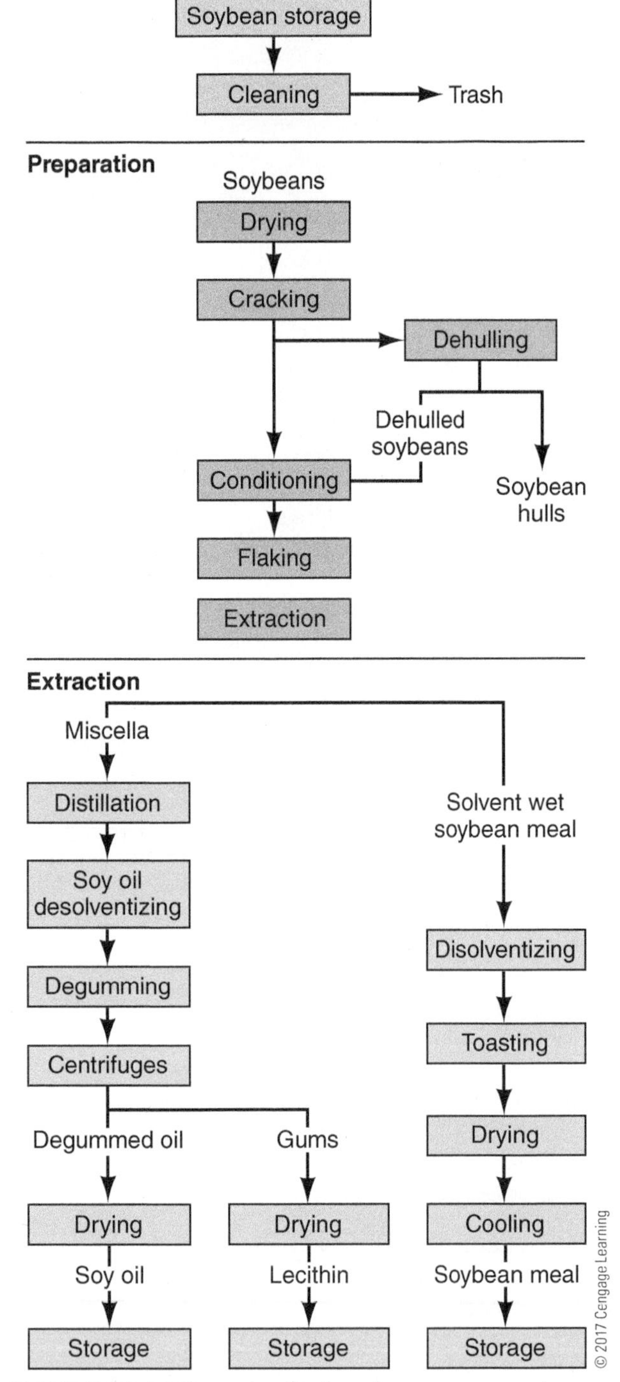

FIGURE 20-10 Flowchart of soybean processing.

Infant Formulas, Soy-Based

Soy-based infant formulas are similar to other infant formulas except that a soy protein isolate powder is used as a base instead of cow's milk. Carbohydrates and fats are added to achieve a fluid similar to breast milk.

Lecithin

Extracted from soybean oil, lecithin is used in food manufacturing as an emulsifier in products high in fats and oils. It also promotes stabilization, antioxidation, and crystallization and helps control spattering. Powdered lecithins can be found in natural and health food stores.

Meat Alternatives (Meat Analogs)

Meat alternatives made from soybeans contain soy protein or tofu and other ingredients mixed together to simulate various kinds of meat. These meat alternatives are sold as frozen, canned, or dried foods. Usually, they can be used the same way as the foods they replace. With so many different meat alternatives available to consumers, the nutritional value of these foods varies considerably. Meat alternatives made from soybeans are excellent sources of protein, iron, and B vitamins.

Natto

Natto is made of fermented, cooked whole soybeans. Because the fermentation process breaks down the beans' complex proteins, natto is more easily digested than whole soybeans. It has a sticky, viscous coating with a cheesy texture. In Asian countries, natto is served as a topping for rice, in miso soups, and with vegetables. Natto can be found in Asian and natural food stores.

Nondairy Soy Frozen Dessert

Nondairy frozen desserts are made from soy milk or soy yogurt. Soy ice cream is one of the most popular desserts made from soybeans and can be found in natural foods stores.

Soy Cheese

Soy cheese is made from soy milk. Its creamy texture makes it an easy substitute for sour cream or cream cheese and can be found in a variety of flavors in natural foods stores. Products made with soy cheese include soy pizza.

Soy Fiber (Okara, Soy Bran, and Soy Isolate Fiber)

There are three basic types of soy fiber: okara, soy bran, and soy isolate fiber. All these products are high-quality, inexpensive sources of dietary fiber. Okara is a pulp fiber by-product of soy milk. It has less protein than whole soybeans, but the protein is of high quality. Okara tastes similar to coconut and can be baked or added as fiber to granola and cookies. Okara also has been made into sausage (see Figure 20-11).

Soy bran is made from hulls (the outer covering of the soybean), which are removed during initial processing. The hulls contain a fibrous material that can be extracted and then refined for use as a food ingredient.

Soy isolate fiber, also known as *structured protein fiber*, is soy protein isolate in a fibrous form.

Soy Flour

Soy flour is made from roasted soybeans ground into a fine powder.

Three kinds of soy flour are available:

1. Natural or full fat, containing the natural oils found in the soybean;
2. Defatted, having the oils removed during processing; and
3. Lecithinated, having lecithin added to it.

All soy flour gives a protein boost to recipes. However, defatted soy flour is an even more concentrated source of protein than full-fat soy flour. Although used mainly by the food industry, soy flour can be found in natural foods stores and some supermarkets. Soy flour is gluten free, so yeast-raised breads made with soy flour are more dense in texture.

Source: USDA, Agricultural Research Service (ARS), photo by Scott Bauer.

FIGURE 20-11 Soybean products.

Soy Grits

Soy grits are similar to soy flour except that the soybeans have been toasted and cracked into coarse pieces rather than the fine powder of soy flour. Soy grits can be used as a substitute for flour in some recipes. High in protein, soy grits can be added to rice and other grains and cooked together.

Soy Protein Concentrate

Soy protein concentrate comes from defatted soy flakes. It contains approximately 70% protein and retains most of the bean's dietary fiber.

Soy Protein Isolates (Isolated Soy Protein)

When protein is removed from defatted flakes, the result is soy protein isolates, the most highly refined soy protein. Containing 92% protein, soy protein isolates possess the greatest amount of protein of all soy products. They are a highly digestible source of amino acids.

Soy Protein, Textured

Textured soy protein usually refers to products made from textured soy flour, although the term can also be applied to textured soy protein concentrates and spun soy fiber.

Textured soy flour is made by running defatted soy flour through an extrusion cooker, which allows for many different forms and sizes. It is widely used as a meat extender. Textured soy flour contains approximately 70% protein and retains most of the bean's dietary fiber. Textured soy flour is sold dried in granules and in chunks.

Soy Sauce

Soy sauce is a dark brown liquid made from soybeans that have been fermented. Soy sauces have a salty taste. Specific types of soy sauce include shoyu, tamari, and teriyaki. Shoyu is a blend of soybeans and wheat. Tamari is made only from soybeans and is a by-product of making miso. Teriyaki sauce can be thicker than other types of soy sauce and includes other ingredients such as sugar, vinegar, and spices.

Soy Yogurt

Soy yogurt is made from soy milk. Its creamy texture makes it an easy substitute for sour cream or cream cheese. Soy yogurt can be found in a variety of flavors in natural foods stores.

Soy Milk and Soy Beverages

When soaked, ground fine, and strained, soybeans produce a fluid called *soybean milk*, which is a substitute for cow's milk. Plain, unfortified soy milk is an excellent source of high-quality protein, and B vitamins. Soy milk is also sold as a powder, which must be mixed with water.

Soy Nut Butter

Soy nut butter is made from roasted, whole soy nuts that are crushed and blended with soy oil and other ingredients. Soy nut butter competes with peanut butter.

Soy Nuts

Roasted soy nuts are whole soybeans that have been soaked in water and then baked until browned. Soy nuts are similar in texture and flavor to peanuts.

Soy Oil and Products

Soy oil is the natural oil extracted from whole soybeans. It is the most widely used oil in the United States. Oil sold in the grocery stores under the generic name *vegetable oil* is usually 100% soy oil or a blend of soy oil and other oils. Soy oil is cholesterol free and high in polyunsaturated fat. Soy oil also is used to make margarine and shortening (see Table 20-2).

Soy Sprouts

Not as popular as mung bean sprouts or alfalfa sprouts, soy sprouts (also called *soybean sprouts*) are an excellent source of protein and vitamin C. Soy sprouts must be cooked quickly at low heat so they do not get mushy. They can also be used raw in salads or soups, or in stir-fried, sautéed, or baked dishes.

Tempeh

Tempeh, a traditional Indonesian food, is a chunky, tender soybean cake. Whole soybeans, sometimes mixed with another grain such as rice or millet, are fermented into a rich cake of soybeans with a smoky or nutty flavor. Tempeh can be marinated and grilled and added to soups, casseroles, and chili.

Tofu and Tofu Products

Also known as *soybean curd*, tofu is a soft cheeselike food made by curdling fresh hot soy milk with a coagulant (refer to Figure 20-11). Tofu is a bland product that easily absorbs the flavors of other ingredients being cooked with it. Tofu is rich in high-quality protein and B vitamins and low in sodium. Firm tofu is dense and solid and can be cubed and served in soups, stir fried, or grilled. Firm tofu is higher in protein, fat, and calcium than other forms of tofu. Silken tofu is a creamy product and can be used as a replacement for sour cream in many dip recipes. Tofu is also available as a powder.

Soy-Based Whipped Toppings

Soy-based whipped toppings are similar to other nondairy whipped toppings, except that hydrogenated soy oil is used instead of other vegetable oils.

Table 20-2 summarizes all of the edible and technical products obtained from the extraction of soybeans.

SCIENCE CONNECTION!

How many soy products can you find in your household? Make a list of as many soy products you can find. This can include edible and nonedible soy products. Are you surprised at the amount of soy-based products?

SUMMARY

Cereal grains and legume seeds, and products from these, are used as food for people throughout the world. Cereal grains provide mainly starch and some protein. Legumes provide mainly protein, oil, and some starch. Starch has unique properties that are used in foods. Seeds of grains and legumes are used to produce flour. These flours are used to produce a variety of other food products.

Corn refining and soybean extraction separate corn seed and soybeans into component parts and convert these to high-value products. These processes are some of the best examples of value-added agriculture and the application of food science.

REVIEW QUESTIONS

Success in any career requires knowledge. Test your knowledge of this chapter by answering these questions or solving these problems.

1. Explain the processes that take place during baking.
2. Describe the various soy products available and how they are made.
3. List three properties of starch.
4. Name five types of wheat flour.
5. List four general uses of legumes.
6. Name the large central portion of the kernel that contains most of the starch.
7. Identify five food products of soybean extraction.
8. What is the purpose of steeping, the first step of corn refining?
9. List four factors that must be controlled when cooking starch.
10. Name four products derived from corn.

STUDENT ACTIVITIES

1. Compare a labeled cross section of a kernel of grain to that of a legume seed. Record the observed differences.
2. Collect the labels on five different ready-to-eat breakfast cereals. Report on the grains or grain products used. Do the same for a type of hot breakfast cereal that requires cooking.
3. Visit a corn milling or soybean-extraction facility on the Web. Develop a visual presentation of the information you discovered.
4. Develop a bread-making demonstration that explains all of the components and processes taking place. Experiment with different types of flours. Use a bread machine if possible.
5. Use a hand mill or a small electric mill to grind some wheat. Separate and screen to produce flour. Develop a report on your results. Include yield and the evaluation of a product produced with the flour.
6. Create a display of soybean-extracted products (Table 20-2) or products from corn refining.
7. Collect, preserve, and display the seeds of cereal grains or legumes.

ADDITIONAL RESOURCES

- Information on Postharvest Operations: http://www.fao.org/inpho/inpho-post-harvest-compendium/cereals-grains/en/
- Grain Science and Industry: http://www.grains.k-state.edu/undergraduate-programs/degree-options/milling-science-and-management.html
- Legumes: http://lpi.oregonstate.edu/mic/food-beverages/legumes
- Iowa State University Soybean Extension and Research Program: http://extension.agron.iastate.edu/soybean/

Please note that internet sites represent a vast resource of information. Those provided in this chapter were vetted by industry experts but you may also wish to explore the topics discussed in this chapter further using a search engine, such as Google. Keywords or phrases may include the following: *wheat, specific types of wheat, flour milling, small grains, grain milling, corn milling, cereal grains, breakfast cereals, corn refining.* In addition, Table B-7 provides a listing of some useful Internet sites that can be used as a starting point.

REFERENCES

Fast, R. B., and E. F. Caldwell. 2000. *Breakfast cereals and how they are made*, 2nd ed. St. Paul, MN: American Association of Cereal Chemists.

Delcour, J.A. and Hoseney, R. C. 2010. *Principles of cereal science and technology.* 3rd ed. St. Paul, MN: American Association of Cereal Chemists.

Lehner, E., and J. Lehner. 1973. *Folklore and odysseys of food and medicinal plants.* New York: Farrar Straus Giroux.

Parker, R. 2009. *Plant and soil science: Fundamentals and applications.* Clifton Park, NY: Delmar Cengage Learning.

Posner, E. S., and A. N. Hibbs. 2004. *Wheat flour milling,* 2nd ed. St. Paul, MN: American Association of Cereal Chemists.

ENDNOTE

1. Ask http://www.ask.com/math/many-bowls-cereal-average-american-eat-annually-b6383df080f94cf2. Last accessed July 8, 2015.

CHAPTER 21

Fruits and Vegetables

OBJECTIVES

After reading this chapter, you should be able to:

- **Identify the parts of a plant considered a vegetable or a fruit**
- **Describe the nutrient composition of a fresh fruit or vegetable**
- **Discuss the structure of a plant cell**
- **Describe plant tissues and their functions**
- **Explain climacteric and nonclimacteric fruits and provide examples**
- **Name one pigment in fruits or vegetables and describe how it responds to heat or pH**
- **List four factors affecting the texture of fruits or vegetables**
- **Name four general compounds that give fruits and vegetables their flavor**
- **Identify the quality grades for fruits and vegetables**

NATIONAL AFNR STANDARD

FPP.03

Select and process food products for storage, distribution, and consumption.

KEY TERMS

Choice
climacteric
dermal
ethylene
grades
nonclimacteric
standards
storage tissue
tubers
U.S. Fancy
vascular

OBJECTIVES (continued)

- **Describe how quality grade determines the use of a fruit or vegetable**
- **List five factors considered during storage**
- **Describe the processing of fruits and vegetables**

Fruits, vegetables, and other plant tissues either directly or indirectly supply all foods to humans. An estimated 270,000 plant species exist. The number of crops that fit into humans' dietary picture is probably between 1,000 and 2,000 species. The fruit of the plant ensures the survival of the plant because the fruit contains the seeds of the plant. Fruits and vegetables take in a wide variety of edible plant parts. Fruits include apples, pears, peaches, apricots, plums, cherries, bananas, blueberries, oranges, tangerines, and grapes. The term *vegetable* includes many different parts of plants, including some fruits.

TYPES OF FRUIT

Fruits grow on trees, bushes or vines. Every fruit has a stem end where it was attached to the tree, bush or vine. Some fruit also have a blossom end which is opposite of the stem where the flower blossomed. Fruits can be categorized in a variety of ways. Some will group fruits by their seed structure and others will utilize the botanical scientific name. Others group fruit by where they are grown.

VEGETABLE PROPERTIES AND STRUCTURAL FEATURES

Leafy vegetables are generally high in water and low in carbohydrates, protein, and fats. They frequently contain the mechanism for photosynthesis. Examples include spinach, lettuce, mustard greens, and cabbage.

Seeds vary in water content and are a source of carbohydrates and protein. They may be fresh and high in water or dried and relatively low in water content. Corn and beans are examples.

Tubers are generally higher in carbohydrates and lower in water content than stem, flower, or leafy vegetables. Tubers are enlarged underground stems. Potatoes are the best example and include all varieties.

The fruit part of the plant can be eaten either as a fruit or as a vegetable. Examples include cantaloupe, eggplant, squash, and peas (snow peas) (Figure 21-1).

Stems are plant portions generally high in water and fiber. They have relatively little nutritional value. Asparagus is an example.

Bulbs are generally higher in carbohydrates and lower in water content than stem, flower, or

FIGURE 21-1 A fruit is generally the fruit of a plant, but vegetables include a plant's fruits, leaves, stems, bulbs, flowers, seeds, roots, and tubers.

leafy vegetables. Bulbs such as onions and garlic are enlargements above the roots.

Broccoli, cauliflower, and artichokes are generally listed as edible flowers. Flowers are generally high in water and low in carbohydrates.

Root vegetables grow underground. Roots such as carrots, parsnips, and radishes are generally higher in carbohydrates and lower in water content than stem, flower, and leafy vegetables. Roots are the part of a plant that grow downward into the soil to anchor the plant and furnish nourishment by absorbing nutrients.

GENERAL COMPOSITION

Fresh as well as canned and frozen vegetables provide a variety of vitamins, minerals, and fiber. They are typically low in fat and carbohydrates.

Fresh fruits and fruit juices contain many vitamins and minerals, are low in fat (except avocados) and sodium, and provide dietary fiber. Whole, unpeeled fruit is higher in fiber than peeled fruit or fruit juice. Canned and frozen fruits and fruit juices contain many vitamins and minerals and are low in fat and sodium.

Table B-8 provides the composition of many fruits and vegetables. Fresh and canned fruits and vegetables contain a high water content. The energy content increases when the fruit is dried. The carbohydrate content in fruit is higher than the protein content. Vegetable nutrient content varies widely. Some fresh and canned vegetables have a high water content with some protein but more carbohydrate. Seeds that are considered vegetables contain high amounts of energy, protein, and carbohydrates.

Fresh Vegetable Labels

Under federal guidelines, retailers must provide nutrition information for the 20 most frequently eaten raw vegetables: potatoes, iceberg lettuce, tomatoes, onions, carrots, celery, sweet corn, broccoli, green cabbage, cucumbers, bell peppers, cauliflower, leaf lettuce, sweet potatoes, mushrooms, green onions, green (snap) beans, radishes, summer squash, and asparagus. Information about other vegetables may also be provided. The nutritional information may appear on posters, brochures, leaflets, or stickers near the vegetable display. It may include serving size; calories per serving; amount of protein, total carbohydrates, total fat, and sodium per serving; and percent of the U.S. Recommended Daily Allowances for iron, calcium, and vitamins A and C per serving.

Fresh Fruit Labels

Under federal guidelines, retailers must provide nutrition information for the 20 most frequently eaten raw fruits: bananas, apples, watermelons, oranges, cantaloupes, grapes, grapefruit, strawberries, peaches, pears, nectarines, honeydew melons, plums, avocados, lemons, pineapples, tangerines, sweet cherries, kiwifruit, and limes. Information about other fruits may also be provided. The nutritional information may appear on posters, brochures, leaflets, or stickers near the fruit display (Figure 21-2). It may include serving size; calories per serving; amount of protein, total

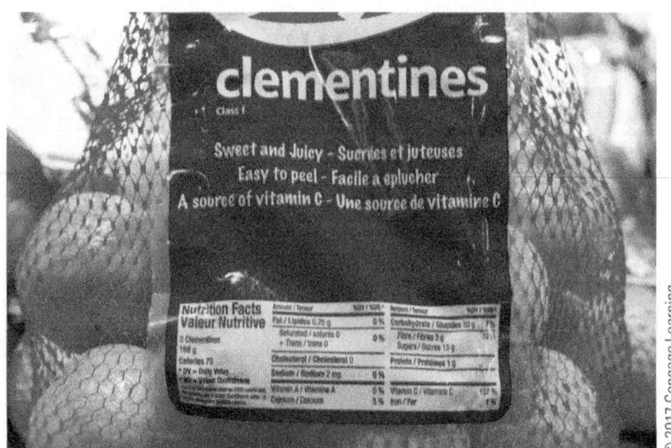

FIGURE 21-2 Under federal guidelines, retailers must provide nutrition information for raw fruits and vegetables.

carbohydrates, total fat, and sodium per serving; and percent of the U.S. Recommended Daily Allowances for iron, calcium, and vitamins A and C per serving.

ACTIVITIES OF LIVING SYSTEMS

Because so many parts of plants are eaten as vegetables or fruits, an understanding of the plant cell and plant tissues is essential to food scientists.

Plant cell structure depends on the role and function of the cell (Figure 21-3). The cell wall consists of a primary wall and a secondary wall. The primary walls of two cells are joined together by a common layer called the *middle lamella*. The cell wall and middle lamella's chief components are cellulose, hemicellulose, and pectic substances.

A vacuole is contained within plant cells. Its size depends on the cell's function. The vacuole is composed of water with dissolved soluble substances. These may include sugars, acids, volatile esters, aldehydes, ketones, and water-soluble pigments, depending on the particular fruit or vegetable.

Energy conversion in the cell is carried out by the chloroplasts and mitochondria. The leucoplasts store starch that is used for energy. The mitochondria are small spheres, rods, or filaments that produce energy for the cell through cellular respiration. They contain fats, proteins, and enzymes.

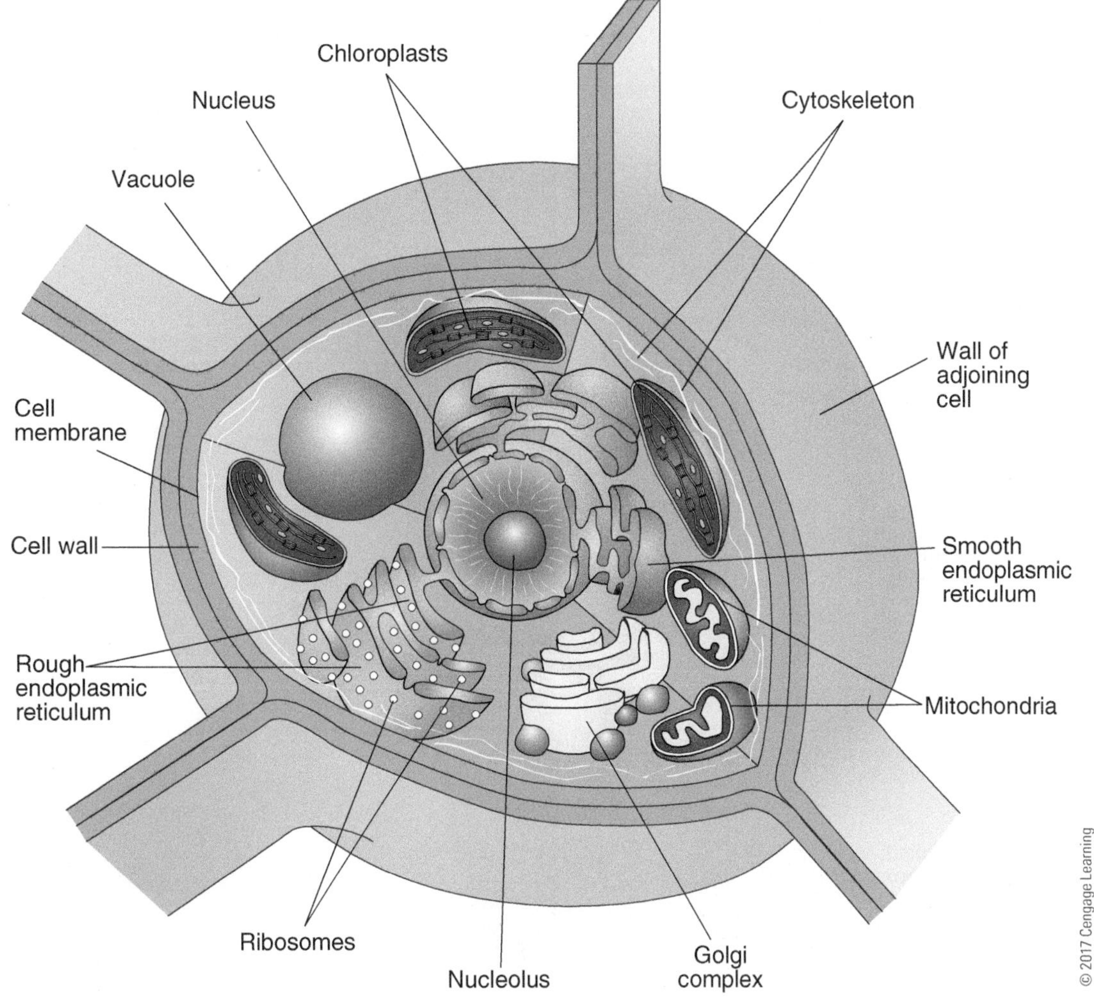

FIGURE 21-3 A plant cell.

The nucleus of the cell is embedded within the cytoplasm. It controls reproduction and protein synthesis. Both the nucleus and mitochondria are needed for the continued life of the cell.

Plant Tissues

Four main types of plant tissues exist in fruits and vegetables (Figure 21-4): **dermal** (epidermis and endodermis), **vascular** (xylem and phloera), supporting, and **storage tissue** (cortex). The dermal tissue generally is a layer of protective tissue. It has the stoma that will penetrate to the interior. The vascular system has two clearly defined structures: the xylem and the phloem. The xylem transports water within the plant, and the phloem transports food. Supporting tissue varies, depending on the particular plant. Generally, storage or parenchymal plant tissues make up most of the edible portion of fruits and vegetables. The leucoplasts, protein bodies, and other food- and energy-containing bodies are located here.

Storage tissue is located in the cytoplasm in leucoplasts. The leucoplasts are colorless plastids and the storage structures of the cell that may accumulate fats and oils (elaioplasts), proteins (aleuroneplasts), and starch (amyloplasts). As the cell matures, lipid, protein, or starch content increases. In storage tissue, plastids may dominant the cell. They are particularly dominant in roots, tubers, bulbs, and seeds.

Protective tissue is made up of modified parenchymal cells that contain suberin, may secrete cutin, and grow tightly together to form an epidermal layer for the cell. The thickness and composition of this protective tissue will vary with the part and type of plants. It is the layer that protects against insects, fungi, microorganisms, and small abrasions. The extent of cutin or suberin will affect the quality of a food, particularly in apples. A shiny apple will usually have a higher content of these components in its outer skin.

Quality characteristics are influenced by water transport. This is critical during the growing

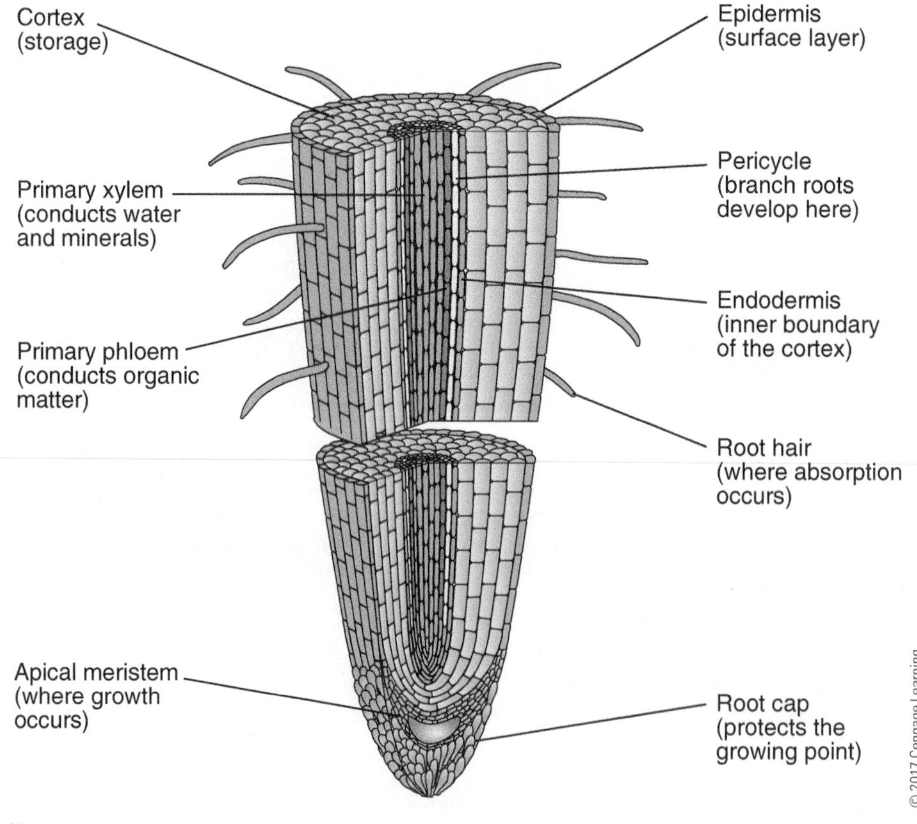

FIGURE 21-4 Plant tissues.

period and also influences the crispness of the final fruit or vegetable. Water movement occurs because of evaporation and controlled loss of water from stomas in the stems and leaves. The more stomas the dermal tissue possesses, the greater the potential for water loss. Dermal tissue is sensitive to light and heat, allowing the stomas to regulate water transport by opening and closing. Supporting tissues are generally cells whose cellulose cell walls have thickened and have become embedded with lignin or pectic substances. The cells themselves are frequently elongated, and these portions of the plant have a tough texture.

Although water is lost through the stomas in the stems and particularly in the leaves, the transport and translocation of water and plant sap from the vacuole occurs in vascular tissue or conducting tissue. One vascular category is the xylem, which chiefly transports water and minerals from the roots up the stem and into the leaves. The transport of carbohydrates, amino acids, and other constituents of the cell sap in the vacuole occurs through the second vascular category, the phloem. The phloem is essentially the nutrient transport. Some overlap exists between it and the xylem.

HARVESTING

Many factors affect production of the vegetable and fruit to the point at which it will be harvested (Figure 21-5). The soil will vary for the different types of vegetables and fruits. Some fruits and vegetables require more water than others, different fertilizers than others, and so forth. The actual days available for a plant to grow and ripen is a critical factor. For that reason, some fruits and vegetables do not have adequate time at the appropriate temperature to reach maturity.

FIGURE 21-5 Proper harvesting of potatoes is important to maintaining quality.

Harvesting by hand or by machine will affect the quality and quantity of fruits and vegetables harvested. Mechanical harvesting machines generally have more impact and damage fruits and vegetables. To offset the greater handling force, new cultivars and varieties have been developed. In many cases, these fruits and vegetables are firmer and higher in solids and with different flavors. For example, horticulturalists and crop scientists have developed a firm, high-solid, and high-pH tomato to meet the impacts of mechanical harvesting. The food processor or preparer must recognize these changes occurring in fruits and vegetables because they influence both safety and quality. In the case of the tomato, no longer is this vegetable necessarily safe if canned only under hot water. In most instances, pressure canning must be used.

The seasonal environment and length are also important. The producer, and to some extent the consumer, is interested in how long it takes a fruit or vegetable to mature. However, urban consumers are more interested in when they can purchase selected fruits and vegetables.

Ripening

Many factors and interactions affect ripeness as well as maturation. Temperature, time, or added gases are of real interest and influence ripening. Each fruit appears to function differently.

One aspect of ripening is the production of **ethylene** gas (C_2H_4), which signals and orchestrates the growth stages of fruits and flowers. Ethylene activates fruit ripening.

TABLE 21-1 Classification of Edible Fruits According to Ripening Patterns

CLIMACTERIC	NONCLIMACTERIC
Apple	Cherry
Apricot	Cucumber
Avocado	Fig
Banana	Grape
Cherimoya	Grapefruit
Feijoa	Lemon
Mango	Melon
Papaya	Orange
Passion fruit	Pineapple
Papaw	Strawberry
Peach	
Pear	
Plum	
Sapote	
Tomato	

Some fruits are **climacteric**, and some are **nonclimacteric** fruits. Climacteric fruits produce ethylene during ripening, and they are ethylene sensitive (Table 21-1).

Appearance

Fruits and vegetables get their characteristic color from pigments. Anthocyanins are purple, blue, and red. Anthoxanthins are white to yellow, and betalains are red. Carotenoids are orange to yellow, and chlorophyll is green. These pigments respond differently to the processing environment, as shown in Table 21-2.

Texture

Texture of plant foods has been difficult to describe in precise terms—for example, a comparison of crisp versus wilted lettuce, a crisp tender carrot versus a crisp tough carrot, a plump, watery strawberry compared to a plump, pithy strawberry. All these comparisons emphasize the two-sided nature of texture. Ultimately, the texture of fruits and vegetables is mainly caused either by the structural components of the plants themselves or by the process of osmosis and diffusion.

The toughness of a fruit or vegetable is the result of changes in cell wall components—pectins, hemicelluloses, and cellulose—during maturation, storage, and processing. Many factors affecting toughness include tissue conditions, pH, enzymes, and salt concentrations.

The crispness of a vegetable is the result of water movement within the plant. Crispness of a plant cell is influenced by turgidity factors such as the concentration of osmotically active substances, the permeability of the protoplasm, and elasticity and toughness. Water movement in plants is a combined effect of capillary action, diffusion, transpiration, and osmosis (Figure 21-6).

TABLE 21-2 Affect of Heat, Acid, and Alkali on Pigments

NAME OF PIGMENT	ACID	ALKALI	PROLONGED HEATING
Anthocyanins	Red	Purple or Blue	Little effect
Anthoxanthins	White	Yellow	Darkens if excessive*
Betalains	Red	Little effect	Fading if slightly acid
Carotenoids	Less intense	Little effect	May be less intense*
Chlorophylls	Olive green	Intensifies green	Olive green

*Heating usually produces little effect.

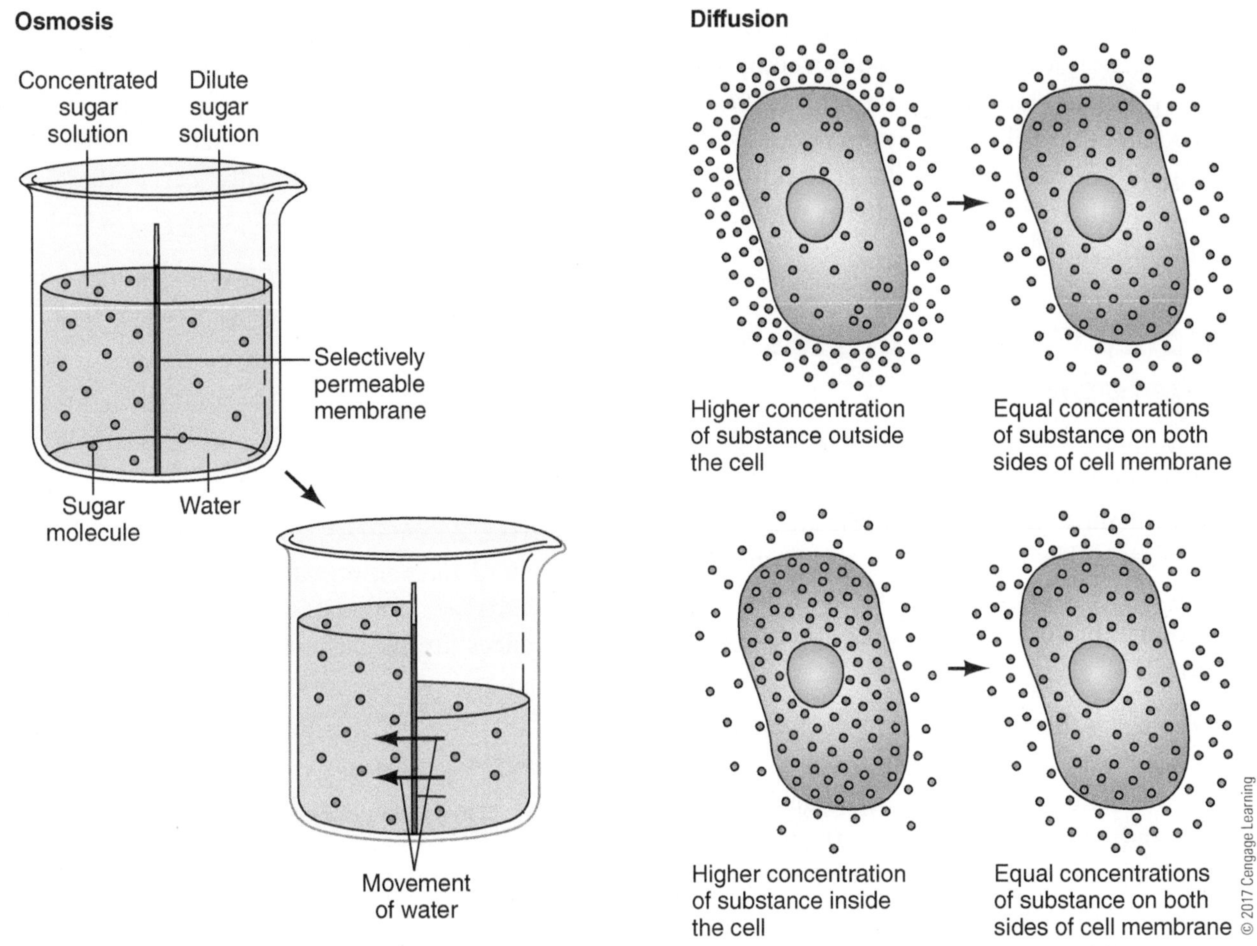

FIGURE 21-6 Osmosis and diffusion, two important processes in plant cells and tissues.

Flavor

Flavors and aromas in fruits and vegetables result from a variety of compounds working together to give unique and distinctive characteristics: aldehydes, alcohols, ketones, organic acids, esters, sulfur compounds, and trace amounts of other chemicals.

Astringency in fruits and vegetables primarily comes from the flavonoid pigments classified as *tannins* or *phenolics*. These flavor components make the mouth pucker. Examples include lemons and chokecherries.

Fruity flavor is extremely complex. It can be partially caused by a combination of esters, alcohols, aldehydes, ketones, and minor compounds.

Sweetness of fruits and vegetables has been the one taste perception that is constantly searched for. Some plants such as sugar cane and sugar beets are grown for sucrose, their sweet component. Other foods are consumed for a combination of sugars and other flavor component interactions. Glucose, fructose, maltose, xylose, and less common sugars are also found. The types of sugars in plants vary considerably. Also, nonsugar sources of plant sweetness include compounds such as glycyrrhizin from licorice.

Acid flavor from fruits and vegetables is formed by many different acids. Although malic and citric acid are the most common, many others can be found in selected plant foods. For example, grapes have considerable tartaric acid, and oxalic acid (rhubarb) and benzoic acid (plums, cranberries) are found in many fruits. These acids give a range of pH values.

The cabbage and onion family give flavors and odors because of a variety of sulfur compounds. Cooking will develop strong flavors from hydrogen sulfide and other volatile sulfur compounds. Also, cutting or shredding across the cell wall releases an enzyme that develops the distinctive flavor of onion, garlic, and leek.

Quality Grades for Fresh Vegetables

Most vegetables are labeled with a U.S. Department of Agriculture (USDA) quality grade. The quality of most fresh vegetables can be judged reasonably well by their external appearance. Vegetables are available year-round from both domestic production and imports from other countries (Figure 21-7).

The USDA has established grade **standards** for most fresh vegetables. The grade is awarded based on the appearance of the vegetable—size, shape, and color. The standards are used extensively as a basis for trading between growers, shippers, wholesalers, and retailers. They are used to a limited extent in sales from retailers to consumers. Use of U.S. grade standards is voluntary in most cases. However, most marketing firms require grading for their products. Large retail chains can demand a certain quality or grade of products they will accept or receive. It is rare to not use grading.

FIGURE 21-7 Besides variety, consumers expect high-quality fruits and vegetables.

However, fruits and vegetables sold directly from so-called U-pick farms and farmers' markets are not usually graded. Grade designations are most often seen on packages of potatoes and onions. Other vegetables occasionally carry the grade name. Grade designations are Fancy, 1, 2, and 3.

U.S. Fancy. **U.S. Fancy** vegetables are of more uniform shape and have fewer defects than U.S. No. 1.

U.S. No. 1. No. 1 vegetables should be tender, appear fresh, have good color, and be relatively free from bruises and decay.

U.S. No. 2 and No. 3. Even though U.S. No. 2 and No. 3 have lower quality requirements than Fancy or No. 1, all **grades** are nutritious. The differences are mainly in appearance, waste, and preference.

Quality Grades for Canned and Frozen Vegetables

The grade standards are used extensively by processors, buyers, and others in wholesale trading to establish the value of a product described by the grades.

U.S. Grade A. Grade A vegetables are carefully selected for color, tenderness, and freedom from blemishes. They are the most tender, succulent, and flavorful vegetables produced. The term *fancy* may appear on the label to reflect the Grade A product.

U.S. Grade B. Grade B vegetables are of excellent quality but not quite as well selected for color and tenderness as Grade A. They are usually slightly more mature and therefore have a slightly different taste than the more succulent vegetables in Grade A.

U.S. Grade C. Grade C vegetables are not so uniform in color and flavor as vegetables in the higher grades, and they are usually more mature. They are a thrifty buy when appearance is not too important. These vegetables can be used as ingredients in soups, stews, and casseroles.

Other names also describe the quality grades of canned or frozen vegetables—Grade A as "Fancy," Grade B as "Extra Standard," and Grade C as "Standard." The brand name of a frozen or canned vegetable may also be an indication of quality. Producers of nationally advertised products spend considerable money and effort to maintain the same quality for their brand labels year after year. Unadvertised brands may also offer an assurance of quality, often at a slightly lower price. Many stores, particularly chain stores, carry two or more qualities under their own name labels (private labels).

Quality Grades for Fresh Fruit

The USDA also has established grade standards for most fresh fruits. The grades are used extensively as a basis for trading among growers, shippers, wholesalers, and retailers. Grade standards are used to a limited extent in sales from retailers to consumers. Although use of U.S. grade standards is voluntary, most packers grade their fruits, and some mark consumer packages with the grade. Some state laws and federal marketing programs require grading and grade labeling of certain fruits. Fruits are graded as Fancy, 1, 2, and 3.

U.S. Grade A or Fancy. Grade A or Fancy means premium quality (Figure 21-8). Only a small percentage of fruits are packed in this grade. These fruits have excellent color, uniform size, weight, and shape. They have the proper ripeness and possess few or no blemishes. Fruits of this grade are used for special purposes where appearance and flavor are important.

FIGURE 21-8 Grade A or fancy apples.

U.S. No. 1. U.S. No. 1 means good quality and is the most commonly used grade for most fruits.

U.S. No. 2 and U.S. No. 3. U.S. No. 2 is noticeably superior to U.S. No. 3, which is the lowest grade practical to pack under normal commercial conditions.

Quality Grades for Canned and Frozen Fruits

Grading and use of fruits for canning and freezing is slightly different. These fruits are graded U.S. Grades A, B, and C.

U.S. Grade A. Grade A fruits are the best, with excellent color and uniform size, weight, and shape. Having the proper ripeness and few or no blemishes, fruits of this grade are excellent to use for special purposes where appearance and flavor are important. This highest grade of fruits is the most flavorful and attractive. Often they are the most expensive. They are excellent to use for special luncheons or dinners, served as dessert, used in fruit plates, or broiled or baked to serve with meat entrees.

U.S. Grade B. Grade B fruits make up much of the fruits that are processed and are of very good quality. Only slightly less perfect than Grade A in color, uniformity, and texture, Grade B fruits have good flavor and are suitable for most uses. Grade B fruits, which are not quite as attractive or tasty as Grade A, are still of good quality. They are used as breakfast fruits; in gelatin molds, fruit cups, and compotes; as topping for ice cream; and as side dishes.

U.S. Grade C. Grade C fruits may contain some broken and uneven pieces. Though flavor may not be as sweet as in higher qualities, these fruits are still good and wholesome. They are useful where color and texture are not of great importance such as in puddings, jams, and frozen desserts. Grade C fruits vary more in taste and appearance than the higher grades, and they cost less. They are useful in many dishes, especially where appearance is not important—for example, in sauces for meats and in cobblers, tarts, upside-down cakes, frozen desserts, jams, or puddings.

Other names are often used to describe the quality grades of canned and frozen fruits: Grade A as Fancy, Grade B as **Choice**, and Grade C as Standard.

Styles. Both canned and frozen fruit are sold in many forms, shapes, or styles. Larger fruit such as pears and peaches may be found in whole, halves, quarters, slices, and diced. Smaller fruits are usually whole, but fruits such as strawberries are sliced and halved. It is best to examine the label for a description of the type or style that will best suit the purpose a customer has in mind when serving the fruit. The grade, style, and syrup or special flavorings in which processed fruits are prepared all affect the cost of the fruits and how they are used.

Country of Origin Labeling

Country of origin labeling (COOL) is a labeling law that requires retailers such as full-line grocery stores, supermarkets, and club warehouse stores to provide their customers information about the source of certain foods, including fresh and frozen fruits and vegetables, peanuts, pecans, macadamia nuts, and ginseng. The final rule for all covered commodities went into effect in March 2009. Agricultural Marketing Services, a division of the USDA, is responsible for administering and enforcing COOL.

DEVELOPMENT OF POTATO CHIPS

In terms of consumption, potatoes rank second worldwide behind rice. Many potatoes are consumed as potato chips. The potato chip is more popular in the United States than in any other part of the world. This favorite snack food is a direct descendant of another popular potato snack, the french fry.

According to one popular story, a dinner guest was dining at Moon's Lake House in Saratoga Springs, New York, in 1853. He sent his french fries back to the kitchen because they were too thick. The chef, a Native American named George Crum, was annoyed at the guest's complaint, so he responded by slicing the potatoes into extremely thin sections, which he fried in oil and salted. The plan backfired, and soon potato chips began appearing on the menu as Saratoga Chips.

For a time, potato chips were available only in the North. In the 1920s, Herman Lay, a traveling salesman in the South, helped popularize the chip from Atlanta to Tennessee. Lay sold potato chips to Southern grocers out of the trunk of his car. Soon he built a business and a name. Lay's potato chips became the first successfully marketed national brand. Today potato chips have evolved into many forms and varieties, including chips of many flavors, fat-free potato chips cooked in high-tech synthetic chemicals, and even artificially shaped chips pressed from potato pulp and sold in cardboard tubes.

For more information about potato chips and other vegetable products, search the Web or visit these Web sites:

- George Crum—Inventor of Potato Chips: http://www.todayifoundout.com/index.php/2014/09/real-story-potato-chip/
- FritoLay: http://www.fritolay.com/snacks
- Potato Chips: http://www.ideafinder.com/history/inventions/potatochips.htm
- How It's Made—Potato Chips: https://youtu.be/LkqBbr7Ewsw

Most processed fruits are available in at least two grades. The grade is not often indicated on processed fruits, but individuals can learn to tell differences in quality by trying different brands. Whole fruits, halves, or slices of similar sizes are more expensive than mixed pieces of various sizes and shapes. Some of the most popular fruits, along with the styles in which they are available, are shown in Figure 21-9.

How to Use Grades and Styles

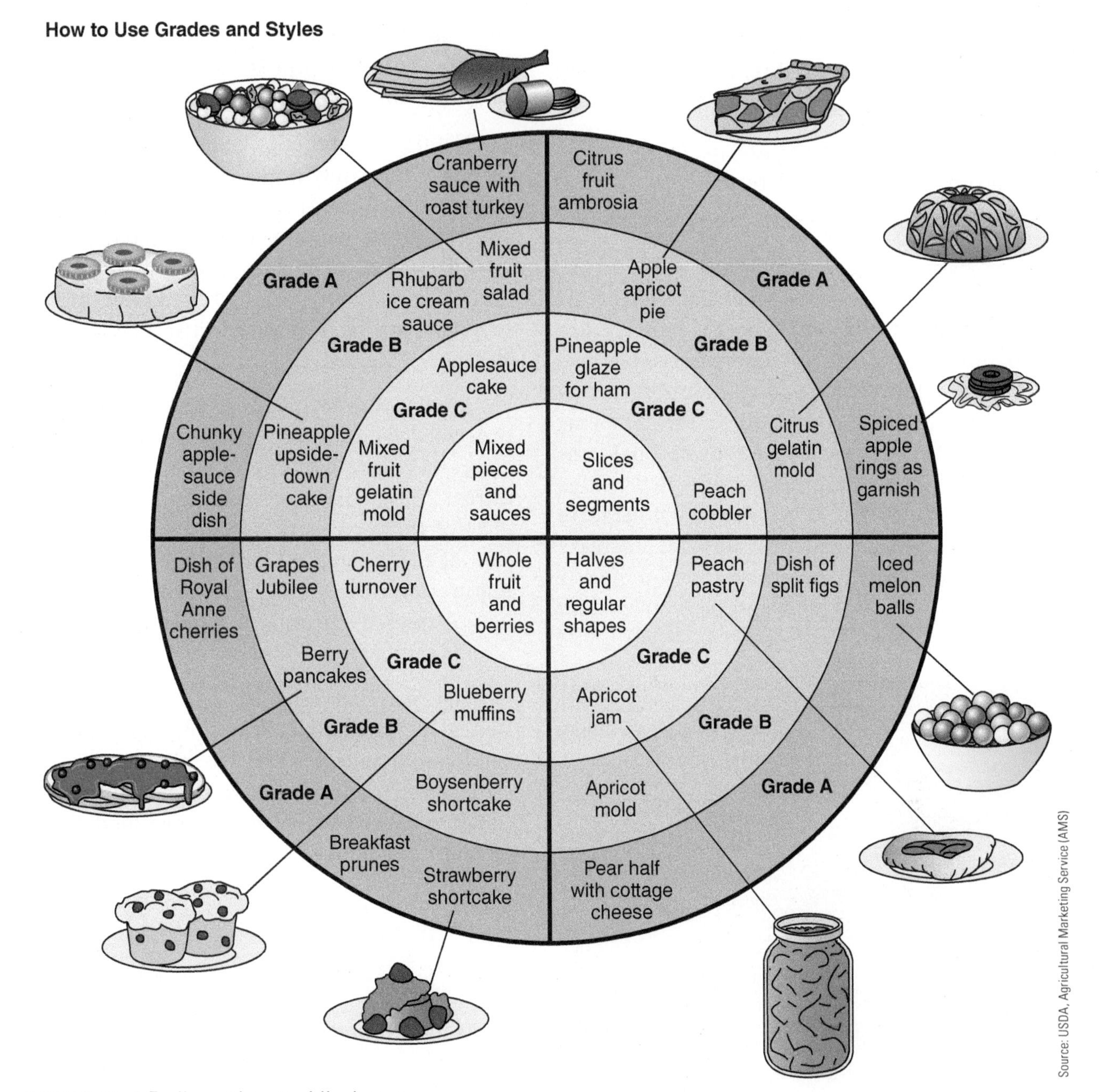

FIGURE 21-9 Fruit grades and their uses.

POST-HARVEST

The quality of harvested fruits and vegetables can be optimal, but postharvest care and storage will affect what the processor, food preparer, or consumer receives (Figure 21-10). Postharvest factors that affect the marketing of quality produce include:

- Shipping and storage at optimal temperature
- Control of carbon dioxide (CO_2) and oxygen (O_2)

- Good humidity control
- Minimal exposure to ethylene gas
- Application of appropriate chemicals
- Good sanitation
- Harvest at the correct maturity stage
- Care in handling during harvest, packing, and shipping

Likely the single most important factor influencing quality during storage is temperature reduction and postharvest maintenance because metabolic activity and spoilage microorganisms within the produce is reduced by low temperature. Each fruit or vegetable has its optimal temperature regime. The generally accepted rule is that for every 50 °F (10 °C) rise in temperature from optimum, metabolic rate will increase two- to threefold. This increase in the temperature sensitivity of fruits and vegetables and subsequent deterioration make fruit and vegetable transportation and storage by the consumer of critical importance.

© Dream79/Shutterstock.com

FIGURE 21-10 Potatoes are an example of fresh produce; chips are an example of a processed form of potatoes.

Natural or synthetic chemicals are used commercially to maintain quality through prolonged storage life. Three primary gases—water vapor (humidity), carbon dioxide (CO_2), and ethylene—strongly affect vegetables and fruits. Modified atmospheres (lower O_2 or higher CO_2) in the form of gases have been most effective in prolonging storage.

Modified atmosphere is done for intact or processed fruit or vegetable through such processes as controlled atmospheric storage procedures. These conditions attempt to control the level of those gases that influence the respiration of the vegetable—for example, ripening or maturation. Generally, CO_2 evolution occurs with increased maturation. Controlling the CO_2 level is critical.

The interrelationship between ethylene gas and storage life is dynamic. Fruits and vegetables may be classified as either ethylene producers or ethylene sensitive (climacteric versus nonclimacteric). (Refer to Table 21-1.) Processors and food preparers must be aware of this in their packaging and storage.

PROCESSING FRUITS

Fruits for canning or freezing are harvested at the proper stage of ripeness so that a good texture and flavor may be preserved. Much of the processing is done by automated equipment, and there is minimal fruit handling by plant workers. Current practices help ensure wholesome, sanitary products with good flavor and quality.

Table 21-3 lists common fruits and provides information about each one.

The initial work in preparing canned or frozen fruits is similar. At the processing plant, fresh

TABLE 21-3 Common Fruits

FRUITS	NOTES
Apples © Africa Studio/Shutterstock.com	The many varieties of apples differ widely in appearance, flesh characteristics, seasonal availability, and suitability for different uses.
Apricots © Nataliya Arzamasova/Shutterstock.com	Apricots develop their flavor and sweetness on the tree and should be mature but firm at the time that they are picked.
Avocados © Nataliya Arzamasova/Shutterstock.com	Avocados vary greatly in shape, size, and color. Most tend to be pear shaped, but some are almost spherical. Some have rough or leathery textured skin, whereas others have smooth skin. The skin color of most varieties is some shade of green, but certain varieties turn maroon, brown, or purplish-black as they ripen.
Bananas © Iurii Kachkovskyi/Shutterstock.com	Bananas develop their best eating quality after they are harvested. This allows bananas to be shipped great distances. Bananas are sensitive to cool temperatures and will be injured in temperatures below 55 °F (12.8 °C), so they should never be kept in the refrigerator. The ideal temperature for ripening bananas is between 60 °F and 70 °F (15.6 °C and 21.1 °C).
Blueberries © Brian A Jackson/Shutterstock.com	Large berries are cultivated varieties, and smaller berries are wild varieties. A dark blue color with a silvery bloom is the best indication of quality. This silvery bloom is a natural and protective waxy coating.
Cherries © Nitr/Shutterstock.com	Good cherries have bright, glossy, plump-looking surfaces and fresh-looking stems. A deep, dark color is the most important indication of good flavor and maturity in sweet cherries.

(*Continues*)

TABLE 21-3 Common Fruits

FRUITS	NOTES
Cranberries  © Gita Kulinitch Studio/Shutterstock.com	These fruits differ considerably in size and color but are not identified by variety names. Plump, firm berries with a lustrous color provide the best quality. Duller varieties should at least have some red color.
Grapefruit  © Valentyn Volkov/Shutterstock.com	Several varieties are marketed, but the principal distinction at retail is between those that are seedless (having few or no seeds) and the seeded type. Another distinction is pink- or red-fleshed fruit and white-fleshed fruit. Grapefruit is picked tree ripe and is ready to eat. Thin-skinned fruits have more juice than coarse-skinned ones.
Grapes  © Su Jianfei/Shutterstock.com	European types are firm fleshed and generally have high sugar content. American-type grapes have softer flesh and are juicier than European types. The outstanding variety for flavor is the Concord, which is blue-black when fully matured. Well-colored, plump grapes are firmly attached to the stem. White or green grapes are sweetest when the color has a yellowish cast or straw color, with a tinge of amber. Red varieties are better when good red predominates on all or most of the berries. Bunches are more likely to hold together if the stems are predominantly green and pliable.
Kiwifruit  © 5 second Studio/Shutterstock.com	A relatively small, ellipsoid-shaped fruit with a bright green, slightly acid-tasting pulp surrounding many small, black, edible seeds, which in turn surround a pale heart. The exterior of the kiwifruit is unappealing to some, being somewhat "furry" and light to medium brown in color. Kiwifruit contains an enzyme, actinidin, similar to papain in papayas that reacts chemically to break down proteins. Actinidin prevents gelatin from setting.
Lemons  © Antonova Anna/Shutterstock.com	Lemons should have a rich yellow color, reasonably smooth-textured skin with a slight gloss, and be firm and heavy. A pale or greenish-yellow color means the freshest fruit with slightly higher acidity. Coarse or rough skin texture is a sign of thick skin and not much flesh.

(Continues)

TABLE 21-3 Common Fruits

FRUITS	NOTES
Limes © Vitalina Rybakova/Shutterstock.com	Limes should have glossy skin and feel heavy for their size.
Cantaloupes © Donal Joski/Shutterstock.com	A cantaloupe might be mature, but not ripe. The stem should be gone, leaving a smooth symmetrical, shallow base called a *full slip*. The netting or veining should be thick, coarse, and corky, and it should stand out in bold relief over some part of the surface. The skin color (ground color) between the netting should have changed from green to yellowish-buff, yellowish-gray, or pale yellow.
Casaba © Teri Virbickis/Shutterstock.com	This sweet, juicy melon is normally pumpkin shaped with a slight tendency to be pointed at the stem end. It is not netted but has shallow, irregular furrows running from the stem end toward the blossom end. The rind is hard with light green or yellow color. The stem does not separate from the melon and must be cut in harvesting.
Crenshaw © rj lerich/Shutterstock.com	Its large size and distinctive shape make this melon easy to identify. It is rounded at the blossom end and tends to be pointed at the stem end. The rind is relatively smooth with only shallow lengthwise furrowing. The flesh is pale orange, juicy, and delicious and is generally considered outstanding in the melon family.
Honeydew © vanillaechoes/Shutterstock.com	The outstanding flavor characteristics of honeydews make them highly prized as a dessert fruit. A soft, velvety texture indicates maturity. The stem does not separate from the fruit and must be cut for harvesting.
Watermelon © topseller/Shutterstock.com	Judging the quality of a watermelon is extremely difficult unless it is cut in half or quartered. The flesh should be firm and juicy with good red color free from white streaks; seeds should be dark brown or black. Seedless watermelons often contain small white, immature seeds, which are normal for this type.

(Continues)

TABLE 21-3 Common Fruits

FRUITS	NOTES
Nectarines © Alexlukin/Shutterstock.com	This fruit combines characteristics of both the peach and the plum. Most varieties have an orange-yellow background color between the red areas, but some varieties have a greenish background color.
Oranges © Jochen Schoenfeld/Shutterstock.com	Leading varieties are the Washington navel and the Valencia, both characterized by a rich orange skin color. The navel orange has a thicker, somewhat more pebbled skin than the Valencia; the skin is more easily removed by hand, and the segments separate more readily. It is ideally suited for eating as a whole fruit or in segments in salads. The Valencia orange is excellent either for juicing or for slicing in salads. Oranges are required by strict state regulations to be mature before being harvested and shipped out of the producing state. Thus, skin color is not a reliable index of quality, and a greenish cast or green spots do not mean that the orange is immature. Often, fully matured oranges will turn greenish (called "regreening") late in the marketing season. Some oranges are artificially colored to improve their appearance. This practice has no effect on eating quality, but artificially colored fruits must be labeled "color added."
Peaches © Zadorozhna Natalia/Shutterstock.com	Peaches fall into two general types: freestone (flesh readily separates from the pit) and clingstone (flesh clings tightly to the pit). Freestones are usually preferred for eating fresh or for freezing; clingstones are used primarily for canning, although they are sometimes sold fresh.
Pears © kuvona/Shutterstock.com	The most popular variety of pear is the Bartlett, both for canning and for sale as a fresh fruit.
Pineapple © JIANG HONGYAN/Shutterstock.com	Pineapples are available all year but are most abundant from March through June. Current marketing practices, including air shipments, allow pineapples to be harvested as nearly ripe as possible. They should have a bright color, fragrant pineapple aroma, and a slight separation of the eyes or pips—the berrylike fruitlets patterned in a spiral on the fruit core.

(Continues)

TABLE 21-3 Common Fruits

FRUITS	NOTES
Plums and prunes © Dionisvera/Shutterstock.com	Quality characteristics for both are similar, and the same buying tips apply to both. Plum varieties differ slightly in appearance and flavor. Only a few varieties of prunes are commonly marketed, and they are all closely similar.
Raspberries, boysenberries, etc. © Vezzani Photography/Shutterstock.com	Blackberries, raspberries, dewberries, loganberries, and youngberries are similar in general structure. They differ from one another in shape or color, but quality factors are about the same for all. Look for berries that are fully ripened with no attached stem caps.
Strawberries © Dionisvera/Shutterstock.com	Berries should have a full red color and bright luster, firm flesh, and the cap stem still attached. The berries should be dry and clean, and usually medium to small strawberries have better eating quality than large ones.
Tangerines © Africa Studio/Shutterstock.com	Deep yellow or orange color and a bright luster is your best sign of fresh, mature, good-flavored tangerines. Because of the typically loose nature of tangerine skins, they will frequently not feel firm to the touch.

fruits are usually sorted into sizes by machine and washed in continuously circulating water or under sprays of water (Figure 21-11). Some fruits—including apples, pears, and pineapples—are mechanically peeled and cored. Next, they are moved on conveyor belts to plant workers who do any additional peeling or cutting necessary. Pits and seeds are removed by automatic equipment, and the fruits are prepared in the various styles (halves, slices, or pieces) by machine. Before the fruits are canned or frozen, plant workers remove any undesirable portions.

© Vaclav Psota/Dreamstime.com

FIGURE 21-11 Washing and grading apples.

SCIENCE CONNECTION!

Research what fruits are grown in your area or region. Why? Does the soil, temperature, or humidity have an influence on what is grown locally? Do you have a local farmers' market? If you wanted to start a garden at your home, what fruits would you select? What do you think will grow best?

Canned Fruits

Cans or glass jars are filled with fruit by semiautomatic machines. Next, the containers are moved to machines that fill them with the correct amount of syrup or liquid and then to equipment that automatically seals them. The sealed containers are cooked under carefully controlled conditions of time and temperature to ensure that the products will keep without refrigeration. After the containers are cooled, they are stored in cool, dry, well-ventilated warehouses until they are shipped to market.

Frozen Fruits

Frozen fruits are most often packed with dry sugar or syrup. After the initial preparation, packages are filled with fruit by semiautomatic equipment, sugar or syrup is added, and the containers are automatically sealed. The packaged fruit is then quickly frozen in special low-temperature chambers and stored at temperatures of 0 °F (−18 °C) or lower.

Fruit Juices

Orange juice is probably the most commonly processed juice. Steps in orange juice processing are also common for the manufacture of other juices. The main steps in the production of most juices include:

- Extraction
- Clarification (clearing)
- Deaeration (removal of air)
- Pasteurization
- Concentration
- Essence add-back (flavors)
- Canning or bottling
- Freezing

Not all juices go through every process. The desired end product will determine whether the juice will be concentrated and frozen.

Blending juices is also a popular processing technique. Some blends include mango and apple, apple and cranberry, and mango and orange. Pear juice is used as a base for many juices because its fruit flavor is strong but not necessarily characteristic. Finally, juices low in vitamin C may be fortified with this vitamin.

PROCESSING VEGETABLES

Vegetables for canning and freezing are grown especially for that purpose, and the processing preserves much of their nutritional value. Both canning and freezing plants are usually located in the vegetable production areas so that the harvested vegetables can be quickly brought to the plant for processing while fresh and at their peak quality.

Table 21-4 lists the common vegetables and provides information about each one.

In today's modern processing plants, most vegetables are canned or frozen by automated equipment, and the procedures used are similar to those used for fruit. The initial work in preparing canned or frozen vegetables is similar. At the processing plant, the fresh product is usually sorted into sizes by machine and washed in

TABLE 21-4 Common Vegetables

VEGETABLES	NOTES
Artichoke © Sabino Parente/Shutterstock.com	The globe artichoke is the large, unopened flower bud of a plant belonging to the thistle family. The many leaflike parts that make up the bud are called *scales*.
Beets © Anna Kucherova/Shutterstock.com	Many beets are sold in bunches with the tops still attached; others are sold with the tops removed. If beets are bunched, you can judge their freshness fairly accurately by the condition of the tops.
Broccoli © mama_mia/Shutterstock.com	A member of the cabbage family and a close relative of cauliflower, it should have a firm, compact cluster of small flower buds, with none opened enough to show the bright-yellow flower. Bud clusters should be dark green or sage green—or even green with a decidedly purplish cast.
Brussel sprouts © Brent Hofacker/Shutterstock.com	Another close relative of the cabbage, Brussels sprouts develop as enlarged buds on a tall stem, one sprout appearing where each main leaf is attached. These sprouts are cut off and, in most cases, are packed in small consumer containers.
Cabbage © Peter Zijlstra/Shutterstock.com	Three major groups of cabbage varieties are available: smooth-leaved green cabbage; crinkly-leaved green Savoy cabbage; and red cabbage. All types are suitable for any use. Cabbage may be sold fresh (called "new" cabbage) or from storage.

(Continues)

TABLE 21-4 Common Vegetables

VEGETABLES	NOTES
Carrots © Donatella Tandelli/ Shutterstock.com	Freshly harvested carrots are available year round. Most are marketed when relatively young, tender, well colored, and mild flavored—an ideal stage for use as raw carrot sticks. Larger carrots are packed separately and used primarily for cooking or shredding.
Cauliflower © Tim UR/Shutterstock.com	The white edible portion is called the *curd*, and the heavy outer leaf covering is called the *jacket leaves*. Cauliflower is generally sold with most of the jacket leaves removed and wrapped in plastic film. A slightly granular or "ricey" texture of the curd will not hurt the eating quality if the surface is compact.
Celery © jiangdi/Shutterstock.com	Most celery is of the so-called Pascal type, which includes thick-branched, green varieties.
Chinese cabbage © Dan loei tam loei/ Shutterstock.com	Primarily a salad vegetable, Chinese cabbage plants are elongated, with some varieties developing a firm head and others an open, leafy form.
Chicory, endive, escarole © Madlen/Shutterstock.com © Lim ChewHow/ Shutterstock.com	Used mainly in salads, are available practically all year round, but primarily in the winter and spring. Chicory or endive has narrow, notched edges, and crinkly leaves resembling the dandelion leaf. Chicory plants often have "blanched" yellowish leaves in the center that are preferred by many people. Escarole leaves are much broader and less crinkly than those of chicory.

(*Continues*)

TABLE 21-4 Common Vegetables

VEGETABLES	NOTES
Corn © taweesak thiprod/Shutterstock.com	Sweet corn is available practically every month of the year, but is most plentiful from early May until mid-September. Yellow-kernel corn is the most popular, but some white-kernel and mixed-color corn is sold. Sweet corn is produced in a large number of states during the spring and summer. For best quality, corn should be refrigerated immediately after being picked. Corn will retain fairly good quality for a number of days, if it has been kept cold and moist since harvesting.
Cucumber © NIPAPORN PANYACHAROEN/Shutterstock.com	Produced at various times of the year in many states, and imported during the colder months. The supply is most plentiful in the summer months. Cucumbers should be a good green color, well developed, but not too large in diameter.
Eggplant © Nattika/Shutterstock.com	Most plentiful during late summer, but is available all year. Although the purple eggplant is more common, white eggplant is occasionally seen in the marketplace.
Greens © Alena Haurylik/Shutterstock.com	A large number of widely differing species of plants are grown for use as "greens." Spinach, kale, collard, turnip, beet, chard, mustard, broccoli leaves, chicory, endive, escarole, dandelion, cress, and sorrel are well-known greens. Many others, some of them wild, are also used to a limited extent as greens.
Lettuce © Svitlana-ua/Shutterstock.com	One of the leading U.S. vegetables, lettuce owes its prominence to the growing popularity of salads in our diets. It is available throughout the year in various seasons from California, Arizona, Florida, New York, New Jersey, and other states. Four types of lettuce are generally sold: iceberg, butter-head, Romaine, and leaf.
Mushrooms © SJ Travel Photo and Video/Shutterstock.com	Grown in houses, cellars, or caves, mushrooms are available year-round in varying amounts. They are mainly produced in Pennsylvania, California, New York, Ohio, and other states. They are described as having a cap (the wide portion on top), gills (the numerous rows of paper-thin tissue seen underneath the cap when it opens), and a stem.

(Continues)

TABLE 21-4 Common Vegetables

VEGETABLES	NOTES
Okra © merc67/Shutterstock.com	Okra is the immature seedpod of the okra plant, generally grown in southern states.
Onions © ORLIO/Shutterstock.com	The many varieties of onions grown commercially fall into three general classes, distinguished by color: yellow, white, and red. Onions are available year-round, either fresh or from storage. Consumers should avoid onions with thick, hollow, woody centers in the neck or with fresh sprouts.
Onions, green and leeks © Binh Thanh Bui/ Shutterstock.com	Sometimes called *scallions*, green onions are similar in appearance but somewhat different in nature. They are ordinary onions harvested very young. They have little or no bulb formation, and their tops are tubular. Leeks have slight bulb formation and broad, flat, dark-green tops.
Parsley © Diana Taliun/Shutterstock .com	Parsley is generally available year-round. It is used both as a decorative garnish and to add its own unique flavor.
Parsnips © margouillat photo/ Shutterstock.com	Although available to some extent throughout the year, parsnips are primarily late-winter vegetables because the flavor becomes sweeter and more desirable after long exposure to cold temperatures below 40 °F (4.4 °C).
Peppers © topseller/Shutterstock .com	Most of the peppers are sweet green peppers, available in varying quantities throughout the year, but most plentiful during late summer. Fully matured peppers of the same type have a bright red color. A variety of colored peppers are also available, including white, yellow, orange, red, and purple.

(*Continues*)

TABLE 21-4 Common Vegetables

VEGETABLES	NOTES
Potatoes © Quang Ho/Shutterstock.com	Potatoes can be put into three groups: new potatoes, general purpose, and baking. Some overlap exists. New potatoes frequently describe potatoes that have been freshly harvested and marketed during the late winter or early spring. The name is also widely used to designate freshly dug potatoes that are not fully matured. The best uses for new potatoes are boiling or creaming. They vary widely in size and shape, depending on their variety, but are likely to be affected by *skinning* or *feathering* of the outer layer of skin. Skinning usually only affects appearance. General-purpose potatoes include the great majority of supplies, both round and long types, offered for sale in markets. With the aid of air-cooled storage, they are available throughout the year. They are used for boiling, frying, and baking. Potatoes grown specifically for their baking quality also are available. Variety and area where grown affect baking quality. The Russet Burbank is commonly used for baking. Potatoes should be free from blemishes and sunburn (a green discoloration under the skin). For cooking purposes, potatoes are sometimes classified as *mealy* or *waxy*. Mealy potatoes (russets, purple) have thick skin and a high starch content but are low in moisture and sugar. Waxy potatoes (red, new) are low in starch with thin skins and a lot of moisture and sugar. Mealy potatoes are best for deep frying and the best choice for mashed potatoes. Waxy potatoes are a good choice for roasting, sautéing, and boiling, but they are not a good choice for frying because of their high moisture content.
Radishes © phloen/Shutterstock.com	Radishes are available year-round, but they are most plentiful from May through July. Medium-size radishes are ¾ to 1 inch in diameter.
Rhubarb © Diana Taliun/Shutterstock.com	Rhubarb is a specialized vegetable used like a fruit in sweetened sauces and pies. Limited supplies are available during most of the year; the best supplies are available from January to June. The petiole should be red, tender, and not fibrous.
Squash (summer) © de2marco/Shutterstock.com	Summer squash includes those varieties that are harvested while still immature and when the entire squash is tender and edible. They include the yellow crookneck, the large straight neck, the greenish-white patty pan, and the slender green zucchini.

(Continues)

TABLE 21-4 Common Vegetables

VEGETABLES	NOTES
Squash (fall and winter) © Svetlana Foote/Shutterstock.com	Winter squash are those varieties that are marketed only when fully mature. Some of the most important varieties are the small, corrugated acorn (available year-round), butternut, buttercup, green and blue Hubbard, green and gold delicious, and banana.
Sweet Potatoes © mama_mia/Shutterstock .com	Available in varying amounts year-round, moist sweet potatoes, sometimes mistakenly called *yams*, are the most common type. They have orange-colored flesh and are especially sweet. (The true yam is the root of a tropical vine that is not grown commercially in the United States.) Dry sweet potatoes have pale-colored flesh and are low in moisture.
Tomatoes © Jiri Vaclavek/Shutterstock .com	Popular and nutritious, tomatoes are in moderate to liberal supply throughout the year. The best flavor usually comes from locally grown tomatoes produced on nearby farms, because the tomato is allowed to ripen completely before being picked. Many areas now ship tomatoes that are picked right after the color has begun to change from green to pink. If tomatoes need further ripening, keep them in a warm place but not in direct sunlight.
Turnips © Jacques PALUT/Shutterstock.com	Popular turnips have white flesh and a purple top (reddish-purple tinting of upper surface). They may be sold "topped" (with leaves removed) or in bunches with tops still on. Rutabagas are distinctly yellow-fleshed, large-sized relatives of turnips.
Watercress © Hawk777/Shutterstock.com	Watercress is a small, round-leaved plant that grows naturally (or it may be cultivated) along the banks of freshwater streams and ponds. It is prized as an ingredient in mixed green salads and as a garnish because of its spicy flavor. Watercress is available in limited supply through most of the year.

© Gemenacom/Dreamstime.com

FIGURE 21-12 Potatoes ready for packaging.

continuously circulating water or sprays of water. Some vegetables—carrots, beets, and potatoes, for example—are mechanically peeled. Next, they are moved onto conveyer belts where plant workers do any additional peeling or cutting in preparation for the various styles—whole, cut, sliced, and so on (Figure 21-12).

Canned Vegetables

Cans or glass jars are filled with vegetables by semiautomatic or automatic machines. Next, the containers are moved to machines that fill them with the correct amount of brine or liquid and then to machines that preheat them before automatically sealing them. The sealed containers are then cooked under carefully controlled conditions of time and temperature to ensure that the product will keep without refrigeration. After the containers are cooled, they are stored in cool, dry, well-ventilated warehouses until they are shipped to market.

Vegetables sold in glass jars with screw-on or vacuum-sealed lids are sealed tightly to preserve the contents.

© 2017 Cengage Learning

FIGURE 21-13 Frozen fruit and vegetable products and packaging.

Frozen Vegetables

Like fruit, vegetables can be frozen for future use (Figure 21-13). After initial preparation, vegetables that are to be frozen are usually blanched or slightly precooked. This precooking process ensures that the frozen vegetables will retain much of their natural appearance and flavor for long periods of time in storage. Without blanching, the product would prematurely turn brown or oxidize before it could be marketed. After freezing, the vegetables are packaged in polyethylene bags of varying sizes or packaged in retail-sized fiber cartons with a labeled overwrap that identifies the product.

BY-PRODUCTS

Processing fruits and vegetables produces many by-products. Often these by-products become feed for livestock. For example, in different areas of the country, milled corn, citrus pulp, processed potato wastes, and grape seeds and skins can be fed to cattle (Figure 21-14).

SCIENCE CONNECTION!

Research which vegetables are grown in your area or region. Why? Does the soil, temperature, or humidity influence what is grown locally? Do you have a local farmers' market in your area? If you wanted to start a garden at your home, what vegetables would you select? What do you think will grow best?

© riaan van den Berg/Shutterstock.com

FIGURE 21-14 Many by-products from fruit and vegetable processing are fed to cattle.

advantage. By doing so, they will be able to design environmentally hardy food-producing plants that are naturally resistant to pests and diseases and capable of growing under extreme conditions of temperature, moisture, and salinity. Or scientists may design an array of fresh fruits and vegetables with excellent flavor, appealing texture, and optimal nutritional content that stay fresh for several weeks. Possibly they could custom design plants with defined structural and functional properties for specific food-processing applications. On the control side, scientists might use biotechnology to produce microsensors that accurately measure the physiological state of plants or temperature-abuse indicators for refrigerated foods, and shelf-life monitors built into food packages. The possibilities are endless.

SCIENCE CONNECTION!

What is a by-product? Select a few fruits or vegetables and see if you can find any by-products made from them! *Hint:* Think of home cleaning products and the scents they use.

BIOTECHNOLOGY

Biotechnology allows researchers to target the genetics of plants, animals, and microorganisms and to manipulate them to our food-production

SUMMARY

Fruits and vegetables include a wide variety of edible parts. They vary widely in carbohydrate and protein content. Many have high water content. Fruits and vegetables are a good source of many vitamins and minerals. Because many parts of plants are eaten as fruits or vegetables, an understanding of plant tissues is critical to food scientists.

The harvesting of fruits and vegetables can be affected by variety, soil type, water, temperature, and season. Climacteric fruits produce ethylene gas during ripening; nonclimacteric fruits do not. Fruits and vegetables get their characteristic color from numerous pigments. In general, water transport in plant tissues influences the texture of fruits and vegetables; flavors and aromas are the result of compounds such as aldehydes, alcohols, ketones, esters, organic acids, and sulfur compounds.

The USDA assigns quality grades to both fruits and vegetables. These quality grades determine the eventual use of the fruit or vegetable. Postharvest care is critical to maintaining the optimal quality of fruits and vegetables. Fruits and vegetables are sold fresh, canned, and frozen. Most of these processes are automated. By-products from fruit and vegetables processing are often used for livestock feed. Biotechnology offers the promise of providing fruits and vegetables to meet new consumer demands.

REVIEW QUESTIONS

Success in any career requires knowledge. Test your knowledge of this chapter by answering these questions or solving these problems.

1. How do fruits and vegetables get their characteristic colors?
2. What are the grade designations for fresh fruits and vegetables and canned fruits and vegetables?
3. What is the difference between climacteric and nonclimacteric fruits?
4. List four compounds that give fruits or vegetables their flavor.
5. What plant material makes up most of the edible portion of fruits and vegetables?
6. The crispness of a vegetable is the result of which plant function?
7. Describe the processing of fruit for canning or freezing.
8. Name the single most important factor that influences quality during the storage of fruits and vegetables.
9. Why are frozen vegetables blanched or precooked?
10. List the steps in orange juice processing.

STUDENT ACTIVITIES

1. Make a list of all vegetables you consume during a week. Determine which part of the plant you are considering to be a vegetable—for example, the flower, the stem, the root, the fruit, and so on. Using Table B-8, list the composition of five of these vegetables.
2. Pick a fruit or vegetable and list all the fresh and processed forms for the fruit or vegetable. Compete with other class members to see who found a vegetable or fruit sold in the most forms. Create a class list to display.
3. Develop a report or presentation on the production, harvesting, and processing of a specific fruit or vegetable listed in Table 21-3 or Table 21-4. Research information provided on the Internet to find facts for this project.
4. Develop an experiment to extract the pigment from a fruit or vegetable such as a carrot, an apple, or spinach. Conduct the experiment and record your findings.
5. Develop an experiment to store a selected vegetable or fruit under different conditions—for example, light or dark, dry or humid, warm or cold, or some combination. Record what happens to the quality of the fruit or vegetable under each set of conditions.

ADDITIONAL RESOURCES

- Structure and Types of Fruit: http://lifeofplant.blogspot.com/2011/04/structure-and-types-of-fruit.html
- Proper Harvesting of Fruits and Vegetables: https://web.extension.illinois.edu/cfiv/homeowners/020824.html

- Guidelines for Harvesting Vegetables: http://www.gardening.cornell.edu/factsheets/vegetables/harvestguide.pdf
- FAO—Fruit and Vegetable Products: http://www.fao.org/wairdocs/x5434e/x5434e05.htm

Internet sites represent a vast resource of information. Although those provided in this chapter were vetted by industry experts, you may also wish to further explore the topics discussed in this chapter using a search engine such as Google. Keywords or phrases may include the following: *a specific fruit or vegetable*; *grading of fruits and vegetables such as U.S. Fancy, U.S. No. 1, 2, 3*; *grades of fruits and vegetables: grade A, B, C*; *processing of specific fruits or vegetables*; *harvesting of specific fruits or vegetables.* In addition, Table B-7 provides a listing of useful Internet sites that can be used as a starting point.

REFERENCES

Corriher, S. O. 1997. *Cookwise: The hows and whys of successful cooking.* New York: William Morrow & Co.

Horn, J., J. Fletcher, and A. Gooch. 1997. *Cooking A to Z: The complete culinary reference source.* Glen Ellen, CA: Cole Publishing Group.

Lehner, E., and J. Lehner. 1973. *Folklore and odysseys of food and medicinal plants.* New York: Farrar Straus Giroux.

Parker, R. 2009. *Plant and soil science: Fundamentals and applications* Clifton Park, NY: Cengage Learning.

Shewfelt, R. L., and V. Bruckner. 2000. *Fruit and vegetable quality: An integrated view.* Boca Raton, FL: CRC Press.

Smith, D. S., J. N. Cash, W. K. Nip, and Y. H. Hui (Eds.). 1997. *Processing vegetables: Science and technology.* Boca Raton, FL:: CRC Press.

Somogyi, L. P., D. M. Barrett, H. Ramaswamy, Y. H. and Hui (Eds.). 2004. *Processing fruits: Science and technology,* 2nd ed. Boca Raton, FL: CRC Press.

Vaclavik, V. A., and E. W. Christina. 2014. *Essentials of food science,* 4th ed. New York: Springer.

Wagner, S. (Ed.). 1999. *The recipe encyclopedia: The complete illustrated guide to cooking.* San Diego: Thunder Bay Press.

CHAPTER 22

Fats and Oils

OBJECTIVES

After reading this chapter, you should be able to:

- Explain saturated and unsaturated, cis, and trans in terms of fatty acids
- Describe fatty acids
- Understand melting point and the structure of fatty acids
- Identify six sources of fats and oils
- List eight functions that fats and oils serve in foods
- Compare the extraction of fats or oils from animals to those of plants
- Describe the process used on oils after extraction
- List five processes in refining and modifying oils and fats after extraction
- Discuss monoglycerides and diglycerides and their uses
- Identify substances that may substitute for fat
- Describe two tests conducted on fats and oils

NATIONAL AFNR STANDARD

FPP.02

Apply principles of nutrition, biology, microbiology, chemistry, and human behavior to the development of food products.

KEY TERMS

bleaching
cis
degumming
deodorization
diglycerides
double bonds
fatty acid
glycerol
hydrogenation
iodine value
monoglycerides
peroxide value
rancidity
rendering
saponification value
saturated
trans
triglyceride
unsaturated
winterization

Plants and animals provide sources of fats and oils. Like proteins and carbohydrates, fats and oils are composed of carbon, hydrogen, and oxygen, yet they contain 2.25 times more energy per unit of weight. In foods and food processing, they contribute unique characteristics. Fats and oils contribute to a food's flavor, texture, and freshness.

EFFECTS OF COMPOSITION ON FAT PROPERTIES

Fats can be classified into simple lipids, compound lipids, composite lipids, spingolipids, and derived lipids. *Lipid* is the chemical name for a group of compounds that include fats, oils, cholesterol, and lecithin. Food science is concerned most with the simple lipids. These are the **triglyceride** lipids. They make up the major components of fat, butter, shortening, and oil.

A triglyceride molecule is made up of three (tri) fatty acid molecules connected to (*esterified*) a **glycerol** molecule. Think of the glycerol as the backbone of the triglyceride. The fatty acids are made of carbon atoms joined like links in a chain. The structure of the **fatty acids** that are esterified to the glycerol determines the properties of fats—for example, whether they are solid or liquid at room temperature.

Fatty acids are chains of 4 to 28 carbon atoms, with the carbons in the chain joined by single or **double bonds**, depending on the number of hydrogen atoms attached. The length of the chain may vary, which influences the fat's ability to dissolve in water. Generally, triglycerides do not dissolve in water. If a fatty acid has all the hydrogens possible attached to the carbons in the chain, it is said to be a **saturated** fatty acid; a saturated fatty acid is fully saturated with hydrogen atoms. If some of the carbons in the chain are joined by double bonds, thus reducing the number of hydrogen atoms, the fatty acid is called **unsaturated**. This means that one or more hydrogen atoms was missing and unsaturated. The acid portion of a fatty acid is represented by —COOH. For example, the following formulas represent a saturated and an unsaturated fatty acid, respectively:

CH3-CH2-CH2-CH2-CH2-CH2-CH2-CH2-CH2-CH2-CH2-CH2-CH2-CH2-CH2-COOH
CH3-CH2-CH2-CH2-CH2-CH2-CH2-CH=CH-CH2-CH2-CH2-CH2-CH2-CH2-COOH

If a fatty acid contains a double bond, the hydrogen atom or other groups attached to the carbon atoms involved in the double bond may have different orientations. In other words, atoms or other groups may exist as **cis** or **trans** forms around the double bond of the fatty acids. These are called *isomers*. Cis and trans forms have the same number of carbon, hydrogen, and oxygen atoms but in a different geometrical arrangement. This also gives a fatty acid different chemical and physical properties. Table 22-1 shows how the number of carbon atoms and saturation influences the melting point of some common fatty acids.

In general, a longer carbon chain increases the melting point; the more double bonds, the lower the melting point; and cis fatty acids have a lower melting point than trans.

In practice, the characteristics of the fatty acids are only a part of the factors that influence the properties of fats. The arrangement of fatty acids on the glycerol backbone (Figure 22-1) will also make a difference. The properties of fats also vary because they are made up of a variety of fatty acids and triglycerides.

TABLE 22-1 Influence of the Size of a Fatty Acid and Saturation on the Melting Point

FATTY ACID	NUMBER OF CARBON ATOMS	MELTING POINT
Saturated Fatty Acids		
Butyric	4	17.8 °F (−7.9 °C)
Caproic	6	25.0 °F (−3.9 °C)
Caprylic	8	61.3 °F (16.3 °C)
Capric	10	88.3 °F (31.3 °C)
Lauric	12	111.2 °F (44.0 °C)
Myristic	14	129.9 °F (54.4 °C)
Palmitic	16	145.0 °F (62.8 °C)
Stearic	18	157.3 °F (69.6 °C)
Arachidic	20	167.7 °F (75.4 °C)
Unsaturated Fatty Acids		
Palmitoleic	16	31.1 °F to 32.9 °F (−0.5 °C to 0.5 °C)
Oleic	18	55.4 °F (13.0 °C)
Linoleic	18	23.0 °F to 10.4 °F (−5 °C to −12 °C)
Linolenic	18	5.9 °F (−14.5 °C)
Arachidonic	20	−57.1 °F (−49.5 °C)

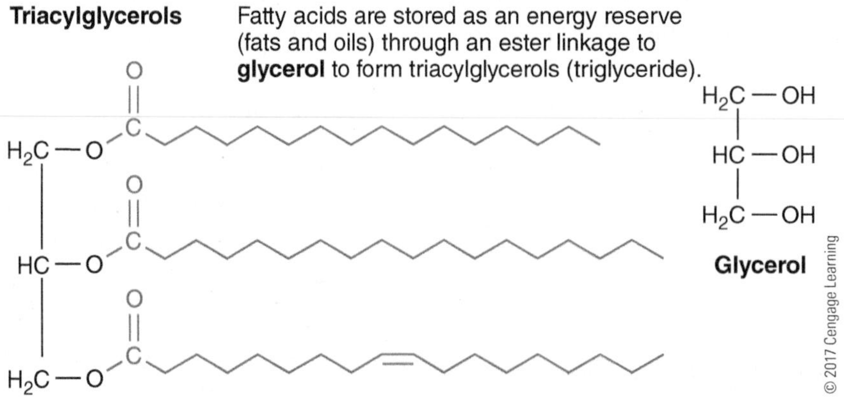

FIGURE 22-1 Fatty acids attached to a glycerol backbone making a triglyceride. (Systemic chemical name is triacylglycerol.)

TRIGLYCERIDES: WHY DO THEY MATTER TO ME?

Having a high level of triglycerides, a type of fat (lipid) in your blood, can increase your risk of heart disease. However, the same lifestyle choices that promote overall health can help lower your triglycerides, too.

Triglycerides are a type of fat (lipid) found in your blood. When you eat, your body converts any calories it does not need to use right away into triglycerides. The triglycerides are stored in your fat cells. Later, hormones release triglycerides for energy between meals. If you regularly eat more calories than you burn, particularly "easy" calories like carbohydrates and fats, then you may have high triglycerides (hypertriglyceridemia).

A simple blood test can reveal whether your triglycerides fall into a healthy range:

- Normal—less than 150 milligrams per deciliter (mg/dL), or less than 1.7 millimoles per liter (mmol/L)
- Borderline high—150 to 199 mg/dL (1.8 to 2.2 mmol/L)
- High—200 to 499 mg/dL (2.3 to 5.6 mmol/L)
- Very high—500 mg/dL or above (5.7 mmol/L or above)

The American Heart Association considers a triglyceride level of 100 mg/dL (1.1 mmol/L) or lower "optimal." For those trying to lower their triglycerides to this level, lifestyle changes such as diet, weight loss, and physical activity are encouraged because triglycerides usually respond well to dietary and lifestyle changes.

Your doctor will usually check for high triglycerides as part of a cholesterol test (sometimes called a *lipid panel* or *lipid profile*). You will have to fast for 9 to 12 hours before blood can be drawn for an accurate triglyceride measurement.

High triglycerides are often a sign of other conditions that increase the risk of heart disease and stroke as well, including obesity and metabolic syndrome.

Sometimes high triglycerides are a sign of poorly controlled type 2 diabetes, low levels of thyroid hormones (hypothyroidism), liver or kidney disease, or rare genetic conditions that affect how your body converts fat to energy. High triglycerides could also be a side effect of certain medications.

The best way to lower triglycerides is to make healthy lifestyle choices:

- Lose weight if you are considered overweight.
- Cut back on calories.
- Avoid sugary and refined foods.
- Limit the cholesterol in your diet.
- Choose healthier fats.
- Substitute fish high in omega-3 fatty acids—such as mackerel and salmon—for red meat.
- Eliminate trans fats.
- Exercise regularly. Aim for at least 30 minutes of physical activity on most or all days of the week.[1]

Adapted. For more information on triglycerides see the Mayo Clinic Web site at http://www.mayoclinic.org/diseases-conditions/high-blood-cholesterol/in-depth/triglycerides/art-20048186?pg=1.

For more information about triglycerides and your health visit:

- Triglycerides and How to Lower Them: http://www.webmd.com/cholesterol-management/lowering-triglyceride-levels
- Cleveland Clinic—Triglycerides: http://my.clevelandclinic.org/services/heart/prevention/risk-factors/cholesterol/triglycerides

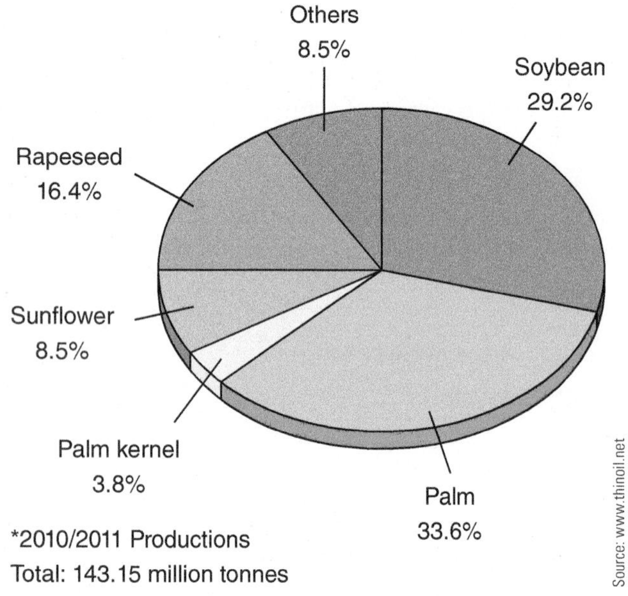

FIGURE 22-2 The world production of fats and oils is dominated by those produced by plants.

SOURCES OF FATS AND OILS

Fats and oils come from plant and animal sources, including fish (Figure 22-2). The plant or vegetable fats include cocoa butter, corn oil, sunflower oil, soybean oil, cottonseed oil, peanut oil, olive oil, canola oil, and many others. Animal fats include lard from pigs, tallow from beef, and butterfat from milk. Fish oils include cod liver oil, oil from menhaden, and whale oil. (The whale is really a mammal, but whale oil is included as a fish or marine oil.) Fats and oils and their composition are listed in Table B-8.

FUNCTIONAL PROPERTIES OF FATS

From a nutritional perspective, fats and oils have a high caloric value, 2.25 times more energy than carbohydrates or proteins. Each gram of fat contains 9 kcal. Fats and oils also carry the fat-soluble vitamins.

The uses of fats in foods continue to expand as they become more healthy and as the industry learns to modify the natural product. In general, functionalities and properties of fats are represented in these six major uses:

1. Textural qualities
2. Emulsions
3. Shortening or tenderizers
4. Medium for transferring heat
5. Aeration and leavening
6. Spray oils

These uses are influenced by the functionality of the particular fat or oil. The functionalities of fat include:

- Producing satiety (fullness after eating)
- Transfering heat
- Adding flavor
- Providing texture (body and mouthfeel)
- Tenderizing
- Decreasing temperature shock in frozen desserts
- Solubilizing flavors and colors
- Dispersing
- Aiding in the incorporation of air (foaming)

PRODUCTION AND PROCESSING METHODS

Food fats and oils are first extracted from both plant and animals to produce the products butter, margarine, lard, and hydrogenated shortening.

Refined oils include soybean oil, cottonseed oil, sunflower oil, peanut oil, olive oil, corn oil, canola oil, safflower oils, coconut oils, palm oil, and palm kernel oils.

Fats and oils are extracted by several different methods. For example, the adipose tissue (fat) of the pig is heated to melt the fat and allow further processing. Melting animal fat to extract it is called **rendering**. Rendering can be accomplished by heating meat scraps in steam or water and then skimming or centrifuging to separate the fat. Dry heat and a vacuum can also be used to render fat. The temperature used can influence the render fat's color and flavor.

The extraction of butterfat was covered in Chapter 16, and butter is made by reversing the oil-in-water emulsion of cream from the dairy cow into a water-in-oil emulsion.

Extracting plant fats requires a more complex method of processing. First, the oil is removed from the plant source by mechanical presses and expellers that squeeze the oil from oilseeds such as soybeans, cottonseed, and peanuts, among others. Often the seeds are ground and cooked before the oil is extracted. Another method of extracting oil from seeds is the solvent extraction method. A nontoxic fat solvent such as hexane is percolated through the cracked seeds. The oil is then distilled from the solvent, and the solvent is reused. Often pressing and solvent methods are combined. The oil-free residue from extraction is called *meal*, and it is used for animal feed.

Oil removed from the seed, pod, or grain is further refined and modified by **degumming**, refining, **bleaching**, **winterization** (fractionation), **hydrogenation**, **deodorization**, or interesterification (Figure 22-3).

SCIENCE CONNECTION!

Margarine is sold in stores with butter, but have you ever compared them? How are butter and margarine different? How are they alike? What is the price comparison between butter and margarine? Which is better for you?

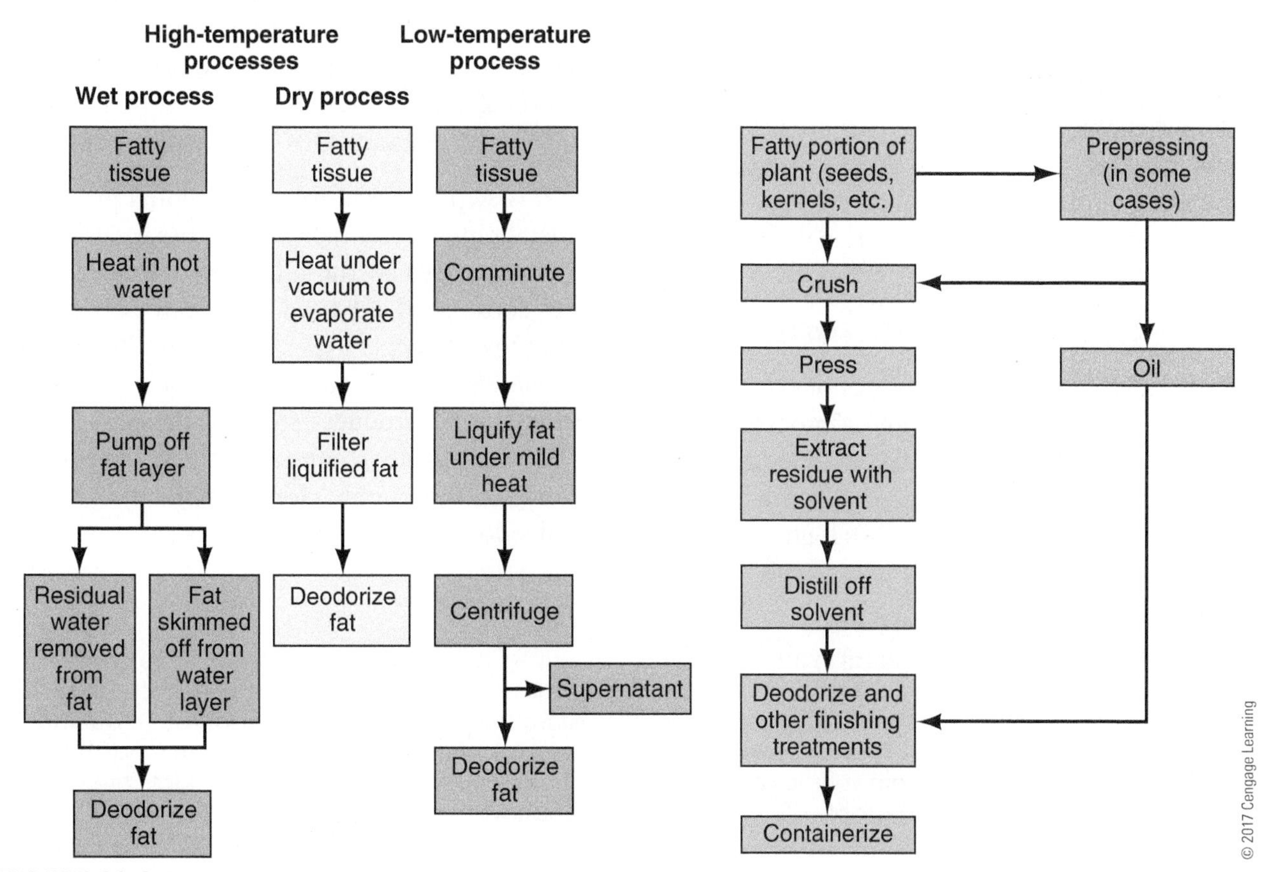

FIGURE 22-3 Flowcharts for the recovery of animal fats and plant oils.

Degumming

The first step in the refining process of many oils is *degumming.* Oils are degummed by mixing them with water. Degumming may be enhanced by adding phosphoric acid, citric acid, or silica gel. Degumming removes valuable emulsifiers such as lecithin. Cottonseed oils are not degummed, but the process is necessary for such oils as soybean and canola.

Alkali Refining

The degummed oil is then treated with an alkali (base) to remove free fatty acids, glycerol, carbohydrates, resins, metals, phosphatides, and protein meal. The oil and alkali are mixed, allowing free fatty acids and alkali to form a soap. The resulting soapstock is removed through centrifuging. Residual soaps are removed with hot water washings.

Bleaching

During the bleaching process, trace metals, color bodies such as chlorophyll, soaps, and oxidation products are removed using bleaching clays, which adsorb the impurities. Bleached oils are nearly colorless and have a **peroxide value** of near zero. Depending on the desired finished product, oils are then subjected to one or more additional processes.

Winterization (Fractionation)

Oils destined for use as salad oils or oils that are to be stored in cool places undergo a process called *winterization* so they will not become cloudy when chilled. The refined, deodorized oils are chilled with gentle agitation, which causes higher melting fractions to precipitate. The fraction that settles out is called *stearin.* Soybean oil does not require winterization, but canola, corn, cottonseed, sunflower, safflower, and peanut oils must be winterized to be clear at cool temperatures.

Hydrogenation

Treatment of fats and oils with hydrogen gas in the presence of a catalyst results in the addition of hydrogen to the carbon–carbon double bond. Hydrogenation produces oil with mouthfeel, stability, melting point, and the lubricating qualities that many manufacturers require. Hydrogenation is a selective process that can be controlled to produce various levels of hardening from slight to almost solid.

Deodorization

Deodorization is a steam distillation process carried out under a vacuum, which removes volatile compounds from the oil. This may be a batch or continuous process. The end product is a bland oil with a low level of free fatty acids and a zero peroxide value. (See "Tests on Fats and Oils" later in this chapter.)

This step also removes any residual pesticides or metabolites present; these are more volatile than the triglycerides in the oil. Some manufacturers favor cottonseed oil because it can be deodorized at lower temperatures, which results in more tocopherols (natural antioxidants) being retained. Deodorization produces some of the purest food products available to consumers. Few other products are so thoroughly clean as refined, bleached, and deodorized oil.

GOOD FAT AND BAD FAT

More than anything, too much fat in the diet is bad, but some types of fats and oils may have good or bad effects on health. Fats contain both saturated and unsaturated (monounsaturated and polyunsaturated) fatty acids. Saturated fat raises blood cholesterol more than other forms of fat. Reducing saturated fat to less than 10% of calories will help lower blood cholesterol level. The fats from meat, milk, and milk products are the main sources of saturated fats in most diets. Many bakery products are also sources of saturated fats. Vegetable oils supply smaller amounts of saturated fat.

(Continues)

GOOD FAT AND BAD FAT (continued)

Olive and canola oils are particularly high in monounsaturated fats; most other vegetable oils, nuts, and high-fat fish are good sources of polyunsaturated fats. Both kinds of unsaturated fats reduce blood cholesterol when they replace saturated fats in the diet. The fats in most fish are low in saturated fatty acids and contain a certain type of polyunsaturated fatty acid: omega-3. This fatty acid is under study because of a possible association with decreased heart disease risks in certain people. Total fat in the diet should be consumed at a moderate level—that is, no more than 30% of calories. Mono- and polyunsaturated fat sources should replace saturated fats within this limit.

Partially hydrogenated vegetable oils such as those used in many margarines and shortenings contain a particular form of unsaturated fat known as *trans-fatty acids* that may raise blood cholesterol levels, although not as much as saturated fat.

For more information about fat intake and the types of fats, search the following Web sites.

- National Agricultural Library: http://fnic.nal.usda.gov/
- U.S. Department of Agriculture's Agricultural Research Service: http://www.ars.usda.gov/main/main.htm
- PUFA: What Is It and Why Should It Be Avoided? http://butterbeliever.com/what-is-pufa/

SCIENCE CONNECTION!

A lot of different types of cooking oils can be found in the grocery store or supermarket. How many different oils can you name? Research different oils and see what health benefits may exist from using different oils.

Interesterification

Interesterification allows fatty acids to be rearranged or redistributed on the glycerol backbone. This is most often accomplished by catalytic methods at low temperatures. The oil is heated, agitated, and mixed with the catalyst at 194 °F (90 °C). Enzymatic systems also may be used for interesterification. The process does not change the degree of saturation or the isomeric state of the fatty acids, but it can improve the functional properties of the oil.

PRODUCTS MADE FROM FATS AND OILS

Butter, of course, is made by churning the butterfat from milk. Modern margarines can be made from any of a wide variety of animal or vegetable fats, mixed with skim milk, salt, and emulsifiers. Margarines and various spreads found in the market can range from 10% to 90% fat. Depending on the final fat content and the margarine's purpose (spreading, cooking, or baking), the level of water and vegetable oils used will slightly vary. Margarine that is comparable to butter is made from vegetable oils that have been fully or partially hydrogenated to solidify for spreading. Some vegetable oils may be mixed with small amounts of animal fat; like butter, legal margarine contains no less than 80% fat. Margarine is produced by heating, blending, and cooling a water-in-oil emulsion. The softer tub margarines are made with less hydrogenated and more liquid oils.

The largest portion of the market for edible oil products includes margarine, spreads, dressings, retail bottled oils, and frying oils. The sources of these lipids have facilitated the selection and development of products that are healthier and more stable. Originally, the use of animal fats and tropical oils (cocoa, palm, and coconut) minimized the problem of stability during food use. These

FIGURE 22-4 Many types of cooking oils are used but soybean, corn, canola, and sunflower, and general vegetable oils are the most common.

products are not currently widely used. The exception may be the increasing use of butter.

In the United States, many types of cooking oils are used, including olive oil, palm oil, soybean oil, canola oil, pumpkin seed oil, corn oil, sunflower oil, safflower oil, peanut oil, grapeseed oil, coconut oil, sesame oil, argan oil, rice bran oil, and other vegetable oils. Animal-based oils such as butter and lard are also used. Soybean, corn, canola, sunflower, and general vegetable oils are the most common (see Figure 22-4). Stability and functionality are obtained by their modification through hydrogenation. Stability and healthfulness have been enhanced recently through development of high–oleic-acid sunflower and safflower oils, low–linolenic-acid soybean oil, and high-oleic and low-linolenic canola oils. In addition, fats with varying degrees of saturation are also being developed. Future food scientists and health professionals need to be constantly aware that the fats and oils they are using may be changing, and as these changes take place, functional properties may change.

MONOGLYCERIDES AND DIGLYCERIDES

Monoglycerides and **diglycerides** are used as emulsifiers in a variety of foods. A monoglyceride is a glycerol molecule with only one fatty acid attached; a diglyceride is a glycerol with two fatty acids attached. Heating a mixture of triglycerides in the presence of a sodium hydroxide catalyst causes some of the fatty acids to disassociate from the glycerol and react with excess glycerol in the mixture. Some groups on the glycerol molecules remain unesterified, producing monoglycerides and diglycerides.

Monoglycerides and diglycerides are hydrophilic (attract water) and hydrophobic (repel water), so they are partially soluble in water and partially soluble in fat, making them excellent emulsifying agents.

FAT SUBSTITUTES

The advantages of low and decreased fat in the diet have been extensively studied and reported, especially in regard to cancer and heart disease risks. The U.S. Surgeon General has recommended the *Dietary Guidelines for Americans*, which describes healthy eating as consuming a variety of nutritious foods and beverages, especially vegetables, fruits, low-fat and fat-free dairy products, and whole grains; limiting the intake of saturated fats, added sugars, and sodium; keeping trans-fat intake as low as possible; and balancing caloric intake with calories burned to manage body weight. When fat was included in the Nutrition Labeling and Education Act, it became advantageous to lower fats in various products. However, the Food and Drug Administration (FDA) allowed several terms to evolve to indicate foods with reduced fat (Figure 22-5).

FIGURE 22-5 Reduced fat and fat-free products.

Unfortunately, fat not only enhances flavor but also plays many other roles. In most instances, carbohydrates and proteins are modified or used directly because they contain less calories per gram (9 kcal versus 4 kcal). The fat industry is attempting to modify fat itself. Approaches to fat reduction in food generally fall into two categories: (1) decreasing fat content and (2) using fat replacers, substitutes, extenders, mimetics, or synesthetic fat.

The use of fat replacers includes various carbohydrate-based, protein-based, and fat-based replacers that perform some of the many functions that fats do in food. For example, protein particles can be used in some foods to give a mouthfeel similar to that of fat at about 40% of fat's caloric content. Carbohydrates, such as some forms of cellulose, can be used to increase the viscosity of foods and mimic oil. The many functions of fats and consumer demands suggest a constant demand by consumers for new and improved fat replacers.

Food technologists are investigating a wide range of ingredients and processes to replace fat in foods and beverages. The following is a list of fat replacers currently in use or that are being researched as possible replacements.

PROTEIN-BASED FAT REPLACERS

The use of protein-based fat replacers is widespread. The various types and their applications are shown in Table 22-2.

CARBOHYDRATE-BASED FAT REPLACERS

Carbhohydrates are also being used as fat replacers, as shown in Table 22-3.

FAT-BASED FAT REPLACERS

As Table 22-4 shows, fat replacers are used in cake mixes, cookies, icings, and numerous vegetable dairy products.

TABLE 22-2 Protein-Based Fat Replacers

TYPE OF PRODUCT	PRODUCT	APPLICATIONS
Microparticulated Protein	Simplesse®	Dairy products, salad dressing, margarine- and mayonnaise-type products, baked goods, coffee creamers, soups, sauces
Modified Whey Protein Concentrate	Dairy-Lo®	Milk and dairy products, baked goods, frostings, salad dressings, mayonnaise-type products
Others	K-Blazer® ULTRA-BAKETM ULTRA-FREEZETM Lita®	Frozen desserts, baked goods

Adapted from Calorie Control Council; see http://www.caloriecontrol.org/articles-and-video/feature-articles/glossary-of-fat-replacers

TABLE 22-3 Carbohydrate-Based Fat Replacers

TYPE OF PRODUCT	PRODUCT	APPLICATIONS
Cellulose	Avicel® Cellulose Gel Methocel™ Solka-Floc®	Dairy-type products, sauces, frozen desserts, salad dressings
Dextrins	Amylum N-Oil®	Salad dressings, puddings, spreads, dairy-type products, frozen desserts
Fiber	Opta™ Oat Fiber Snowite Ultracel™ Z-Trim	Baked goods, meats, spreads, and extruded products.
Gums	KELCOGEL® KELTROL® Slendid™	Reduced-calorie, fat-free salad dressings; reduced fat content in other formulated foods, including desserts and processed meats
Inulin	Raftiline® Fruitafit® Fibruline®	Yogurt, cheese, frozen desserts, baked goods, icings, fillings, whipped cream, dairy products, fiber supplements, processed meats
Maltodextrins	CrystaLean® Lorelite Lycadex® MALTRIN® Paselli®D-LITE Paselli®EXCEL Paselli®SA2 STAR-DRI®	Baked goods, dairy products, salad dressings, spreads, sauces, frostings, fillings, processed meat, frozen desserts, extruded products and beverages
Nu-Trim		Foods and beverages such as baked goods, milk, cheese and ice cream, yielding products that are both reduced fat and high in beta-glucan, which has been cited as responsible for the beneficial reduction in cardiovascular risk factors
Oatrim (hydrolyzed oat flour	Beta-Trim™ TrimChoice	Baked goods, fillings and frostings, frozen desserts, dairy beverages, cheese, salad dressings, processed meats, confections
Polydextrose	Litesse® Sta-Lite™	Baked goods, chewing gums, confections, salad dressings, frozen dairy desserts, gelatins, puddings

(Continues)

TABLE 22-3 Carbohydrate-Based Fat Replacers

TYPE OF PRODUCT	PRODUCT	APPLICATIONS
Polyols	Many brands available	This group of sweeteners provides the bulk of sugar, without as many calories; also may be used to replace the bulk of fat in reduced-fat and fat-free products
Starch and modified food starch	Amalean® I & II Fairnex™ VA15 VA20 Instant Stellar™ N-Lite OptaGrade®# Perfectamyl™ AC AX-1 and AX-2 PURE-GEL® STA-SLIM™	Processed meats, salad dressings, baked goods, fillings and frostings, sauces, condiments, frozen desserts, dairy products
Z-Trim		Baked goods (to replace part of the flour), burgers, hot dogs, cheese, ice cream, yogurt

Adapted from Calorie Control Council; see http://www.caloriecontrol.org/articles-and-video/feature-articles/glossary-of-fat-replacers

TABLE 22-4 Fat-Based Fat Replacers

TYPE OF PRODUCT	PRODUCT	APPLICATIONS
Emulsifiers	Dur-Lo® EC^T-25	Cake mixes, cookies, icings, numerous vegetable dairy products
Salatrim	Benefat™	Confections, baked goods, dairy items, and other applications

Adapted from Calorie Control Council; see http://www.caloriecontrol.org/articles-and-video/feature-articles/glossary-of-fat-replacers

LIPID (FAT AND OIL) ANALOGS

These reduced-calorie replacers also have a wide variety of consumer and commercial applications, as shown in Table 22-5.

TABLE 22-5 Lipid (Fat and Oil) Analogs

PRODUCT	APPLICATIONS
Esterified Propoxylated Glycerol (EPG)*	Reduced-calorie fat replacer may partially or fully replace fats and oils in all typical consumer and commercial applications, including formulated products and in baking and frying.
Olestra (Olean®)	This product was approved by the FDA to replace the fat used in making salty snacks and crackers.
Sorbestrin*	Suitable for use in all vegetable oil applications including fried foods, salad dressing, mayonnaise and baked goods

*May require FDA approval

Adapted from Calorie Control Council; see http://www.caloriecontrol.org/articles-and-video/feature-articles/glossary-of-fat-replacers

For more information on fat replacers, see http://www.caloriecontrol.org/articles-and-video/feature-articles/glossary-of-fat-replacers

The average molecular weight of the fatty acids in a fat influences the firmness of the fat, the flavor, and the odor. A **saponification value** indicates the average molecular weight of the fatty acids in a fat. This value represents the number of milligrams of potassium hydroxide needed to saponify (convert to soap) one gram of fat. The saponification value increases and decreases inversely (opposite of) with the average molecular weight.

SCIENCE CONNECTION!

Research three different fat replacers listed above. Are they in use today? What foods can you find them in? Are these foods sold in your area?

MATH CONNECTION!

How do these fat replacers help make food healthier? How do the nutritional facts for the food items differ from regular fat items? Which version are you more likely to eat? Why?

TESTS ON FATS AND OILS

Fats and oils are tested to obtain information about their function in foods, measure any degree of deterioration (stability), and check for misrepresentation or adulteration. Chemical tests can determine the degree of unsaturation of the fatty acids in a fat. This is expressed as the **iodine value** of the fat. The test is based on the amount of iodine absorbed by a fat per 100 grams. The higher the iodine value, the greater the degree of unsaturation. Another chemical test yields the peroxide value. This indicates the degree of oxidation that has taken place in a fat or oil. The test is based on the amount of peroxides that form at the site of double bonds. These peroxides release iodine from potassium iodide when it is added to the system.

Hydrolytic **rancidity** refers to the rancidity that occurs under conditions of moisture, high temperature, and natural lipolytic enzymes. The *acid value* refers to a measure of free fatty acids present in a fat that were released during hydrolysis.

One of the most common physical tests performed on fats is a determination of the melting point. Fats and oils used in frying are subjected to measurements of their smoke point, flash point, and fire point (Table 22-6). Other physical determinations include color and specific gravity.

TABLE 22-6 Smoke, Flash, and Fire Points of Oils (°C)

OIL OR FAT	SMOKE	FLASH	FIRE
Castor, refined	200	298	335
Corn, crude	178	294	346
Corn, refined	227	326	359
Olive, virgin	175–199	322	361

SUMMARY

Fats and oils contain 2.25 times more energy than carbohydrates or proteins. Food science is concerned most with simple lipids such as the triglycerides that make up fat, butter, shortening, and oil. Some foods such as butter, cooking oils, and shortening are pure fat. Other foods make use of the properties of fats. Chemical characteristics of fats influence their properties. The function and properties of fats in foods are important. For example, fats can be used for textural qualities, emulsions, tenderizers, heat transfer, aeration, flavor, and dispersion.

Fats and oils are extracted from plants and animals. Rendering extracts fats from animal products. Expellers and solvents extract fats from plants. Extracted fats undergo additional processing such as degumming, alkali refining, bleaching, winterization, hydrogenation, deodorization, and interesterification. Chemical tests such as iodine value, rancidity, melting point, smoke point, and saponification provide information about function and stability.

Because fats contain more calories per pound than carbohydrates or proteins, the food industry continues to reduce the fat content of foods and to search for fat substitutes.

REVIEW QUESTIONS

Success in any career requires knowledge. Test your knowledge of this chapter by answering these questions or solving these problems.

1. List eight functions fats and oils serve in foods.
2. What influences the melting point of some common fatty acids?
3. Describe fatty acids and how they relate to saturated and unsaturated fats.
4. Name the two methods of extracting oil from plants.
5. List the eight processes used on oils after extraction.
6. How are monoglycerides and diglycerides used in food production?
7. Explain the difference between cis and trans-fatty acids.
8. Name and discuss the two chemical tests on fats and oils.
9. List six sources of fats and oils.
10. What is Olestra?

STUDENT ACTIVITIES

1. Make a diagram or a three-dimensional model to explain cis and trans.
2. Create a list of foods where the flavor provided by the fat makes the unique "taste" of the food. Create a class list to discuss and display.
3. Identify four food items that you consumed in the last 24 hours. Determine and report the fat content of these foods. Use Table B-8.

4. Find labels that list some substances used to prevent the oxidation of fats. (Refer to Chapter 14.) Present these labels to the class.
5. With a partner demonstrate that some fats are liquid at room temperature and some are solid at room temperature; discuss the reasons for this. For example, compare lard, shortening, vegetable oil, olive oil, and margarine or butter.
6. Use chemical models to demonstrate the concept of single and double bonds and saturation and unsaturation in fatty acids.
7. Use a fat or oil and make a digital video demonstrating how to make soap.

ADDITIONAL RESOURCES

- ChooseMyPlate.gov: http://www.choosemyplate.gov/food-groups/oils.html
- The Truth About Fats: The Good, The Bad, and the In-Between: http://www.health.harvard.edu/staying-healthy/the-truth-about-fats-bad-and-good
- FAO—Selection and Grading of Raw Materials for Meat Processing: http://www.fao.org/docrep/010/ai407e/ai407e05.htm
- University of Maine—Overview of the Rendering Industry: http://umaine.edu/byproducts-symposium/files/2011/10/Overview-of-the-Rendering-Industry-pdf.pdf
- Amrita University—Estimation of Iodine Value of Fats and Oils: https://www.youtube.com/watch?v=_5ObG6fIAdQ
- Fat Replacers: http://www.ift.org/knowledge-center/read-ift-publications/science-reports/scientific-status-summaries/fat-replacers.aspx

Internet sites represent a vast resource of information. Although those provided in this chapter were vetted by industry experts, you may wish to further explore the topics discussed in this chapter using a search engine such as Google. Keywords or phrases may include the following: *fat substitutes*; *saturated fats*; *unsaturated fats*; *fatty acids*; *production/processing of fats or oils*; *names of oils*—for example, *soybean oil*, *olive oil*, and the like; *monoglycerides*; *diglycerides*; *fat replacers*. In addition, Table B-7 provides a listing of useful Internet sites that can be used as a starting point.

REFERENCES

Rolland, J. 2014. *The cook's essential kitchen dictionary: A complete culinary resource*, 2nd ed. Toronto: Robert Rose.

Corriher, S. O. 1997. *Cookwise: The hows and whys of successful cooking.* New York: William Morrow & Co.

Cremer, M. L. 1998. *Quality food in quantity: Management and science.* Berkeley, CA: McCutchan Publishing.

Ensminger, A. H., M. E. Ensminger, J. E. Konlande, and J. R. Robson. 1994. *Foods and nutrition encyclopedia* (2 vols.). Boca Raton, FL: CRC Press.

Horn, J., J. Fletcher, and A. Gooch. 1997. *Cooking A to Z: The complete culinary reference source.* Glen Ellen, CA: Cole Publishing Group.

Moore, J. T. 2011. *Chemistry for dummies*, 2nd ed. Hoboken, NJ: Wiley.

Potter, N. N., and J. H. Hotchkiss. 1999. *Food science,* 5th ed. New York: Springer.

Vaclavik, V. A., and E. W. Christina. 2014. *Essentials of food science*, 4th ed. New York: Springer.

ENDNOTES

1. Mayo Foundation for Medical Education and Research: http://www.mayoclinic.org/diseases-conditions/high-blood-cholesterol/in-depth/triglycerides/art-20048186?pg=1. Last Accessed August 6, 2015.
2. Calorie Control Council: http://www.caloriecontrol.org/articles-and-video/feature-articles/glossary-of-fat-replacers. Last accessed August 6, 2015.

CHAPTER 23

Candy and Confectionery

OBJECTIVES

After reading this chapter, you should be able to:

- **Identify three crystalline and three noncrystalline candies**
- **Describe the relationship between sugar concentration and the boiling point**
- **Understand the common components of candies and confectioneries**
- **Identify two ways to produce invert sugar**
- **Explain caramelization in candy making**
- **Name four sugar-based sweeteners developed from cornstarch**
- **Describe uses of high-fructose corn syrup in the food supply**
- **Define cocoa**
- **Explain conching**

KEY TERMS

caramelization
conching
crystalline
Dutch processing
enrober
high-fructose corn syrup (HFCS)
hydrolyze
interfering agent
invert sugar
nib
noncrystalline
polymerize
substrate
viscosity

NATIONAL AFNR STANDARD

FPP.02

Apply principles of nutrition, biology, microbiology, chemistry, and human behavior to the development of food products.

OBJECTIVES (continued)

- **Describe modern candy and confectionery manufacturing**
- **List four sugar alcohols and four high-intensity sweeteners**
- **Discuss the labeling information and requirements for candy**

The word *candy* covers more than pure-sugar concoctions; it also includes an array of tasty confectioneries that combine sugar or similar substances with other compatible ingredients such as fruits, nuts, and chocolate. Some people may not like the idea, but candy is a food. Some candies such as those containing milk or nuts actually offer beneficial food values.

SOURCES OF SUGAR

Table sugar or sucrose is extracted from plant sources. The most important sugar crops are sugarcane (*Saccharum* spp.) and sugar beets (*Beta vulgaris*). Other minor commercial sugar crops include the date palm (*Phoenixdactylifera*), sorghum (*Sorghum vulgare*), and the sugar maple (*Acre saccharum*). Most sweeteners used today are extracted from these plants, which are high in sugar content. Sweet syrups are extracted, impurities are removed, and then all or some of water is removed. Honey is manufactured by bees rather than extracted from plants. Types of sugars used commonly as food ingredients include granulated sugar, brown sugar, confectioners' sugar, honey, corn syrup, molasses, and maple syrup.[1]

- Molasses, granulated sugar, brown sugar, and confectioners' sugar are all refined from sugar beets or cane sugar. Molasses is a viscous by-product of the refining of sugarcane or sugar beets into sugar.
- Brown sugar is usually regular refined sucrose with some of the molasses brought back in. Light brown sugar is usually 3.5% molasses, and dark brown sugar is usually 6.5% molasses.
- Confectionary sugar, also referred to as *powdered sugar* or *icing sugar*, is refined sucrose that is ground into a fine powder. Most confectioners' sugar has cornstarch added to prevent lumps. Confectioners' sugar comes in 4 varieties: 10XX, 10X, 6X, and 4X. The X refers to how finely ground the sugar is.[2]
- Honey was the first sweetener used in food and is also the only product with no expiration date or shelf life. Honey takes on the aromatic flavor of the flowers used for pollination. Popular types of honey include clover, orange blossom, buckwheat, blueberry, and wildflower.
- Corn syrup is a thick sweet syrup processed from cornstarch. It comes in both light and dark varieties. It is used often in making candies and pies.
- Sorghum is a grass crop and is grown both for sorghum syrup and as a gluten-free grain for flour.
- Maple syrup is the concentrated sap of sugar maple trees. This is an expensive process because it takes more than 40 gallons of sap to produce 1 gallon of maple syrup.

SCIENCE CONNECTION!

What is the difference between corn syrup and high-fructose corn syrup? Why has there been so much controversy in the news about high-fructose corn syrup? Research and find sweet food items you eat that do and do not contain high-fructose corn syrup.

SUGAR-BASED CONFECTIONERY

Honey, sorghum, molasses, maple syrup, and selected fruit juices and pulps serve as sweetener substitutes for cane sugar and sugar beet sugar. A variety of sweeteners have been developed, including sugar-based sweeteners, which have been developed from cornstarch.

Confectionery (candy) can be divided into those in which sugar is the main ingredient and those that are based on chocolate. In sugar-based candies, the sugar is manipulated to achieve some desired changes in the texture of the candy. For example, the sugar can be **crystalline**, and these crystals can be large or small, or it can be **noncrystalline** and glasslike. If the sugar is crystalline or noncrystalline, the sugar structure can be made hard or soft, modified by the amount of moisture, the air that is whipped in, or other ingredients. Crystallinity and moisture in a finished candy are determined by the ingredients, the heat used in cooking, cooling, and stirring.

Candies based on a crystalline sugar include rock candy, fondant, and fudge (Figure 23-1). Noncrystalline sugar candies include hard candies, brittles, chewy candies, and gummy candies.

© Bogdan Hoda/Dreamstime.com

FIGURE 23-1 Chocolate production line in an industrial factory.

Composition

Sugars and sugary foods provide a valuable and inexpensive source of energy. Other than energy, however, some sugary foods provide no real nutrition. Table B-8 shows the nutrient composition for a variety of common candies such as caramels, chocolates, chocolate, fondant, fudge, gum drops, hard candy, and jelly beans. Sugar and honey are also listed in Table B-8.

Ingredients

The principal ingredient of candies, including chocolate, is the sweetener. The most common sweetener used in candies and chocolates is sucrose—sugar from cane and sugar beets. At room temperature, a concentrated sucrose solution can be created by adding two parts sucrose to one part water. When cooled, this solution becomes supersaturated. Agitation during cooling causes crystallization, and the addition of a small sugar crystal speeds crystallization.

Heating the water before adding sucrose increases the sucrose concentration and increases the boiling point. The relationship between the sucrose concentration and the boiling point is used in making candy. It determines the percentage of water and sugar in the final product. When a boiling syrup (sugar and water) reaches a specific temperature, it will have the desired sugar concentration, as shown in Table 23-1. The numbers in Table 23-1 do not apply to mixed sugar solutions such as sucrose solutions containing corn syrup, glucose, or molasses.

Depending on the type of product being created, the concentration of sugar in a sugar syrup can be measured by a reading on a thermometer or by its appearance when dropped into ice water, as indicated in Table 23-2. The temperature and concentration then lead to a predicted behavior of the syrup to form the product.

Candies are made of sugar (sucrose), water or other liquid, and usually an **interfering agent**.

TABLE 23-1 Concentration of Sucrose, Water, and the Boiling Point

PERCENT SUCROSE	PERCENT WATER	BOILING TEMPERATURE °F/°C
30.0	70.0	212/100
50.0	50.0	216/102
70.0	30.0	223/106
90.0	10.0	253/123
95.0	5.0	284/140
97.0	3.0	304/151
99.5	0.5	331/166
99.6	0.4	340/171

As their name suggests, these agents interfere with the formation of sucrose crystals and provide some secondary properties to the candies. Butter, milk, starch, egg white, gelatin, fats, pectins, gums, cocoa, and corn syrup are commonly used as crystal interfering agents. The secondary properties these agents bring to a candy include thickening, chewiness, whipping, flavoring, tenderizing, and lubricating.

Invert Sugar

Heat and acid will **hydrolyze** and invert the sugar into the component monosaccharides—fructose and glucose. This is particularly true if the product

TABLE 23-2 Sugar Concentration, Temperature, and Behavior for Various Candies and Confections

CANDY OR CONFECTION	TEMPERATURE OF SYRUP AT SEA LEVEL	STAGE OF CONCENTRATION DESIRED	BEHAVIOR OF SYRUP STATE DESIRED
Syrup	230 °F to 234 °F (110 °C to 112 °C)	Thread	Spins a 2-in. thread when dropped from fork or spoon
Fondant Fudge Penuchi	234 °F to 240 °F (112 °C to 115 °C)	Soft ball	When dropped into ice cold water, forms a soft ball that flattens on removal
Caramels	244 °F to 248 °F (118 °C to 120 °C)	Firm ball	When dropped into ice cold water, forms a firm ball that does not flatten on removal
Divinity Marshmallows Nougat Popcorn balls Saltwater taffy	250 °F to 265 °F (121 °C to 130 °C)	Hard ball	When dropped into ice cold water, forms a ball that is hard enough to hold its shape yet remains plastic
Butterscotch Taffies	270 °F to 290 °F (132 °C to 143 °C)	Soft crack	When dropped into ice cold water, separates into threads that are hard but not brittle
Brittle Glace	300 °F to 310 °F (149 °C to 154 °C)	Hard crack	When dropped into ice cold water, separates into threads that are hard and brittle
Barley sugar	320 °F (160 °C)	Clear liquid	Sugar liquefies
Caramel	338 °F (170 °C)	Brown liquid	Melted sugar becomes brown

is heated. Fructose and glucose production create changes in a product. Unlike sucrose, they are reducing sugars, which means they will enhance browning. They are more soluble and more hygroscopic than sucrose. The confectionery industry refers to glucose as *dextrose* and fructose as *levulose.* Hydrolysis of sucrose is shown in the following reaction:

$$C_{12}H_{22}O_{11} + H_2O \longrightarrow C_6H_{12}O_6 + C_6H_{12}O_6$$
$$\text{Sucrose} + \text{Water} \longrightarrow \text{Glucose} + \text{Fructose}$$

(*Note:* Glucose and fructose have the same chemical formula, but the atoms are arranged differently from one another.)

Cream of tartar (an acid salt) as an added ingredient in a candy formula serves indirectly to decrease the rate of crystallization as well as crystal size. It does this through its ability to hydrolyze sucrose into its **invert sugar**.

combination is much more soluble than the sucrose crystals, and so the consumer (eater) perceives a syrup that is extremely sweet. The reason for the increased sweetness is that fructose is 40% to 70% sweeter.

Another category of inversion is the development of **high-fructose corn syrup (HFCS)**. HFCS is manufactured from cornstarch. Cornstarch is hydrolyzed by acid or enzyme, and the resulting glucose is then inverted into fructose by an enzyme. The percentage will vary. This is one of the new processing methods in foods, particularly in the sweetener area.

Caramelization

Caramelization is the application of heat to the point that the sugars dehydrate, break down, and **polymerize**. Although a relatively complex reaction, it can be simply done. Once the melting point is reached, sugars will caramelize.

SCIENCE CONNECTION!

Have you ever tried to pour honey? Honey is probably more difficult to pour than other liquids because of its high **viscosity** or thickness. What does that mean? Compare the viscosities of water, milk, soda, corn syrup, and molasses. What effect does viscosity have on these liquids?

The classic example of using invert sugar in foods is chocolate-covered cherries. These cherries are made by adding invertase, an enzyme, to fondant. As a solid crystalline candy, fondant can be placed around the cherries and then coated with chocolate. The cherries are allowed to sit, and the invertase inside hydrolyzes the sucrose into fructose and glucose. The fructose and glucose

The sugar actually comes apart chemically and forms smaller sugars, some of which join together again to form different smaller sugars. During the caramelizing process, new sugars and similar compounds are constantly being formed. By the time a dark caramel is formed, more than 128 different sugars and related compounds have been formed.

CANDY CANE HISTORY

In 1670, the choirmaster at the Cologne Cathedral gave sugar sticks to his young singers to keep them quiet during the long living creche ceremony. In honor of the occasion, he had the candies bent into shepherds' crooks. In 1847, a German-Swedish immigrant named August Imgard of Wooster, Ohio, decorated a small pine tree with paper ornaments and candy canes. At the beginning of the 1900s, red and white stripes and peppermint flavors became the norm for candy canes. The white and red stripes are said to be symbolic to Christians.

CANDY CANE HISTORY (continued)

In the 1920s, Bob McCormack began making candy canes as Christmas treats for his children, friends, and local shopkeepers in Albany, Georgia. It was a difficult process—pulling, twisting, cutting, and bending the candy by hand. It could only be done for local sales.

In the 1950s, Bob's brother-in-law, Gregory Keller, a Catholic priest, invented a machine to automate candy cane production. Packaging innovations by other members of the McCormack family made it possible to transport the delicate canes. This transformed Bobs Candies, Inc., into the largest producer of candy canes in the world.

Modern technology has made candy canes accessible and plentiful, and a Christmas tradition for many people.

For more information on candy canes, search the Web or visit the following Web sites.

- Ferrara Candy Company: http://www.ferrarausa.com/brands/bobs/
- National Confectioners Association: http://www.candyusa.com/

Each sugar has its own caramelization temperature. Caramel from sucrose forms at 338 °F (170 °C) or above. Galactose and glucose caramelize at about the same temperature as sucrose. Fructose caramelizes at 230 °F (110 °C) and maltose at about 356 °F (180 °C). Caramel is brown. It has a characteristic pungent taste, often bitter, less sweet, and it is noncrystalline.

A good example of caramelization is in the making of peanut brittle (Figure 23-2). Sugar is slowly and carefully heated in a skillet. This slow heating allows for uniform *unsaturated polymer formation* to occur. Brown, flavorful peanut brittle is a result of this caramelization process. The presence of sugar acids produced during this process is evident when foamy peanut brittle is made. This is produced by taking the heated brown mixture, while still hot, and adding a small amount of baking soda. The reaction of the baking soda with the sugar acids produces carbon dioxide gas, which causes the foaming.

FIGURE 23-2 Peanut brittle is a good example of what can result from the caramelization process.

Corn Syrups and Other Sweeteners

Sugar-based sweeteners are developed from cornstarch. The development of the various types of corn syrups, maltodextrins, and HFCS from cornstarch sources could be called one of the greatest changes in the sugar and sweetener industry over several centuries. In the late 1800s, it was found that cornstarch could be hydrolyzed and a sugar formed. In the 1970s, it became a major commercial product and brought major changes to the food industry.

These sweeteners are processed and refined using a series of steeping, separating, and grinding processes. Finally, the products are converted and fermented. From this processing, the five classes of corn sweeteners are formed:

1. Corn syrup (glucose syrup)
2. Dried corn syrup (dried glucose syrup)
3. Maltodextrin

4. Dextrose monohydrate
5. Dextrose anhydrous

This does not include the conversion of corn syrup into high-fructose corn syrup.

If these products are steamed, however, they can become gummy. A fast-food hamburger business that precooks and wraps its products may prefer the firmer product.

MATH CONNECTION!

Is high-fructose corn syrup good or bad? Research and defend your opinion. Compare the costs of food products and the costs to consumers as well as health concerns.

Fructose and Fructose Products

Fructose is a monosaccharide that is approximately 75% sweeter than sucrose. For this reason, fructose and fructose products are frequently substituted for sucrose. High-fructose corn syrup is often used.

The HFCS story is one of the most revolutionary in food science in the last decade. Consumption has increased since its inception. The products themselves are made up of hydrolyzed cornstarch. The cornstarch is hydrolyzed, and the resulting corn syrup has an invertase that changes glucose into fructose.

HFCSs retain moisture that prevents products from drying out. They also control crystallization, produce an osmotic pressure that is higher than those of sucrose and medium invert sugar, help control microbiological growth, and help penetrate cell membranes. HFCSs provide a ready yeast-fermentable **substrate**. HFCSs provide a controllable substrate for browning and the Maillard reaction. They impart a degree of sweetness that is essentially the same as invert liquid sugar. Less HFCS is required because it is sweeter than liquid sucrose and corn syrup blends. HFCSs blend easily with sweeteners, acids, and flavorings.

These attributes are advantages in many instances but can be disadvantages in others. High-fructose corn syrup is extremely soluble and hygroscopic. Generally, baked products made with HFCS will be softer than those made with sucrose.

CHOCOLATE AND COCOA PRODUCTS

Cocoa is made from finely pulverized, defatted, and roasted cacao kernels to which natural and artificial spices and flavors may be added. It is manufactured by pumping hot chocolate liquor (semiliquid ground cacao kernels) into hydraulic cage presses. Under extreme pressure, part of the fat, or cocoa butter, is removed. The fat content of cocoa varies from less than 10% to 22%. Cocoa may be **Dutch processed** with a mild alkali treatment that darkens the cocoa and improves its flavor. Cocoa is the flavoring ingredient in many confectioneries, baked goods, ice creams, puddings, and beverages.

Federal standards define several kinds of chocolate products. Bitter chocolate, or chocolate liquor, is the roasted ground kernel (**nib**) of the cacao bean; it is commonly known as baker's or baking chocolate. A minimum of 15% liquor mixed with sugar and cocoa butter is sweet chocolate. When the amount of chocolate liquor is greater than 35%, the product is bittersweet chocolate. A combination of at least 12% dry whole-milk solids, sugar, cocoa butter, and at least 10% chocolate liquor produces milk chocolate.

Cocoa

Cocoa is made by removing some of the cocoa butter. Eating chocolate is made by adding cocoa butter. This is true of all eating chocolate, whether it is dark, bittersweet, or milk chocolate. Besides enhancing the flavor, the added cocoa butter serves to make the chocolate more fluid.

One example of eating chocolate is sweet chocolate, a combination of unsweetened chocolate, sugar, cocoa butter, and sometimes a little vanilla. Production involves melting and combining these ingredients in a large mixing machine until the mass has the consistency of dough.

Milk Chocolate

Milk chocolate, the most common form of eating chocolate, goes through essentially the same mixing process, except that it involves using less unsweetened chocolate and adding milk (Figure 23-3). Whatever ingredients are used, the mixture then travels through a series of heavy rollers set one atop the other. During the grinding that takes place here, the mixture is refined into a smooth paste ready for **conching**.

Conching is a flavor-development process that puts the chocolate through a kneading action and takes its name from the shell-like shape of the containers originally used. The conches, as the machines are called, are equipped with heavy rollers that plow back and forth through the chocolate mass anywhere from a few hours to several days. Under regulated speeds, these rollers can produce different degrees of agitation and aeration in developing and modifying the chocolate flavors.

In some manufacturing setups, an emulsifying operation either takes the place of conching or supplements it. This operation is carried out by a machine that works like an eggbeater to break up sugar crystals and other particles in the chocolate mixture to give it a fine, velvety smoothness.

After the emulsifying or conching machines, the mixture goes through a tempering interval—heating, cooling, and reheating—and then finally into molds to be formed into the shape of the completed product. Molds take a variety of shapes and sizes, from the popular individual-size bars available to consumers to 10-pound blocks used by confectionery manufacturers.

When the molded chocolate reaches the cooling chamber, cooling proceeds at a fixed rate that keeps hard-earned flavor intact. Bars are then removed from the molds and passed along to wrapping machines to be packed for shipment.

For convenience, chocolate is frequently shipped in a liquid state when intended for use by other food manufacturers. In addition, a portion of the U.S. total chocolate output goes into coatings, powders, and flavorings that add zest to our foods in a thousand different ways.

SCIENCE CONNECTION!

How are milk chocolate, dark chocolate, white chocolate, and semisweet chocolate different? How are they the same? What are different uses for the different chocolates?

CONFECTIONERY MANUFACTURING PRACTICES

During modern confectionery manufacturing, many batch or continuous processes are used to produce the basic fondants, taffies, brittles, and hard candies. Then specializing machines extrude, divide, mold, glaze, and enrobe the candies during final production stages. A Web site called Candy USA (http://www.candyusa.com/CST/) features links to Web sites for everyone's favorite candies

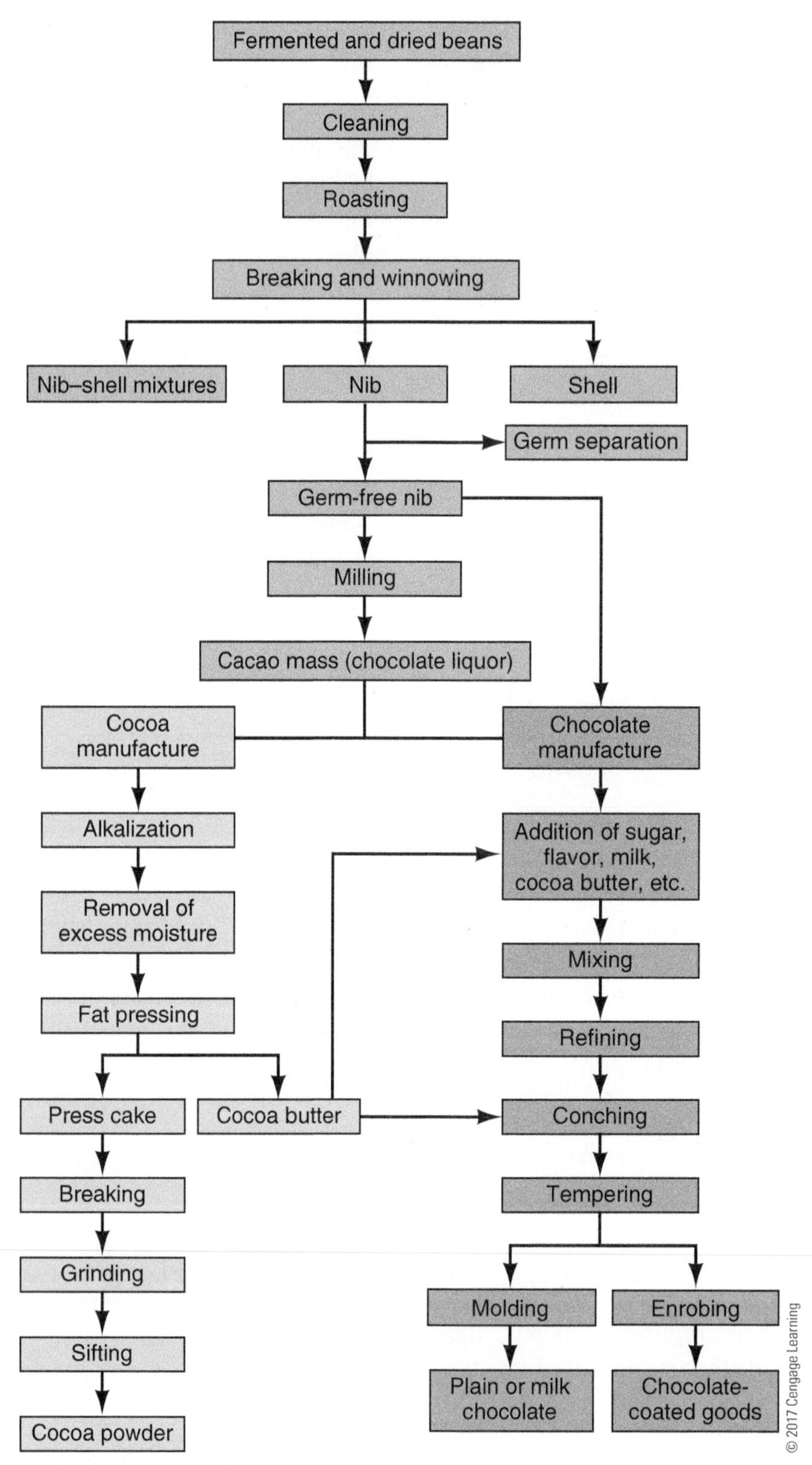

FIGURE 23-3 Flowchart of cocoa and chocolate manufacturing.

from a variety of companies, including American Licorice Company, Ben Myerson Candy Co., Clark Bar America Inc., Certs, Elmer Candy Corporation, Fannie May Candies, Favorite Brands International, Georgia Nut Co. Inc., Gimbal's Candy, Hershey Foods Corp., M&M's Brand Chocolate Candies, Nestles' Foods, See's Candies, Snickers, and Willy Wonka Candy Factory.

In a chocolate factory, precision instruments regulate temperatures, stabilize the moisture content of the air, and control the time intervals of manufacturing operations and other items necessary to achieve quality results.

The industry employs many machines that shape and package chocolate into familiar forms. Some of the shaping machines perform at amazing speeds, squirting jets of liquid chocolate product that solidify into special shapes at the rate of several hundred per minute. Other machines wrap and package the finish products at high speed.

Enrobing

The **enrober** is used by many candy manufactures in the creation of assorted chocolates. The enrober receives lines of assorted centers (nuts, nougats, fruit, etc.) and showers them with a waterfall of liquid chocolate. This generally covers and surrounds each center with a blanket of chocolate. Other confectionery machines create a molded hollow shell of chocolate that is then filled with a soft or liquid center before the bottom is sealed with chocolate.

Standards

Mechanization of the entire chocolate-making process contributes to the industry's high standards of hygiene and sanitation. Chocolate factories constantly run quality tests to determine whether the process is running within the strict limitations designed for each product. These tests examine the viscosity of chocolate, the cocoa butter content, acidity, the fineness of a product, and the purity and taste of the desired finished product.

All chocolate manufacturers must meet standards as set forth in Food and Drug Administration (FDA) rules and regulations. These govern manufacturing formulas, even to the extent of specifying the minimum content of chocolate liquor and milk used. They also impose strict rules regarding flavorings and other ingredients that may be used.

FDA standards for cacao products were updated and published in 1993. Those rules are highly technical, down to prescribing analytic techniques and specifying approved processing methods. Specifications for cacao nibs say they may contain "not more than 1.75% by weight" of residual shell. The standards provide definitions of intermediate and end products, including chocolate liquor—for example, chocolate liquor "contains not less than 50% nor more than 60% by weight of cacao fat." The FDA also sets standards for breakfast cocoa, sweet chocolate, semisweet or bittersweet chocolate, milk chocolate, skim milk chocolate, and so on.

MATH CONNECTION!

Americans sure do love candy! Which holiday do you think sells the most candy each year? What holidays make up the top five in candy sales each year? Research and find evidence to see if you are correct in your hypothesis!

SUGAR SUBSTITUTES

Sugar alcohols are made by chemically reducing a sugar to an alcohol that is less sweet than sugar (see Table 23-3). These are not fermentable by bacteria in the mouth, so they do not contribute to tooth decay. Products using these sugar alcohols are often labeled "sugar free," but this does not mean

TABLE 23-3 Relative Sweetness of Sugar Alcohols Compared to a Sucrose Value of 100

SUGAR ALCOHOL	SWEETNESS
Xylitol	90
Sorbitol	63
Galactitol	58
Malitol	68
Lactitol	35

that they are calorie free. Sugar alcohols contain 4 kcal of energy per gram, just like sugar.

High-intensity sweeteners are used in confectioneries to reduce their caloric content. The calories are reduced because less sweetener is required or the sweetener cannot be metabolized by the human body, so they do not contribute calories. Examples of these sweeteners and their relative sweetness are listed in Table 23-4.

Not all the sweeteners in Table 23-4 are approved for use at this time. The FDA has approved the use of the following nonnutritive sweeteners: acesulfame K, aspartame, neotame, saccharin, sucralose and stevia. Also, high-intensity sweeteners often do not provide the same functional properties as sugar. Additional additives are used to provide bulking, mouthfeel, and other desirable characteristics.

TABLE 23-4 Approximate Sweetness of Sugar Substitutes Compared to a Sucrose Value of 1

SUBSTITUTE	SWEETNESS
Acesulfame K	200
Aspartame	180
Cyclamate	30
Dihydrochalcones	300 to 2,000
Glycyrrhizin	50 to 100
Monellin	1,500 to 2,000
Saccharin	300
Stevioside	300
Talin	2,000 to 3,000

LABELING

By law, all packaged foods must bear a label that lists ingredients in order of predominance. Candy is no exception, and every package of hard candies and every chocolate bar must include such a listing. As part of the food labeling rules under the Nutrition Labeling and Education Act of 1990, manufacturers must include substantially more nutrition information on labels than they did in the past.

For example, the chocolate bar shown in Figure 23-4 includes the information that it provides 280 calories, 5% of the daily value for carbohydrate, and a significant 95% of the daily value for saturated fat.

A representative FDA-required label listing of ingredients reads sugar, corn syrup, citric acid, artificial flavor, Red No. 40, Yellow No. 5, Yellow No. 6, Blue No. 1. The last four are specific FDA-sanctioned food colorings, synthetic additives that a few people are allergic to but that otherwise pose no threat to human health.

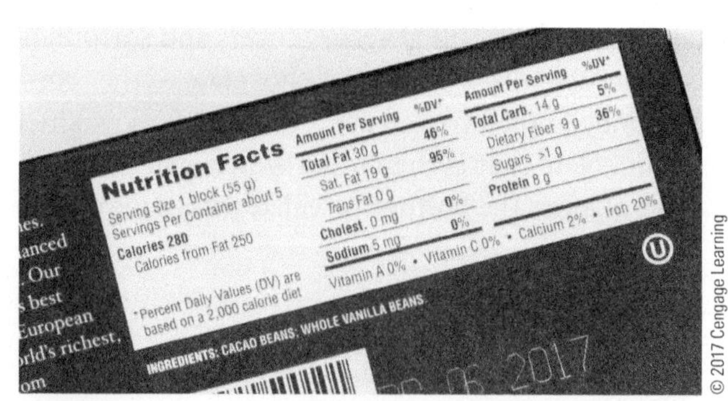

FIGURE 23-4 Nutrition label on the back of a chocolate candy bar.

SUMMARY

Candy and confectionery are divided into those in which sugar is the main ingredient and those based on chocolate. Further, candies can be classified as crystalline or noncrystalline. The processes of creating an invert sugar and caramelization produce unique flavors and characteristics. The relationship between sucrose concentration and the boiling point determines the percentage of water and sugar in the final product. Interfering agents help prevent the formation of sucrose crystals and provide secondary properties to candies.

Sugar-based sweeteners developed from cornstarch brought about major changes in the food industry. These sweeteners include corn syrup, maltodextrin, dextrose monohydrate, and dextrose anhydrous. High-fructose corn syrups are produced when cornstarch is hydrolyzed by an invertase. Again, high-fructose corn syrup revolutionized the food industry.

Cocoa is made from ground, defatted, and roasted cacao kernels. Federal standards define cocoa and other chocolate products, including milk chocolate, dark chocolate, and baking chocolate. Much of candy manufacturing, including chocolate manufacturing, relies on automation and precision instrumentation.

Sugar substitutes and high-intensity sweeteners are often used to reduce the calories in candy and confectionery. Labeling requirements for candy and confectionery are the same as for other foods.

REVIEW QUESTIONS

Success in any career requires knowledge. Test your knowledge of this chapter by answering these questions or solving these problems.

1. Define conching.
2. What are sugar-based sweeteners developed from?
3. Identify three candy products based on a crystalline sugar.
4. Explain how an enrober works.
5. Why are fructose and high-fructose products frequently substituted for sucrose?
6. What is the most common sugar in candies?
7. What labeling information is required on candies and confectioneries?
8. Define cocoa.
9. What substance is added to the ingredients of candy that serves indirectly to decrease the rate of crystallization as well as crystal size?
10. Which process applies heat to the point that sugars dehydrate and break down and polymerize?

STUDENT ACTIVITIES

1. Visit the Candy USA Web site (http://www.candyusa.com/) and link with the manufacturer of your favorite candy. Report your findings on the manufacturing process with a visual presentation.
2. Identify a candy recipe that relies on the formation of an invert sugar from sucrose. Explain the recipe to the class.

3. Develop a report or presentation on products using some of the sugar alcohols listed in Table 23-3.
4. Make a list of chocolate products used in cooking.
5. Visit the Hershey Chocolate Web site (www.hersheys.com/) and report on the historical and current characteristics of the company.
6. Develop a report or presentation on of the use of the sugar substitutes in Table 23-3.
7. Use a simple recipe for a candy and make it in class. Discuss the changes in the sugar as the final product is reached.
8. Conduct a taste test of a variety of candies and classify them as crystalline or noncrystalline. Record the results of the taste test.

ADDITIONAL RESOURCES

- Resident Course in Technology, University Of Wisconsin–Madison: http://foodsci.wisc.edu/extension/candy/
- Cargill Cocoa and Chocolate: http://www.cargillcocoachocolate.com/
- Sweeteners—Sugar Substitutes: http://umm.edu/health/medical/ency/articles/sweeteners-artificial
- Sugar Substitutes Not So Super Sweet After All: http://news.psu.edu/story/325074/2014/09/04/research/sugar-substitutes-not-so-super-sweet-after-all
- FDA's Policy on Declaring Small Amounts of Nutrients and Dietary Ingredients on Nutrition Labels: http://www.fda.gov/Food/GuidanceRegulation/GuidanceDocumentsRegulatoryInformation/ucm456062.htm

Internet sites represent a vast resource of information. Although those provided in this chapter were vetted by industry experts, you may wish to further explore the topics discussed in this chapter using a search engine such as Google. Keywords or phrases may include the following: *candies*; *confectionery*; *specific candy brand names*; *conching*; *enrober*; types of sugars—*sucrose*, *fructose*, *corn syrup*, and the like; *cocoa products*; *sugar substitutes*. In addition, Table B-7 provides a listing of useful Internet sites that can be used as a starting point.

REFERENCES

Alexander, R. J. 1998. *Sweeteners: Nutritive*. St. Paul, MN: Eagan Press.

Bartlett, J. 1996. *The cook's dictionary and culinary reference*. Chicago: Contemporary Books.

Corriher, S. O. 1997. *Cookwise: The hows and whys of successful cooking*. New York: William Morrow & Co.

Horn, J., J. Fletcher, and A. Gooch. 1997. *Cooking A to Z: The complete culinary reference source*. Glen Ellen, CA: Cole Publishing Group.

McGee, H. 2004. *On food and cooking: The science and lore of the kitchen*. New York: Scribner.

ENDNOTES

1. "Roses Sugar Bible," Food Art http://www.foodarts.com/news/classics/20964/roses-sugar-bible. Last accessed July 9, 2015.
2. Hive and Honey Apiary; http://www.hiveandhoneyapiary.com/honeyvarieties.html. Last accessed July 9, 2015.

CHAPTER 24

Beverages

OBJECTIVES

After reading this chapter, you should be able to:

- Describe how carbonated nonalcoholic beverages are manufactured
- List the steps in the production of beer
- Compare the production of wine to vinegar
- Indicate how fermentation plays a role in the production of coffee
- Explain six ways enzymes are used in the production of beverages
- Discuss how beverages meet the demand for a healthful drink
- Identify the fastest-growing segment of the beverage industry
- Name five herbs used in beverages
- Identify the plants that produce coffee and tea
- Describe how to produce a coffee substitute
- Compare tea to herbal teas

KEY TERMS

black tea
carbonated
carbonator
green tea
lagering
malt
mashing
nonnutritive
nutritionally enhanced
oolong
steeping
synchrometer
vitamin-fortified
vinification
wort

NATIONAL AFNR STANDARD

FPP.02
Apply principles of nutrition, biology, microbiology, chemistry, and human behavior to the development of food products.

For thousands of years, water was the only beverage choice for people. With the domestication of animals, milk was added to the diet. Through history, beer, wine, chocolate, coffee, and tea all became important beverages. Many alternate forms of these beverages were experimented with because some were too costly to make or import. In recent years, soft drinks, juice drinks, energy drinks, sports drinks, and others have hit the beverage market. Today, consumers have hundreds of different drink beverages to choose from. Because there are many beverage choices, most consumers drink a variety of beverages. Some beverages provide nutrition and vitamins, while others provide high sugar with little nutritional value.

FIGURE 24-2 Today's consumer can choose from a wide variety of soft drinks in plastic bottles and aluminum cans.

CARBONATED NONALCOHOLIC BEVERAGES

Carbonated soft drinks are the most popular beverage (Figure 24-1). Carbonated nonalcoholic beverages are usually sweetened, flavored, acidified, colored, artificially carbonated, and chemically preserved. Their origin is in the ancient Greek and Roman use of naturally occurring mineral waters. In 1767, British chemist Joseph Priestley discovered how to artificially carbonate water. Soda fountains were popular in parts of drugstores in the 1800s and offered widely enjoyed carbonated beverage flavors, including orange and grape. Many of the most well- known brands of carbonated beverages were invented between 1860 and 1900. Carbonated beverages became known as "pop" in some places because of the sound the bubbles produced when they exploded. Ever since Priestley's discovery, many flavors and forms of carbonated soft drinks have been developed for consumers (Figure 24-2).

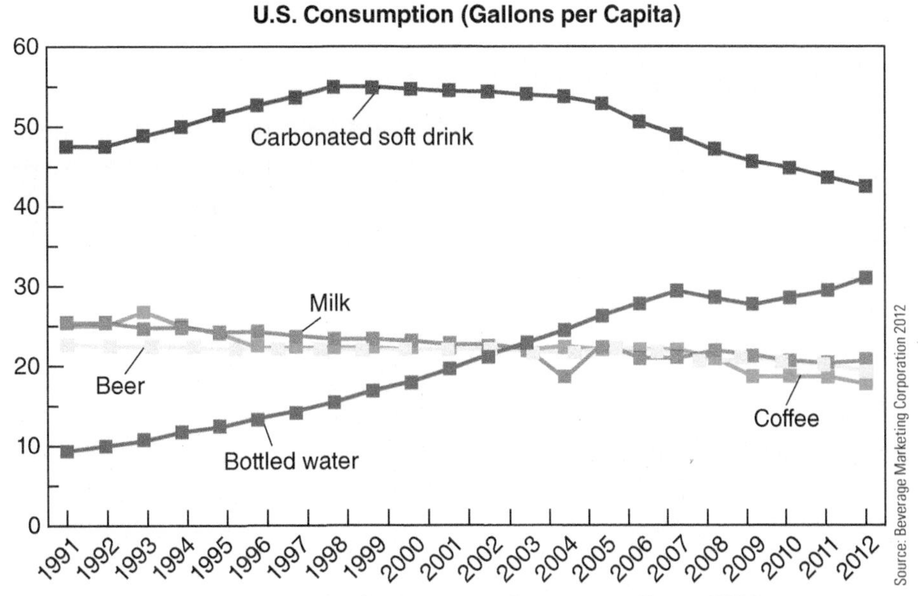

FIGURE 24-1 U.S. Beverage Consumption in Gallons per Consumer Since 1990.

Sweeteners

Syrup made from sucrose was commonly used in soft drinks, but now the most common sugar used in soft drinks is high-fructose corn syrup (HFCS). High-fructose sugars are sweeter than sucrose. The final concentration of sugar in a soft drink is 8% to 14%. Besides contributing sweetness, sugar in a soft drink contributes body and mouthfeel.

Soft drinks with no calories or reduced calories have become extremely popular. **Nonnutritive** sweeteners (no calories) that have been used in soft drinks include saccharin and cyclamate. Currently, reduced-calorie soft drinks use a sweetener called *aspartame* (trademark NutraSweet®). This sweetener is 150 to 200 times sweeter than sucrose, so only tiny amounts are needed even though it contains the same number of calories per gram as sugar (4 kcal/g). Another fairly new nonnutritive sweetener is stevia. This sugar substitute is extracted from the leaves of the plant species *Stevia rebaudiana*. Extracts are 150 times sweeter than sugar. When these nonnutritive or reduced-calorie sweeteners are used, nonnutritive carbohydrates such as carboxymethyl cellulose or a pectin are sometimes added to give the soft drink a desired mouthfeel.

Flavors

Flavors used in soft drinks include synthetic flavors (refer to Chapter 14), natural flavor extracts, and fruit juice concentrates. Synthetic flavors are highly complex and may easily contain hundreds of distinct compounds. Some cola and other flavors are complex, and their secret formulas are carefully guarded.

Colors

Important coloring agents for soft drinks are synthetic and are U.S. certified food colors approved by the Food and Drug Administration (refer to Chapter 14). These colors meet strict requirements for purity. Heated sugar produces a caramel color used in darker beverages. This natural color is preferred because of its coloring power and color stability. When natural fruit juices or extracts are used in soft drinks, their colors are often supplemented with synthetic colors.

Acid

Carbon dioxide in soft drinks contributes to the acidity, but the main acids used are phosphoric, citric, fumaric, tartaric, and malic acids. Besides acting as preservatives, the acids enhance flavors. Of course, acids in the soft drink lower the pH, as shown in Table 24-1.

Water

Water is the major ingredient of carbonated soft drinks. By volume, it makes up as much as 92%. Pure water—or nearly chemically pure water—is essential because traces of impurities may react with chemicals in a soft drink. Alkalinity, iron, and manganese must be low. Chlorine cannot be present. Obviously, the water must be colorless, and it must not have any odor, taste, or organic matter. Bottling plants condition the water they use so it will meet their high standards.

TABLE 24-1 The pH of Common Soft Drinks

FLAVOR	pH
Colas	2.6
Cherry	3.7
Lemon-lime	3.0
Grape	3.0
Root beer	4.0

Carbon Dioxide

Carbon dioxide (CO_2) gas provides the zest and sparkle of a carbonated drink. Sources of carbon dioxide gas include carbonates, limestone, the burning of fuels, and industrial fermentation. Soft drink manufacturers use carbon dioxide from high-pressure cylinders purchased from manufacturers who produce the gas under strict food purity regulations. The amount of carbon dioxide gas in each beverage is different and is measured out in terms of volumes of gas per volume of liquid. Most beverages are carbonated in the range of 1.5 to 4 volumes of carbon dioxide. The equipment that carbonates a soft drink is called a **carbonator**. Carbonation of the soft drink takes place at a lowered temperature because the solubility of the carbon dioxide is greater at lower temperatures.

Mixing

Operations producing soft drinks include mixing, carbonating, and bottling. Water that has been treated and deaerated (air has been removed) and the flavored syrup are pumped to a **synchrometer**. This metering device measures the syrup and water in fixed proportion to the carbonator. Next, the mix is cooled and sent to the carbonator. After carbonation, the soft drink is placed in cans or bottles and sealed. In restaurants, fast-food establishments, and other businesses, soft drinks are served directly into a container, the syrup from the manufacturer being held in one pressurized tank and carbon dioxide in another. The syrup and carbon dioxide are mixed as the drink is being drawn.

NONCARBONATED HERBAL AND HEALTHFUL BEVERAGES

Not long ago, carbonated beverages dominated the shelves and cooler space in markets, but newer competition has arrived in the form of fortified, healthful, or natural beverages. The world's first **vitamin-fortified** fruit drinks appeared in 1948. Other herbal and healthful beverages followed, and new ones continue to be developed and marketed. Four examples are Hi-C®, Gatorade®, SoBe®, and Snapple®.

Hi-C®

Hi-C was conceived as a result of a weather-related shortage of orange juice in the United States. In 50 years, Hi-C has become the world's largest brand of vitamin-fortified fruit drink, and the fifth largest-selling trademark of the Coca-Cola® Company. Hi-C fruit drinks contain 100% of the recommended daily intake of vitamin C per serving. In 1987, The Minute Maid® Company (a Coca-Cola subsidiary) was the first to introduce a calcium-fortified orange juice and the first and only company to offer drink boxes containing juices and juice drinks that are fortified with calcium and vitamin C. Providing a vitamin-fortified beverage is also important in other countries. For example, Kapo® fruit-flavored beverages fortified with vitamins were originally developed in 1975 under a Chilean government request to develop a low-cost way to provide children with a nutritious drink.

MATH CONNECTION!

What is your favorite carbonated beverage? What is the history of the drink product? Is it marketed locally, regionally, nationally, or internationally? How much of this beverage is sold annually?

The Web site for Hi-C and other Minute Maid products is www. minutemaid.com.

Gatorade®

In the 1960s, a team of researchers at the University of Florida began a project to develop a product for rapid fluid replacement for human bodies to help prevent severe dehydration and loss of body salts brought about by physical exertion in high temperatures. By 1965, the group developed a formula that was ready for testing. Because football players experience tremendous fluid losses during practices and games, the formula was tested on 10 members of the University of Florida football team (the Gators). This beverage became known as *Gatorade*. The name was later trademarked. A coach's statement that "Gatorade made the difference" in the team's success was published by *Sports Illustrated*, and this marked the beginning of the Gatorade phenomenon.

In May 1967, Stokely-Van Camp, then a leading processor and marketer of fruits and vegetables, acquired the rights to produce and sell Gatorade throughout the United States. In 1983, the Quaker Oats Company acquired Stokely-Van Camp, including the Gatorade brand. The Quaker Oats Company marketed the brand nationally.

Gatorade provides the human body with fluids, minerals, and energy during exercise. It remains the number-one sports drink in the United States even with increasing competition. The Gatorade sports drink is scientifically formulated to quickly replace the fluids and minerals the body loses during exercise or other strenuous physical activities. Gatorade is formulated to stimulate rapid fluid absorption, ensure rapid rehydration, provide carbohydrate energy to working muscles, and stimulate an individual to drink more. It is composed of water, sucrose, glucose–fructose syrup, citric acid, natural flavors, salt, sodium citrate, and monopotassium phosphate. Gatorade also has created several varieties of sports drinks and other items for consumers.

The Web site for Gatorade is http://www.gatorade.com/.

SoBe®

SoBe is a registered trademark of South Beach Beverage Company, makers of **nutritionally enhanced** refreshment beverages. The company started in 1995 with the introduction of South Beach brand iced teas and fruit drinks. The four partners who started SoBe were into health and fitness as a way of life. The first product introduced was SoBe **black tea** 3g with ginseng, ginkgo, and guarana. South Beach has since introduced the SoBe brand, a line of energizing wellness beverages targeted at the mass market. The line includes exotic teas, juice blends, elixirs, and "effect" beverages such as Energy, Power, and Wisdom. All of these are fortified with herbs, minerals, vitamins, and other nutrient enhancers. SoBe's package designs feature dual lizards representing the Asian concept of the yin and yang of life. The bottles are sculpted glass with the lizard embossed on the neck (Figure 24-3).

More information about SoBe and the claims for their beverages can be found at their Web site at http://www.sobe.com/#!/home/coconut.

FIGURE 24-3 Products such as SoBe compete for the market of traditional carbonated beverages.

THE OLDEST BRAND OF CARBONATED DRINK

The oldest continuously commercially marketed carbonated drink is Moxie, which became available in apothecaries as a medical tonic in 1876. Made by the Nerve Food Company of Lowell, Massachusetts, Moxie was first sold as a beverage in 1884. Although Hires Root Beer was developed in 1876, eight years before Moxie, it was periodically pulled off the market.

By the early 20th century, the so-called Nerve Food was carbonated and brilliantly merchandised and had become a household name. Despite the claim restrictions placed on Moxie by the Food and Drug Act, many ads from this explosive growth period touted the "healthful" and alleged medicinal benefits of the tonic. An early newspaper ad had Professor Allyn, "Food Authority," giving his esteemed testimonial.

Bottlers were opened all over the country. Frank Archer, who started with the company as a clerk, continued to brilliantly promote Moxie using every promotional gimmick known at the time. In its glory days, the beverage was strongly associated with amusement parks, dance halls, and East Coast resorts. These places were synonymous with good times and the "vigorous" life that drinking Moxie was supposed to sustain.

Moxie has an odd flavor that has been described as a combination of cola, root beer, and licorice. The first version contained wintergreen as well as root gentian, a bittersweet herb. Even those who like it say it is an acquired taste.

Moxie was also the first carbonated beverage to offer a sugar-free version. The potent brew with its distinctive orange labels is still available in Maine, where it is considered a local classic.

The Monarch Corporation purchased the Moxie brand in 1966 but continued production and marketing in New England. In 2007, Monarch sold Moxie to its current owner, Cornucopia Beverages of Bedford, New Hampshire, which is owned by the Coca-Cola Bottling Company of Northern New England, a subsidiary of the Kirin Company based in Tokyo.

"Moxie" is a proper name that is now in the dictionary as a noun synonymous with having "spunk." It is still common to hear of someone as having "a lot of moxie."

To read more about the Moxie phenomenon or about root beer, visit the following Web sites.

- Moxie: http://www.drinkmoxie.com/history.php
- Moxie in Maine: https://www.youtube.com/watch?v=D6heTaAxOEY&feature=related
- Moxie World: http://www.moxiecongress.org/archives.htm
- History of Root Beer: http://www.cascience.org/csta/pdf/CCS/History_of_Root_Beer.pdf

Snapple®

Snapple beverages claim to be all natural, without preservatives, artificial flavorings, or chemical dyes. Of Snapple's 50 different flavors, the top 10 flavors are named:

1. Lemon Tea
2. Kiwi Strawberry
3. Diet Peach Tea
4. Raspberry Tea
5. Pink Lemonade
6. Mango Madness
7. Diet Raspberry Tea
8. Fruit Punch
9. Diet Cranberry Raspberry
10. Diet Lemon Tea

The Web site for Snapple is www.snapple.com

MATH CONNECTION!

What is your favorite bottled juice drink? What are the nutritional facts for this beverage? How many servings of your favorite drink do you drink a day? Each week? How many calories do you consume from your favorite drink?

BOTTLED WATER

Bottled water started out being fashionable, but now it has become a basic staple in many American households. In 1998, bottled water sales in the United States were $4.3 billion, according to the Beverage Marketing Corporation. In 2012, total U.S. bottled water consumption was 9.67 billion gallons, having increased by 6.7% to total $11.8 billion. The consumption of bottled water in 2012 amounted to every person in the United States drinking an average of 30.8 gallons of bottled water. This increase in volume is larger than any other beverage category in the United States.[1] In fact, bottled water represents the fastest-growing segment of the beverage industry when compared to fruit beverages, other soft drinks, and beer.

This fastest-growing segment of the beverage industry is the result of several trends. Baby boomers are maturing, and both their tastes and waistlines are guiding them toward more natural and less caloric beverages. America's passion with fitness has encouraged consumers to drink beverage alternatives. The deteriorating taste and quality of tap water and fear of unknown contaminants have seemed to make bottled water the perfect solution.

Companies leading the bottled water business in terms of estimated dollar sales (at wholesale), according to the Beverage Marketing Corporation, include Perrier Group of America, Suntory, McKesson Water Products Company, Danone International, Crystal Geyser, and U.S. Filter. All other companies in the bottled water industry make up the remaining 45% and represent more than 900 brands.

The top 10 leading brands are:

1. Poland Spring®
2. Arrowhead®
3. Evian®
4. Sparkletts®
5. Hinckley & Schmitt®
6. Zephyrills®
7. Ozarka®
8. Deer Park®
9. Crystal Geyser®
10. Crystal Springs®

These leading brands make up approximately 45% of the market share. All other brands make up the remaining 55% of the market share.

Within the bottled water business, two distinct industries and segments are represented. By volume, the largest is the five-gallon or returnable container water business. Companies such as Arrowhead, Sparkletts, and Hinckley & Schmitt are the leaders in this field. Often associated with the office water cooler, bottlers also use 2½-gallon as well as 1-gallon containers for supermarket distribution. This type of bottled water is sold as an alternative to tap water. Premium bottled waters such as Evian, Vittel, and Perrier are sold as soft drink and alcohol alternatives. Packaging ranges from 6 fluid ounces to 2 liters and from custom glass and polyethylene plastic to aluminum cans (refer to Chapter 15). More and more bottled water producers have switched from glass to polycarbonate because the public has increasingly associated quality with this kind of packaging.

In the United States, waters labeled "spring water" must come from a spring source or from

© Pressmaster/Shutterstock.com

FIGURE 24-4 Water is water, but many brands are available.

a borehole adjacent to a spring. Famous spring waters in the United States include Mountain Valley in Arkansas, Belmont Springs in Massachusetts, Saratoga in New York, and Poland Spring in Maine. Artesian water is water from a well that taps a confined underground aquifer where the water level stands above the natural water table. Kentwood Springs in Louisiana is a well-known artesian water. Some bottled water is filtered and reprocessed water from municipal sources.

Bottled water is a profitable industry (Figure 24-4). A bottle of water similar in size to a carbonated drink can sell for as much as or more than the carbonated drink. Current statistics on the bottled water industry are maintained at the International Bottled Water Association (http://www.bottledwater.org/economics/industry-statistics).

ALCOHOLIC BEVERAGES

Fermenting a carbohydrate source such as corn, rye, rice, molasses, agave, wheat, potatoes, or barley will create an alcoholic beverage. Whether it is beer, whiskey, sake, vodka, rum, or tequila depends on the carbohydrate source. Fermenting the sugar-containing juices from grapes creates wine or brandy if it is further refined. The production of beer and wine is discussed in the sections that follow.

Beer

Beer and ale are produced from **malt**, hops, yeast, malt adjuncts, and water. Malt is prepared from barley that has germinated. It is dried, and the sprouts are removed. Malt adjuncts are starch- or sugar-containing materials such as corn, potato, rice, wheat, and barley.

Malting. Using the process of **steeping**, barley grains are soaked at 50 °F to 60 °F (10 °C to 15.6 °C), germinated at 60 °F to 70 °F (15.6 °C to 21.1 °C) for 5 to 7 days and then dried. Most of the sprouts are removed (Figure 24-5). The malt will then serve as a source of amylases (enzymes) that will break down starch into sugar that yeast can consume and thus ferment a liquid.

Mashing. Mashing involves mixing the ground malt with a previously boiled malt adjunct

SCIENCE CONNECTION!

Do you drink bottled water? Do you have a favorite brand? Why or why not? Different brands can carry trace minerals that offer different tastes. Do you prefer bottle water to fountain or tap water? Why or why not? Does convenience play a factor in your choice?

MATH CONNECTION!

How many plastic bottles of soda, juice drinks, or water do you drink a day? Estimate how many you consume in a week. Do you recycle these bottles? How many plastic bottles end up in landfills each year? Research and find statistics on plastic bottles in landfills. Are you surprised by your findings? How can you reduce your use of plastic bottles? What could your chapter or school do to help with this issue?

© Iurii Osadchi/Dreamstime.com

FIGURE 24-5 Barley is sprouted during the malting process.

at a temperature of 65 °F to 70 °F (18.3 °C to 21.1 °C) (see Figure 24-6). The enzymes in the malt digest the starch in the adjunct and release sugar. The mix is heated to 75 °F (24.9 °C) to denature the enzymes and is then filtered. The clear filtrate is known as **wort**.

Boiling. Hops, natural flowers that offer flavoring and stability to a distillation, are added to the wort and boiled for about 2.5 hours and then filtered. Boiling concentrates the solids, kills microorganisms, inactivates enzymes, coagulates proteins, and caramelizes sugars.

Fermentation. Wort is inoculated with a beer yeast, *Saccharomyces carlsbergenis*. The temperature is held between 38 °F and 57 °F (3.3 °C to 14 °C). Fermentation is complete in 8 to 14 days.

© Anna Penigina/Dreamstime.com

FIGURE 24-6 Copper kettles for cooking mash and adding hops during beer making at a brewery.

At this time, the alcohol content of the wort is approximately 4.6% by volume and pH around 4.0. Bacterial growth is minimized as much as is possible. As soon as fermentation is complete, the beer is quickly chilled to 32 °F (0 °C) and passed through a series of filters to remove yeast and other suspended materials before storage and completion.

Completion. Beer is aged at 32 °F (0 °C) for from several weeks to several months. This storage time is known as **lagering**. During storage, the beer is carbonated to a CO_2 content of 0.45% to 0.52%.

After storage, the beer is given a final filtration to remove traces of suspended materials and leave a crystal clear appearance.

Beer in cans is pasteurized by heating to about 140 °F (60 °C) to remove organisms. Draft beer is kept under refrigeration and does not need pasteurization. Many consumers say the beer has a better flavor because it has not been pasteurized. Cold pasteurization or filtration removes leftover yeast and bacteria. This stabilizes the beer without conventional pasteurization's heat.

Wine

Wine is an alcoholic beverage made from fermented grape juice. It can be made from many fruits and berries, but the grape is the most popular. Growing grapes is a major feature of the economy of many wine-producing countries. Wines can be red, white, or rose as well as dry, medium, or sweet. They fall into three basic categories:

1. Natural or table wines with an alcohol content of 8% to 14% and generally consumed with meals
2. Sparkling wines such as champagne, which contains carbon dioxide
3. Fortified wines, with alcohol contents of 15% to 24% and with varying levels of sweetness

The various types of fortified wines include port, sherry, and aromatic wines and bitters such as vermouth.

The quality and quantity of any vineyard's grapes depends on its geographical, geological, and climatic conditions as much as the grape variety and methods of cultivation. Viniculture is the science and art of growing grapes for wine, **vinification** is the production of wine from grapes, and viticulture is the science and art of growing grapes.

Harvesting. Crops are harvested in the autumn when the grapes contain the optimal balance of sugar and acidity. For the sweet white wines of France and Germany, picking is delayed until the grapes are affected by a beneficial mold, *Botrytis cinerea,* which concentrates the juice by dehydrating each grape.

Vinification. For red wine, the grapes are crushed immediately after picking, and the stems generally removed. The yeasts present on the skins come into contact with the grape sugars, and fermentation begins naturally. Cultures of the yeast *Saccharomyces ellipsoideus* are sometimes added, and the grapes are often treated with sulfur dioxide (SO_2) to control the growth of undesirable microorganisms, especially bacteria. During fermentation, the sugars are converted by yeast to ethyl alcohol (ethanol) and carbon dioxide (Figure 24-7). The alcohol extracts color from the skins of the grapes; the longer the vatting period, the deeper the color. Glycerol and some esters, aldehydes, and acids that contribute to a wine's character, bouquet (aroma), and taste are by-products of fermentation.

© Ene/Dreamstime.com

FIGURE 24-7 Wine fermentation tanks.

Traditional maturation of red wine takes as long as 2 years in 50-gallon oak casks. During this time, the wine is racked—drawn off to leave sediment behind—three or four times into fresh casks to avoid bacterial spoilage. Further aging typically takes place after bottling.

The juice of most grape varieties is colorless. Grapes for white wine are also pressed immediately after picking. Fermentation can proceed until it is completed, which will make a dry white wine. Or it can be stopped to make a sweeter wine. Maturation of white Burgundy and some California chardonnays still takes place in oak casks, but vintners tend now to use large stainless steel tanks. Minimal contact with the air helps keep the grapes fresh.

To make rose wines, the fermenting grape juice is left in contact with the skins just long enough for the alcohol to extract the required degree of color. Vinification then proceeds as for white wine.

The best and most expensive sparkling wines are made by the champagne method. Cultured yeasts and sugar are added to the base wine to induce a second fermentation but this time in the bottle. The resulting carbon dioxide is retained in the wine. Sparkling wines are also made by carbonation.

The alcohol content of fortified wines is raised by adding alcohol. With port, madeira, and brandy, spirit (alcohol) is added during fermentation to kill off the yeast, stop fermentation, and leave the desired degree of natural grape sugar in the wine. Sherry is made by adding spirit to the fully fermented wine. Its color, strength, and sweetness are then adjusted to the required style before bottling.

In the United States, the U.S. Department of the Treasury's Bureau of Alcohol, Tobacco, Firearms, and Explosives issues regulations governing the labeling and taxing of wines and other alcoholic beverages. The labels must also meet the requirements of the Food, Drug, and Cosmetic Act.

COFFEE

Coffee beans or cherries come from a small tree (shrub) of the genus *Coffea*. Three important species are *C. arabica, C. canephora,* and *C. liberica.* After harvesting, the coffee cherries may be dried and the pulp around the beans removed. In wet climates or for particular types of coffee, the harvested cherries may be washed and then pulped to separate the beans. The dry and wet methods of preparation produce distinctive flavors in the beans and, along with the differences between varieties, account for the subtle flavor distinctions between beans from various growing areas. The pulpy berry is removed from the skin by bacteria that break down pectin (they are *pectinolytic*). This is followed by an acid fermentation by lactic acid bacteria. After fermentation, beans are dried, hulled, and roasted.

The flavor of coffee is determined not only by the variety but also by the length of time the green beans are roasted (Figure 24-8). In continuous roasting, hot air that is 400 °F to 500 °F (200 °C to 260 °C) is forced through small quantities of beans for 5 minutes. In batch roasting, much larger quantities of beans are roasted for longer. Dark-roasted coffees (French and espresso roasts) are stronger and mellower than lightly roasted beans.

FIGURE 24-8 A wide variety of specialty coffees are available on the market today.

After roasting, the beans are usually ground and vacuum packed in cans. The flavor of coffee deteriorates rapidly after it is ground and after a sealed can is opened.

Instant coffee is prepared by forcing an atomized spray of extremely strong coffee extract through a jet of hot air. This evaporates the water in the extract and leaves dried coffee particles, which are packaged as instant, or soluble, coffee. Another method of producing instant coffee is freeze-drying.

To make decaffeinated coffee, the green bean is processed in a bath of methylene chloride, which removes the caffeine, and then steam to remove the methylene chloride. In a newer method, the caffeine is removed using steam only.

COFFEE SUBSTITUTES

An aromatic beverage similar to coffee can be created by roasting certain grains and grain products in combination with other flavor sources. These cereal beverages are both commercially available and can be made at home. One popular product, Postum®, was developed by C. W. Post, and it helped form the foundation for the General Foods Corporation. Other commercial products include Cafix®, Soyava®, and Pero®.

These commercial products are generally prepared from barley, wheat, rye, malt, and bran with additional flavoring from items such as molasses, dandelion, chicory, carob, cassia bark, allspice, and star anise, depending on the manufacturer. Some of these cereal beverages can even be percolated like coffee. A homemade roasted grain drink also can be prepared by baking a combination of wheat bran, eggs, cornmeal, and molasses.

SCIENCE CONNECTION!

Why would someone consume a coffee substitute? What reasons might lead someone to make this a beverage of choice?

TEA

Tea ranks first as the most popular beverage in the world. It is made when the processed leaves of the tea plant are infused with boiling water. The tea plant (*Camellia sinesis*) is native to Southeast Asia. Its dark green leaves contain the chemicals caffeine and tannin.

Types of tea include fermented (black), unfermented (green), and partially fermented (**oolong**). Production of tea is not really a fermentation, only the action of enzymes contained within the tea leaves.

Processing

The leaves are hand plucked by experienced workers. Only the smallest, youngest leaves are used to produce tea.

To make black tea, harvested leaves are spread on withering racks to dry. The leaves become soft and pliable and are then roller crushed to break the cell walls and release an enzyme. This process gives the tea its flavor. After rolling, the lumps of tea are broken and spread in a fermentation room to oxidize, which turns the leaves to a copper color. The leaves are finally hot-air dried in a process that stops fermentation and turns the leaves black. After the tea is processed, it is sieved to produce tea leaves of a uniform size and to facilitate blending and packing.

Leaf-grade sizes run from *pekoe*, the coarsest size, to *flowery orange pekoe*, the smallest. *Tippy golden flowery orange pekoe* indicates a tea containing the golden-colored tip obtained from the bud. The broken teas have leaves that were broken during processing. These include broken orange pekoe, broken pekoe, fannings, pekoe fannings, and dust.

Oolong tea begins like black tea. The aroma develops more quickly. When the leaf is fired or dried, a coppery color forms around the edge of the leaf while the center remains green. The oolong flavor is fruity and pungent.

Green tea is produced much like the others, except that the leaf is heated before rolling in order to destroy the enzyme. The leaf then remains green throughout further manufacturing, and the aroma characteristic of black tea does not develop. Green tea is graded by age and style.

Blended and Unblended Varieties

Various blended and unblended teas achieved fame for their characteristic flavors, including Assam, Darjeeling, and Keemun, which is also known as English Breakfast tea. Popular blended teas include Irish Breakfast, Russian style, and Earl Grey.

Instant Tea and Bottled Tea

Instant tea is manufactured in a process similar to that of instant coffee. First the tea is extracted from the tea leaves using hot water, 140 °F to 212 °F (60 °C to 100 °C). This extract is concentrated in low-temperature evaporators and is dried using spray driers in a low-temperature vacuum. Just before concentration, the aromatics (flavors) are distilled with flavor-recovery equipment. The flavor is added back later. Some teas may be freeze-dried instead of spray dried.

Recently, bottled and canned teas have gained popularity. The manufacture of these products is similar to carbonated beverages except that they begin with a brewed product. If their pH is above 4.6, retort processing is required.

HERBAL TEA

Tea also includes herbal teas. These teas are made from a wide variety of plants and use not only leaves but also flowers, roots, bark, and seeds. Unlike the limited flavor variations of black tea, herb tea exists in a variety of distinctively different flavors, colors, and aromas. Blending the flavors of different herbs results in an infinite variety of taste sensations. Most herbal teas contain no caffeine. Herbal tea consumption was widespread in Europe long before the arrival of black tea. Some of the lasting favorites include such flavors as chamomile, peppermint, and rose hips.

The parts of the plant picked depend on the type of plant and the intended use. Leaves, flowers, roots, bark, and seed are all potential herbal tea ingredients. After harvest, the herbs are dried by either spreading them on large screens or by tying them in bundles and hanging them upside down. This is done indoors or outdoors in the shade, but it must be done quickly to retain the plants' natural oils and color. Oven drying is less effective than natural drying in terms of preserving the natural oils and flavor. After drying, the herbs are bundled into large sacks and wooden chests for shipment to the herbal tea maker. The next processes include cleaning, milling, sifting, and blending the herbs into the desired flavor combinations.

The most successful herbal tea company is Celestial Seasonings® in Boulder, Colorado. Celestial Seasonings began in 1969 in Aspen, Colorado, where 19-year-old Mo Siegel and a friend, Wyck Hay, gathered wild herbs in the forests and canyons of the Rocky Mountains and made them into herbal teas. Their first tea was produced and packaged with the help of wives and friends and sold in local health food stores.

Now Celestial Seasonings purchases more than 100 different varieties of herbs, spices, and fruits from more than 35 different countries. Each herb and spice has its own flavor and taste characteristics—for example, tangy hibiscus, citruslike rose hips, applelike chamomile, soothing peppermint, fruity blackberry leaves, sweet cloves, and exotic cinnamon. These create a variety of teas when blended.

Celestial's herbalists sample and select entire crops for purchase. Once the herbs and spices arrive in Boulder, they are inspected and tested again. The approved herbs are cleaned and inspected once again and then sent to the milling room to be cut into different degrees of fineness demanded for individual types of teas. Finally, they are sifted to achieve uniformity of texture and purity of content necessary for consistent flavor.

Next, the herbs are blended to match a standard recipe. The batch is compared with previous blends and readjusted until it is absolutely perfect in terms of consistency, quality, and flavor. It is then carefully packaged to protect the flavor.

The most popular herbs for tea are those such as hibiscus flowers, lemon grass, and peppermint. These herbal teas are recommended by many physicians and nutrition advocates as beneficial alternatives to such beverages as coffee, black tea, and sweetened, carbonated drinks.

Kraft Inc. bought Celestial Seasonings in 1984. The company's marketing expertise brought Celestial Seasonings to the attention of new consumers, strengthening its lead in the herbal tea industry while also introducing a gourmet line of traditional, or black, teas. In September 1988, Kraft sold Celestial Seasonings back to its management, returning the tea company to independent ownership. The company remains headquartered in Boulder in a new corporate facility on Sleepytime Drive. Its Web site is located at www.celestialseasonings.com/.

Today, the company claims to serve more than 1.2 billion cups of tea per year. It is the largest herbal tea manufacturer in North America and is expanding internationally. Among the herbal teas being produced now are the following:

Almond Sunset
Antioxidant Green Tea
Authentic Green Tea
Bengal Spice
Black Raspberry
Caffeine-Free Tea
Ceylon Apricot Ginger
Chamomile
Cinnamon Apple Spice
Country Peach Passion
Cranberry Cove
Detox A.M. Herb Tea
Diet Partner Herb Tea
Echinacea Cold Season
Echinacea Herb Tea
Emerald Gardens Green Tea
Emperor's Choice
English Toffee
Estate Blend
Fast Lane
Fruit Sampler
GingerEase Herb Tea
GinkgoSharp Herb Tea
Ginseng Energy
Grandma's Tummy Mint
Green Tea Sampler
Harvest Chamomile
Heart Health
Honey Lemon Ginseng Green Tea
LaxaTea Herb Tea
Lemon Berry Zinger
Lemon Zinger
Mandarin Orange Spice
Mint Magic
Misty Jasmine Green Tea
Misty Mango Oolong
Mood Mender
Morning Thunder
Nutcracker Sweet
Orange Mango Zinger
Peppermint
Raspberry Zinger
Red Zinger
Roastaroma
Sleepytime
Strawberry Kiwi
Sugar Plum Spice
Sunburst C
Tension Tamer
Throat Soothers Tea
Vanilla Hazelnut
Vanilla Maple
Wild Berry Zinger
Wild Cherry Blackberry

In 1998, the company announced a strategic transition to leverage the Celestial brand. It launched a line of herbal dietary supplements under the Celestial Seasonings Mountain Chai® label. The company also introduced a line of "wellness" teas and organic teas and reintroduced a line of green teas. It also announced the return of Red Zinger.

SCIENCE CONNECTION!

Tea has a prominent history in the United States. Create a timeline of the historical importance of tea in the United States.

SUMMARY

The food industry creates a wide variety of beverages. These include carbonated, nonalcoholic beverages with different flavors and with nutritive and nonnutritive sweeteners. Carbonation in beverages provides their zest and sparkle. Recently, the popularity of carbonated beverages has given way to new products created by the food industry such as vitamin-fortified fruit drinks, scientifically formulated sports drinks, nutritionally enhanced beverages, and water. Bottled water represents a profitable, fast-growing segment of the beverage industry.

Fermentation of carbohydrates from corn, rye, rice, molasses, wheat, potatoes, barley, agave, and fruit juices creates alcoholic beverages such as beer, wine, whiskey, vodka, rum, sake, and tequila. Fermentation by yeast changes sugar to alcohol; the taste of the beverage depends on the carbohydrate source and the process.

Coffee comes from roasted beans of a small variety of tree. Coffee flavor is determined by the variety and length of roasting time. Coffee substitutes are produced from roasted grains and grain products in combination with other flavors. Tea comes from the processed leaves of a plant. The type of tea produced depends on the enzymatic action in the harvested tea leaves. Both tea and coffee have been produced in an instant form. Tea has been bottled as a beverage.

Herbal teas come from many plants and not just the leaves: flowers, roots, bark, and seeds of herbs all can be used in a blend for herbal tea. Herbal teas are enjoyed because of their wide ranging tastes, and they are often recommended as beneficial alternatives to coffee, tea, and carbonated drinks.

REVIEW QUESTIONS

Success in any career requires knowledge. Test your knowledge of this chapter by answering these questions or solving these problems.

1. List the five reasons why hops and wort are boiled in the beer-making process.
2. Name the three categories of wines.
3. What is the difference between viticulture and vinification?
4. How is the flavor of coffee determined?
5. What two plants are coffee and tea made from?
6. What is the fastest-growing segment of the beverage industry?
7. List five herbs used in beverages.
8. What is the difference between what plain teas and herbal teas are made from?
9. What are coffee substitutes made from?
10. What drink provides minerals, vitamins, and energy during exercise?

STUDENT ACTIVITIES

1. List the ingredients on a can or bottle of a beverage and explain the function of each ingredient. Bring the label of the bottle to class and compare ingredients with other students.

2. Identify some of the new healthful beverages and make a list of their claims. Conduct a taste test to determine which ingredients consumers can identify. Record your findings.
3. Develop and conduct a survey to determine the favorite beverage enjoyed by people in your community. Report your findings to the class.
4. Develop a report or presentation on the physiological action of caffeine and beverages containing caffeine.
5. Conduct a taste test (or contest) to see if students can differentiate between brand name colas and a generic cola.
6. Make one of the coffee substitutes or herbal teas in class and conduct a taste test.

ADDITIONAL RESOURCES

- A Guide to the Nonalcoholic Beverage Industry: http://marketrealist.com/2014/11/guide-non-alcoholic-beverage-industry/
- Healthy Beverage Guidelines: https://www.hsph.harvard.edu/nutritionsource/healthy-drinks-full-story/
- Types of Beverages: http://www.beverageinstitute.org/article/types-of-beverages/
- University of Michigan Healthy Beverage Program: http://www.uofmhealth.org/drink

Internet sites represent a vast resource of information. Although those provided in this chapter were vetted by industry experts, you may wish to further explore the topics discussed in this chapter using a search engine such as Google. Keywords or phrases may include the following: *nonalcoholic beverages, teas (specific brand names) and types, colas (specific brand names), wine (specific brand names), beer (specific brand names), malting, mashing, vinegar, cold pasteurization, carbonated soft drinks, diet drinks, carbonation*. In addition, Table B-7 provides a listing of useful Internet sites that can be used as a starting point.

REFERENCES

Alexander, R. J. 1998. *Sweeteners: Nutritive*. St. Paul, MN: Eagan Press.

Bartlett, J. 1996. *The cook's dictionary and culinary reference*. Chicago: Contemporary Books.

Bennion, M., and B. Scheule. 2014. *Introductory foods*, 14th ed. Upper Saddle River, NJ: Prentice Hall.

Conforti, F.D. 2008. *Food selection and preparation: A laboratory manual*, 2nd ed. Hoboken, NJ: Wiley-Blackwell.

Delgado-Vargas, F., and O. Paredes-Lopez. 2002. *Natural colorants for food and nutraceutical uses*. Boca Raton, FL: CRC Press.

Horn, J., J. Fletcher, and A. Gooch. 1997. *Cooking A to Z: The complete culinary reference source*. Glen Ellen, CA: Cole Publishing Group.

Lehner, E., and J. Lehner. 1973. *Folklore and odysseys of food and medicinal plants.* New York: Tudor Publishing.

McGee, H. 2004. *On food and cooking. The science and lore of the kitchen.* New York: Scribner.

ENDNOTE

1. International Bottled Water Association http://www.bottledwater.org/us-consumption-bottled-water-shows-continued-growth-increasing-62-percent-2012-sales-67-percent. Last accessed July 13, 2015.

SECTION *Four*

Related Issues

CHAPTER 25

Environmental Concerns and Processing

OBJECTIVES

After reading this chapter, you should be able to:

- Describe the properties and the requirements of water used in food processing
- Understand four methods that food-processing industries use to dispose of solid wastes
- Explain how water becomes wastewater during food processing
- Relate the level of solids in wastewater to biological oxygen demand (BOD)
- Describe wastewater treatments to lower BOD
- List eight products that could be in wastewater
- Identify four methods of separating water from solid wastes
- Name four ways food processors reduce the amount of solid wastes and water discharge
- Explain five methods of conserving water during food processing

NATIONAL AFNR STANDARD

FPP.01

Develop and implement procedures to ensure safety, sanitation, and quality in food products and processing facilities.

KEY TERMS

alkaline

biological oxygen demand (BOD)

caustic

flocculation

landfilling

potable

tramp material

turbidity

wastewater

wet scrubber

Advances in food processing and food packaging play a primary role in keeping the American food supply the safest in the world. However, packing technology must balance food protection with other issues, including material costs, environmental consciousness, the regulation of pollutants, and the disposal of solid waste. Solid wastes and wastewater from food processing are environmental concerns. The food-processing industry uses technology to reduce the quantity of high-moisture-content solid wastes generated by washing, cleaning, extracting, and separating undesirable solids from fruit and vegetable processing. Research in this area continues to develop ways to reduce the volume and water content of solid wastes, increase the value of solid wastes sold as animal feed, and convert solid wastes into other by-products.

WATER IN FOOD PRODUCTION

In the production of quality food, the role of water in processing and preparation is almost infinite. Water serves as a universal solvent. The ability of water to serve as a solvent has tremendous effects on many of its other roles. It affects the osmotic pressure of a solution and often determines a food's structure—for example, the use of crystalline water in frozen desserts. Water also serves as a heat-transfer medium and serves as a cleansing agent. Managing the use and disposal of water during food processing continues to be an important issue for the industry.

PROPERTIES AND REQUIREMENTS OF PROCESSING WATERS

Converting raw products into finished food products requires large quantities of water. Incoming water quality is important to a food-manufacturing operation, and the water quality exiting a food manufacturing operation is important to everyone. All water is part of the water (hydrologic) cycle (see Figure 25-1). Food scientists are always searching for ways to prevent or eliminate contamination of water by food processing.

Food-production and food-processing industries concern themselves with three aspects of water:

1. Microbiological and chemical purity and safety
2. Suitability for processing use
3. Decontamination after use

Water entering a food-processing plant must meet health standards for drinking (**potable**) water. National Primary Drinking Water Regulations have been issued by the federal Environmental Protection Agency. These regulations, which can be found in the Code of Federal Regulations, set limits on such primary characteristics as:

- Inorganic chemicals (e.g., As, Ba, Cd, Cr, Pb, Hg, NO_3, Se, Ag, F)
- Organic chemicals (e.g., Endrin®, Lindane®, Methoxychlor®, Toxaphene®, 2,4-D®, 2,4,5-TP Silvex®)
- **Turbidity** (cloudiness)
- Coliform bacteria

Strict environmental regulations govern the discharge of polluted water from processing plants (Figure 25-2). Contaminated water and the difficulties of disposing of **wastewater** increase the overall cost of manufacturing food.

ENVIRONMENTAL CONCERNS

For many food-processing plants, a large fraction of the solid waste produced at the plant comes from the separation of desired food components

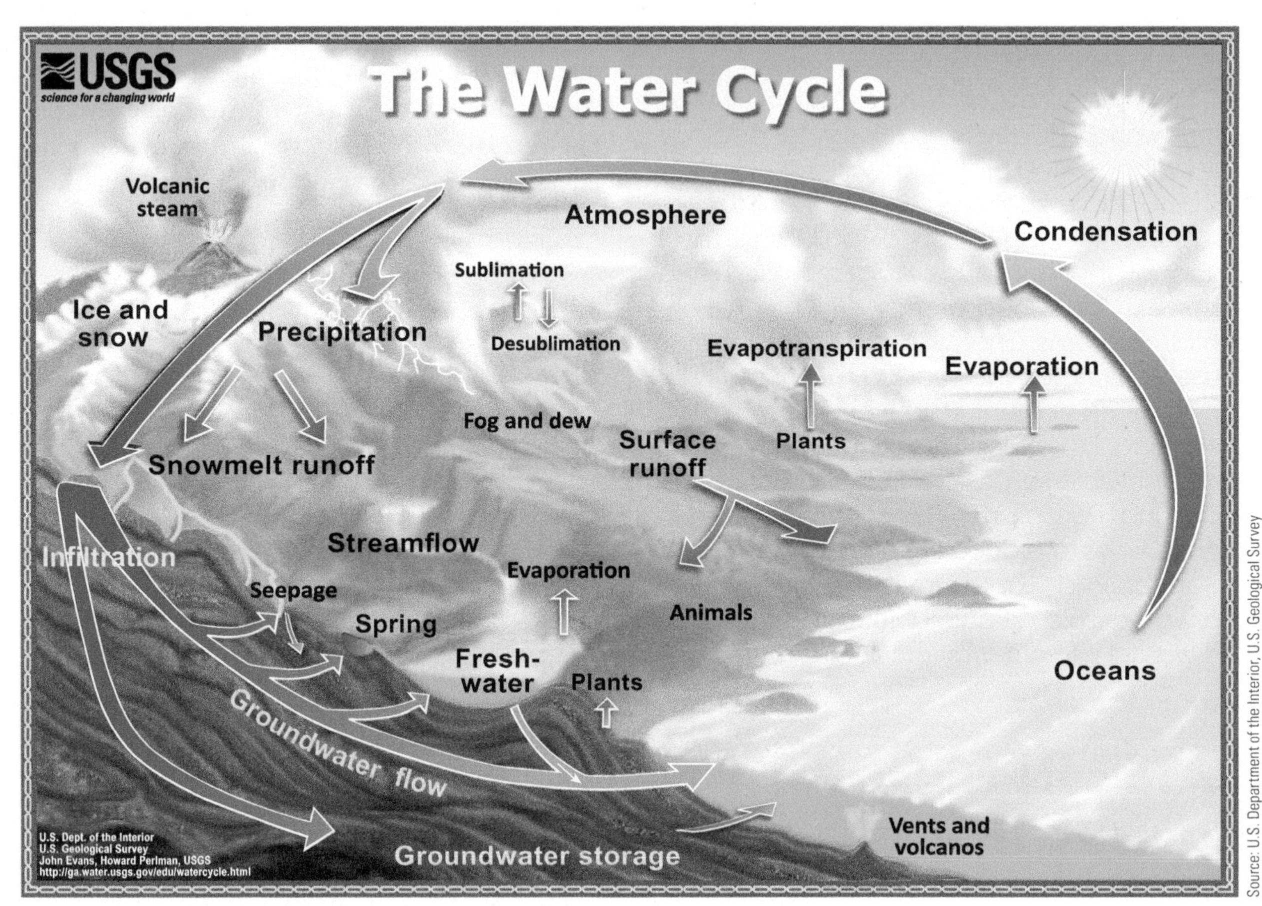

Source: U.S. Department of the Interior, U.S. Geological Survey

FIGURE 25-1 The water (hydrology) cycle.

from undesired components in the early stages of processing. Undesirable components include:

- **Tramp material** (soil and extraneous material)
- Spoiled food stocks
- Fruit and vegetable trimmings, peels, pits, seeds, and pulp

© John Dorado/Dreamstime.com

FIGURE 25-2 Freshwater storage tanks are important to any community.

In some food-processing plants, **caustic** peeling is used to remove skins from soft fruit and vegetables such as tomatoes. This operation produces a highly **alkaline** or salty solid waste, depending on whether the alkalinity is neutralized. High-moisture solid waste materials can also be generated by water cleanup and reuse operations in which the dissolved or suspended solids are concentrated and separated from wastewater streams.

DISPOSAL OF SOLID WASTES

Currently, food-processing solid waste can be disposed of by sewer or by **landfilling**, and chemical production or converted into animal feed or fuel. The effective recovery of the energy value in the solid waste is often determined by the moisture content and the processing method—gasification, bioconversion, incineration. A final use for solid wastes is composting or some type of land application with limited soil-amendment value.

Disposal of solid wastes as animal feed is limited by several factors, including cost of shipping, spoilage during storage and transport, and the presence of undesirable components such as alkaline elements or salt. For example, distillers' dried grains and solubles have been fed to animals for decades. Water content is a major contributor to shipping costs and to some extent the spoilage rate. Spoilage itself reduces the value of solid wastes and limits the animal feeding option to local animal herds.

Incineration and use of solid wastes as fuel are options in certain cases, but the solid wastes must have relatively low water content and be further dried with ease. The moisture content of suitable fuels is approximately 10% or less.

Composting is an option for disposal, but odor and leaching of soluble constituents are limiting factors (see Figure 25-3). Composted material is valued as a soil amendment or potting soil, but its widespread use and marketability are limited by shipping cost. Composition of the composting materials needs to be controlled to obtain the correct physical mix that will allow natural aerobic bioprocesses to proceed. A full range of food-processing wastes can be composted, including fruit and vegetable wastes such as peelings, skins, pumice, cores, leaves, and twigs; fish processing waste such as bones, heads, fins, tails, skin, whole fish, and fish offal; meat processing wastes such as paunch contents, blood, fats, intestines, and manure; and grain processing wastes such as chaff, hulls, pods, stems, and weeds.

FIGURE 25-3 Composting could potentially solve some food-processing waste issues.

Disposing solid wastes from food processing in domestic sewers is becoming less favorable because of increased sewer rates and the hesitation of municipal sewage-treatment plants to accept waste streams that have high **biological oxygen demand (BOD)** and, in some cases, high salt content.

The practice of landfilling is becoming less favorable in general because of the generation of foul odors as communities expand and build housing near food-processing plants. Leaching of undesirable constituents (salts, soluble organics) into the soil and groundwater is also an important concern when groundwater is used by communities or migrates into nearby streams.

Currently, solid wastes disposed by landfilling or composting are minimally treated using dewatering screens, centrifugal screens, or strainers to separate free liquids from solids. In the case of solids from juice extractors and sorting operations to remove blemished or spoiled fruits and vegetables, the solid waste is not treated before disposal. Similarly, low-moisture-content solid wastes that are potentially suitable as fuels or for incineration undergo minimal processing such as size reduction and minimal drying.

Solid wastes sent to a sewer undergo size reduction and are mixed with water or other liquid waste streams to produce acceptable flow properties and BOD loading.

Some solid wastes disposed of as animal feeds are not further treated but are fed to local livestock, particularly dairy and beef cattle. For example, solid wastes from orange juice processors and distilleries can be dried and sold as livestock feed.

SCIENCE CONNECTION!

How can we reduce processing waste and reduce landfilling? Research and strategize ways to reduce waste.

HOH—WATER

Water—the stuff of life—covers three-fourths of Earth's surface and represents a major component in the bodies of plants and animals. A 200-lb (91-kg) human would weigh only about 90 lbs (41 kg) if all the water was removed from his body. This miracle liquid is formed from two gases—hydrogen and oxygen. Two atoms of hydrogen (H) and one atom of oxygen (O) combine to form water, making its chemical formula H_2O or HOH. The water molecule is bipolar, having charged poles like a magnet, which provides water with unique properties.

Depending on the temperature, water exists in three forms. It is a liquid between 32 °F and 212 °F (0 °C and 100 °C). It is a gas or vapor at temperatures above 212 °F (100 °C). It becomes a solid (ice) at temperatures below 32 °F (0 °C).

Of all naturally occurring substances, water has the highest specific heat. This makes it a good coolant in biological systems and makes it resistant to rapid temperature changes. Specific heat is the amount of heat required to raise the temperature of a substance by 1 °C.

Water is the universal solvent, dissolving almost everything. It is powerful enough to dissolve rocks yet gentle enough to hold an enzyme in a fragile cell. As a solvent, it acts as a medium for biochemical reactions and carries waste products and nutrients.

This miracle liquid supports life.

To learn more about water and water quality, visit the following Web sites.

- Encyclopedia Brittanica: www.britannica.com/
- U.S. Environmental Protection Agency—Learn About Water: http://water.epa.gov/

PROPERTIES OF WASTEWATERS

Some food-processing plants may be washing profits down the drain. This industry typically uses a large volume of water to process food products and clean plant equipment, yielding large amounts of wastewater that must be treated. Excessive water use and wastewater production add financial and ecological burdens to the industry and the environment. However, food processors can take actions that will dramatically reduce water use, wastewater production, and the high costs associated with these problems.

Cleanup water in food-processing plants flushes loose meat, blood, soluble protein, inorganic particles, and other food waste into the sewer. Some of these raw materials could be recovered and sold to other industries but instead are lost. Also, most of this waste adds a high level of biological oxygen demand to the wastewater.

Wastewater-treatment plants (Figure 25-4) use BOD levels to gauge the amount of waste present in water: The higher the BOD level, the more treatment is required. Sewer plants add surcharges for each pound of BOD that exceeds a set limit. These charges can cost a company hundreds of thousands of dollars each year.

FIGURE 25-4 Treatment of wastewater.

© Jaroslow Domanski/Dreamstime.com

Food-processing companies use current methods and interventions that can assist in effectively managing their water resources. Without knowledge and use of these wastewater management techniques, companies lose money through water use charges, raw material losses, sewage surcharges, and possible fines from environmental agencies. With the public emphasis on environmental quality, the food industry has further incentive to reduce its water usage and its wastewater production.

WASTEWATER TREATMENT

New dewatering schemes are needed to make food-processing solid wastes more economical for many applications. Common dewatering processes use a variety of mechanical means such as screw presses, belt presses, vacuum filters, and so on. Separations involving other mechanical forces such as ultrasound and nonmechanical forces such as electric and magnetic fields have been used in a few specialized sectors.

Primary treatments such as screening, filtering, centrifuging, skimming, settling, and coagulation or **flocculation** lower the wastewater BOD and the total solids in the wastewater. Secondary treatment often involves the use of trickling filters, activated sludge tanks, and various ponds. Sometimes the secondary treatments are preceded by anaerobic digestion.

Lowering Discharge Volumes

Plant surveys have determined where water use occurs and where wastes are generated in food-processing plants. The results show that more than half the waste load results from wet cleanup practices. Waste in the form of bits of food was being flushed down the drains.

Where cleanup practices (Figure 25-5) are involved in increasing the wastewater, specialists suggest techniques for dry cleanup that would

FIGURE 25-5 Cleaning requires large quantities of water, as this high-pressure washer used in food-processing plants shows.

© Endostock/Dreamstime.com

reduce wastewater production. Dry cleanup uses methods to capture all nonliquid waste and prevent it from entering the wastewater. If most of the waste that comes out of a plant consists of carbohydrates or proteins, dry cleanup allows much of the waste to be reclaimed and put to secondary use—for example, as animal feed. Some of the remaining wastes in an animal-processing facility can be sold to a rendering plant.

A grease trap, a solids-recovery basin, and an activated sludge system with provisions for pH control are part of an overall wastewater-management system. Where odors are part of the problem, they can be eliminated by passing a building's exhaust air through a **wet scrubber**.

A concept as simple as keeping wastes off floors and out of drains will save a company money and reduce the strain on a city's sewage-treatment plant. Most of the changes made to reduce water use and waste cost a company little or nothing. Carelessness can be prevented by managerial emphasis and making employees aware of the problem.

Commonsense approaches to cleanup—using trays beneath machines to catch spillage, picking up spillage before hosing down floors, and placing screens over drains—were used at little cost. Awareness of the serious problems caused by overuse and product waste cost a company only the time needed to educate employees thoroughly. A successful pollution-prevention program requires frequent retraining to keep employees focused and careful.

Any food-processing plant can conserve water by following these conservation guidelines:

- Always treat water as a raw material with a real cost.
- Set water conservation goals for the plant.
- Make water conservation a management priority.
- Install water meters and monitor water use.
- Train employees how to use water efficiently.
- Use automatic shut-off nozzles on all water hoses.
- Use high-pressure, low-volume cleaning systems.
- Do not let people use water hoses as brooms.
- Reuse water where possible.
- Minimize spills of ingredients and raw and finished products on the floor.
- Always clean up spills before washing.

RESPONSIBILITY

Water conservation and waste reduction are important. Water costs and sewer charges are rising. Water quality and availability are threatened by increased consumption and pollution in many areas of the country. Pollution is being attacked aggressively by public agencies and the public at large. Future regulations may require water conservation and the elimination of pollutant discharges. A food processor's image can be tarnished and sales hurt if a plant is perceived to be harming the environment. Enforcement actions are becoming more severe. Lawsuits, fines, and even prison terms may face those who do not fully comply with environmental laws.

SUMMARY

Food processing produces solid wastes. To remain environmentally friendly, food processors seek ways to dispose of or use solid wastes. Because food processing requires large quantities of potable water and produces large quantities of wastewater, other environmental issues must be addressed. Food processors not only must clean up wastewater but also work to reduce the amount of water used in processing and thus lower wastewater production. Responsibility for disposing of solid wastes and treating and reducing wastewater resides with the food processor in accordance with environmental standards, laws, and restrictions.

REVIEW QUESTIONS

Success in any career requires knowledge. Test your knowledge of this chapter by answering these questions or solving these problems.

1. What main use does water serve?
2. List five methods to conserve water during food processing.
3. Name the three aspects of water that concern food-production and food-processing industries.
4. Define BOD.
5. Explain how water becomes wastewater during food processing.
6. When BOD levels are high, what is required?
7. List four methods for separating solids from wastes.
8. Name four ways food processors can reduce solid wastes and water discharge.
9. Explain dry cleanup.
10. Name one way dry cleanup can be used.

STUDENT ACTIVITIES

1. Visit a wastewater-treatment facility and develop a report or presentation on the operation.
2. Diagram the water (hydrologic) cycle and show how using water and disposing of wastewater by food processors fits into the cycle.
3. Collect two current articles related to water quality. Summarize these and report on them to the class.
4. Demonstrate the value of clear, clean water. Start with a clear, clean quart bottle of water. Take a drink. Then add a tablespoon of soil, some bread crumbs, a few pieces of hair, a little catsup, some crumbs of hamburger, and a tablespoon of oil. Shake the mixture. This mixture represents wastewater. Devise an experiment to show how this wastewater could be cleaned up.
5. Invite a wastewater-treatment specialist to visit the classroom and discuss methods used to treat and clean water. Prepare questions for the guest speaker before the visit.

ADDITIONAL RESOURCES

- Food Processing—Background Reading: http://www.jhsph.edu/research/centers-and-institutes/teaching-the-food-system/curriculum/_pdf/Food_Processing-Background.pdf
- UNESCO—Advanced Water Management for Food Production: https://www.unesco-ihe.org/msc-programmes/specialization/advanced-water-management-food-production
- Water Use in the Food Industry: http://pods.dasnr.okstate.edu/docushare/dsweb/Get/Document-8508/FAPC-180web.pdf
- Food Processing Wastewater Treatment: http://fabe.osu.edu/research/food-processing-wastewater-treatment

Internet sites represent a vast resource of information. Although those provided in this chapter were vetted by industry experts, you may wish to further explore the topics discussed in this chapter using a search engine such as Google. Keywords or phrases may include the following: *wastewater, wastewater treatment, solid wastes, potable water, processing waters, National Primary Drinking Water Regulations, composting, landfilling, sludge system, water conservation, waste reduction, Environmental Protection Agency, Department of Environmental Quality.* In addition, Table B-7 provides a listing of some useful Internet sites that can be used as a starting point.

REFERENCES

Council for Agricultural Science and Technology (CAST). 1995. *Waste management and utilization in food production and processing* (Task force report No. 125). Ames, IA: Author.

Cremer, M. L. 1998. *Quality food in quantity: Management and science.* Berkeley, CA: McCutchan Publishing.

Kreith, F., and T. Tchobanoglous. 2002. *Handbook of solid waste management,* 2nd ed. New York: McGraw Hill.

Potter, N. N., and J. H. Hotchkiss. 1999. *Food science,* 5th ed. New York: Springer.

Tchobanoglous, G., F. L. Burton, and D. H. Stensel. 2002. *Wastewater engineering: Treatmentand reuse,* 4th ed. New York: McGraw Hill.

CHAPTER 26

Food Safety

OBJECTIVES

After reading this chapter, you should be able to:

- List three categories of food safety
- Name four factors that contribute to the development of a food-borne disease
- List four types of microorganisms that can cause food-borne illness
- List five factors that affect microbial growth
- Identify the microorganisms that provide an index of food sanitation
- Discuss the role of sanitation and cleaning during processing in food safety
- Determine the correct order of sanitizing or cleaning a food contact surface
- Name three types of food soils
- Explain two types of sanitization
- Identify agencies involved in food-safety regulation
- Describe the role of HACCP in food safety

NATIONAL AFNR STANDARD

FPP.01

Develop and implement procedures to ensure safety, sanitation and quality in food products and processing facilities.

KEY TERMS

biofilms
biotechnology
clean out of place (COP)
cross-contamination
food soil
gastroenteritis
generation time
genetically modified organism (GMO)
Good Manufacturing Practices (GMPs)
heating, ventilating, and air-conditioning (HVAC)
lag time
psychrotrophic
sanitization
standard plate count (SPC)
thermotrophic
toxins
transgenic crop
transgenic organism

Food safety is a broad topic. Pesticides, herbicides, chemical additives, and spoilage are all concerns in the food industry. However, food scientists, food processors, and consumers focus most on microbiological quality. Microorganisms pose a challenge to the food industry, and most food processes are designed with microbial quality in mind. Microorganisms are too small to be seen with the unaided eye and have the ability to reproduce rapidly. Many of them produce toxins and can cause infections. For these reasons, the microbiological quality of food is scrutinized closely.

SAFETY, HAZARDS, AND RISKS

The United States has the safest food supply in the world. However, people still become ill from eating contaminated foods. Foods can become contaminated in a variety of ways, although contaminant fall into three basic categories of hazard: physical, chemical, and biological. Physical hazards are unwanted substances—bones, metal, glass, plastic, dirt, insects, wood—become part of a food product. Natural occurring objects can be contaminants left in food such as fruit pits and bones. Chemical hazards can contaminate foods when there is improper storage of materials such as cleaners, sanitizers, polishes, hand lotions, and pesticides. Although physical and chemical hazards can and do cause injuries, biological hazards are the largest concern in food safety.

Food safety hazards include all microorganisms, chemicals, and foreign materials that could cause injury or harm if consumed. The following three conditions directly result in food-safety problems.

1. *Illness caused through transmission of disease germs.* Pathogenic germs are passed from person to person through soiled objects such as money, doorknobs, railings, common drinking cups, and the like. Food can serve as a mere vehicle of disease transmission. Transmission of animal pathogens to humans by way of food is also possible.
2. *Food poisonings and food infections caused by bacteria.* The terms *food poisoning* and *food infections* refer to a violent illness of the stomach and intestinal tract (known as **gastroenteritis**) after an offending food has been eaten. If the offending food contains high numbers of bacteria, a victim has a high likelihood of suffering gastroenteritis. Some pathogens are able to release **toxins** into food itself and directly cause illness (an intoxication). Other pathogens do not act until swallowed. These can cause an infection of the gastrointestinal tract.
3. *Food poisoning caused by agents other than microorganisms.* An offending food will contain poisonous chemicals or be a poisoned plant or animal product. Two examples are (a) tuna and mercury and (b) apples and alar.

Recent developments in diagnosing and tracking reported illnesses have helped the public become more aware that certain types of illness might be related to food consumed prior to illness.

FOOD-RELATED HAZARDS

Today, food-borne illness is of serious concern. Its frequency is not known because a great majority of cases go unreported. Reporting food-borne diseases to public health authorities is not required in the United States. Estimates claim as many as 200 million cases in the United States per year. Only a small percentage of these are hospitalized. Most are passed off as traveler's diarrhea, 24-hour flu, or upset stomach. Symptoms of food-borne illness vary depending on the illness, but diarrhea, vomiting, fever, nausea, cramps and jaundice are all common symptoms. How quickly symptoms develop, or the onset time varies as well. Onsite time can range from 30 minutes to six weeks depending on the food-borne illness.

TABLE 26-1 Food-Borne Disease Outbreaks, 2011–2012

Outbreaks reported	1,632
Cases of illness	29,112
Hospitalizations	1,750
Deaths	68

Source: Foodborne Disease Outbreak Surveillance System, 2011–2012, the most recent years for which outbreak data are finalized.

Most food-borne illness can be avoided if food is handled properly. Statistics from the Centers for Disease Control show that the most commonly reported food-preparation practice that contributed to food-borne disease was improper holding temperatures, followed by poor personal hygiene, inadequate cooking, contaminated equipment, and food from an unsafe source, as shown in Table 26-1.

Cross-Contamination

Cross-contamination is the transportation of harmful substances to food by the following means.

- Hands touch raw foods such as uncooked chicken and then touch food that will not be cooked—salad ingredients, for example.
- Surfaces such as cutting boards or cleaning cloths that touch raw foods are not cleaned and sanitized before they touch ready-to-eat food.
- Raw or contaminated foods touch or drip fluids on cooked or ready-to-eat foods.
- Raw and ready-to-eat foods are improperly stored in a refrigerator.

High-Risk Foods and Individuals

The U.S. Public Health Service classifies moist, high-protein, and low-acid foods as potentially hazardous. High-protein foods consist, in whole or in part, of milk or milk products, shell eggs, meats, poultry, fish, shellfish, and edible crustaceans (shrimp, lobster, crab). Baked or boiled potatoes, tofu, soy protein foods, plant foods that have been heat-treated, and raw seed sprouts (such as alfalfa or bean sprouts) also pose a hazard. These foods can support rapid growth of infectious or disease-causing microorganisms.

The immune system helps fight infection, but the immune systems of very young children, pregnant women, the elderly, and chronically ill people are at greatest risk to develop food-borne infections. Infants and children in particular produce less acid in their stomachs, making it easier for them to get sick. For pregnant women, the fetus is at risk because it does not have a fully developed immune system. For elderly individuals, poor nutrition, lack of protein in the diet, and poor blood circulation may result in a weakened immune system. Those with immunocompromised systems from diabetes, cancer, and AIDS patients as well as those on antibiotics are at greater risk.

Microorganisms are everywhere—on the body, in the air, on kitchen counters and utensils, and in food. The main microorganisms are viruses, parasites, fungi, and bacteria. Table 26-2 describes some of the common food-borne diseases.

SCIENCE CONNECTION!

How can you reduce cross-contamination in your kitchen at home? Come up with five ways to cut down on cross-contamination and the possible spread of food-borne diseases.

TABLE 26-2 Microorganisms in Food

DISEASES AND ORGANISMS THAT CAUSES IT	SOURCE OF ILLNESS	SYMPTOMS
Botulinum toxin (produced by *Clostridium botulinum* bacteria)	Although the spores of these bacteria are widespread, they only produce toxins in anaerobic (oxygenless) environments that have little acidity. They are found in a considerable variety of canned foods, including corn, green beans, soups, beets, asparagus, mushrooms, tuna, and liver pate. They can also be found in luncheon meats, ham sausage, stuffed eggplant, lobster, and smoked and salted fish.	Onset: Generally 4 to 36 hours after eating. Neurotoxic symptoms, including double vision, inability to swallow, speech difficulty, a progressive paralysis of the respiratory system. *Get medical help immediately. Botulism can be fatal.*
Campylobacteriosis (*Campylobacter jejuni*)	Bacteria on poultry, cattle, and sheep can contaminate the meat and milk of these animals. Chief food sources: raw poultry, meat, and unpasteurized milk.	Onset: Generally 2 to 5 days after eating. Diarrhea, abdominal cramping, fever, and sometimes bloody stools. Lasts 7 to 10 days.
Listerosis (*Listeria monocytogens*)	Found in soft cheese, unpasteurized milk, imported seafood products, frozen cooked crabmeat, cooked shrimp, and cooked surimi (imitation shellfish). The *Listeria* bacteria resist heat, salt, nitrite, and acidity better than many other microorganisms. They survive and grow at low temperatures.	Onset: From 7 to 30 days after eating, but most symptoms are reported 48 hours to 72 hours after consumption of contaminated food. Fever, headache, nausea, and vomiting. Primarily affects pregnant women and their fetuses, newborns, the elderly, people with cancer, and those with impaired immune systems. Can cause fetal and infant death.
Perfringen food poisoning (*Clostridium perfringens*)	In most instances, caused by failure to keep food hot. A few organisms are often present after cooking and multiply to toxic levels during cool down and storage of prepared foods. Meats and meat products are the foods most frequently implicated. These organisms grow better than other bacteria between 120 °F and 130 °F (49 °C and 54 °C). Gravies and stuffing must be kept above 140 °F (60 °C).	Onset: Generally 8 to 12 hours after eating. Abdominal pain, diarrhea, and sometimes nausea and vomiting. Symptoms last a day or less and are usually mild. Can be more serious in older or debilitated people.
Salmonellosis (*Salmonella enteritis*) Shigellosis (bacillary dysentery) (*Shigella* bacteria)	Raw meats, poultry, milk and other dairy products, shrimp, frog legs, yeast, coconut, pasta, and chocolate are most frequently involved.	Onset: Generally 6 to 48 hours after eating. Nausea, abdominal cramps, diarrhea, fever, and headache. All age groups are susceptible, but symptoms are most severe for the elderly, infants, and the infirm.

(*Continues*)

TABLE 26-2 Microorganisms in Food

DISEASES AND ORGANISMS THAT CAUSES IT	SOURCE OF ILLNESS	SYMPTOMS
Staphylococcal food poisoning Staphylococcal enterotoxin (produced by *Staphylococcus aureus* bacteria)	Found in milk and dairy products, poultry, and potato salad. Food becomes contaminated when human carriers do not wash hands and then handle liquid or moist food that is not cooked thoroughly afterward. Organisms multiply in food left at room temperature.	Onset: 1 to 7 days after eating. Abdominal cramps, diarrhea, fever, sometimes vomiting, and blood, pus, or mucus in stools.
Vibrio infection (*Vibrio vulnificus*)	Toxins produced when food contaminated with the bacteria is left too long at room temperature. Meats, poultry, egg products, tuna, potato and macaroni salads, and cream-filled pastries are good environments for these bacteria to produce toxins. The bacteria live in coastal waters and can infect humans either through open wounds or through consumption of contaminated seafood. The bacteria are most numerous in warm weather.	Onset: Generally 30 minutes to 8 hours after eating. Diarrhea, vomiting, nausea, abdominal pain, cramps, and prostration. Lasts 24 to 48 hours. Rarely fatal. Onset: Abrupt. Chills, fever, or prostration. People with liver conditions, low gastric (stomach) acid, and weakened immune systems are most at risk.
Amebiasis (*Entamoeba histolytics*)	This protozoan exists in the human intestinal tract and is expelled in feces. Protozoans are present in polluted water, and vegetables grown in polluted soil spread the infection.	Onset: 3 to 10 days after exposure. Severe cramping pain, tenderness over the colon or liver, loose morning stools, recurrent diarrhea, loss of weight, fatigue, and sometimes anemia.
Giardiasis (*Giardia lamblia*)	Most frequently associated with consumption of contaminated water. May be transmitted by uncooked foods that become contaminated while growing or after cooking by infected food handlers. Cool, moist conditions favor the organism's survival.	Onset: 1 to 3 days. Sudden onset of explosive watery stools, abdominal cramps, anorexia, nausea, and vomiting. Especially infects hikers, children, travelers, and institutionalized patients.
Hepatitis A virus	Mollusks (oysters, clams, mussels, scallops, and cockles) become carriers when their beds are polluted by untreated sewage. Raw shellfish are especially potent carriers—and cooking does not always kill the virus.	Onset: Begins with malaise, appetite loss, nausea, vomiting, and fever. After 3 to 10 days patient develops jaundice with darkened urine. Severe cases can cause liver damage and death.
Enterohemorrhagic and shiga toxin-producing *Escherichia coli* (E. coli)	*E. coli* can be found in the intestines of cattle. The bacteria can contaminate the meat during slaughter. Linked most with undercooked ground beef	Onset: Begins with fever, vomiting and diarrhea. It is contagious and can be spread by infected fecal matter.
Norovirus	Often linked to ready-to-eat foods and contaminated water. Often transferred from infected food handler to foods. Outbreaks have occurred annually on cruise ships.	Onset: vomiting and diarrhea. Infected staff must be removed from work while sick because this disease is highly contagious.

MICROORGANISMS

Food safety requires an understanding of the basics of food microbiology. Harmful microorganisms are called *pathogens*. Some pathogens can sicken someone who eats them. Others produce poisons or toxins that make people sick. Important organisms to be aware of include: bacteria, fungi (yeasts and molds), viruses, and parasites.

Although not living organisms, viruses are generally included as biological agents of concern. These materials are combinations of proteins and nucleic acids that can take over cellular functions. Harmful effects of microorganisms include:

- Spoilage of foods
- Food-borne toxins
- Food-borne infections
- Viral borne infections

Viruses

Viruses are the tiniest, and probably the simplest, organisms. They are not able to reproduce by themselves outside living cells. Once they enter a cell, they force it to make more viruses. Some viruses are extremely resistant to heat and cold. They do not need potentially hazardous food to survive, and they do not multiply once they are in food, which serves mainly as a transportation device to get from one host to another.

Parasites

Parasites need to live on or in a host to survive. Examples of parasites that may contaminate food are *Trichinella spiralis* (trichinosis), which affects pork; and *Anisakis roundworm*, which affects fish. The most important way to prevent parasites from causing food-borne illness is to purchase food from approved and reputable suppliers. Parasites are commonly associated with seafood or foods processed with contaminated water.

Fungi

Fungi can be microscopic or as big as a giant mushroom. Fungi are found in air, soil, plants, animals, water, and some foods. Molds and yeast are both fungi. Some molds and mushroom produce toxins that cause food-borne illness. Some fungi are beneficial. Harmful mushrooms are difficult to recognize, so all mushrooms should be purchased from approved and reputable suppliers.

Bacteria

Of all the microorganisms, bacteria are the greatest threat to food safety. Bacteria are single-celled, living organisms that can grow quickly at favorable temperatures. Some bacteria are useful, and we use them to make foods such as cheese, buttermilk, sauerkraut, pickles, and yogurt. Other bacteria are infectious disease-causing agents called *pathogens* that use the nutrients found in potentially hazardous foods to multiply.

Some bacteria are not infectious on their own, but when they multiply in potentially hazardous food, they release toxins that poison humans when the contaminated food is eaten.

Factors Affecting Microbial Growth

The effect of organisms on the safety of foods is dependent on the initial numbers of organisms present, processing to eliminate the organisms, control of the environment to prevent growth, and sanitation.

The major factors that influence the growth of microorganisms in food are:

- Food
- Acidity (pH)
- Temperature
- Time
- Oxygen
- Moisture

Note the mnemonic created from the first letters of these six factors: They spell out FAT TOM, an easy way to remember the conditions that create danger.

The food industry depends on minimizing microbial populations in the food as well as controlling the environment.

FOOD

Most bacteria need nutrients to survive. Foods that are not time and temperature controlled for safety will support the growth of bacteria better than other types of foods.

ACIDITY

Bacteria grow best in foods that contain little or no acid. The measurement of acidity, pH, uses a scale running from 0 to 14, with 0 highly acidic and 14 highly basic; a value of 7 is neutral. Generally, microorganisms, especially pathogens, cannot grow at pH levels below 4.0. These are termed *acid foods* and depend on the low pH to prevent or minimize growth. These include foods such as fruit beverages and salad dressings.

TEMPERATURE

Bacteria grow rapidly between 41 °F and 135 °F. This range is known as the *temperature danger zone*. Bacterial growth is limited when food is held above or below this zone.

TIME

Bacteria need time to grow. The longer foods remain in the temperature danger zone, the higher the risk of bacterial growth.

OXYGEN

Organisms can be classified as *aerobic* or *anaerobic*. Aerobic organisms require air (oxygen) for growth and will not grow without air. These include yeasts and molds and many bacteria. A common practice is to hot-fill foods and seal them so that a slight vacuum is formed in the package. Many foods have instructions to "Refrigerate After Opening" because yeasts and molds can grow once the product is opened and is exposed to air.

MOISTURE

Bacteria grow well in foods with high levels of moisture. The amount of moisture available in food is called *water activity* (A_w). The A_w scale is 0.0 to 1.0. The more moisture is available in a food, the higher the value. For example, water has a water activity of 1.0.

FAT TOM

You can help keep food safe by controlling FAT TOM. The two factors that we can control the most are time and temperature. Time and temperature are connected and go hand in hand. Controlling time and temperature means limiting how long foods stay in the temperature danger zone.[1]

BACTERIUM OR VIRUS: WHAT'S THE DIFFERENCE?

A bacterium is a microscopic single-celled organism and distinctly different from a virus. The plural of *bacterium* is *bacteria*. Bacteria occur everywhere life exists. They possess tough, rigid outer cell walls through which they absorb their food. Some bacteria have slimy outer capsules, and some may have whiplike flagella that propel them through liquids. If flagella are positioned all around a bacterium, it is called *peritrichous*; if the flagella are at each end, the bacterium is called *lophotrichous*. Some bacteria simply drift in air or water currents.

(Continues)

BACTERIUM OR VIRUS: WHAT'S THE DIFFERENCE? (continued)

Bacteria generally reproduce by splitting into two. This is called *binary fission*, and it may occur once every 15 minutes to 30 minutes. Under favorable conditions, one bacterium could form more than 150 trillion bacteria in 24 hours, although this rarely happens. Bacteria are numerous and tough. A pinch of soil contains millions, and some bacteria can survive freezing, intense heat, drying, and some disinfectants. To survive adverse conditions, bacteria form spores that can remain active for years.

Bacteria can be classified by their shapes. Those shaped like spheres are called *cocci*. Those shaped like rods are called *bacilli*. *Spirillum* have spiral shapes. *Vibrios* are comma shaped. *Mycobacteria* are incredibly small rods. *Flexibacter* form long, thin rods.

Many bacteria perform useful functions for humans. For example, helpful bacteria include those responsible for decay, sewage treatment, cheese and yogurt production, and the nitrification process. Pathogenic bacteria cause disease. Bacterial infections can be treated with antibiotics or similar drugs, and some illnesses can be prevented by vaccination.

A virus is smaller and simpler than a bacterium. In fact, a virus is so small and simple that it is considered less a living organism than an inanimate particle. Viruses are so small they cannot be seen with ordinary light microscopes and must be examined through electron microscopes.

Viruses live and reproduce inside living cells. They depend on these cells to reproduce, but some can survive for quite some time outside cells and even through freezing and drying. Because viruses live inside the cells of plants and animals, chemical treatment is often out of the question because this would kill the host cell. Some drugs relieve the symptoms that viruses produce, but the only effective way to control a viral infection is to remove the infected individual. The affected individual's own immune system must produce antibodies to counteract invading viruses and stop the infection. Vaccination provides one way to prevent viral infections.

To learn more about bacteria and viruses, visit the following Web sites.

- CELLS Alive! www.cellsalive.com
- Virus or Bacterium? http://www.microbeworld.org/what-is-a-microbe/virus-or-bacterium
- How to Know the Difference Between Bacteria and Viruses: http://www.wikihow.com/Know-the-Difference-Between-Bacteria-and-Viruses
- Boston University: www.bu.edu
- National Institute of Allergy and Infections Diseases Microbes: http://www.niaid.nih.gov/topics/microbes/Pages/default.aspx

Anaerobic organisms only grow in the absence of air. One such organism is *Clostridium botulinum*, which produces a toxin that can be lethal if ingested. Where foods are packed under a vacuum, the protection against *C. botulinum* growth is to heat-treat the food (sterilization) to a time and temperature where any organisms present are destroyed.

Moisture Availability. Organisms need free moisture in order to grow. Drying foods removes the available moisture and prevents growth of bacteria, yeasts, and molds. Control of water activity is a common method of preventing the outgrowth of microorganisms. Water activity is a measure of free (unbound) water available for chemical and biological activity.

The variable A_w is the vapor pressure of a food product at a specified temperature as well as a measure of the relative humidity of the food where

$$A_w = 1.0 = 100\% \text{ relative humidity}$$

$$A_w = 0.0 = 0\% \text{ relative humidity}$$

Foods generally have water activities that range between 0.1 for dried foods to 0.96 for fluid foods.

Different organisms require different levels of free water to grow, with bacteria requiring more free water than yeasts and molds. A rule of thumb is that bacteria do not grow at an a_w of less than 0.85 and yeasts and molds do not grow at an a_w of less than 0.65. But some exceptions exist.

Intermediate moisture foods depend on the addition of large quantities of sugars to reduce the a_w to below 0.85 and on packaging to prevent the growth of yeasts and molds.

Nutrient Availability. Most foods contain adequate nutrients to support the growth of microorganisms, especially foods that contain both a fermentable carbohydrate and a protein source. Sugar is the most common carbon source. In starch-based foods, the action of amylases will frequently increase the available sugar source.

Storage Temperature. Organisms can be classified on the basis of their ability to grow at different temperatures:

- *Psychrophiles* grow best at temperatures <50 °F (<10 °C).
- *Mesophiles* grow best at ambient temperatures 77 °F to 95 °F (26 °C to 35 °C)
- *Thermophiles* grow best at temperatures >104 °F (<40 °C).
- **Psychrotrophic** tolerate low temperatures and can grow under refrigeration
- ***Thermotrophic*** tolerate high temperatures and can grow at 131 °F to 140 °F (55 °C to 60 °C)

Most pathogenic bacteria are mesophilic. A few, including *Listeria monocytogenes* and *C. botulinum* type E, can grow under refrigeration conditions.

Organisms that spoil refrigerated products are generally psychrotrophic and frequently belong to the genus *Pseudomonas*.

Lag Time, Generation Time, and Numbers. The amount of time required for an organism to reach the log-growth phase is termed the **lag time**, and the time required to double the population of the organisms is termed the **generation time**.

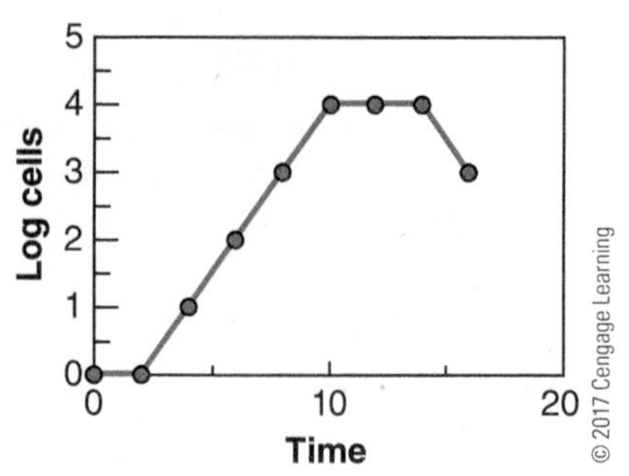

FIGURE 26-1 Logarithmic growth of bacteria over time.

The amount of time required for an organism to reach a specific number is dependent upon the initial population, the lag time, and the generation time. Under ideal growth conditions, microorganisms can double their number in about 30 minutes. As conditions move away from their optimum, the generation time is decreased—until eventually no growth occurs (Figure 26-1).

In processed foods, surviving organisms have generally been stressed and commonly exhibit lag times of 2 to 4 days. In fermented foods, where cultures are added, the organisms are added in an active-growth phase, and there is little or no lag time.

A rule of thumb for the numbers of organisms required to produce toxins or to produce desired or undesired flavors is 1 million per gram. For foodborne toxins such as caused by *Clostridium botulinum*, *Staphylococcus aureus*, and *Bacillus cereus*, there must be a large number produced. Similarly, the numbers of organisms required for fermented foods (such as yogurt) is also enormous.

In the case of organisms that cause infectious disease—for example, *Escherichia coli* 0157, *Listeria monocytogenes*, and *Salmonella*—the numbers of organisms that can cause disease can be tiny. The Food and Drug Administration (FDA) generally requires that no organisms can be recovered from 100 g of food after suitable incubation.

Under ideal growth conditions, it can take a relatively short period of time to increase from a single organism to 1 million per gram (20 generations). At four generations per hour, the food can reach more than 1 million in just 5 hours. If the initial load is 1,000 per gram, then the time required to reach 1 million would be 2.5 hours (10 generations).

MICROBIOLOGICAL METHODOLOGY

Testing foods for the presence of pathogenic microorganisms is critically important. Although 100% of the food cannot be tested, it can be deemed safe through proper auditing of food supplies. In many instances, the pathogenic microorganisms are present in minute numbers, but that is all some of these pathogens need to transmit disease or illness. For that reason, the presence of other microorganisms is monitored. These microorganisms provide an index of the sanitary quality of the product and may serve as an indicator of potential for the presence of pathogenic species. *E. coli* is commonly employed as an indicator microorganism. Because *E. coli* is a coliform bacteria common to the intestinal tract of humans and animals, its relationship to intestinal food-borne pathogens is high.

Total counts of microorganisms are also an indication of the sanitary quality of a food. Referred to as the **standard plate count (SPC)**, this total count of viable microbes reflects the handling history, state of decomposition, and degree of freshness of a food. Total counts may be taken to indicate the type of sanitary control exercised in its production, transport, and storage. Most foods have standards or limits for total counts. This is especially true for milk.

In adopting microbiological standards to milk, the first two concerns are product safety and shelf life. The following bacterial counts are standards for milk as recommended by the U.S. Public Health Service.

- Grade A raw milk for pasteurization: Not to exceed 100,000 bacteria per milliliter (ml) before commingling with other produced milk, and not to exceed 300,000 per milliliter as commingled milk before pasteurization.
- Grade A pasteurized milk: Not more than 20,000 bacteria per milliliter, and not more than 10 coliforms per milliliter.

The objective of pasteurization is to reduce the total microbial load, or SPC. In addition, pasteurization must destroy all pathogens that may be carried in the milk from the cow, particularly undulant fever, tuberculosis, Q fever, and other diseases transmittable to humans. This is accomplished by setting the time and temperature of the heat treatment so that certain heat-resistant pathogens—specifically, *Mycobacterium tuberculosis* and *Coxiella burnetii* (causative agents of Q fever and tuberculosis, respectively)—would be destroyed if present. Milk pasteurization temperatures are sufficient to destroy all yeasts, molds, and many of the spoilage bacteria.

A low SPC does not always represent a safe product. It is possible to have a low count of SPC in foods in which toxin-producing organisms have grown. These organisms produce toxins that remain stable under conditions that may not favor the survival of the microbial cell. In adopting microbiological standards, the first concern is product safety followed by shelf life.

Standard plating methods can take several days. New tests are being developed that will detect specific types of microorganisms in a matter of hours. These new tests are based on being able to detect a specific type of DNA from the microorganisms of interest.

PROCESSING AND HANDLING

Keeping microbial loads at minimal levels is essential to provide safe food of high quality. This requires care in food handling and minimizing microorganisms in the product during processing. One key to this goal is preventing food contamination by contact with equipment (or food contact) surface. Cleaning and sanitizing are important steps in any food plant's operations and will become more important in the future as the industry deals with new and emerging microorganisms such as *E. coli* 0157. That bacteria first became associated with food-borne infections in the 1980s.

Microbial destruction is achieved by heat or by chemicals. The most common method of making foods safe is to use thermal processing to eliminate pathogens and then use good sanitation practices to prevent them from reentering the food afterward. Spores are more resistant than vegetative cells and thus require more heat to kill. Two general heat processes are used:

- *Pasteurization*, or heating to a specific temperature for a specific time to kill the most heat-resistant vegetative pathogen
- *Sterilization*, or heating to a specific temperature for a specific time to kill the most heat-resistant spore-forming organism

Neither process kills all organisms in food, and nonpathogenic spoilage organisms can survive to some degree. Because pasteurization does not kill spores, pasteurized products are kept under refrigeration to control the growth of surviving spore formers that grow well at ambient temperatures.

foods are more popular than ever before. Today, consumers are more likely to purchase foods that need little or no preparation or cooking before consumption. This trend means that if these foods are contaminated with harmful microorganisms, there may be no consumer preparation step that will reduce or eliminate the hazard. Because of this, it is critically important to control food-borne pathogens during the manufacturing and storage of foods, especially for ready-to-eat foods.

Along with changes in food manufacturing and marketing, the growth in scientific understanding of food-borne illness has been even more significant. The significance of pathogens such as *E. coli* O157:H7, *Campylobacter jejuni*, *Cryptosporidium parvum*, *Cyclospora cayetanensis,* and *Norovirus* were not as well understood in the 1980s as they are today. In addition to these new pathogens, familiar pathogens such as *Salmonella* continue to present a challenge. Modern good manufacturing practices can play a role in reducing the risk of these pathogens as well.

Along with good agricultural practices, good laboratory practices, and good clinical practices, good manufacturing practices are overseen by the FDA in the United States. Good manufacturing practice guidelines provide guidance for manufacturing, testing, and quality assurance to ensure that a food or drug product is safe for human consumption.

All guidelines observe the same basic principles.

- Hygiene: Food manufacturing facility must be clean and hygienic.
- Controlled environmental conditions must prevent the cross-contamination of food.
- Manufacturing processes must be clearly defined and controlled.

SCIENCE CONNECTION!

How many foods get pasteurized? Which foods do you think are? Hypothesize and research your answers. How many did you guess correctly?

GOOD MANUFACTURING PRACTICES

The food industry has undergone considerable change in the past 30 years since **Good Manufacturing Practices (GMPs)** regulations were instituted. Ready-to-eat foods now represent a larger and growing portion of the American diet. Ready-to-eat fresh produce salads are a popular replacement for salads prepared in the home. Refrigerated foods and heat-and-serve

- Any changes in manufacturing processes must be evaluated.
- Instructions and procedures must be written in clear and unambiguous language (good documentation practices).
- Operators must be trained to carry out and document procedures.
- The distribution of a food or drug must minimize any risks to their quality.
- A system must be available to recall any product from sale or supply.

RODENTS, BIRDS, AND INSECTS

Rodents carry many diseases and parasites that can be transmitted to humans. These diseases and parasites include leptospirosis, salmonellosis, tapeworms, and trichinosis. Rodents also deposit excreta, urine, and other filth on food products in and around food facilities. They will also gnaw on materials that they will use to build nests. Rodents contaminate much more than they eat.

Some rodents can walk along telephone wires or leap horizontally 18 feet. They can squeeze through gaps the width of a pencil or drop 50 feet without being killed. Their instinct for survival is high. They are extremely prolific.

Birds also carry diseases and parasites potentially hazardous to people. They are capable of flying through any open window, door, or other gap in a building, and, like rodents, will leave droppings that can contaminate a plant and its food products.

Insects seek heat, moisture, and darkness and can be even more elusive than rodents or birds. They leave trails in the dust and can be spotted around likely insect hideouts such as holes, damp places, boxes, and seams in bags and folds of paper. Some insects—cockroaches, for example—have a highly developed survival instinct and are extremely adaptable. They can develop resistance to poisons within a few insect generations. Insects are even more prolific than rodents. With their hairy legs, they spread dirt, debris, and bacteria.

A reputable pest control company or exterminator is generally the most cost-effective method to deal with invasions of rodents, birds, or insects. When professional pest control is used, the owner or manager of a food-industry facility should frequently monitor the activities of the pest control representative and know not only what poisons are being used but also the locations of bait stations and traps. Unfortunately, no all-purpose pesticide for the food industry yet exists.

Investing in building and grounds maintenance remains the best practice to solve pest problems. Extermination is a poor second choice for pest control and it will cost as much if not more. The following guidelines will help prevent a rodent, bird, or insect problem at a food-industry site:

- Keep all grounds surrounding buildings clear of weeds, grass, brush, and standing water.
- Seal all windows and doors tightly.
- Fit windows with fine mesh screens.
- Fill all holes and cracks in the building.
- Clean the building and equipment on a regular basis.
- Fix any leaks in the roof.
- Eliminate areas of food buildup.
- Eliminate empty space in or around equipment.
- Make sure trash, debris, and clutter are picked up frequently.
- Regularly inspect using a detailed checklist.
- Cover all garbage containers.
- Keep building humidity as low as possible.
- Maintain proper building temperature.
- When storing all foods, follow FDA guidelines.

Preventive and control measures avoid potentially costly, mandated adjustments that might arise when an FDA investigator visits. This also ensures that only quality, safe food products find their way to the consumers.

CLEANING AND SANITIZING

For a food-processing plant, cleaning and sanitizing may be the most important aspects of a hygiene program. Sufficient time should be given to outline proper procedures and parameters. The detailed procedures must be developed for all food-product contact surfaces (equipment, utensils, and so on) as well as for nonproduct surfaces such as nonproduct portions of equipment, overhead structures, shields, walls, ceilings, lighting devices, refrigeration units, and **heating, ventilation, and air-conditioning (HVAC)** systems. In short, anything that could affect food safety should be sanitized on a regular basis.

Cleaning frequency must be clearly defined for each process line—for example, daily, after production runs, or more often if necessary. The type of cleaning required must also be identified.

The objective of cleaning and sanitizing food-contact surfaces is to remove food (nutrients) that bacteria need to grow and to kill any bacteria that are present. The clean, sanitized equipment and surfaces must drain dry and be stored dry to prevent bacterial growth. Necessary equipment (brushes and so on) must also be clean and stored in a clean, sanitary manner.

Cleaning and sanitizing procedures must be evaluated for adequacy through evaluation and inspection procedures. How well employees adhere to prescribed written procedures should be continuously monitored by inspection, swab testing, and direct observation, and records should be maintained to evaluate long-term compliance. The correct order of events for cleaning or sanitizing food-product contact surfaces is rinse, clean, rinse, sanitize.

Cleaning

Cleaning is the complete removal of **food soil** using appropriate detergent chemicals under recommended conditions. People employed in food production need to have a working understanding of the nature of the different types of food soil and the chemistry of their removal.

Equipment can be categorized with regard to the cleaning method as follows.

- Mechanical cleaning: Often referred to as clean in place, this requires no disassembly or partial disassembly.
- **Clean out of place (COP)**: Items can be partially disassembled and cleaned in specialized COP pressure tanks.
- Manual cleaning: Total disassembly is required for both cleaning and inspection.

Food soil is generally defined as unwanted matter on food-contact surfaces. Soil is visible or invisible. The primary source of soil is from the food product being handled. However, minerals from water residue and residues from cleaning compounds contribute to films left on surfaces. Microbiological **biofilms** also contribute to soil buildup on surfaces.

Because soils vary widely in composition, no single detergent is capable of removing all types. Many complex films contain combinations of food components, surface oil or dust, insoluble cleaner components, and insoluble hard-water salts. Their solubility properties depend on such factors as heat effect, age, dryness, and time, among others. Soils may be classified as:

- Soluble in water (sugars, some starches, most salts)
- Soluble in acid (limestone and most mineral deposits)
- Soluble in alkali (protein, fat emulsions)
- Soluble in water, alkali, or acid

Sanitation

Appropriate and approved **sanitization** procedures are processes. The duration or time as well as the chemical conditions must be described. The official definition of sanitizing food product contact surfaces from the Association of Official Analytical Chemists is a process that reduces contamination by 99.999% (5 logs) in 30 seconds.

Two general types of sanitization are *thermal* (heat) and *chemical*. Thermal sanitization uses hot water or steam for a specified temperature and contact time. Chemical sanitization uses an approved chemical sanitizer at a specified concentration and contact time. As with any heat treatment, the effectiveness of thermal sanitizing depends on many factors, including initial contamination load, humidity, pH, temperature, and time.

The use of steam as a sanitizing process has limited application. It is generally more expensive than alternatives, and regulating and monitoring contact temperature and time is difficult. Further, the principal by-product of steam—condensation—can complicate other cleaning operations.

Hot-water sanitizing through immersion (small parts, knives, etc.), spray (dishwashers), and circulating systems—is commonly used. The time required is determined by the temperature of the water (Figure 26-2). When using hot water in

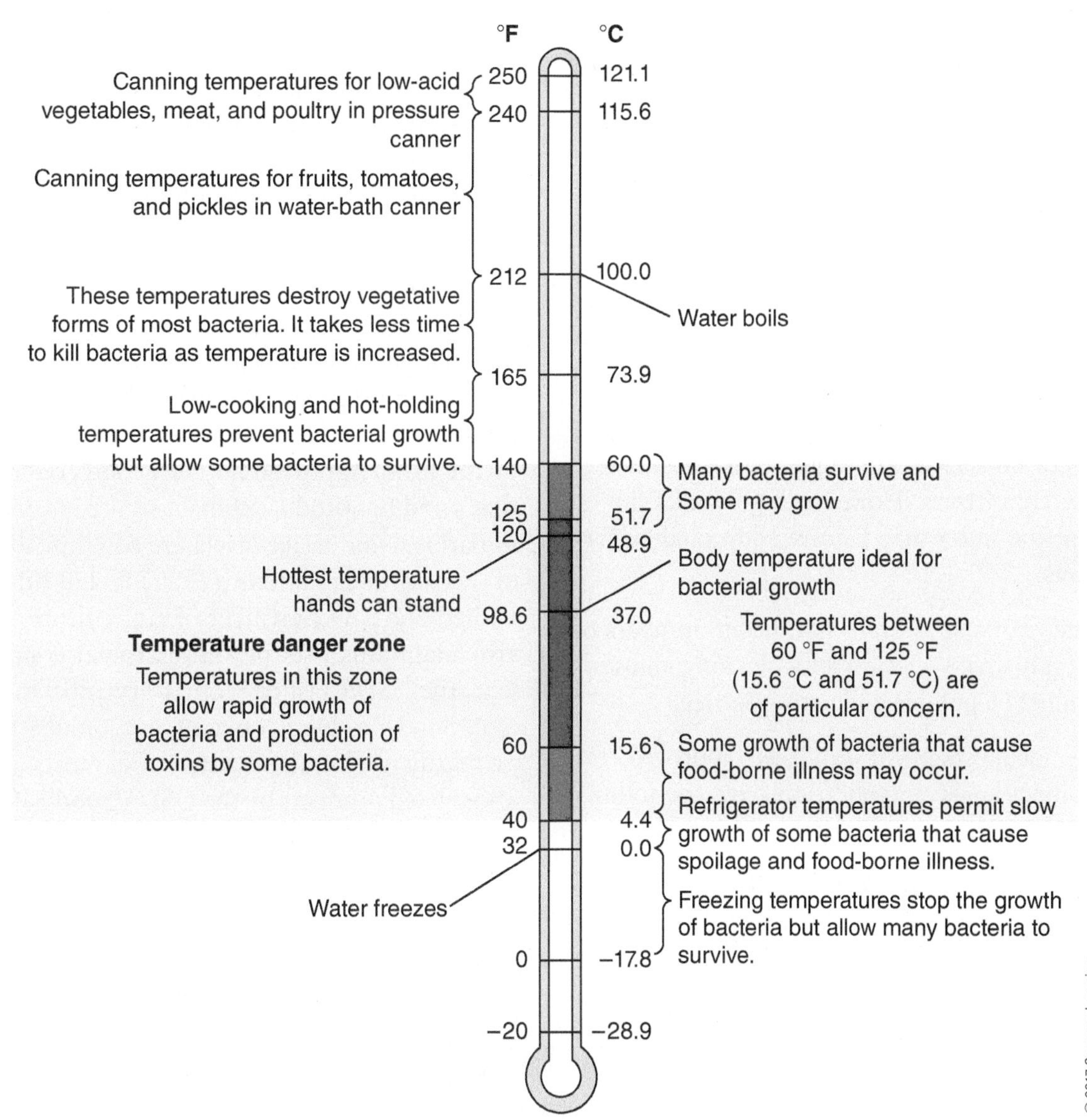

FIGURE 26-2 Relationships between temperature and microbial growth.

dishwashing and utensil-sanitizing applications, typical regulatory requirements (FDA Food Code, 1995) specify immersion for at least 30 seconds at 170 °F (77 °C) for manual operations and a final rinse temperature of 165 °F (74 °C) in single-tank, single-temperature machines and 180 °F (82 °C) for other machines.

Many state regulations require a utensil surface temperature of 160 °F (71 °C) as measured by an irreversibly registering temperature indicator in utensil-washing machines.

Water comprises approximately 95% to 99% of cleaning and sanitizing solutions. Water functions to carry the detergent or sanitizer to the surface and carry soils or contamination from the surface.

Impurities in water can drastically alter the effectiveness of a detergent or a sanitizer. Water hardness is the most important chemical property with a direct effect on cleaning and sanitizing efficiency. (Other impurities can affect food-contact surfaces or affect soil deposit properties or film formation.)

Water pH ranges generally from 5.0 to 8.5. This range is of no serious consequence to most detergents and sanitizers. However, highly alkaline or highly acidic water may require additional buffering agents.

Water can also contain significant numbers of microorganisms. Water used for cleaning and sanitizing must be potable and pathogen free.

The ideal chemical sanitizer approved for food-contact surface application have the following characteristics or properties:

- A wide range or scope of activity
- The ability to destroy microorganisms rapidly
- Stability under all types of conditions
- Tolerance of a broad range of environmental conditions
- Ready solubility with some detergency
- Low toxicity and corrosiveness
- Low cost

No sanitizer on the market meets all of these criteria. For this reason, every chemical sanitizer must be evaluated for its properties, advantages, and disadvantages for each specific application.

Regulatory Considerations

The regulatory concerns for chemical sanitizers are their antimicrobial activity or efficacy, the safety of residues on food-contact surfaces, and environmental safety. Users must follow regulations that apply for each chemical usage situation. The registration of chemical sanitizers and antimicrobial agents for use on food and food-product contact surfaces as well as nonproduct contact surfaces is through the U.S. Environmental Protection Agency (EPA). Before approval and registration, the EPA reviews efficacy and safety data and product labeling information.

The FDA is primarily involved in evaluating residues from sanitizer use that could enter the food supply. Thus, any antimicrobial agent and its maximum usage level for direct use on food or on food-product contact surfaces must be approved by the FDA. Approved no-rinse food contact sanitizers and nonproduct contact sanitizers, their formulations, and usage levels are listed in the Code of Federal Regulations (21 CFR 178.1010). The United States Department of Agriculture (USDA) also maintains lists of antimicrobial compounds (e.g., the USDA List of Proprietary Substances and Non Food Product Contact Compounds) that are primarily used in the regulation of meats, poultry, and related products by the USDA Food Safety and Inspection Service (Figure 26-3).

HACCP AND FOOD SAFETY

The Food and Drug Administration has adapted a food-safety program developed in the 1960s for astronauts for much of the U.S. food supply. The

Safe Handling Instructions

This product was inspected for your safety. Some animal products may contain bacteria that could cause illness if the product is mishandled or cooked improperly. For your protection, follow these safe handling instructions.

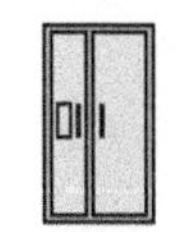

Keep refrigerated or frozen.
Thaw in refrigerator or microwave.

Keep raw (meats or poultry) separate from other foods. Wash working surfaces (including cutting boards), utensils and hands after touching raw (meat or poultry).

Cook thoroughly.

Refrigerate leftovers within 2 hours.

FIGURE 26-3 Safe handling instructions on many meat and poultry products.

program for astronauts focused on preventing hazards that could cause food-borne illnesses by applying science-based controls from raw material to finished products. FDA new system would do the same.

Traditionally, industry and regulators have depended on spot checks of manufacturing conditions and random sampling of final products to ensure safe food. This system, however, tends to be reactive rather than preventative and can be less efficient than the new system.

As noted in Chapter 19, the new system became known as Hazard Analysis and Critical Control Point, or HACCP (pronounced "hass-sip"). Many of its principles already are in place in the FDA-regulated low-acid canned food industry and have been incorporated into FDA Food Code. The Food Code serves as model legislation for state and territorial agencies that license and inspect food services, retail food stores, and food vending operations in the United States.

The USDA also has established HACCP for the meat and poultry industry. Larger establishments were required to start using HACCP by late January 1998. Smaller companies had until late January 1999, and the smallest plants had until late January 2000. (The USDA regulates meat and poultry; FDA all other foods.)

The FDA is considering the adoption of HACCP regulations as a standard throughout much of the rest of the U.S. food supply. The regulations would cover both domestic and imported foods.

To help determine the degree to which such regulations would be feasible, the agency is now conducting a pilot HACCP program with volunteer food companies that make cheese, frozen dough, breakfast cereals, salad dressings, and other products.

HACCP uses the following seven steps.

1. *Analyze hazards.* Potential hazards associated with a food and measures to control those hazards are identified. The hazard could be biological (e.g., microbial), chemical (e.g., a pesticide), or physical (e.g., ground glass or metal fragments).
2. *Identify critical control points.* These are points in a food's production—from its raw state through processing and shipping to consumption by the consumer—at which a potential hazard can be controlled or eliminated. Examples are cooking, cooling, packaging, and metal detection.
3. *Establish preventative measures with critical limits for each control point.* For a cooked food, for example, this might include setting the minimum cooking temperature and time required to ensure the elimination of any microbes.
4. *Establish procedures to monitor the critical control points.* Such procedures might include determining how cooking time and temperature should be monitored and by whom.

5. *Establish corrective actions to be taken when monitoring shows that a critical limit has not been met.* For example, if the minimum cooking temperature has not been met, food should be reprocessed if possible or thrown away.
6. *Establish procedures to verify that the system is working properly.* For example, testing time-and-temperature recording devices must verify that a cooking unit is working properly.
7. *Establish effective recordkeeping to document the HACCP system.* This would include recording hazards and their control methods as well as monitoring safety requirements and actions taken to correct potential problems.

Each step would have to be backed by sound scientific knowledge—for example, published microbiological studies.

New challenges to the U.S. food supply have prompted FDA to consider adopting a HACCP-based food-safety system. One of the most important challenges is the increasing number of new food pathogens. For example, between 1973 and 1988, bacteria not previously recognized as important causes of food-borne illness—such as *E. coli* O157:H7 and *Salmonella enteritidis*—became more widespread (Figure 26-4).

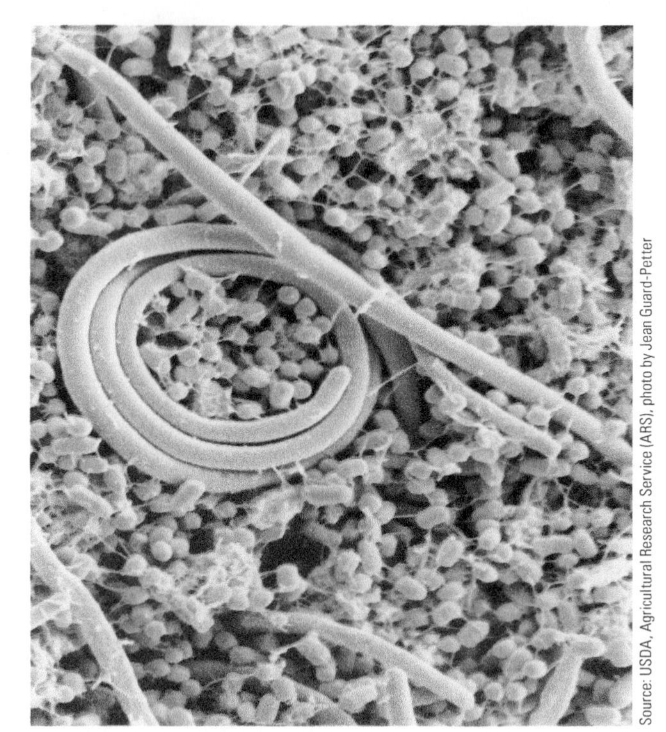

Source: USDA, Agricultural Research Service (ARS), photo by Jean Guard-Petter

FIGURE 26-4 Microscopic view of the bacterium *Salmonella enteritis.*

Public health officials and agencies are also increasingly concerned about chemical contamination of food—for example, the effects of lead on the nervous system.

E. COLI O157:H7 AND JACK IN THE BOX RESTAURANTS

In 1993, 623 people in the western United States fell ill with a little-known bacteria called *E. coli* O157:H7. Ultimately, four children died from their infections, and many others suffered long-term medical complications. The culprit was later traced to undercooked hamburger served at Jack in the Box restaurants. This outbreak thrust food-borne illness onto the national stage as a real and present threat, sparking a change in the way Americans and the government treat this issue.

Michael Taylor had been the USDA new Food Safety and Inspection Service Administrator for just 6 weeks when he took to the podium at the American Meat Institute's annual convention in San Francisco on September 29, 1994, and changed the beef industry forever.

Every point in the beef chain, from slaughterhouse to marketplace, Taylor said, required scrutiny to protect consumers from dangerous microbes.

Between the efforts of industry leaders and the baseline government regulations, action inspired by the Jack in the Box outbreak have resulted in major reductions in *E. coli* O157:H7 infections in the past 20 years.

(*Continues*)

E. COLI O157:H7 AND JACK IN THE BOX RESTAURANTS (continued)

Dave Theno, PhD, the man hired to take over the company's safety system after the outbreak, introduced to Jack in the Box the systematic approach to food safety known as HACCP. Theno's plan targeted four major components of the business: suppliers, transportation, restaurant staff, and corporate infrastructure.

Jack in the Box now required random *E. coli* sampling of all beef supplies. The company went from having no microbial testing at all to conducting a random test every 15 minutes.

Jack in the Box now also had a checklist for slaughterhouses. During transportation, patties were to be strictly held at safe temperatures from plant to burger grill, at which point they would be heated to an internal temperature of 155 °F by employees versed in food-safety training. Corporate incentives were structured around food-safety priorities.

For the record, the company has not had an outbreak since 1993.

Granted, beef still causes *E. coli* outbreaks, and product recalls still occur every year. For the most part, however, multimillion-pound recalls are a thing of the past.

Adapted from Food Safety News, "Jack in the Box and the Decline of *E. coli*," by J. Andrews. For the full article, visit http://www.foodsafetynews.com/2013/02/jack-in-the-box-and-the-decline-of-e-coli/#.VaRSVI9VjF8.

Another important factor when implementing a HACCP system is that the size of the food industry and the diversity of its products and processes have grown tremendously—in the amount of domestic food manufactured and the number and kinds of foods imported. At the same time, the FDA and state and local agencies have the same limited numbers of resources to ensure food safety.

SCIENCE CONNECTION!

Research food-borne illness outbreaks. What do you find? Have there been any recent outbreaks? What diseases caused the outbreaks? What foods were associated with them? What was done about them?

HACCP offers a number of advantages. Some of the most prominent are the following:

- Focus on identifying and preventing hazards from contaminating food (see Figures 26-5, 26-6, and 26-7)
- Basis in sound science
- Permits more efficient and effective government oversight, primarily because the recordkeeping allows investigators to see how well a firm is complying with food-safety laws over a period of time rather than how well it is doing on any given day
- Places responsibility for ensuring food safety appropriately on the food manufacturer or distributor
- Helps food companies compete more effectively in world markets

BIOTECHNOLOGY

Over the last several decades, research and development in the field of **biotechnology** has increased. Biotechnology is not new. Civilization has cultivated plants and bred animals to reach desired traits for thousands of years. Biotechnology methods span a variety of techniques. There are different ways of moving genes to produce desirable traits, for example. A more traditional method for both plants

© Sebastian Czaphnik/Dreamstime.com

FIGURE 26-5 Wearing and using proper equipment—an important part of food safety.

and animals is selective breeding in which a plant or animal with a desired trait is chosen and bred to produce more plants or animals with desired traits. With the advancement of technology, genes that express the desired trait is physically moved or added to the new plant in a laboratory. Often this process is performed on crops to produce insect- or herbicide-resistant plants. A **transgenic organism** is an organism whose genes have been artificially altered through genetic engineering. A **transgenic crop** is also known as a **genetically modified organism (GMO)**.[2]

Genetically Modified Foods

Some people use the term *genetically modified organism* and others use the term *genetic engineering.* Regardless of its name, genetic engineering is not a new science, but it is now making profound genetic changes more easily. Insulin, for example, has now been genetically engineered using the insulin gene from the intestines of pigs as inserts into the DNA of bacteria. After the bacteria grows and produces insulin, the insulin is purified and used for medical uses.

This genetic engineering also extends into food crop production. In 1901, for example, *Bacillus thuringiensis* (Bt) was first discovered as a natural pesticide. Research and development over the next 50 years led to the discovery of how well Bt could specifically control lepidoteran insects (the moth family) because of the toxic patasporal crystal. Bt became a registered pesticide with the EPA in 1961. In the 1980s, the use of Bt increased as insects became resistant to synthetic insecticides. Because Bt is organic and targets specific insects and does not remain in the environment, industries began earnest research on ways to adopt Bt for other commercial uses. With the advancements of molecular biology in the 1990s, it became feasible to move the genes that encode the toxic crystals into a plant. Bt corn, the first genetically engineered plant, was registered with the EPA in 1995.

Today, there are genetically modified crops of corn, cotton, soybeans, canola, sugar beet, and alfalfa planted and grown throughout the world. Potatoes, apples, and rice research plots are currently being tested. Scientists continue to research ways to provide a safe and wholesome food supply. Growing more food on less land is a growing expectation as the world's population continues to grow. Hopefully, the use of biotechnology will continue to reduce pesticide usage and increase food production.[3]

Regulations. The FDA regulates GM foods the same way it regulates traditional foods. Foods from genetically engineered plants must meet the same requirements, including safety requirements, as foods from traditionally bred plants. FDA has a consultation process that encourages developers of these plants to consult with FDA before marketing. This process helps developers determine the necessary steps to ensure their food products are

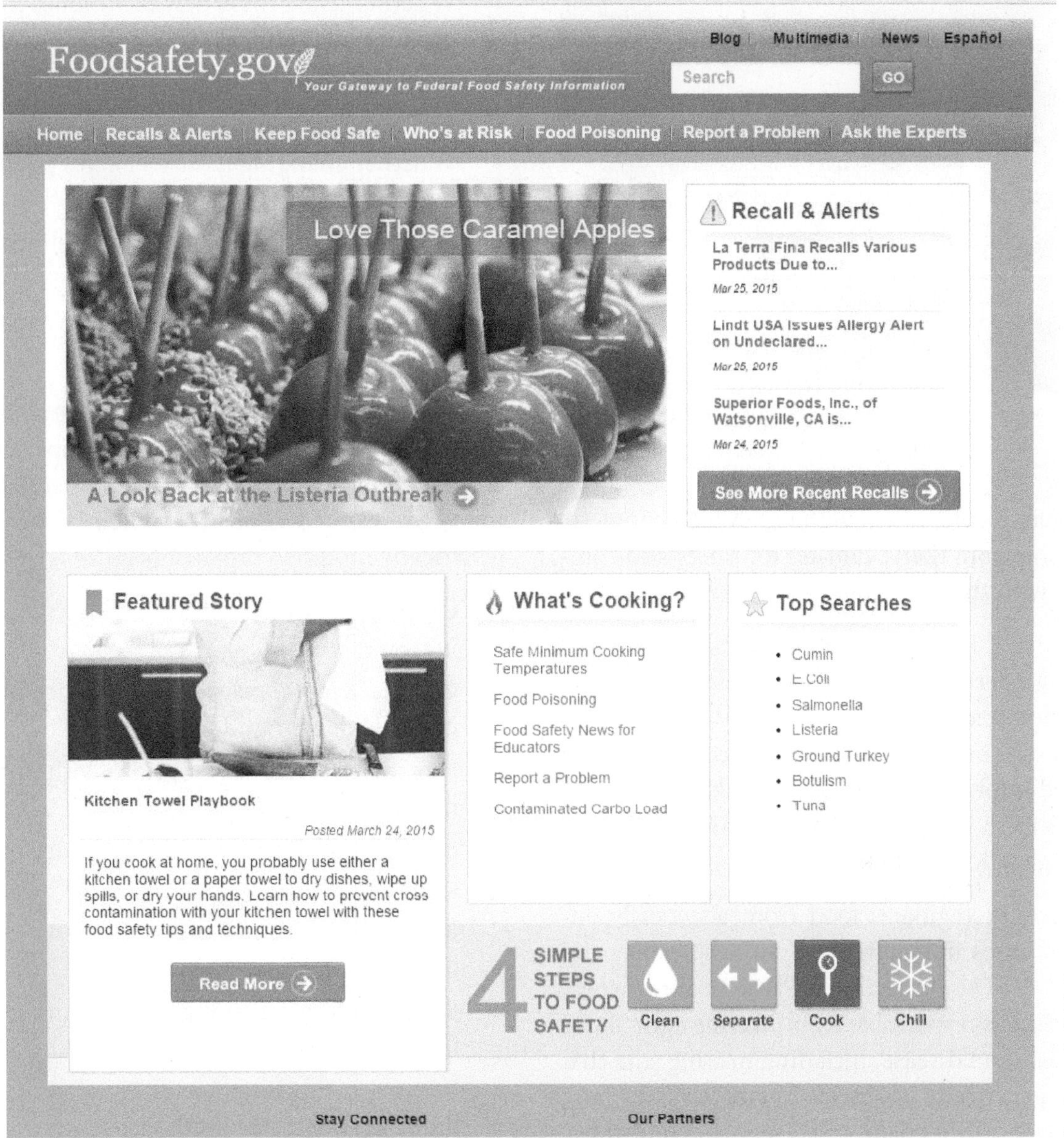

Source: www.foodsafety.gov

FIGURE 26-6 The U.S. government maintains a Web site promoting food safety.

safe and lawful. The goal of the consultation process is to ensure that any safety or other regulatory issues related to a food product are resolved before commercial distribution. No food from a genetically engineered plant grown in the United States and evaluated by the FDA through its consultation process has gone on the market without FDA questions about safety being resolved.

Evaluating the safety of food from a genetically engineered plant is a comprehensive process that includes several steps. Generally, the developer identifies the distinguishing attributes of new genetic traits and assesses whether any new material that a person consumed in food made from the new plants could be toxic or allergenic. The developer also compares the levels of nutrients in the new genetically engineered plant to traditionally bred plants. This typically includes assessing such nutrients as fiber, protein, fat, vitamins, and minerals. The developer includes this information in

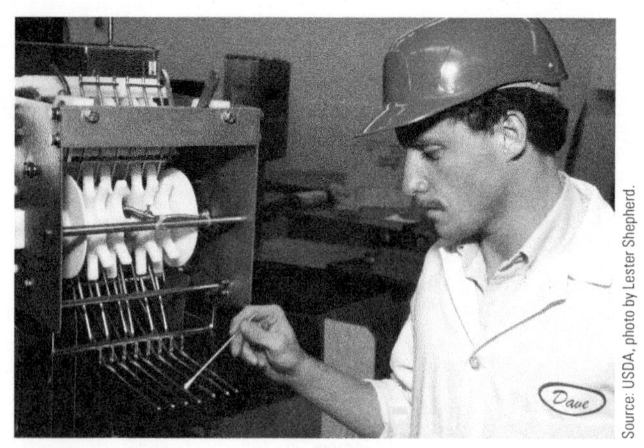
Source: USDA, photo by Lester Shepherd.

FIGURE 26-7 Food-processing equipment being inspected after cleaning.

a safety assessment, which the FDA biotechnology evaluation team then evaluates for safety and compliance with the law.

Teams of FDA scientists knowledgeable in genetic engineering, toxicology, chemistry, nutrition, and other scientific areas carefully evaluate the safety assessments, taking into account relevant data and information. It typically takes developers 13 years to 15 years and more than $125 million to bring a genetically modified food to market.

The FDA regulates food from genetically engineered crops in conjunction with the USDA and the EPA. USDA Animal and Plant Health Inspection Service is responsible for protecting agriculture from pests and disease, including making sure that all new genetically engineered plant varieties pose no pest risk to other plants. EPA regulates pesticides, including those bioengineered into food crops, to make sure that pesticides are safe for human and animal consumption and do not pose unreasonable risks to human health or the environment.[4]

Concerns of GMOs. Some consumers are concerned about the safety and environmental effects of genetically modified foods. The following list shows some of these concerns:

- Not enough research has been done on the long-term effect of genetically modified organisms.
- Native species will be contaminated because of cross pollination with transgenic crops.
- Whether using these new techniques in food production is ethical.
- Whether an unhealthy or harmful food will be created.
- Whether changing nature violates people's religious beliefs.
- Will genetically modified species eliminate natural species.
- Genetically modified seeds are too expensive for many farmers to plant.

Benefits of Genetically Modified Organisms. Just as some consumer groups are concerned about using genetically engineered crops, many are excited about the potential benefits of genetically modified organisms (GMOs):

- Crops are disease resistant and have increased yields.
- Crops are insect resistant and reduce the use of pesticides sprayed on crops.
- Provide healthier food options with increased nutritional value.
- Increased shelf life of foods to reduce food spoilage.
- Reduce bruising and enzymatic browning of foods.
- Reduce or remove allergens from the food product.
- Increase yields for plants that are difficult to breed in traditional methods.

Even though biotechnology and the use of gene transfers has been around for 20 years, there is still much to learn about the technology. Continued research and regulation will be key to genetic engineering's future. Consumers, lawmakers, food scientists, and corporate interests will continue to work on viable options. Labeling GMO foods is just one issue currently under consideration. Media hype over GMO foods may not always portray accurate scientific information, so consumers will need to continue working to find information so they can make informed decisions about their food choices.[5]

SCIENCE CONNECTION!

There is a lot of propaganda in the news about GMOs. Find two articles that support GMO use and find two articles against it. After reading all the articles, what are your thoughts? Create a list of pros and cons for GMOs and then defend your position.

FFA FOOD SCIENCE AND TECHNOLOGY CAREER DEVELOPMENT EVENT

The FFA Food Science and Technology Career Development Event (CDE) includes practicums in which participants must perform sensory evaluations such as triangle tests and flavor and aroma identifications. It also challenges participants with food safety and quality practicums. Use the FDA Web site for more information while preparing for the Food Science FFA Career Development Event: www.fda.gov.

SUMMARY

Food safety concerns include pesticides, additives, and spoilage, but most food scientists and consumers focus on microorganisms in food. These microorganisms cause food-borne illnesses that vary in severity. Many people experience food-borne illnesses as something like a 24-hour flu, an upset stomach, or just simple diarrhea. For the youngest and oldest, however, these food-borne illnesses can be life-threatening. Microorganisms of concern include viruses, bacteria, parasites, and fungi.

To control microorganisms, food scientists must understand the factors that affect microbial growth. These include pH, oxygen availability, nutrient availability, moisture availability, storage temperature, lag time, and generation time. Because foods cannot be tested for all microorganisms, the presence of *E. coli* serves as an indicator that a food is not safe to consumer. *E. coli* is detected by the standard plate count.

The goal of proper processing and handling is to keep microbial loads to a minimum to provide safe, high-quality food. During processing, microbial destruction and control are achieved by heat or chemicals. Cleaning before sanitization is an important part of maintaining safe food in any processing environment. Depending on the type of processing, appropriate and approved sanitation procedures must be followed. The USDA, FDA, and EPA are all involved in regulatory considerations of food safety. HACCP is a food-safety program developed for the astronauts in the 1960s and 1970s. HACCP involves seven steps and is now being used to monitor and control food-borne diseases. One major area of focus of biotechnology is genetically modified organisms. Although several commercial crops have been genetically modified, the public is still uncertain about this use of technology.

REVIEW QUESTIONS

Success in any career requires knowledge. Test your knowledge of this chapter by answering these questions or solving these problems.

1. List three types of food soils.
2. Briefly list the seven steps involved in HACCP.

3. List four food-borne diseases.
4. What is a rule of thumb used to determine the numbers of organisms required to produce toxins?
5. Name three ways cross-contamination occurs.
6. Describe the four main microorganisms that concern the food industry.
7. Name the two general heat processes for microbial destruction.
8. Name the correct order of events for cleaning and sanitizing food-product contact surfaces.
9. What are the two types of sanitization used in food preparation?
10. List the five most commonly reported food preparation practices from the Centers for Disease Control that contribute to food-borne diseases.
11. Name three current GMO crops in production.

STUDENT ACTIVITIES

1. Choose one of the food-borne diseases in Table 26-2. Develop a short presentation on the disease and include an example of a disease outbreak.
2. Collect two current articles that relate to food safety issues. Summarize and report on what the articles described.
3. Develop a report or presentation on commonly used chemical sanitizers. Bring a label from the sanitizer to share with the class.
4. Invite a guest speaker from an industry to discuss the use of the HACCP process. Prepare questions for the guest speaker before the visit.

ADDITIONAL RESOURCES

- IFT: http://www.ift.org
- USDA Food Safety Inspection Service: http://www.fsis.usda.gov
- Pennsylvania State University—Food Science for Kids of All Ages! http://foodscience.psu.edu/public/kitchen-chemistry
- Foodsafety.gov: http://www.foodsafety.gov
- USDA Food Safety Inspection Service—For Kids and Teens: http://www.fsis.usda.gov/wps/portal/fsis/topics/food-safety-education/teach-others/download-materials/for-kids-and-teens/for-kids-and-teens

Internet sites represent a vast resource of information. Although those provided in this chapter were vetted by industry experts, you may wish to further explore the topics discussed in this chapter using a search engine such as Google. Keywords or phrases may include the following: *HACCP, food toxins, sanitize, biofilms, psychrophiles, mesophiles, thermophiles, psychrotrophic, thermotrophic, food safety, food poisoning, food hazards, food-borne illness, cross-contamination, U.S. Public Health Service, food microorganisms, food-borne infections, viral borne infections, log-bacteria growth, pasteurization, sanitization, heat sterilization, Environmental Protection Agency, Food Safety and Inspection Service, Food and Drug Administration*. In addition, Table B-7 provides a listing of useful Internet sites that can be used as a starting point.

REFERENCES

Cremer, M. L. 1998. *Quality food in quantity: Management and science.* Berkeley, CA: McCutchan Publishing.

National Council for Agricultural Education. 1993. *Food science, safety, and nutrition.* Madison, WI: National FFA Foundation.

Potter, N. N., and J. H. Hotchkiss. 1999. *Food science,* 5th ed. New York: Springer.

Seperich, G. J. 1998. *Food science and safety.* Danville, IL: Interstate Printers.

Vaclavik, V. A., and E. W. Christina. 2014. *Essentials of food science*, 4th ed. New York: Springer.

ENDNOTES

1. James Andrews, "Jack in the Box and the Decline of E. coli": http://www.foodsafetynews.com/2013/02/jack-in-the-box-and-the-decline-of-e-coli/#.VaRSVl9VjF8. Last accessed August 6, 2015.
2. Restaurant Management 101: http://restaurantmgmt101.com/2011/05/fattom-and-foodborne-illness/. Last accessed July 13, 2015.
3. University of California at San Diego, *Bacillus thuringiensis:* http://www.bt.ucsd.edu/index.html. Last accessed July 15, 2015.
4. FDA, Genetic Engineering: http://www.fda.gov/Food/FoodScienceResearch/Biotechnology/ucm346030.htm. Last accessed July 15, 2015.
5. MIT Technology Review, "Why We Will Need Genetically Modified Foods": http://www.technologyreview.com/featuredstory/522596/why-we-will-need-genetically-modified-foods/. Last accessed July 16, 2015.

CHAPTER 27

Regulation and Labeling

OBJECTIVES

After reading this chapter, you should be able to:

- **Identify the agencies and laws that regulate foods and labeling**
- **Describe the functions of a quality-assurance department**
- **Discuss the history of food labels**
- **List five features of new labels**
- **Name two general categories of food exempt from food labels**
- **Explain six components found on the nutritional panel**
- **Describe the format of the nutritional panel**
- **Discuss the use of Daily Reference Values (DRVs)**
- **Identify the appropriate use of the following words: free, low, high, less, light, and more**
- **List two health-claim relationships that can be listed on a food package**

KEY TERMS

coliform
compliance
enhancers
Food, Drug, and Cosmetic Act
food labeling
hydrolysates
Infant Health Formula Act
Meat Inspection Act
nutrition labeling
Nutrition Labeling and Education Act
standard plate count (SPC)
quality assurance (QA)

NATIONAL AFNR STANDARD

FPP.04

Explain the scope of the food industry and the historical and current developments of food products and processing.

The Food and Drug Administration (FDA), the U.S. Department of Agriculture (USDA), and various federal legislative acts regulate foods and the labeling of foods and food laws to protect consumers. In addition, many states and cities have food laws. Food labeling provides basic information about the ingredients in and the nutritional value of, food products so that consumers can make informed choices in the marketplace. Recent changes in food labeling provide the consumer with more complete, useful, and accurate information than ever before.

FEDERAL FOOD, DRUG, AND COSMETIC ACT

The FDA, operating under the federal **Food, Drug, and Cosmetic Act**, regulates the labeling of all foods other than meat and poultry (Figure 27-1). Meat and poultry products are regulated by the USDA under the federal **Meat Inspection Act**.

ADDITIONAL FOOD LAWS

Beside the Food, Drug, and Cosmetic Act, additional federal laws cover specific foods. The Federal Meat Inspection Act of 1906 mandated inspection of all meat animals, slaughtering conditions, and meat-processing facilities. The USDA Food Safety and Inspection Service (FSIS) enforces the act.

The Federal Poultry Products Inspection Act of 1957 is similar to the Meat Inspection Act but applies to poultry and poultry products.

To protect the public and the food industry against false advertising, the Federal Trade Commission Act was amended for food in 1938.

For infants, the **Infant Health Formula Act** of 1980 provides that manufactured formulas contain the known essential nutrients at the correct levels.

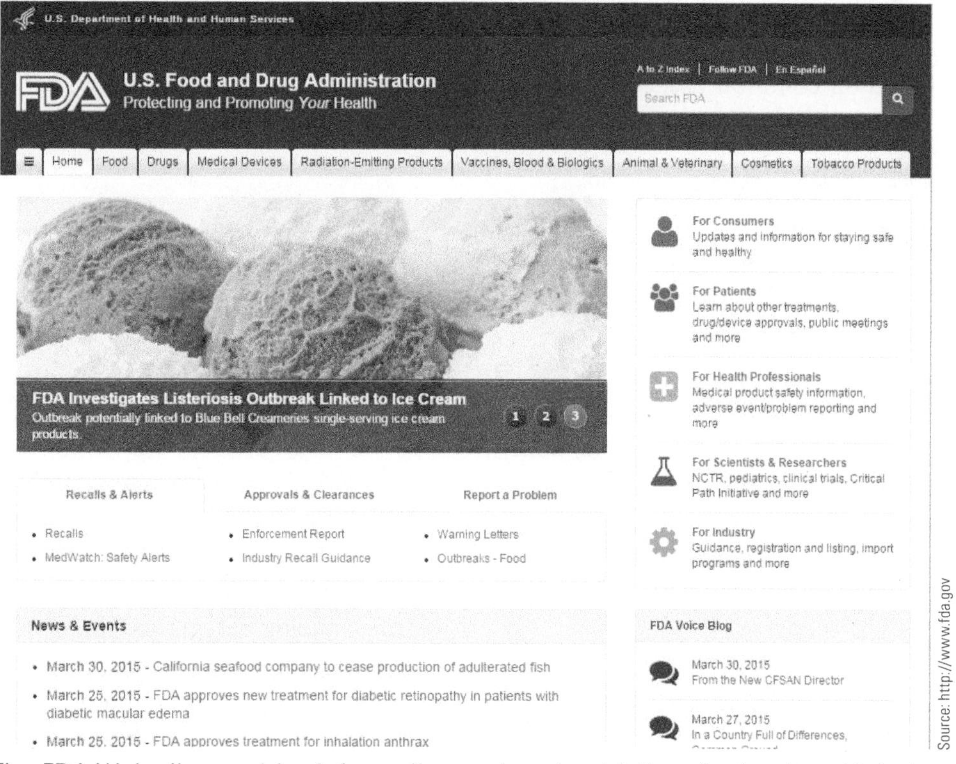

FIGURE 27-1 The FDA Web site provides information on legal guidelines for food and labels.

The **Nutrition Labeling and Education Act** of 1990 protects consumers against partial truths, mixed messages, and fraud regarding nutrition information.

Federal grade standards have already been discussed in Chapters 17, 18, 19, 20, and 21. These are standards of quality that help producers, dealers, wholesalers, retailers, and consumers market and purchase food. Standards and inspection come under the Agricultural Marketing Service of the Department of Agriculture (http://www.ams.usda.gov/AMSv1.0/standards).

Besides federal laws, all states and many cities also have their own food laws to further protect consumers from unsanitary conditions and deception. These laws govern retail food outlets and eating establishments.

LEGAL CATEGORIES OF FOOD SUBSTANCES

In the United States, substances that become parts of foods can be legally divided into several categories. Substances added to foods that have a history of being safe based on common usage in food are call *generally recognized as safe* (GRAS). Substances such as common spices, natural seasonings, many flavorings, baking powders, citric acid, malic acid, phosphoric acid, various gums, many emulsifiers, and numerous other substances are included in the list of GRAS substances.

Food additives are a very specific group of substances that are added intentionally and directly to foods. These are regulated and approved by the FDA. Scientific data must show that the additive is harmless in the intended food application and at the intended level of use. Food additives fall into one of the following categories:

- Preservatives
- Antioxidants
- Flavoring agents
- Sweeteners
- Emulsifiers, stabilizers, and thickeners
- Leavening agents
- Anticaking agents
- Humectants
- Coloring agents
- Bleaches
- Acids, bases, and buffers
- Nutrients

Pesticide residues may become a small part of a food product rather unintentionally through the application of a pesticide to a crop that has a minute amount of carryover to the food product. Most toxicologists do not believe that the small residues present a significant risk to human health. Even so, federal and some state laws regulate pesticide uses. For additional information on additives, refer to Chapter 14.

SCIENCE CONNECTION!

What laws regulate foods in your local community? What effects do local regulations have on those foods and food-related jobs and service? Does this have any effect on you personally? Why or why not?

TESTING FOR SAFETY

Food safety is a broad topic, and one covered in Chapter 26. Pesticides, herbicides, chemical additives, and spoilage are all of concern, but food scientists, food processors, and consumers focus most on microbiological quality. Microorganisms pose a challenge to the food industry, and most food processes are designed with microbial quality in mind. Microorganisms are too small to be seen with the unaided eye and have the ability to reproduce rapidly. Many of them produce toxins and can cause infections. For these reasons, the microbiological quality of the food we eat is scrutinized closely. *Escherichia coli* (*E. coli*) is commonly employed as an indicator microorganism. Because *E. coli* is a **coliform** bacterium common to the intestinal tract of humans and animals, its relationship to intestinal food-borne pathogens is high.

Total counts of microorganisms are also an indication of the sanitary quality of a food. Referred to as the **standard plate count (SPC)**, this total count of viable microbes reflects the handling history, state of decomposition, or degree of freshness of a food. Total counts may be taken to indicate the type of sanitary control exercised in the food's production, transport, and storage. Most foods have standards or limits for total counts. A low SPC does not always represent a safe product, however. A low-count food can still contain toxin-producing organisms. These toxins can remain stable even under conditions that do not favor the survival of the microbial cell.

QUALITY ASSURANCE

Of all functions in the food industry, **quality assurance (QA)** requires more diverse technical and analytical skills than most jobs. QA personnel continually monitor incoming raw milk and finished milk products to ensure **compliance** with compositional standards, microbiological standards, and various government regulations. A QA manager can halt production, refuse to accept raw materials, and stop shipments if specifications for products or processes are not met. Normally, this department has no control over a product unless something has gone wrong.

The major functions of the QA department include:

- Compliance with specifications—legal requirements, industry standards, internal company standards, shelf-life tests, customers' specifications

FIGURE 27-2 Modern laboratories and technicians monitor the safety of the food supply.

- Test procedures—testing of raw materials, finished products, and in-process tests
- Sampling schedules—use a suitable sampling schedule to maximize the probability of detection while minimizing workload
- Records and reporting—maintain all QA records so that customer complaints and legal problems can be dealt with
- Troubleshooting—solve various problems caused by poor-quality raw materials, erratic supplies and malfunctioning process equipment, and investigate reasons for poor-quality product to avoid repetition
- Special problems—customer complaints, production problems, personnel training, short courses, and so on

A typical QA department may have a chemistry lab, a raw materials inspection lab, a sensory lab, and a microbiology lab (Figure 27-2). All of these disciplines work together to ensure that the food we consume is of the highest quality. After all, quality is what makes a customer return.

FOOD LABELING

Food labeling for most of the food products sold in the United States must show the product's name, the manufacturer's name and address, the amount of product in the package, and the product's ingredients. The ingredients are listed in descending order based on their weight. Under current laws, fresh fruits, vegetables, and meat are exempt from these labeling requirements.

In 1973, the FDA established **nutrition labeling** or guidelines for labeling the nutrient and caloric content of food products. Nutrition labeling is mandatory only for those foods that have nutrients added or make a nutritional claim. Manufacturers are encouraged, but not required, to provide nutrition labeling on other food products.

In a 1990 Food Marketing Institute survey, more than 70% of food shoppers identified taste, nutrition, and product safety as being critically important factors in making food purchases. In the same survey, 36% of shoppers reported that they always read the ingredient and nutrition labels; another 45% said they sometimes read nutrition labels.

The growing importance of the role of nutrition in promoting health and preventing disease, as well as consumer demand for clearer and easier-to-understand information, led to the passage of the Nutrition Labeling and Education Act (NLEA) of 1990. Federal regulations that detail the format and content of food labels have been in effect for more than 20 years.

Under these regulations from the FDA, the Department of Health and Human Services, and the USDA's FSIS, the contemporary food label offers more complete, useful, and accurate nutrition information than did previous labels (see Figure 27-3).

Key features of the food label include:

- Nutrition labeling for almost all foods
- Distinctive, easy-to-read format
- Information on the amount per serving of saturated fat, cholesterol, dietary fiber, and other nutrients
- Nutrient reference values, expressed as percentage daily values
- Uniform definitions for terms that describe a food's nutrient content such as *light*, *low-fat*, and *high-fiber*
- Claims about the relationship between a nutrient or food and a disease or health-related condition
- Standardized serving sizes
- Declaration of total percentage of juice in juice drinks
- Voluntary nutritional information for many raw foods

Examples of nutrition labels are shown in Figure 27-4.

Programs for parents, youth, seniors, and restaurants and retail establishments have been developed by the FDA to assist consumers and food-industry entrepreneurs implement and interpret the nutrition fact label (Figure 27-5). Visit The FDA Nutrition Facts Label Programs and Materials (http://www.fda.gov/Food/IngredientsPackagingLabeling/LabelingNutrition/ucm20026097.htm) for more information and many resources.

Foods Affected

The FDA regulations call for nutrition labeling for most foods. In addition, they set up voluntary programs for nutrition information for many raw foods—the 20 most frequently eaten raw fruits, vegetables, and fish and the 45 best-selling cuts of meat—under the FDA voluntary point-of-purchase nutrition information program. Labels for meat and poultry products are regulated by the FSIS.

Exemptions

Some foods are exempt from nutrition labeling:

- Food served for immediate consumption, such as that served in hospital cafeterias and airplanes, by food-service vendors such as mall cookie counters, sidewalk vendors, and food trucks
- Vending machines operators who operate less than 20 machines

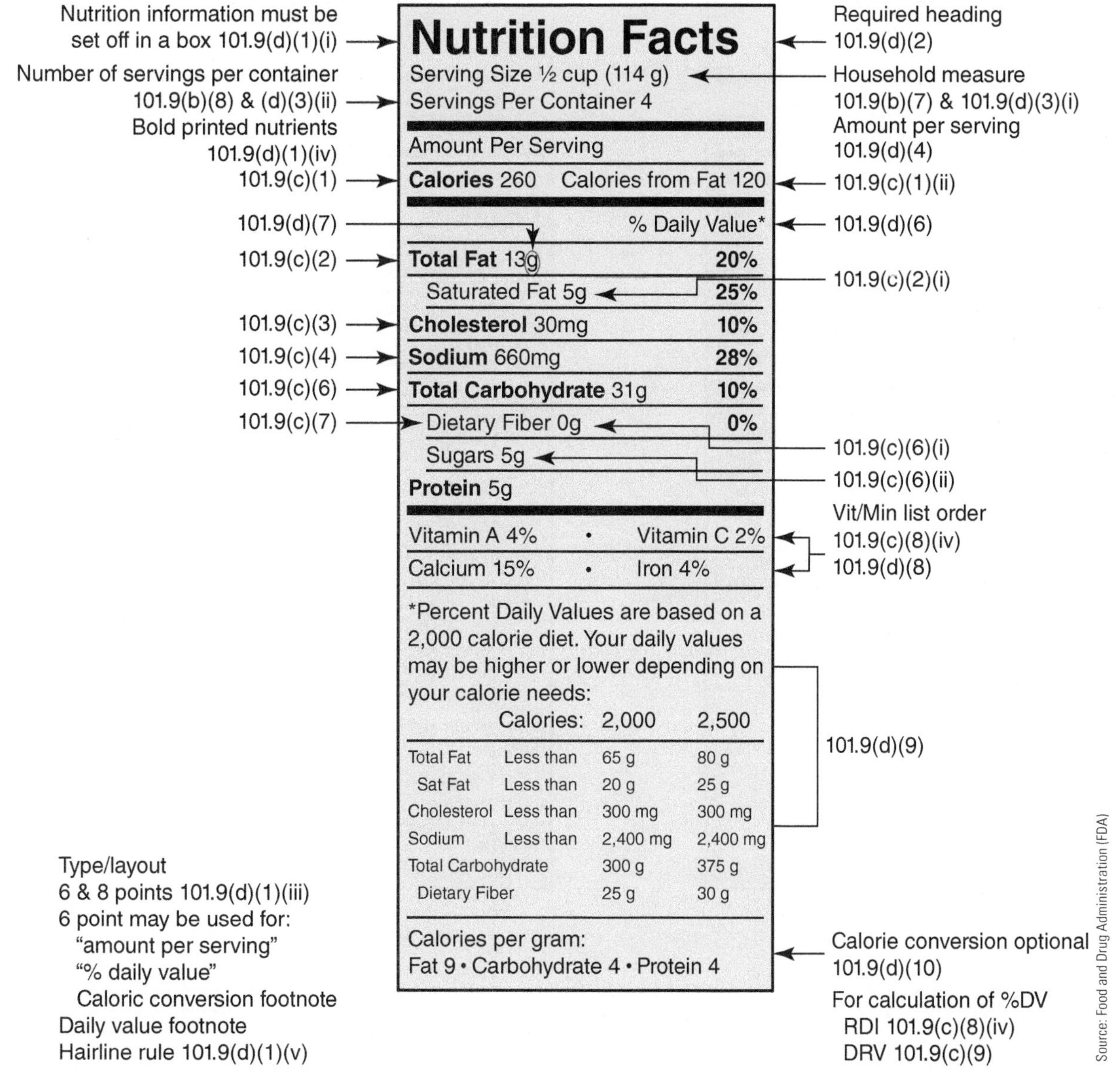

FIGURE 27-3 Under regulations from the FDA, the Department of Health and Human Services, and the USDA's Food Safety and Inspection Service, the food label now offers more complete, useful, and accurate nutrition information.

- Ready-to-eat food that is not for immediate consumption but is prepared primarily on-site—for example, bakery, deli, and candy store items
- Food shipped in bulk, as long as it is not for sale in that form to consumers
- Medical foods, such as those used to address the nutritional needs of patients with certain diseases
- Plain coffee and tea, some spices, and other foods that contain no significant amounts of any nutrient
- Food produced by small businesses—based on the number of people a company employs

Although these foods are exempt, they are free to carry nutrition information, when appropriate—as long as it complies with regulations. Also, they will lose their exemption if their labels carry a

Nutrition Facts

Serving Size 1 cup (228 g)
Servings Per Container 2

Amount Per Serving	
Calories 260	Calories from Fat 120
	% Daily Value*
Total Fat 13g	**20%**
Saturated Fat 5g	**25%**
Trans Fat 2g	
Cholesterol 30mg	**10%**
Sodium 660mg	**28%**
Total Carbohydrate 31g	**10%**
Dietary Fiber 0g	**0%**
Sugars 5g	
Protein 5g	

Vitamin A 4%	•	Vitamin C 2%
Calcium 15%	•	Iron 4%

*Percent Daily Values are based on a 2,000 calorie diet. Your Daily Values may be higher or lower depending on your calorie needs:

		2,000	2,500
	Calories:	2,000	2,500
Total Fat	Less than	65 g	80 g
Sat Fat	Less than	20 g	25 g
Cholesterol	Less than	300 mg	300 mg
Sodium	Less than	2,400 mg	2,400 mg
Total Carbohydrate		300 g	375 g
Dietary Fiber		25 g	30 g

Calories per gram:
Fat 9 • Carbohydrate 4 • Protein 4

Nutrition Facts

Serving Size 1 cup (245 g)
Servings Per Container 2

Amount Per Serving	
Calories 60	Calories from Fat 10
	% Daily Value*
Total Fat 1g	**2%**
Sodium 800mg	**33%**
Total Carbohydrate 10g	**3%**
Dietary Fiber 0g	**4%**
Protein 2g	

Vitamin A 20% • Vitamin C 4% • Iron 2%

Not a significant source of saturated fat, *trans* fat, cholesterol, sugars, or calcium.

*Percent Daily Values are based on a 2,000 calorie diet. Your daily values may be higher or lower depending on your calorie needs:

		2,000	2,500
	Calories:	2,000	2,500
Total Fat	Less than	65 g	80 g
Sat Fat	Less than	20 g	25 g
Cholesterol	Less than	300 mg	300 mg
Sodium	Less than	2,400 mg	2,400 mg
Total Carbohydrate		300 g	375 g
Dietary Fiber		25 g	30 g

Calories per gram:
Fat 9 • Carbohydrate 4 • Protein 4

FRUIT DESSERT FOR CHILDREN LESS THAN 2 YEARS OLD

Nutrition Facts

Serving Size 1 jar (110g)

Amount Per Serving	
Calories 110	
Amount Per Serving	
Total Fat	0g
Trans Fat	0g
Sodium	10mg
Total Carbohydrate	27g
Dietary Fiber	4g
Sugars	0g
Protein	5g

% Daily Value

Protein 0%	• Vitamin A 6%
Vitamin C 45% •	Calcium 2%
Iron 2%	

21 CFR 101.9(j)(5)(i)

Nutrition Facts

Serving Size 1/3 cup (56g)
Servings about 3
Calories 90
Fat Cal. 20

*Percent Daily Values (DV) are based on a 2,000 calorie diet

Amount/serving	%DV*	Amount/serving	%DV*
Total Fat 2g	3%	**Total Carb.** 0g	0%
Sat. Fat 1g	5%	Fiber 0g	0%
Trans Fat 0.5g		Sugars 0g	
Cholest. 10mg	3%	**Protein** 17g	
Sodium 200mg	8%		

Vitamin A 0% • Vitamin C 0% • Calcium 0% • Iron 6%

Nutrition Facts Serv. Size: 1 package, Amount Per Serving: **Calories** 45, Fat Cal. 10, **Total Fat** 1g (2% DV), Sat. Fat 0.5g (3% DV), *Trans* Fat 0.5g, **Cholest.** 0mg (0% DV), **Sodium** 50mg (2% DV), **Total Carb.** 8g (3% DV), Fiber 1g (4% DV), Sugars 4g, **Protein** 1g, Vitamin A (8% DV), Vitamin C (8% DV), Calcium (0% DV), Iron (2% DV). Percent Daily Values (DV) are based on a 2,000 calorie diet.

Source: Food and Drug Administration (FDA)

FIGURE 27-4 Though the format may vary, all nutrition labels carry the same basic nutritional information.

Source: Food and Drug Administration (FDA)/www.fda.gov/nutritioneducation

FIGURE 27-5 This poster demonstrates a program put in place by the USDA to educate youth on how to interpret and understand the importance of the nutrition fact label.

nutrient content or health claim or any other type of nutritional information.

Nutritional information about game meats—such as deer, bison, rabbit, quail, wild turkey, and ostrich—is not required on individual packages. Instead, it can be given on counter cards, signs, or other point-of-purchase materials. Because few nutrient data exist for these foods, the FDA believes that allowing this option will enable producers of game meats to give first priority to collecting appropriate data and make it easier for them to update the information as it becomes available.

The FDA is considering changes to the nutrition facts label. For example, in July 2015, the FDA issued a supplemental proposed rule that, among other things, would (1) require declaration of the percent daily value (% DV) for added sugars and (2) change the current footnote on the Nutrition Facts label to help consumers understand the percent daily value concept.

Restaurant Nutritional Labeling

Americans eat and drink about one-third of their calories away from home. Making calorie information available on chain restaurant menus will help consumers make informed choices for themselves and their families.

The FDA new regulations that require calorie information on restaurant menus and menu boards as well as on vending machines should be helpful for consumers. Some states, localities, and large restaurant chains were already doing their own forms of menu labeling, but this information was not consistent across all areas even when implemented. Calorie information will now be required on menus and menu boards in chain restaurants (and other places selling restaurant-type food) and on certain vending machines. This new calorie labeling will be consistent nationwide and should provide easy-to-understand nutrition information in a direct and accessible manner. Calorie labeling on restaurant menus and menu boards and on vending machines is set to begin December 1, 2016.

Calorie labeling is required for restaurants and similar retail food establishments that are part of a chain of 20 or more locations. For standard menu items, calories will be listed clearly and prominently on menus and menu boards and next to the name or price of the food or beverage.

For self-service foods—buffets and salad bars, for example—calories will be shown on signs set near the foods. Calories are not required to be listed for condiments, daily specials, custom orders, or temporary and seasonal menu items.

On vending machines, calorie labeling is required for machine operators who own or operate 20 or more vending machines. Calories will be shown on a sign (such as a small placard, sticker, or poster) or on electronic or digital displays near the food item or selection button on vending machines and bulk vending machines (e.g., gumball machines and mixed nut machines) unless calories are already visible on the actual food packages before purchase.

In addition to calorie information, restaurants are also required to provide written nutrition information on their menu items, including total fat, calories from fat, saturated fat, trans-fat, cholesterol, sodium, total carbohydrates, dietary fiber, sugars, and protein. This information maybe on posters, tray liners, signs, counter cards, handouts, booklets, computers, or kiosks. If consumers do not see such information, they can ask restaurant personnel for nutrition information.

With this new ruling from the FDA, calorie information will be on the following:

- Meals or snacks from sit-down and fast-food restaurants, bakeries, coffee shops, and ice cream stores
- Foods purchased at drive-through windows
- Take-out and delivery foods such as pizza
- Foods such as sandwiches ordered from a menu or menu board at a grocery store, convenience store, or delicatessen
- Self-serve foods from a salad or hot-food bar at a restaurant or grocery store
- Foods such as popcorn purchased at a movie theater or amusement park
- Alcoholic drinks such as cocktails when they are listed on menus

No calorie information is required for posting on the following items:

- Foods sold at deli counters and typically intended for more than one person
- Foods purchased in bulk in grocery stores such as loaves of bread from the bakery section
- Bottles of liquor displayed behind a bar
- Food in transportation vehicles such as food trucks, airplanes, and trains
- Food on menus in elementary, middle, and high schools that are part of the USDA National School Lunch Program (although vending machines in these locations are not exempt from displaying nutrition information)

Nutrition Panel Title

The food label features a nutrition panel (Figure 27-6) with the title "Nutrition Facts." There are also requirements on type size, style, spacing, and contrast to ensure more distinctive and easy-to-read labels.

Serving Sizes

The serving size remains the basis for reporting each food's nutrient content. However, unlike in the past when serving sizes were left to food manufacturers' discretion, serving sizes are now more uniform and reflect the amounts people actually eat. They also must be expressed in both common household and metric measures.

Among the household measures allowed by the FDA are the cup, tablespoon, teaspoon, piece, slice, fraction (such as "¼ pizza"), and common household containers used to package food products (such as a jar or tray). Ounces may be used but only if a common household unit is not applicable and an

INGREDIENTS

Water, flour, beef (beef, beef broth and salt), **vegetable oil, tomato paste, seasoning blend** (salt, spices), **seasoning blend** (spices, salt, powdered onion, granulated garlic), **hydroxipropyl methylcellulose** (a stabilizer), **parsley, red pepper.**

Prepared & Packaged by Food Company, Anytown, USA, 55555

For questions or comments call:
1 800 555-1234
1 800 555-7685

Nutrition Facts

Serving Size 1 (126 g)
Servings Per Container 15

Amount per Serving
Calories 240 Calories from Fat 100

	% Daily Value*
Total Fat 12g	**18%**
Saturated Fat 2g	**9%**
Cholesterol 15mg	**5%**
Sodium 710mg	**30%**
Total Carbohydrate 23g	**8%**
Dietary Fiber 3g	**13%**
Sugars 0g	
Protein 11g	**22%**

Vitamin A 8% • Vitamin C 4%
Calcium 4% • Iron 15%

*Percent Daily Values are based on 2,000 calorie diet. Your daily values may be higher or lower based on your calorie needs:

	Calories:	2,000	2,500
Total Fat	Less than	65 g	80 g
Sat Fat	Less than	20 g	25 g
Cholesterol	Less than	300 mg	300 mg
Sodium	Less than	2,400 mg	2,400 mg
Total Carbohydrates		300 g	375 g
Dietary Fiber		25 g	30 g

Calories per gram:
Fat 9 • Carbohydrates 4 • Protein 4

FIGURE 27-6 Nutrition panels provide a variety of information to consumers.

appropriate visual unit is given—for example, 1 oz. (28 g, "about ½ pickle"). Grams (g) and milliliters (ml) are the metric units that are used in serving size statements. NLEA defines serving size as the amount of food customarily eaten at one time. The serving sizes that appear on food labels are based on FDA-established lists of "Reference Amounts Customarily Consumed Per Eating Occasion."

The serving sizes of products that come in discrete units such as cookies, candy bars, and sliced products is the number of whole units that most closely approximates the reference amount. Cookies are an example. Under the "bakery products" category, cookies have a reference amount of 30 g. The household measure closest to that amount is the number of cookies that come closest to weighing 30 g. Thus, the serving size on the label of a package of cookies in which each cookie weighs 13 g would read "2 cookies (27 g)."

Nutrition Information

Dietary components on the nutrition panel include the mandatory (boldface) components and the voluntary components. The following is the order in which they must appear on the label:

- Total calories
- Calories from fat
- Calories from saturated fat
- Total fat
- Saturated fat
- Polyunsaturated fat
- Monounsaturated fat
- Cholesterol
- Sodium
- Potassium
- Total carbohydrate
- Dietary fiber
- Soluble fiber
- Insoluble fiber
- Sugars
- Sugar alcohol (sugar substitutes—xylitol, mannitol, and sorbitol)
- Other carbohydrates (the difference between total carbohydrate and the sum of dietary fiber, sugars, and sugar alcohol if declared)
- Protein
- Vitamin A
- Percent of vitamin A present as beta-carotene
- Vitamin C
- Calcium
- Iron
- Other essential vitamins and minerals

If a claim is made about any of the optional components or if a food is fortified or enriched with any optional component, then the nutritional information for these components becomes mandatory.

These mandatory and voluntary components are the only ones allowed on the nutrition panel. Listings of single amino acids, maltodextrin, calories from polyunsaturated fat, and calories from carbohydrates, for example, may not appear as part of the nutrition facts on the label.

The required nutrients were selected because they address today's health concerns. The order in which they must appear reflects the priority of current dietary recommendations. Thiamin, riboflavin, and niacin are no longer required in nutrition labeling because deficiencies of each are no longer considered of public health significance. However, they may be listed voluntarily.

Nutrition Panel Format

All nutrients must be declared as percentages of the Daily Values—the food label reference values (Figure 27-7). The amount, in grams or milligrams, of macronutrients (such as fat, cholesterol, sodium, carbohydrates, and protein) still must be listed to the immediate right of the names of each nutrient. For the first time, a column headed "Percent Daily Value" now appears.

Nutrition Facts
Serving Size 1/2 cup (114g)
Servings Per Container 4

Amount per Serving	
Calories 90 Calories from Fat 30	
	% Daily Value*
Total Fat 3g	**5%**
Saturated Fat 0g	**0%**
Cholesterol 0mg	**0%**
Sodium 300mg	**13%**
Total Carbohydrate 13g	**4%**
Dietary Fiber 3g	**12%**
Sugars 3g	
Protein 3g	
Vitamin A 80% • Vitamin C 60%	
Calcium 4% • Iron 4%	

*Percent Daily Values are based on a 2,000 calorie diet. Your daily values may be higher or lower based on your calorie needs:

		Calories: 2,000	2,500
Total Fat	Less than	65 g	80 g
Sat Fat	Less than	20 g	25 g
Cholesterol	Less than	300 mg	300 mg
Sodium	Less than	2,400 mg	2,400 mg
Total Carbohydrates		300 g	375 g
Dietary Fiber		25 g	30 g

Calories per gram:
Fat 9 • Carbohydrates 4 • Protein 4

FIGURE 27-7 The ideal nutrition label.

Requiring nutrients to be declared as a percentage of the Daily Values is intended to prevent misinterpretations that arise with quantitative values. For example, a food with 140 milligrams (mg) of sodium could be mistaken for a high-sodium food because 140 is a relatively large number. In actuality, the amount represents less than 6% of the Daily Value for sodium, which is 2,400 mg.

On the other hand, a food with 5 g of saturated fat could be construed as being low in that nutrient. In fact, that food would provide 25% of the total Daily Value because 20 g is the Daily Value for saturated fat based on a 2,000-calorie diet.

Nutrition Panel Footnote. The Percent (%) Daily Value listing carries a footnote saying that the percentages are based on a 2,000-calorie diet. Some nutrition labels—at least those on larger packages—have these additional footnotes:

- A sentence noting that a person's individual nutrient goals are based on his or her calorie needs
- Lists of the daily values for selected nutrients for a 2,000-calorie and a 2,500-calorie diet
- An optional footnote for packages of any size is the number of calories per gram of fat (9) and carbohydrate and protein (4).

SNAKE OIL

In the late 19th and early 20th centuries, White Eagle's Indian Rattle Snake Oil Liniment was sold as a cure-all for any kind of pain. Of course it provided no such relief, and today the term *snake oil* is used to describe a liquid concoction of questionable medical value sold as an all-purpose curative, especially by traveling hucksters. Snake oil also is used to describe any type of questionable cure-all. The term *nostrum* is also used to describe a medicine sold with false or exaggerated claims—a quack medicine.

Nostrums permeated American society by late 19th century. These products appealed to exotica, the supposed medical knowledge of Native Americans, death, religion, patriotism, mythology, and new developments in science in particular. There was nothing to stop patent medicine makers from making any outrageous claim or putting anything in their products. For example, Lydia E. Pinkham's Vegetable Compound was first marketed in 1875 as the "female complaint" nostrum. It was widely advertised in the backs of newspapers and women's magazines. After many years and successful sales, people discovered that Lydia E. Pinkham's Vegetable Compound contained 15% to 20% alcohol! This is typical of many of the early nostrums. They contained varying levels of alcohol and sometimes morphine.

(*Continues*)

SNAKE OIL (continued)

The rise of advertising in America paralleled the rise of nostrums. At the same time, the biomedical sciences in this country were still in their infancy, and medicine was ill equipped to deal with most diseases. Enterprising individuals were prepared to step in and alleviate potential customers' suffering with such products as:

- Ayer's Sarsaparilla: a "blood purifier"
- Dr. Morse's Indian Root Pills: curing everything from tapeworm to skin eruptions
- Pond's Extract: relieving pain of every kind inside and outside the body
- Rumford Yeast Powder: restoring the phosphorus lost by using flour
- Fat Off: an obesity-curing cream
- Dr. Bonker's Egyptian Oil: curing pains and ills in adults, children, and horses
- Dr. Lindley's Epilepsy Remedy: curing epilepsy, fits, spasms, and convulsions
- Eckman's Alterative: curing all throat and lung diseases, including tuberculosis
- Rite Wate Vegetable Compound: reducing body fat
- Liquozone: killing disease germs (first bottle free)
- Kickapoo Sagwa: claiming to be a "renovator"
- Anti-Morbific Liver and Kidney Medicine: preventing all diseases
- Dr. Shreves' Anti-Gallstone: treating gallstones and kidney stones
- Mixer's Cancer and Scrofula Syrup: curing cancer, piles, ulcers, tumors, abscesses, and all blood diseases

As ridiculous as these claims seem now, similar products are still sold today. Their makers rely on people who are gullible and looking for easy solutions to complex problems.

For more information about nostrums of the past, visit these Web sites or search the FDA Web site for "snake oil."

- FDA—Drugs: www.fda.gov/cder/
- FDA: A Brief History of the Center for Drug Evaluation and Research: http://www.fda.gov/aboutfda/whatwedo/history/forgshistory/cder/ucm320822.htm
- The Great American Fraud: www.mc.vanderbilt.edu/biolib/hc/nostrums/
- FDA—Protecting and Promoting Your Health: www.fda.gov/
- FDA—Office of Generic Drugs: http://www.fda.gov/AboutFDA/CentersOffices/OfficeofMedicalProductsandTobacco/CDER/ucm119100.htm
- Quacks and Nostrums: http://exhibits.hsl.virginia.edu/caricatures/en4-quacks/

Format Modifications. In limited circumstances, variations in the format of the nutrition panel are allowed. Some formats are mandatory.

For example, the labels of foods for children under 2 (except infant formula, which has special labeling rules under the Infant Formula Act of 1980) may not carry information about saturated fat, polyunsaturated fat, monounsaturated fat, cholesterol, calories from fat, or calories from saturated fat. The reason is to prevent parents from wrongly assuming that infants and toddlers should restrict their fat intake when, in fact, they should not. Fat is important during these years to ensure adequate growth and development.

Some format exceptions exist for small and medium-sized packages. Packages with less than 12 square inches of available labeling space (about the size of a package of chewing gum) do not have to carry nutrition information unless a nutrient content or health claim is made for the product. However, they must provide an address or telephone number for consumers to obtain the required nutrition information.

Daily Values—DRVs

Daily Value, the food label reference value, comprises two sets of dietary standards: Daily Reference Values (DRVs) and Reference Daily Intakes (RDIs). Only Daily Value appears on the label, though, to make label reading less confusing (Figure 27-8).

DRVs have been established for macronutrients that are sources of energy: fat, carbohydrate (including fiber), and protein; and for cholesterol, sodium, and potassium, which do not contribute calories.

DRVs for the energy-producing nutrients are based on the number of calories consumed per day. A daily intake of 2,000 calories has been established as the reference. This level was chosen, in part, because it approximates the caloric requirements for postmenopausal women. This group has the highest risk of excessive calorie and fat intake. DRVs for the energy-producing nutrients are calculated as follows:

- Fat—based on 30% of calories
- Saturated fat—based on 10% of calories
- Carbohydrate—based on 60% of calories
- Protein—based on 10% of calories (the DRV for protein applies only to adults and children 4 and older)
- Fiber—based on 11.5 g of fiber per 1,000 calories

Because of current public health recommendations, DRVs for some nutrients represent the uppermost limit that is considered desirable. The DRVs for fats and sodium are:

- Total fat—less than 65 g
- Saturated fat—less than 20 g
- Cholesterol—less than 300 mg
- Sodium—less than 2,400 mg

Nutrient Content Descriptions

The regulations also spell out what terms may be used to describe the level of a nutrient in a food and how they can be used. These include *free*, *low*, *lean* and *extra lean*, *high*, *good source*, *reduced*, *less*, and *light*.

Free. *Free* means that a product contains no amount of, or only trivial or "physiologically inconsequential" amounts of, one or more of these components: fat, saturated fat, cholesterol, sodium, sugars, and calories. Synonyms for *free* include *without*, *no*, and *zero*.

Low. The term *low* can be used on foods that can be eaten frequently without exceeding dietary guidelines for one or more of these components: fat, saturated fat, cholesterol, sodium, and calories, for example:

- Low fat—3 g or less per serving
- Low saturated fat—1 g or less per serving
- Low sodium—140 mg or less per serving
- Very low sodium—35 mg or less per serving
- Low cholesterol—20 mg or less and 2 g or less of saturated fat per serving
- Low calorie—40 calories or less per serving

Synonyms for *low* include *little*, *few*, and *low source of*.

Lean and Extra Lean. Lean and extra lean can be used to describe the fat content of meat, poultry, seafood, and game meats.

- Lean—less than 10 g fat, 4.5 g or less saturated fat, and less than 95 mg cholesterol per serving and per 100 g
- Extra lean—less than 5 g fat, less than 2 g saturated fat, and less than 95 mg cholesterol per serving and per 100 g

High. *High* can be used if the food contains 20% or more of the Daily Value for a particular nutrient in a serving.

Good Source. *Good source* means that one serving of a food contains 10% to 19% of the Daily Value for a particular nutrient.

Reference Values for Nutrition Labeling[1] (Based on a 2,000 Calorie Intake; for Adults and Children 4 or More Years of Age)		
Nutrient[2]	**Unit of Measure**	**Daily Values**
Total fat	grams (g)	65
Saturated fatty acids	grams (g)	20
Cholesterol	milligrams (mg)	300
Sodium	milligrams (mg)	2,400
Potassium	milligrams (mg)	3,500
Total carbohydrate	grams (g)	300
Fiber	grams (g)	25
Protein	grams (g)	50
Vitamin A	International Unit (IU)	5,000
Vitamin C	milligrams (mg)	60
Calcium	milligrams (mg)	1,000
Iron	milligrams (mg)	18
Vitamin D	International Unit (IU)	400
Vitamin E	International Unit (IU)	30
Vitamin K	micrograms (μg)	80
Thiamin	milligrams (mg)	1.5
Riboflavin	milligrams (mg)	1.7
Niacin	milligrams (mg)	20
Vitamin B_6	milligrams (mg)	2.0
Folate	micrograms (μg)	400
Vitamin B_{12}	micrograms (μg)	6.0
Biotin	micrograms (μg)	300
Pantothenic acid	milligrams (mg)	10
Phosphorus	milligrams (mg)	1,000
Iodine	micrograms (μg)	150
Magnesium	milligrams (mg)	400
Zinc	milligrams (mg)	15
Selenium	micrograms (μg)	70
Copper	milligrams (mg)	2.0
Manganese	milligrams (mg)	2.0
Chromium	micrograms (μg)	120
Molybdenum	micrograms (μg)	75
Chloride	milligrams (mg)	3,400

Source: Food and Drug Administration (FDA)

[1] *Source:* U.S. Food and Drug Administration (FDA), Center for Food Safety and Applied Nutrition. Revised January 30, 1998.
[2] Nutrients in this table are listed in the order in which they are required to appear on a label. This list includes only those nutrients for which a Daily Reference Value (DRV) has been established or a Reference Daily Intake (RDI).

FIGURE 27-8 Table of Reference Values for nutrition labeling.

Reduced. *Reduced* means that a nutritionally altered product contains at least 25% less of a nutrient or calories than the regular, or reference, product. However, a *reduced* claim cannot be made on a product if its reference food already meets the requirement for a *low* claim.

Less. *Less* means that a food, altered or not, contains 25% less of a nutrient or calories than the reference food. For example, pretzels that have 25% less fat than potato chips could carry a *less* claim. *Fewer* is an acceptable synonym.

Light. *Light* can mean two things. First, a nutritionally altered product that contains one-third fewer calories or half the fat of the reference food can use *light*. If the food derives 50% or more of its calories from fat, then the reduction must be 50% of the fat. Second, the sodium content of a low-calorie, low-fat food that has been reduced by 50% can also use *light*. In addition, *light in sodium* may be used on foods in which the sodium content has been reduced by at least 50%.

The term *light* still can be used to describe such properties as texture and color, as long as the label explains the intent—for example, "light brown sugar" and "light and fluffy."

More. *More* means that a serving of food, altered or not, contains a nutrient that is at least 10% of the Daily Value more than the reference food. The 10% of Daily Value also applies to *fortified*, *enriched*, and *added* claims, but in those cases the food must be altered (Figure 27-9).

Other Definitions

The label regulations also address other claims such as *percent fat free*, *implied*, *healthy*, and *fresh*.

Percent Fat Free. A product bearing this claim must be a low-fat or fat-free product. In addition, the claim must accurately reflect the amount of fat present in 100 grams of the food. Thus, if a food contains 2.5 grams fat per 50 grams, the claim must be "95% fat free."

Implied. These types of claims are prohibited when they wrongfully imply that a food contains or does not contain a meaningful level of a nutrient. For example, a product claiming to be made with an ingredient known to be a source of fiber (such as "made with oat bran") is not allowed unless the product contains enough of that ingredient (e.g., oat bran) to meet the definition for "good source" of fiber.

Healthy. A *healthy* food must be low in fat and saturated fat and contain limited amounts of cholesterol and sodium. In addition, if it is a single-item food, it must provide at least 10% of one or more of vitamins A or C, iron, calcium, protein, or fiber. If it is a meal-type product, such as a frozen entree or multicourse frozen dinner, then it must provide 10% of two or three of these vitamins or minerals or of protein or fiber in addition to meeting the other criteria.

Fresh. Although not mandated by NLEA, the FDA issued a regulation for the term *fresh*. The

Country Home

Old fashioned enriched white bread

Ingredients wheat flour, (enriched with niacin, iron, thiamine mononitrate, riboflavin, and folic acid), water, sugar, salt, potato flour, all vegetable shortening (partially hydrogenated soybean and cottonseed oils), yeast, gluten, cultured whey, vinegar, lecithin.

Net wt 1.5 Lb (680 g)

Nutrition Facts

Serving Size 1 slice (40g)
Servings Per Container 17

Amount per Serving
Calories 100 Calories from Fat 10

	% Daily Value*
Total Fat 1g	2%
Saturated Fat 0g	0%
Cholesterol 0mg	0%
Sodium 230mg	10%
Total Carbohydrate 21g	7%
Dietary Fiber 0g	0%
Sugars 3g	
Protein 2g	

Vitamin A 0% • Vitamin C 0%
Calcium 0% • Iron 4%

*Percent Daily Values are based on a 2,000 calorie diet. Your daily values may be higher or lower based on your calorie needs:

	Calories:	2,000	2,500
Total Fat	Less than	65g	80g
Sat Fat	Less than	20g	25g
Cholesterol	Less than	300mg	300mg
Sodium	Less than	2,400mg	2,400mg
Total Carbohydrates		300g	375g
Dietary Fiber		25g	30g

Calories per gram:
Fat 9 • Carbohydrates 4 • Protein 4

The Bread Shop Inc.

FIGURE 27-9 Label from enriched or fortified products.

regulation was issued because of concern over the term's possible misuse on some food labels.

The regulation defines the term *fresh* when it is used to suggest that a food is raw or unprocessed. *Fresh* can be used only on a food that is raw, has never been frozen or heated, and contains no preservatives, but irradiation at low levels is allowed. *Fresh frozen*, *frozen fresh*, and *freshly frozen* can be used for foods that are quickly frozen while still fresh. Blanching (brief scalding before freezing to prevent nutrient breakdown) is allowed. Other uses of the term *fresh*, such as in "fresh milk" or "freshly baked bread," are not affected.

- Update daily values for nutrients such as sodium, dietary fiber and vitamin D. Daily values are used to calculate the Percent Daily Value listed on the label, which helps consumers understand the nutrition information in the context of a total daily diet.
- Require manufacturers to declare the amount of potassium and vitamin D on the label because they are new "nutrients of public health significance." Calcium and iron would continue to be required, and vitamins A and C could be included on a voluntary basis.

SCIENCE CONNECTION!

How familiar are you with food nutritional labels? Gather food labels from several different food items you consume on a regular basis. Read and compare the labels. Are you aware of the nutritional information on the foods you eat? Are you surprised by anything on the label?

PROPOSED REVISIONS TO THE NUTRITIONAL FACTS LABEL

In March 2014, the FDA proposed updates to the Nutrition Facts label found on most food packages in the United States. In July 2015, supplemental proposed changes were also included. The FDA is currently deliberating these changes, and rulings are expected in the next year (see Figure 27-10). The Nutrition Facts label introduced 20 years ago helps consumers make informed food choices and maintain healthy dietary practices. If adopted, the proposed changes would include the following:

1. Greater Understanding of Nutrition Science
    - Require information about "added sugars." Many experts recommend consuming fewer calories from added sugar because they can decrease the intake of nutrient-rich foods while increasing calorie intake.
    - While continuing to require "Total Fat," "Saturated Fat," and "Trans Fat" on the label, "Calories from Fat" would be removed because research shows the type of fat is more important than the amount.
2. Updated Serving Size Requirements and New Labeling Requirements for Certain Package Sizes
    - Change the serving size requirements to reflect how people eat and drink today, which has changed since serving sizes were first established in the 1990s. By law, the label information on serving sizes must be based on what people actually eat, not on what they "should" be eating.
    - Require that packaged foods, including drinks, that are typically eaten in one sitting be labeled as a single serving and that calorie and nutrient information be declared for the entire package. For example, a 20-ounce bottle of soda,

Nutrition Facts
Serving Size 2/3 cup (55g)
Servings Per Container About 8

Amount Per Serving	
Calories 230	Calories from Fat 72
	% Daily Value*
Total Fat 8g	**12%**
Saturated Fat 1g	**5%**
Trans Fat 0g	
Cholesterol 0mg	**0%**
Sodium 160mg	**7%**
Total Carbohydrate 37g	**12%**
Dietary Fiber 4g	**16%**
Sugars 1g	
Protein 3g	
Vitamin A	10%
Vitamin C	8%
Calcium	20%
Iron	45%

*Percent Daily Values are based on a 2,000 calorie diet. Your daily value may be higher or lower depending on your calorie needs.

	Calories:	2,000	2,500
Total Fat	Less than	65 g	80 g
Sat Fat	Less than	20 g	25 g
Cholesterol	Less than	300 mg	300 mg
Sodium	Less than	2,400 mg	2,400 mg
Total Carbohydrate		300 g	375 g
Dietary Fiber		25 g	30 g

(a) Current Label

Nutrition Facts
8 servings per container
Serving size 2/3 cup (55g)

Amount per 2/3 cup
Calories 230

% DV*	
12%	**Total Fat** 8g
5%	**Saturated Fat** 1g
	***Trans* Fat** 0g
0%	**Cholesterol** 0mg
7%	**Sodium** 160 mg
12%	**Total Carbs** 37g
14%	Dietary Fiber 4g
	Sugars 1g
	Added Sugars 0g
	Protein 3g
10%	**Vitamin D** 2 mcg
20%	**Calcium** 260 mg
45%	**Iron** 8 mg
5%	**Potassium** 235 mg

*Footnote on Daily Values (DV) and calories reference to be inserted here.

Source: Food and Drug Administration (FDA)

(b) Proposed Label

FIGURE 27-10 A comparison of the current nutrition fact label with the new label proposed in 2014–2015.

typically consumed in a single sitting, would be labeled as one serving rather than more than one serving.

- For certain packages that are larger and could be consumed in one sitting or multiple sittings, manufacturers would have to provide "dual column" labels to indicate both "per serving" and "per package" calories and nutrient information. Examples would be a 24-ounce bottle of soda or a pint of ice cream. This will help people more easily understand how many calories and nutrients they are getting if they consume the entire package at one time.

3. Refreshed Design
   - Make calories and serving sizes more prominent to emphasize parts of the label that are important in addressing current public health concerns such as obesity, diabetes, and cardiovascular disease.
   - Shift the Percent Daily Value to the left of the label, so it comes first. This is important because the Percent Daily

Value tells consumers how much of certain nutrients they are getting from particular foods in the context of a total daily diet.
- Change the footnote to more clearly explain the meaning of the Percent Daily Value.

Changes were proposed in March 2014 with supplemental changes proposed in July 2015. If these changes are approved or changed, manufactures will have two years to make the necessary changes to package labeling. Readers are encouraged to visit the FDA site for current information on implementation of the proposed changes.[3]

USDA's Meat Grading Program

USDA has quality grades for beef, veal, lamb, yearling mutton, and mutton. It also has yield grades for beef, pork, and lamb. Although there are USDA quality grades for pork, these do not carry through to the retail level as do the grades for other kinds of meat graders, who are routinely checked by supervisors who travel throughout the country to make sure that all graders are interpreting and applying the standards in a uniform manner. A USDA Choice rib roast, for example, must have met the same grade criteria no matter where or when a consumer buys it.

When meat is graded, a shield-shaped purple mark is stamped on the carcass. With today's close trimming at the retail level, however, you may not see the USDA grade shield on meat cuts at the store. Instead, retailers put stickers with the USDA grade shield on individual packages of meat. In addition, grade shields and inspection legends may appear on bags containing larger wholesale cuts (Figure 27-11).

Health Claims

Claims for eight relationships between a nutrient or a food and the risk of a disease or health-related condition are now allowed (Figure 27-12). They can be made in several ways: through third-party references, such as the National Cancer Institute; statements; symbols, such as a heart; and vignettes or descriptions. Whatever the case, the claim must meet the requirements for authorized health claims; for example, it cannot state the degree of risk reduction and can only use *may* or *might* in discussing the nutrient or food–disease relationship. It must also state that other factors play a role in that disease.

The claims also must be phrased so that consumers can understand the relationship between the nutrient and the disease and the nutrient's importance in relationship to a daily diet. An example of an appropriate claim is: "While many factors affect heart disease, diets low in saturated fat and cholesterol may reduce the risk of this disease."

The allowed nutrient–disease relationship claims include:

- Calcium and osteoporosis
- Fat and cancer
- Saturated fat and cholesterol and coronary heart disease (CHD)
- Fiber-containing grain products, fruits, and vegetables and cancer
- Fruits, vegetables, and grain products that contain fiber and the risk of CHD
- Sodium and hypertension (high blood pressure)
- Fruits and vegetables and cancer
- Folic acid and neural tube defects

Ingredient Labeling

The list of ingredients has undergone some changes, too. Chief among them is a requirement for full ingredient labeling on so-called standardized foods, which were exempt previously. Ingredient declaration is now required on all foods that have more than one ingredient. Also, the ingredient list includes, when appropriate:

- FDA-certified color additives by name, such as FD&C Blue No. 1

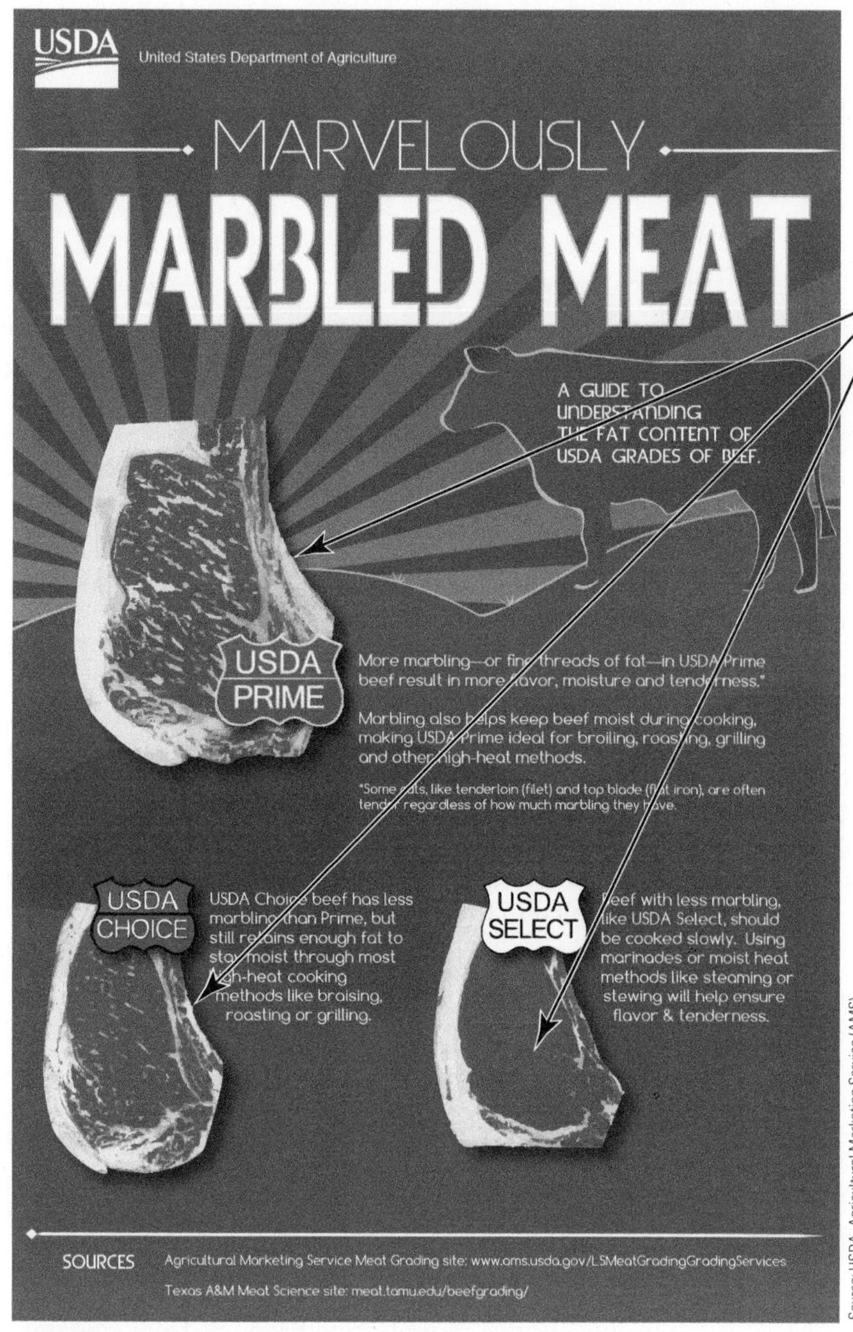

The degree of marbling, or fine threads of fat, is an indicator of the quality of meat. The higher the degree of marbling, the higher quality of meat. Here, the USDA distinguishes between USDA PRIME (best), USDA CHOICE (better), and USDA SELECT (acceptable). These grade shields communicate to consumers the quality of the beef that they are purchasing.

FIGURE 27-11 The degree of marbling in a cut of beef determines the grade shield it will receive from the USDA.

- Sources of protein **hydrolysates**, which are used in many foods as flavors and flavor **enhancers**
- Declaration of caseinate as a milk derivative in the ingredient list of foods that claim to be nondairy such as coffee whiteners

The main reason for these new requirements is that some people may be allergic to such additives and now have a better chance of avoiding them.

As required by NLEA, beverages that claim to contain juice now must declare the total percentage of juice on the information panel. In addition,

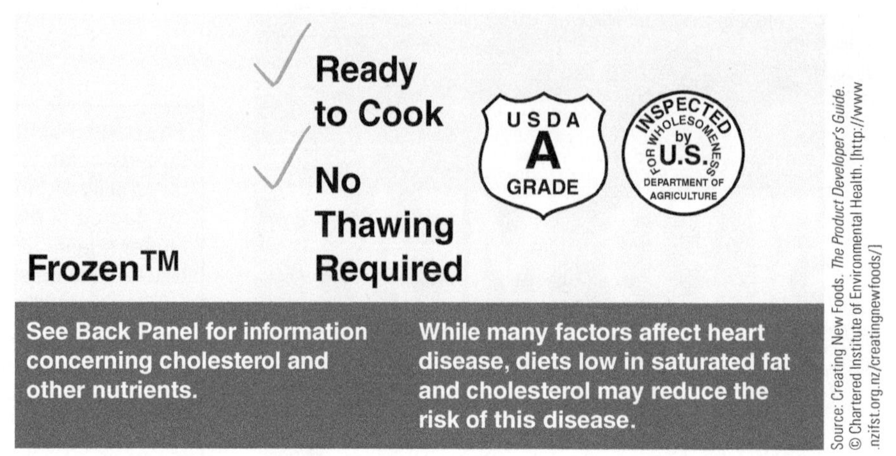

FIGURE 27-12 Label making a health claim.

the FDA regulation establishes criteria for naming juice beverages. For example, when the label of a multi-juice beverage states one or more—but not all—of the juices present, and the predominantly named juice is present in minor amounts, the product's name must state that the beverage is flavored with that juice or declare the amount of the juice in a 5% range—for example, "raspberry-flavored juice blend" or "juice blend, 2% to 7% raspberry juice."

FORMULATION AND COSTING

Many food products are made by combining raw materials in specific proportions in a formulation, and research on the effects of various formulations on product qualities is common in product design. Product formulation is a calculation to control the cost of the final product.

Basic formulation steps ensure that a proper mixture includes:

1. Setting product quality requirements
2. Finding information about the raw material such as composition, quality, and costs
3. Determining the limits of the raw materials and the processing variables
4. Using quantitative techniques such as linear programming or experimental designs
5. Using product profile tests and technical tests to make changes in formulations.[4]

The important properties of the raw materials in relation to the product qualities are recognized in the product design. An example of using the Pearson's Square in product formulation is highlighted in Chapter 17.

Factors in Packaging Design

Many sizes, designs, and materials are used to preserve and market food products. The packaging process must ensure that the product stays in a usable, edible, attractive, and safe state from packaging until consumption. Table 27-1 describes the various factors that packaging must account for to be effective in protecting products from contamination and damage of any kind.

Packaging materials include films, cardboard, metal, glass, and solid plastics. Packaging type includes bottles, cartons, and cans. The packaging size will determine the size of packaging equipment and materials needed. Designers do not have a great deal of room for creativity in food packaging except with regard to the graphic design,

TABLE 27-1 Factors Packaging Must Account for to Be Effective in Protecting Products from Contamination and Damage

FACTOR	PRINCIPAL CONSIDERATIONS
Consumer	Buying, transporting, storing, using, eating, disposing
Product	Containment Protection in external environment, distribution Presentation for communication, promotion, selling Use by consumer: convenient, dispensable, ergonomic, information Legal requirements
Process	Preservation of food, processing ability, interaction with processing Product packaging quality Machine ability in making, forming, filling, closing
Distribution	Outer packing, unitization, transport, storage conditions Retailer needs Storage, display, communication, bar coding, tamper-proofing
Environment	Resources used: energy, raw materials Waste: reuse, recycle, or disposable

Adapted from *Creating New Foods: The Product Developer's Guide.* (c) Chartered Inst. of Environmental Health. Web Edition published by NZIFST (Inc.)

but there are still examples of creative packaging designs on grocery store shelves everywhere. The use of computer design has advanced the industry packaging design in many ways.

FFA FOOD SCIENCE CAREER DEVELOPMENT EVENT

The FFA Food Science Career Development Event (CDE) includes a product-development activity in which participants are required to use skills in product formulation, costing, and packaging design while responding to various industry scenarios. The following are examples of scenario products from past FFA events:

- Ready-to-eat breakfast cereal for retail
- Refrigerated frozen cookie dough for wholesale
- Yogurt parfait for convenience stores
- Refrigerated, heat-and-serve pizza for retail
- Shelf-stable, dried fruit snack mix for retail

More information, contest rules, and preparation suggestions can be found on the national FFA Web site, www.ffa.org.

SUMMARY

Operating under the federal Food, Drug, and Cosmetic Act, the FDA regulates the labeling for all foods except meat and poultry. Under the federal Meat Inspection Act, the USDA regulates meat and poultry products. Additional federal acts cover specific foods.

Recent changes in the food label may provide consumers with more accurate and more useful nutritional information. This information is provided on the nutrition panel. Some of the components of the food label are mandatory, and some are voluntary. DRVs establish dietary standards for labels. Guidelines for labels promote the uniform definitions for words such as *free, low, lean, high, reduced,* and *less*. In addition, label regulations address how and when certain health claims can be made for foods. Finally, food-labeling requirements also address ingredient labeling.

REVIEW QUESTIONS

Success in any career requires knowledge. Test your knowledge of this chapter by answering these questions or solving these problems.

1. What are DRVs?
2. What is the new requirement on ingredient labels?
3. Name three agencies that regulate foods and labeling.
4. List three foods exempt from food labels.
5. Name four functions of quality-assurance programs.
6. A nutrition label reads "While many factors affect heart disease, diets low in saturated fat and cholesterol may reduce the risk of this disease." What type of claim is this?
7. List six components found on the nutritional panel.
8. What does it mean when a product is labeled as "light"?
9. Name the two terms used to describe the fat content of meat, poultry, seafood, and game meats.
10. What are the DRVs for fats and sodium?

STUDENT ACTIVITIES

1. Visit the FDA Web site and select one of the food guidance (good manufacturing) documents and develop a short report using available information. The address for the Web site is http://www.fda.gov/food/guidanceregulation/cgmp/default.htm.
2. Find an article in a newspaper or magazine or on the Internet that promotes the healthy effects of some food—for example, an article discussing foods low in cholesterol or foods with the ability to reduce cholesterol or reduce the risk of cancer or heart disease. Summarize this article and report to the class. As an additional part of the assignment, find the food described in the article and see if the label on the food makes the same claim.
3. Contribute five different labels from foods around your house to a classroom bulletin board displaying labels of all kinds. When the bulletin board display is complete, make a list of any

claims of "low," "free," "light," "reduced," "good source," or "high" and any health claims made on any of the labels.

4. Cut the nutrition label off of a familiar food. Make photocopies for the class and then go over each part of the label point by point. Lead students in discovering what they can learn about the food from its label.

ADDITIONAL RESOURCES

- Task—Ground Beef: http://www.achieve.org/files/CCSS-CTE-Ground-Beef-FINAL.pdf
- National FFA Career Development Events: https://www.ffa.org/SiteCollectionDocuments/cde_fst_2012.pdf
- Profitability of Ground Meat: http://www.canadabeef.ca/pdf/profitability.pdf
- Food Processing: http://www.foodprocessing.com/resource-centers/ingredients-formulation/
- Cargill Product Development: http://www.cargillfoods.com/na/en/product-development/

Internet sites represent a vast resource of information. Although those provided in this chapter were vetted by industry experts, you may wish to further explore the topics discussed in this chapter using a search engine such as Google. Keywords or phrases may include the following: *food labeling, nutrition panel title, serving size, nutrition information, Infant Health Formula Act, Nutrition Labeling and Education Act, quality assurance (QA), food laws, food substances, Daily Reference Values, nutrient panel descriptions, percentage daily values.* In addition, Table B-7 provides a listing of useful Internet sites that can be used as a starting point.

REFERENCES

Fortin, N. D. 2015. *Food regulation: Law, science, policy and practice*, 2nd ed. Hoboken, NJ: Wiley.

Potter, N. N., and J. H. Hotchkiss. 1999. *Food science*, 5th ed. New York: Springer.

Sanchez, M. 2015. *Food law and regulation for non-lawyers: A U.S. perspective.* New York: Springer

Vaclavik, V. A., and E. W. Christina. 2014. *Essentials of food science*, 2nd ed. New York: Springer.

ENDNOTES

1. Food and Drug Administration, Food Labeling: http://www.fda.gov/Food/IngredientsPackagingLabeling/LabelingNutrition/ucm248732.htm. Last accessed August 6, 2015.
2. Food and Drug Administration, Food Labeling: http://www.fda.gov/Food/IngredientsPackagingLabeling/LabelingNutrition/ucm248731.htm. Last accessed August 6, 2015.
3. Food and Drug Administration, Nutritional Labeling: http://www.fda.gov/Food/GuidanceRegulation/GuidanceDocumentsRegulatoryInformation/LabelingNutrition/ucm385663.htm. Last accessed August 6, 2015.
4. Wiriyacharee, P. (1990) The systematic development of a controlled fermentation process for Nham, a Thai semi-dry sausage, Ph.D. thesis, Massey University, New Zealand.

CHAPTER 28

World Food Needs

OBJECTIVES

After reading this chapter, you should be able to:

- **Discuss the effects of hunger and malnutrition**
- **Describe the impact of hunger worldwide**
- **Discuss possible causes of world hunger**
- **List seven steps identified by the United Nations for eliminating hunger**
- **Explain the role of technology in eliminating hunger**
- **Identify agencies and organizations involved in preventing and eliminating hunger**
- **Discuss the Plan of Action developed at the World Food Summit**
- **Recognize agencies and organizations concerned with eliminating hunger**

KEY TERMS

famine
food security
foreign aid
hunger
integrated pest management (IPM)
malnutrition
stunting
sustainable
undernutrition
underweight
wasting

NATIONAL AFNR STANDARD

FPP.04

Explain the scope of the food industry and the historical and current developments of food products and processing.

World hunger is a serious problem with no simple solution. Awareness is often the first step in solving any problem. In this case, approximately 805 million people are hungry. With an estimated world population of 7.3 billion people, this means one 1 of 9 suffers from chronic undernourishment. In fact, if all the world's undernourished people were gathered in one place, their population would be greater than every continent except Asia. Each year, people die from hunger or problems caused by hunger, including many children under 5 years of age.

WORLD FOOD HUNGER AND MALNUTRITION

What is **hunger**? Hunger means different things to different people. World hunger usually means **malnutrition**, **undernutrition**, or **famine**. *Undernutrition* means a person does not get enough food to have a healthy life. People can go short periods of time without eating and not suffer any permanent damage. Over long periods of time, however, hunger can slow physical and mental development in children. Because hunger weakens the body, otherwise survivable diseases often can cause death.

Malnutrition implies that a person eats but does not receive the amounts of nutrients needed to keep his or her body healthy. Because of this, a malnourished person does not always feel hunger. The malnourished just do not get enough of the right things to eat.

Hunger (malnutrition and undernutrition) not only kills but also causes serious physical injury to the body, including brain damage. Children suffer the most. The effects of hunger on children lead to **stunting**, being **underweight**, and **wasting**. A high proportion of children in the developing world suffer from undernutrition that results from a combination of inadequate food intake and diseases such as diarrhea that prevent the proper digestion of food.

Hunger can be a famine (a great shortage of food), but more often it is a highly limited diet. Often undernutrition occurs on a seasonal basis. In developing countries, the time before harvest is difficult. People have used all of their food and economic resources. They are counting on the new harvest to bring them food and money. If just one of many natural disasters occurs (e.g., drought, insects, freeze), then people will be without food or money to buy food. If they are lucky and harvest their crops, they will have food. Each year this problem happens during harvest somewhere in the world.

Many of the people who suffer from hunger are people who live in rural areas in developing countries (Figure 28-1). Strangely, the people in the country who grow the food are hungry. However, when there is a shortage of food or a crisis, these people are the hardest to get food to (Figure 28-2).

Another group of people who suffer from hunger are those who live in slums. A slum is an area, usually in a large city, that is typically overcrowded. Living conditions are unhealthy, poverty is high, and most people are without jobs (Figure 28-3).

Supplying food for a city of 10 million people is a massive undertaking that stretches the resources of a region. A city of this size requires the daily importation of at least 6,000 tons (12 million pounds) of food!

Worldwide, more people are living in large cities than ever before (Figure 28-4), and these large cities have slums associated with them. This means more and more people live in slums, where the availability of food is limited (Figure 28-5).

Most of the world's malnourished people live in Southeast and South Asia, Latin America, the Caribbean, the Near East and North Africa, and Sub-Saharan Africa (Figure 28-6). Poverty is common.

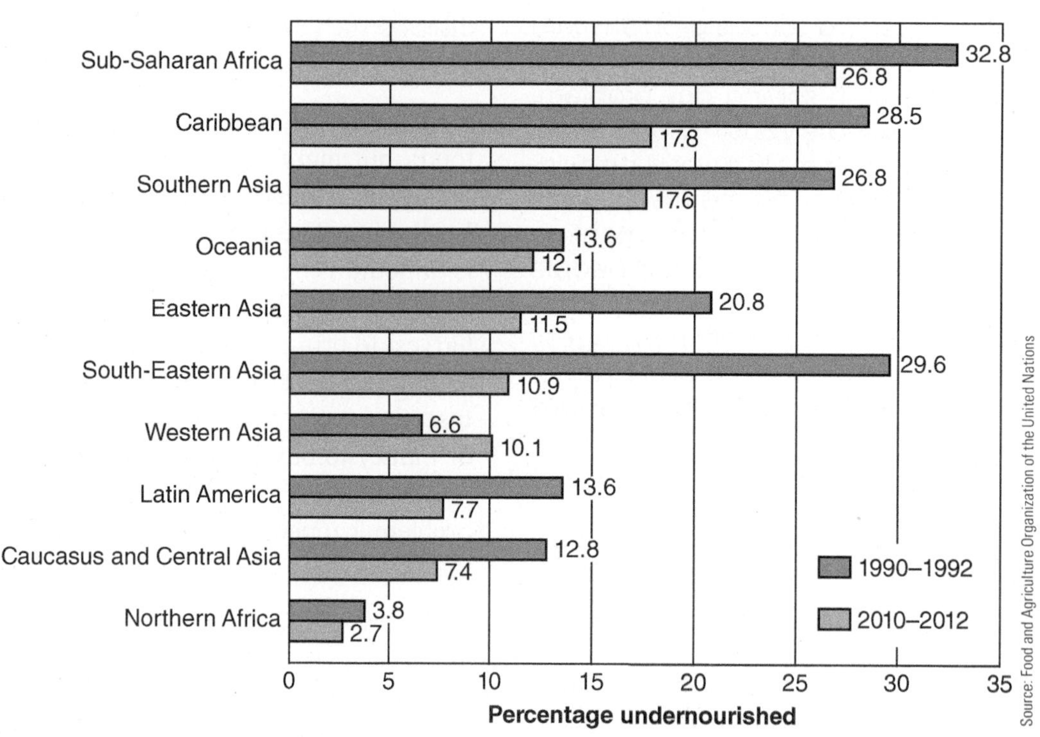

FIGURE 28-1 Percent of undernourished in various regions around the world.

FIGURE 28-2 Rural hunger—a problem in many developing countries.

FIGURE 28-3 The slums of large cities often have many hungry people.

Causes of Hunger

The four most common misconceptions about the causes of hunger are:

1. Not enough food is available to feed everyone.
2. The population is too large.
3. Governments cause hunger.
4. Foreign aid helps eliminate hunger.

Not Enough Food Available to Feed Everyone. Currently, enough food to feed everyone in the world is produced. Concern is growing, however, because the world's population is expected to reach 9 billion people by 2050. Currently, other conditions affect how food is produced or distributed. If the food produced in the world could be divided so that everyone would

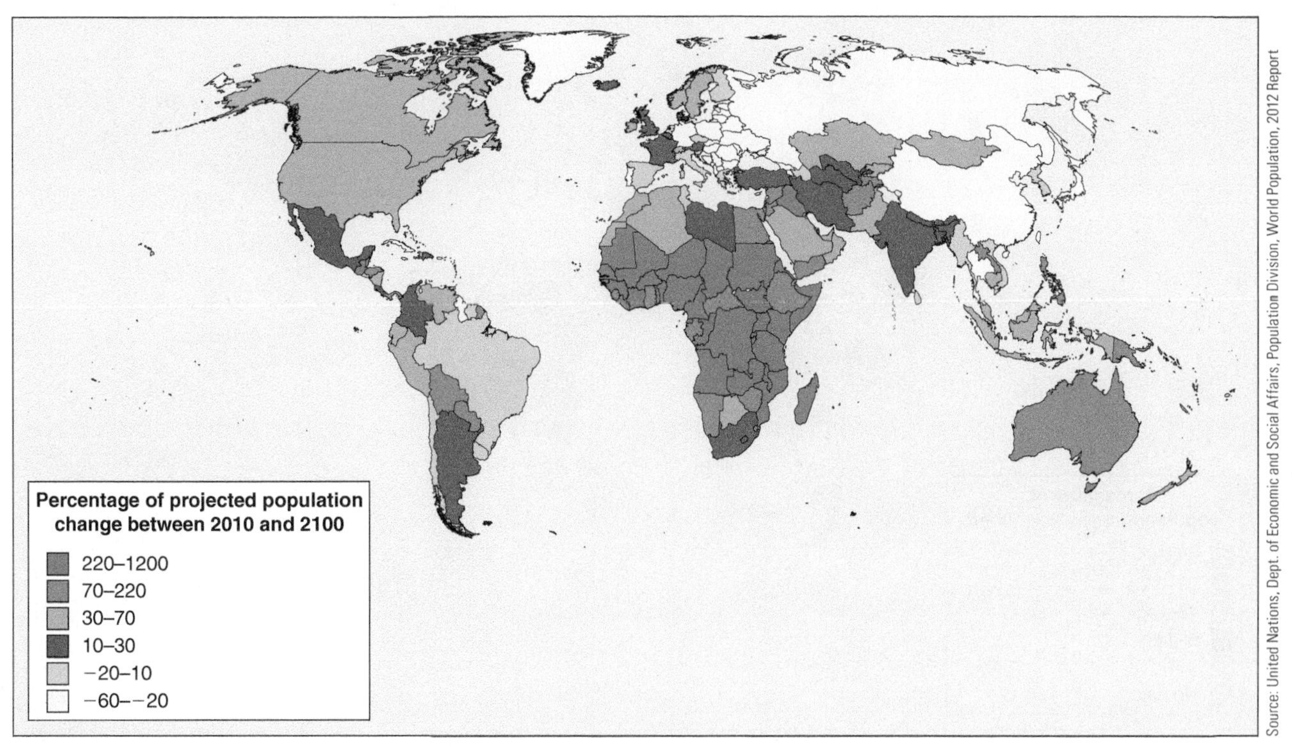

FIGURE 28-4 World map: Projected population growth, 2010–2100.

FIGURE 28-5 Garbage dumps outside large cities become a source of food for some people.

get the same amount, some 10% of food produced would be left over. Enough grain is produced in the world to give every man, woman, and child 2 pounds each day. This does not include all the beans, potatoes, fruits, and vegetables that are produced. Two pounds of grain can provide 3,000 calories. The average recommended daily minimum is 2,300 calories.

Almost every country in the world has the resources needed to feed its people and rid the country of hunger. In many areas where food is produced, harvested food is sent to other areas where people can pay more. Food production is not the issue, poverty is. If people had more money to pay for the harvested food, they would stay in or move to an area.

The cause of poverty includes a lack of resources for poor people because of extremely unequal income distributions in the world and conflict within specific countries. As of 2015, the World Bank has estimated that slightly more than 1 billion poor people in developing countries live on $1.25 a day or less. This is better than 1.91 billion in 1990 and 1.93 billion in 1981. Even though we are moving in the right direction, this still means that 17% of people in the developing world live at or below $1.25 a day and in extreme poverty.

The United Nations asked questions of 83 countries and found that 3% of the landlords control almost 80% of the land. This means that most

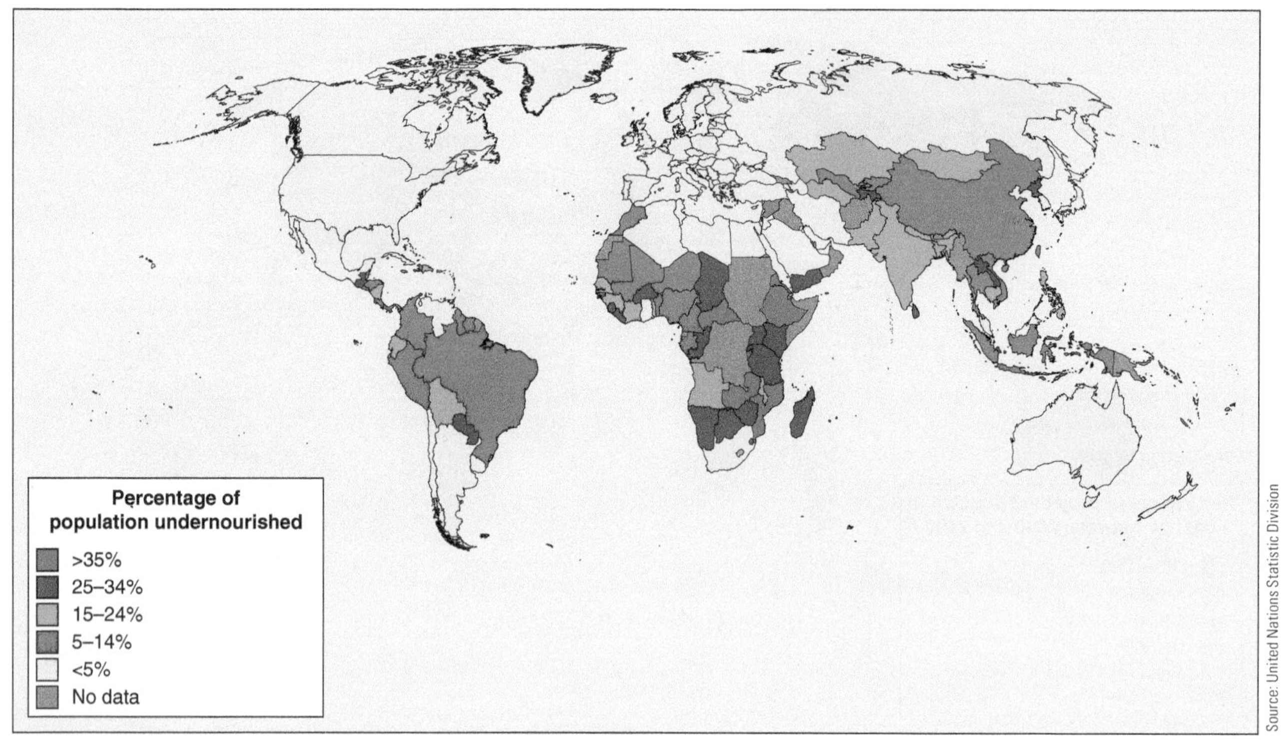

FIGURE 28-6 Percentage of the population affected by undernourishment by country.

people live on another person's land. Usually the food is produced for the people who own the land.

If people do not have the money to buy food, they also cannot borrow the money they need. In most countries, only 5% to 20% of the producers are able to borrow money from institutions such as banks. The others must go to landlords or moneylenders who charge interest rates of up to 200%. An interest rate of 200% on a loan of $100 means someone must pay back $300. With large interest rates, people only sink deeper into poverty.

Sometimes poor weather conditions (Figure 28-7) destroy an entire season's harvest. If these bad weather conditions happen in a small or poor country, a localized famine can result. If several countries have a famine at the same time, food becomes hard to get and can even effect the price of food worldwide.

FIGURE 28-7 Hay bales float in floodwaters in the middle of rows of corn.

The Population Is Too Large. Total world population is not as important as where people live. People in poor countries tend to have many children. Large numbers of children can drain a country's food supply. Children, however, are critically important to countries that are still developing. In many poor areas, people have a lot of children because of poor health care and nutrition and because many children do not survive to adulthood (Figure 28-8). In developing countries, children usually help with work, bringing

© africa924/Shutterstock.com

FIGURE 28-8 Nutritional assessment helps determine to what extent children in underdeveloped countries suffer from undernourishment.

in extra money to their families. Most of the time, these people depend on their children to take care of them when they get old. In countries like the United States and Canada, on the other hand, many people are able to save money for retirement, but this is not always true in developing countries. Although children may use a lot of a country's resources, they are necessary to the country's economy and well-being.

Still, a dense population does not always mean that the people suffer from hunger. Surveys have shown that countries that have more people per acre (population density—the average number of people per space unit) are not always the most hungry.

Governments Cause Hunger. A government is just one element that controls a country. Other controlling agencies or organizations are multinational corporations (businesses that have operations in more than one country), international agencies, and other governments. Altogether they form a group of people who lives differently than the country's rural population. A country's government may add to the hunger of its people, but it is seldom if ever the only cause.

Foreign Aid Helps Eliminate Hunger. Every year the United States and other countries send large quantities of food to hungry people in many countries. Much of the food that is supposed to get to the hungry never reaches them. But even if it did, it would only be treating the symptoms and not the cause. It would only provide temporary relief. It also puts more food in the foreign market, causing food prices to drop. When food prices go down, rural farmers get less money for what they produce.

When food is free and available, people start to depend on it. When the food is gone, the people are at least as badly off as when the food first arrived because they often do not try to get food on their own. Foreign aid needs to help fix the problems that cause hunger. If the problem of hunger is to be solved, the most important goal should be growing and preserving more food in a region. This approach has been part of what has been called the Green Revolution. Its purpose is to teach countries to produce enough food to meet their own needs.

FIGHTING THE PROBLEM

No easy answers exist to fix the problem of world hunger. Sending food to countries where there is an emergency can help, but such **foreign aid** is only a temporary solution.

The Food and Agriculture Organization (FAO) is an agency of the United Nations that leads international efforts to defeat hunger. Serving both developed and developing countries, the FAO acts as a neutral forum where all nations meet as equals to negotiate agreements and debate policy. FAO is also a source of knowledge and information, and it helps developing countries and countries in transition modernize and improve their agricultural, forestry and fisheries practices, thus ensuring good nutrition and food security for all.

The FAO has outlined the following priorities in its fight against hunger:

- *Help eliminate hunger, food insecurity, and malnutrition* by contributing to the eradication of hunger with policies and political commitments that support food security and by making sure that up-to-date information about hunger and nutritional challenges and solutions is available and accessible.

- *Make agriculture, forestry, and fisheries more productive and sustainable* by promoting evidence-based policies and practices that support highly productive agricultural sectors (crops, livestock, forests, and fisheries) while ensuring that the natural resource base does not suffer in the process.
- *Reduce rural poverty* by helping the rural poor gain access to the resources and services they need, including rural employment and social protection, to forge a path out of poverty.
- *Enable inclusive and efficient agricultural and food systems* by helping to build safe and efficient food systems that support smallholder agriculture and reduce poverty and hunger in rural areas.
- *Increase the resiliencies of livelihoods from disasters* by helping countries prepare for natural and human-caused disasters, reducing their risks, and enhancing the stability of their food and agricultural systems.

The FAO (www.fao.org) created a seven-step plan against hunger that provides starting points for a complex problem.

Step 1: More Self-Sufficiency

Many countries suffering from hunger purchase many of the items they need from other countries. Many need to learn to do more for themselves so they will not need to purchase as much from other countries.

Step 2: Check Farming Regulations

Many developing countries have rules and regulations for their farmers. These rules must ensure that farmers will want to produce food and will get a fair amount of money for their food. For the farmers to do their part, they need training to use their land and water wisely.

THE MOST IMPORTANT GRAIN

For more than 60% of people in the world, rice is their main food. Although the production of wheat is greater in absolute tonnage, rice directly supports more people than any other crop. The Asia-Pacific region, where more than 56% of the world's population live, adds 51 million more rice consumers every year. As a result, the thin line of rice self-sufficiency experienced by many countries is disappearing quickly. Whether the current 524 million metric tonnes of rice produced annually can be increased to 700 million metric tonnes by 2025 with less land, fewer people, less water, and fewer pesticides is a big question. The task will be enormous: substantially increasing current levels of production by putting more area under cultivation with modern varieties and more fertilizer use. Irrigated rice areas currently occupy some 56% of the total area and contribute 76% to total production. Increasing the size of this area is difficult because of problems with soil salinity, the high cost of development, water scarcity, alternative and competing uses of water, and environmental concerns. Increased productivity on a tight time scale will need to use more advanced technologies.

Rice contributes to a wide variety of products other than edible grain—beer, wine, straw, and paper. This tropical grass originated in Southeast Asia and was first cultivated at least 10,000 years ago. Today, it is grown in every tropical area around the world and is also successfully cultivated in areas where the climate is far cooler, such as mountain terraces in Nepal.

In most rice-growing areas, young rice seedlings are planted in standing water in fields called *paddies*. The water remains until the seed heads emerge, after which the paddies are drained while the crop ripens.

(*Continues*)

THE MOST IMPORTANT GRAIN (continued)

For more information about rice and culture, visit the following Web sites:

- Rice Production in the Asia-Pacific Region: http://www.fao.org/docrep/003/x6905e/x6905e04.htm
- Rice: http://www.ers.usda.gov/topics/crops/rice.aspx
- Rice Anatomy: http://www-plb.ucdavis.edu/labs/rost/Rice/introduction/intro.html
- Grain Genes: http://wheat.pw.usda.gov/ggpages/maps.shtml
- Everything You Need to Know About the History of Rice: https://www.ucl.ac.uk/rice/historyofrice

Step 3: Proper Storage

After the food is harvested, it must be properly stored until needed. When the food is needed, it has to be taken where and when it is needed most. Both the storage and transportation of food need improvement.

Step 4: Check Food Aid

Food aid has to be checked to ensure it gets to the hungry. The country's production then must be checked. No country should be allowed to produce less food for itself because it receives food aid.

Step 5: Work Together

Developed and developing countries need to work together more. This makes trade easier and can stabilize food prices.

Step 6: Prevent Waste

Countries need not only to keep from using too much food but also from wasting food.

Step 7: Pay Off Debt

Many developing countries owe a lot of money. These countries need to pay off their debts. This can be done by paying back money they make on exports (things that they sell to other countries). Unfortunately, countries that owe a lot of money have a lot of trouble borrowing more money.

A visit to the FAO Web site (http://www.fao.org/) provides links to the actions, countries, themes, publications, statistics, and partnerships of the FAO. The site is a great resource for anyone who wants to know and do more about world hunger.

ROLES OF TECHNOLOGY

To alleviate hunger, technology research in the following areas should take high priority:

- Improving the application of technology to natural resource management
- Protecting crops without relying heavily on pesticides
- Genetically improving key crops
- Taking global action to advance scientific knowledge and its application

Resource Management

Using available technology, a country's agricultural resource base must be characterized and evaluated—both in relation to existing cultivated areas and to land currently under forest or pasture. This assessment should be done using such modern tools and techniques as remote sensing, aerial photography, sonic devices, and geographic information systems.

Even with an evaluation of the resource base, three factors determine the future success of resource management:

1. The extent to which cultivars and cropping systems can be adapted to fit the resource rather than trying to modify the resource itself
2. The degree to which the local community can be involved in the planning and management of the resource and feel ownership of it
3. The willingness of governments to adjust their policies to encourage efficient and **sustainable** resource management (Refer to Chapter 2, Food System and Sustainability.)

Using this information, a country could manage soil fertility for higher productivity by improving soil and water management in rain-fed farming and irrigated land. Techniques such as drainage and canal lining could reduce salinity and waterlogging.

Protection of Crops

Important progress is being made to protect crops. Less toxic chemicals and more efficient methods of applying them have been developed. Plant breeding allows the introduction of host-plant resistance. This is being accelerated by gene mapping and genetic engineering techniques to identify sources of genetic resistance in crops and their wild relatives. Biological control through the use of predators and parasites is also being used on some crops.

Integrated pest management (IPM) combines biological and crop-management practices to develop cost-effective controls with reduced dependence on pesticides. This method could be used on a wider scale worldwide, but it requires highly skilled farmers. IPM may require community cooperation and crop consulting to help producers monitor pests. Because pests and diseases do not respect national boundaries, crop protection merits high priority for international cooperation to better understand and monitor its status, improve diagnosis, maintain databases, and identify natural enemies and sources of resistance.

Genetic Improvement

Plant breeding is the cornerstone of yield-increasing technology, and it also plays a key role in preventing yield losses from disease, weather, and insects (Figure 28-9). Raising yields in regions where they are already high—for example, wheat in Western Europe and irrigated rice in East Asia—may not be realistic. Plant breeding must also improve the nutritional quality of a food, especially nutrients such as vitamins. Plant breeding may also help crops overcome stresses such as drought, extreme temperatures, soil acidity, and other nutrient problems that keep yields low in many other areas.

Source: USDA, Agricultural Research Service (ARS), photo by Peggy Greb.

FIGURE 28-9 Research in genetic improvement of plants and animals offers hope for feeding more hungry people.

Biotechnology offers hope of finding solutions to these problems. It already contributes to genetic improvements in cereals, root crops, vegetables, and industrial crops as well as to animal health. Biotechnology will also improve diagnostic and asexual propagation techniques in raising the nutritional density of crops and increasing resistance to disease, weather, and insects. The evolution of new human-made species to complement existing cereals and improvement of tolerance to stresses will eventually be a reality.

Global Action

Whether world food needs in 2020 can be met both in amount and quality likely will depend on the successful enlistment of scientific resources for research and on improving farmers' skills to manage their resources. Sustainability will require knowledge distribution, training, and accessibility of inputs (seed, fertilizer, money, and so on). Four hopeful signs in this area are:

1. Strong international support for seed and tissue (germplasm) collection, conservation, and evaluation. In many countries, an adequate supply of quality seed is still a problem at the farm level.
2. Increasing international cooperation in resource assessment and monitoring.
3. Development and application of the potential for expanding knowledge created by modern information networks, videos, computers, and related technology.
4. Expansion of regional research and technology transfer networks among developing countries.

WORLD FOOD SUMMIT

A worldwide effort to address poverty and world hunger needs started with the 1974 World Food Conference. In 1996, member attendees of the World Food Summit pledged to achieve food security and reduce the number of undernourished people by half by 2015. In 2002, World Food Summit members called for establishing an intergovernmental working group to prepare guidelines on implementing the rights of all people to food. This led to drafting of the Right to Food Guidelines.

In 2009, the World Summit on Food Security was held in Rome. Some 60 heads of government attended and unanimously adopted a new declaration, pledging their governments to eradicate hunger from the world at the earliest possible date. The entire declaration and proceedings can be read at the Web site for the World Food Summit: www.fao.org/wsfs/world-summit.

World Food Summit Plan of Action

Food security exists when all people at all times have access—physical and economic—to sufficient, safe, and nutritious food to meet their dietary needs and food preferences for active and healthy lives.

Eradicating poverty is essential to improving access to food. Most of those who are undernourished cannot produce or cannot afford to buy enough food. They have inadequate access to means of production such as land, water, inputs, improved seeds and plants, appropriate technologies, and farm credit. In addition, wars, civil strife, natural disasters, climate-related ecological

SCIENCE CONNECTION!

Global warming has been a term used for several years, but what does it mean? Will climate changes around the world affect food production? How do weather and climate currently affect food production?

changes, and environmental degradation have adversely affected hundreds of millions of people. Although food assistance may ease their plight, it is not a long-term solution to the underlying causes of food insecurity.

There can also be pockets of food insecurity across a nation. For example, although the United States is not an impoverished country, some areas within the country are food insecure. An estimated 14.3% of American households were food insecure at least some time during 2013, meaning they lacked access to enough food for active healthy lives for all household members.[1]

A peaceful and stable environment in every country is a fundamental condition for realizing and making food security sustainable. Governments should create environments that facilitate private and group initiatives to allocate their skills, efforts, and resources, especially to meet the common goal of food for all. Farmers, fishers, foresters, and other food producers and providers have critical roles in achieving food security, and their full involvement and ability are crucial for success.

The goal of enough food for all can be reached. The 7.3 billion people in the world today have, on average, 15% more food per person than the global population of 4 billion people had 30 years ago.[2] Large increases in world food production, through the sustainable management of natural resources, are required to feed a growing population and to achieve improved diets. Increased production, including traditional crops and their products—in combination with food imports, reserves, and international trade—can strengthen food security and address regional disparities.

The follow-up to the World Food Summit includes actions at the national, intergovernmental, and interagency levels. The international community and the UN system, including the Food and Agriculture Organization and other agencies, has important contributions to the implementation of the World Food Summit Plan of Action. The FAO Committee on World Food Security will have responsibility to monitor the implementation of the Plan of Action, which should allow people to enjoy their human rights, including the right to development. The plan should also recognize various religious and ethical values, cultural backgrounds, and philosophical convictions of individuals and their communities.

MATH CONNECTION!

What do you think can be done to end world hunger? How can it be done? How much will it cost? Who will be responsible for paying that cost? Brainstorm some ideas that you think could help end hunger. What can *you* do?

HUNGER AGENCIES AND ORGANIZATIONS

Table 28-1 lists some of the agencies and organizations involved in hunger-related projects. These organizations can serve as useful sources of information and educational materials on fighting hunger.

SCIENCE CONNECTION!

Choose and research one of the hunger organizations listed in Table 28-1. What is it doing to help end world hunger? Is there a way you can become involved?

TABLE 28-1 Hunger Organizations

NAME OF ORGANIZATION AND WEB SITE	ADDRESS
Action Against Hunger www.actionagainsthunger.org	247 West 37th Street 10th Floor New York, NY 10018
Bread for the World www.bread.org	425 3rd Street SW Ste 1200 Washington, DC 20024
CARE International www.care-international.org	777 First Avenue 5th Floor NY 10017 New York
Center for Science in the Public Interest www.cspinet.org	1220 L St. NW Suite 300 Washington, DC 20005
Children's Hunger Fund childrenshungerfund.org	13931 Balboa Blvd. Sylmar, CA 91342
Committee for Economic Development www.ced.org	2000 L Street NW Suite 700 Washington, DC 20036
Committee for UNICEF www.unicef.org	125 Maiden Lane 11th Floor New York, NY 10038
Church World Service www.cwsglobal.org	475 Riverside Dr. Suite 700 New York, NY 10115
Cross International www.crossinternational.org	600 SW Third Street Suite 2201 Pompano Beach, FL 33060
Feed the Children www.feedthechildren.org	333 N. Meridian Oklahoma City, OK 73107
Food and Agriculture Organization of the United Nations www.fao.org	One United Nations Plaza Suite DC1 - 1125 New York, NY 10017
Food for the Hungry www.fh.org	1224 E. Washington Street Phoenix, AZ 85034-1102
Institute for Food and Development Policy foodfirst.org	398 60th Street Oakland, CA 94618
Freedom from Hunger www.freedomfromhunger.org	1644 DaVinci Court Davis, CA 95618

(Continues)

TABLE 28-1 Hunger Organizations

NAME OF ORGANIZATION AND WEB SITE	ADDRESS
Hope Link www.hope-link.org	10675 Willows Road NE Willows Creek Corporate Center Suite 275 Redmond, WA 98052
The Hunger Project www.thp.org	5 Union Square West 7th Floor New York, NY 10003
Institute for Policy Studies www.ips-dc.org	1112 16th St. NW Suite 600 Washington, DC 20036
International Bank for Reconstruction and Development (IBRD) (World Bank) www.worldbank.org	1818 H Street, NW Washington, DC 20433
International Institute for Environment and Development www.iied.org	80-86 Gray's Inn Road London WC1X 8NH, UK
National Academy of Sciences National Academy of Engineering National Research Council Institute of Medicine www.nas.edu	500 Fifth St., N.W. Washington, DC 20001
Oxfam America www.oxfamamerica.org	226 Causeway Street 5th Floor Boston, MA 02114-2206
Pan American Health Organization www.paho.org	525 23rd Street NW Washington, DC 20037
Project Open Hand www.openhand.org	730 Polk Street San Francisco, CA 94109
Resources for the Future www.rff.org	1616 P St. NW Washington, DC 20036
Stop Hunger Now www.stophungernow.org	615 Hillsborough St. Suite 200 Raleigh, NC 27603
United Nations Development Program (UNDP) www.undp.org	One United Nations Plaza New York, NY 10017
World Food Program www.wfp.org	1725 I Street NW Suite 510 Washington, DC 20006

(*Continues*)

TABLE 28-1 Hunger Organizations

NAME OF ORGANIZATION AND WEB SITE	ADDRESS
World Health Organization (WHO) www.who.int	525 23rd Street NW Washington, DC 20037
World Hunger Education Service www.worldhunger.org	P.O. Box 29056 Washington, DC 20017
World Hunger Year (WHY) www.whyhunger.org	505 Eighth Avenue Suite 2100 New York, NY 10018
World Watch Institute www.worldwatch.org	1400 16th St. NW Ste. 430 Washington, DC 20036

SUMMARY

One word explains hunger—poverty. Poverty is the state of living without enough money, material possessions, or other resources. Enough food is produced around the world to feed everyone, however. Most countries have the resources they need to eliminate hunger, although people often do not have the money they need to buy the food their countries produce.

In developing countries, poverty, hunger, and malnutrition are among the principal causes of accelerated migrations from rural to urban areas. The largest population shift of all time is now underway.

Harmful seasonal variations and midyear instability of food supplies can be reduced. Making progress in addressing food security should include minimizing people's vulnerabilities to climate fluctuations, agricultural pests, and diseases.

Unless national governments and the international community address the many causes of food insecurity, the number of hungry and malnourished people will remain high in developing countries, particularly in Africa south of the Sahara. Sustainable food security will not be achieved. The resources required for investment in food security will be generated mostly from domestic, private, and public organizations and groups.

REVIEW QUESTIONS

Success in any career requires knowledge. Test your knowledge of this chapter by answering these questions or solving these problems.

1. What are the three effects of hunger on children?
2. List five agencies involved in tackling world food hunger.
3. In developing countries, where do many of those who suffer from hunger live?
4. List the four ways technology research can help with hunger.
5. What three things does world hunger usually mean?

6. List the seven steps identified by the United Nations for eliminating hunger.
7. Define food security.
8. List the four common misconceptions about the causes of hunger.
9. What is an essential challenge to improving access to food?
10. What is the difference between famine and undernourishment?

STUDENT ACTIVITIES

1. Develop a report or presentation about hunger and malnutrition in an area of the world.
2. Visit the Web site for the FAO and develop a report on the type of information it provides. Determine how the information is helpful in the fight against hunger.
3. A city of 10 million people requires at least 6,000 tons of food each day. Find different ways to express this need. For example, how many pounds per year is this? How many semitruck loads are required each week or month? Share your examples with the class.
4. Contact one of the agencies in Table 28-1 and report on the information it provides.
5. Develop a visual presentation on a new food or technology that has the potential to help eliminate hunger in one area of the world.
6. Conduct a class debate on the causes of world hunger and the possible solutions.

ADDITIONAL RESOURCES

- Food Needs and Population: http://www.fao.org/docrep/x0262e/x0262e23.htm
- Can We Meet the World's Growing Demand for Food? http://www.agmrc.org/renewable_energy/renewable_energy/can-we-meet-the-worlds-growing-demand-for-food/
- World Food Program: http://www.wfp.org/
- World Health Organization—Food Security: http://www.who.int/trade/glossary/story028/en/

Internet sites represent a vast resource of information. Although those provided in this chapter were vetted by industry experts, you may wish to further explore the topics discussed in this chapter using a search engine such as Google. Keywords or phrases may include the following: *hunger organizations, famine, malnutrition, food security,* and *foreign aid*. In addition, Table B-7 provides a listing of useful Internet sites that can be used as a starting point.

REFERENCES

Food and Agriculture Organization (FAO). 2015. *The state of food and agriculture 2015 in brief (SOFA).* Available at the FAO Web site: www.fao.org.

Food and Agriculture Organization (FAO). 2016. Enhancing early warning capabilities and capacities for food safety. Available at the FAO Web site: www.fao.org.

Food and Agriculture Organization (FAO). 2016. Achieving zero hunger: The critical role of investments in social protection and agriculture, 2nd ed. Available at the FAO Web site: www.fao.org.

Janssen, S. (Ed.). 2016. *The world almanac and book of facts 2016*. New York: World Almanac.

Worldwatch Institute. 2000. *State of the world 2016: Confronting hidden threats to sustainability.* Washington, DC: Island Press.

ENDNOTES

1. USDA http://www.ers.usda.gov/publications/err-economic-research-report/err173.aspx. Last accessed July 14, 2015.
2. World Hunger Educational Service: http://www.worldhunger.org/articles/Learn/world%20hunger %20facts%202002.htm. Last accessed July 14, 2015.

CHAPTER 29

Food and Health

OBJECTIVES

After reading this chapter, you should be able to:

- Describe the relationship between diet and health
- Recognize health issues associated with being overweight or obese
- Define and use the body mass index (BMI)
- Determine suggested caloric intake for a specific age, gender, and activity
- Identify the relationship between free radicals and antioxidants
- Explain type 2 diabetes
- Recommend foods for those with type 2 diabetes
- Identify and describe 10 digestive disorders
- List dietary treatments for lactose intolerance
- Describe possible dietary causes of diarrhea
- Recognize the dietary treatment for celiac disease

NATIONAL AFNR STANDARD

FPP.04FPP.02

Apply principles of nutrition, biology, microbiology, chemistry, and human behavior to the development of food products.

KEY TERMS

anorexia nervosa
antioxidants
binge eating
body mass index (BMI)
bulimia nervosa
calorie
celiac disease
constipation
Crohn's disease
dental caries
diabetes
diarrhea
diverticular disease
fiber
flatulence
food allergy
free radicals
gastroesophageal reflux disease (GERD)
gluten
Helicobacter pylori
herbal supplements

OBJECTIVES (continued)

- List good sources of fiber in the diet
- Name three eating disorders
- Recognize the cause of dental caries
- Describe a food allergy and give two foods that can cause allergic reactions in some people
- Define and give an example of a phytonutrient
- Explain nutrigenomics
- Discuss cautions when taking herbs or herbal supplements
- Identify the roles of fruits, vegetables, whole grains, protein, carbohydrates and fats in maintaining health
- Recommend foods to help maintain health

KEY TERMS (continued)

insulin
lactase
lactose
lactose intolerance
monounsaturated fatty acid (MUFA)
nutraceutical
nutrigenomics
obesity
pancreatitis
peptic ulcer
phytonutrients
polyunsaturated fatty acid (PUFA)
saturated fatty acids
sleep apnea
trans-fatty acid

Trendy diets and nutrition research aimed at healthy living seem to change almost daily. Still, amidst all the hype, study after study shows that good food choices have a positive impact on health, and poor diets have negative long-term effects. A healthy diet provides the body the nutrients it needs to perform physically, maintain wellness, and fight disease.

FOOD AND DISEASE

In general, a poor diet in combination with a sedentary lifestyle, large portion sizes, and high stress, is blamed for the increase in obesity and associated diseases in the United States over the last three decades. According to the Centers for Disease Control and Prevention, more than one-third of U.S. adults are obese. Diseases associated with obesity include type 2 diabetes, high blood pressure, coronary heart disease, stroke, gallbladder disease, osteoarthritis, sleep apnea, respiratory problems, and certain cancers, including breast cancer in women.

Other food-related health issues include digestive disorders such as constipation, diarrhea, celiac disease, gallstones, heartburn, food allergies, lactose intolerance, ulcers, osteoporosis, and Crohn's disease. Also included are eating disorders such as anorexia nervosa, bulimia nervosa, and binge-eating disorders. Finally, tooth decay—dental caries—is a food-related disease.

FIGURE 29-1 Good food choices for meals influences the maintenance of health and the prevention, alleviation, or cure of disease.

What individuals choose to eat is central to their health, Figure 29-1. Food acts as medicine to maintain health, as well as prevent and treat disease. Food influences the maintenance of health and the prevention, alleviation, or cure of disease.

Obesity

Millions of Americans and people around the world are overweight or obese. Being overweight or obese puts an individual at risk for many health problems. The more body fat an individual has and the more an individual weighs, the more he or she is likely to develop:

- Coronary heart disease
- High blood pressure
- Type 2 diabetes
- Gallstones
- Breathing problems
- Certain cancers

The terms *overweight* and *obesity* refer to body weight that is greater than is considered healthy for a certain height. The most useful measure of overweight and **obesity** is **body mass index (BMI)**. BMI is calculated from a person's height and weight (weight in kilograms divided by the square of height in meters). The higher a person's BMI, the higher the risk for certain diseases such as heart disease, high blood pressure, type 2 diabetes, gallstones, breathing problems, and certain cancers.

The BMI applies to most adult men and women age 20 and older. Although BMI can be used for most men and women, it does have limits. It may overestimate body fat in athletes and others who have muscular builds, and it may underestimate body fat in older persons and others who have lost muscle. For children age 2 and older, BMI percentile is the best assessment of body fat.

BMI calculators can be found on a number of Web sites (e.g., http://www.nhlbi.nih.gov/health/educational/lose_wt/BMI/bmicalc.htm), and BMI calculators can be downloaded as smartphone apps. These calculators take an individual's height and weight and then calculate the BMI. Tables such as Table 29-1 can also help determine BMI, and Figure 29-2 is a visual guide to BMI.

BMI categories include:

- Underweight <18.5
- Normal weight = 18.5 to 24.9
- Overweight = 25 to 29.9
- Obesity = 30 and more

An individual's weight is the result of many factors, including environment, family history and genetics, metabolism (the way the body changes food and oxygen into energy), and behavior or habits. Some factors—family history, for example—cannot be changed. However, other

TABLE 29-1 Body Mass Index Table

	NORMAL						OVERWEIGHT					OBESE										EXTREME OBESITY														
BMI	**19**	**20**	**21**	**22**	**23**	**24**	**25**	**26**	**27**	**28**	**29**	**30**	**31**	**32**	**33**	**34**	**35**	**36**	**37**	**38**	**39**	**40**	**41**	**42**	**43**	**44**	**45**	**46**	**47**	**48**	**49**	**50**	**51**	**52**	**53**	**54**
Height (inches)																Body Weight (pounds)																					
58	91	96	100	105	110	115	119	124	129	134	138	143	148	153	158	162	167	172	177	181	186	191	196	201	205	210	215	220	224	229	234	239	244	248	253	258
59	94	99	104	109	114	119	124	128	133	138	143	148	153	158	163	168	173	178	183	188	193	198	203	208	212	217	222	227	232	237	242	247	252	257	262	267
60	97	102	107	112	118	123	128	133	138	143	148	153	158	163	168	174	179	184	189	194	199	204	209	215	220	225	230	235	240	245	250	255	261	266	271	276
61	100	106	111	116	122	127	132	137	143	148	153	158	164	169	174	180	185	190	195	201	206	211	217	222	227	232	238	243	248	254	259	264	269	275	280	285
62	104	109	115	120	126	131	136	142	147	153	158	164	169	175	180	186	191	196	202	207	213	218	224	229	235	240	246	251	256	262	267	273	278	284	289	295
63	107	113	118	124	130	135	141	146	152	158	163	169	175	180	186	191	197	203	208	214	220	225	231	237	242	248	254	259	265	270	278	282	287	293	299	304
64	110	116	122	128	134	140	145	151	157	163	169	174	180	186	192	197	204	209	215	221	227	232	238	244	250	256	262	267	273	279	285	291	296	302	308	314
65	114	120	126	132	138	144	150	156	162	168	174	180	186	192	198	204	210	216	222	228	234	240	246	252	258	264	270	276	282	288	294	300	306	312	318	324
66	118	124	130	136	142	148	155	161	167	173	179	186	192	198	204	210	216	223	229	235	241	247	253	260	266	272	278	284	291	297	303	309	315	322	328	334
67	121	127	134	140	146	153	159	166	172	178	185	191	198	204	211	217	223	230	236	242	249	255	261	268	274	280	287	293	299	306	312	319	325	331	338	344
68	125	131	138	144	151	158	164	171	177	184	190	197	203	210	216	223	230	236	243	249	256	262	269	276	282	289	295	302	308	315	322	328	335	341	348	354
69	128	135	142	149	155	162	169	176	182	189	196	203	209	216	223	230	236	243	250	257	263	270	277	284	291	297	304	311	318	324	331	338	345	351	358	365
70	132	139	146	153	160	167	174	181	188	195	202	209	216	222	229	236	243	250	257	264	271	278	285	292	299	306	313	320	327	334	341	348	355	362	369	376
71	136	143	150	157	165	172	179	186	193	200	208	215	222	229	236	243	250	257	265	272	279	286	293	301	308	315	322	329	338	343	351	358	365	372	379	386
72	140	147	154	162	169	177	184	191	199	206	213	221	228	235	242	250	258	265	272	279	287	294	302	309	316	324	331	338	346	353	361	368	375	383	390	397
73	144	151	159	166	174	182	189	197	204	212	219	227	235	242	250	257	265	272	280	288	295	302	310	318	325	333	340	348	355	363	371	378	386	393	401	408
74	148	155	163	171	179	186	194	202	210	218	225	233	241	249	256	264	272	280	287	295	303	311	319	326	334	342	350	358	365	373	381	389	396	404	412	420
75	152	160	168	176	184	192	200	208	216	224	232	240	248	256	264	272	279	287	295	303	311	319	327	335	343	351	359	367	375	383	391	399	407	415	423	431
76	156	164	172	180	189	197	205	213	221	230	238	246	254	263	271	279	287	295	304	312	320	328	336	344	353	361	369	377	385	394	402	410	418	426	435	443

Source: Adapted from *Clinical Guidelines on the Identification, Evaluation, and Treatment of Overweight and Obesity in Adults: The Evidence Report.*

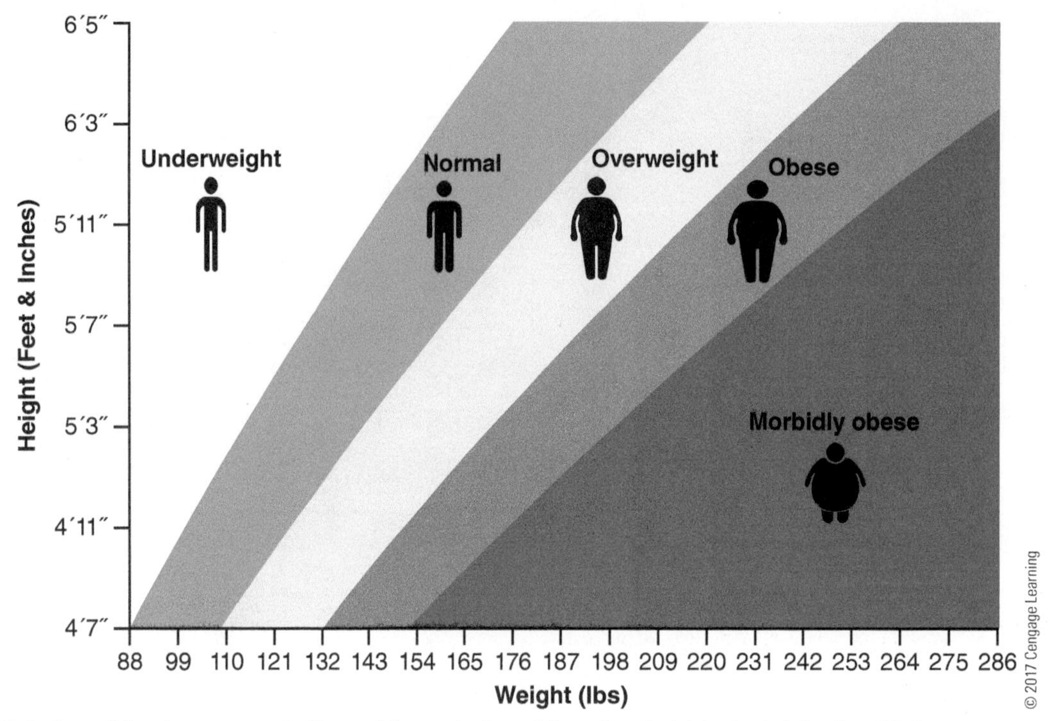

FIGURE 29-2 Graphical representation of the relationship of weight to height—the BMI.

factors, such as lifestyle habits and eating choices can be changed. For example, individuals can follow a healthy eating plan, keep caloric needs in mind, and be physically active, limiting the amount of inactive time. Although caloric needs vary with age, gender, activity, and genetic background, there are resources (such as Table 29-2 here) that provide suggested caloric intake levels.

TABLE 29-2 Estimated Calorie Needs per Day by Age, Gender, and Physical Activity Level

Estimated amounts of calories[a] needed to maintain calorie balance for various gender and age groups at three different levels of physical activity. The estimates are rounded to the nearest 200 calories for assignment to a U.S. Department of Agriculture (USDA) Food Pattern. An individual's calories needs may be higher or lower than these average estimates.

ACTIVITY LEVEL[B]	MALE			FEMALE[C]		
	SEDENTARY	MODERATELY ACTIVE	ACTIVE	SEDENTARY	MODERATELY ACTIVE	ACTIVE
Age (years)						
2	1,000	1,000	1,000	1,000	1,000	1,000
3	1,200	1,400	1,400	1,000	1,200	1,400
4	1,200	1,400	1,600	1,200	1,400	1,400
5	1,200	1,400	1,600	1,200	1,400	1,600
6	1,400	1,600	1,800	1,200	1,400	1,600
7	1,400	1,600	1,800	1,200	1,600	1,800

(Continues)

TABLE 29-2 Estimated Calorie Needs per Day by Age, Gender, and Physical Activity Level

ACTIVITY LEVEL[b]	MALE			FEMALE[c]		
	SEDENTARY	MODERATELY ACTIVE	ACTIVE	SEDENTARY	MODERATELY ACTIVE	ACTIVE
8	1,400	1,600	2,000	1,400	1,600	1,800
9	1,600	1,800	2,000	1,400	1,600	1,800
10	1,600	1,800	2,200	1,400	1,800	2,000
11	1,800	2,000	2,200	1,600	1,800	2,000
12	1,800	2,200	2,400	1,600	2,000	2,200
13	2,000	2,200	2,600	1,600	2,000	2,200
14	2,000	2,400	2,800	1,800	2,000	2,400
15	2,200	2,600	3,000	1,800	2,000	2,400
16	2,400	2,800	3,200	1,800	2,000	2,400
17	2,400	2,800	3,200	1,800	2,000	2,400
18	2,400	2,800	3,200	1,800	2,000	2,400
19–20	2,600	2,800	3,000	2,000	2,200	2,400
21–25	2,400	2,800	3,000	2,000	2,200	2,400
26–30	2,400	2,600	3,000	1,800	2,000	2,400
31–35	2,400	2,600	3,000	1,800	2,000	2,200
36–40	2,400	2,600	2,800	1,800	2,000	2,200
41–45	2,200	2,600	2,800	1,800	2,000	2,200
46–50	2,200	2,400	2,800	1,800	2,000	2,200
51–55	2,200	2,400	2,800	1,600	1,800	2,200
56–60	2,200	2,400	2,600	1,600	1,800	2,200
61–65	2,000	2,400	2,600	1,600	1,800	2,000
66–70	2,000	2,200	2,600	1,600	1,800	2,000
71–75	2,000	2,200	2,600	1,600	1,800	2,000
76+	2,000	2,200	2,400	1,600	1,800	2,000

Source: http://www.cnpp.usda.gov/sites/default/files/usda_food_patterns/EstimatedCalorieNeedsPerDayTable.pdf])

a. Based on Estimated Energy Requirements (EER) equations, using reference heights (average) and reference weights (healthy) for each age-gender group. For children and adolescents, reference height and weight vary. For adults, the reference man is 5 feet 10 inches tall and weights 154 pounds. The reference woman is 5 feet 4 inches tall and weights 126 pounds. EER equations are from the institute of Medicine. Dietary Reference intakes for Energy, Carbohydrate, Fiber, Fat, Fatty Acids, Cholesterol, Protein, and Amino Acids. Washington (DC): The National Academies Press, 2002.

b. Sedentary means a lifestyle that includes only the light physical activity associated with typical day-to-day life. Moderately active means a lifestyle that includes physical activity equivalent to walking about 1.5 to 3 miles per day at 3 to 4 miles per hour, in addition to the light physical activity associated with typical day-to-day life. Active means a lifestyle that includes physical activity equivalent to walking more than 3 miles per day at 3 to 4 miles per hour, in addition to the light physical activity associated with typical day-to-day life.

c. Estimates for females do not include women who are pregnant or breastfeeding.

To maintain or lose weight, many people count their daily caloric intake. A variety of Web sites are helpful for doing this, and smartphone apps are available to estimate calories expended during exercise as well as track overall daily calories. One-size-fits-all **calorie** recommendations do not work. Any recommendation must be customized to an individual.

Nutrition labels on foods list the calories per serving, and many restaurants now provide the caloric content of their foods. When the caloric value is not listed, other sources can provide estimated or guideline caloric values for foods.

Doctor-prescribed weight-loss medicines and certain surgeries are also options for some people if lifestyle changes are not enough.

Reaching and staying at a healthy weight is a long-term challenge for people who are overweight or obese. But it also is a chance to lower risk for other serious health problems. With the right treatment and motivation, it is possible to lose weight and lower long-term disease risk.

Possible Cancer Prevention

Free radicals are highly reactive chemicals that have the potential to harm cells. They are created when an atom or a molecule (a chemical that has two or more atoms) either gains or loses an electron (a small negatively charged particle found in atoms). Free radicals are formed naturally in the body and play an important role in many normal cellular processes. At high concentrations, however, free radicals can be hazardous and damage all major components of cells, including DNA, proteins, and cell membranes. The damage to cells caused by free radicals, especially the damage to DNA, may play a role in the development of cancer and other health conditions.

Abnormally high concentrations of free radicals in the body can be caused by exposure to ionizing radiation and other environmental toxins. When ionizing radiation hits an atom or a molecule in a cell, an electron may be lost, leading to the formation of a free radical. The production of abnormally high levels of free radicals is the mechanism by which ionizing radiation kills cells. Moreover, some environmental toxins—cigarette smoke, some metals, and high-oxygen atmospheres—may contain large amounts of free radicals or stimulate the body's cells to produce more free radicals. Free radicals that contain the element oxygen are the most common type of free radicals produced in living tissue. Another name for them is *reactive oxygen species.*

WHAT IS A CALORIE?

Calories are used to express the energy content of foods. We count calories. We check calories, and we know that too many calories can cause us to gain weight. We reduce calories to lose weight. But what is a calorie?

Calories are one of two units of heat energy. Specifically, a calorie (lowercase "c") is the energy needed to raise the temperature of 1 gram of water by 1 °C, which is also defined as 4.1868 joules (happily, there is no movement to use joules on nutrition labels). A big Calorie (capital "C") is the energy needed to raise the temperature of 1 kilogram of water by 1 °C. It is equal to 1,000 small calories and is often used to measure the energy value of foods.

- A Calorie Is a Calorie, or Is It? http://www.health.gov/dietaryguidelines/dga2005/healthieryou/html/chapter5.html
- ChooseMyPlate: http://www.choosemyplate.gov/weight-management-calories/calories.html
- What Is a Calorie? http://ed.ted.com/lessons/what-is-a-calorie-emma-bryce
- Calories Count: http://www.dining.ucla.edu/housing_site/dining/SNAC_pdf/CaloriesCount.pdf

Antioxidants. Antioxidants are chemicals that interact with and neutralize free radicals, preventing them from causing damage. Antioxidants are also known as *free radical scavengers.* The body makes some of the antioxidants it uses to neutralize free radicals; these are called *endogenous antioxidants.* However, the body relies on external (exogenous) sources, primarily through diet, to obtain the rest of the antioxidants it needs. These exogenous antioxidants are commonly called *dietary antioxidants.*

Fruits, vegetables, and grains are rich sources of dietary antioxidants. Some dietary antioxidants are also available as dietary supplements. Examples of dietary antioxidants include beta-carotene, lycopene, and vitamins A, C, and E (alpha-tocopherol). The mineral element selenium is often thought to be a dietary antioxidant, but the antioxidant effects of selenium are most likely the result of proteins' antioxidant activities of selenium in the proteins and not selenium itself.

In laboratory and animal studies, the presence of increased levels of exogenous antioxidants has been shown to prevent the types of free radical damage that have been associated with cancer development. For this reason, researchers have investigated whether taking dietary antioxidant supplements can help lower the risk of developing or dying from cancer in humans.

Many observational studies, including case-control studies and group studies, have been conducted to investigate whether dietary antioxidant supplement use is associated with reduced cancer risks in humans. These studies have yielded mixed results. Because observational studies cannot adequately control for biases that might influence study outcomes, the results of any individual observational study must be viewed with caution.

Randomized trials are considered to provide the strongest and most reliable evidence of the benefit or harm of a health-related intervention. Randomized controlled trials of dietary antioxidant supplements for cancer prevention have been conducted worldwide. Many of the trials were sponsored by the National Cancer Institute. The following are brief summaries of the results of some of these trials.

- In a large-scale trial, healthy Chinese men and women at increased risk of developing esophageal cancer and gastric cancer were randomly assigned to (1) take a combination of 15 milligrams (mg) beta-carotene, 30 mg alpha tocopherol, and 50 micrograms (μg) selenium daily for 5 years; or (2) to take no antioxidant supplements. The initial results of the trial showed that people who took the antioxidant supplements had a lower risk of death from gastric cancer but not from esophageal cancer. However, their risks of developing gastric cancer or esophageal cancer were not affected by antioxidant supplementation. In recently updated results, a reduced risk of death from gastric cancer was no longer found for those who took antioxidant supplements compared with those who did not.
- An alpha-tocopherol and beta-carotene cancer-prevention study investigated whether the use of alpha tocopherol or beta-carotene supplements (or both) for 5 to 8 years could help reduce the incidence of lung and other cancers in middle-aged male smokers in Finland. Initial results of the trial, reported in 1994, showed an increase in the incidence of lung cancer among participants who took beta-carotene supplements (20 mg per day); in contrast, alpha-tocopherol supplementation (50 mg per day) had no effect on lung cancer incidence. Later results showed no effect of beta-carotene or alpha-tocopherol supplementation on the incidence of cancers of the bladder, ureter, renal pelvis, pancreas, colon and rectum, kidney, pharynx, esophagus, or larynx.
- A carotene and retinol efficacy trial in the U.S. trial examined the effects of daily supplementation with beta-carotene and retinol (vitamin A) on the incidence of lung cancer, other cancers, and death among people who were at high risk of lung cancer because of a history of smoking or exposure

to asbestos. The trial began in 1983 and ended in late 1995, 2 years earlier than originally planned. Results reported in 1996 showed that daily supplementation with both 15 mg beta-carotene and 25,000 International Units (IUs) retinol was associated with increased lung cancer and increased death from all causes. A 2004 report showed that these adverse effects persisted up to 6 years after supplementation ended. Additional results reported in 2009 showed that beta-carotene and retinol supplementation had no effect on the incidence of prostate cancer.

- A trial called the Physicians' Health Study I examined the effects of long-term beta-carotene supplementation on cancer incidence, cancer mortality, and all-cause mortality among U.S. male physicians. The results of the study reported in 1996 showed that beta-carotene supplementation (50 mg every other day for 12 years) had no effect on any of these outcomes in smokers or nonsmokers.
- The Women's Health Study trial investigated the effects of beta-carotene supplementation (50 mg every other day), vitamin E supplementation (600 IU every other day), and aspirin (100 mg every other day) on the incidence of cancer and cardiovascular disease in U.S. women ages 45 and older. The results reported in 1999 showed no benefit or harm associated with 2 years of beta-carotene supplementation. In 2005, similar results were reported for vitamin E supplementation.
- The Heart Outcomes Prevention Evaluation—The Ongoing Outcomes Study was an international trial that examined the effects of alpha-tocopherol supplementation on cancer incidence, death from cancer, and the incidence of major cardiovascular events (heart attack, stroke, or death from heart disease) in people diagnosed with cardiovascular disease or diabetes. The results reported in 2005 showed no effect of daily supplementation with alpha tocopherol (400 IU) for a median of 7 years on any of the outcomes.
- The Physicians' Health Study II trial examined whether supplementation with vitamin E, vitamin C, or both would reduce the incidence of cancer in male U.S. physicians ages 50 years and older. The results reported in 2009 showed that the use of these supplements (400 IU vitamin E every other day, 500 mg vitamin C every day, or a combination of the two) for a median of 7.6 years did not reduce the incidence of prostate cancer or other cancers, including lymphoma, leukemia, melanoma, and cancers of the lung, bladder, pancreas, and colon and rectum.

Overall, randomized controlled clinical trials did not provide evidence that dietary antioxidant supplements are beneficial in primary cancer prevention. In addition, a systematic review of the available evidence regarding the use of vitamin and mineral supplements for the prevention of chronic diseases, including cancer, conducted for the United States Preventive Services Task Force likewise found no clear evidence of benefit in preventing cancer.

Possibly, the lack of benefit in clinical studies can be explained by differences in the effects of the tested antioxidants when they are consumed as purified chemicals as opposed to when they are consumed in foods, which contain complex mixtures of antioxidants, vitamins, and minerals. Acquiring a more complete understanding of the antioxidant content of individual foods, how the various antioxidants and other substances in foods interact with one another, and factors that influence the uptake and distribution of food-derived antioxidants in the body are active areas of ongoing cancer-prevention research.

Additional large randomized controlled trials are needed to provide clear scientific evidence about the potential benefits or harms of taking antioxidant supplements during cancer treatment. Until more is known about the effects of antioxidant supplements in cancer patients, these supplements should be used with caution. Cancer patients should inform their doctors about their use of any dietary supplement.

SCIENCE CONNECTION!

Have you seen commercials or advertisements for dietary supplements? What do you really know about them? Are they really healthy? Research and record your findings.

Type 2 Diabetes

Diabetes is a disorder of metabolism, or the way the body uses digested food for growth and energy. In diabetes, the pancreas either produces little or no **insulin** (a hormone that helps get glucose, the body's main source of fuel, into cells) or the cells do not respond appropriately to the insulin that is produced. The three main types of diabetes are type 1, type 2, and gestational diabetes. Approximately 90% to 95% of people with diabetes have type 2. This form of diabetes is most often associated with older age, obesity, family history of diabetes, previous history of gestational diabetes, physical inactivity, and certain ethnicities. Some 80% of people with type 2 diabetes are overweight. Prediabetes, also called *impaired fasting glucose* or *impaired glucose tolerance*, is a state in which blood glucose levels are higher than normal but not high enough to be called diabetes.

People can develop type 2 diabetes at any age, even during childhood. However, this type of diabetes develops most often in middle-aged and older people. People who are overweight and inactive are also more likely to develop type 2 diabetes.

In type 2 and other types of diabetes, individuals have too much glucose, also called *sugar*, in their blood. People with diabetes have problems converting food to energy. Normally, after a meal, food is broken down into glucose, which is carried by the blood to cells throughout the body. With the help of the hormone insulin, cells absorb blood glucose and use it for energy. Insulin is made in the pancreas, an organ located behind the stomach.

Type 2 diabetes usually begins with insulin resistance, a condition linked to excess weight in which the body's cells do not use insulin properly. As a result, the body needs more insulin to help glucose enter cells. At first, the pancreas keeps up with the added demand by producing more insulin. But in time, the pancreas loses its ability to produce enough insulin, and blood glucose levels rise (Figure 29-3).

Over time, high blood glucose damages nerves and blood vessels, leading to problems such as heart disease, stroke, kidney disease, blindness, dental disease, and amputations. Other problems of diabetes may include increased risk of getting other diseases, loss of mobility with aging, depression, and pregnancy problems.

Treatment includes taking diabetes medicines, making wise food choices, being physically active on a regular basis, controlling blood pressure and cholesterol, and, for some people, taking aspirin daily.

Prevention. The results of the Diabetes Prevention Program (DPP) proved that weight loss through moderate diet changes and physical activity can delay or prevent type 2 diabetes. The DPP was a federally funded study of 3,234 people at high risk for diabetes. This study showed that a 5% to 7% weight loss—10 to 14 pounds for a 200-pound person—slowed development of type 2 diabetes.

DPP study participants were overweight and had higher than normal levels of blood glucose, a condition called *prediabetes*. Many had family members with type 2 diabetes. Prediabetes, obesity, and a family history of diabetes are strong risk factors for type 2 diabetes. Approximately half of the DPP participants were from minority groups with high rates of diabetes, including African Americans, Alaskan Natives, American Indians, Asian Americans, Hispanics or Latinos, and Pacific Islander Americans.

Others at high risk for developing type 2 diabetes were also included, such as women with histories of gestational diabetes and people ages 60 years and older.

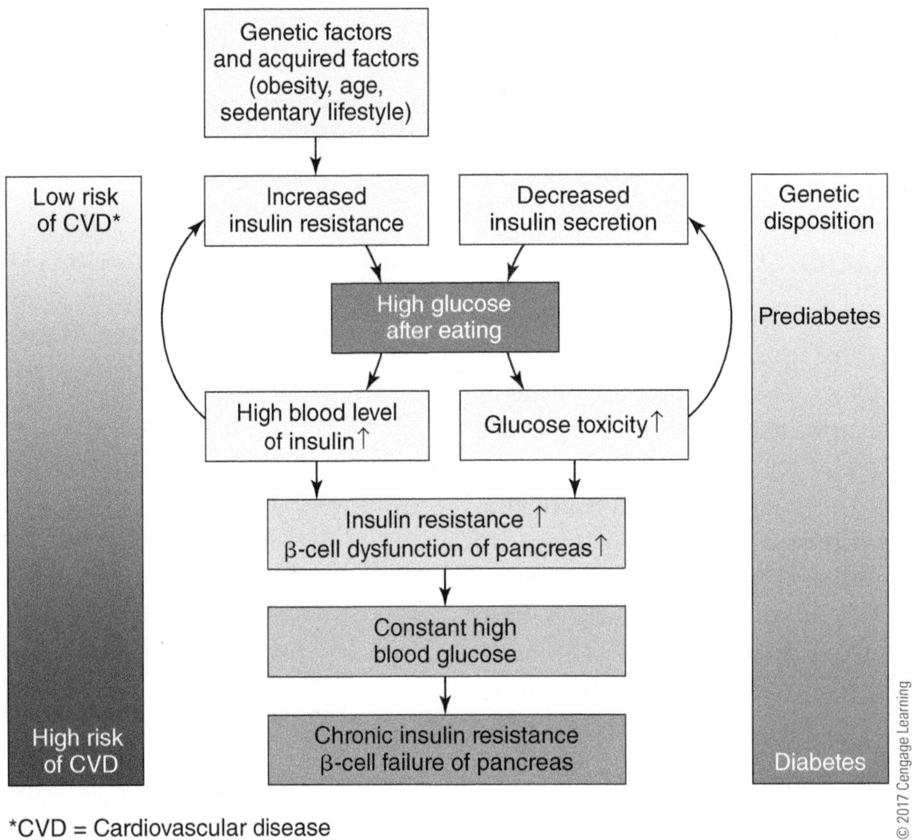

FIGURE 29-3 A flow chart of the development of type 2 diabetes.

More information about the DPP, funded by the National Institutes of Health (NIH), is available at www.bsc.gwu.edu/dpp.

Other Types of Diabetes. In addition to type 2, the other main types of diabetes are type 1 diabetes and gestational diabetes. Type 1 diabetes, formerly called *juvenile diabetes*, is usually first diagnosed in children, teenagers, and young adults. In this type of diabetes, the pancreas can no longer make insulin because the body's immune system has attacked and destroyed the cells that make it. Treatment for type 1 diabetes includes taking insulin shots or using an insulin pump, making wise food choices, being physically active on a regular basis, controlling blood pressure and cholesterol, and, for some people, taking aspirin daily.

Gestational diabetes is a type of diabetes that develops only during pregnancy. Hormones produced by the placenta and other pregnancy-related factors contribute to insulin resistance, which occurs in all women during late pregnancy. Insulin resistance increases the amount of insulin needed to control blood glucose levels. If the pancreas cannot produce enough insulin, gestational diabetes occurs.

As with type 2 diabetes, excess weight is linked to gestational diabetes. Overweight or obese women are at particularly high risk for gestational diabetes because they start pregnancy with a higher need for insulin because of insulin resistance. Excessive weight gain during pregnancy may also increase the risk. Gestational diabetes occurs more often in some ethnic groups and among women with family histories of diabetes.

Although gestational diabetes usually goes away after the baby is born, a woman who has had gestational diabetes is more likely to develop type 2 diabetes later in life. Babies born to mothers who had gestational diabetes are also more likely to develop obesity and type 2 diabetes as they grow up.

Signs and Symptoms of Type 2. The signs and symptoms of type 2 diabetes can be so mild that a person might not even notice them. Nearly 7 million people in the United States have type 2 diabetes and do not know they have the disease. Many have no signs or symptoms. Some people have symptoms but do not suspect diabetes. Symptoms include:

- Increased thirst
- Increased hunger
- Fatigue
- Increased urination, especially at night
- Unexplained weight loss
- Blurred vision
- Numbness or tingling in the feet or hands
- Sores that do not heal

Many people do not find out they have the disease until they have diabetes problems such as blurred vision or heart trouble. If diabetes is discovered early, treatments can be received that will prevent damage to the body.

Individuals can reduce the risk of getting diabetes and even return blood glucose levels to normal by losing a little weight through healthy eating and being more physically active.

Sleep Problems

Studies show that untreated sleep problems, especially sleep apnea, can increase the risk of type 2 diabetes. **Sleep apnea** is a common disorder in which individuals have pauses in breathing or shallow breaths while they sleep. Most people who have sleep apnea do not know they have it, and it often goes undiagnosed. Night shift workers who have problems with sleepiness may also be at increased risk for obesity and type 2 diabetes.

Weight and Type 2. Individuals who are overweight should reach and maintain a reasonable body weight. Even a 10- to 15-pound weight loss makes a big difference. They should also make wise food choices most of the time and be

FIGURE 29-4 Physical activity, even walking, contributes to good healthy and complements a healthy diet.

physically active every day. If high blood pressure is also a problem, then reducing sodium and alcohol intakes can also help. If cholesterol or triglyceride levels are too high, then these steps plus a medication to control cholesterol levels may be needed.

Be Physically Active Every Day. Regular physical activity tackles several risk factors at once (Figure 29-4). Activity helps an individual lose weight; keeps blood glucose, blood pressure, and cholesterol under control; and helps the body use insulin.

Prescribed Medicines. Some people need medicine to help control their blood pressure or cholesterol levels. These should be taken as directed. One medicine, metformin, maybe prescribed to prevent type 2 diabetes. Metformin makes insulin work better and can reduce the risk of type 2 diabetes.

Eating, Diet, and Nutrition. An individual's eating, diet, and nutrition choices play an important role in preventing or delaying diabetes. Some suggestions that can be used to reach and maintain a reasonable weight and make wise food choices most of the time include:

- Change habits and be patient
- Get help from a dietitian or join a weight-loss program for support
- Know your BMI
- Reduce serving sizes of main courses, meat, desserts, and other foods high in fat

- Increase the amount of fruits and vegetables eaten at meals
- Limit fat intake to about 25% of total calories (check food labels for fat content)
- Limit sodium intake to less than 2,300 milligrams—about 1 teaspoon of salt—each day
- Limit alcohol intake to one drink for women or two drinks for men per day
- Reduce the number of calories eaten each day
- Keep a food and physical activity log

Additional help can be found on many reliable Web sites—for example, the USDA Web site at www.choosemyplate.gov.

Dietary Supplements. Vitamin D studies show a link between people's ability to maintain healthy blood glucose levels and having enough vitamin D in their blood. However, studies to determine the proper vitamin D levels for people with diabetes and for preventing diabetes are ongoing, and there no special recommendations yet about vitamin D levels or supplements for people with diabetes. Currently, the Institute of Medicine, the agency that recommends supplementation levels based on current science, provides the following guidelines for daily vitamin D intake:

- People ages 1 to 70 years may require 600 International Units (IUs)
- People age 71 and older may require as much as 800 IUs
- No more than 4,000 IUs of vitamin D should be taken per day

To help ensure coordinated and safe care, individuals should discuss their use of complementary and alternative medicine practices, including the use of dietary supplements, with a doctor.

More information about using dietary supplements to help with diabetes is available from the National Diabetes Information Clearinghouse at www.diabetes.niddk.nih.gov/dm/pubs/alternativetherapies/index.aspx.

DIABETIC EATING

Individuals with diabetes must change their eating habits. They eat smaller portions and learn what a serving size is for different foods and how many servings are needed in a meal. They eat less fat by choosing fewer high-fat foods and using less fat for cooking. They especially want to limit foods that are high in saturated fats or trans-fats such as:

- Fatty cuts of meat
- Fried foods
- Whole milk and dairy products made from whole milk
- Cakes, candy, cookies, crackers, and pies
- Salad dressings
- Lard, shortening, stick margarine, and nondairy creamers

Individuals with diabetes eat more fiber by eating more whole-grain foods such as:

- Breakfast cereals made with 100% whole grains
- Oatmeal
- Whole grain rice
- Whole-wheat bread, bagels, pita bread, and tortillas

These individuals eat a variety of fruits and vegetables every day (Figure 29-5), choosing fresh, frozen, canned, or dried fruit and 100% fruit juices most of the time. They also eat plenty of vegetables such as:

- Dark green vegetables—for example, broccoli, spinach, and Brussels sprouts
- Orange vegetables such as carrots, sweet potatoes, pumpkin, and winter squash

- Beans and peas such as black beans, garbanzo beans, kidney beans, pinto beans, split peas, and lentils

Those who have diabetes should eat fewer foods that are high in sugar, including:

- Fruit-flavored drinks
- Sodas
- Tea or coffee sweetened with sugar

Finally, individuals with diabetes should use less salt in cooking and at the table and eat fewer foods that are high in salt such as:

- Canned and package soups
- Canned vegetables
- Pickles
- Processed meats

Source: USDA, Agricultural Research Service (ARS), photo by Peggy Greb

FIGURE 29-5 A wide variety of fresh fruits and vegetables are part of a healthy diet.

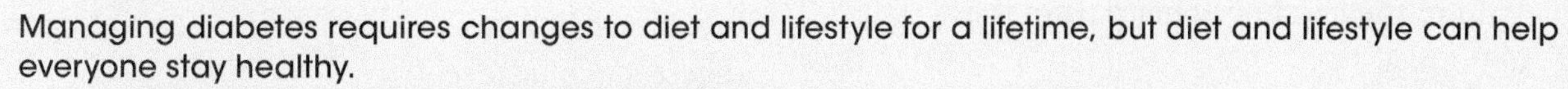

Managing diabetes requires changes to diet and lifestyle for a lifetime, but diet and lifestyle can help everyone stay healthy.

- Diabetes Diet: http://umm.edu/health/medical/reports/articles/diabetes-diet
- Four Steps to Manage Your Diabetes for Life: http://ndep.nih.gov/publications/publicationdetail.aspx?pubid=4
- Newcastle Study—600 Calorie Diet: http://www.diabetes.co.uk/diet/newcastle-study-600-calorie-diet.html
- Diabetes Management: http://www.mayoclinic.org/diseases-conditions/diabetes/in-depth/diabetes-management/art-20047963

DIGESTIVE DISORDERS

Digestive disorders include gas, heartburn, indigestion, lactose intolerance, peptic ulcer disease, diarrhea, constipation, diverticular disease, celiac disease, Crohn's disease, and pancreatitis. Some are caused by foods, and some are managed by the food eaten.

Gas

Gas in the digestive tract is usually caused by swallowing air and by the breakdown of certain foods in the large intestine by bacteria. Most foods that contain carbohydrates can cause gas. In contrast, fats and proteins cause little gas. Foods that produce gas in one person may not cause gas in another, depending on how well individuals digest carbohydrates and the types of bacteria present in the intestines. Gas-producing foods include:

- Beans
- Vegetables such as broccoli, cauliflower, cabbage, Brussels sprouts, onions, mushrooms, artichokes, and asparagus
- Fruits such as pears, apples, and peaches
- Whole grains such as whole wheat and bran
- Sodas; fruit drinks, especially apple juice and pear juice; and other drinks that contain high-fructose corn syrup, a sweetener made from corn
- Milk and milk products such as cheese, ice cream, and yogurt
- Packaged foods such as bread, cereal, and salad dressing that contain small amounts of lactose, a sugar found in milk and foods made with milk
- Sugar-free candies and gums that contain sugar alcohols such as sorbitol, mannitol, and xylitol

The most common symptoms of gas (**flatulence**) are burping, passing gas, bloating, and abdominal pain or discomfort.

Heartburn

The most common symptom of **gastroesophageal reflux disease (GERD)** is regular heartburn—a painful, burning feeling in the middle of the chest, behind the breastbone, and in the middle of the abdomen. Other common GERD symptoms include:

- Bad breath
- Nausea
- Pain in the chest or upper part of the abdomen
- Problems swallowing or painful swallowing
- Respiratory problems
- Vomiting
- Erosion of the teeth

GERD happens when the lower esophageal sphincter becomes weak or relaxes when it should not, causing stomach contents to rise up into the esophagus. The lower esophageal sphincter becomes weak or relaxes because of certain conditions such as increased pressure on the abdomen from being overweight, obese, or pregnant and certain medicines.

Control of GERD may include some of the following:

- Avoiding items that may cause it such as greasy or spicy foods and alcoholic drinks
- Not overeating
- Not eating 2 to 3 hours before bedtime
- Losing weight if overweight or obese
- Quitting smoking and avoiding secondhand smoke
- Taking over-the-counter medicines

Depending on the severity of GERD, doctors may recommend additional lifestyle changes, medicines, or surgery.

Lactose Intolerance

Lactose is a sugar found in milk and milk products. The small intestine, the organ where most food digestion and nutrient absorption take place, produces an enzyme called *lactase*. Lactase breaks down lactose into two simpler forms of sugar: glucose and galactose. The body then absorbs these simpler sugars into the bloodstream.

Lactose intolerance is a condition in which people have digestive symptoms such as bloating, diarrhea, and gas after eating or drinking milk or milk products. Lactase deficiency and lactose malabsorption may lead to lactose intolerance. In people who have a lactase deficiency, the small intestine produces low levels of lactase and cannot digest much lactose.

Lactase deficiency may cause lactose malabsorption in which undigested lactose passes to the colon. As part of the large intestine, the colon absorbs water from stool and changes it from liquid to solid. Here bacteria break down undigested lactose and create fluid and gas. Not all people with lactase deficiency and lactose malabsorption have digestive symptoms.

Most people with lactose intolerance can eat or drink some amount of lactose without having digestive symptoms. Individuals vary in the amount of lactose they can tolerate.

People sometimes confuse lactose intolerance with a milk allergy. Although lactose intolerance is a digestive system disorder, a milk allergy is a reaction by the body's immune system to one or more milk proteins. An allergic reaction to milk can be life threatening even if the person eats or drinks only a small amount of milk or milk product. A milk allergy most commonly occurs in the first year of life, whereas lactose intolerance occurs more often during adolescence or adulthood.

Many people can manage the symptoms of lactose intolerance by changing their diet. Some people may only need to limit the amount of lactose they eat or drink. Others may need to avoid lactose altogether. Using **lactase** products can help some people manage their symptoms. A dietary plan can help people manage the symptoms of lactose intolerance and make sure they get enough nutrients.

Gradually introducing small amounts of milk or milk products may help some people adapt to them with fewer symptoms. People often can better tolerate milk or milk products by having them with meals, such as milk with cereal or cheese with crackers. People with lactose intolerance are generally more likely to tolerate hard cheeses, such as cheddar or Swiss, than a glass of milk. A 1.5-ounce serving of low-fat hard cheese has less than 1 gram of lactose, while a 1-cup serving of low-fat milk has approximately 11 to 13 grams of lactose. An interesting fact is that people with lactose intolerance are more likely to tolerate yogurt than milk, even though yogurt and milk have similar amounts of lactose.

Lactose-free and lactose-reduced milk and milk products are available at most supermarkets and are identical nutritionally to regular milk and milk products (Figure 29-6). Manufacturers treat lactose-free milk with the lactase enzyme, which breaks down the lactose in the milk. Lactose-free milk remains fresh for about the same length of time or, if it is ultrapasteurized, longer than regular milk. Lactose-free milk may have a slightly sweeter taste than regular milk.

People can use lactase tablets and drops when they eat or drink milk products. The lactase enzyme digests the lactose in the food and therefore reduces the chances of developing digestive symptoms. People should check with a health-care provider before using these products because some groups—young children and pregnant and breastfeeding women—may not be able to use them.

Peptic Ulcer Disease

A **peptic ulcer** is a sore on the lining of the stomach or duodenum—a section of the small intestine. Causes of peptic ulcers include:

FIGURE 29-6 Lactose-free milk, usually available wherever milk is sold.

- Long-term use of nonsteroidal anti-inflammatory drugs (NSAIDs) such as aspirin and ibuprofen
- An infection with the bacteria ***Helicobacter pylori*** (*H. pylori*)
- Rare cancerous and noncancerous tumors in the stomach, duodenum, or pancreas, known as Zollinger-Ellison syndrome

Approximately 30% to 40% of people in the United States get an *H. pylori* infection at some point in their lives. In most cases, the infection remains dormant or quiet without signs or symptoms for years. Most people get an *H. pylori* infection as a child but rarely develop peptic ulcers. Adults who have an *H. pylori* infection may get a peptic ulcer, also called an *H. pylori-induced peptic ulcer*. However, most people with an *H. pylori* infection never develop a peptic ulcer.

H. pylori are spiral-shaped bacteria that can damage the lining of the stomach and duodenum and cause peptic ulcer disease. Researchers are not certain how *H. pylori* spreads, but it could be spread through unclean food, unclean water, unclean eating utensils, and contact with an infected person's saliva and other bodily fluids.

Treatment for peptic ulcer disease varies, depending on the severity and cause. Researchers have not found that diet and nutrition play an important role in causing or preventing peptic ulcers. Before acid blocking drugs became available, milk was used to treat ulcers. However, milk is not an effective way to prevent or relieve a peptic ulcer. Alcohol and smoking both contribute to ulcers and should be avoided.

Diarrhea

Diarrhea is loose, watery stools. Having diarrhea means passing loose stools three or more times a day. Acute diarrhea is a common problem that usually lasts 1 or 2 days and goes away on its own. Diarrhea lasting more than 2 days may be a sign of a more serious problem. Chronic diarrhea—diarrhea that lasts at least 4 weeks—may be a symptom of a chronic disease. Chronic diarrhea symptoms may be continual or come and go.

Diarrhea of any duration may cause dehydration, which means the body lacks enough fluid and electrolytes—which are chemicals in salts, including sodium, potassium, and chloride—to function properly. Loose stools contain more fluid, and electrolytes and weigh more than solid stools.

People of all ages can get diarrhea. In the United States, adults average one bout of acute diarrhea each year, and young children have an average of two episodes of acute diarrhea each year.

Acute diarrhea is usually caused by a bacterial, viral, or parasitic infection. Chronic diarrhea is usually related to a functional disorder such as irritable bowel syndrome or an intestinal disease such as Crohn's disease.

Bacterial infections from several types of bacteria consumed through contaminated food or water are one well-known cause of diarrhea. Common culprits include *Campylobacter, Salmonella, Shigella,* and *Escherichia coli.*

Many viruses cause diarrhea, including rotavirus, norovirus, cytomegalovirus, herpes simplex virus, and viral hepatitis. Infection with the rotavirus is the most common cause of acute diarrhea in children. Rotavirus diarrhea usually resolves in 3 to 7 days but can cause problems digesting lactose for up to a month or longer.

Parasites cause diarrhea by entering the body through food or water and settling in the digestive system. Parasites that cause diarrhea include *Giardia lamblia, Entamoeba histolytica,* and *Cryptosporidium.*

Diarrhea can be a symptom of irritable bowel syndrome, and inflammatory bowel disease, ulcerative colitis, Crohn's disease, and celiac disease often lead to diarrhea.

Food intolerances and sensitivities such as lactose and excessive quantities of certain types of sugar substitutes can cause diarrhea, as can antibiotics, cancer drugs, and antacids containing magnesium.

In most cases of diarrhea, the only treatment necessary is replacing lost fluids and electrolytes to prevent dehydration. Some over-the-counter medicines maybe helpful, too. Until diarrhea subsides, avoiding caffeine and foods that are greasy, high in fiber, or sweet may lessen the symptoms. These foods can aggravate diarrhea. Some people also have problems digesting lactose during or after a bout of diarrhea. Yogurt, which has less lactose than milk, is often better tolerated. Yogurt with active, live bacterial cultures may even help people recover from diarrhea more quickly. As symptoms improve, soft, bland foods can be added to the diet, including bananas, plain rice, boiled potatoes, toast, crackers, cooked carrots, and baked chicken without the skin or fat.

Constipation

Constipation is characterized as fewer than three bowel movements a week and bowel movements with stools that are hard, dry, and small, making them painful or difficult to pass. Some people think they are constipated if they do not have a bowel movement every day. However, people have different bowel movement patterns. Somc people may have three movements a day, and others only three movements a week.

Constipation most often lasts for only a short time and is not dangerous. Some steps can be taken to prevent or relieve constipation. Changes in eating, diet, and nutrition can treat constipation. These changes include drinking liquids throughout the day, eating more fruits and vegetables, and eating more fiber. Some food are to be avoided if an individual is constipated. These include:

- Cheese
- Chips
- Fast food
- Ice cream
- Meat
- Prepared foods such as some frozen meals and snack foods
- Processed foods such as hot dogs or some microwavable dinners

Exercising every day may help prevent and relieve constipation. Also some over-the-counter medicines may also be helpful. As always, consult a health-care provider.

Diverticular Disease, Celiac Disease, Crohn's Disease, and Pancreatitis

Diverticular disease is a condition that occurs when a person has problems from small pouches or sacs that have formed and pushed outward through weak spots in the colon wall. Each pouch is called a *diverticulum*. Multiple pouches are called *diverticula*. Diverticula are most common in the sigmoid colon, the lower part of the colon (Figure 29-7).

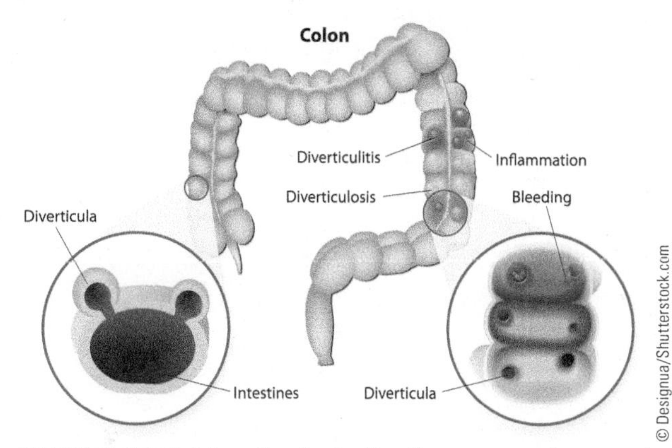

FIGURE 29-7 Diverticula in the human colon and the effects of diverticulitis and diverticulosis.

Problems that occur with diverticular disease include diverticulitis and diverticular bleeding. Diverticulitis occurs when the diverticula become inflamed, irritated and swollen, and infected. Diverticular bleeding occurs when a small blood vessel within the wall of a diverticulum bursts.

Studies have shown that a high-fiber diet (Table 29-3) can help prevent diverticular disease in people who already have diverticulosis.

Fiber supplements such as methylcellulose (Citrucel) or psyllium (Metamucil) may be recommended. These products are available as powders, pills, and wafers and provide 0.5 grams to 3.5 grams of fiber per dose. Fiber products should be taken with at least 8 ounces of water.

Medications and probiotics may be used to reduce the symptoms of diverticulosis. Although more research is needed, probiotics may help treat the symptoms of diverticulosis, prevent the onset of diverticulitis, and reduce the chance of

TABLE 29-3 Fiber Rich Foods

BEANS, CEREALS, AND BREADS	AMOUNT OF FIBER
1/2 cup of navy beans	9.5 grams
1/2 cup of kidney beans	8.2 grams
1/2 cup of black beans	7.5 grams
WHOLE-GRAIN CEREAL, COLD	
1/2 cup of All-Bran	9.6 grams
3/4 cup of Total	2.4 grams
3/4 cup of Post Bran Flakes	5.3 grams
1 packet of whole-grain cereal, hot (oatmeal, Wheatena)	3.0 grams
1 whole wheat English muffin	4.4 grams
FRUITS	
1 medium apple, with skin	3.3 grams
1 medium pear, with skin	4.3 grams
1/2 cup of raspberries	4.0 grams
1/2 cup of stewed prunes	3.8 grams
VEGETABLES	
1/2 cup of winter squash	2.9 grams
1 medium sweet potato, with skin	4.8 grams
1/2 cup of green peas	4.4 grams
1 medium potato, with skin	3.8 grams
1/2 cup of mixed vegetables	4.0 grams
1 cup of cauliflower	2.5 grams
1/2 cup of spinach	3.5 grams
1/2 cup of turnip greens	2.5 grams

recurrent symptoms. Probiotics are live bacteria, like those normally found in the gastrointestinal (GI) tract. Probiotics can be found in dietary supplements, capsules, tablets, and powders, as well as some foods such as yogurt.

SCIENCE CONNECTION!

Fiber is an important part of the diet. Search online and find three high-fiber recipes using any of the foods listed in Table 29-3. Would you eat the food recipe you found? Why or why not?

Scientists now think that people with diverticular disease do not need to eliminate certain foods from their diet. In the past, health-care providers recommended that people with diverticular disease avoid nuts, popcorn, and seeds—sunflower, pumpkin, caraway, and sesame—because they thought food particles could enter, block, or irritate the diverticula. Recent data suggest that these foods are not harmful. The seeds in tomatoes, zucchini, cucumbers, strawberries, and raspberries, as well as poppy seeds, are also fine to eat. Still, people with diverticular disease may differ in the amounts and types of foods that worsen their symptoms.

Celiac Disease. Celiac disease is a digestive disease that damages the small intestine and interferes with absorption of nutrients from food. People who have celiac disease cannot tolerate **gluten**, a protein in wheat, rye, and barley. Gluten is found in many food products and also in everyday products such as medicines, vitamins, and lip balms.

When people with celiac disease eat foods or use products containing gluten, their immune system responds by damaging or destroying villi, the tiny, fingerlike protrusions lining the small intestine (Figure 29-8). Villi normally allow nutrients from food to be absorbed through the walls of the small intestine into the bloodstream. Without healthy villi, a person becomes malnourished, no matter how much food he or she eats.

Celiac disease is both a disease of malabsorption, meaning nutrients are not absorbed properly, and an abnormal immune reaction to gluten. Celiac disease is also known as *celiac sprue*, *nontropical sprue*, and *gluten-sensitive enteropathy*. Celiac disease is genetic—that is, it runs in families. Sometimes the disease is triggered, or becomes active for the first time, after surgery, pregnancy, childbirth, viral infection, or severe emotional stress.

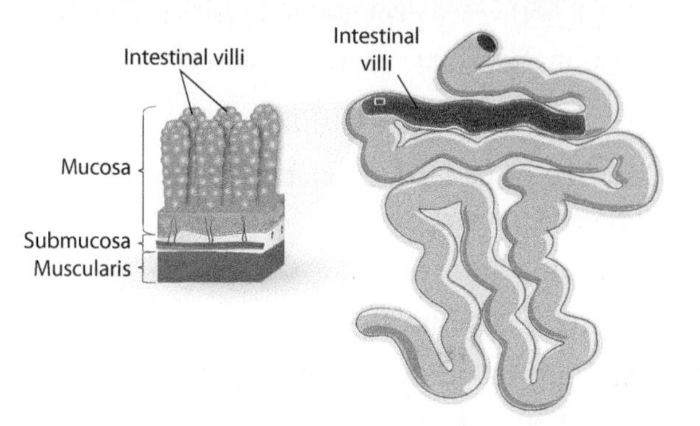

FIGURE 29-8 Villi inside the small intestine.

The only treatment for celiac disease is a gluten-free diet. A person with celiac disease should not eat most grains, pasta, cereal, and many processed foods. Despite these restrictions, by making informed choices people with celiac disease can eat a well-balanced diet with a variety of foods.

Crohn's Disease. Crohn's disease is a chronic, or long-lasting, disease that causes inflammation, irritation or swelling, in the GI tract. Most commonly, Crohn's affects the small intestine and the beginning of the large intestine, but the disease can affect any part of the GI tract from mouth to anus. It is also called *inflammatory bowel disease*.

Crohn's disease most often begins gradually and worsens over time. Most people have periods of remission, times when symptoms disappear, that can last for weeks or years. Some people with Crohn's disease receive care from a gastroenterologist, a doctor who specializes in digestive diseases.

The exact cause of Crohn's disease is unknown. Researchers believe various factors play a role in causing Crohn's disease, including autoimmune reactions, genes, and the environment.

Researchers have not found that eating, diet, and nutrition cause Crohn's disease symptoms, but good nutrition is important in managing Crohn's disease. Dietary changes can help reduce symptoms. A health-care provider may recommend that a person make dietary changes such as:

- Avoiding carbonated drinks
- Avoiding popcorn, vegetable skins, nuts, and other high-fiber foods
- Drinking more liquids
- Eating smaller meals more often
- Keeping a food diary to help identify troublesome foods

Health-care providers may also recommend nutritional supplements and vitamins for people who do not absorb enough nutrients.

Pancreatitis. Pancreatitis is inflammation of the pancreas. The pancreas is a large gland behind the stomach and close to the duodenum, the first part of the small intestine. The pancreas secretes digestive juices, or enzymes, into the duodenum through a tube called the *pancreatic duct.* Pancreatic enzymes join with bile, a liquid produced in the liver and stored in the gallbladder, to digest food. The pancreas also releases the hormones insulin and glucagon into the bloodstream. These hormones help the body regulate the glucose it takes from food for energy. Normally, digestive enzymes secreted by the pancreas do not become active until they reach the small intestine. But when the pancreas is inflamed, the enzymes inside it attack and damage the tissues that produce them. Pancreatitis can be acute or chronic.

Treatment for acute pancreatitis requires a few days' stay in the hospital for intravenous fluids, antibiotics, and medication to relieve pain. The person cannot eat or drink so the pancreas can rest. If vomiting occurs, a tube may be placed through the nose and into the stomach to remove fluid and air. Unless complications arise, acute pancreatitis usually resolves in a few days. In severe cases, the person may require nasogastric feeding, a special liquid given in a long, thin tube inserted through the nose and throat and into the stomach for several weeks while the pancreas heals. Before leaving the hospital, the person will be advised not to smoke, drink alcoholic beverages, or eat fatty meals. In some cases, the cause of the pancreatitis is clear, but in others more tests are needed after the person is discharged and the pancreas is healed.

EATING DISORDERS

The eating disorders **anorexia nervosa, bulimia nervosa**, and **binge-eating disorder**, and their variants, all feature serious disturbances in eating behavior and weight regulation. They are associated with a wide range of adverse psychological, physical, and social consequences. A person with an eating disorder may start out just eating smaller or larger amounts of food, but at some point, his or her urge to eat less or more spirals out of control. Severe distress or concern about body weight or shape or extreme efforts to manage weight or food intake also may characterize an eating disorder.

Eating disorders are real, treatable medical illnesses. They frequently coexist with other illnesses such as depression, substance abuse, and anxiety disorders. Other symptoms can become life-threatening if a person does not receive treatment. Anorexia is associated with the highest mortality rate of any psychiatric disorder.

Eating disorders affect both genders, although rates among women and girls are two and one-half times greater than for men and boys. Eating disorders frequently appear during the teen years or young adulthood but also may develop during childhood or later in life.

Anorexia Nervosa

Individuals with anorexia nervosa see themselves as overweight, even when they are clearly underweight. Eating, food, and weight control become obsessions. People with anorexia nervosa typically weigh themselves repeatedly, portion food carefully, and eat tiny quantities of only certain foods. Some people with anorexia nervosa also may engage in binge eating followed by extreme dieting, excessive exercise, self-induced vomiting, or misuse of laxatives, diuretics, or enemas.

Some who have anorexia nervosa recover with treatment after only one episode. Others get well but have relapses. Still others have a more chronic, or long-lasting, form of anorexia nervosa, in which their health declines as they battle the illness.

Other symptoms and medical complications may develop over time, including:

- Thinning of the bones (osteopenia or osteoporosis)
- Brittle hair and nails
- Dry and yellowish skin
- Growth of fine hair all over the body (lanugo)
- Mild anemia, muscle wasting, and weakness
- Severe constipation
- Low blood pressure, or slowed breathing and pulse
- Damage to the structure and function of the heart
- Brain damage
- Multiple-organ failure
- Drop in internal body temperature, causing a person to feel cold all the time
- Lethargy, sluggishness, or feeling tired all the time
- Infertility

Bulimia Nervosa

People with bulimia nervosa have recurrent and frequent episodes of eating unusually large amounts of food and feel a lack of control over these episodes. This binge eating is followed by behavior that compensates for the overeating such as forced vomiting, excessive use of laxatives or diuretics, fasting, excessive exercise, or a combination of these behaviors.

Unlike anorexia nervosa, people with bulimia nervosa usually maintain what is considered a healthy or normal weight, while some are slightly overweight. Like people with anorexia nervosa, however, they often fear gaining weight, want desperately to lose weight, and are intensely unhappy with their body size and shape. Usually, bulimic behavior is done secretly because it is often accompanied by feelings of disgust or shame. The binge eating and purging cycle can happen anywhere from several times a week to many times a day. Other symptoms include:

- Chronically inflamed and sore throat
- Swollen salivary glands in the neck and jaw area
- Worn tooth enamel and increasingly sensitive and decaying teeth as a result of exposure to stomach acid
- Acid reflux disorder and other gastrointestinal problems
- Intestinal distress and irritation from laxative abuse
- Severe dehydration from purging of fluids
- Electrolyte imbalance—too low or too high levels of sodium, calcium, potassium, and other minerals that can lead to a heart attack or stroke.

Binge-Eating Disorder

People with binge-eating disorder lose control over their eating. Unlike bulimia nervosa, periods of binge eating are not followed by offsetting behaviors such as purging, excessive exercise, or fasting. As a result, people with binge-eating disorder often are overweight or obese and at higher risk of developing cardiovascular disease and high blood pressure. They also experience guilt, shame, and distress about their binge eating, which can lead to more binge eating.

FOOD ALLERGIES

A **food allergy** is an abnormal response to a food triggered by the body's immune system. There are several types of immune responses to food. The body produces a specific type of antibody called immunoglobulin E (IgE).The binding of IgE to specific molecules present in a food triggers the immune response. Sometimes, reaction to food is not an allergy at all but another type of reaction called *food intolerance.*

An allergic reaction to food usually takes place within a few minutes to several hours after exposure to the allergen. The process of eating and digesting food and the location of mast cells both affect the timing and location of the reaction. All or some of the following symptoms maybe experienced:

- Itching in the mouth
- Swelling of lips and tongue

Source: USDA, Agricultural Research Service (ARS), photo by Scott Bauer

FIGURE 29-9 A common food allergy in both children and adults is peanuts and peanut products.

SCIENCE CONNECTION!

People have varying levels of allergic reactions. Research one of the major foods that people most often have allergic reactions to—citrus fruit, nuts, fish, peanuts, shellfish, and wheat—and find out what happens when a person has a reaction to this food item. Why do reaction levels vary? Can medicines help with the allergic reactions? Record your findings.

- GI symptoms, such as vomiting, diarrhea, or abdominal cramps and pain
- Hives
- Worsening of eczema
- Tightening of the throat or trouble breathing
- Drop in blood pressure

The response may be mild, but in rare cases it can produce a severe and life-threatening reaction called *anaphylaxis*. People with a food allergy must work with a health-care professional to learn which foods cause an allergic reaction.

Among children, most allergic reactions to food are to peanuts (Figure 29-9), milk, soybeans, nuts from trees, eggs, and wheat. The majority of children stop being allergic to foods early in childhood. Allergic adults typically react to citrus fruit, nuts, fish, peanuts, shellfish, and wheat.

DENTAL CARIES OR TOOTH DECAY

Dental caries—tooth decay—is a major oral health problem in most industrialized countries, affecting 60% to 90% of schoolchildren and the vast majority of adults. The early sign of the caries process is a small patch of demineralized (softened) enamel at the tooth surface, often hidden from sight in the fissures (grooves) of teeth or in between the teeth. The destruction spreads into the softer, sensitive part of the tooth beneath the enamel (dentine). The weakened enamel then collapses to form a cavity, and the tooth is progressively destroyed. Caries can also attack the roots of teeth if they become exposed by gum recession. This is more common in older adults.

Dental caries is caused by the action of acids on the enamel surface. The acid is produced when sugars (mainly sucrose) in foods or drinks react with bacteria present in the dental biofilm (plaque) on the tooth surface. The acid produced leads to a loss of calcium and phosphate from the enamel; this process is called *demineralization.*

Prevention includes reduced consumption of sugars, good oral health (brush and floss), the use of fluorides, and sometimes the application of sealants.

PHYTONUTRIENTS

In addition to vitamins and minerals, plants contain compounds called **phytonutrients** (sometimes referred to as *phytochemicals*). Essentially, these compounds are the plants' protection. A plant cannot fight or flee, so it is equipped with *phyto* or plant nutrients that defend against disease, blight, radiation, weather, insects, and anything else that may threaten the plant's survival.

When we eat the plants, we not only benefit from the vitamin and mineral content of the plant but also from the protection these phytonutreints provide. Phytonutrients are considered anti-inflammatory and have been shown to possess anticancer properties, to repair DNA damage, to aid detoxification, to enhance immunity, and to influence insulin–glucose balance.

Hundreds of phytonutrients have been discovered thus far. Because fruits and vegetables contain different amounts of these beneficial compounds, it is best to eat a variety of plants. Sources of beneficial phytonutrients include:

- Cabbage, Brussels sprouts, broccoli, kale, cauliflower, and turnips, which all contain the phytonutrient indole-3-carbinol
- Oranges, tangerines, lemons, and limes, which all contain a flavonoid called *limonene*
- Grapes, apples, cherries, blueberries, and raspberries, which contain anthocyanins
- Onions and garlic, which contain the phytonutrient quercitin
- Tomatoes, red pepper, watermelon, and radishes, which all contain lycopene
- Peaches, carrots, apricots, pumpkin, and squash, which contain the phytonutrient carotinoids
- Swiss chard, kale, and parsley, which contain the phytonutrient lutein

Phytonutrient content is categorized by color (dark green, light green, red, orange, and purple.) An individual who gets at least one food from each of these color groups every day not only will be getting a variety of beneficial phytonutrients but also meeting the recommended minimum of five or more servings of fruits or vegetables a day. (*Note:* Color indicates the predominant phytonutrient content, but most fruits and vegetables contain multiple phytonutrients.)

NUTRACEUTICALS

The word *nutraceutical* is a combination of the words *nutrition* and *pharmaceutical.* It was coined in 1989 by Stephen L. DeFelice, founder and chairman of the Foundation of Innovation Medicine. A **nutraceutical** is a food or part of a food that allegedly provides medicinal or health benefits, including the prevention and treatment of disease. A nutraceutical may be a naturally nutrient-rich or medicinally active food, such as garlic or soybeans, or it may be a specific component of a food, such as the omega-3 fish oil that can be derived from salmon and other cold-water fish. The term is applied to products that range from isolated nutrients, dietary supplements and herbal products, specific diets, and processed foods such as cereals, soups, and beverages.

HERBS

Many people take **herbal supplements** in an effort to be well and stay healthy. Hundreds of herbal supplements are available in grocery stores, health food stores, and pharmacies as well as for sale on the Internet (Figure 29-10). There are many claims about their health benefits. Deciding what is effective and safe is difficult. Often the use of herbs is based on their historical uses or current folk or traditional uses. Some have scientific evidence about effectiveness and cautions about side effects. To help ensure coordinated and safe care, individuals should notify their health-care providers about any complementary health approaches they are using, including herbal supplements.

A 2007 survey found that more than 38 million U.S. adults reported using a natural product, such as herbs, for health purposes. Among the top 10 natural products used were several botanicals: echinacea, flaxseed, ginseng, ginkgo, and garlic. Some other common herbs are listed in Table 29-4.

People have used herbs as medicine since ancient times. For example, the use of aloe vera can be traced back to early Egypt, where the plant was depicted on stone carvings. Known as the "plant of immortality," it was presented as a burial gift to deceased pharaohs. Lavender, native to the Mediterranean region, was used in ancient Egypt as part of the process for mummifying bodies. Chasteberry, the fruit of the chaste tree, has been used for thousands of years by women to ease menstrual problems and to stimulate the production of breast milk. Historically, cat's claw, which grows wild in Central and South America, especially in the Amazon rainforest, has been used for centuries in South America to prevent and treat disease. Hoodia, a flowering, cactus-like plant native to the Kalahari Desert in southern Africa, was used by Kalahari Bushmen to reduce hunger and thirst during long hunts.

FIGURE 29-10 The number and variety of over-the-counter herbal supplements is as large as the industry providing them.

What Are Herbs?

An herb (also called a *botanical*) is a plant or plant part used for its scent, flavor, or health-related properties. An herbal supplement is a type of dietary supplement that contains herbs alone or in mixtures.

Herbs still play a part in the health practices of many countries and cultures. Ayurvedic medicine, which originated in India, uses herbs, plants, oils, common spices (such as ginger and turmeric), and other naturally occurring substances. Traditional Chinese medicine uses herbs such as astragalus, bitter orange, and ginkgo for various health conditions. Herbs are also an important part of Native American healing traditions. Dandelion and goldenseal are examples of herbs used by Native Americans for different health conditions.

Research on Herbs

Although millions of Americans use herbal supplements, much remains to be learned about their safety and effectiveness. The National Center for Complementary and Alternative Medicine at the National Institutes of Health is the federal government's lead agency for studying all types of complementary health approaches, including herbal supplements. This research covers a wide range of studies from laboratory-based research studying

TABLE 29-4 **Sixty-Five Common Herbs**

COMMON NAME	SCIENTIFIC NAME
All-heal (or self-heal)	*Prunella vulgaris L.*
Aloe vera gel	*Aloe vera*
American ginseng	*Panax quinquefolius*
Angelica root	*Angelica archangelica*
Anise seed	*Pimpinella anisum*
Apple	*Malus domestica*
Arnica	*Arnica montana*
Basil	*Ocimum basilicum*
Bilberry	*Vaccinium myrtillus*
Black cohosh root	*Cimicifuga racemosa*
Black pepper	*Piper nigrum*
Black walnut	*Juglans nigra*
Cacao	*Theobroma cacao L.*
Calendula	*Calendula officinalis*
Caraway seed	*Carum carvi*
Cardamom	*Elettaria cardamomum*
Carob	*Ceratonia siliqua*
Castor oil	*Ricinus communis*
Cayenne pepper	*Capsicum minimum*
Celery seed	*Apium graveolens*
Chamomile	*Matricaria recutita*
Cilantro	*Coriandrum sativum*
Cinnamon	*Cinnamomum zeylanicum, C. cassia*
Clove oil	*Syzygium aromaticum*
Coconut oil	*Cocos nucifera*
Coffee	*Coffea arabica*
Comfrey leaf and root	*Symphytum officinale*
Cranberry	*Vaccinium macrocarpon*

(Continues)

TABLE 29-4 Sixty-Five Common Herbs

COMMON NAME	SCIENTIFIC NAME
Dandelion root	*Taraxacum officinale*
Dill	*Anethum graveolens*
Echinacea	*Echinacea angustifolia*
Eucalyptus	*Eucalyptus globulus*
Fennel seed	*Foeniculum vulgare*
Garlic	*Allium sativum*
Ginger root	*Zingiber officinale*
Ginseng root, American	*Panax quinquefolius*
Hawthorn berry	*Crataegus monogyna*
Hibiscus	*Hibiscus sabdariffa*
Horseradish root	*Armoracia rusticana*
Juniper berries	*Juniperus communis*
Lavender	*Lavandula spp.*
Lemon	*Citrus limonum*
Lemon balm	*Melissa officinalis*
Lemongrass	*Cymopogon citratus, C.flexuosus*
Mint	*Mentha piperita*
Mustard	*Brassica nigra*
Oats	*Avena sativa L.*
Olive oil	*Olea europea*
Orange	*Citrus sinensis, Citrus* spp.
Oregano	*Origanum vulgare*
Peppermint	*Mentha piperita*
Pine	*Pinus sylvestris*
Pumpkin seed	*Cucurbita pepo*
Rose	*Rosa spp*
Rosemary	*Rosmarinus officinalis*
Sage	*Salvia officinalis*

(Continues)

TABLE 29-4 Sixty-Five Common Herbs

COMMON NAME	SCIENTIFIC NAME
Sesame	*Sesamum indicum*
Skullcap	*Scutellaria lateriflora*
St. John's wort	*Hypericum perforatum*
Sunflower	*Helianthus annuus*
Tea	*Camellia sinensis*
Tea tree oil	*Melaleuca alternifolia*
Thyme	*Thymus vulgaris*
Turmeric	*Curcuma longa L.*
Valerian root	*Valeriana officinalis*
White willow bark	*Salix alba*
Witch hazel	*Hamamelis virginiana*

how herbs might affect the body to large clinical trials testing their use in people, such as studying ginkgo's effects on memory in older adults or whether St. John's wort may help people with minor depression.

Exploring how and why botanicals act in the body is an important step in evaluating their safety and effectiveness. Although herbs have been used for thousands of years as natural medicines, "natural" does not always mean "safe." Herbs can act in the body in ways similar to prescription drugs, and herbs may have side effects. They may also affect how the body responds to prescription drugs or over-the-counter medicines, possibly decreasing or increasing their effects.

Regulation of Herbal Supplements

The U.S. Food and Drug Administration (FDA) regulates herbal and other dietary supplements differently from conventional medicines. The standards of safety and effectiveness that prescription and over-the-counter medicines have to meet before they are marketed do not apply to supplements. The standards for supplements are found in the Dietary Supplement Health and Education Act (DSHEA), a federal law that defines dietary supplements and sets product-labeling standards and health-claim limits. More information about DSHEA can be found on the FDA Web site.

Using Caution

Anyone considering or using an herbal supplement should consider the following:

- Some herbal supplements are known to interact with medicines (both prescription and over-the-counter). For example, St. John's wort can interact with birth control pills.
- Research has shown that what is listed on the label of an herbal supplement may not be what is in the bottle.
- Individuals may be getting less or more of an ingredient than the label indicates. A manufacturer's use of the terms *standardized*,

certified, or *verified* does not necessarily guarantee product quality or consistency.

- Many factors, including manufacturing and storage methods, can affect the contents of an herbal product.
- Some herbal supplements have been found to be contaminated with metals, unlabeled prescription drugs, microorganisms, and other substances.
- Not all herbal claims can be validated.

Individuals who use herbal supplements should do so under the guidance of a medical professional who has been properly trained in herbal medicine. This is especially important for herbs that are part of a whole medical system such as traditional Chinese medicine or Ayurvedic medicine. Women who are pregnant or nursing should be especially cautious about using herbal supplements. This caution also applies to giving children herbal supplements. For more information, see "Using Dietary Supplements Wisely" (www.nccam.nih.gov/health/supplements/wiseuse.htm).

Individuals who take herbs or supplements should give their health-care providers a full picture of everything that is being done to manage health, including all complementary health approaches. This will help ensure coordinated and safe care. It is especially important for anyone taking any prescription or over-the-counter medication that could interact with an herbal supplement.

Many Web sites and books tout the uses of herbs and the symptoms of disease that they will relieve. Individuals should study the scientific research on a specific herb, which will allow them to better make informed decisions.

NUTRIGENOMICS

One of the breakthrough concepts from the Human Genome Project, a decades-long effort that led to the complete mapping of human DNA, is that "genes in and of themselves do not create disease." Only when they are exposed to a harmful environment unique to the individual do they create the outcome of disease.

An advancing area of study called **nutrigenomics** looks at how different foods may interact with specific genes in the hopes of modifying the risk of common chronic diseases such as type 2 diabetes, obesity, heart disease, stroke, and certain cancers. Nutrigenomics also seeks to identify the molecules in the diet that affect health by altering the expression of genes.

For example, one study showed that participants who consumed a diet of whole rye (low insulin response) experienced changes in their gene expression that reduced their risk of developing diabetes. Participants who consumed an oat–wheat–potato (high insulin response) diet experienced the opposite—a change in their gene expression that increased their risk. Perhaps the age old wisdom "to use food as medicine" is not far off.

Humans cannot change their genes, but they can change the environment that affects how the genes manifest. One important component of this environment is food. Research in nutrigenomics represents a highly promising area of discovery (Figure 29-11).

SCIENCE CONNECTION!

Research one of the herbs in Table 29-4. Where is it grown? How is it harvested? How is it used in the food industry?

© 2017 Cengage Learning

FIGURE 29-11 Nutrigenomics looks at how different foods may interact with specific genes to modify the risk of common chronic diseases.

COMPONENTS OF FOOD

This section covers how some of the components of the foods we eat contribute to our health. It can help identify the type of diet we should eat.

Fruits and Vegetables

The health-related benefits of a diet rich in fruits and vegetables are known to most Americans, but the scientific literature since 2005 has increasingly pointed out the influence of these food groups on a variety of diseases. Several studies show that the higher the consumption of fruit and vegetables, the lower the incidence of cardiovascular disease, including stroke. Fruits and vegetables provide a wide variety of benefits to the body.

Vitamins and minerals (including antioxidants, such as vitamins A, C, E and selium) are found in all foods, but fruits and vegetables are a particularly good source for some of them.

Fiber, which assists digestion, slows carbohydrate absorption and promotes satiety.

Phytonutrients that are considered anti-inflammatory have been shown to possess anticancer properties, repair DNA damage, aid detoxification, enhance immunity, and influence insulin–glucose balance.

Source: USDA, Agricultural Research Service (ARS), photo by Peggy Greb

FIGURE 29-12 Whole grain products are part of a healthy diet.

Whole Grains

Whole grains contain beneficial nutrients vital to a healthy diet (see Figure 29-12). Whole grains include vitamins and minerals, which are stripped out during food processing and may or may not be replaced by manufacturers in the final product. Whole grains also include fiber, which is not replaced when refined but offers many essential benefits. Whole grains provide information and materials to help the body do the following:

- Regulate blood sugar (because complex carbohydrates are metabolized more slowly)
- Aid digestion by producing good bacteria in the gut

- Control appetite (because the fiber in the grains signals satiety-the sense of being full)
- Reduce cholesterol
- Remove toxins (because fiber binds to toxins in the gut and removes them during elimination)
- Improve digestive system function
- Synthesize neurotransmitters (the chemical messengers made by the body such as serotonin for sleep and mood)

Protein

Meats, fish, and beans are key sources of protein in the diet. Less is known about protein and its relationship to health and disease than fats or carbohydrates. Protein is abundant in the body and regulates multiple messengers that keep us functioning. Protein provides the body with amino acids, which are needed daily because the body does not store them. Proteins provide information to help the body do the following:

- Regulate blood sugar and insulin balance
- Produce hormones that regulate mood and sleep
- Detoxify (during the second phase of detoxification in the liver, protein attaches to waste molecules and escorts them out of the body)
- Make connective tissue for skin, cartilage, and bone
- Build muscle
- Promote wound healing
- Aid adrenal and thyroid function
- Produce and maintain a feeling of satiety (feeling full)

Fats and Oils

The popularity of low-fat diets in our culture leads many people to assume that eating any fat is bad, but our bodies require some fat to be healthy. As scientific and public opinion of fats is slowly shifting, the emerging consensus is that eating the right kind of fats is important to health and the prevention of disease. Fats provide information to help the body do the following:

- Provide insulation for the organs
- Transport fat-soluble vitamins (A, D, E, K)
- Provide materials critical to the integrity of cellular membranes
- Lubricate mucous membranes and skin
- Provide materials used to make hormones
- Use glucose more effectively
- Contribute to healthy joints
- Enjoy efficient gut health
- Facilitate immune system function
- Increase or decrease inflammation (depending on the type of fat).

Generally, individuals should eat fats that decrease inflammation such as oils from plants, nuts, and seeds and fats from fish whose diets are made up of algae (these all contain a predominance of omega-3 fatty acids).

More on Fats. A **monounsaturated fatty acid (MUFA)** has one double bond. Plant sources that are rich in MUFAs include nuts and vegetable oils that are liquid at room temperature (e.g., canola oil, olive oil, and high-oleic safflower and sunflower oils). A **polyunsaturated fatty acid (PUFA)** has two or more double bonds and may be one of two types based on the position of the first double bond.

Linoleic acid is required but cannot be synthesized by humans and, therefore, is considered essential in the diet. Primary sources are liquid vegetable oils, including soybean oil, corn oil, and safflower oil. Also called n-6 fatty acids, or omega-6 fatty acids.

Alpha-linolenic acid is an n-3 fatty acid (omega-3 fatty acid). that is required because it is not synthesized by humans and is thus considered essential in their diets. It is obtained from plant sources, including soybean oil, canola oil, walnuts,

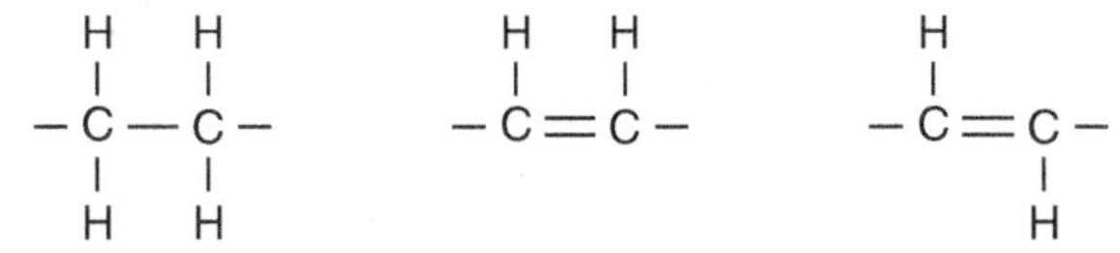

```
 H   H          H   H          H
 |   |          |   |          |
-C — C-        -C = C-        -C = C-
 |   |                             |
 H   H                             H
```

Saturated fat Unsaturated fat Trans-fat

FIGURE 29-13 A trans-fatty acid.

and flaxseed. Eicosapentaenoic acid and docosahexaenoic acid are long-chain n-3 fatty acids that are contained in fish and shellfish.

Saturated fatty acids have no double bonds. Examples include the fatty acids found in animal products such as meat, milk and milk products, hydrogenated shortening, and coconut or palm oils. In general, foods with relatively high amounts of saturated fatty acids are solid at room temperature.

Unsaturated fatty acids contain one or more isolated double bonds in a trans configuration produced by chemical hydrogenation (adding hydrogen). Sources of trans-fatty acids (see Figure 29-13) include hydrogenated or partially hydrogenated vegetable oils that are used to make shortening and commercially prepared baked goods, snack foods, fried foods, and margarine. **Trans-fatty acids** also are present in foods that come from ruminant animals such as cattle and sheep. These foods include dairy products, beef, and lamb.

BAD DIETS, GOOD DIETS

New diets and weight-loss schemes are frequently promoted in various media. People seem to want an easy method or pill that will help them lose weight. Although many people around the world struggle to find enough to eat, many Americans seem obsessed with what to eat and how to lose weight. Some diets are fads, and some are reasonable. Examples include:

- Alkaline Diet
- Atkins Diet
- Baby Food Diet
- Cookie Diet
- Five-Bite Diet
- Grapefruit Diet
- HCG (human chorionic gonadotropin) Diet
- Macrobiotic Diet
- Master Cleanse–Lemonade Diet
- Mediterranean Diet
- NutriSystem
- Paleo Diet
- Raw Food Diet
- South Beach Diet
- Tapeworm Diet
- The Zone Diet
- Volumetrics
- Weight Watchers
- Werewolf Diet

These diets are often expensive and highly restrictive—and sometimes dangerous. Many are designed to sell books that promote a corresponding diet or program of weight loss.

What to Eat

Because of their individual nutritional needs, nutritional statuses, health conditions, and genetic differences what people should eat to maintain their health varies widely. Overall, proper nutrition should maintain caloric balance over time to achieve and sustain a healthy body weight. People who are most successful at achieving and maintaining a healthy weight do so through continued attention to consuming only enough calories from foods and beverages to meet their needs and by being physically active. To curb the obesity epidemic and improve their health, many Americans should should focus on consuming nutrient-dense foods and beverages, decreasing the calories they consume, and increasing the calories they expend through physical activity.

Americans currently consume too much sodium and too many calories from solid fats, added sugars, and refined grains. These replace nutrient-dense foods and beverages and make it difficult for people to achieve recommended nutrient intake while controlling caloric intake. A healthy eating pattern limits he intake of sodium, solid fats, added sugars, and refined grains and emphasizes nutrient-dense foods and beverages such as vegetables, fruits, whole grains, fat-free or low-fat milk and milk products, seafood, lean meats and poultry, eggs, beans and peas, and nuts and seeds.

ChooseMyPlate.gov is an approach designed to help Americans make healthy food choices. MyPlate is based on the *Dietary Guidelines for Americans.* In January 2016, the Secretary of Health and Human Services and Secretary of Agriculture released the 2015–2020 Dietary Guidelines for Americans (http://health.gov/dietaryguidelines/2015/). This is the 8th edition of the Dietary Guidelines since 1980. Besides focusing on making healthy food choices this new edition reaffirms guidance about the core building blocks of a healthy lifestyle. Also, the 2015 edition includes updated guidance on topics such as added sugars, sodium, and cholesterol and new information on caffeine.

A healthy eating pattern promotes health, helps decrease the risk of chronic diseases, and helps prevents food-borne illness. Four basic food safety principles work together to reduce the risk of food-borne illnesses:

1. Clean
2. Separate
3. Cook
4. Chill

In addition, foods that pose high risk for food-borne illness should be avoided, including unpasteurized milks, cheeses, and juices and undercooked animal foods.

Balancing Calories

Controlling total caloric intake manages body weight. For people who are overweight or obese, this will mean consuming fewer calories from foods and beverages, increasing physical activity, and reducing time spent in sedentary behaviors. These individuals also need to maintain appropriate caloric balance during each stage of life—from childhood, adolescence, adulthood, and pregnancy and breastfeeding to older age.

Individuals need to select an eating pattern that meets their nutrient needs over time at an appropriate calorie level, account for all foods and beverages they consume, and assess how they fit within a total healthy eating pattern. Healthy eating also means following food-safety recommendations when preparing and eating foods to reduce the risk of food-borne illnesses.

SUMMARY

Trendy diets and nutrition research aimed at healthy living seem to change almost daily. Still, amidst all the hype, study after study shows that good food choices have positive effects on health, and poor diets have negative long-term effects. A healthy diet provides the human body the nutrients it needs to perform physically, to maintain wellness, and to fight disease.

In general, a poor diet in combination with a sedentary lifestyle, large portion sizes, and high stress is blamed for the increase in obesity and associated diseases in the United States; according to the Centers for Disease Control and Prevention, more than one-third of American adults are obese). Diseases associated with obesity include type 2 diabetes, high blood pressure, coronary heart disease, stroke, gallbladder disease, osteoarthritis, sleep apnea, respiratory problems, and certain cancers, including breast cancer in women.

Other food-related health issues include digestive disorders such as constipation, diarrhea, celiac disease, gallstones, heartburn, food allergies, lactose intolerance, ulcers, osteoporosis, and Crohn's disease. Also included are eating disorders such as anorexia nervosa, bulimia nervosa, and binge-eating disorders. Finally, dental caries—tooth decay—can also be the product of poor nutrition.

What individuals choose to eat is central to their health. Food acts as medicine to maintain health, as well as prevent and treat disease. Food influences the maintenance of health and the prevention or alleviation—and even cure—of some diseases. Millions of Americans and people worldwide are overweight or obese. Being overweight or obese puts an individual at risk for many health problems. The more body fat an individual has, or the more and individual weighs, the more likely he or she is to develop health problems.

The most useful measure of overweight and obesity is BMI. BMI is calculated from a person's height and weight (weight in kilograms divided by the square of height in meters).

Overall, randomized controlled clinical trials did not provide evidence that dietary antioxidant supplements are beneficial in primary cancer prevention.

An individual's eating, diet, and nutrition choices play an important role in preventing or delaying diabetes. Some suggestions that can be used to reach and maintain a reasonable weight and make wise food choices most of the time include digestive disorders, including gas, heartburn, indigestion, lactose intolerance, peptic ulcer disease, diarrhea, constipation, diverticular disease, celiac disease, Crohn's disease, and pancreatitis. Some are caused by foods, and some are managed by the food eaten.

Eating disorders include anorexia nervosa, bulimia nervosa, binge-eating disorder, and their variants. A food allergy is an abnormal response to a food triggered by the body's immune system. There are several types of immune responses to food.

Dental caries is caused by the action of acids on a tooth's enamel surface. The acid is produced when sugars (mainly sucrose) in foods or drinks react with bacteria present in the dental biofilm (plaque) on the tooth surface.

Phytonutrients are considered anti-inflammatory and have been shown to possess anticancer properties, repair DNA damage, aid detoxification, enhance immunity, and influence insulin-glucose balance.

A nutraceutical is a food or part of a food that allegedly provides medicinal or health benefits, including the prevention and treatment of disease.

Many Web sites and books tout the uses of herbs to relieve symptoms of disease. Individuals should study scientific research into a specific herb to make an informed decision.

An advancing area of study called *nutrigenomics* looks at how different foods may interact with specific genes to modify the risk of common chronic diseases such as type 2 diabetes, obesity, heart disease, stroke, and certain cancers.

A healthy eating pattern promotes health, helps decrease the risk of chronic diseases, and prevents food-borne illness. The 8th edition of the Dietary Guidelines for Americans, released in 2015, is a governmental effort to promote healthy food choices.

REVIEW QUESTIONS

Success in any career requires knowledge. Test your knowledge of this chapter by answering these questions or solving these problems.

1. Find suggested daily caloric intakes for a highly active 22-year-old male and a highly active female.
2. What is the BMI for male who is 5 feet 11 inches tall and weighs 230 pounds?
3. Identify 10 digestive disorders.
4. Name three eating disorders.
5. Define and give an example of a phytonutrient.
6. What are the health issues associated with being overweight or obese?
7. Discuss the possible dietary causes of diarrhea.
8. Explain type 2 diabetes.
9. List five good dietary sources of fiber and explain why fiber is important in dealing with some dietary diseases.
10. Discuss the role of fruits, vegetables, whole grains, protein, carbohydrates, and fats in maintaining health.

STUDENT ACTIVITIES

1. Keep a log of the foods you eat during a week, including the number of calories. Then evaluate the healthfulness of your diet.
2. Calculate your BMI and determine what it says about your health.
3. Develop a presentation on one of the eating disorders, including treatments.
4. Report on information available through one of the Internet resources listed under "Additional Resources."
5. Evaluate the nutrient density and caloric content of favorite snack foods and report to the class.
6. Develop a weekly diet for someone with type 2 diabetes.
7. Research one of the common food allergies for adults and give a visual class presentation.
8. Create an informational poster on lactose-free milk, including the action of lactase. Determine why lactose-free milk has a slightly sweeter taste.
9. Develop a presentation on a common herbal supplement and its actions. Support or disprove its actions using a search of scientific studies.

ADDITIONAL RESOURCES

- CDC—Healthy Weight: http://www.cdc.gov/healthyweight/index.html
- Choose My Plate: http://www.choosemyplate.gov/
- How Does Food Impact Health? http://www.takingcharge.csh.umn.edu/explore-healing-practices/food-medicine/how-does-food-impact-health
- Mayo Clinic—Nutrition and Healthy Eating: http://www.mayoclinic.org/healthy-lifestyle/nutrition-and-healthy-eating/basics/nutrition-basics/hlv-20049477
- National Institute of Diabetes and Digestive and Kidney Diseases: http://www.niddk.nih.gov/health-information/health-topics/Pages/default.aspx
- Nutrition.gov—Nutrition and Health Issues: http://www.nutrition.gov/nutrition-and-health-issues
- Nutrition.gov—What's in Food: http://www.nutrition.gov/whats-food

REFERENCES

Dole Food Company, Mayo Clinic, and UCLA Center for Healthy Policy Research. (2002). *Encyclopedia of foods: A guide to healthy nutrition*. San Diego: Academic Press.

McWilliams, J. E. (2009). *Just food: Where locavores get it wrong and how we can truly eat responsibly.* New York: Little, Brown & Co.

Kowalchik, C., and W. H. Hylton. (1987). *Rodale's illustrated encyclopedia of herbs*. Emmaus, PA: Rodale Press.

Haas, E. M., and B. Levin. (2006). *Staying healthy with nutrition: The complete guide to diet and nutritional medicine* (rev. ed.). Berkeley, CA: Celestial Arts.

Margen, S. (2002). *Wellness foods A to Z: An indispensable guide for health-conscious food lovers.* New York: Rebus.

Duyff, R. L. (2012). *American dietetic association complete food and nutrition guide* (4th ed., revised and updated). Boston: Houghton Mifflin Harcourt.

CHAPTER 30

Careers in Food Science

OBJECTIVES

After reading this chapter, you should be able to:

- Understand the basic skills and knowledge needed for successful employment and job advancement
- Describe the thinking skills needed for the workplace of today
- Recognize the traits of an entrepreneur
- Identify six occupational areas of the food industry
- Identify the careers that require a science background
- Describe the general duties of the occupations in six areas of the food industry
- Explain the education and experience needed to enter six areas of the food industry
- List six general competencies needed in the workplace
- List eight guidelines for choosing a job

NATIONAL AFNR STANDARD

FPP.01

Develop and implement procedures to ensure safety, sanitation, and quality in food products and processing facilities.

KEY TERMS

competencies
creative thinking
cultural diversity
data sheet
demographic
entrepreneur
follow-up letter
letter of application
letter of inquiry
résumé

OBJECTIVES (continued)

- **Identify 10 guidelines for filling out an application form**
- **Understand the appropriate content of a letter of inquiry or application**
- **Describe the elements of a résumé or data sheet**
- **Determine 10 reasons an interview may fail**
- **Discuss what research studies indicate about basic skills and thinking skills for the workplace**

In terms of the value of shipments, food processing is the largest manufacturing industry in the United States. The major technological support of the food-processing industry comes from food scientists, technicians, and other industry employees who use their training and experience to convert raw foods into quality products quickly, efficiently, and with a minimum of waste. They are directly concerned with the industry's high standards of quality, new manufacturing methods, new preservation techniques, and new packaging materials. A knowledge of chemistry, microbiology, engineering, and other basic and applied sciences plays an important part in maintaining the flavor, color, texture, nutritional value, and safety of our food.

GENERAL SKILLS AND KNOWLEDGE

Over the past few years, research study after research study indicated that potential employees never receive some basic skills and knowledge. Without these, the specific skills and knowledge needed for employment in the food industry are of little value. The new workplace, in fact, demands a better-prepared individual than in the past. Finally, those individuals working for themselves must develop a trait called *entrepreneurship*. This may also be a good trait for any employee.

Basic Skills

Success in the workplace requires that individuals possess skills in reading, writing, mathematics, listening, and speaking at levels identified by employers nationwide as well as a certain level of proficiency using computers (Figure 30-1).

Reading. An individual who is ready for today's workplace and the future demonstrates reading with the following **competencies**:

- Locates, understands, and interprets written information, including manuals, graphs, and schedules to perform job tasks
- Learns from text by determining the main idea or essential message
- Identifies relevant details, facts, and specifications

Source: USDA, Agricultural Research Service (ARS), photo by Stephen Ausmus

FIGURE 30-1 Using computers and interpreting data are important tasks for jobs throughout the food industry.

- Infers or locates the meaning of unknown or technical vocabulary
- Judges the accuracy, appropriateness, style, and plausibility of reports, proposals, or theories of other writers

Employee reading skills in the food industry are needed so they can keep up with new information and read directions and instructions as part of any job.

Writing. An individual ready for today's workplace and the future demonstrates writing abilities with the following competencies:

- Communicates thoughts, ideas, information, and messages
- Records information completely and accurately
- Composes and creates documents such as letters, directions, manuals, reports, proposals, graphs, and flowcharts with appropriate language, grammar, style, organization, and format
- Checks, edits, and revises for correct information, emphasis, form, grammar, spelling, and punctuation

In the food industry, writing skills are needed to perform such tasks as keeping records, making reports, and communicating with co-workers, as well as others inside and outside the industry.

Mathematics. The workplace of today and the future require individuals who are competent with certain areas of mathematics. Mathematics is the science of computing with numbers by addition, subtraction, multiplication, and division. Inn the food industry, these skills are used especially to figure out conversions and ratios. These important competencies are:

- Performing basic computations
- Using numerical concepts such as whole numbers, fractions, and percentages in practical situations
- Making reasonable estimates of mathematic results without a calculator
- Using tables, graphs, diagrams, and charts to obtain or convey information
- Approaching practical problems by choosing from a variety of mathematical techniques
- Using quantitative data to construct logical explanations of real-world situations
- Expressing mathematical ideas and concepts orally and in writing
- Understanding the role of chance in the occurrence and prediction of events

Listening. Individuals working today and in the future must demonstrate an ability to really listen. This means to receive, interpret, and to respond to oral messages and other cues such as body language. Real listening means that an individual comprehends, learns, evaluates, appreciates, or supports a speaker without interruption.

Speaking. Finally, an individual successful in the workplace of today and the future must demonstrate these speaking competencies:

- Organize ideas and communicate oral messages appropriate to listeners and situations
- Participate in conversation, discussion, and group presentations
- Use verbal language, body language, style, tone, and level of complexity appropriate for each audience and occasion
- Speak clearly and communicate the message
- Understand and respond to listener feedback
- Ask and answer questions when needed
- Use appropriate telephone etiquette, especially by being pleasant and courteous

Thinking Skills

Contrary to the old workplace, employers in the new workplace want workers who can think, according to many research studies. Employers search for individuals who show competencies in these areas: **creative thinking**, decision making, problem solving, mental visualization, knowing how to learn, and reasoning (Figure 30-2).

© Marin Bulat/Dreamstime.com

FIGURE 30-2 The modern workplace encourages employees to use thinking skills.

Creative Thinking. Creative thinkers generate new ideas by making nonlinear or unusual connections or by changing or reshaping goals to imagine new possibilities. These individuals use imagination to freely combine ideas and information in new ways. Creative thinkers may also use computer graphics and other forms of multimedia to enhance presentations and ideas.

Decision Making. Individuals who use thinking skills to make decisions are able to specify goals and limitations to a problem. They generate alternatives and consider risks for all of them before choosing the best one.

Problem Solving. As obvious as it sounds, the first step to problem solving is recognizing that a problem exists. After this, individuals with problem-solving skills identify possible reasons for the problem and then devise and begin a plan of action to resolve it. As the problem is being solved, problem solvers monitor the progress and fine-tune the plan. Being able to recognize the need for a new product and looking for solutions are good examples of problem solving in food science.

Mental Visualization. This thinking skill requires that an individual see things in his or her mind's eye by and organize and process symbols, pictures, graphs, objects, and other information.

Knowing How to Learn. Of all thinking skills, knowing how to learn is perhaps the most important given the rapid changes in available technology. An individual able to employ this skill can recognize and use learning techniques to apply and adjust existing and new knowledge and skills in familiar and changing situations. Knowing how to learn means awareness of personal learning styles—formal and informal learning strategies.

Reasoning. The individual who uses reasoning discovers the rule or principle connecting two or more objects and applies this to solving a problem. For example, chemistry teaches the theory of pH measurements, but the reasoning individual is able to use this information in understanding pH in food chemistry.

General Workplace Competencies

Besides basic skills and thinking skills, the workplace of today and the future demands general competencies in the use of resources, interpersonal skills, information use, systems, and technology.

Resources. Resources for any business include time, money, materials, facilities, and people. Individuals in the food-industry workplace must know how to manage the following:

- Time by using goals, priorities, and schedules
- Money when budgeting and forecasting

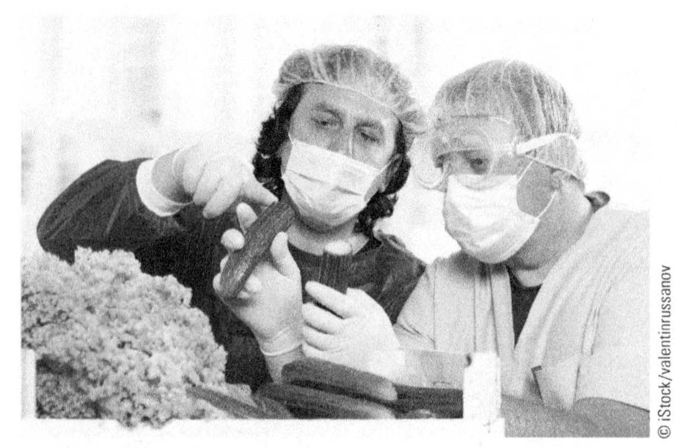

FIGURE 30-3 Working in teams—an important skill for employees.

FIGURE 30-4 Colleges and universities provide training in computer skills.

- Material and facility resources such as parts, equipment, space, and products
- Human resources by determining knowledge, skills, and performance levels

Interpersonal Skills. Although solitary individual jobs exist in the job market, most people are members of teams in which they contribute to their work groups (Figure 30-3). They teach others in their workplace when new knowledge or skills are needed. More than ever and at all levels, individuals must remember to serve customers and satisfy their expectations. Through teams, individuals frequently exercise leadership to communicate, justify, encourage, persuade, and motivate individuals or groups. As part of employment teams, individuals negotiate resources or interests to arrive at a decision. The United States has a culturally diverse workforce made up of people from around the world, so respecting **cultural diversity** is important and necessary in this growing global market. Many nonverbal communications have different meanings in different countries and cultures.

Information Technology

This is the information technology age. Individuals in the workplace must excel at using information technology. Successful individuals will identify the need for certain information and evaluate the information as it relates to a specific job and then use the necessary technology to make it successful. With the computer, individuals in the workplace must gather, organize, and process information in a systematic way (Figure 30-4). Also, given the considerable information available on the World Wide Web, individuals must be able to interpret and communicate information to others using oral, written, and graphical methods. Computer skills are key to managing production information. Information technology in today's workplace includes videoconferencing, online auctions, virtual meetings, webinars, social media. and more.[1]

Systems. No longer can any aspect of a business or industry be viewed as a part that stands alone. Every part is belongs to a system, and individuals now seek to understand systems, whether they are

SCIENCE CONNECTION!

What is the meaning of a handshake? Are there different meanings to a thumbs-up hand gesture? Research nonverbal communications and find at least three different meanings for one of these examples. What effects can this have on a workplace and global business communications?

social, organizational, technological, or biological. With an understanding of the systems in a business, trends can be determined and predictions can be made. Individuals can then modify a system to improve a product or service. For example, successful development and marketing of a new product requires understanding the entire food system.

Personal Qualities

Even after training in basic skills, thinking skills, and general workplace competencies, individuals can still fail because they lack certain personal qualities or values. These include responsibility, self-esteem, sociability, self-management, integrity, and honesty. These qualities together describe the term *work ethics.*

Responsible individuals work hard at tasks even when a task is unpleasant. Responsibility shows itself in meeting high standards of attendance, being punctual, being enthusiastic, performing jobs with vitality, and being optimistic in starting and finishing tasks.

Those who have good self-esteem believe in themselves and maintain a positive view of themselves even in high-pressure, demanding situations. These individuals know their skills, abilities, and emotional capacities. In short, they feel good about themselves.

Successful individuals demonstrate understanding, friendliness, adaptability, empathy, and politeness when interacting with other people. These skills are demonstrated in both familiar and unfamiliar social situations. The best examples are those individuals who take sincere interest in what others say and do.

Along with self-esteem is self-management. Individuals who are successful in business can accurately assess their own knowledge, skills, and abilities while setting well-defined and realistic personal goals. Once goals are set, they manage themselves by monitoring their progress and motivating themselves to achieve desired goals. Self-management also means a person exhibits self-control and responds to feedback unemotionally and nondefensively.

Finally, to be successful in the food industry, an employee or **entrepreneur** requires good old-fashioned honesty and integrity. Good work ethics will probably always be a part of good business.

ENTREPRENEURSHIP

The most common view of an entrepreneur is one who takes risks and starts a new business (Figure 30-5). Although this may be true for some in the food industry, certain entrepreneurial traits are desirable at most levels of employment. Within any organization, an entrepreneur may need to do the following:

- Find a better or higher use for resources
- Apply technology in a new way
- Develop a new market for an existing product
- Develop a new product to reach a specific market
- Improve an existing product in a new way
- Use technology to develop a new approach to serving an existing market
- Develop a new idea that creates a new business or diversifies an existing business

Most people can be entrepreneurs. It is more of an attitude—but one that incorporates many

FIGURE 30-5 Success of new products and businesses requires entrepreneurship.

desired traits. The attitude of an entrepreneur includes the following:

- Risk taking with clear understanding of the odds
- Focus on opportunities and not problems
- Seeking constant improvement
- Being impressed with productivity and not appearances
- Recognizing the importance of examples
- Keeping things simple
- Providing an open door and personal contact leadership
- Focusing on the customer
- Encouraging flexibility
- Being purposeful and communicating a vision

Entrepreneurs are ready for the unexpected, differences, new needs, change, **demographic** shifts, changes in perception, and new knowledge. Entrepreneurs are good employees and good employers. Entrepreneurs keep the food industry growing.

JOBS AND COURSES IN THE FOOD INDUSTRY

Opportunities open to graduates in the food industry and allied industries include research, development, and production work; technical sales within the food industry or in closely related areas such as the container and equipment manufacturing fields; extension work; research work in experimental stations or in other branches of government; food consulting; and promotional work with public or private utilities.

Food-industry skills and knowledge allow an individual to do the following:

- Enjoy a career in a dynamic, multibillion-dollar business
- Develop new food products and food-processing technologies
- Improve nutritional quality of foods and ensure that food is safe and wholesome
- Manage food-processing companies
- Pursue an advanced graduate or professional degree

Food science is a broad scientific field. Food processing is the largest manufacturing industry in the world. Food scientists are employed by the industry itself as well as by government and universities. Employment opportunities with excellent pay scales are plentiful. The following is a general list of opportunities in food-related positions:

- Production manager or supervisor—manages food-production and food-processing facilities
- Product development technologist—assists in designing, researching, and developing new food products
- Food engineer—involved in the design and manufacture of machinery needed to produce processed food that is safe and nutritious
- Flavor chemist—develop new food flavors in food or beverage items
- Food microbiologist—involved with microbiological safety of foods, including using microorganisms to produce new kinds of foods or improve existing foods
- Quality-control scientist—inspects and determines the quality of food in one or more stages of manufacturing
- Sensory scientist—tests food items based on the senses (appearance, smell, taste, feel, and sound)
- Research technician—helps government and university scientists perform food-related research
- Technical representative—provides technical support to salespeople in food-manufacturing or equipment-manufacturing plants

In addition, thousands of careers specialize in the food industries, including accounting, administrative, customer service, legal, logistics, marketing, packaging, research and development, and sales.

SCIENCE CONNECTION!

Think about some of your favorite food items. What jobs do you think it took to create that food and bring it to you? Brainstorm some of the jobs that helped create that food item. Pick two job ideas to further research. Would one of these interest you as a career?

EDUCATION AND EXPERIENCE

Education and experience vary widely for jobs and careers in the food industry. Some require only a high school education with on-the-job training provided (Figure 30-6). Others require college training in a technical area; and still others require a bachelor's, a master's, or doctorate degrees. Some careers have their own certification programs, and many require experience before advancing through the career choices. In the sections that follow, the requirements, education, and experience for jobs and careers are suggested.

© iStock/michaeljung

FIGURE 30-6 Most jobs require on-the-job training.

IDENTIFYING A JOB

More specific careers or jobs in the food industry include the areas of food safety and inspection, retailing and wholesaling, research and development, food science, marketing, and communication. Specific jobs and educational requirements are described below.

Food Inspection and Safety

With the ongoing demand for food quality and safety, high standards must be maintained. Individuals who work in this field monitor the safety and quality of foods, verify the quantity and composition of food products, and verify the accuracy of labels (see Figure 30-7). They also help businesses develop inspection systems that focus

© Photographerlondon/Dreamstime.com

FIGURE 30-7 Large processing plants employ many inspectors.

on preventing contamination in the entire food-manufacturing process. Some of the job titles in this area are:

- Crop certification inspector
- Dairy products inspector
- Fish and fish products inspector
- Fruit and vegetable inspector
- Grain inspector
- Livestock inspector
- Plant protection inspector
- Poultry inspector
- Public health and restaurant inspector

Education and Experience. A bachelor's degree or college diploma in agriculture, biology, or food-processing technology is required with several years of experience in agricultural production or fish processing. Individuals often are required to complete in-house training courses.

Job Description. In general, food inspectors check agricultural and fish products to ensure they conform to prescribed standards for production, storage, and transportation. Depending on the area of specialization, inspectors look for a variety of different things.

Fruit and vegetable inspectors inspect both fresh and frozen fruit and vegetables and prepare reports on crop production and market conditions. Grain inspectors inspect and grade all classes of grain at terminal elevators, monitor the fumigation of infested grain, and monitor storage, handling, transportation, and equipment to ensure that sanitary procedures are followed. Meat inspectors monitor the operations and sanitary conditions of slaughtering or meat-processing plants and inspect carcasses to ensure they are fit for human consumption. Plant protection inspectors certify seed crops; oversee the quarantine, treatment, or destruction of plants and plant products; and monitor the fumigation of plants and plant product imports and exports. Many of these individuals work for various levels of government, food-processing plants, and slaughterhouses.

Public health inspectors inspect workplaces to ensure that equipment, materials, and production processes do not present a health or safety hazard to food-service employees or the general public. Restaurant inspectors inspect the sanitary conditions of restaurants, hotels, schools, hospitals, and public facilities or institutions and investigate the outbreaks of diseases and poisonings resulting from spoiled food supplies. These individuals often work in municipal, state, or federal government positions.

Food-Service Industry

Individuals in the food-service industry area prepare quick meals for the lunchtime crowd, or bake breads and pastries at a local or industrial bakery. Careers in the food-service industry include bakers, butchers and meat cutters, chefs, cooks, food-service counter attendants, food preparers or kitchen helpers, food-service supervisors, and restaurant and food-service managers.

Bakers. Job titles associated with bakers include *baker, baker apprentice, bakery supervisor,* and *head baker.*

Education. Generally, bakers require a high school diploma, a three- or four-year apprenticeship, or a college or other program for bakers. Sometimes on-the-job training is provided.

Job Description. Bakers prepare pies, breads, rolls, muffins, cookies, cakes, icings and frostings, and many other foods, depending on where they work. In local bakeries and the hospitality industries, bakers will perform the following:

- Experiment with new recipes and ingredients
- Prepare their own creations and special customer orders
- Draw up production schedules to determine the types and quantities of goods to produce
- Order or purchase baking supplies
- Market and sell baked goods

In commercial bakeries or supermarkets, bakers will perform the following tasks:

- Prepare dough for pies, bread and rolls, and sweet goods and prepare batters for muffins, cookies, cakes, icings, and frostings according to recipes
- Frost and decorate cakes or other baked goods
- Hire and train baking personnel
- Most bakers will work in local and commercial bakeries, supermarkets, hotels and resorts, restaurants, clubs, ships, or their own bakeries.

Butchers and Meat Cutters. Job titles as butchers and meat cutters could include *butcher, butcher apprentice, head butcher, meat cutter,* and *supermarket meat cutter.*

Education. A high school diploma may be required. College or other programs in meat cutting may be required. On-the-job training in food stores is usually provided. Frequently, trade certification is available after working in the industry.

Job Description. Retail and wholesale butchers and meat cutters prepare those standard cuts of meat, poultry, fish, or shellfish that customers buy at wholesalers, in supermarkets, or at specialty shops in their towns or cities. They also perform the following duties:

- Cut, trim, and otherwise prepare food for sale at self-serve counters or according to customers' special orders
- Grind meats and slice cooked meats using powered grinders and slicing machines
- Shape, lace, and tie roasts and other meats, poultry, or fish
- Prepare special displays
- Supervise other butchers or meat cutters

Butchers and meat cutters work in supermarkets (Figure 30-8), grocery stores, butcher shops, and fish stores, or they own their own businesses.

FIGURE 30-8 Meat cutter.

Chefs. Job titles associated with career choices surrounding chefs include *chef, chef de cuisine, chef de partie, corporate chef, executive chef, executive sous chef, garde-manger chef, head chef, master chef, pastry chef, saucier,* and *specialist chef.*

Education. A high school diploma and 3-year cook's apprenticeship program is required; some individuals take formal training abroad. Some executive chef positions require several years of experience in commercial food preparation, including 2 years in a supervisory capacity and experience as a sous chef, specialist chef, or chef. Sous chefs, specialist chefs, and chefs usually require several years of experience in commercial food preparation. The American Chefs' Federation provides certification.

Job Description. Chefs who prepare delicious meals are revered around the world. There are even culinary olympics to show off chefs' talents. Anyone interested in becoming a chef can certainly specialize because considerable variety exists in chef positions.

Executive chefs often plan and direct the food preparation and cooking activities of several restaurants in a hotel, restaurant chain, hospital, or other establishment with food services. They plan menus and ensure that the food meets quality standards; estimate food requirements and may estimate food and labor costs; supervise the work of sous chefs, specialist chefs, chefs, and cooks; and they may even prepare and cook food on a regular basis or for special guests and functions.

Sous chefs usually supervise the activities of specialist chefs, chefs, cooks, and other kitchen workers. They may demonstrate new cooking techniques and new equipment to cooking staffs and plan menus and requisition food and kitchen supplies. They may also prepare and cook meals or specialty foods.

Chefs and specialist chefs prepare and cook complete meals, banquets, and specialty foods such as pastries, sauces, soups, salads, vegetables, and meat, poultry, and fish dishes. They may also create decorative food displays such as artistic ice carvings. Chefs and specialist chefs often instruct cooks in preparing, cooking, garnishing, and presenting food; plan menus; and requisition food and kitchen supplies.

Most chefs of all types work in restaurants, hotels and resorts, hospitals and other health-care institutions, central food commissaries, clubs and similar establishments, and on ships. Executive chefs may progress to managerial positions in food-preparation businesses (see Figure 30-9).

FIGURE 30-9 Chef and chef's assistant begin daily meal production.

Cooks. *Apprentice cook, cook, dietary cook, grill cook, hospital cook, institutional cook, journeyman* or *journeywoman cook, licensed cook, second cook,* and *short-order cook* are all titles for cooks that can be part of anyone's career.

Education. A high school diploma is usually required with an apprenticeship program for cooks. Some individuals attend a college or other program in cooking, or they gain several years of commercial cooking experience. Frequently, trade certification is available for work experience.

Job Description. Cooks perform some or all of the following duties:

- Prepare and cook complete meals or individual dishes and foods
- Prepare and cook special meals for patients as instructed by a dietitian or chef
- Plan menus, determine the size of food portions, estimate food requirements and costs
- Monitor and order supplies
- Supervise kitchen helpers in the handling of food

Cooks work in restaurants, hotels and resorts, hospitals and other health-care institutions, central food commissaries, educational institutions, construction or logging camp sites, and even ships.

Food-Service Counter Attendants, Food Preparers, and Kitchen Helpers. Food-service titles include *cafeteria counter attendant, fast-food preparer, food preparer, ice cream counter attendant, salad bar attendant,* and *sandwich maker.*

Education. Some high school may be required, but most of these workers learn through on-the-job training provided by an organization or company. Some college programs are available.

Job Description. Food-service counter attendants and food preparers take customer orders. They prepare foods such as sandwiches, hamburgers, salads, milk shakes, and ice cream dishes and take the money for the food purchased. Some may also serve customers at counters or buffet tables.

With additional training and experience, an individual can move into other occupations in food preparation and service such as cook or waiter. Most of these people work in cafeterias, fast-food outlets, restaurants, hotels, hospitals, and other establishments.

Food-Service Supervisors. Job titles for food-service supervisors include *cafeteria supervisor, catering supervisor, canteen supervisor,* and *food-service supervisor.*

Education. A high school diploma is often required. Some community colleges provide programs in food-service administration, hotel and restaurant management, and related disciplines. Also, several years of experience in food preparation or service can lead to a job in food-service supervision.

Job Description. A good supervisor has must be able to deal effectively with people. Food-service supervisors coordinate and schedule the activities of staff who prepare and portion food. They may estimate and order ingredients and supplies and prepare food-order summaries for chefs according to requests from dietitians, patients in hospitals, or other customers. They supervise and check the assembly of regular and special diet trays and the delivery of food trolleys in hospitals to ensure that both food and service meet quality standards.

They work in hospitals and other health-care establishments, cafeterias, catering companies, and other food-service establishments (Figure 30-10).

Source: USDA, photo by Lance Cheung

FIGURE 30-10 School service worker in a school lunch program.

Restaurant and Food-Service Managers. Titles for restaurant and food-service managers include *assistant manager, manager, banquet manager, bar manager, cafeteria manager, catering service manager, dining room manager, food-services manager, hotel food and beverage service,* and *restaurateur.*

Education. Most of these jobs require college or some other program related to hospitality or food and beverage management. Also, several years of experience in the food-service sector, including supervisory experience, are usually necessary.

Job Description. These individuals manage restaurants and other food-service establishments. They plan, organize, direct, and control the operations of restaurants, bars, cafeterias, and other food or beverage services, and often they interact with customers. Restaurant and food-services managers also perform some or all of the following duties:

- Determine the type of services to be offered and implement operational procedures
- Recruit staff and oversee staff training
- Set staff work schedules and monitor staff performance
- Control costs and inventories, monitor revenues, and modify procedures and prices
- Negotiate arrangements with suppliers for food and other supplies

- Negotiate arrangements with clients for catering or the use of facilities for banquets or receptions
- Resolve customer complaints
- Ensure that health and safety regulations are followed

Restaurant and food-service managers work in food and beverage service establishments or be self-employed.

Food Retail and Wholesale Industry

Much of the food industry relies on buyers and sales representatives. These individuals work for organizations that buy and sell food products for customers. The general categories of careers include *retail buyers*, *wholesale buyers*, and *sales representatives.*

Job possibilities as retail and wholesale buyers could include *food buyer, beverage taster and buyer,* and *produce buyer.*

Education. Jobs as retail and wholesale buyers require a high school diploma. A university degree or college diploma in business, marketing, or a related program is usually required. Also, experience as a sales supervisor or sales representative in an occupation related to the product usually is typically a must. Supervisors and senior buyers require experience.

Job Description. Retail and wholesale buyers decide on what merchandise grocery stores and local cooperatives will carry. Experienced buyers specialize in particular product lines. They study market reports, trade periodicals and promotional materials, and visit trade shows and factories. Once they see what is available, they select products that best fit their company's needs, interview suppliers, and negotiate prices, discounts, credit terms, and transportation arrangements. They oversee the distribution of the products to different outlets and maintain adequate stock levels. More and more, these careers will require the use of computers linked to suppliers to place orders.

FIGURE 30-11 Maintaining product displays—a sales position in many stores.

Retail and wholesale buyers work for food wholesalers, supermarket chains, and other retail outlets.

Sales Representative. For wholesale trade sales representatives possible job titles include: account executive, food products sales rep (representative), and liquor sales rep (Figure 30-11).

Education. High school diploma and a university degree or college diploma in business administration, marketing, or a related program are usually required. Experience as a sales supervisor or sales representative in an occupation related to the product is usually required. Supervisors and senior buyers require experience.

Many universities and colleges offer business administration and marketing programs.

Job Description. Sales representatives in the wholesale trade sell products and services to retail, wholesale, commercial, industrial, and professional customers. They look for new customers and take good care of existing ones. Sales reps must be able to convince customers of the benefits of their products and provide cost estimates, credit terms, warranties, and delivery dates. They prepare or oversee the preparation of sales contracts. They consult with customers after the sale and provide ongoing support, and they review information about product innovations, competitors, and market conditions. All of this means almost constant contact with people. Sales reps may also travel a lot.

Sales representatives work for companies that produce or provide products and services, including food, beverages, and tobacco products.

Research and Development

Research and development provides a career for those who are interested in developing the next food craze or breaking down the bacteria behind animal and plant diseases that cost farmers millions of dollars every year. This type of career in the food industry is steeped in science—biology, chemistry, or biotechnology. The possibilities for scientists in the food industry include *applied chemical technician, applied chemical technologist, food bacteriological technician, food scientist,* and *food chemist.*

Applied Chemical Technician. An applied chemical technician could be a *formulation technician, laboratory technician,* or *food-processing quality-control technician* (Figure 30-12).

Education. These careers require the completion of a 1- or 2-year college program in chemical or biochemical technology. Some chemical technicians and technologists are university graduates.

Job Description. Generally, applied chemical technicians work under chemists and technologists, helping them with their research. Applied chemical technicians typically will:

- Assist in setting up and conducting chemical experiments
- Operate and maintain laboratory equipment and prepare solutions, formulations, and so on
- Compile records for analytical studies
- Carry out a limited range of other technical functions
- Assist in the design and fabrication of experimental apparatus

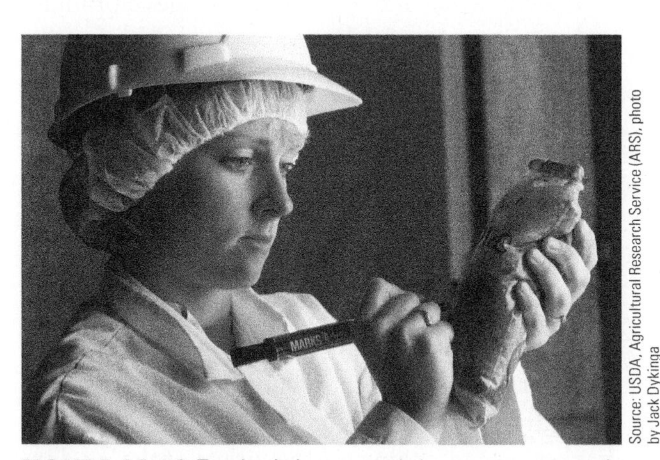

Source: USDA, Agricultural Research Service (ARS), photo by Jack Dykinga

FIGURE 30-12 Technicians work in many aspects of the food industry.

Applied chemical technicians work in laboratories, food-processing industries, and health, education, and government establishments, or they may be self-employed.

Applied Chemical Technologist. An applied chemical technologist can be called a *food technologist.* This career requires the completion of a 1- to 3-year college program in chemistry, biochemistry, or a closely related discipline. Many colleges offer chemistry programs. Certification programs are available and may be required by an employer.

Applied chemical technologists or food technologists will conduct chemical experiments, tests, and analyses. They operate and maintain laboratory equipment and prepare solutions, reagents, and sample formulations, and they compile records and interpret results.

Applied chemical technologists work in laboratories, food-processing industries, and health, education, and government establishments, or they may be self-employed.

Food Bacteriological Technician. A career as a food bacteriological technician requires the completion of a 1- to 2-year college program in a related field. Many types of technicians rely on a certification program.

Generally, bacteriological technicians provide technical support to scientists, engineers, and other professionals working in the fields of agriculture, plant and animal biology, microbiology, and cell and molecular biology. Biological technicians assist in conducting biological, microbiological, and biochemical tests and laboratory analyses. They perform a limited range of technical functions in support of agriculture, plant breeding, animal husbandry, and biology. Technicians assist in conducting field research and surveys to collect data and samples of water, soil, and plant and animal populations, and they assist in analyzing data and preparing reports.

Technicians work in laboratory and field settings for governments; for manufacturers of food products and pharmaceuticals; for biotechnology companies; for health, research, and educational institutions; and for environmental consulting companies, They may also be self-employed.

Food Scientist and Related Scientists

Food science is the discipline in which biology, physical sciences, and engineering are used to study the nature of foods, the causes of their deterioration, and the principles underlying food processing.

Education. A food science career requires a bachelor's degree in a related discipline. A master's or doctorate degree is necessary for employment as a research scientist. Postdoctoral research experience is usually required before employment in academic departments and research institutions.

Job Description. Individuals can specialize in the particular discipline that most interests them. Generally, scientists conduct basic and applied research to develop new practices and products related to food and agriculture.

Food scientists work in laboratory and field settings for governments, pharmaceutical and biotechnology companies, as well as for health and educational institutions (Figure 30-13).

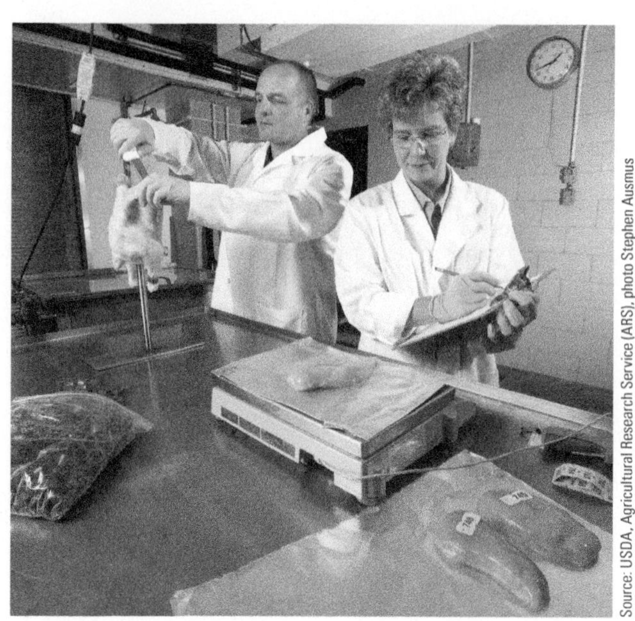

Source: USDA, Agricultural Research Service (ARS), photo Stephen Ausmus

FIGURE 30-13 Food scientists constantly research new products and new processing methods.

Food Chemist. Chemists play an important role in the development of new foods and nonfood uses such as the development of cosmetics made from milk ingredients. A bachelor's degree in chemistry, biochemistry, or a related discipline is necessary. A master's or doctorate degree is usually required to work as a research chemist.

Chemists analyze, synthesize, purify, modify, and characterize chemical or biochemical compounds. They conduct research to develop new chemical formulations and processes, and they research the synthesis and properties of chemical compounds and the mechanisms of chemical reactions. Chemists participate in interdisciplinary research and development projects working with biologists, microbiologists, agronomists, and other professionals, and they act as technical consultants in particular fields of expertise.

Chemists work in research, development, and quality-control laboratories; food-processing and biotechnology companies; and governmental and educational institutions.

Marketing and Communications

Marketing products locally or worldwide and communicating globally provide yet other career areas for individuals in the food industry. Successful individuals could end up preparing big-time marketing campaigns for the giants of the food industry. *Advertising specialist, marketing specialist,* and *writing and public relations professionals* are titles for these careers.

Education. Advertising and marketing specialists or managers require a university degree or college diploma in business administration with sales or marketing specialization, public relations, communications, marketing, journalism, or a related field. Managers also require several years' experience in advertising, public relations, or communications.

Job Description. Advertising and marketing specialists analyze advertising needs and current marketing strategies and provide advice on advertising and marketing strategies. They may also plan, develop, and implement advertising campaigns for print or electronic media.

Advertising and marketing specialists work for commercial, industrial, and wholesale organizations; marketing and public relations consulting companies; and government departments.

Writing and Public Relations Professionals. Writing and public relations professionals are called *writers*, *editors*, or *journalists*, and they can be specialists in public relations, communications, or lobbying.

Education. Educational requirements differ, depending on the specialization. Usually a university degree or college diploma in English, French, journalism, marketing, communications, or another discipline is required. To be hired, most writers require talent and ability as demonstrated by a work portfolio. Editors must have several years' experience in journalism, writing, publishing, or a related field.

Job Description. Writers and public relations professionals are usually responsible for writing copy for and preparing events that help promote the food industries. Writers research and write books, speeches, manuals, specifications, and other nonjournalistic articles. Copywriters study and determine the selling features of products and services and write text for advertisements and commercials. Writers may specialize in a particular subject or type of writing.

Editors review, evaluate, and edit manuscripts, articles, news reports, and other material for publication and broadcast as well as coordinate staff activities. Journalists research and write for newspapers, television, radio, and other media. Specialists in public relations and communications develop and implement communications strategies and information programs, publicize activities and events, and maintain media relations on behalf of clients. They may also work as lobbyists on behalf of the food industry or related associations.

Writers and public relations professionals work for governments, nongovernment associations, or private industry, or they are self-employed.

Others

Various other occupational areas may also have food industry-related jobs, among them:

- Environmental agriculture
- Exporting
- Livestock or crop production
- Horticulture
- Policy development and analysis
- Veterinary and animal health

As individuals learn their abilities, capabilities, and areas of interest, they can find food-industry careers in these occupational areas.

SCIENCE CONNECTION!

Use the Careers in Food Web site (http://www.careersinfood.com) to find three different careers to explore. What did you learn? What interests you in these careers?

FOOD-INDUSTRY SUPERVISED AGRICULTURAL EXPERIENCE

A Supervised Agricultural Experience (SAE) is designed to provide students the opportunity to gain experience in agricultural areas based on their interests. An SAE represents the actual, planned application of concepts and principles learned in agricultural education. Students experience and apply what is learned in the classroom to real-life situations. Students are supervised by agriculture teachers in cooperation with parents or guardians, employers, and other adults who assist them in the development and achievement of their educational goals. The purpose is to help students develop skills and abilities that lead toward a career.

Planning and conducting an SAE for food science could include areas of interest such as food processing, food chemistry, nutrition, food packaging, food commodities, and food regulations. Students should work with their instructors to do the following:

- Identify an appropriate SAE opportunity in the community
- Ensure that the SAE represents meaningful learning activities that benefit the student, the agriculture education program, and the community
- Obtain classroom and individual instruction on SAE
- Adopt a suitable record-keeping system
- Plan the SAE and acquire needed resources
- Coordinate release time and visits to SAE
- Sign a training agreement along with the employer, teacher, and parent or guardian
- Report on and evaluate the SAE and records resulting from it

Additional help and ideas for planning and conducting an SAE can be found through the National FFA Organization Web site (https://www.ffa.org).

GETTING A JOB

After identifying job possibilities, your real work begins. Getting the job is difficult and requires preparation. Again, whole books, videos, and seminars can teach you how to get a job. A few tips follow.

After a job is identified, do a little research on the company and the job before applying. Know these things about the job and the company:

- Its full name
- Name of the personnel manager
- Company address and phone number
- Position available
- Requirements for the position
- Geographic scope of the company—local, county, state, regional, national
- Company's product(s)
- Recent company developments
- Responsibilities of the position
- Demand for the company's product(s)

Before you get too far along in the application process, be certain the job is what you want to pursue. Money is not everything in a job. Compare the requirements and demands of the occupation with the characteristics you possess.

Application Forms

If the company requires an application form, remember that you are selling yourself with the information you provide. If possible, review the entire application form before you begin to fill it out. Pay particular attention to any special instructions to print or write in your own handwriting.

Many companies have online applications. It is a good idea to type out answers in a Word document to check for spelling and grammar before entering information online.[2] When answering ads that require potential employees to apply in person, be prepared to complete an application form on the spot. Take an ink pen. Prepare a list of information you will need to complete the application form: the addresses of schools you have attended; names, phone numbers, and addresses of previous employers and supervisors; and other important names, phone numbers, and addresses of references. The following guidelines will provide you with some direction when completing application forms:

1. Follow all instructions carefully and exactly.
2. If an application is to be handwritten, rather than typed, write neatly and legibly. Handwritten answers should be printed unless otherwise directed.
3. Application forms should be written in ink unless otherwise requested. If you make a mistake, mark through it with one neat line.
4. Be honest and realistic.
5. Give all the facts for each question.
6. Keep answers brief.
7. Fill in all the blanks. If the question does not pertain to you, write "not applicable" or "N/A." If there is no answer, write "none" or draw a short line through the blank.
8. Many application forms ask what salary you expect. If you are not sure what is appropriate, write "negotiable," "open," or "scale" in the blank. Before applying, try to find out what the going rate is for similar work at other locations. Give a salary range rather than an exact figure.

Letters of Inquiry and Application

The purpose of a **letter of inquiry** is to obtain information about possible job vacancies. The purpose of a **letter of application** is to apply for a specific position that has been publicly advertised. Both letters indicate your interest in working for a particular company, acquaint an employer with your qualifications, and encourage the employer to invite you for a job interview.

Both types of letter represent you. They should be accurate, informative, and attractive. Your written communications should present a strong, positive, professional image as both a job seeker and future employee. The following list should be used as a guide when writing letters of inquiry and application:

1. Be short and specific, one or two pages (details left to résumé). Use 8½ × 11-inch white typing paper, not personal or fancy paper.
2. Type neatly and make sure there are no errors.
3. Keep the form attractive and free of smudges.
4. Write to a specific person. Use "To Whom It May Concern" if answering a blind ad.
5. Make sure paragraphs are logical, organized, and to the point.
6. Carefully construct sentences and make sure they are free of spelling and grammatical errors.
7. Keep a positive tone.
8. Express ideas in a clear, concise, and direct manner.
9. Avoid slang words and expressions.
10. Avoid excessive use of the word "I."
11. Avoid mentioning salary and fringe benefits.
12. Write a first draft and then make revisions.
13. Proofread the final letter yourself and then have someone else proofread it.
14. Address and sign correctly. Type envelope addresses.

This information should be included in a letter of inquiry:

1. Specify the reasons why you are interested in working for the company and ask whether any positions are available now or are expected in the near future.

2. Express your interest in being considered a candidate for a position when one becomes available.
3. Because you are not applying for a particular position, you cannot relate your qualifications directly to job requirements. (You can explain how your personal qualifications and work experience would help meet the needs of the company.)
4. Mention and include your résumé.
5. State your willingness to meet with a company representative to discuss your background and qualifications. (Include your address and a phone number where you can be reached.)
6. Address letters of inquiry to the "Personnel Manager" unless you know his or her name.

A letter of application should include:

1. Your source for the job lead.
2. The particular job you are applying for and the reason why you are interested in the position and the company.
3. How your personal qualifications meet the needs of the employer.
4. How your work experience relates to job requirements.
5. Your résumé.
6. A request for an interview and a statement of your willingness. (Include your address and a phone number where you can be reached.)

Résumé or Data Sheet

Some jobs require a **résumé** (Figure 30-14) or **data sheet**. The following information should be considered when writing a résumé or data sheet:

- Name, address, and phone number
- Brief, specific statement of career objective
- Educational background—names of schools, dates, major field of study, degrees or diplomas—listed in reverse chronological order
- Leadership activities, honors, and accomplishments
- Work experience listed in reverse chronological order
- Special technical skills and interests related to job
- References
- One page if possible
- Neatly typed and error free
- Logically organized
- Honestly listed qualifications and experiences

Employers look for a quick overview of who you are and how you fit into their business. On the first reading, the employer will spend 10 to 15 seconds reading a résumé. Be sure to present relevant information clearly and concisely in an eye-catching format.

The Interview

The next step in the job-hunting process is the interview. While many do's and don'ts of an interview are available, perhaps the best advice comes from the interviewer's side of the desk. This list provides common reasons interviewers give for not being able to place applicants in a job.

1. Poor attitude
2. Unstable work record
3. Bad references
4. Lack of self-selling ability
5. Lack of skill and experience
6. Not really anxious to work
7. "Bad mouthing" former employers
8. Too demanding (wanting too much money or to work only under certain conditions)
9. Unable to be available for interviews or canceling out
10. Poor appearance
11. Lack of manners and personal courtesy
12. Chewing gum, smoking, fidgeting

RÉSUMÉ

Susan Smith

CURRENT ADDRESS

PO Box 1238 Anywhere, ID 00000
Telephone: 000/888-8888
E-mail address: sasmith3@gmail.com

EDUCATION

- Local High School, Anywhere, ID: Graduated 2010. 3.8/4.0 GPA
- College of Southern Idaho, Twin Falls, ID, 2010–2014: Bachelor of Science. Food Science and Nutrition with a Minor in Marketing

CAREER OBJECTIVE

To pursue a career in the food science industry that provides advancement opportunities during my career.

ACTIVITIES AND HONORS

- Active member of 4-H Club for three years. Learned food preparation.
- Member FFA for four years and was elected President during senior year.
    - Member of the Food Science CDE team 2 years
- Member Postsecondary Agricultural Student (PAS) organization 2012–2014
- Advisor to local 4-H Club 2011 to present.

EMPLOYMENT AND WORK EXPERIENCES

- **January 2010 to Present:** Sunrise Bakery, Arco, ID; general help; prepare breads and donuts, work some in sales.
- **July 2008 to December 2010:** ABC Grocery, McCall, ID; restocked shelves; boxed groceries; worked into checker position.

REFERENCES

Lane Johnson; Sunrise Bakery 000/777-7777
Jared Proctor; FFA Advisor 000/999-9999

FIGURE 30-14 Neat, complete resumes—a necessary component for career development.

13. No attempt to establish rapport; not looking the interviewer in the eye
14. Being interested only in the salary and benefits of the job
15. Lack of confidence; being evasive
16. Poor grammar, use of slang
17. Not having any direction or goals

Follow-Up Letters

Follow-up letters are sent immediately after an interview. The follow-up letter demonstrates your knowledge of business etiquette and protocol. Always send a follow-up letter regardless of whether or not you had a good interviewing experience and regardless of whether you are interested in the position.

When employers do not receive follow-up letters from job candidates, they often assume that the candidate is not aware of the professional courtesy and protocol they will need to demonstrate on the job.

The major purpose of a follow-up letter is to thank those individuals who participated in your interview. In addition, a follow-up letter reinforces your name, application, and qualifications to the employer and indicates whether you are still interested in the job position.

OCCUPATIONAL SAFETY

Employees in the food industry should expect a safe and healthful workplace. Still, individuals may encounter such hazards as:

- Toxic chemicals in cleaning products
- Slippery floors
- Hot cooking equipment
- Sharp objects
- Heavy lifting
- Stress
- Harassment
- Poor workstation designs

To prevent or minimize exposure to occupational hazards, employers are expected to provide employees with safety and health training, including providing information on chemicals that could be harmful to an individual's health. If an employee is injured or becomes ill because of a job, many employers pay for medical care and lost wages are sometimes provided.

Not only are food-industry employers responsible for creating and maintaining a safe workplace, but also employees must do their part, including:

- Following all safety rules and instructions
- Using safety equipment and protective clothing when needed
- Looking out for co-workers
- Keeping work areas clean and neat
- Knowing what to do in an emergency
- Reporting any health and safety hazard to the supervisor

A JOB IS MORE THAN MONEY

Before taking a job, be certain it is what you want. Although the salary or the wage is important, job satisfaction is something quite different and highly important. Jobs quickly become routine and mundane. For example, a job with little fulfillment and challenge for some people can easily become a chore just to go to. Before taking a job or even while looking for a job, answer these questions:

1. Does the job description fit your interests?
2. Is this the level of occupation in which you wish to engage?
3. Does this type of work appeal to your interests?
4. Are the working conditions suitable to you?
5. Will you be satisfied with the salaries and benefits offered?
6. Can you advance in this occupation as rapidly as you would like? What are the advancement opportunities?
7. Does the future outlook satisfy you?
8. Is the occupation in demand now and in the foreseeable future?
9. Do you have or can you get the education needed for the occupation?
10. What type of training is available after taking the job?
11. Can you get the finances needed to get into the occupation?

(Continues)

A JOB IS MORE THAN MONEY (continued)

12. Can you meet the health and physical requirements?
13. Will you be able to meet the entry requirements?
14. Do you know of any other reasons you might not be able to enter this occupation?
15. Is the occupation available locally or are you willing to move to a part of the country where it is available?

Also, before taking a job or looking for a job, do a little personality inventory of yourself. Consider the following:

1. Do I like to be alone or with people?
2. Am I mechanical or artistic?
3. Would I rather work independently or work under supervision?
4. Would I like to think or be active?
5. Could I take authority and responsibility for others?
6. Must I have freedom to express creativity?
7. What things do I like to do? Make a list.
8. At what time of day can I work best?
9. Can I work under pressure or stress?
10. Make a list of your strong points. Consider skills, hobbies, and leisure time activities you can offer an employer.

Do your research and your job will be more rewarding and you will feel better about yourself.

For more information about building a career, search the Web or visit these Web sites:

- What Do Employers Really Want? http://www.quintcareers.com/job_skills_values.html
- Career Planning: http://www.educationplanner.org/students/career-planning/
- Explore Your Career Options: http://www.actstudent.org/career/
- CareersNet: www.careersnet.com
- AgCareers: http://www.agcareers.com/
- Forbes—10 Interview Questions You Should Ask: http://www.forbes.com/sites/nextavenue/2014/06/18/10-job-interview-questions-you-should-ask/
- Career Placement: http://foodscience.psu.edu/students/career-placement
- What Is Food Science? http://www.worldwidelearn.com/online-education-guide/health-medical/nutrition-food-science-major.htm
- Major—Food Science: https://bigfuture.collegeboard.org/majors/agriculture-related-sciences-food-science

For additional information about personal and occupational safety practices in the workplace, contact the Occupational Safety and Health Administration (OSHA) on the Web at www.osha.gov, the National Institute for Occupational Safety and Health (NIOSH) on the Web at www.cdc.gov/niosh/homepage.html, or the Department of Labor at www.dol.gov.

SUMMARY

The primary goal of education and training is to become employable and stay employable—to get and keep a job or career or to run a successful business. The world of work requires people who can read, write, do math, and communicate. Rapidly changing technology has made this even more critical. Also, the modern workplace looks for people who possess thinking skills. Even with a solid set of basic skills, future employees also need to be able to relate to other people, be able to use information, need to understand the concept of systems, and know how to use technology. Old-fashioned ideas such as responsibility, self-esteem, sociability, self-management, and integrity are still relevant.

Jobs or careers in the food industry range from those closely tied to the industry to those that support the food industry. In general, potential job or career areas include production management, product development, food engineering, microbiology, quality control, research, technical representation, sales, and service. Education and training in the food industry vary from on-the-job training to high school and college degrees and certificates.

After training and education, finding and getting the right job or career may still be a challenge. Good resources exist for locating a job. Still, one of the best resources is personal contact. Well-written letters of inquiry, application forms, a clear and eye-catching résumé, and being prepared for the job interview are the best ways to secure a job.

REVIEW QUESTIONS

Success in any career requires knowledge. Test your knowledge of this chapter by answering these questions or solving these problems.

1. What are the six basic skills required in the workplace?
2. Define *entrepreneur.*
3. List the parts of a résumé.
4. What is the difference between a letter of application and a letter of inquiry?
5. Name 10 reasons why an interview may fail.
6. What is the largest manufacturing industry in the world?
7. List the five thinking skills needed in the new workplace.
8. What background or education is generally required to obtain a job in research and development in any area of the food industry?
9. What education is required for a chef certification?
10. Identify four jobs in the food safety and inspection area of the food industry.

STUDENT ACTIVITIES

1. Develop your own résumé or data sheet. Ask your instructor to help edit your résumé.
2. Collect position announcements and classified ads for jobs in the food industry. Write a letter of job inquiry and a letter of job application for a selected job using this information.

3. Develop a list of questions frequently asked during an interview. Use the questions in role-playing job interviews and videotape the interviews.
4. Organize a field trip to a public or private placement office. Following the field trip, discuss the office's policies and how they affect job searchers and employers. Alternatively, invite a representative from a state employment agency to explain how employment agencies can help students gain employment.
5. Hold a food-industry career field day. Invite individuals currently employed in the food industry to present a panel discussion on career opportunities. For example, invite representatives from research, education, and government.
6. Select one career in the food industry of interest and prepare a research paper on the career using a computer and word-processing software. The paper should identify the knowledge and skills required and the employment opportunities.
7. Collect pictures or photographs of people engaged in various food-related careers. Use them to prepare a poster for the People in Agriculture FFA CDE.
8. Invite a resource person such as a business owner or personnel manager to discuss what he or she looks for in résumés, application letters, and forms as well as during interviews.
9. Invite a panel of local businesspeople to discuss the importance of employee work habits, basic skills, and attitudes and how they affect the entire business.
10. Prepare for and participate in the FFA Job Interview CDE.

ADDITIONAL RESOURCES

- Careers in Food Science and Human Nutrition: http://academics.aces.illinois.edu/career-services/linking-careers/fshn-careers
- What Kind of Careers Do Food Scientists Have? http://sfs.wsu.edu/prospective-students/faq/food-science-careers/
- Cornell University—Food Science: http://courses.cornell.edu/preview_program.php?catoid=12&poid=3348
- Food Products and Processing Systems—Interest in a Supervised Agricultural Experience (SAE): http://www.exploresae.com/docs/Resources/FoodProductsandProcessing.pdf
- Occupational Outlook Handbook—Agricultural and Food Scientists: http://www.bls.gov/ooh/life-physical-and-social-science/agricultural-and-food-scientists.htm

Internet sites represent a vast resource of information. Although those provided in this chapter were vetted by industry experts, you may wish to further explore the topics discussed in this chapter using a search engine such as Google. Keywords or phrases may include the following: *résumé writing, data sheet, food science careers, entrepreneur, competencies (reading, writing, listening, speaking), letter of inquiry, letter of application, job interview, any specific food science career listing.* In addition, Table B-7 provides a listing of useful Internet sites that can be used as a starting point.

REFERENCES

Aslett, D. 1993. *Everything I needed to know about business I learned in the barnyard.* Pocatello, ID: Marsh Creek Press.

Bolles, R. N. 2015. *What color is your parachute? 2016: A practical manual for job hunters and career changers.* Berkeley, CA: Ten Speed Press.

Business Council for Effective Literacy Bulletin. (1987). *Job-related basic skills: A guide for planners of employee programs.* New York: BCEL.

Carnegie, D. 1998. *How to win friends and influence people.* New York: Pocket Books

Ricketts, C. 2011. *Leadership: Personal development and career success.* Clifton Park, NY: Cengage Learning.

U.S. Department of Education. (1991). *America 2000: An education strategy.* Sourcebook. Washington, DC: United States Department of Education.

Ziglar, Z. 2000. *See you at the top: 25th anniversary edition.* Gretna, LA: Pelican Publishing Co.

ENDNOTES

1. Houston Chronicle: http://smallbusiness.chron.com/importance-technology-workplace-10607.html. Last accessed July 14, 2015.
2. Careers in Foods: http://www.careersinfood.com/. Last accessed July 14, 2015.

APPENDIX A

Review of Chemistry

The study of food science requires an understanding of simple chemistry and organic chemistry. Carbohydrates, lipids, and proteins are all important to food science, and they are composed of smaller building blocks. The information found in this appendix includes a review of important chemical interactions and concepts encountered in food science, which provides a foundation for the discussions within the text.

ELEMENTS

The atom is the smallest unit of an element that still exhibits the properties of that element. Atomic structure is shown in Figure A-1. Atoms consist of a nucleus containing protons (1 mass unit, positive charge) and neutrons (1 mass unit, no charge). Surrounding the nucleus are one or more electrons, depending on the element. The number of electrons equals the number of protons, and the atom is electrically neutral. Electrons travel around the nucleus at extremely high speed, each one traveling in one of several possible energy levels (also called *shells*, *orbits*, or *orbitals*). Each level has a maximum number of electrons—two in the first level, eight in the second level, and so on. The energy level nearest the nucleus is the one with the lowest energy electrons. Orbitals further from the nucleus contain higher energy electrons.

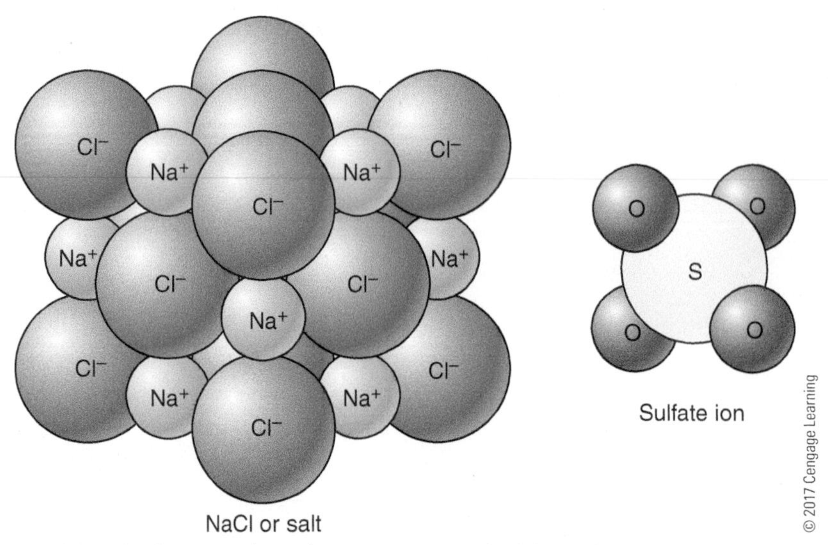

FIGURE A-1 Atoms combine to form molecules as represented here by nucleus and electron cloud.

1 H																	2 He
3 Li	4 Be											5 B	6 C	7 N	8 O	9 F	10 Ne
11 Na	12 Mg											13 Al	14 Si	15 P	16 S	17 Cl	18 Ar
19 K	20 Ca	21 Sc	22 Ti	23 V	24 Cr	25 Mn	26 Fe	27 Co	28 Ni	29 Cu	30 Zn	31 Ga	32 Ge	33 As	34 Se	35 Br	36 Kr
37 Rb	38 Sr	39 Y	40 Zr	41 Nb	42 Mo	43 Tc	44 Ru	45 Rh	46 Pd	47 Ag	48 Cd	49 In	50 Sn	51 Sb	52 Te	53 I	54 Xe
55 Cs	56 Ba	57 La	72 Hf	73 Ta	74 W	75 Re	76 Os	77 Ir	78 Pt	79 Au	80 Hg	81 Tl	82 Pb	83 Bi	84 Po	85 At	86 Rn
87 Fr	88 Ra	89 Ac	104 Rf	105 Db	106 Sg	107 Bh	108 Hs	109 Mt	110 Uun								

58 Ce	59 Pr	60 Nd	61 Pm	62 Sm	63 Eu	64 Gd	65 Tb	66 Dy	67 Ho	68 Er	69 Tm	70 Yb	71 Lu
90 Th	91 Pa	92 U	93 Np	94 Pu	95 Am	96 Cm	97 Bk	98 Cf	99 Es	100 Fm	101 Md	102 No	103 Lr

FIGURE A-2 The Periodic Table of the Elements provides information about every known atom.

The *atomic number* of an atom is the total number of protons. The *atomic weight* of an atom is the total number of protons plus neutrons. The periodic table shows that columns in the table contain elements with the same number of electrons in their outermost energy levels or have full energy levels. The elements important to life include carbon, hydrogen, nitrogen, and oxygen (Figure A-2).

Every element is determined by the number of protons in the atom. Isotopes (different forms) of elements are determined by the number of neutrons in the atom.

Chemical properties of an element are determined by the number of electrons in the outermost energy level of an atom. These outermost electrons interact with the electrons of other atoms when two atoms come together. Elements in vertical columns of the periodic table contain the same number of electrons in their outermost levels, and they share similar chemical properties. Arsenic occurs right below phosphorus and is a poison. Silicon appears right below carbon, but life as we know it is based on carbon, not silicon.

CHEMICAL BONDS

This section provides a quick review of chemical bonds. Emphasis is placed on bonds between the six major elements found in biological systems: H, C, N, O, P, and S.

Covalent Bonds

Covalent bonds are the strongest chemical bonds and are formed by the sharing of a pair of electrons (Figure A-3). Once formed, covalent bonds rarely break spontaneously because of simple energetic

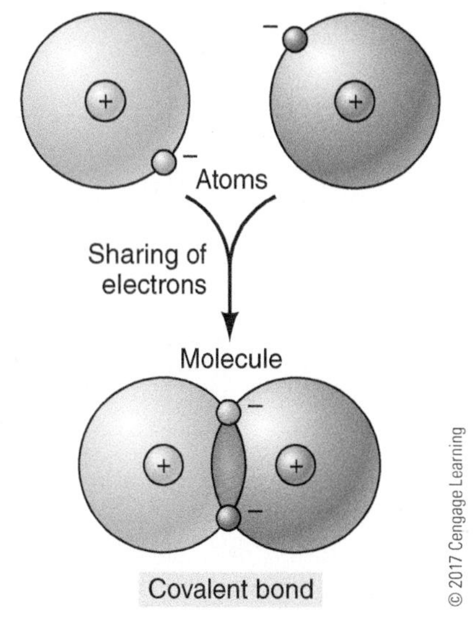

FIGURE A-3 Sharing of electrons in a covalent bond.

considerations: The thermal energy of a molecule at room temperature is much lower than the energy required to break a covalent bond. Covalent bonds can be single, double, or triple.

Carbon–carbon bonds are unusually strong and stable covalent bonds.

The major organic elements have standard bonding capabilities, as shown in Figure A-4.

Covalent bonds can also have partial charges when the atoms involved have different electronegativities. Water is perhaps the most obvious example of a molecule with partial charges. The symbols

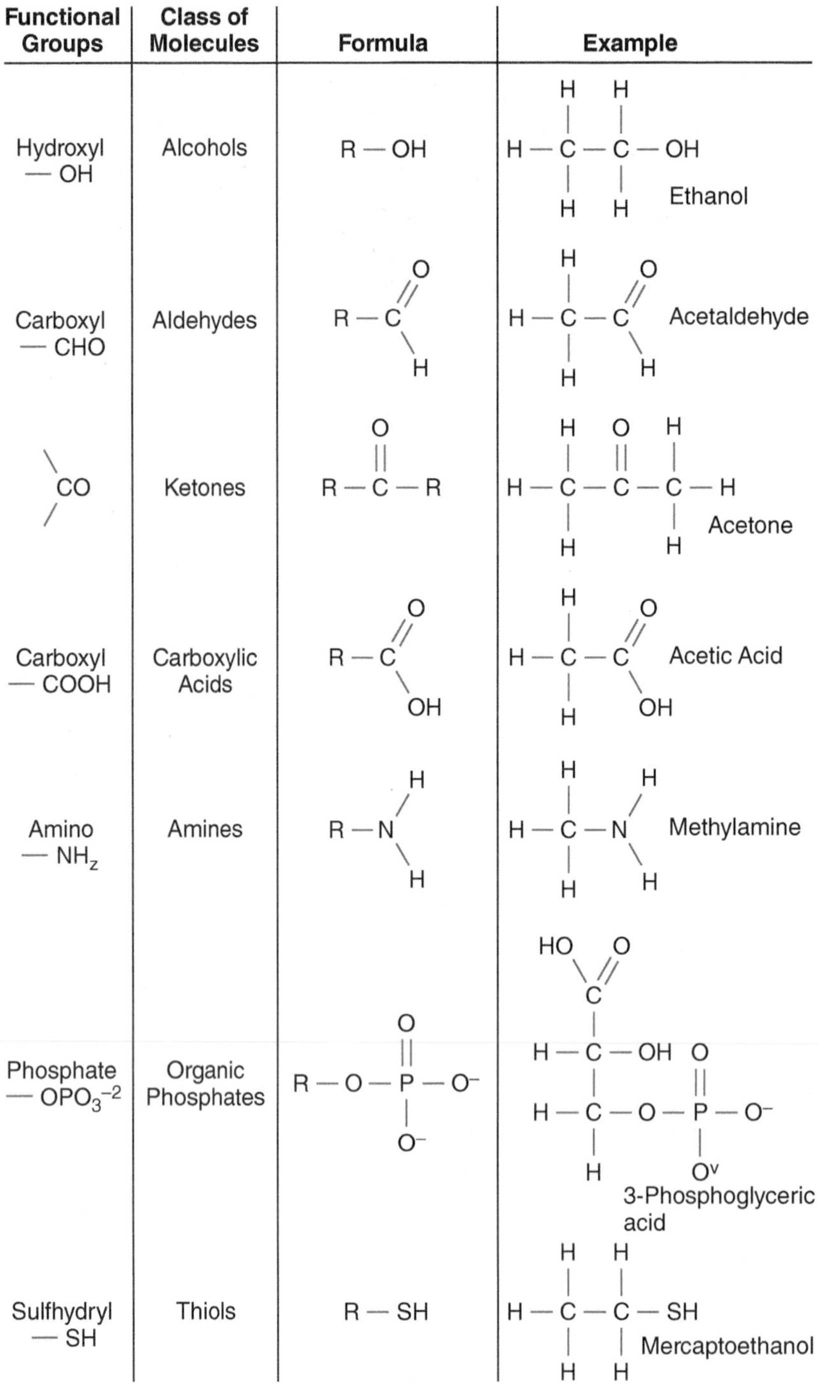

Functional Groups	Class of Molecules	Formula	Example
Hydroxyl — OH	Alcohols	R — OH	H — C(H)(H) — C(H)(H) — OH Ethanol
Carboxyl — CHO	Aldehydes	R — C(=O) — H	H — C(H)(H) — C(=O) — H Acetaldehyde
\CO/	Ketones	R — C(=O) — R	H — C(H)(H) — C(=O) — C(H)(H) — H Acetone
Carboxyl — COOH	Carboxylic Acids	R — C(=O) — OH	H — C(H)(H) — C(=O) — OH Acetic Acid
Amino — NH_z	Amines	R — N(H) — H	H — C(H)(H) — N(H) — H Methylamine
Phosphate — OPO_3^{-2}	Organic Phosphates	R — O — P(=O)(O⁻) — O⁻	HO — C(=O) — C(H)(OH) — C(H)(H) — O — P(=O)(O⁻) — O⁻ 3-Phosphoglyceric acid
Sulfhydryl — SH	Thiols	R — SH	H — C(H)(H) — C(H)(H) — SH Mercaptoethanol

FIGURE A-4 Representation of functional groups found in organic molecules.

delta plus (δ^+) and delta minus (δ^-) are used to indicate partial charges.

Because of its high electronegativity, oxygen attracts the electrons away from the hydrogen atoms, resulting in a partial negative charge on the oxygen and a partial positive charge on each hydrogen atom.

The possibility of hydrogen bonds (H-bonds) is a consequence of partial charges.

Hydrogen Bonds

Hydrogen bonds are formed when a hydrogen atom is shared between two molecules (Figure A-5).

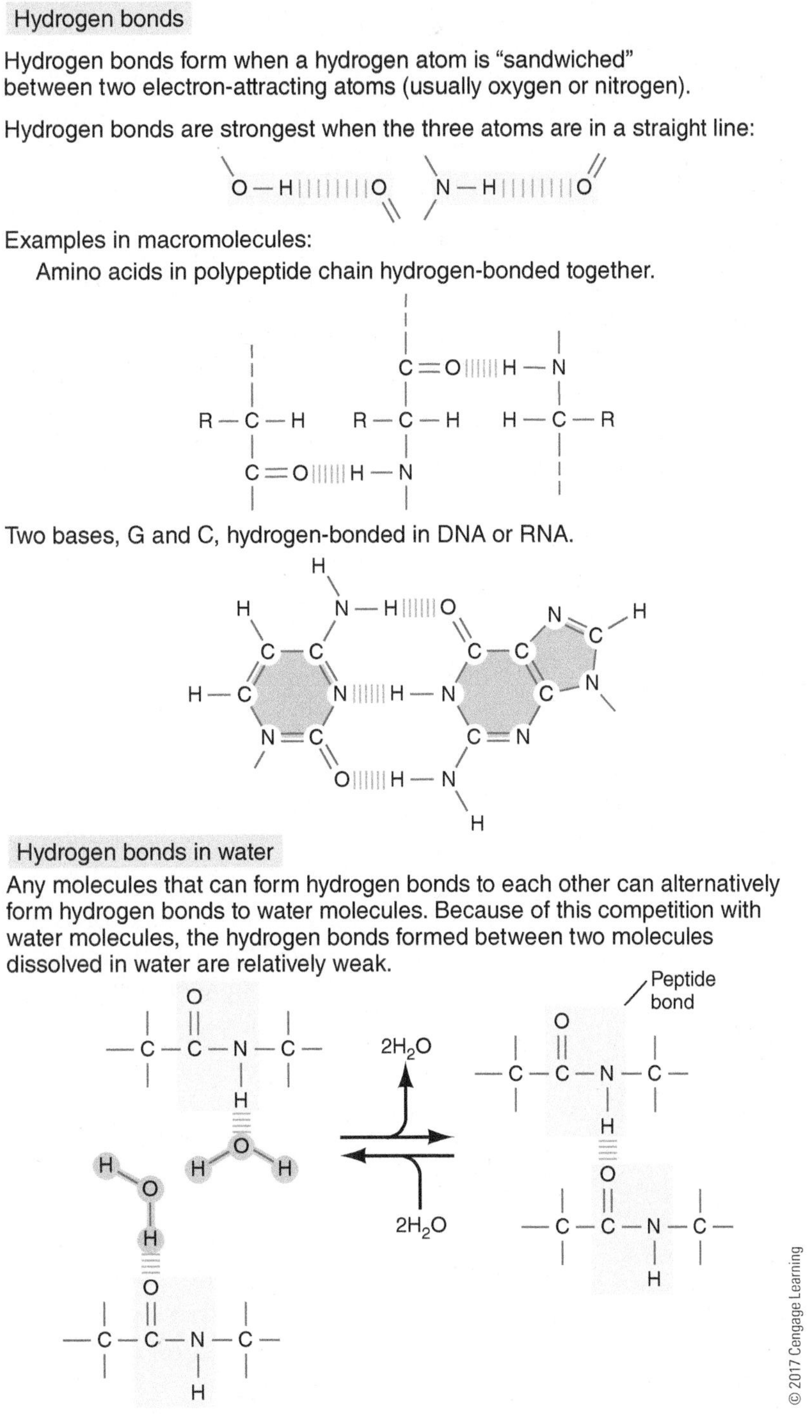

Hydrogen bonds

Hydrogen bonds form when a hydrogen atom is "sandwiched" between two electron-attracting atoms (usually oxygen or nitrogen).

Hydrogen bonds are strongest when the three atoms are in a straight line:

Examples in macromolecules:

Amino acids in polypeptide chain hydrogen-bonded together.

Two bases, G and C, hydrogen-bonded in DNA or RNA.

Hydrogen bonds in water

Any molecules that can form hydrogen bonds to each other can alternatively form hydrogen bonds to water molecules. Because of this competition with water molecules, the hydrogen bonds formed between two molecules dissolved in water are relatively weak.

Peptide bond

$2H_2O$

$2H_2O$

FIGURE A-5 Description of a hydrogen bond.

Hydrogen bonds have polarity. A hydrogen atom covalently attached to a highly electronegative atom (N, O, or P) shares its partial positive charge with a second electronegative atom (N, O, or P). One common example is the hydrogen bonding between water molecules.

Hydrogen bonds are frequently found in proteins and nucleic acids (as in DNA); by reinforcing each other, they serve to keep the protein (or nucleic acid) structure secure. Because the hydrogen atoms in the protein could also hydrogen bond to the surrounding water, the relative strength of protein–protein hydrogen bonds versus protein–water bonds is smaller.

Ionic Bonds

Ionic bonds are formed when electrons are completely transferred from one atom to another; the result is two ions, one positively charged and one negatively charged (Figure A-6). For example, when a sodium atom (Na) donates the one electron in its outer valence shell to a chlorine (Cl) atom, which needs one electron to fill its outer valence shell, NaCl (table salt) results. The symbol for sodium chloride is Na^+Cl^-. Ionic bonds are often 4 kcal/mol to 7 in strength.

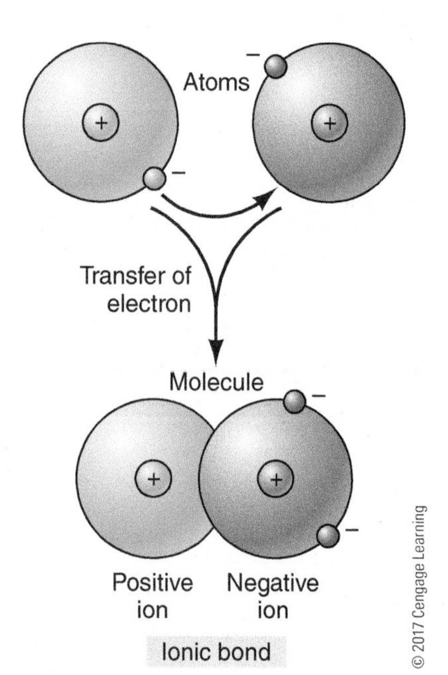

FIGURE A-6 Representation of an ionic bond.

Van der Waals Bonds

Van der Waals bonds or interactions are extremely weak bonds (generally no greater than 1 kcal/mol) that form between nonpolar molecules or nonpolar parts of a molecule (Figure A-7). The weak bond is created because a C–H bond can have a transient dipole and induce a transient dipole in another C–H bond.

MOLECULES

Molecules are the smallest identifiable unit into which a pure substance can be divided and still retain the composition and chemical properties of that substance. The division of a sample of a substance into progressively smaller parts produces no change in either its composition or its chemical properties.

Molecules are held together by shared electron pairs—that is, covalent bonds. Such bonds are directional, meaning that the atoms adopt specific positions relative to one another to maximize bond strengths.

The molecular weight of a molecule is the sum of the atomic weights of its component atoms. If a substance has a molecular weight of 32, then 32 grams of the substance is termed *1 mole*. For example, the molecular weight of NaCl is 58.5 (23 + 35.5), and 1 mole of NaCl weighs 58.5 grams.

REACTIONS

A few examples of a common and important type of chemical reaction are seen in the rusting of metals, the process involved in traditional photography, the way living systems produce and use energy, and the operation of a car battery. These chemical

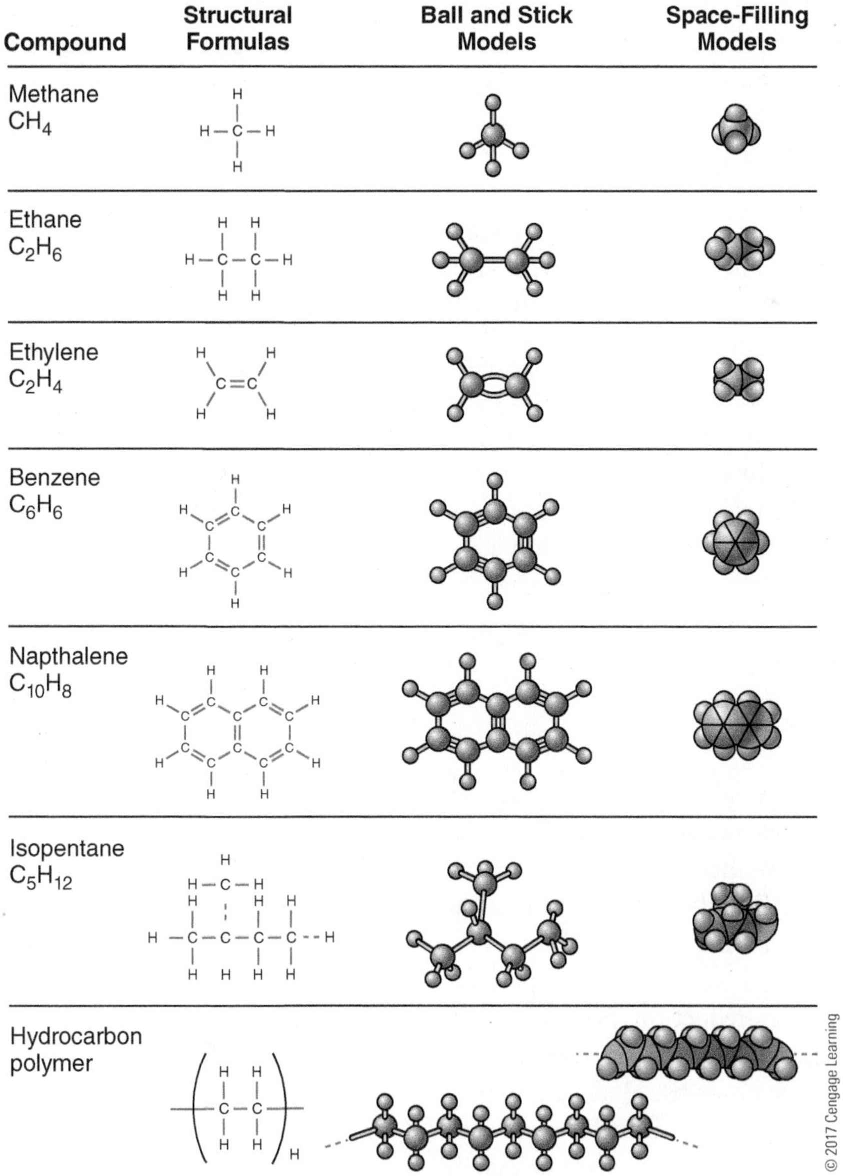

Compound	Structural Formulas	Ball and Stick Models	Space-Filling Models
Methane CH_4			
Ethane C_2H_6			
Ethylene C_2H_4			
Benzene C_6H_6			
Napthalene $C_{10}H_8$			
Isopentane C_5H_{12}			
Hydrocarbon polymer			

FIGURE A-7 Four different methods of describing organic molecules.

changes are all classified as electron-transfer or oxidation–reduction reactions.

The term *oxidation* was derived from the observation that almost all elements reacted with oxygen to form compounds called *oxides*. A typical example is the corrosion or rusting of iron, which is described by the following chemical equation:

$$4Fe + 3O_2 \longrightarrow 2Fe_2O_3$$

Reduction was the term originally used to describe the removal of oxygen from metal ores, which reduced the metal ore to pure metal, as shown in this chemical reaction:

$$2Fe_2O_3 + 3C \longrightarrow 3CO_2 + 4Fe$$

Based on these two examples, oxidation can be simply defined as the addition of oxygen. Reduction can be defined as the removal of oxygen.

The logical starting point in the discussion of oxidation–reduction reactions is the atom and the terms and conventions used by chemists to describe this phenomenon. All atoms are electrically neutral even though they comprise charged, subatomic particles. The terms *oxidation state* and *oxidation number* have been developed to describe this "electrical state" of the atom. The oxidation state or oxidation number of an atom is simply defined as the sum of the negative and positive charges in an atom. Because every atom contains equal numbers of positive and negative charges, the oxidation state or oxidation number of any atom is always zero.

Oxidation–reduction reactions always involve a change in the oxidation state of the atoms or ions involved. This change in oxidation state is due to the loss or gain of electrons. The loss of electrons from an atom produces a positive oxidation state, whereas the gain of electrons results in negative oxidation states.

METABOLISM

Metabolism is the general term used to for all chemical reactions that occur in a living system. Metabolism can be divided into two processes: (1) *anabolism*, or reactions that are involved in the synthesis of compounds, and (2) *catabolism*, or reactions that involve the breakdown of compounds. In terms of oxidation–reduction principles, anabolic reactions are primarily characterized by reduction reactions, such as the dark reaction in photosynthesis where carbon dioxide is reduced to form glucose. Catabolic reactions are primarily oxidation reactions. Although catabolism involves many separate reactions, an example of such a process can be described by the oxidation of glucose, as shown here:

$$C_6H_{12}O_6 + 6O_2 \longrightarrow 6CO_2 + 6H_2O + \text{Energy}$$

In this reaction, the carbon atoms in glucose are oxidized, undergoing an increase in oxidation state (each carbon loses two electrons) as they are converted to form carbon dioxide. At the same time, each oxygen atom is reduced by gaining two electrons when it is converted to water. Part of the energy is released as heat, and the remainder is stored in the chemical bonds of energetic compounds such as adenosine triphosphate (ATP) and nicotinamide adenine dinucleotide (NADH).

Catabolic reactions can be divided into many different groups of reactions called *catabolic pathways*. In these pathways, glycolysis, the citric acid cycle, and electron transport (Figure A-8), the carbon atoms are slowly oxidized by a series of reactions that gradually modify the carbon skeleton of the compound as well as the oxidation state of carbon. Coupled to these reactions are other reversible oxidation–reduction reactions designed to capture the energy released and temporarily store it within the chemical bonds of ATP and NADH. These compounds are then use to provide the energy that drives the cellular machinery.

ORGANIC CHEMISTRY

Nutrients are chemicals and understanding their function requires some knowledge of organic chemistry. Table A-1 indicates some of the basics required for such an understanding.

Organic chemistry involves carbon-containing molecules, and all carbon atoms have four bonds to account for. In carbohydrates, fats, and proteins, each carbon can connect to:

- Another carbon
- A hydroxyl
- A hydrogen
- An amino group
- An oxygen (double bond)

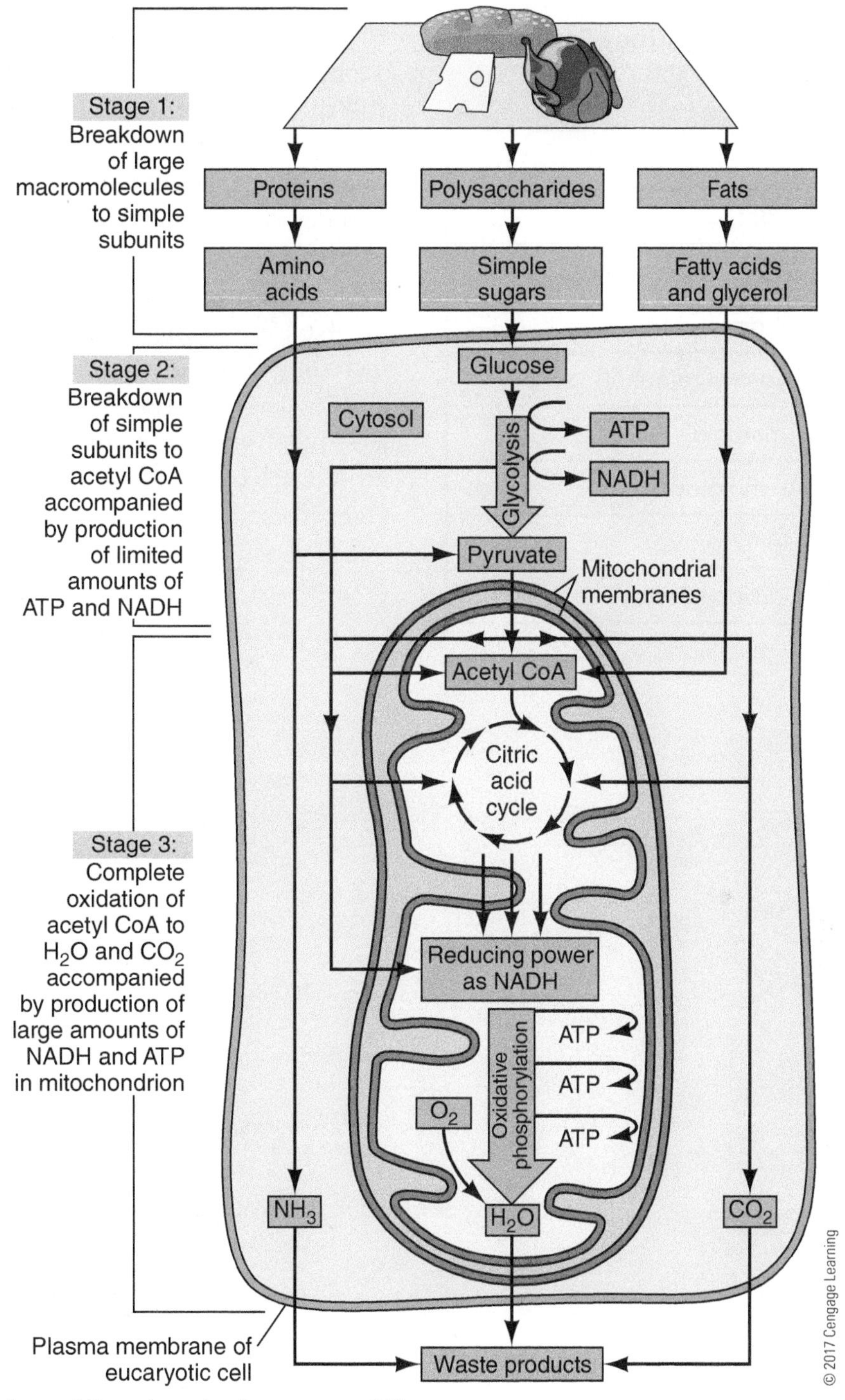

FIGURE A-8 Overview of the chemical process of life.

TABLE A-1
Chemical Symbols and What They Represent

SYMBOL	REPRESENTS
C	Carbon atom
H	Hydrogen atom
O	Oxygen atom
N	Nitrogen atom
OH	Hydroxyl (alcohol)
NH_3	Ammonia
NH_2	Amino group
CH_3	Methyl group
COOH	Carboxyl (acid)

SUMMARY

Atoms are the smallest unit of an element that still shows properties of the element. Atoms bond with other atoms in chemical bonds such as covalent bonds, hydrogen bonds, ionic bonds, and Van der Waals bonds. Molecules are the smallest identifiable units of pure substances. These molecules are formed by chemical reactions or become involved in chemical reactions. These reactions, including those involved in metabolism, can be classified as oxidation–reduction. Organic chemistry involves carbon-containing molecules; such groups are hydroxyl, amino, ammonia, methyl, and carboxyl.

APPENDIX B

Reference Tables

The information in this appendix includes a variety of useful conversions, conversion factors, measurement standards, and common measures. This appendix also includes a table listing of some Internet sites (by URL) that lead to many additional sources of data and information. Finally, the last table in this appendix is a food-composition table that lists many common foods.

By making full use of this appendix, the reader can understand more, plan better, do more, and learn more.

TABLE B-1
Conversion Tables for Common Weights and Measures

NONMETRIC UNIT	METRIC EQUIVALENT
1 pound	454 grams
2.2 pounds	1 kilogram
1 quart	1 liter
15.43 grains	1 gram
2,205 pounds	1 metric ton
1 inch	2.54 centimeters
0.39 inch	10 millimeters or 1 centimeter
39.37 inches	1 meter
1 acre	406 hectares

TABLE B-2
Weight Conversions

MEASUREMENTS	EQUIVALENT AMOUNT
8 tablespoons	1/2 cup
3 teaspoons	1 tablespoon
1 pint	2 cups
2 pints	1 quart
4 quarts	1 gallon or 8 pounds of water
2,000 pounds	1 ton
16 ounces	1 pound
27 cubic feet	1 cubic yard
1 peck	8 quarts
1 bushel	4 pecks
OTHER CONVERSIONS	
1%	.01 10,000 ppm
1 Megacalorie (M-cal)	1,000 calories
1 calorie (big calorie)	1,000 calories (small calorie)
1 M-cal	1 therm

TABLE B-3 Common Measures and Approximate Equivalents

1 liquid teaspoon	5 milliliters (ml)
3 liquid teaspoons	1 liquid tablespoon = 15 ml
2 liquid tablespoons	1 liquid ounce = 30 ml
8 liquid ounces	1 liquid cup = 0.24 liter
2 liquid cups	1 liquid pint = 0.47 liter
2 liquid pints	1 liquid quart = 0.9463 liter
4 liquid quarts	1 liquid gallon (U.S.) = 3.7854 liters

TABLE B-4 Fahrenheit to Centigrade Temperature Conversions

°F	°C	°F	°C	°F	°C
100	37.8	77	25.0	54	12.2
99	37.2	76	24.4	53	11.7
98	36.7	75	23.9	52	11.1
97	36.1	74	23.3	51	10.6
96	35.6	73	22.8	50	10.0
95	35.0	72	22.2	49	9.4
94	34.4	71	21.7	48	8.9
93	33.9	70	21.1	47	8.3
92	33.3	69	20.6	46	7.8
91	32.8	68	20.0	45	7.2
90	32.2	67	19.4	44	6.7
89	31.7	66	18.9	43	6.1
88	31.1	65	18.3	42	5.6
87	30.6	64	17.8	41	5.0
86	30.0	63	17.2	40	4.4
85	29.4	62	16.7	39	3.9
84	28.9	61	16.1	38	3.3
83	28.3	60	15.6	37	2.8
82	27.8	59	15.0	36	2.2
81	27.2	58	14.4	35	1.7
80	26.7	57	13.9	34	1.1
79	26.1	56	13.3	33	0.6
78	25.6	55	12.8	32	0.0

Formulas used: °C = (°F − 32) × 5/9 *or* °F = (°C × 9/5) + 32

TABLE B-5 Conversion Factors for English and Metric Measurements

TO CONVERT THE ENGLISH	TO THE METRIC MULTIPLY BY	TO CONVERT METRIC	MULTIPLY BY	TO GET ENGLISH
acres	0.4047	hectares	2.47	Acres
acres	4047	m^2	0.000247	Acres
BTU	1055	joules	0.000948	BTU
BTU	0.0002928	kwh	3415.301	BTU
BTU/hr	0.2931	watts	3.411805	BTU/hr
bu	0.03524	m^3	28.37684	bu
bu	35.24	L	0.028377	bu
ft^3	0.02832	m^3	35.31073	ft^3
ft^3	28.32	L	0.035311	ft^3
in^3	16.39	cm^3	0.061013	in^3
in^3	1.639×10^{-5}	m^3	61012.81	in^3
in^3	0.01639	L	61.01281	in^3
yd^3	0.7646	m^3	1.307873	yd^3
yd^3	764.6	L	0.001308	yd^3
ft	30.48	cm	0.032808	ft
ft	0.3048	m	3.28084	ft
ft/min	0.508	cm/sec	1.968504	ft/min
ft/sec	30.48	cm/sec	0.032808	ft/sec
gal	3785	cm^3	0.000264	gal
gal	0.003785	m^3	264.2008	gal
gal	3.785	L	0.264201	gal
gal/min	0.06308	L/sec	15.85289	
in	2.54	cm	0.393701	in
in	0.0254	m	39.37008	in
mi	1.609	km	0.621504	mi
mph	26.82	m/min	0.037286	Mph
oz	28.349	gm	0.035275	oz

(Continues)

TABLE B-5 Conversion Factors for English and Metric Measurements

TO CONVERT THE ENGLISH	TO THE METRIC MULTIPLY BY	TO CONVERT METRIC	MULTIPLY BY	TO GET ENGLISH
fl oz	0.02947	L	33.93281	fl oz
liq pt	0.4732	L	2.113271	liq pt
lb	453.59	gm	0.002205	lb
qt	0.9463	L	1.056747	qt
ft^2	0.0929	m^2	10.76426	ft^2
yd^2	0.8361	m^2	1.196029	yd^2
tons	0.9078	tonnes	1.101564	tons
yd	0.0009144	km	1093.613	yd
yd	0.9144	m	1.093613	yd

TABLE B-6 More Conversion Factors for Metric and English Units

LENGTH
1 mile = 1.609 kilometers; 1 kilometer = 0.621 mile 1 yard = 0.914 meter; 1 meter = 1.094 yards 1 inch = 2.54 centimeters; 1 centimeter = 0.394 inch
AREA
1 square mile = 2.59 square kilometers; 1 square kilometer = 0.386 square mile 1 acre = 0.00405 square kilometer; 1 square kilometer = 247. 1 acres 1 acre = 0.405 hectare; 1 hectare = 2.471 acres
VOLUME
1 acre/inch = 102.8 cubic meters; 1 cubic meter = 0.00973 acre/inch 1 quart = 0.946 liter; 1 liter = 1.057 quarts 1 bushel = 0.352 hectoliter; 1 hectoliter = 2.838 bushels

(Continues)

TABLE B-6 More Conversion Factors for Metric and English Units

WEIGHT
1 pound = 0.454 kilogram; 1 kilogram = 2.205 pounds
1 pound = 0.00454 quintal; 1 quintal = 220.5 pounds
1 ton = 0.9072 metric ton; 1 metric ton = 1.102 tons
YIELD OR RATE
1 pound/acre = 1.121 kilograms/acre; 1 kilogram/acre = 0.892 pound/acre
1 ton/acre = 2.242 tons/hectare; 1 ton/hectare = 0.446 ton/acre
1 bushel/acre = 1.121 quintals/hectare; 1 quintal/hectare = 0.892 bushel/acre
1 bushel/acre = (60#) = 0.6726 quintal/hectare; 1 quintal/hectare = 1.487 bushel/acre (60#)
1 bushel/acre = (56#) = 0.6278 quintal/acre; 1 quintal/acre = 1.597 bushels/acre (56#)
TEMPERATURE
To convert Fahrenheit (F) to Celsius (C): 0.555 × (F − 32)
To convert Celsius (C) to Fahrenheit (F): 1.8 × (C + 32)

TABLE B-7 Food Science Resources on the Internet

NAME/TOPIC	URL
Agricultural Marketing Resource Center	http://www.agmrc.org/markets_industries/food/
Agriculture and Agri-Food Canada	http://www.agr.gc.ca/
American Association of Cereal Chemists	http://www.aaccnet.org/Pages/default.aspx
Agriculture News Network	http://www.agriculturenewsnetwork.com/
American Culinary Federation	http://www.acfchefs.org/
American Dairy Science Association	http://www.adsa.org/
Academy of Nutrition and Dietetics	http://www.eatright.org/
American Egg Board	http://www.aeb.org/
American Institute of Baking	http://www.aibonline.org/aibOnline/en/home.aspx
North American Meat Institute	https://www.meatinstitute.org/
American Soybean Association	https://soygrowers.com/

(Continues)

TABLE B-7 Food Science Resources on the Internet

NAME/TOPIC	URL
Beef.Org	www.beef.org
National Cattlemen's Beef Organization	http://www.beefusa.org/
Burger King	www.bk.com
Calorie Counter (USDA)	http://www.newcaloriecounter.com/articles/goverment/usda/usda_national_nutrient_database_for_standard_reference.html
Canning Basics for Food Preservation	http://www.canning-food-recipes.com/canning.htm
Center for Disease Control and Prevention	http://www.cdc.gov/
Center for Food Safety and Applied Nutrition	http://www.fda.gov/Food/
Calculators and Counters (USDA)	https://fnic.nal.usda.gov/dietary-guidance/interactive-tools/calculators-and-counters
Campbell's	http://www.campbellsoup.com/
ChooseMyPlate.gov	http://www.choosemyplate.gov/
Coca-Cola	http://us.coca-cola.com/home/
ConAgra Foods	http://www.conagrafoods.com/
Cornell University Food Science and Technology	https://foodscience.cals.cornell.edu/
Council for Agricultural Science and Technology	http://www.cast-science.org/
USDA Current Research Information System	http://cris.nifa.usda.gov/
Dairy Farmers of America	http://www.dfamilk.com/
Dairy Network.com	http://www.dairynetwork.com/
Dietary Guidance/Food and Nutrition Information Center	https://fnic.nal.usda.gov/dietary-guidance
Feeder's Digest	http://www.feedersdigest.net/
FDA Bad Bug Book (Second Edition)	http://www.fda.gov/Food/FoodborneIllnessContaminants/CausesOfIllnessBadBugBook/
Food and Agriculture Organization of the United Nations	http://www.fao.org/about/en/
Food and Drug Administration	http://www.fda.gov/
Food and Nutrition Information Center	https://fnic.nal.usda.gov/

(*Continues*)

TABLE B-7 Food Science Resources on the Internet

NAME/TOPIC	URL
Food and Nutrition (USDA)	http://www.usda.gov/wps/portal/usda/usdahome?navid=food-nutrition
FoodCom	http://www.food.com/
Food Industry (USDA)	http://www.fda.gov/Food/ResourcesForYou/Industry/
Food Institute: U.S. National & Regional Associations on the Internet	https://www.foodinstitute.com/assn_natl
Food Institute, the	https://www.foodinstitute.com/
Food Labeling Guide (FDA)	http://www.fda.gov/Food/GuidanceRegulation/GuidanceDocumentsRegulatoryInformation/LabelingNutrition/ucm2006828.htm
Food Manufacturing Resources—Careers in Food	http://www.careersinfood.com/resources.cfm
Food Marketing Institute/Industry	http://www.fmi.org/
Food Network	http://www.foodnetwork.com/
Food Preservation	http://nchfp.uga.edu/ http://extension.oregonstate.edu/fch/food-preservation http://extension.psu.edu/food/preservation http://food.unl.edu/preservation http://science.howstuffworks.com/innovation/edible-innovations/food-preservation.htm http://www.dummies.com/how-to/content/food-preservation-methods-canning-freezing-and-dry.html
Food Safety and Inspection Service(USDA)	http://www.fsis.usda.gov/wps/portal/fsis/home http://www.epa.gov/radiation/sources/food_safety.html http://www.fda.gov/Food/ResourcesForYou/Consumers/ucm261680.htm
Foster Farms	http://www.fosterfarms.com/
Incredible Egg	http://www.incredibleegg.org/
Institute of Food Technologists	https://www.ift.org/
International Food Information Council Foundation	http://www.foodinsight.org/
Fruit Net	http://www.fruitnet.com/

(*Continues*)

TABLE B-7 Food Science Resources on the Internet

NAME/TOPIC	URL
General Mills	http://www.generalmills.com/
Global Meat News	http://www.globalmeatnews.com/
Government Food Safety Information	http://www.foodsafety.gov/
Grocery Manufacturers of America	http://www.gmaonline.org/
Hershey's	http://www.hersheys.com/
Horizon Organic Dairy	https://www.horizondairy.com/
Hormel Foods	http://www.hormelfoods.com/
International Organization for Standard (ISO)	http://www.iso.org/iso/home.html
I0I Loders Croklaan (Oils & Fats)	http://northamerica.croklaan.com/
Institute of Child Nutrition	http://www.theicn.org//
Iowa Beef Processors	http://www.tyson.com/
JustFood.com	http://www.just-food.com/
KFC	http://www.kfc.com/
Kraft	http://www.kraftrecipes.com/
Kroger	https://www.kroger.com/
McDonalds	http://www.mcdonalds.com/us/en/home.html
Meat and Poultry Online	http://www.meatpoultry.com/
Minute Maid	http://www.minutemaid.com/content/minutemaid/en/home/
NASCO	http://www.enasco.com/
Food Safety Database (WHO)	http://www.who.int/foodsafety/en//
National Meat Association/North American Meat Processors Association	http://beefmagazine.com/allied-industry/national-meat-association-north-american-meat-processors-association-consolidate
National Pork Producers Council	http://www.nppc.org/
National Restaurant Association	http://www.restaurant.org/Home
Nutrition and Food Curriculum Guide (USDA)	https://fnic.nal.usda.gov/professional-and-career-resources/nutrition-education/curricula-and-lesson-plans

(*Continues*)

TABLE B-7 Food Science Resources on the Internet

NAME/TOPIC	URL
Nutrition	http://www.nutrition.gov/
Nutrient Database (USDA)	http://ndb.nal.usda.gov/
Pastry Chef Central	http://www.pastrychef.com/
PennState: Food Science Resources for Teachers	http://foodscience.psu.edu/youth/educators
Pepsi	http://www.pepsi.com/en-us/d
PepsiCo	http://www.pepsico.com/
Perdue Farms	http://www.perdue.com/
Performance Food Group	http://www.pfgc.com/Pages/default.aspx
Pizza Hut	https://order.pizzahut.com/home?
Produce Marketing Association	http://www.pma.com/
RecipeLink	http://www.recipelink.com/index.html
SafeFood	http://www.safefood.eu/Home.aspx
Safeway	http://www.safeway.com/
Seafood Education (National Marine Fisheries Service)	http://www.afsc.noaa.gov/education/activities /seafood_ed.htm
Seafood Network Information Center (Oregon State University)	http://seafood.oregonstate.edu/
Seafood Products Association	http://www.spa-food.org/
Smithfield Foods	http://smithfieldfoods.com/
Soyfoods Association of North America	http://www.soyfoods.org/
Soyfoods Council, The	http://thesoyfoodscouncil.com/
Soyfoods Directory	http://www.soyfoods.com/
Taco Bell	http://www.tacobell.com/
Tyson Foods, Inc.	http://www.tysonfoods.com/
U.S. Meat Animal Research Center	http://www.ars.usda.gov/main/site_main .htm?modecode=30-40-05-00
U.S. Poultry & Egg Association	http://www.uspoultry.org/
Wendy's	https://www.wendys.com/

TABLE B-8 Food Composition Table

								FATTY ACIDS										VITAMIN A					
FOOD NO.	FOOD DESCRIPTION	MEASURE OF EDIBLE PORTION	WEIGHT (G)	WATER (%)	CALORIES (KCAL)	PROTEIN (G)	TOTAL FAT (G)	SAT (G)	MONO (G)	POLY (G)	CHOLEST (MG)	CARB (G)	TOTAL D F(G)	CA (MG)	FE (MG)	K (MG)	NA (MG)	(IU)	(RE)	THMN (MG)	RIBOFL (MG)	NIACIN (MG)	ASCORBIC ACID (MG)
Beverages																							
	Alcoholic																						
	Beer																						
1	Regular	12 fl oz	355	92	146	1	0	0.0	0.0	0.0	0	13	0.7	18	0.1	89	18	0	0	0.02	0.09	1.6	0
2	Light	12 fl oz	354	95	99	1	0	0.0	0.0	0.0	0	5	0.0	18	0.1	64	11	0	0	0.03	0.11	1.4	0
	Gin, rum, vodka, whiskey																						
3	80 proof	1.5 fl oz	42	67	97	0	0	0.0	0.0	0.0	0	0	0.0	0	Tr	1	Tr	0	0	Tr	Tr	Tr	0
4	86 proof	1.5 fl oz	42	64	105	0	0	0.0	0.0	0.0	0	Tr	0.0	0	Tr	1	Tr	0	0	Tr	Tr	Tr	0
5	90 proof	1.5 fl oz	42	62	110	0	0	0.0	0.0	0.0	0	0	0.0	0	Tr	1	Tr	0	0	Tr	Tr	Tr	0
6	Liqueur, coffee, 53 proof	1.5 fl oz	52	31	175	Tr	Tr	0.1	Tr	0.1	0	24	0.0	1	Tr	16	4	0	0	Tr	0.01	0.1	0
	Mixed drinks, prepared from recipe																						
7	Daiquiri	2 fl oz	60	70	112	Tr	Tr	Tr	Tr	Tr	0	4	0.0	2	0.1	13	3	2	0	0.01	Tr	Tr	1
8	Pina colada	4.5 fl oz	141	65	262	1	3	1.2	0.2	0.5	0	40	0.8	11	0.3	100	8	3	0	0.04	0.02	0.2	7
	Wine																						
	Dessert																						
9	Dry	3.5 fl oz	103	80	130	Tr	0	0.0	0.0	0.0	0	4	0.0	8	0.2	95	9	0	0	0.02	0.02	0.2	0
10	Sweet	3.5 fl oz	103	73	158	Tr	0	0.0	0.0	0.0	0	12	0.0	8	0.2	95	9	0	0	0.02	0.02	0.2	0
	Table																						
11	Red	3.5 fl oz	103	89	74	Tr	0	0.0	0.0	0.0	0	2	0.0	8	0.4	115	5	0	0	0.01	0.03	0.1	0
12	White	3.5 fl oz	103	90	70	Tr	0	0.0	0.0	0.0	0	1	0.0	9	0.3	82	5	0	0	Tr	0.01	0.1	0
	Carbonated*																						
13	Club soda	12 fl oz	355	100	0	0	0	0.0	0.0	0.0	0	0	0.0	18	Tr	7	75	0	0	0.00	0.00	0.0	0
14	Cola type	12 fl oz	370	89	152	0	0	0.0	0.0	0.0	0	38	0.0	11	0.1	4	15	0	0	0.00	0.00	0.0	0
	Diet, sweetened with aspartame																						
15	Cola	12 fl oz	355	100	4	Tr	0	0.0	0.0	0.0	0	Tr	0.0	14	0.1	0	21	0	0	0.02	0.08	0.0	0
16	Other than cola or pepper type	12 fl oz	355	100	0	Tr	0	0.0	0.0	0.0	0	0	0.0	14	0.1	7	21	0	0	0.00	0.00	0.0	0
17	Ginger ale	12 fl oz	366	91	124	0	0	0.0	0.0	0.0	0	32	0.0	11	0.7	4	26	0	0	0.00	0.00	0.0	0
18	Grape	12 fl oz	372	89	160	0	0	0.0	0.0	0.0	0	42	0.0	11	0.3	4	56	0	0	0.00	0.00	0.0	0
19	Lemon lime	12 fl oz	368	90	147	0	0	0.0	0.0	0.0	0	38	0.0	7	0.3	4	40	0	0	0.00	0.00	0.1	0

*Mineral content varies depending on water source.

Source: USDA, Agricultural Research Services (ARS), Nutritive Value of Foods

(Continues)

TABLE B-8 Food Composition Table

FOOD NO.	FOOD DESCRIPTION	MEASURE OF EDIBLE PORTION	WEIGHT (G)	WATER (%)	CALORIES (KCAL)	PROTEIN (G)	TOTAL FAT (G)	FATTY ACIDS SAT (G)	MONO (G)	POLY (G)	CHOLEST (MG)	CARB (G)	TOTAL D F(G)	CA (MG)	FE (MG)	K (MG)	NA (MG)	VITAMIN A (IU)	(RE)	THMN (MG)	RIBOFL (MG)	NIACIN (MG)	ASCORBIC ACID (MG)
Beverages (*Continued*)																							
	Carbonated (*Continued*)																						
20	Orange	12 fl oz	372	88	179	0	0	0.0	0.0	0.0	0	46	0.0	19	0.2	7	45	0	0	0.00	0.00	0.0	0
21	Pepper type	12 fl oz	368	89	151	0	Tr	0.3	0.0	0.0	0	38	0.0	11	0.1	4	37	0	0	0.00	0.00	0.0	0
22	Root beer	12 fl oz	370	89	152	0	0	0.0	0.0	0.0	0	39	0.0	19	0.2	4	48	0	0	0.00	0.00	0.0	0
	Chocolate flavored beverage mix																						
23	Powder	2-3 heap-ing tsp	22	1	75	1	1	0.4	0.2	Tr	0	20	1.3	8	0.7	128	45	4	Tr	0.01	0.03	0.1	Tr
24	Prepared with milk	1 cup	266	81	226	9	9	5.5	2.6	0.3	32	31	1.3	301	0.8	497	165	311	77	0.10	0.43	0.3	2
	Cocoa																						
	Powder containing nonfat dry milk																						
25	Powder	3 heaping tsp	28	2	102	3	1	0.7	0.4	Tr	1	22	0.3	92	0.3	202	143	4	1	0.03	0.16	0.2	1
26	Prepared (6 oz water plus 1 oz powder)	1 serving	206	86	103	3	1	0.7	0.4	Tr	2	22	2.5	97	0.4	202	148	4	0	0.03	0.16	0.2	Tr
	Powder containing nonfat dry milk and aspartame																						
27	Powder	½-oz envelope	15	3	48	4	Tr	0.3	0.1	Tr	1	9	0.4	86	0.7	405	168	5	1	0.04	0.21	0.2	0
28	Prepared (6 oz water plus 1 envelope mix)	1 serving	192	92	48	4	Tr	0.3	0.1	Tr	2	8	0.4	90	0.7	405	173	4	0	0.04	0.21	0.2	0
	Coffee																						
29	Brewed	6 fl oz	178	99	4	Tr	0	Tr	0.0	Tr	0	1	0.0	4	0.1	96	4	0	0	0.00	0.00	0.4	0
30	Espresso	2 fl oz	60	98	5	Tr	Tr	0.1	0.0	0.1	0	1	0.0	1	0.1	69	8	0	0	Tr	0.11	3.1	Tr
31	Instant, prepared (1 rounded tsp powder plus 6 fl oz water)	6 fl oz	179	99	4	Tr	0	Tr	0.0	Tr	0	1	0.0	5	0.1	64	5	0	0	0.00	Tr	0.5	0
	Fruit drinks, noncarbonated, canned or bottled, with added ascorbic acid																						
32	Cranberry juice cocktail	8 fl oz	253	86	144	0	Tr	Tr	Tr	0.1	0	36	0.3	8	0.4	46	5	10	0	0.02	0.02	0.1	90
33	Fruit punch drink	8 fl oz	248	88	117	0	0	Tr	Tr	Tr	0	30	0.2	20	0.5	62	55	35	2	0.05	0.06	0.1	73
34	Grape drink	8 fl oz	250	88	113	0	0	Tr	0.0	Tr	0	29	0.0	8	0.4	13	15	3	0	0.01	0.01	0.1	85

(Continues)

TABLE B-8 Food Composition Table

FOOD NO.	FOOD DESCRIPTION	MEASURE OF EDIBLE PORTION	WEIGHT (G)	WATER (%)	CALORIES (KCAL)	PROTEIN (G)	TOTAL FAT (G)	FATTY ACIDS SAT (G)	FATTY ACIDS MONO (G)	FATTY ACIDS POLY (G)	CHOLEST (MG)	CARB (G)	TOTAL D F(G)	CA (MG)	FE (MG)	K (MG)	NA (MG)	VITAMIN A (IU)	VITAMIN A (RE)	THMN (MG)	RIBOFL (MG)	NIACIN (MG)	ASCORBIC ACID (MG)
	Beverages (*Continued*)																						
	Fruit drinks, noncarbonated, canned or bottled, with added ascorbic acid (*Continued*)																						
35	Pineapple grapefruit juice drink	8 fl oz	250	88	118	1	Tr	Tr	Tr	0.1	0	29	0.3	18	0.8	153	35	88	10	0.08	0.04	0.7	115
36	Pineapple orange juice drink	8 fl oz	250	87	125	3	0	0.0	0.0	0.0	0	30	0.3	13	0.7	115	8	1,328	133	0.08	0.05	0.5	56
	Lemonade																						
37	Frozen concentrate, prepared	8 fl oz	248	89	99	Tr	0	Tr	Tr	Tr	0	26	0.2	7	0.4	37	7	52	5	0.01	0.05	Tr	10
	Powder, prepared with water																						
38	Regular	8 fl oz	266	89	112	0	0	Tr	Tr	Tr	0	29	0.0	29	0.1	3	19	0	0	0.00	Tr	0.0	34
39	Low calorie, sweetened with aspartame	8 fl oz	237	99	5	0	0	0.0	0.0	0.0	0	1	0.0	50	0.1	0	7	0	0	0.00	0.00	0.0	6
	Malted milk, with added nutrients																						
	Chocolate																						
40	Powder	3 heaping tsp	21	3	75	1	1	0.4	0.2	0.1	1	18	0.2	93	3.6	251	125	2,751	824	0.64	0.86	10.7	32
41	Prepared	1 cup	265	81	225	9	9	5.5	2.6	0.4	34	29	0.3	384	3.8	620	244	3,058	901	0.73	1.26	10.9	34
	Natural																						
42	Powder	4-5 heaping tsp	21	3	80	2	1	0.3	0.2	0.1	4	17	0.1	79	3.5	203	85	2,222	668	0.62	0.75	10.2	27
43	Prepared	1 cup	265	81	231	10	9	5.4	2.5	0.4	34	28	0.0	371	3.6	572	204	2,531	742	0.71	1.14	10.4	29
	Milk and milk beverages. See Dairy Products.																						
44	Rice beverage, canned (RICE DREAM)	1 cup	245	89	120	Tr	2	0.2	1.3	0.3	0	25	0.0	20	0.2	69	86	5	0	0.08	0.01	1.9	1
	Soy milk. See Legumes, Nuts, and Seeds.																						
	Tea																						
	Brewed																						
45	Black	6 fl oz	178	100	2	0	0	Tr	Tr	Tr	0	1	0.0	0	Tr	66	5	0	0	0.00	0.02	0.0	0
	Herb																						

(*Continues*)

TABLE B-8 Food Composition Table

FOOD NO.	FOOD DESCRIPTION	MEASURE OF EDIBLE PORTION	WEIGHT (G)	WATER (%)	CALORIES (KCAL)	PROTEIN (G)	TOTAL FAT (G)	FATTY ACIDS SAT (G)	FATTY ACIDS MONO (G)	FATTY ACIDS POLY (G)	CHOLEST (MG)	CARB (G)	TOTAL D F(G)	CA (MG)	FE (MG)	K (MG)	NA (MG)	VITAMIN A (IU)	VITAMIN A (RE)	THMN (MG)	RIBOFL (MG)	NIACIN (MG)	ASCORBIC ACID (MG)
Beverages (*Continued*)																							
	Tea (*Continued*)																						
46	Chamomile	6 fl oz	178	100	2	0	0	Tr	Tr	Tr	0	Tr	0.0	4	0.1	16	2	36	4	0.02	0.01	0.0	0
47	Other than chamomile	6 fl oz	178	100	2	0	0	Tr	Tr	Tr	0	Tr	0.0	4	0.1	16	2	0	0	0.02	0.01	0.0	0
	Instant, powder, prepared																						
48	Unsweetened	8 fl oz	237	100	2	0	0	0.0	0.0	0.0	0	Tr	0.0	5	Tr	47	7	0	0	0.00	Tr	0.1	0
49	Sweetened, lemon flavor	8 fl oz	259	91	88	Tr	0	Tr	Tr	Tr	0	22	0.0	5	0.1	49	8	0	0	0.00	0.05	0.1	0
50	Sweetened with saccharin, lemon flavor	8 fl oz	237	99	5	0	0	0.0	0.0	Tr	0	1	0.0	5	0.1	40	24	0	0	0.00	0.01	0.1	0
51	Water, tap	8 fl oz	237	100	0	0	0	0.0	0.0	0.0	0	0	0.0	5	Tr	0	7	0	0	0.00	0.00	0.0	0
Dairy Products																							
	Butter. See Fats and Oils.																						
	Cheese																						
	Natural																						
52	Blue	1 oz	28	42	100	6	8	5.3	2.2	0.2	21	1	0.0	150	0.1	73	396	204	65	0.01	0.11	0.3	0
53	Camembert (3 wedges per 4-oz container)	1 wedge	38	52	114	8	9	5.8	2.7	0.3	27	Tr	0.0	147	0.1	71	320	351	96	0.01	0.19	0.2	0
	Cheddar																						
54	Cut pieces	1 oz	28	37	114	7	9	6.0	2.7	0.3	30	Tr	0.0	204	0.2	28	176	300	79	0.01	0.11	Tr	0
55		1 cubic inch	17	37	68	4	6	3.6	1.6	0.2	18	Tr	0.0	123	0.1	17	105	180	47	Tr	0.06	Tr	0
56	Shredded	1 cup	113	37	455	28	37	23.8	10.6	1.1	119	1	0.0	815	0.8	111	701	1,197	314	0.03	0.42	0.1	0
	Cottage																						
	Creamed (4% fat)																						
57	Large curd	1 cup	225	79	233	28	10	6.4	2.9	0.3	34	6	0.0	135	0.3	190	911	367	108	0.05	0.37	0.3	0
58	Small curd	1 cup	210	79	217	26	9	6.0	2.7	0.3	31	6	0.0	126	0.3	177	850	342	101	0.04	0.34	0.3	0
59	With fruit	1 cup	226	72	279	22	8	4.9	2.2	0.2	25	30	0.0	108	0.2	151	915	278	81	0.04	0.29	0.2	0
60	Low fat (2%)	1 cup	226	79	203	31	4	2.8	1.2	0.1	19	8	0.0	155	0.4	217	918	158	45	0.05	0.42	0.3	0
61	Low fat (1%)	1 cup	226	82	164	28	2	1.5	0.7	0.1	10	6	0.0	138	0.3	193	918	84	25	0.05	0.37	0.3	0
62	Uncreamed (dry curd, less than ½% fat)	1 cup	145	80	123	25	1	0.4	0.2	Tr	10	3	0.0	46	0.3	47	19	44	12	0.04	0.21	0.2	0
	Cream																						
63	Regular	1 oz	28	54	99	2	10	6.2	2.8	0.4	31	1	0.0	23	0.3	34	84	405	108	Tr	0.06	Tr	0

(Continues)

TABLE B-8 Food Composition Table

FOOD NO.	FOOD DESCRIPTION	MEASURE OF EDIBLE PORTION	WEIGHT (G)	WATER (%)	CALORIES (KCAL)	PROTEIN (G)	TOTAL FAT (G)	FATTY ACIDS			CHOLEST (MG)	CARB (G)	TOTAL D F(G)	CA (MG)	FE (MG)	K (MG)	NA (MG)	VITAMIN A		THMN (MG)	RIBOFL (MG)	NIACIN (MG)	ASCORBIC ACID (MG)
								SAT (G)	MONO (G)	POLY (G)								(IU)	(RE)				
Dairy Products (*Continued*)																							
	Butter. See Fats and Oils. (*Continued*)																						
64		1 tbsp	15	54	51	1	5	3.2	1.4	0.2	16	Tr	0.0	12	0.2	17	43	207	55	Tr	0.03	Tr	0
65	Low fat	1 tbsp	15	64	35	2	3	1.7	0.7	0.1	8	1	0.0	17	0.3	25	44	108	33	Tr	0.04	Tr	0
66	Fat free	1 tbsp	16	76	15	2	Tr	0.1	0.1	Tr	1	1	0.0	29	Tr	25	85	145	44	0.01	0.03	Tr	0
67	Feta	1 oz	28	55	75	4	6	4.2	1.3	0.2	25	1	0.0	140	0.2	18	316	127	36	0.04	0.24	0.3	0
68	Low fat, cheddar or colby	1 oz	28	63	49	7	2	1.2	0.6	0.1	6	1	0.0	118	0.1	19	174	66	18	Tr	0.06	Tr	0
	Mozzarella, made with																						
69	Whole milk	1 oz	28	54	80	6	6	3.7	1.9	0.2	22	1	0.0	147	0.1	19	106	225	68	Tr	0.07	Tr	0
70	Part skim milk (low moisture)	1 oz	28	49	79	8	5	3.1	1.4	0.1	15	1	0.0	207	0.1	27	150	199	54	0.01	0.10	Tr	0
71	Muenster	1 oz	28	42	104	7	9	5.4	2.5	0.2	27	Tr	0.0	203	0.1	38	178	318	90	Tr	0.09	Tr	0
72	Neufchatel	1 oz	28	62	74	3	7	4.2	1.9	0.2	22	1	0.0	21	0.1	32	113	321	85	Tr	0.06	Tr	0
73	Parmesan, grated	1 cup	100	18	456	42	30	19.1	8.7	0.7	79	4	0.0	1,376	1.0	107	1,862	701	173	0.05	0.39	0.3	0
74		1 tbsp	5	18	23	2	2	1.0	0.4	Tr	4	Tr	0.0	69	Tr	5	93	35	9	Tr	0.02	Tr	0
75		1 oz	28	18	129	12	9	5.4	2.5	0.2	22	1	0.0	390	0.3	30	528	199	49	0.01	0.11	0.1	0
76	Provolone	1 oz	28	41	100	7	8	4.8	2.1	0.2	20	1	0.0	214	0.1	39	248	231	75	0.01	0.09	Tr	0
	Ricotta, made with																						
77	Whole milk	1 cup	246	72	428	28	32	20.4	8.9	0.9	124	7	0.0	509	0.9	257	207	1,205	330	0.03	0.48	0.3	0
78	Part skim milk	1 cup	246	74	340	28	19	12.1	5.7	0.6	76	13	0.0	669	1.1	308	307	1,063	278	0.05	0.46	0.2	0
79	Swiss	1 oz	28	37	107	8	8	5.0	2.1	0.3	26	1	0.0	272	Tr	31	74	240	72	0.01	0.10	Tr	0
	Pasteurized process cheese American																						
80	Regular	1 oz	28	39	106	6	9	5.6	2.5	0.3	27	Tr	0.0	174	0.1	46	406	343	82	0.01	0.10	Tr	0
81	Fat free	1 slice	21	57	31	5	Tr	0.1	Tr	Tr	2	3	0.0	145	0.1	60	321	308	92	0.01	0.10	Tr	0
82	Swiss	1 oz	28	42	95	7	7	4.5	2.0	0.2	24	1	0.0	219	0.2	61	388	229	65	Tr	0.08	Tr	0
83	Pasteurized process cheese food, American	1 oz	28	43	93	6	7	4.4	2.0	0.2	18	2	0.0	163	0.2	79	337	259	62	0.01	0.13	Tr	0
84	Pasteurized process cheese spread, American	1 oz	28	48	82	5	6	3.8	1.8	0.2	16	2	0.0	159	0.1	69	381	223	54	0.01	0.12	Tr	0
	Cream, sweet																						
85	Half and half (cream and milk)	1 cup	242	81	315	7	28	17.3	8.0	1.0	89	10	0.0	254	0.2	314	98	1,050	259	0.08	0.36	0.2	2
86		1 tbsp	15	81	20	Tr	2	1.1	0.5	0.1	6	1	0.0	16	Tr	19	6	65	16	0.01	0.02	Tr	Tr

(*Continues*)

TABLE B-8 Food Composition Table

FOOD NO.	FOOD DESCRIPTION	MEASURE OF EDIBLE PORTION	WEIGHT (G)	WATER (%)	CALORIES (KCAL)	PROTEIN (G)	TOTAL FAT (G)	FATTY ACIDS			CHOLEST (MG)	CARB (G)	TOTAL D F(G)	CA (MG)	FE (MG)	K (MG)	NA (MG)	VITAMIN A		THMN (MG)	RIBOFL (MG)	NIACIN (MG)	ASCORBIC ACID (MG)
								SAT (G)	MONO (G)	POLY (G)								(IU)	(RE)				
Dairy Products (*Continued*)																							
	Cream, sweet (*Continued*)																						
87	Light, coffee, or table	1 cup	240	74	469	6	46	28.8	13.4	1.7	159	9	0.0	231	0.1	292	95	1,519	437	0.08	0.36	0.1	2
88		1 tbsp	15	74	29	Tr	3	1.8	0.8	0.1	10	1	0.0	14	Tr	18	6	95	27	Tr	0.02	Tr	Tr
	Whipping, unwhipped (volume about double when whipped)																						
89	Light	1 cup	239	64	699	5	74	46.2	21.7	2.1	265	7	0.0	166	0.1	231	82	2,694	705	0.06	0.30	0.1	1
90		1 tbsp	15	64	44	Tr	5	2.9	1.4	0.1	17	Tr	0.0	10	Tr	15	5	169	44	Tr	0.02	Tr	Tr
91	Heavy	1 cup	238	58	821	5	88	54.8	25.4	3.3	326	7	0.0	154	0.1	179	89	3,499	1,002	0.05	0.26	0.1	1
92		1 tbsp	15	58	52	Tr	6	3.5	1.6	0.2	21	Tr	0.0	10	Tr	11	6	221	63	Tr	0.02	Tr	Tr
93	Whipped topping (pressurized)	1 cup	60	61	154	2	13	8.3	3.9	0.5	46	7	0.0	61	Tr	88	78	506	124	0.02	0.04	Tr	0
94		1 tbsp	3	61	8	Tr	1	0.4	0.2	Tr	2	Tr	0.0	3	Tr	4	4	25	6	Tr	Tr	Tr	0
	Cream, sour																						
95	Regular	1 cup	230	71	493	7	48	30.0	13.9	1.8	102	10	0.0	268	0.1	331	123	1,817	449	0.08	0.34	0.2	2
96		1 tbsp	12	71	26	Tr	3	1.6	0.7	0.1	5	1	0.0	14	Tr	17	6	95	23	Tr	0.02	Tr	Tr
97	Reduced fat	1 tbsp	15	80	20	Tr	2	1.1	0.5	0.1	6	1	0.0	16	Tr	19	6	68	17	0.01	0.02	Tr	Tr
98	Fat free	1 tbsp	16	81	12	Tr	0	0.0	0.0	0.0	1	2	0.0	20	0.0	21	23	100	13	0.01	0.02	Tr	0
	Cream product, imitation (made with vegetable fat)																						
	Sweet																						
	Creamer																						
99	Liquid (frozen)	1 tbsp	15	77	20	Tr	1	0.3	1.1	Tr	0	2	0.0	1	Tr	29	12	13*	1*	0.00	0.00	0.0	0
100	Powdered	1 tsp	2	2	11	Tr	1	0.7	Tr	Tr	0	1	0.0	Tr	Tr	16	4	4	Tr	0.00	Tr	0.0	0
	Whipped topping																						
101	Frozen	1 cup	75	50	239	1	19	16.3	1.2	0.4	0	17	0.0	5	0.1	14	19	646*	65*	0.00	0.00	0.0	0
102		1 tbsp	4	50	13	Tr	1	0.9	0.1	Tr	0	1	0.0	Tr	Tr	1	1	34*	3*	0.00	0.00	0.0	0
103	Powdered, prepared with whole milk	1 cup	80	67	151	3	10	8.5	0.7	0.2	8	13	0.0	72	Tr	121	53	289*	39*	0.02	0.09	Tr	1
104		1 tbsp	4	67	8	Tr	Tr	0.4	Tr	Tr	Tr	1	0.0	4	Tr	6	3	14*	2*	Tr	Tr	Tr	Tr
105	Pressurized	1 cup	70	60	184	1	16	13.2	1.3	0.2	0	11	0.0	4	Tr	13	43	331*	33*	0.00	0.00	0.0	0
106		1 tbsp	4	60	11	Tr	1	0.8	0.1	Tr	0	1	0.0	Tr	Tr	1	2	19*	2*	0.00	0.00	0.0	0

*The vitamin A values listed for imitation sweet cream products are mostly from beta-carotene added for coloring.

(Continues)

TABLE B-8 Food Composition Table

FOOD NO.	FOOD DESCRIPTION	MEASURE OF EDIBLE PORTION	WEIGHT (G)	WATER (%)	CALORIES (KCAL)	PROTEIN (G)	TOTAL FAT (G)	FATTY ACIDS SAT (G)	FATTY ACIDS MONO (G)	FATTY ACIDS POLY (G)	CHOLEST (MG)	CARB (G)	TOTAL D F(G)	CA (MG)	FE (MG)	K (MG)	NA (MG)	VITAMIN A (IU)	VITAMIN A (RE)	THMN (MG)	RIBOFL (MG)	NIACIN (MG)	ASCORBIC ACID (MG)
Dairy Products (*Continued*)																							
	Cream product, imitation (made with vegetable fat) (*Continued*)																						
107	Sour dressing (filled cream type, nonbutterfat)	1 cup	235	75	417	8	39	31.2	4.6	1.1	13	11	0.0	266	0.1	380	113	24	5	0.09	0.38	0.2	2
108		1 tbsp	12	75	21	Tr	2	1.6	0.2	0.1	1	1	0.0	14	Tr	19	6	1	Tr	Tr	0.02	Tr	Tr
	Frozen dessert																						
	Frozen yogurt, soft serve																						
109	Chocolate	½ cup	72	64	115	3	4	2.6	1.3	0.2	4	18	1.6	106	0.9	188	71	115	31	0.03	0.15	0.2	Tr
110	Vanilla	½ cup	72	65	114	3	4	2.5	1.1	0.2	1	17	0.0	103	0.2	152	63	153	41	0.03	0.16	0.2	1
	Ice cream																						
	Regular																						
111	Chocolate	½ cup	66	56	143	3	7	4.5	2.1	0.3	22	19	0.8	72	0.6	164	50	275	79	0.03	0.13	0.1	Tr
112	Vanilla	½ cup	66	61	133	2	7	4.5	2.1	0.3	29	16	0.0	84	0.1	131	53	270	77	0.03	0.16	0.1	Tr
113	Light (50% reduced fat), vanilla	½ cup	66	68	92	3	3	1.7	0.8	0.1	9	15	0.0	92	0.1	139	56	109	31	0.04	0.17	0.1	1
114	Premium low fat, chocolate	½ cup	72	61	113	3	2	1.0	0.6	0.1	7	22	0.7	107	0.4	179	50	163	47	0.02	0.13	0.1	1
115	Rich, vanilla	½ cup	74	57	178	3	12	7.4	3.4	0.4	45	17	0.0	87	Tr	118	41	476	136	0.03	0.12	0.1	1
116	Soft serve, french vanilla	½ cup	86	60	185	4	11	6.4	3.0	0.4	78	19	0.0	113	0.2	152	52	464	132	0.04	0.16	0.1	1
117	Sherbet, orange	½ cup	74	66	102	1	1	0.9	0.4	0.1	4	22	0.0	40	0.1	71	34	56	10	0.02	0.06	Tr	2
	Milk																						
	Fluid, no milk solids added																						
118	Whole (3.3% fat)	1 cup	244	88	150	8	8	5.1	2.4	0.3	33	11	0.0	291	0.1	370	120	307	76	0.09	0.40	0.2	2
119	Reduced fat (2%)	1 cup	244	89	121	8	5	2.9	1.4	0.2	18	12	0.0	297	0.1	377	122	500	139	0.10	0.40	0.2	2
120	Lowfat (1%)	1 cup	244	90	102	8	3	1.6	0.7	0.1	10	12	0.0	300	0.1	381	123	500	144	0.10	0.41	0.2	2
121	Nonfat (skim)	1 cup	245	91	86	8	Tr	0.3	0.1	Tr	4	12	0.0	302	0.1	406	126	500	149	0.09	0.34	0.2	2
122	Buttermilk	1 cup	245	90	99	8	2	1.3	0.6	0.1	9	12	0.0	285	0.1	371	257	81	20	0.08	0.38	0.1	2
	Canned																						
123	Condensed, sweetened	1 cup	306	27	982	24	27	16.8	7.4	1.0	104	166	0.0	868	0.6	1,136	389	1,004	248	0.28	1.27	0.6	8
	Evaporated																						
124	Whole milk	1 cup	252	74	339	17	19	11.6	5.9	0.6	74	25	0.0	657	0.5	764	267	612	136	0.12	0.80	0.5	5
125	Skim milk	1 cup	256	79	199	19	1	0.3	0.2	Tr	9	29	0.0	741	0.7	849	294	1,004	300	0.12	0.79	0.4	3

(Continues)

TABLE B-8 Food Composition Table

FOOD NO.	FOOD DESCRIPTION	MEASURE OF EDIBLE PORTION	WEIGHT (G)	WATER (%)	CALORIES (KCAL)	PROTEIN (G)	TOTAL FAT (G)	FATTY ACIDS SAT (G)	FATTY ACIDS MONO (G)	FATTY ACIDS POLY (G)	CHOLEST (MG)	CARB (G)	TOTAL D F(G)	CA (MG)	FE (MG)	K (MG)	NA (MG)	VITAMIN A (IU)	VITAMIN A (RE)	THMN (MG)	RIBOFL (MG)	NIACIN (MG)	ASCORBIC ACID (MG)
	Dairy Products (*Continued*)																						
	Milk (*Continued*)																						
	Dried																						
126	Buttermilk	1 cup	120	3	464	41	7	4.3	2.0	0.3	83	59	0.0	1,421	0.4	1,910	621	262	65	0.47	1.89	1.1	7
127	Nonfat, instant, with added vitamin A	1 cup	68	4	244	24	Tr	0.3	0.1	Tr	12	35	0.0	837	0.2	1,160	373	1,612	483	0.28	1.19	0.6	4
	Milk beverage																						
	Chocolate milk (commercial)																						
128	Whole	1 cup	250	82	208	8	8	5.3	2.5	0.3	31	26	2.0	280	0.6	417	149	303	73	0.09	0.41	0.3	2
129	Reduced fat (2%)	1 cup	250	84	179	8	5	3.1	1.5	0.2	17	26	1.3	284	0.6	422	151	500	143	0.09	0.41	0.3	2
130	Lowfat (1%)	1 cup	250	85	158	8	3	1.5	0.8	0.1	7	26	1.3	287	0.6	426	152	500	148	0.10	0.42	0.3	2
131	Eggnog (commercial)	1 cup	254	74	342	10	19	11.3	5.7	0.9	149	34	0.0	330	0.5	420	138	894	203	0.09	0.48	0.3	4
	Milk shake, thick																						
132	Chocolate	10.6 fl oz	300	72	356	9	8	5.0	2.3	0.3	32	63	0.9	396	0.9	672	333	258	63	0.14	0.67	0.4	0
133	Vanilla	11 fl oz	313	74	350	12	9	5.9	2.7	0.4	37	56	0.0	457	0.3	572	299	357	88	0.09	0.61	0.5	0
	Sherbet. See Dairy Products, frozen dessert.																						
	Yogurt																						
	With added milk solids																						
	Made with lowfat milk																						
134	Fruit flavored	8-oz container	227	74	231	10	2	1.6	0.7	0.1	10	43	0.0	345	0.2	442	133	104	25	0.08	0.40	0.2	1
135	Plain	8-oz container	227	85	144	12	4	2.3	1.0	0.1	14	16	0.0	415	0.2	531	159	150	36	0.10	0.49	0.3	2
	Made with nonfat milk																						
136	Fruit flavored	8-oz container	227	75	213	10	Tr	0.3	0.1	Tr	5	43	0.0	345	0.2	440	132	16	5	0.09	0.41	0.2	2
137	Plain	8-oz container	227	85	127	13	Tr	0.3	0.1	Tr	4	17	0.0	452	0.2	579	174	16	5	0.11	0.53	0.3	2
	Without added milk solids																						
138	Made with whole milk, plain	8-oz container	227	88	139	8	7	4.8	2.0	0.2	29	11	0.0	274	0.1	351	105	279	68	0.07	0.32	0.2	1
139	Made with nonfat milk, low calorie sweetener, vanilla or lemon flavor	8-oz container	227	87	98	9	Tr	0.3	0.1	Tr	5	17	0.0	325	0.3	402	134	0	0	0.08	0.37	0.2	2

(*Continues*)

TABLE B-8 Food Composition Table

FOOD NO.	FOOD DESCRIPTION	MEASURE OF EDIBLE PORTION	WEIGHT (G)	WATER (%)	CALORIES (KCAL)	PROTEIN (G)	TOTAL FAT (G)	FATTY ACIDS			CHOLEST (MG)	CARB (G)	TOTAL D F(G)	CA (MG)	FE (MG)	K (MG)	NA (MG)	VITAMIN A		THMN (MG)	RIBOFL (MG)	NIACIN (MG)	ASCORBIC ACID (MG)
								SAT (G)	MONO (G)	POLY (G)								(IU)	(RE)				
Eggs																							
	Egg																						
	Raw																						
140	Whole	1 medium	44	75	66	5	4	1.4	1.7	0.6	187	1	0.0	22	0.6	53	55	279	84	0.03	0.22	Tr	0
141		1 large	50	75	75	6	5	1.6	1.9	0.7	213	1	0.0	25	0.7	61	63	318	96	0.03	0.25	Tr	0
142		1 extra large	58	75	86	7	6	1.8	2.2	0.8	247	1	0.0	28	0.8	70	73	368	111	0.04	0.29	Tr	0
143	White	1 large	33	88	17	4	0	0.0	0.0	0.0	0	Tr	0.0	2	Tr	48	55	0	0	Tr	0.15	Tr	0
144	Yolk	1 large	17	49	59	3	5	1.6	1.9	0.7	213	Tr	0.0	23	0.6	16	7	323	97	0.03	0.11	Tr	0
	Cooked, whole																						
145	Fried, in margarine, with salt	1 large	46	69	92	6	7	1.9	2.7	1.3	211	1	0.0	25	0.7	61	162	394	114	0.03	0.24	Tr	0
146	Hard cooked, shell removed	1 large	50	75	78	6	5	1.6	2.0	0.7	212	1	0.0	25	0.6	63	62	280	84	0.03	0.26	Tr	0
147		1 cup, chopped	136	75	211	17	14	4.4	5.5	1.9	577	2	0.0	68	1.6	171	169	762	228	0.09	0.70	0.1	0
148	Poached, with salt	1 large	50	75	75	6	5	1.5	1.9	0.7	212	1	0.0	25	0.7	60	140	316	95	0.02	0.22	Tr	0
149	Scrambled, in margarine, with whole milk, salt	1 large	61	73	101	7	7	2.2	2.9	1.3	215	1	0.0	43	0.7	84	171	416	119	0.03	0.27	Tr	Tr
150	Egg substitute, liquid	¼ cup	63	83	53	8	2	0.4	0.6	1.0	1	Tr	0.0	33	1.3	208	112	1,361	136	0.07	0.19	0.1	0
Fats and Oils																							
	Butter (4 sticks per lb)																						
151	Salted	1 stick	113	16	813	1	92	57.3	26.6	3.4	248	Tr	0.0	27	0.2	29	937	3,468	855	0.01	0.04	Tr	0
152		1 tbsp	14	16	102	Tr	12	7.2	3.3	0.4	31	Tr	0.0	3	Tr	4	117	434	107	Tr	Tr	Tr	0
153		1 tsp	5	16	36	Tr	4	2.5	1.2	0.2	11	Tr	0.0	1	Tr	1	41	153	38	Tr	Tr	Tr	0
154	Unsalted	1 stick	113	18	813	1	92	57.3	26.6	3.4	248	Tr	0.0	27	0.2	29	12	3,468	855	0.01	0.04	Tr	0
155	Lard	1 cup	205	0	1,849	0	205	80.4	92.5	23.0	195	0	0.0	Tr	0.0	Tr	Tr	0	0	0.00	0.00	0.0	0
156		1 tbsp	13	0	115	0	13	5.0	5.8	1.4	12	0	0.0	Tr	0.0	Tr	Tr	0	0	0.00	0.00	0.0	0
	Margarine, vitamin A-fortified, salt added																						
	Regular (about 80% fat)																						
157	Hard (4 sticks per lb)	1 stick	113	16	815	1	91	17.9	40.6	28.8	0	1	0.0	34	0.1	48	1,070	4,050	906	0.01	0.04	Tr	Tr
158		1 tbsp	14	16	101	Tr	11	2.2	5.0	3.6	0	Tr	0.0	4	Tr	6	132	500	112	Tr	0.01	Tr	Tr
159		1 tsp	5	16	34	Tr	4	0.7	1.7	1.2	0	Tr	0.0	1	Tr	2	44	168	38	Tr	Tr	Tr	Tr
160	Soft	1 cup	227	16	1,626	2	183	31.3	64.7	78.5	0	1	0.0	60	0.0	86	2,449	8,106	1,814	0.02	0.07	Tr	Tr

(Continues)

TABLE B-8 Food Composition Table

								FATTY ACIDS										VITAMIN A					
FOOD NO.	FOOD DESCRIPTION	MEASURE OF EDIBLE PORTION	WEIGHT (G)	WATER (%)	CALORIES (KCAL)	PROTEIN (G)	TOTAL FAT (G)	SAT (G)	MONO (G)	POLY (G)	CHOLEST (MG)	CARB (G)	TOTAL D F(G)	CA (MG)	FE (MG)	K (MG)	NA (MG)	(IU)	(RE)	THMN (MG)	RIBOFL (MG)	NIACIN (MG)	ASCORBIC ACID (MG)
Fats and Oils (*Continued*)																							
	Margarine, vitamin A-fortified, salt added (*Continued*)																						
161		1 tsp	5	16	34	Tr	4	0.6	1.3	1.6	0	Tr	0.0	1	0.0	2	51	168	38	Tr	Tr	Tr	Tr
	Spread (about 60% fat)																						
162	Hard (4 sticks per lb)	1 stick	115	37	621	1	70	16.2	29.9	20.8	0	0	0.0	24	0.0	34	1,143	4,107	919	0.01	0.03	Tr	Tr
163		1 tbsp	14	37	76	Tr	9	2.0	3.6	2.5	0	0	0.0	3	0.0	4	139	500	112	Tr	Tr	Tr	Tr
164		1 tsp	5	37	26	Tr	3	0.7	1.2	0.9	0	0	0.0	1	0.0	1	48	171	38	Tr	Tr	Tr	Tr
165	Soft	1 cup	229	37	1,236	1	139	29.3	72.1	31.6	0	0	0.0	48	0.0	68	2,276	8,178	1,830	0.02	0.06	Tr	Tr
166		1 tsp	5	37	26	Tr	3	0.6	1.5	0.7	0	0	0.0	1	0.0	1	48	171	38	Tr	Tr	Tr	Tr
167	Spread (about 40% fat)	1 cup	232	58	801	1	90	17.9	36.4	32.0	0	1	0.0	41	0.0	59	2,226	8,285	1,854	0.01	0.05	Tr	Tr
168		1 tsp	5	58	17	Tr	2	0.4	0.8	0.7	0	Tr	0.0	1	0.0	1	46	171	38	Tr	Tr	Tr	Tr
169	Margarine butter blend	1 stick	113	16	811	1	91	32.1	37.0	18.0	99	1	0.0	32	0.1	41	1,014	4,035	903	0.01	0.04	Tr	Tr
170		1 tbsp	14	16	102	Tr	11	4.0	4.7	2.3	12	Tr	0.0	4	Tr	5	127	507	113	Tr	Tr	Tr	Tr
	Oils, salad or cooking																						
171	Canola	1 cup	218	0	1,927	0	218	15.5	128.4	64.5	0	0	0.0	0	0.0	0	0	0	0	0.00	0.00	0.0	0
172		1 tbsp	14	0	124	0	14	1.0	8.2	4.1	0	0	0.0	0	0.0	0	0	0	0	0.00	0.00	0.0	0
173	Corn	1 cup	218	0	1,927	0	218	27.7	52.8	128.0	0	0	0.0	0	0.0	0	0	0	0	0.00	0.00	0.0	0
174		1 tbsp	14	0	120	0	14	1.7	3.3	8.0	0	0	0.0	0	0.0	0	0	0	0	0.00	0.00	0.0	0
175	Olive	1 cup	216	0	1,909	0	216	29.2	159.2	18.1	0	0	0.0	Tr	0.8	0	Tr	0	0	0.00	0.00	0.0	0
176		1 tbsp	14	0	119	0	14	1.8	9.9	1.1	0	0	0.0	Tr	0.1	0	Tr	0	0	0.00	0.00	0.0	0
177	Peanut	1 cup	216	0	1,909	0	216	36.5	99.8	69.1	0	0	0.0	Tr	0.1	Tr	Tr	0	0	0.00	0.00	0.0	0
178		1 tbsp	14	0	119	0	14	2.3	6.2	4.3	0	0	0.0	Tr	Tr	Tr	Tr	0	0	0.00	0.00	0.0	0
179	Safflower, high oleic	1 cup	218	0	1,927	0	218	13.5	162.7	31.3	0	0	0.0	0	0.0	0	0	0	0	0.00	0.00	0.0	0
180		1 tbsp	14	0	120	0	14	0.8	10.2	2.0	0	0	0.0	0	0.0	0	0	0	0	0.00	0.00	0.0	0
181	Sesame	1 cup	218	0	1,927	0	218	31.0	86.5	90.9	0	0	0.0	0	0.0	0	0	0	0	0.00	0.00	0.0	0
182		1 tbsp	14	0	120	0	14	1.9	5.4	5.7	0	0	0.0	0	0.0	0	0	0	0	0.00	0.00	0.0	0
183	Soybean, hydrogenated	1 cup	218	0	1,927	0	218	32.5	93.7	82.0	0	0	0.0	0	0.0	0	0	0	0	0.00	0.00	0.0	0
184		1 tbsp	14	0	120	0	14	2.0	5.8	5.1	0	0	0.0	0	0.0	0	0	0	0	0.00	0.00	0.0	0
185	Soybean, hydrogenated and cottonseed oil blend	1 cup	218	0	1,927	0	218	39.2	64.3	104.9	0	0	0.0	0	0.0	0	0	0	0	0.00	0.00	0.0	0
186		1 tbsp	14	0	120	0	14	2.4	4.0	6.5	0	0	0.0	0	0.0	0	0	0	0	0.00	0.00	0.0	0
187	Sunflower	1 cup	218	0	1,927	0	218	22.5	42.5	143.2	0	0	0.0	0	0.0	0	0	0	0	0.00	0.00	0.0	0

(*Continues*)

TABLE B-8 Food Composition Table

FOOD NO.	FOOD DESCRIPTION	MEASURE OF EDIBLE PORTION	WEIGHT (G)	WATER (%)	CALORIES (KCAL)	PROTEIN (G)	TOTAL FAT (G)	FATTY ACIDS			CHOLEST (MG)	CARB (G)	TOTAL D F(G)	CA (MG)	FE (MG)	K (MG)	NA (MG)	VITAMIN A		THMN (MG)	RIBOFL (MG)	NIACIN (MG)	ASCORBIC ACID (MG)
								SAT (G)	MONO (G)	POLY (G)								(IU)	(RE)				
Fats and Oils (*Continued*)																							
	Oils, salad or cooking (*Continued*)																						
188		1 tbsp	14	0	120	0	14	1.4	2.7	8.9	0	0	0.0	0	0.0	0	0	0	0	0.00	0.00	0.0	0
	Salad dressings																						
	Commercial																						
	Blue cheese																						
189	Regular	1 tbsp	15	32	77	1	8	1.5	1.9	4.3	3	1	0.0	12	Tr	6	167	32	10	Tr	0.02	Tr	Tr
190	Low calorie	1 tbsp	15	80	15	1	1	0.4	0.3	0.4	Tr	Tr	0.0	14	0.1	1	184	2	Tr	Tr	0.02	Tr	Tr
	Caesar																						
191	Regular	1 tbsp	15	34	78	Tr	8	1.3	2.0	4.8	Tr	Tr	Tr	4	Tr	4	158	3	Tr	Tr	Tr	Tr	0
192	Low calorie	1 tbsp	15	73	17	Tr	1	0.1	0.2	0.4	Tr	3	Tr	4	Tr	4	162	3	Tr	Tr	Tr	Tr	0
	French																						
193	Regular	1 tbsp	16	38	67	Tr	6	1.5	1.2	3.4	0	3	0.0	2	0.1	12	214	203	20	Tr	Tr	Tr	0
194	Low calorie	1 tbsp	16	69	22	Tr	1	0.1	0.2	0.6	0	4	0.0	2	0.1	13	128	212	21	0.00	0.00	0.0	0
	Italian																						
195	Regular	1 tbsp	15	38	69	Tr	7	1.0	1.6	4.1	0	1	0.0	1	Tr	2	116	11	4	Tr	Tr	Tr	0
196	Low calorie	1 tbsp	15	82	16	Tr	1	0.2	0.3	0.9	1	1	Tr	Tr	Tr	2	118	0	0	0.00	0.00	0.0	0
	Mayonnaise																						
197	Regular	1 tbsp	14	15	99	Tr	11	1.6	3.1	5.7	8	Tr	0.0	2	0.1	5	78	39	12	0.00	0.00	Tr	0
198	Light, cholesterol free	1 tbsp	15	56	49	Tr	5	0.7	1.1	2.8	0	1	0.0	0	0.0	10	107	18	2	0.00	0.00	0.0	0
199	Fat free	1 tbsp	16	84	12	0	Tr	0.1	0.1	0.2	0	2	0.6	0	0.0	15	190	0	0	0.00	0.00	0.0	0
	Russian																						
200	Regular	1 tbsp	15	35	76	Tr	8	1.1	1.8	4.5	3	2	0.0	3	0.1	24	133	106	32	0.01	0.01	0.1	1
201	Low calorie	1 tbsp	16	65	23	Tr	1	0.1	0.1	0.4	1	4	Tr	3	0.1	26	141	9	3	Tr	Tr	Tr	1
	Thousand island																						
202	Regular	1 tbsp	16	46	59	Tr	6	0.9	1.3	3.1	4	2	0.0	2	0.1	18	109	50	15	Tr	Tr	Tr	0
203	Low calorie	1 tbsp	15	69	24	Tr	2	0.2	0.4	0.9	2	2	0.2	2	0.1	17	153	49	15	Tr	Tr	Tr	0
	Prepared from home recipe																						
204	Cooked, made with margarine	1 tbsp	16	69	25	1	2	0.5	0.6	0.3	9	2	0.0	13	0.1	19	117	66	20	0.01	0.02	Tr	Tr
205	French	1 tbsp	14	24	88	Tr	10	1.8	2.9	4.7	0	Tr	0.0	1	Tr	3	92	72	22	Tr	Tr	Tr	Tr
206	Vinegar and oil	1 tbsp	16	47	70	0	8	1.4	2.3	3.8	0	Tr	0.0	0	0.0	1	Tr	0	0	0.00	0.00	0.0	0

(*Continues*)

TABLE B-8 Food Composition Table

FOOD NO.	FOOD DESCRIPTION	MEASURE OF EDIBLE PORTION	WEIGHT (G)	WATER (%)	CALORIES (KCAL)	PROTEIN (G)	TOTAL FAT (G)	FATTY ACIDS SAT (G)	MONO (G)	POLY (G)	CHOLEST (MG)	CARB (G)	TOTAL D F(G)	CA (MG)	FE (MG)	K (MG)	NA (MG)	VITAMIN A (IU)	(RE)	THMN (MG)	RIBOFL (MG)	NIACIN (MG)	ASCORBIC ACID (MG)
Fats and Oils (*Continued*)																							
	Salad dressings (*Continued*)																						
207	Shortening (hydrogenated soybean and cottonseed oils)	1 cup	205	0	1,812	0	205	51.3	91.2	53.5	0	0	0.0	0	0.0	0	0	0	0	0.00	0.00	0.0	0
208		1 tbsp	13	0	113	0	13	3.2	5.7	3.3	0	0	0.0	0	0.0	0	0	0	0	0.00	0.00	0.0	0
Fish and Shellfish																							
209	Catfish, breaded, fried	3 oz	85	59	195	15	11	2.8	4.8	2.8	69	7	0.6	37	1.2	289	238	24	7	0.06	0.11	1.9	0
	Clam																						
210	Raw, meat only	3 oz	85	82	63	11	1	0.1	0.1	0.2	29	2	0.0	39	11.9	267	48	255	77	0.07	0.18	1.5	11
211		1 medium	15	82	11	2	Tr	Tr	Tr	Tr	5	Tr	0.0	7	2.0	46	8	44	13	0.01	0.03	0.3	2
212	Breaded, fried	¾ cup	115	29	451	13	26	6.6	11.4	6.8	87	39	0.3	21	3.0	266	834	122	37	0.21	0.26	2.9	0
213	Canned, drained solids	3 oz	85	64	126	22	2	0.2	0.1	0.5	57	4	0.0	78	23.8	534	95	485	145	0.13	0.36	2.9	19
214		1 cup	160	64	237	41	3	0.3	0.3	0.9	107	8	0.0	147	44.7	1,005	179	912	274	0.24	0.68	5.4	35
	Cod																						
215	Baked or broiled	3 oz	85	76	89	20	1	0.1	0.1	0.3	40	0	0.0	8	0.3	439	77	27	9	0.02	0.04	2.1	3
216		1 fillet	90	76	95	21	1	0.1	0.1	0.3	42	0	0.0	8	0.3	465	82	29	9	0.02	0.05	2.2	3
217	Canned, solids and liquid	3 oz	85	76	89	19	1	0.1	0.1	0.2	47	0	0.0	18	0.4	449	185	39	12	0.07	0.07	2.1	1
	Crab																						
	Alaska king																						
218	Steamed	1 leg	134	78	130	26	2	0.2	0.2	0.7	71	0	0.0	79	1.0	351	1,436	39	12	0.07	0.07	1.8	10
219		3 oz	85	78	82	16	1	0.1	0.2	0.5	45	0	0.0	50	0.6	223	911	25	8	0.05	0.05	1.1	6
220	Imitation, from surimi	3 oz	85	74	87	10	1	0.2	0.2	0.6	17	9	0.0	11	0.3	77	715	56	17	0.03	0.02	0.2	0
	Blue																						
221	Steamed	3 oz	85	77	87	17	2	0.2	0.2	0.6	85	0	0.0	88	0.8	275	237	5	2	0.09	0.04	2.8	3
222	Canned crabmeat	1 cup	135	76	134	28	2	0.3	0.3	0.6	120	0	0.0	136	1.1	505	450	7	3	0.11	0.11	1.8	4
223	Crab cake, with egg, onion, fried in margarine	1 cake	60	71	93	12	5	0.9	1.7	1.4	90	Tr	0.0	63	0.6	194	198	151	49	0.05	0.05	1.7	2
224	Fish fillet, battered or breaded, fried	1 fillet	91	54	211	13	11	2.6	2.3	5.7	31	15	0.5	16	1.9	291	484	35	11	0.10	0.10	1.9	0
225	Fish stick and portion, breaded, frozen, reheated	1 stick (4" × 1" × ½")	28	46	76	4	3	0.9	1.4	0.9	31	7	0.0	6	0.2	73	163	30	9	0.04	0.05	0.6	0

(*Continues*)

TABLE B-8 Food Composition Table

FOOD NO.	FOOD DESCRIPTION	MEASURE OF EDIBLE PORTION	WEIGHT (G)	WATER (%)	CALORIES (KCAL)	PROTEIN (G)	TOTAL FAT (G)	FATTY ACIDS SAT (G)	MONO (G)	POLY (G)	CHOLEST (MG)	CARB (G)	TOTAL D F(G)	CA (MG)	FE (MG)	K (MG)	NA (MG)	VITAMIN A (IU)	(RE)	THMN (MG)	RIBOFL (MG)	NIACIN (MG)	ASCORBIC ACID (MG)
Fish and Shellfish (*Continued*)																							
	Fish stick and portion, breaded, frozen, reheated (*Continued*)																						
226		1 portion (4" × 2" × ½")	57	46	155	9	7	1.8	2.9	1.8	64	14	0.0	11	0.4	149	332	60	18	0.07	0.10	1.2	0
227	Flounder or sole, baked or broiled	3 oz	85	73	99	21	1	0.3	0.2	0.5	58	0	0.0	15	0.3	292	89	32	9	0.07	0.10	1.9	0
228		1 fillet	127	73	149	31	2	0.5	0.3	0.8	86	0	0.0	23	0.4	437	133	48	14	0.10	0.14	2.8	0
229	Haddock, baked or broiled	3 oz	85	74	95	21	1	0.1	0.1	0.3	63	0	0.0	36	1.1	339	74	54	16	0.03	0.04	3.9	0
230		1 fillet	150	74	168	36	1	0.3	0.2	0.5	111	0	0.0	63	2.0	599	131	95	29	0.06	0.07	6.9	0
231	Halibut, baked or broiled	3 oz	85	72	119	23	2	0.4	0.8	0.8	35	0	0.0	51	0.9	490	59	152	46	0.06	0.08	6.1	0
232		½ fillet	159	72	223	42	5	0.7	1.5	1.5	65	0	0.0	95	1.7	916	110	285	86	0.11	0.14	11.3	0
233	Herring, pickled	3 oz	85	55	223	12	15	2.0	10.2	1.4	11	8	0.0	65	1.0	59	740	732	219	0.03	0.12	2.8	0
234	Lobster, steamed	3 oz	85	76	83	17	1	0.1	0.1	0.1	61	1	0.0	52	0.3	299	323	74	22	0.01	0.06	0.9	0
235	Ocean perch, baked or broiled	3 oz	85	73	103	20	2	0.3	0.7	0.5	46	0	0.0	116	1.0	298	82	39	12	0.11	0.11	2.1	1
236		1 fillet	50	73	61	12	1	0.2	0.4	0.3	27	0	0.0	69	0.6	175	48	23	7	0.07	0.07	1.2	Tr
	Oyster																						
237	Raw, meat only	1 cup	248	85	169	17	6	1.9	0.8	2.4	131	10	0.0	112	16.5	387	523	248	74	0.25	0.24	3.4	9
238		6 medium	84	85	57	6	2	0.6	0.3	0.8	45	3	0.0	38	5.6	131	177	84	25	0.08	0.08	1.2	3
239	Breaded, fried	3 oz	85	65	167	7	11	2.7	4.0	2.8	69	10	0.2	53	5.9	207	354	257	77	0.13	0.17	1.4	3
240	Pollock, baked or broiled	3 oz	85	74	96	20	1	0.2	0.1	0.4	82	0	0.0	5	0.2	329	99	65	20	0.06	0.06	1.4	0
241		1 fillet	60	74	68	14	1	0.1	0.1	0.3	58	0	0.0	4	0.2	232	70	46	14	0.04	0.05	1.0	0
242	Rockfish, baked or broiled	3 oz	85	73	103	20	2	0.4	0.4	0.5	37	0	0.0	10	0.5	442	65	186	56	0.04	0.07	3.3	0
243		1 fillet	149	73	180	36	3	0.7	0.7	0.9	66	0	0.0	18	0.8	775	115	326	98	0.07	0.13	5.8	0
244	Roughy, orange, baked or broiled	3 oz	85	69	76	16	1	Tr	0.5	Tr	22	0	0.0	32	0.2	327	69	69	20	0.10	0.16	3.1	0
	Salmon																						
245	Baked or broiled (red)	3 oz	85	62	184	23	9	1.6	4.5	2.0	74	0	0.0	6	0.5	319	56	178	54	0.18	0.15	5.7	0
246		½ fillet	155	62	335	42	17	3.0	8.2	3.7	135	0	0.0	11	0.9	581	102	324	98	0.33	0.27	10.3	0
247	Canned (pink), solids and liquid (includes bones)	3 oz	85	69	118	17	5	1.3	1.5	1.7	47	0	0.0	181	0.7	277	471	47	14	0.02	0.16	5.6	0
248	Smoked (chinook)	3 oz	85	72	99	16	4	0.8	1.7	0.8	20	0	0.0	9	0.7	149	666	75	22	0.02	0.09	4.0	0

(*Continues*)

TABLE B-8 Food Composition Table

FOOD NO.	FOOD DESCRIPTION	MEASURE OF EDIBLE PORTION	WEIGHT (G)	WATER (%)	CALORIES (KCAL)	PROTEIN (G)	TOTAL FAT (G)	FATTY ACIDS SAT (G)	MONO (G)	POLY (G)	CHOLEST (MG)	CARB (G)	TOTAL D F(G)	CA (MG)	FE (MG)	K (MG)	NA (MG)	VITAMIN A (IU)	(RE)	THMN (MG)	RIBOFL (MG)	NIACIN (MG)	ASCORBIC ACID (MG)
Fish and Shellfish (*Continued*)																							
	Salmon (*Continued*)																						
249	Sardine, Atlantic, canned in oil, drained solids (includes bones)	3 oz	85	60	177	21	10	1.3	3.3	4.4	121	0	0.0	325	2.5	337	429	190	57	0.07	0.19	4.5	0
	Scallop, cooked																						
250	Breaded, fried	6 large	93	58	200	17	10	2.5	4.2	2.7	57	9	0.2	39	0.8	310	432	70	20	0.04	0.10	1.4	2
251	Steamed	3 oz	85	73	95	20	1	0.1	0.1	0.4	45	3	0.0	98	2.6	405	225	85	26	0.09	0.05	1.1	0
	Shrimp																						
252	Breaded, fried	3 oz	85	53	206	18	10	1.8	3.2	4.3	150	10	0.3	57	1.1	191	292	161	48	0.11	0.12	2.6	1
253		6 large	45	53	109	10	6	0.9	1.7	2.3	80	5	0.2	30	0.6	101	155	85	25	0.06	0.06	1.4	1
254	Canned, drained solids	3 oz	85	73	102	20	2	0.3	0.2	0.6	147	1	0.0	50	2.3	179	144	51	15	0.02	0.03	2.3	2
255	Swordfish, baked or broiled	3 oz	85	69	132	22	4	1.2	1.7	1.0	43	0	0.0	5	0.9	314	98	116	35	0.04	0.10	10.0	1
256		1 piece	106	69	164	27	5	1.5	2.1	1.3	53	0	0.0	6	1.1	391	122	145	43	0.05	0.12	12.5	1
257	Trout, baked or broiled	3 oz	85	68	144	21	6	1.8	1.8	2.0	58	0	0.0	73	0.3	375	36	244	73	0.20	0.07	7.5	3
258		1 fillet	71	68	120	17	5	1.5	1.5	1.7	48	0	0.0	61	0.2	313	30	204	61	0.17	0.06	6.2	2
	Tuna																						
259	Baked or broiled	3 oz	85	63	118	25	1	0.3	0.2	0.3	49	0	0.0	18	0.8	484	40	58	17	0.43	0.05	10.1	1
	Canned, drained solids																						
260	Oil pack, chunk light	3 oz	85	60	168	25	7	1.3	2.5	2.5	15	0	0.0	11	1.2	176	301	66	20	0.03	0.10	10.5	0
261	Water pack, chunk light	3 oz	85	75	99	22	1	0.2	0.1	0.3	26	0	0.0	9	1.3	201	287	48	14	0.03	0.06	11.3	0
262	Water pack, solid white	3 oz	85	73	109	20	3	0.7	0.7	0.9	36	0	0.0	12	0.8	201	320	16	5	0.01	0.04	4.9	0
263	Tuna salad: light tuna in oil, pickle relish, mayo type salad dressing	1 cup	205	63	383	33	19	3.2	5.9	8.5	27	19	0.0	35	2.1	365	824	199	55	0.06	0.14	13.7	5
Fruits and Fruit Juices																							
	Apples																						
	Raw																						
264	Unpeeled, 2¾" dia (about 3 per lb)	1 apple	138	84	81	Tr	Tr	0.1	Tr	0.1	0	21	3.7	10	0.2	159	0	73	7	0.02	0.02	0.1	8
265	Peeled, sliced	1 cup	110	84	63	Tr	Tr	0.1	Tr	0.1	0	16	2.1	4	0.1	124	0	48	4	0.02	0.01	0.1	4
266	Dried (sodium bisulfite used to preserve color)	5 rings	32	32	78	Tr	Tr	Tr	Tr	Tr	0	21	2.8	4	0.4	144	28	0	0	0.00	0.05	0.3	1

(*Continues*)

TABLE B-8 Food Composition Table

								FATTY ACIDS										VITAMIN A					
FOOD NO.	FOOD DESCRIPTION	MEASURE OF EDIBLE PORTION	WEIGHT (G)	WATER (%)	CALORIES (KCAL)	PROTEIN (G)	TOTAL FAT (G)	SAT (G)	MONO (G)	POLY (G)	CHOLEST (MG)	CARB (G)	TOTAL D F(G)	CA (MG)	FE (MG)	K (MG)	NA (MG)	(IU)	(RE)	THMN (MG)	RIBOFL (MG)	NIACIN (MG)	ASCORBIC ACID (MG)
Fruits and Fruit Juices (*Continued*)																							
	Apples (*Continued*)																						
267	Apple juice, bottled or canned	1 cup	248	88	117	Tr	Tr	Tr	Tr	0.1	0	29	0.2	17	0.9	295	7	2	0	0.05	0.04	0.2	2
268	Apple pie filling, canned	1/8 of 21-oz can	74	73	75	Tr	Tr	Tr	0.0	Tr	0	19	0.7	3	0.2	33	33	10	1	0.01	0.01	Tr	1
	Applesauce, canned																						
269	Sweetened	1 cup	255	80	194	Tr	Tr	0.1	Tr	0.1	0	51	3.1	10	0.9	156	8	28	3	0.03	0.07	0.5	4
270	Unsweetened	1 cup	244	88	105	Tr	Tr	Tr	Tr	Tr	0	28	2.9	7	0.3	183	5	71	7	0.03	0.06	0.5	3
	Apricots																						
271	Raw, without pits (about 12 per lb with pits)	1 apricot	35	86	17	Tr	Tr	Tr	0.1	Tr	0	4	0.8	5	0.2	104	Tr	914	91	0.01	0.01	0.2	4
	Canned, halves, fruit and liquid																						
272	Heavy syrup pack	1 cup	258	78	214	1	Tr	Tr	0.1	Tr	0	55	4.1	23	0.8	361	10	3,173	317	0.05	0.06	1.0	8
273	Juice pack	1 cup	244	87	117	2	Tr	Tr	Tr	Tr	0	30	3.9	29	0.7	403	10	4,126	412	0.04	0.05	0.8	12
274	Dried, sulfured	10 halves	35	31	83	1	Tr	Tr	0.1	Tr	0	22	3.2	16	1.6	482	4	2,534	253	Tr	0.05	1.0	1
275	Apricot nectar, canned, with added ascorbic acid	1 cup	251	85	141	1	Tr	Tr	0.1	Tr	0	36	1.5	18	1.0	286	8	3,303	331	0.02	0.04	0.7	137
	Asian pear, raw																						
276	2¼" high × 2½" dia	1 pear	122	88	51	1	Tr	Tr	0.1	0.1	0	13	4.4	5	0.0	148	0	0	0	0.01	0.01	0.3	5
277	3 3/8" high × 3" dia	1 pear	275	88	116	1	1	Tr	0.1	0.2	0	29	9.9	11	0.0	333	0	0	0	0.02	0.03	0.6	10
	Avocados, raw, without skin and seed																						
278	California (about 1/5 whole)	1 oz	28	73	50	1	5	0.7	3.2	0.6	0	2	1.4	3	0.3	180	3	174	17	0.03	0.03	0.5	2
279	Florida (about 1/10 whole)	1 oz	28	80	32	Tr	3	0.5	1.4	0.4	0	3	1.5	3	0.2	138	1	174	17	0.03	0.03	0.5	2
	Bananas, raw																						
280	Whole, medium (7" to 7 7/8" long)	1 banana	118	74	109	1	1	0.2	Tr	0.1	0	28	2.8	7	0.4	467	1	96	9	0.05	0.12	0.6	11
281	Sliced	1 cup	150	74	138	2	1	0.3	0.1	0.1	0	35	3.6	9	0.5	594	2	122	12	0.07	0.15	0.8	14
282	Blackberries, raw	1 cup	144	86	75	1	1	Tr	0.1	0.3	0	18	7.6	46	0.8	282	0	238	23	0.04	0.06	0.6	30
	Blueberries																						
283	Raw	1 cup	145	85	81	1	1	Tr	0.1	0.2	0	20	3.9	9	0.2	129	9	145	15	0.07	0.07	0.5	19
284	Frozen, sweetened, thawed	1 cup	230	77	186	1	Tr	Tr	Tr	0.1	0	50	4.8	14	0.9	138	2	101	9	0.05	0.12	0.6	2

(*Continues*)

TABLE B-8 Food Composition Table

FOOD NO.	FOOD DESCRIPTION	MEASURE OF EDIBLE PORTION	WEIGHT (G)	WATER (%)	CALORIES (KCAL)	PROTEIN (G)	TOTAL FAT (G)	FATTY ACIDS SAT (G)	MONO (G)	POLY (G)	CHOLEST (MG)	CARB (G)	TOTAL D F(G)	CA (MG)	FE (MG)	K (MG)	NA (MG)	VITAMIN A (IU)	(RE)	THMN (MG)	RIBOFL (MG)	NIACIN (MG)	ASCORBIC ACID (MG)
Fruits and Fruit Juices (*Continued*)																							
	Cantaloupe. See Melons.																						
	Carambola (starfruit), raw																						
285	Whole (3⁵/₈" long)	1 fruit	91	91	30	Tr	Tr	Tr	Tr	0.2	0	7	2.5	4	0.2	148	2	449	45	0.03	0.02	0.4	19
286	Sliced	1 cup	108	91	36	1	Tr	Tr	Tr	0.2	0	8	2.9	4	0.3	176	2	532	53	0.03	0.03	0.4	23
	Cherries																						
287	Sour, red, pitted, canned, water pack	1 cup	244	90	88	2	Tr	0.1	0.1	0.1	0	22	2.7	27	3.3	239	17	1,840	183	0.04	0.10	0.4	5
288	Sweet, raw, without pits and stems	10 cherries	68	81	49	1	1	0.1	0.2	0.2	0	11	1.6	10	0.3	152	0	146	14	0.03	0.04	0.3	5
289	Cherry pie filling, canned	¹/₅ of 21-oz can	74	71	85	Tr	Tr	Tr	Tr	Tr	0	21	0.4	8	0.2	78	13	152	16	0.02	0.01	0.1	3
290	Cranberries, dried, sweetened	¼ cup	28	12	92	Tr	Tr	Tr	Tr	0.1	0	24	2.5	5	0.1	24	1	0	0	0.01	0.03	Tr	Tr
291	Cranberry sauce, sweetened, canned (about 8 slices per can)	1 slice	57	61	86	Tr	Tr	Tr	Tr	Tr	0	22	0.6	2	0.1	15	17	11	1	0.01	0.01	0.1	1
	Dates, without pits																						
292	Whole	5 dates	42	23	116	1	Tr	0.1	0.1	Tr	0	31	3.2	13	0.5	274	1	21	2	0.04	0.04	0.9	0
293	Chopped	1 cup	178	23	490	4	1	0.3	0.3	0.1	0	131	13.4	57	2.0	1,161	5	89	9	0.16	0.18	3.9	0
294	Figs, dried	2 figs	38	28	97	1	Tr	0.1	0.1	0.2	0	25	4.6	55	0.8	271	4	51	5	0.03	0.03	0.3	Tr
	Fruit cocktail, canned, fruit and liquid																						
295	Heavy syrup pack	1 cup	248	80	181	1	Tr	Tr	Tr	0.1	0	47	2.5	15	0.7	218	15	508	50	0.04	0.05	0.9	5
296	Juice pack	1 cup	237	87	109	1	Tr	Tr	Tr	Tr	0	28	2.4	19	0.5	225	9	723	73	0.03	0.04	1.0	6
	Grapefruit																						
	Raw, without peel, membrane and seeds (3¾" dia)																						
297	Pink or red	½ grapefruit	123	91	37	1	Tr	Tr	Tr	Tr	0	9	1.4	14	0.1	159	0	319	32	0.04	0.02	0.2	47
298	White	½ grapefruit	118	90	39	1	Tr	Tr	Tr	Tr	0	10	1.3	14	0.1	175	0	12	1	0.04	0.02	0.3	39
299	Canned, sections with light syrup	1 cup	254	84	152	1	Tr	Tr	Tr	0.1	0	39	1.0	36	1.0	328	5	0	0	0.10	0.05	0.6	54

(*Continues*)

TABLE B-8 Food Composition Table

FOOD NO.	FOOD DESCRIPTION	MEASURE OF EDIBLE PORTION	WEIGHT (G)	WATER (%)	CALORIES (KCAL)	PROTEIN (G)	TOTAL FAT (G)	FATTY ACIDS SAT (G)	FATTY ACIDS MONO (G)	FATTY ACIDS POLY (G)	CHOLEST (MG)	CARB (G)	TOTAL D F(G)	CA (MG)	FE (MG)	K (MG)	NA (MG)	VITAMIN A (IU)	VITAMIN A (RE)	THMN (MG)	RIBOFL (MG)	NIACIN (MG)	ASCORBIC ACID (MG)
	Fruits and Fruit Juices (*Continued*)																						
	Grapefruit (*Continued*)																						
	Grapefruit juice																						
	Raw																						
300	Pink	1 cup	247	90	96	1	Tr	Tr	Tr	0.1	0	23	0.2	22	0.5	400	2	1,087	109	0.10	0.05	0.5	94
301	White	1 cup	247	90	96	1	Tr	Tr	Tr	0.1	0	23	0.2	22	0.5	400	2	25	2	0.10	0.05	0.5	94
	Canned																						
302	Unsweetened	1 cup	247	90	94	1	Tr	Tr	Tr	0.1	0	22	0.2	17	0.5	378	2	17	2	0.10	0.05	0.6	72
303	Sweetened	1 cup	250	87	115	1	Tr	Tr	Tr	0.1	0	28	0.3	20	0.9	405	5	0	0	0.10	0.06	0.8	67
	Frozen concentrate, unsweetened																						
304	Undiluted	6-fl-oz can	207	62	302	4	1	0.1	0.1	0.2	0	72	0.8	56	1.0	1,002	6	64	6	0.30	0.16	1.6	248
305	Diluted with three parts water by volume	1 cup	247	89	101	1	Tr	Tr	Tr	0.1	0	24	0.2	20	0.3	336	2	22	2	0.10	0.05	0.5	83
306	Grapes, seedless, raw	10 grapes	50	81	36	Tr	Tr	0.1	Tr	0.1	0	9	0.5	6	0.1	93	1	37	4	0.05	0.03	0.2	5
307		1 cup	160	81	114	1	1	0.3	Tr	0.3	0	28	1.6	18	0.4	296	3	117	11	0.15	0.09	0.5	17
	Grape juice																						
308	Canned or bottled	1 cup	253	84	154	1	Tr	0.1	Tr	0.1	0	38	0.3	23	0.6	334	8	20	3	0.07	0.09	0.7	Tr
	Frozen concentrate, sweetened, with added vitamin C																						
309	Undiluted	6-fl-oz can	216	54	387	1	1	0.2	Tr	0.2	0	96	0.6	28	0.8	160	15	58	6	0.11	0.20	0.9	179
310	Diluted with three parts water by volume	1 cup	250	87	128	Tr	Tr	0.1	Tr	0.1	0	32	0.3	10	0.3	53	5	20	3	0.04	0.07	0.3	60
311	Kiwi fruit, raw, without skin (about 5 per lb with skin)	1 medium	76	83	46	1	Tr	Tr	Tr	0.2	0	11	2.6	20	0.3	252	4	133	14	0.02	0.04	0.4	74
312	Lemons, raw, without peel (2 1/8" dia with peel)	1 lemon	58	89	17	1	Tr	Tr	Tr	0.1	0	5	1.6	15	0.3	80	1	17	2	0.02	0.01	0.1	31
	Lemon juice																						
313	Raw (from 2 1/8"-dia lemon) juice of	1 lemon	47	91	12	Tr	0	0.0	0.0	0.0	0	4	0.2	3	Tr	58	Tr	9	1	0.01	Tr	Tr	22
314	Canned or bottled, unsweetened	1 cup	244	92	51	1	1	0.1	Tr	0.2	0	16	1.0	27	0.3	249	51*	37	5	0.10	0.02	0.5	61
315		1 tbsp	15	92	3	Tr	Tr	Tr	Tr	Tr	0	1	0.1	2	Tr	16	3*	2	Tr	0.01	Tr	Tr	4

(*Continues*)

TABLE B-8 Food Composition Table

								Fatty Acids										Vitamin A					
Food No.	Food Description	Measure of Edible Portion	Weight (g)	Water (%)	Calories (kcal)	Protein (g)	Total Fat (g)	Sat (g)	Mono (g)	Poly (g)	Cholest (mg)	Carb (g)	Total D F(g)	Ca (mg)	Fe (mg)	K (mg)	Na (mg)	(IU)	(RE)	Thmn (mg)	Ribofl (mg)	Niacin (mg)	Ascorbic Acid (mg)
Fruits and Fruit Juices (*Continued*)																							
	Lime juice																						
316	Raw (from 2"-dia lime) juice of	1 lime	38	90	10	Tr	Tr	Tr	Tr	Tr	0	3	0.2	3	Tr	41	Tr	4	Tr	0.01	Tr	Tr	11
317	Canned, unsweetened	1 cup	246	93	52	1	1	0.1	0.1	0.2	0	16	1.0	30	0.6	185	39*	39	5	0.08	0.01	0.4	16
318		1 tbsp	15	93	3	Tr	Tr	Tr	Tr	Tr	0	1	0.1	2	Tr	11	2*	2	Tr	Tr	Tr	Tr	1
	Mangos, raw, without skin and seed (about 1½ per lb with skin and seed)																						
319	Whole	1 mango	207	82	135	1	1	0.1	0.2	0.1	0	35	3.7	21	0.3	323	4	8,061	805	0.12	0.12	1.2	57
320	Sliced	1 cup	165	82	107	1	Tr	0.1	0.2	0.1	0	28	3.0	17	0.2	257	3	6,425	642	0.10	0.09	1.0	46
	Melons, raw, without rind and cavity contents																						
	Cantaloupe (5" dia)																						
321	Wedge	1/8 melon	69	90	24	1	Tr	Tr	Tr	0.1	0	6	0.6	8	0.1	213	6	2,225	222	0.02	0.01	0.4	29
322	Cubes	1 cup	160	90	56	1	Tr	0.1	Tr	0.2	0	13	1.3	18	0.3	494	14	5,158	515	0.06	0.03	0.9	68
	Honeydew (6"-7" dia)																						
323	Wedge	1/8 melon	160	90	56	1	Tr	Tr	Tr	0.1	0	15	1.0	10	0.1	434	16	64	6	0.12	0.03	1.0	40
324	Diced (about 20 pieces per cup)	1 cup	170	90	60	1	Tr	Tr	Tr	0.1	0	16	1.0	10	0.1	461	17	68	7	0.13	0.03	1.0	42
325	Mixed fruit, frozen, sweetened, thawed (peach, cherry, raspberry, grape and boysenberry	1 cup	250	74	245	4	Tr	0.1	0.1	0.2	0	61	4.8	18	0.7	328	8	805	80	0.04	0.09	1.0	188
326	Nectarines, raw (2½" dia)	1 nectarine	136	86	67	1	1	0.1	0.2	0.3	0	16	2.2	7	0.2	288	0	1,001	101	0.02	0.06	1.3	7
	Oranges, raw																						
327	Whole, without peel and seeds (2⅝" dia)	1 orange	131	87	62	1	Tr	Tr	Tr	Tr	0	15	3.1	52	0.1	237	0	269	28	0.11	0.05	0.4	70
328	Sections without membranes	1 cup	180	87	85	2	Tr	Tr	Tr	Tr	0	21	4.3	72	0.2	326	0	369	38	0.16	0.07	0.5	96
	Orange juice																						
329	Raw, all varieties	1 cup	248	88	112	2	Tr	0.1	0.1	0.1	0	26	0.5	27	0.5	496	2	496	50	0.22	0.07	1.0	124
330		juice from 1 orange	86	88	39	1	Tr	Tr	Tr	Tr	0	9	0.2	9	0.2	172	1	172	17	0.08	0.03	0.3	43
331	Canned, unsweetened	1 cup	249	89	105	1	Tr	Tr	0.1	0.1	0	25	0.5	20	1.1	436	5	436	45	0.15	0.07	0.8	86

*Sodium benzoate and sodium bisulfite added as preservatives.

(Continues)

TABLE B-8 Food Composition Table

FOOD NO.	FOOD DESCRIPTION	MEASURE OF EDIBLE PORTION	WEIGHT (G)	WATER (%)	CALORIES (KCAL)	PROTEIN (G)	TOTAL FAT (G)	FATTY ACIDS			CHOLEST (MG)	CARB (G)	TOTAL D F(G)	CA (MG)	FE (MG)	K (MG)	NA (MG)	VITAMIN A		THMN (MG)	RIBOFL (MG)	NIACIN (MG)	ASCORBIC ACID (MG)
								SAT (G)	MONO (G)	POLY (G)								(IU)	(RE)				
	Fruits and Fruit Juices (*Continued*)																						
	Orange juice (*Continued*)																						
332	Chilled (refrigerator case)	1 cup	249	88	110	2	1	0.1	0.1	0.2	0	25	0.5	25	0.4	473	2	194	20	0.28	0.05	0.7	82
	Frozen concentrate																						
333	Undiluted	6-fl-oz can	213	58	339	5	Tr	0.1	0.1	0.1	0	81	1.7	68	0.7	1,436	6	588	60	0.60	0.14	1.5	294
334	Diluted with 3 parts water by volume	1 cup	249	88	112	2	Tr	Tr	Tr	Tr	0	27	0.5	22	0.2	473	2	194	20	0.20	0.04	0.5	97
	Papayas, raw																						
335	½" cubes	1 cup	140	89	55	1	Tr	0.1	0.1	Tr	0	14	2.5	34	0.1	360	4	398	39	0.04	0.04	0.5	87
336	Whole (5⅛" long × 3" dia)	1 papaya	304	89	119	2	Tr	0.1	0.1	0.1	0	30	5.5	73	0.3	781	9	863	85	0.08	0.10	1.0	188
	Peaches																						
	Raw																						
337	Whole, 2½" dia, pitted (about 4 per lb)	1 peach	98	88	42	1	Tr	Tr	Tr	Tr	0	11	2.0	5	0.1	193	0	524	53	0.02	0.04	1.0	6
338	Sliced	1 cup	170	88	73	1	Tr	Tr	0.1	0.1	0	19	3.4	9	0.2	335	0	910	92	0.03	0.07	1.7	11
	Canned, fruit and liquid																						
339	Heavy syrup pack	1 cup	262	79	194	1	Tr	Tr	0.1	0.1	0	52	3.4	8	0.7	241	16	870	86	0.03	0.06	1.6	7
340		1 half	98	79	73	Tr	Tr	Tr	Tr	Tr	0	20	1.3	3	0.3	90	6	325	32	0.01	0.02	0.6	3
341	Juice pack	1 cup	248	87	109	2	Tr	Tr	Tr	Tr	0	29	3.2	15	0.7	317	10	945	94	0.02	0.04	1.4	9
342		1 half	98	87	43	1	Tr	Tr	Tr	Tr	0	11	1.3	6	0.3	125	4	373	37	0.01	0.02	0.6	4
343	Dried, sulfured	3 halves	39	32	93	1	Tr	Tr	0.1	0.1	0	24	3.2	11	1.6	388	3	844	84	Tr	0.08	1.7	2
344	Frozen, sliced, sweetened, with added ascorbic acid, thawed	1 cup	250	75	235	2	Tr	Tr	0.1	0.2	0	60	4.5	8	0.9	325	15	710	70	0.03	0.09	1.6	236
	Pears																						
345	Raw, with skin, cored, 2½" dia	1 pear	166	84	98	1	1	Tr	0.1	0.2	0	25	4.0	18	0.4	208	0	33	3	0.03	0.07	0.2	7
	Canned, fruit and liquid																						
346	Heavy syrup pack	1 cup	266	80	197	1	Tr	Tr	0.1	0.1	0	51	4.3	13	0.6	173	13	0	0	0.03	0.06	0.6	3
347		1 half	76	80	56	Tr	Tr	Tr	Tr	Tr	0	15	1.2	4	0.2	49	4	0	0	0.01	0.02	0.2	1
348	Juice pack	1 cup	248	86	124	1	Tr	Tr	Tr	Tr	0	32	4.0	22	0.7	238	10	15	2	0.03	0.03	0.5	4
349		1 half	76	86	38	Tr	Tr	Tr	Tr	Tr	0	10	1.2	7	0.2	73	3	5	1	0.01	0.01	0.2	1
	Pineapple																						
350	Raw, diced	1 cup	155	87	76	1	1	Tr	0.1	0.2	0	19	1.9	11	0.6	175	2	36	3	0.14	0.06	0.7	24

(*Continues*)

TABLE B-8 Food Composition Table

FOOD NO.	FOOD DESCRIPTION	MEASURE OF EDIBLE PORTION	WEIGHT (G)	WATER (%)	CALORIES (KCAL)	PROTEIN (G)	TOTAL FAT (G)	FATTY ACIDS			CHOLEST (MG)	CARB (G)	TOTAL D F(G)	CA (MG)	FE (MG)	K (MG)	NA (MG)	VITAMIN A		THMN (MG)	RIBOFL (MG)	NIACIN (MG)	ASCORBIC ACID (MG)
								SAT (G)	MONO (G)	POLY (G)								(IU)	(RE)				
Fruits and Fruit Juices (*Continued*)																							
	Pineapple (*Continued*)																						
	Canned, fruit and liquid																						
	Heavy syrup pack																						
351	Crushed, sliced, or chunks	1 cup	254	79	198	1	Tr	Tr	Tr	0.1	0	51	2.0	36	1.0	264	3	36	3	0.23	0.06	0.7	19
352	Slices (3" dia)	1 slice	49	79	38	Tr	Tr	Tr	Tr	Tr	0	10	0.4	7	0.2	51	Tr	7	Tr	0.04	0.01	0.1	4
	Juice pack																						
353	Crushed, sliced, or chunks	1 cup	249	84	149	1	Tr	Tr	Tr	0.1	0	39	2.0	35	0.7	304	2	95	10	0.24	0.05	0.7	24
354	Slice (3" dia)	1 slice	47	84	28	Tr	Tr	Tr	Tr	Tr	0	7	0.4	7	0.1	57	Tr	18	2	0.04	0.01	0.1	4
355	Pineapple juice, unsweetened, canned	1 cup	250	86	140	1	Tr	Tr	Tr	0.1	0	34	0.5	43	0.7	335	3	13	0	0.14	0.06	0.6	27
	Plantain, without peel																						
356	Raw	1 medium	179	65	218	2	1	0.3	0.1	0.1	0	57	4.1	5	1.1	893	7	2,017	202	0.09	0.10	1.2	33
357	Cooked, slices	1 cup	154	67	179	1	Tr	0.1	Tr	0.1	0	48	3.5	3	0.9	716	8	1,400	140	0.07	0.08	1.2	17
	Plums																						
358	Raw (2$^{1}/_{8}$" dia)	1 plum	66	85	36	1	Tr	Tr	0.3	0.1	0	9	1.0	3	0.1	114	0	213	21	0.03	0.06	0.3	6
	Canned, purple, fruit and liquid																						
359	Heavy syrup pack	1 cup	258	76	230	1	Tr	Tr	0.2	0.1	0	60	2.6	23	2.2	235	49	668	67	0.04	0.10	0.8	1
360		1 plum	46	76	41	Tr	Tr	Tr	Tr	Tr	0	11	0.5	4	0.4	42	9	119	12	0.01	0.02	0.1	Tr
361	Juice pack	1 cup	252	84	146	1	Tr	Tr	Tr	Tr	0	38	2.5	25	0.9	388	3	2,543	255	0.06	0.15	1.2	7
362		1 plum	46	84	27	Tr	Tr	Tr	Tr	Tr	0	7	0.5	5	0.2	71	Tr	464	46	0.01	0.03	0.2	1
	Prunes, dried, pitted																						
363	Uncooked	5 prunes	42	32	100	1	Tr	Tr	0.1	Tr	0	26	3.0	21	1.0	313	2	835	84	0.03	0.07	0.8	1
364	Stewed, unsweetened, fruit and liquid	1 cup	248	70	265	3	1	Tr	0.4	0.1	0	70	16.4	57	2.8	828	5	759	77	0.06	0.25	1.8	7
365	Prune juice, canned or bottled	1 cup	256	81	182	2	Tr	Tr	0.1	Tr	0	45	2.6	31	3.0	707	10	8	0	0.04	0.18	2.0	10
	Raisins, seedless																						
366	Cup, not packed	1 cup	145	15	435	5	1	0.2	Tr	0.2	0	115	5.8	71	3.0	1,089	17	12	1	0.23	0.13	1.2	5
367	Packet, ½ oz (1½ tbsp)	1 packet	14	15	42	Tr	Tr	Tr	Tr	Tr	0	11	0.6	7	0.3	105	2	1	Tr	0.02	0.01	0.1	Tr
	Raspberries																						
368	Raw	1 cup	123	87	60	1	1	Tr	0.1	0.4	0	14	8.4	27	0.7	187	0	160	16	0.04	0.11	1.1	31
369	Frozen, sweetened, thawed	1 cup	250	73	258	2	Tr	Tr	Tr	0.2	0	65	11.0	38	1.6	285	3	150	15	0.05	0.11	0.6	41

(*Continues*)

TABLE B-8 Food Composition Table

FOOD NO.	FOOD DESCRIPTION	MEASURE OF EDIBLE PORTION	WEIGHT (G)	WATER (%)	CALORIES (KCAL)	PROTEIN (G)	TOTAL FAT (G)	FATTY ACIDS SAT (G)	MONO (G)	POLY (G)	CHOLEST (MG)	CARB (G)	TOTAL D F(G)	CA (MG)	FE (MG)	K (MG)	NA (MG)	VITAMIN A (IU)	(RE)	THMN (MG)	RIBOFL (MG)	NIACIN (MG)	ASCORBIC ACID (MG)
Fruits and Fruit Juices (*Continued*)																							
	Raspberries (*Continued*)																						
370	Rhubarb, frozen, cooked, with sugar	1 cup	240	68	278	1	Tr	Tr	Tr	0.1	0	75	4.8	348	0.5	230	2	166	17	0.04	0.06	0.5	8
	Strawberries																						
	Raw, capped																						
371	Large (1¹/₈" dia)	1 strawberry	18	92	5	Tr	Tr	Tr	Tr	Tr	0	1	0.4	3	0.1	30	Tr	5	1	Tr	0.01	Tr	10
372	Medium (1¼" dia)	1 strawberry	12	92	4	Tr	Tr	Tr	Tr	Tr	0	1	0.3	2	Tr	20	Tr	3	Tr	Tr	0.01	Tr	7
373	Sliced	1 cup	166	92	50	1	1	Tr	0.1	0.3	0	12	3.8	23	0.6	276	2	45	5	0.03	0.11	0.4	94
374	Frozen, sweetened, sliced, thawed	1 cup	255	73	245	1	Tr	Tr	Tr	0.2	0	66	4.8	28	1.5	250	8	61	5	0.04	0.13	1.0	106
	Tangerines																						
375	Raw, without peel and seeds (2³/₈" dia)	1 tangerine	84	88	37	1	Tr	Tr	Tr	Tr	0	9	1.9	12	0.1	132	1	773	77	0.09	0.02	0.1	26
376	Canned (mandarin oranges), light syrup, fruit and liquid	1 cup	252	83	154	1	Tr	Tr	Tr	0.1	0	41	1.8	18	0.9	197	15	2,117	212	0.13	0.11	1.1	50
377	Tangerine juice, canned, sweetened	1 cup	249	87	125	1	Tr	Tr	Tr	0.1	0	30	0.5	45	0.5	443	2	1,046	105	0.15	0.05	0.2	55
	Watermelon, raw (15" long × 7½" dia)																						
378	Wedge (about ¹/₁₆ of melon)	1 wedge	286	92	92	2	1	0.1	0.3	0.4	0	21	1.4	23	0.5	332	6	1,047	106	0.23	0.06	0.6	27
379	Diced	1 cup	152	92	49	1	1	0.1	0.2	0.2	0	11	0.8	12	0.3	176	3	556	56	0.12	0.03	0.3	15
Grain Products																							
	Bagels, enriched																						
380	Plain	3½" bagel	71	33	195	7	1	0.2	0.1	0.5	0	38	1.6	53	2.5	72	379	0	0	0.38	0.22	3.2	0
381		4" bagel	89	33	245	9	1	0.2	0.1	0.6	0	48	2.0	66	3.2	90	475	0	0	0.48	0.28	4.1	0
382	Cinnamon raisin	3½" bagel	71	32	195	7	1	0.2	0.1	0.5	0	39	1.6	13	2.7	105	229	52	0	0.27	0.20	2.2	Tr
383		4" bagel	89	32	244	9	2	0.2	0.2	0.6	0	49	2.0	17	3.4	132	287	65	0	0.34	0.25	2.7	1
384	Egg	3½" bagel	71	33	197	8	1	0.3	0.3	0.5	17	38	1.6	9	2.8	48	359	77	23	0.38	0.17	2.4	Tr
385		4" bagel	89	33	247	9	2	0.4	0.4	0.6	21	47	2.0	12	3.5	61	449	97	29	0.48	0.21	3.1	1
386	Banana bread, prepared from recipe, with margarine	1 slice	60	29	196	3	6	1.3	2.7	1.9	26	33	0.7	13	0.8	80	181	278	72	0.10	0.12	0.9	1

(*Continues*)

TABLE B-8 Food Composition Table

FOOD NO.	FOOD DESCRIPTION	MEASURE OF EDIBLE PORTION	WEIGHT (G)	WATER (%)	CALORIES (KCAL)	PROTEIN (G)	TOTAL FAT (G)	FATTY ACIDS SAT (G)	MONO (G)	POLY (G)	CHOLEST (MG)	CARB (G)	TOTAL D F(G)	CA (MG)	FE (MG)	K (MG)	NA (MG)	VITAMIN A (IU)	(RE)	THMN (MG)	RIBOFL (MG)	NIACIN (MG)	ASCORBIC ACID (MG)
Grain Products (*Continued*)																							
	Barley, pearled																						
387	Uncooked	1 cup	200	10	704	20	2	0.5	0.3	1.1	0	155	31.2	58	5.0	560	18	44	4	0.38	0.23	9.2	0
388	Cooked	1 cup	157	69	193	4	1	0.1	0.1	0.3	0	44	6.0	17	2.1	146	5	11	2	0.13	0.10	3.2	0
	Biscuits, plain or buttermilk, enriched																						
389	Prepared from recipe, with 2% milk	2½" biscuit	60	29	212	4	10	2.6	4.2	2.5	2	27	0.9	141	1.7	73	348	49	14	0.21	0.19	1.8	Tr
390		4" biscuit	101	29	358	7	16	4.4	7.0	4.2	3	45	1.5	237	2.9	122	586	83	23	0.36	0.31	3.0	Tr
	Refrigerated dough, baked																						
391	Regular	2½" biscuit	27	28	93	2	4	1.0	2.2	0.5	0	13	0.4	5	0.7	42	325	0	0	0.09	0.06	0.8	0
392	Lower fat	2¼" biscuit	21	28	63	2	1	0.3	0.6	0.2	0	12	0.4	4	0.6	39	305	0	0	0.09	0.05	0.7	0
	Breads, enriched																						
393	Cracked wheat	1 slice	25	36	65	2	1	0.2	0.5	0.2	0	12	1.4	11	0.7	44	135	0	0	0.09	0.06	0.9	0
394	Egg bread (challah)	½" slice	40	35	115	4	2	0.6	0.9	0.4	20	19	0.9	37	1.2	46	197	30	9	0.18	0.17	1.9	0
395	French or vienna (includes sourdough)	½" slice	25	34	69	2	1	0.2	0.3	0.2	0	13	0.8	19	0.6	28	152	0	0	0.13	0.08	1.2	0
396	Indian fry (navajo) bread	5" bread	90	27	296	6	9	2.1	3.6	2.3	0	48	1.6	210	3.2	67	626	0	0	0.39	0.27	3.3	0
397		10½" bread	160	27	526	11	15	3.7	6.4	4.1	0	85	2.9	373	5.8	118	1,112	0	0	0.69	0.49	5.8	0
398	Italian	1 slice	20	36	54	2	1	0.2	0.2	0.3	0	10	0.5	16	0.6	22	117	0	0	0.09	0.06	0.9	0
	Mixed grain																						
399	Untoasted	1 slice	26	38	65	3	1	0.2	0.4	0.2	0	12	1.7	24	0.9	53	127	0	0	0.11	0.09	1.1	Tr
400	Toasted	1 slice	24	32	65	3	1	0.2	0.4	0.2	0	12	1.6	24	0.9	53	127	0	0	0.08	0.08	1.0	Tr
	Oatmeal																						
401	Untoasted	1 slice	27	37	73	2	1	0.2	0.4	0.5	0	13	1.1	18	0.7	38	162	4	1	0.11	0.06	0.8	0
402	Toasted	1 slice	25	31	73	2	1	0.2	0.4	0.5	0	13	1.1	18	0.7	39	163	4	1	0.09	0.06	0.8	Tr
403	Pita	4" pita	28	32	77	3	Tr	Tr	Tr	0.1	0	16	0.6	24	0.7	34	150	0	0	0.17	0.09	1.3	0
404		6½" pita	60	32	165	5	1	0.1	0.1	0.3	0	33	1.3	52	1.6	72	322	0	0	0.36	0.20	2.8	0
	Pumpernickel																						
405	Untoasted	1 slice	32	38	80	3	1	0.1	0.3	0.4	0	15	2.1	22	0.9	67	215	0	0	0.10	0.10	1.0	0

(*Continues*)

TABLE B-8 Food Composition Table

FOOD NO.	FOOD DESCRIPTION	MEASURE OF EDIBLE PORTION	WEIGHT (G)	WATER (%)	CALORIES (KCAL)	PROTEIN (G)	TOTAL FAT (G)	FATTY ACIDS SAT (G)	MONO (G)	POLY (G)	CHOLEST (MG)	CARB (G)	TOTAL D F(G)	CA (MG)	FE (MG)	K (MG)	NA (MG)	VITAMIN A (IU)	(RE)	THMN (MG)	RIBOFL (MG)	NIACIN (MG)	ASCORBIC ACID (MG)
Grain Products (*Continued*)																							
	Breads, enriched (*Continued*)																						
406	Toasted	1 slice	29	32	80	3	1	0.1	0.3	0.4	0	15	2.1	21	0.9	66	214	0	0	0.08	0.09	0.9	0
	Raisin																						
407	Untoasted	1 slice	26	34	71	2	1	0.3	0.6	0.2	0	14	1.1	17	0.8	59	101	0	0	0.09	0.10	0.9	Tr
408	Toasted	1 slice	24	28	71	2	1	0.3	0.6	0.2	0	14	1.1	17	0.8	59	102	Tr	0	0.07	0.09	0.8	Tr
	Rye																						
409	Untoasted	1 slice	32	37	83	3	1	0.2	0.4	0.3	0	15	1.9	23	0.9	53	211	2	Tr	0.14	0.11	1.2	Tr
410	Toasted	1 slice	24	31	68	2	1	0.2	0.3	0.2	0	13	1.5	19	0.7	44	174	1	0	0.09	0.08	0.9	Tr
411	Rye, reduced calorie	1 slice	23	46	47	2	1	0.1	0.2	0.2	0	9	2.8	17	0.7	23	93	1	0	0.08	0.06	0.6	Tr
	Wheat																						
412	Untoasted	1 slice	25	37	65	2	1	0.2	0.4	0.2	0	12	1.1	26	0.8	50	133	0	0	0.10	0.07	1.0	0
413	Toasted	1 slice	23	32	65	2	1	0.2	0.4	0.2	0	12	1.2	26	0.8	50	132	0	0	0.08	0.06	0.9	0
414	Wheat, reduced calorie	1 slice	23	43	46	2	1	0.1	0.1	0.2	0	10	2.8	18	0.7	28	118	0	0	0.10	0.07	0.9	Tr
	White																						
415	Untoasted	1 slice	25	37	67	2	1	0.1	0.2	0.5	Tr	12	0.6	27	0.8	30	135	0	0	0.12	0.09	1.0	0
416	Toasted	1 slice	22	30	64	2	1	0.1	0.2	0.5	Tr	12	0.6	26	0.7	29	130	0	0	0.09	0.07	0.9	0
417	Soft crumbs	1 cup	45	37	120	4	2	0.2	0.3	0.9	Tr	22	1.0	49	1.4	54	242	0	0	0.21	0.15	1.8	0
418	White, reduced calorie	1 slice	23	43	48	2	1	0.1	0.2	0.1	0	10	2.2	22	0.7	17	104	1	Tr	0.09	0.07	0.8	Tr
	Bread, whole wheat																						
419	Untoasted	1 slice	28	38	69	3	1	0.3	0.5	0.3	0	13	1.9	20	0.9	71	148	0	0	0.10	0.06	1.1	0
420	Toasted	1 slice	25	30	69	3	1	0.3	0.5	0.3	0	13	1.9	20	0.9	71	148	0	0	0.08	0.05	1.0	0
	Bread crumbs, dry, grated																						
421	Plain, enriched	1 cup	108	6	427	14	6	1.3	2.6	1.2	0	78	2.6	245	6.6	239	931	1	0	0.83	0.47	7.4	0
422		1 oz	28	6	112	4	2	0.3	0.7	0.3	0	21	0.7	64	1.7	63	244	Tr	0	0.22	0.12	1.9	0
423	Seasoned, unenriched	1 cup	120	6	440	17	3	0.9	1.2	0.8	1	84	5.0	119	3.8	324	3,180	16	4	0.19	0.20	3.3	Tr
	Bread crumbs, soft. See White bread.																						
424	Bread stuffing, prepared from dry mix	½ cup	100	65	178	3	9	1.7	3.8	2.6	0	22	2.9	32	1.1	74	543	313	81	0.14	0.11	1.5	0
425	Breakfast bar, cereal crust withfruit filling, fat free	1 bar	37	14	121	2	Tr	Tr	Tr	0.1	Tr	28	0.8	49	4.5	92	203	1,249	125	1.01	0.42	5.0	1

(*Continues*)

TABLE B-8 Food Composition Table

FOOD NO.	FOOD DESCRIPTION	MEASURE OF EDIBLE PORTION	WEIGHT (G)	WATER (%)	CALORIES (KCAL)	PROTEIN (G)	TOTAL FAT (G)	FATTY ACIDS			CHOLEST (MG)	CARB (G)	TOTAL D F(G)	CA (MG)	FE (MG)	K (MG)	NA (MG)	VITAMIN A		THMN (MG)	RIBOFL (MG)	NIACIN (MG)	ASCORBIC ACID (MG)
								SAT (G)	MONO (G)	POLY (G)								(IU)	(RE)				
Grain Products (*Continued*)																							
	Breakfast Cereals																						
	Hot type, cooked																						
	Corn (hominy) grits																						
	Regular or quick, enriched																						
426	White	1 cup	242	85	145	3	Tr	0.1	0.1	0.2	0	31	0.5	0	1.5	53	0	0	0	0.24	0.15	2.0	0
427	Yellow	1 cup	242	85	145	3	Tr	0.1	0.1	0.2	0	31	0.5	0	1.5	53	0	145	15	0.24	0.15	2.0	0
428	Instant, plain	1 packet	137	82	89	2	Tr	Tr	Tr	0.1	0	21	1.2	8	8.2	38	289	0	0	0.15	0.08	1.4	0
	CREAM OF WHEAT																						
429	Regular	1 cup	251	87	133	4	1	0.1	0.1	0.3	0	28	1.8	50	10.3	43	3	0	0	0.25	0.00	1.5	0
430	Quick	1 cup	239	87	129	4	Tr	0.1	0.1	0.3	0	27	1.2	50	10.3	45	139	0	0	0.24	0.00	1.4	0
431	Mix'n Eat, plain	1 packet	142	82	102	3	Tr	Tr	Tr	0.2	0	21	0.4	20	8.1	38	241	1,252	376	0.43	0.28	5.0	0
432	MALT O MEAL	1 cup	240	88	122	4	Tr	0.1	0.1	Tr	0	26	1.0	5	9.6	31	2	0	0	0.48	0.24	5.8	0
	Oatmeal																						
433	Regular, quick or instant, plain, nonfortified	1 cup	234	85	145	6	2	0.4	0.7	0.9	0	25	4.0	19	1.6	131	2	37	5	0.26	0.05	0.3	0
434	Instant, fortified, plain	1 packet	177	86	104	4	2	0.3	0.6	0.7	0	18	3.0	163	6.3	99	285	1,510	453	0.53	0.28	5.5	0
	QUAKER instant																						
435	Apples and cinnamon	1 packet	149	79	125	3	1	0.3	0.5	0.6	0	26	2.5	104	3.9	106	121	1,019	305	0.30	0.35	4.1	Tr
436	Maple and brown sugar	1 packet	155	75	153	4	2	0.4	0.6	0.7	0	31	2.6	105	3.9	112	234	1,008	302	0.30	0.34	4.0	0
437	WHEATENA	1cup	243	85	136	5	1	0.2	0.2	0.6	0	29	6.6	10	1.4	187	5	0	0	0.02	0.05	1.3	0
	Ready to eat																						
438	ALL BRAN	½ cup	30	3	79	4	1	0.2	0.2	0.5	0	23	9.7	106	4.5	342	61	750	225	0.39	0.42	5.0	15
439	APPLE CINNAMON CHEERIOS	¾ cup	30	3	118	2	2	0.3	0.6	0.2	0	25	1.6	35	4.5	60	150	750	225	0.38	0.43	5.0	15
440	APPLE JACKS	1 cup	30	3	116	1	Tr	0.1	0.1	0.2	0	27	0.6	3	4.5	32	134	750	225	0.39	0.42	5.0	15
441	BASIC 4	1 cup	55	7	201	4	3	0.4	1.0	1.1	0	42	3.4	310	4.5	162	323	1,250	375	0.37	0.42	5.0	15
442	BERRY BERRY KIX	¾ cup	30	2	120	1	1	0.2	0.5	0.1	0	26	0.2	66	4.5	24	185	750	225	0.38	0.43	5.0	15
443	CAP'N CRUNCH	¾ cup	27	2	107	1	1	0.4	0.3	0.2	0	23	0.9	5	4.5	35	208	36	4	0.38	0.42	5.0	0
444	CAP'N CRUNCH'S CRUNCHBERRIES	¾ cup	26	2	104	1	1	0.3	0.3	0.2	0	22	0.6	7	4.5	37	190	33	5	0.37	0.42	5.0	Tr

(*Continues*)

TABLE B-8 Food Composition Table

FOOD NO.	FOOD DESCRIPTION	MEASURE OF EDIBLE PORTION	WEIGHT (G)	WATER (%)	CALORIES (KCAL)	PROTEIN (G)	TOTAL FAT (G)	FATTY ACIDS			CHOLEST (MG)	CARB (G)	TOTAL D F(G)	CA (MG)	FE (MG)	K (MG)	NA (MG)	VITAMIN A		THMN (MG)	RIBOFL (MG)	NIACIN (MG)	ASCORBIC ACID (MG)
								SAT (G)	MONO (G)	POLY (G)								(IU)	(RE)				
Grain Products (*Continued*)																							
	Breakfast Cereals (*Continued*)																						
445	CAP'N CRUNCH'S PEANUT BUTTER CRUNCH	¾ cup	27	2	112	2	2	0.5	0.8	0.5	0	22	0.8	3	4.5	62	204	37	4	0.38	0.42	5.0	0
446	CHEERIOS	1 cup	30	3	110	3	2	0.4	0.6	0.2	0	23	2.6	55	8.1	89	284	1,250	375	0.38	0.43	5.0	15
	CHEX																						
447	Corn	1 cup	30	3	113	2	Tr	0.1	0.1	0.2	0	26	0.5	100	9.0	32	289	0	0	0.38	0.00	5.0	6
448	Honey nut	¾ cup	30	2	117	2	1	0.1	0.4	0.2	0	26	0.4	102	9.0	27	224	0	0	0.38	0.44	5.0	6
449	Multi bran	1 cup	49	3	165	4	1	0.2	0.3	0.5	0	41	6.4	95	13.7	191	325	0	0	0.32	0.00	4.4	5
450	Rice	1¼ cup	31	3	117	2	Tr	Tr	Tr	Tr	0	27	0.3	104	9.0	36	291	0	0	0.38	0.02	5.0	6
451	Wheat	1 cup	30	3	104	3	1	0.1	0.1	0.3	0	24	3.3	60	9.0	116	269	0	0	0.23	0.04	3.0	4
452	CINNAMON LIFE	1 cup	50	4	190	4	2	0.3	0.6	0.8	0	40	3.0	135	7.5	113	220	16	2	0.63	0.71	8.4	Tr
453	CINNAMON TOAST CRUNCH	¾ cup	30	2	124	2	3	0.5	0.9	0.5	0	24	1.5	42	4.5	44	210	750	225	0.38	0.43	5.0	15
454	COCOA KRISPIES	¾ cup	31	2	120	2	1	0.6	0.1	0.1	0	27	0.4	4	1.8	60	210	750	225	0.37	0.43	5.0	15
455	COCOA PUFFS	1 cup	30	2	119	1	1	0.2	0.3	Tr	0	27	0.2	33	4.5	52	181	0	0	0.38	0.43	5.0	15
	Corn Flakes																						
456	GENERAL MILLS, TOTAL	1⅓ cup	30	3	112	2	Tr	0.2	0.1	Tr	0	26	0.8	237	18.0	34	203	1,250	375	1.50	1.70	20.1	60
457	KELLOGG'S	1 cup	28	3	102	2	Tr	0.1	Tr	0.1	0	24	0.8	1	8.7	25	298	700	210	0.36	0.39	4.7	14
458	CORN POPS	1 cup	31	3	118	1	Tr	0.1	0.1	Tr	0	28	0.4	2	1.9	23	123	775	233	0.40	0.43	5.2	16
459	CRISPIX	1 cup	29	3	108	2	Tr	0.1	0.1	0.1	0	25	0.6	3	1.8	35	240	750	225	0.38	0.44	5.0	15
460	Complete Wheat Bran Flakes	¾ cup	29	4	95	3	1	0.1	0.1	0.4	0	23	4.6	14	8.1	175	226	1,208	363	0.38	0.44	5.0	15
461	FROOT LOOPS	1 cup	30	2	117	1	1	0.4	0.2	0.3	0	26	0.6	3	4.2	32	141	703	211	0.39	0.42	5.0	14
462	FROSTED FLAKES	¾ cup	31	3	119	1	Tr	0.1	Tr	0.1	0	28	0.6	1	4.5	20	200	750	225	0.37	0.43	5.0	15
	FROSTED MINI WHEATS																						
463	Regular	1 cup	51	5	173	5	1	0.2	0.1	0.6	0	42	5.5	18	14.3	170	2	0	0	0.36	0.41	5.0	0
464	Bite size	1 cup	55	5	187	5	1	0.2	0.2	0.6	0	45	5.9	0	15.4	186	2	0	0	0.33	0.39	4.7	0
465	GOLDEN GRAHAMS	¾ cup	30	3	116	2	1	0.2	0.3	0.2	0	26	0.9	14	4.5	53	275	750	225	0.38	0.43	5.0	15
466	HONEY FROSTED WHEATIES	¾ cup	30	3	110	2	Tr	0.1	Tr	Tr	0	26	1.5	8	4.5	56	211	750	225	0.38	0.43	5.0	15
467	HONEY NUT CHEERIOS	1 cup	30	2	115	3	1	0.2	0.5	0.2	0	24	1.6	20	4.5	85	259	750	225	0.38	0.43	5.0	15
468	HONEY NUT CLUSTERS	1 cup	55	3	213	5	3	0.4	1.8	0.4	0	43	4.2	72	4.5	171	239	0	0	0.37	0.42	5.0	9
469	KIX	1⅓ cup	30	2	114	2	1	0.2	0.1	Tr	0	26	0.8	44	8.1	41	263	1,250	375	0.38	0.43	5.0	15
470	LIFE	¾ cup	32	4	121	3	1	0.2	0.4	0.6	0	25	2.0	98	9.0	79	174	12	1	0.40	0.45	5.3	0
471	LUCKY CHARMS	1 cup	30	2	116	2	1	0.2	0.4	0.2	0	25	1.2	32	4.5	54	203	750	225	0.38	0.43	5.0	15

(*Continues*)

TABLE B-8 Food Composition Table

FOOD NO.	FOOD DESCRIPTION	MEASURE OF EDIBLE PORTION	WEIGHT (G)	WATER (%)	CALORIES (KCAL)	PROTEIN (G)	TOTAL FAT (G)	FATTY ACIDS SAT (G)	MONO (G)	POLY (G)	CHOLEST (MG)	CARB (G)	TOTAL D F(G)	CA (MG)	FE (MG)	K (MG)	NA (MG)	VITAMIN A (IU)	(RE)	THMN (MG)	RIBOFL (MG)	NIACIN (MG)	ASCORBIC ACID (MG)
Grain Products (*Continued*)																							
	Breakfast Cereals (*Continued*)																						
472	NATURE VALLEY Granola	¾ cup	55	4	248	6	10	1.3	6.5	1.9	0	36	3.5	41	1.7	183	89	0	0	0.17	0.06	0.6	0
	100% Natural Cereal																						
473	With oats, honey, and raisins	½ cup	51	4	218	5	7	3.2	3.2	0.8	1	36	3.7	39	1.7	214	11	4	1	0.14	0.09	0.8	Tr
474	With raisins, low fat	½ cup	50	4	195	4	3	0.8	1.3	0.5	1	40	3.0	30	1.3	169	129	9	1	0.15	0.06	0.9	Tr
475	PRODUCT 19	1 cup	30	3	110	3	Tr	Tr	0.2	0.2	0	25	1.0	3	18.0	41	216	750	225	1.50	1.71	20.0	60
476	Puffed Rice	1 cup	14	3	56	1	Tr	Tr	Tr	Tr	0	13	0.2	1	4.4	16	Tr	0	0	0.36	0.25	4.9	0
477	Puffed Wheat	1 cup	12	3	44	2	Tr	Tr	Tr	Tr	0	10	0.5	3	3.8	42	Tr	0	0	0.31	0.22	4.2	0
	Raisin Bran																						
478	GENERAL MILLS, TOTAL	1 cup	55	9	178	4	1	0.2	0.2	0.2	0	43	5.0	238	18.0	287	240	1,250	375	1.50	1.70	20.0	0
479	KELLOGG'S	1 cup	61	8	186	6	1	0.0	0.2	0.8	0	47	8.2	35	5.0	437	354	832	250	0.43	0.49	5.6	0
480	RAISIN NUT BRAN	1 cup	55	5	209	5	4	0.7	1.9	0.5	0	41	5.1	74	4.5	218	246	0	0	0.37	0.42	5.0	0
481	REESE'S PEANUT BUTTER PUFFS	¾ cup	30	2	129	3	3	0.6	1.4	0.6	0	23	0.4	21	4.5	62	177	750	225	0.38	0.43	5.0	15
482	RICE KRISPIES	1¼ cup	33	3	124	2	Tr	0.1	0.1	0.2	0	29	0.4	3	2.0	42	354	825	248	0.43	0.46	5.5	17
483	RICE KRISPIES TREATS cereal	¾ cup	30	4	120	1	2	0.4	1.0	0.2	0	26	0.3	2	1.8	19	190	750	225	0.39	0.42	5.0	15
484	SHREDDED WHEAT	2 biscuits	46	4	156	5	1	0.1	NA	NA	0	38	5.3	20	1.4	196	3	0	NA	0.12	0.05	2.6	0
485	SMACKS	¾ cup	27	3	103	2	1	0.3	0.1	0.2	0	24	0.9	3	1.8	42	51	750	225	0.38	0.43	5.0	15
486	SPECIAL K	1 cup	31	3	115	6	Tr	0.0	0.0	0.2	0	22	1.0	5	8.7	55	250	750	225	0.53	0.59	7.0	15
487	QUAKER Toasted Oatmeal, Honey Nut	1 cup	49	3	191	5	3	0.5	1.2	0.7	Tr	39	3.3	27	4.5	185	166	500	150	0.37	0.42	5.0	6
488	TOTAL, Whole Grain	¾ cup	30	3	105	3	1	0.2	0.1	0.1	0	24	2.6	258	18.0	97	199	1,250	375	1.50	1.70	20.1	60
489	TRIX	1 cup	30	2	122	1	2	0.4	0.9	0.3	0	26	0.7	32	4.5	18	197	750	225	0.38	0.43	5.0	15
490	WHEATIES	1 cup	30	3	110	3	1	0.2	0.2	0.2	0	24	2.1	55	8.1	104	222	750	225	0.38	0.43	5.0	15
	Brownies, without icing																						
	Commercially prepared																						
491	Regular, large (2¾" sq × ⅞")	1 brownie	56	14	227	3	9	2.4	5.0	1.3	10	36	1.2	16	1.3	83	175	39	3	0.14	0.12	1.0	0
492	Fat free, 2" sq	1 brownie	28	12	89	1	Tr	0.2	0.1	Tr	0	22	1.0	17	0.7	89	90	1	Tr	0.03	0.04	0.3	Tr
493	Prepared from dry mix, reduced calorie, 2" sq	1 brownie	22	13	84	1	2	1.1	1.0	0.2	0	16	0.8	3	0.3	69	21	0	0	0.02	0.03	0.2	0
494	Buckwheat flour, whole groat	1 cup	120	11	402	15	4	0.8	1.1	1.1	0	85	12.0	49	4.9	692	13	0	0	0.50	0.23	7.4	0
495	Buckwheat groats, roasted (kasha), cooked	1 cup	168	76	155	6	1	0.2	0.3	0.3	0	33	4.5	12	1.3	148	7	0	0	0.07	0.07	1.6	0

(*Continues*)

TABLE B-8 Food Composition Table

FOOD NO.	FOOD DESCRIPTION	MEASURE OF EDIBLE PORTION	WEIGHT (G)	WATER (%)	CALORIES (KCAL)	PROTEIN (G)	TOTAL FAT (G)	FATTY ACIDS SAT (G)	MONO (G)	POLY (G)	CHOLEST (MG)	CARB (G)	TOTAL D F(G)	CA (MG)	FE (MG)	K (MG)	NA (MG)	VITAMIN A (IU)	(RE)	THMN (MG)	RIBOFL (MG)	NIACIN (MG)	ASCORBIC ACID (MG)
Grain Products (*Continued*)																							
	Bulgur																						
496	Uncooked	1 cup	140	9	479	17	2	0.3	0.2	0.8	0	106	25.6	49	3.4	574	24	0	0	0.32	0.16	7.2	0
497	Cooked	1 cup	182	78	151	6	Tr	0.1	0.1	0.2	0	34	8.2	18	1.7	124	9	0	0	0.10	0.05	1.8	0
	Cakes, prepared from dry mix																						
498	Angelfood (1/12 of 10" dia)	1 piece	50	33	129	3	Tr	Tr	Tr	0.1	0	29	0.1	42	0.1	68	255	0	0	0.05	0.10	0.1	0
499	Yellow, light, with water, egg whites, no frosting (1/12 of 9" dia)	1 piece	69	37	181	3	2	1.1	0.9	0.2	0	37	0.6	69	0.6	41	279	6	1	0.06	0.12	0.6	0
	Cakes, prepared from recipe																						
500	Chocolate, without frosting (1/12 of 9" dia)	1 piece	95	24	340	5	14	5.2	5.7	2.6	55	51	1.5	57	1.5	133	299	133	38	0.13	0.20	1.1	Tr
501	Gingerbread (1/9 of 8" square)	1 piece	74	28	263	3	12	3.1	5.3	3.1	24	36	0.7	53	2.1	325	242	36	10	0.14	0.12	1.3	Tr
502	Pineapple upside down (1/9 of 8" square)	1 piece	115	32	367	4	14	3.4	6.0	3.8	25	58	0.9	138	1.7	129	367	291	75	0.18	0.18	1.4	1
503	Shortcake, biscuit type (about 3" dia)	1 shortcake	65	28	225	4	9	2.5	3.9	2.4	2	32	0.8	133	1.7	69	329	47	12	0.20	0.18	1.7	Tr
504	Sponge (1/12 of 16-oz cake)	1 piece	63	29	187	5	3	0.8	1.0	0.4	107	36	0.4	26	1.0	89	144	163	49	0.10	0.19	0.8	0
	White																						
505	With coconut frosting (1/12 of 9" dia)	1 piece	112	21	399	5	12	4.4	4.1	2.4	1	71	1.1	101	1.3	111	318	43	12	0.14	0.21	1.2	Tr
506	Without frosting (1/12 of 9" dia)	1 piece	74	23	264	4	9	2.4	3.9	2.3	1	42	0.6	96	1.1	70	242	41	12	0.14	0.18	1.1	Tr
	Cakes, commercially prepared																						
507	Angelfood (1/12 of 12-oz cake)	1 piece	28	33	72	2	Tr	Tr	Tr	0.1	0	16	0.4	39	0.1	26	210	0	0	0.03	0.14	0.2	0
508	Boston cream (1/6 of pie)	1 piece	92	45	232	2	8	2.2	4.2	0.9	34	39	1.3	21	0.3	36	132	74	21	0.38	0.25	0.2	Tr
509	Chocolate with chocolate frosting (1/8 of 18-oz cake)	1 piece	64	23	235	3	10	3.1	5.6	1.2	27	35	1.8	28	1.4	128	214	54	16	0.02	0.09	0.4	Tr
510	Coffeecake, crumb (1/9 of 20-oz cake)	1 piece	63	22	263	4	15	3.7	8.2	2.0	20	29	1.3	34	1.2	77	221	70	21	0.13	0.14	1.1	Tr
511	Fruitcake	1 piece	43	25	139	1	4	0.5	1.8	1.4	2	26	1.6	14	0.9	66	116	9	2	0.02	0.04	0.3	Tr
	Pound																						
512	Butter (1/12 of 12-oz cake)	1 piece	28	25	109	2	6	3.2	1.7	0.3	62	14	0.1	10	0.4	33	111	170	44	0.04	0.06	0.4	0

(Continues)

TABLE B-8 Food Composition Table

FOOD NO.	FOOD DESCRIPTION	MEASURE OF EDIBLE PORTION	WEIGHT (G)	WATER (%)	CALORIES (KCAL)	PROTEIN (G)	TOTAL FAT (G)	FATTY ACIDS SAT (G)	MONO (G)	POLY (G)	CHOLEST (MG)	CARB (G)	TOTAL D F(G)	CA (MG)	FE (MG)	K (MG)	NA (MG)	VITAMIN A (IU)	(RE)	THMN (MG)	RIBOFL (MG)	NIACIN (MG)	ASCORBIC ACID (MG)
Grain Products (*Continued*)																							
	Cakes, commercially prepared (*Continued*)																						
513	Fat free (3¼" × 2¾" × ⅝" slice)	1 slice	28	31	79	2	Tr	0.1	Tr	0.1	0	17	0.3	12	0.6	31	95	27	8	0.04	0.08	0.2	0
	Snack cakes																						
514	Chocolate, creme filled, with frosting	1 cupcake	50	20	188	2	7	1.4	2.8	2.6	9	30	0.4	37	1.7	61	213	9	3	0.11	0.15	1.2	0
515	Chocolate, with frosting, low fat	1 cupcake	43	23	131	2	2	0.5	0.8	0.2	0	29	1.8	15	0.7	96	178	0	0	0.02	0.06	0.3	0
516	Sponge, creme filled	1 cake	43	20	155	1	5	1.1	1.7	1.4	7	27	0.2	19	0.5	37	155	7	2	0.07	0.06	0.5	Tr
517	Sponge, individual shortcake	1 shortcake	30	30	87	2	1	0.2	0.3	0.1	31	18	0.2	21	0.8	30	73	46	14	0.07	0.08	0.6	0
	Yellow																						
518	With chocolate frosting	1 piece	64	22	243	2	11	3.0	6.1	1.4	35	35	1.2	24	1.3	114	216	70	21	0.08	0.10	0.8	0
519	With vanilla frosting	1 piece	64	22	239	2	9	1.5	3.9	3.3	35	38	0.2	40	0.7	34	220	40	12	0.06	0.04	0.3	0
520	Cheesecake (⅙ of 17-oz cake)	1 piece	80	46	257	4	18	7.9	6.9	1.3	44	20	0.3	41	0.5	72	166	438	117	0.02	0.15	0.2	Tr
521	Cheese flavor puffs or twists	1 oz	28	2	157	2	10	1.9	5.7	1.3	1	15	0.3	16	0.7	47	298	75	10	0.07	0.10	0.9	Tr
522	CHEX mix	1 oz (about ⅔ cup)	28	4	120	3	5	1.6	NA	NA	0	18	1.6	10	7.0	76	288	41	4	0.44	0.14	4.8	13
	Cookies																						
523	Butter, commercially prepared	1 cookie	5	5	23	Tr	1	0.6	0.3	Tr	6	3	Tr	1	0.1	6	18	34	8	0.02	0.02	0.2	0
	Chocolate chip, medium (2¼"-2½" dia)																						
	Commercially prepared																						
524	Regular	1 cookie	10	4	48	1	2	0.7	1.2	0.2	0	7	0.3	3	0.3	14	32	Tr	0	0.02	0.03	0.3	0
525	Reduced fat	1 cookie	10	4	45	1	2	0.4	0.6	0.5	0	7	0.4	2	0.3	12	38	Tr	0	0.03	0.03	0.3	0
526	From refrigerated dough (spooned from roll)	1 cookie	26	3	128	1	6	2.0	2.9	0.6	7	18	0.4	7	0.7	52	60	15	4	0.04	0.05	0.5	0
527	Prepared from recipe, with margarine	1 cookie	16	6	78	1	5	1.3	1.7	1.3	5	9	0.4	6	0.4	36	58	102	26	0.03	0.03	0.2	Tr
528	Devil's food, commercially prepared, fat free	1 cookie	16	18	49	1	Tr	0.1	Tr	Tr	0	12	0.3	5	0.4	18	28	Tr	NA	0.01	0.03	0.2	Tr
529	Fig bar	1 cookie	16	17	56	1	1	0.2	0.5	0.4	0	11	0.7	10	0.5	33	56	5	1	0.03	0.03	0.3	Tr

(*Continues*)

TABLE B-8 Food Composition Table

FOOD NO.	FOOD DESCRIPTION	MEASURE OF EDIBLE PORTION	WEIGHT (G)	WATER (%)	CALORIES (KCAL)	PROTEIN (G)	TOTAL FAT (G)	FATTY ACIDS SAT (G)	FATTY ACIDS MONO (G)	FATTY ACIDS POLY (G)	CHOLEST (MG)	CARB (G)	TOTAL D F(G)	CA (MG)	FE (MG)	K (MG)	NA (MG)	VITAMIN A (IU)	VITAMIN A (RE)	THMN (MG)	RIBOFL (MG)	NIACIN (MG)	ASCORBIC ACID (MG)
Grain Products (*Continued*)																							
	Cookies (*Continued*)																						
	Molasses																						
530	Medium	1 cookie	15	6	65	1	2	0.5	1.1	0.3	0	11	0.1	11	1.0	52	69	0	0	0.05	0.04	0.5	0
531	Large (3½"-4" dia)	1 cookie	32	6	138	2	4	1.0	2.3	0.6	0	24	0.3	24	2.1	111	147	0	0	0.11	0.08	1.0	0
	Oatmeal																						
	Commercially prepared, with or without raisins																						
532	Regular, large	1 cookie	25	6	113	2	5	1.1	2.5	0.6	0	17	0.7	9	0.6	36	96	5	1	0.07	0.06	0.6	Tr
533	Soft type	1 cookie	15	11	61	1	2	0.5	1.2	0.3	1	10	0.4	14	0.4	20	52	5	1	0.03	0.03	0.3	Tr
534	Fat free	1 cookie	11	13	36	1	Tr	Tr	Tr	0.1	0	9	0.8	4	0.2	23	33	0	0	0.02	0.03	0.1	0
535	Prepared from recipe, with raisins (2⅝" dia)	1 cookie	15	6	65	1	2	0.5	1.0	0.8	5	10	0.5	15	0.4	36	81	96	25	0.04	0.02	0.2	Tr
	Peanut butter																						
536	Commercially prepared	1 cookie	15	6	72	1	4	0.7	1.9	0.8	Tr	9	0.3	5	0.4	25	62	1	Tr	0.03	0.03	0.6	0
537	Prepared from recipe, with margarine (3" dia)	1 cookie	20	6	95	2	5	0.9	2.2	1.4	6	12	0.4	8	0.4	46	104	120	31	0.04	0.04	0.7	Tr
	Sandwich type, with creme filling																						
538	Chocolate cookie	1 cookie	10	2	47	Tr	2	0.4	0.9	0.7	0	7	0.3	3	0.4	18	60	Tr	0	0.01	0.02	0.2	0
	Vanilla cookie																						
539	Oval	1 cookie	15	2	72	1	3	0.4	1.3	1.1	0	11	0.2	4	0.3	14	52	0	0	0.04	0.04	0.4	0
540	Round	1 cookie	10	2	48	Tr	2	0.3	0.8	0.8	0	7	0.2	3	0.2	9	35	0	0	0.03	0.02	0.3	0
	Shortbread, commercially prepared																						
541	Plain (1⅝" sq)	1 cookie	8	4	40	Tr	2	0.5	1.1	0.3	2	5	0.1	3	0.2	8	36	7	1	0.03	0.03	0.3	0
	Pecan																						
542	Regular (2" dia)	1 cookie	14	3	76	1	5	1.1	2.6	0.6	5	8	0.3	4	0.3	10	39	Tr	Tr	0.04	0.03	0.3	0
543	Reduced fat	1 cookie	16	5	73	1	3	0.6	1.6	0.4	0	11	0.2	8	0.5	15	55	1	Tr	0.05	0.03	0.4	Tr
	Sugar																						
544	Commercially prepared	1 cookie	15	5	72	1	3	0.8	1.8	0.4	8	10	0.1	3	0.3	9	54	14	4	0.03	0.03	0.4	Tr
545	From refrigerated dough	1 cookie	15	5	73	1	3	0.9	2.0	0.4	5	10	0.1	14	0.3	24	70	6	2	0.03	0.02	0.4	0

(*Continues*)

TABLE B-8 Food Composition Table

FOOD NO.	FOOD DESCRIPTION	MEASURE OF EDIBLE PORTION	WEIGHT (G)	WATER (%)	CALORIES (KCAL)	PROTEIN (G)	TOTAL FAT (G)	FATTY ACIDS SAT (G)	MONO (G)	POLY (G)	CHOLEST (MG)	CARB (G)	TOTAL D F(G)	CA (MG)	FE (MG)	K (MG)	NA (MG)	VITAMIN A (IU)	(RE)	THMN (MG)	RIBOFL (MG)	NIACIN (MG)	ASCORBIC ACID (MG)
Grain Products (*Continued*)																							
	Cookies (*Continued*)																						
546	Prepared from recipe, with margarine (3" dia)	1 cookie	14	9	66	1	3	0.7	1.4	1.0	4	8	0.2	10	0.3	11	69	135	35	0.04	0.04	0.3	Tr
547	Vanilla wafer, lower fat, medium size	1 cookie	4	5	18	Tr	1	0.2	0.3	0.2	2	3	0.1	2	0.1	4	12	1	Tr	0.01	0.01	0.1	0
	Corn chips																						
548	Plain	1 oz	28	1	153	2	9	1.3	2.7	4.7	0	16	1.4	36	0.4	40	179	27	3	0.01	0.04	0.3	0
549	Barbecue flavor	1 oz	28	1	148	2	9	1.3	2.7	4.6	0	16	1.5	37	0.4	67	216	173	17	0.02	0.06	0.5	Tr
	Cornbread																						
550	Prepared from mix, piece 3¾" × 2½" × ¾"	1 piece	60	32	188	4	6	1.6	3.1	0.7	37	29	1.4	44	1.1	77	467	123	26	0.15	0.16	1.2	Tr
551	Prepared from recipe, with 2% milk, piece 2½" sq × 1½"	1 piece	65	39	173	4	5	1.0	1.2	2.1	26	28	1.9	162	1.6	96	428	180	35	0.19	0.19	1.5	Tr
	Cornmeal, yellow, dry form																						
552	Whole grain	1 cup	122	10	442	10	4	0.6	1.2	2.0	0	94	8.9	7	4.2	350	43	572	57	0.47	0.25	4.4	0
553	Degermed, enriched	1 cup	138	12	505	12	2	0.3	0.6	1.0	0	107	10.2	7	5.7	224	4	570	57	0.99	0.56	6.9	0
554	Self rising, degermed, enriched	1 cup	138	10	490	12	2	0.3	0.6	1.0	0	103	9.8	483	6.5	235	1,860	570	57	0.94	0.53	6.3	0
555	Cornstarch	1 tbsp	8	8	30	Tr	Tr	Tr	Tr	Tr	0	7	0.1	Tr	Tr	Tr	1	0	0	0.00	0.00	0.0	0
	Couscous																						
556	Uncooked	1 cup	173	9	650	22	1	0.2	0.2	0.4	0	134	8.7	42	1.9	287	17	0	0	0.28	0.13	6.0	0
557	Cooked	1 cup	157	73	176	6	Tr	Tr	Tr	0.1	0	36	2.2	13	0.6	91	8	0	0	0.10	0.04	1.5	0
	Crackers																						
558	Cheese, 1" sq	10 crackers	10	3	50	1	3	0.9	1.2	0.2	1	6	0.2	15	0.5	15	100	16	3	0.06	0.04	0.5	0
	Graham, plain																						
559	2½" sq	2 squares	14	4	59	1	1	0.2	0.6	0.5	0	11	0.4	3	0.5	19	85	0	0	0.03	0.04	0.6	0
560	Crushed	1 cup	84	4	355	6	8	1.3	3.4	3.2	0	65	2.4	20	3.1	113	508	0	0	0.19	0.26	3.5	0
561	Melba toast, plain	4 pieces	20	5	78	2	1	0.1	0.2	0.3	0	15	1.3	19	0.7	40	166	0	0	0.08	0.05	0.8	0
562	Rye wafer, whole grain, plain	1 wafer	11	5	37	1	Tr	Tr	Tr	Tr	0	9	2.5	4	0.7	54	87	1	0	0.05	0.03	0.2	Tr
	Saltine																						
563	Square	4 crackers	12	4	52	1	1	0.4	0.8	0.2	0	9	0.4	14	0.6	15	156	0	0	0.07	0.06	0.6	0
564	Oyster type	1 cup	45	4	195	4	5	1.3	2.9	0.8	0	32	1.4	54	2.4	58	586	0	0	0.25	0.21	2.4	0

(*Continues*)

TABLE B-8 Food Composition Table

FOOD NO.	FOOD DESCRIPTION	MEASURE OF EDIBLE PORTION	WEIGHT (G)	WATER (%)	CALORIES (KCAL)	PROTEIN (G)	TOTAL FAT (G)	FATTY ACIDS SAT (G)	MONO (G)	POLY (G)	CHOLEST (MG)	CARB (G)	TOTAL D F(G)	CA (MG)	FE (MG)	K (MG)	NA (MG)	VITAMIN A (IU)	(RE)	THMN (MG)	RIBOFL (MG)	NIACIN (MG)	ASCORBIC ACID (MG)
Grain Products (*Continued*)																							
	Crackers (*Continued*)																						
	Sandwich type																						
565	Wheat with cheese	1 sandwich	7	4	33	1	1	0.4	0.8	0.2	Tr	4	0.1	18	0.2	30	98	5	1	0.03	0.05	0.3	Tr
566	Cheese with peanut butter	1 sandwich	7	4	34	1	2	0.4	0.8	0.3	Tr	4	0.2	6	0.2	17	69	22	2	0.03	0.02	0.5	Tr
	Standard snack type																						
567	Bite size	1 cup	62	4	311	5	16	2.3	6.6	5.9	0	38	1.0	74	2.2	82	525	0	0	0.25	0.21	2.5	0
568	Round	4 crackers	12	4	60	1	3	0.5	1.3	1.1	0	7	0.2	14	0.4	16	102	0	0	0.05	0.04	0.5	0
569	Wheat, thin square	4 crackers	8	3	38	1	2	0.4	0.9	0.2	0	5	0.4	4	0.4	15	64	0	0	0.04	0.03	0.4	0
570	Whole wheat	4 crackers	16	3	71	1	3	0.5	0.9	1.1	0	11	1.7	8	0.5	48	105	0	0	0.03	0.02	0.7	0
571	Croissant, butter	1 croissant	57	23	231	5	12	6.6	3.1	0.6	38	26	1.5	21	1.2	67	424	424	106	0.22	0.14	1.2	Tr
572	Croutons, seasoned	1 cup	40	4	186	4	7	2.1	3.8	0.9	3	25	2.0	38	1.1	72	495	16	4	0.20	0.17	1.9	0
	Danish pastry, enriched																						
573	Cheese filled	1 danish	71	31	266	6	16	4.8	8.0	1.8	11	26	0.7	25	1.1	70	320	104	32	0.13	0.18	1.4	Tr
574	Fruit filled	1 danish	71	27	263	4	13	3.5	7.1	1.7	81	34	1.3	33	1.3	59	251	53	16	0.19	0.16	1.4	3
	Doughnuts																						
575	Cake type	1 hole	14	21	59	1	3	0.5	1.3	1.1	5	7	0.2	6	0.3	18	76	8	2	0.03	0.03	0.3	Tr
576		1 medium	47	21	198	2	11	1.7	4.4	3.7	17	23	0.7	21	0.9	60	257	27	8	0.10	0.11	0.9	Tr
577	Yeast leavened, glazed	1 hole	13	25	52	1	3	0.8	1.7	0.4	1	6	0.2	6	0.3	14	44	2	1	0.05	0.03	0.4	Tr
578		1 medium	60	25	242	4	14	3.5	7.7	1.7	4	27	0.7	26	1.2	65	205	8	2	0.22	0.13	1.7	Tr
579	Eclair, prepared from recipe, 5" × 2" × 1¾"	1 eclair	100	52	262	6	16	4.1	6.5	3.9	127	24	0.6	63	1.2	117	337	718	191	0.12	0.27	0.8	Tr
	English muffin, plain, enriched																						
580	Untoasted	1 muffin	57	42	134	4	1	0.1	0.2	0.5	0	26	1.5	99	1.4	75	264	0	0	0.25	0.16	2.2	0
581	Toasted	1 muffin	52	37	133	4	1	0.1	0.2	0.5	0	26	1.5	98	1.4	74	262	0	0	0.20	0.14	2.0	Tr
	French toast																						
582	Prepared from recipe, with 2% milk, fried in margarine	1 slice	65	55	149	5	7	1.8	2.9	1.7	75	16	0.7	65	1.1	87	311	315	86	0.13	0.21	1.1	Tr
583	Frozen, ready to heat	1 slice	59	53	126	4	4	0.9	1.2	0.7	48	19	0.7	63	1.3	79	292	110	32	0.16	0.22	1.6	Tr
	Granola bar																						
584	Hard, plain	1 bar	28	4	134	3	6	0.7	1.2	3.4	0	18	1.5	17	0.8	95	83	43	4	0.07	0.03	0.4	Tr

(Continues)

TABLE B-8 Food Composition Table

FOOD NO.	FOOD DESCRIPTION	MEASURE OF EDIBLE PORTION	WEIGHT (G)	WATER (%)	CALORIES (KCAL)	PROTEIN (G)	TOTAL FAT (G)	FATTY ACIDS SAT (G)	MONO (G)	POLY (G)	CHOLEST (MG)	CARB (G)	TOTAL D F(G)	CA (MG)	FE (MG)	K (MG)	NA (MG)	VITAMIN A (IU)	(RE)	THMN (MG)	RIBOFL (MG)	NIACIN (MG)	ASCORBIC ACID (MG)
Grain Products (*Continued*)																							
	Granola bar (*Continued*)																						
	Soft, uncoated																						
585	Chocolate chip	1 bar	28	5	119	2	5	2.9	1.0	0.6	Tr	20	1.4	26	0.7	96	77	12	1	0.06	0.04	0.3	0
586	Raisin	1 bar	28	6	127	2	5	2.7	0.8	0.9	Tr	19	1.2	29	0.7	103	80	0	0	0.07	0.05	0.3	0
587	Soft, chocolate-coated, peanut butter	1 bar	28	3	144	3	9	4.8	1.9	0.5	3	15	0.8	31	0.4	96	55	37	10	0.03	0.06	0.9	Tr
588	Macaroni (elbows), enriched, cooked	1 cup	140	66	197	7	1	0.1	0.1	0.4	0	40	1.8	10	2.0	43	1	0	0	0.29	0.14	2.3	0
589	Matzo, plain	1 matzo	28	4	112	3	Tr	0.1	Tr	0.2	0	24	0.9	4	0.9	32	1	0	0	0.11	0.08	1.1	0
	Muffins																						
	Blueberry																						
590	Commercially prepared (2¾" dia × 2")	1 muffin	57	38	158	3	4	0.8	1.1	1.4	17	27	1.5	32	0.9	70	255	19	5	0.08	0.07	0.6	1
591	Prepared from mix (2¼" dia × 1¾v")	1 muffin	50	36	150	3	4	0.7	1.8	1.5	23	24	0.6	13	0.6	39	219	39	11	0.07	0.16	1.1	1
592	Prepared from recipe, with 2% milk	1 muffin	57	40	162	4	6	1.2	1.5	3.1	21	23	1.1	108	1.3	70	251	80	22	0.16	0.16	1.3	1
593	Bran with raisins, toaster type, toasted	1 muffin	34	27	106	2	3	0.5	0.8	1.7	3	19	2.8	13	1.0	60	179	58	16	0.07	0.10	0.8	0
	Corn																						
594	Commercially prepared (2½" dia × 2¾")	1 muffin	57	33	174	3	5	0.8	1.2	1.8	15	29	1.9	42	1.6	39	297	119	21	0.16	0.19	1.2	0
595	Prepared from mix (2¼" dia × 1½")	1 muffin	50	31	161	4	5	1.4	2.6	0.6	31	25	1.2	38	1.0	66	398	105	23	0.12	0.14	1.1	Tr
596	Oat bran, commercially prepared (2½" dia × 2¼")	1 muffin	57	35	154	4	4	0.6	1.0	2.4	0	28	2.6	36	2.4	289	224	0	0	0.15	0.05	0.2	0
597	Noodles, chow mein, canned	1 cup	45	1	237	4	14	2.0	3.5	7.8	0	26	1.8	9	2.1	54	198	38	4	0.26	0.19	2.7	0
	Noodles (egg noodles), enriched, cooked																						
598	Regular	1 cup	160	69	213	8	2	0.5	0.7	0.7	53	40	1.8	19	2.5	45	11	32	10	0.30	0.13	2.4	0
599	Spinach	1 cup	160	69	211	8	3	0.6	0.8	0.6	53	39	3.7	30	1.7	59	19	165	22	0.39	0.20	2.4	0
600	NUTRI GRAIN Cereal Bar, fruit filled	1 bar	37	15	136	2	3	0.6	1.9	0.3	0	27	0.8	15	1.8	73	110	750	227	0.37	0.41	5.0	0

(Continues)

TABLE B-8 Food Composition Table

FOOD NO.	FOOD DESCRIPTION	MEASURE OF EDIBLE PORTION	WEIGHT (G)	WATER (%)	CALORIES (KCAL)	PROTEIN (G)	TOTAL FAT (G)	FATTY ACIDS SAT (G)	FATTY ACIDS MONO (G)	FATTY ACIDS POLY (G)	CHOLEST (MG)	CARB (G)	TOTAL D F(G)	CA (MG)	FE (MG)	K (MG)	NA (MG)	VITAMIN A (IU)	VITAMIN A (RE)	THMN (MG)	RIBOFL (MG)	NIACIN (MG)	ASCORBIC ACID (MG)
Grain Products (*Continued*)																							
	Oat bran																						
601	Uncooked	1 cup	94	7	231	16	7	1.2	2.2	2.6	0	62	14.5	55	5.1	532	4	0	0	1.10	0.21	0.9	0
602	Cooked	1 cup	219	84	88	7	2	0.4	0.6	0.7	0	25	5.7	22	1.9	201	2	0	0	0.35	0.07	0.3	0
603	Oriental snack mix	1 oz (about ¼ cup)	28	3	156	5	7	1.1	2.8	3.0	0	15	3.7	15	0.7	93	117	1	0	0.09	0.04	0.9	Tr
	Pancakes, plain (4" dia)																						
604	Frozen, ready to heat	1 pancake	36	45	82	2	1	0.3	0.4	0.3	3	16	0.6	22	1.3	26	183	36	10	0.14	0.17	1.4	Tr
605	Prepared from complete mix	1 pancake	38	53	74	2	1	0.2	0.3	0.3	5	14	0.5	48	0.6	67	239	12	3	0.08	0.08	0.7	Tr
606	Prepared from incomplete mix, with 2% milk, egg and oil	1 pancake	38	53	83	3	3	0.8	0.8	1.1	27	11	0.7	82	0.5	76	192	95	27	0.08	0.12	0.5	Tr
	Pie crust, baked																						
	Standard type																						
607	From recipe	1 pie shell	180	10	949	12	62	15.5	27.3	16.4	0	86	3.0	18	5.2	121	976	0	0	0.70	0.50	6.0	0
608	From frozen	1 pie shell	126	11	648	6	41	13.3	19.8	5.1	0	62	1.3	26	2.8	139	815	0	0	0.35	0.48	3.1	0
609	Graham cracker	1 pie shell	239	4	1,181	10	60	12.4	27.2	16.5	0	156	3.6	50	5.2	210	1,365	1,876	483	0.25	0.42	5.1	0
	Pies																						
	Commercially prepared (1/6 of 8" dia)																						
610	Apple	1 piece	117	52	277	2	13	4.4	5.1	2.6	0	40	1.9	13	0.5	76	311	145	35	0.03	0.03	0.3	4
611	Blueberry	1 piece	117	53	271	2	12	2.0	5.0	4.1	0	41	1.2	9	0.4	59	380	164	40	0.01	0.04	0.4	3
612	Cherry	1 piece	117	46	304	2	13	3.0	6.8	2.4	0	47	0.9	14	0.6	95	288	329	63	0.03	0.03	0.2	1
613	Chocolate creme	1 piece	113	44	344	3	22	5.6	12.6	2.7	6	38	2.3	41	1.2	144	154	0	0	0.04	0.12	0.8	0
614	Coconut custard	1 piece	104	49	270	6	14	6.1	5.7	1.2	36	31	1.9	84	0.8	182	348	114	28	0.09	0.15	0.4	1
615	Lemon meringue	1 piece	113	42	303	2	10	2.0	3.0	4.1	51	53	1.4	63	0.7	101	165	198	59	0.07	0.24	0.7	4
616	Pecan	1 piece	113	19	452	5	21	4.0	12.1	3.6	36	65	4.0	19	1.2	84	479	198	53	0.10	0.14	0.3	1
617	Pumpkin	1 piece	109	58	229	4	10	1.9	4.4	3.4	22	30	2.9	65	0.9	168	307	3,743	405	0.06	0.17	0.2	1
	Prepared from recipe (1/8 of 9" dia)																						
618	Apple	1 piece	155	47	411	4	19	4.7	8.4	5.2	0	58	3.6	11	1.7	122	327	90	19	0.23	0.17	1.9	3
619	Blueberry	1 piece	147	51	360	4	17	4.3	7.5	4.5	0	49	3.6	10	1.8	74	272	62	6	0.22	0.19	1.8	1
620	Cherry	1 piece	180	46	486	5	22	5.4	9.6	5.8	0	69	3.5	18	3.3	139	344	736	86	0.27	0.23	2.3	2
621	Lemon meringue	1 piece	127	43	362	5	16	4.0	7.1	4.2	67	50	0.7	15	1.3	83	307	203	56	0.15	0.20	1.2	4

(*Continues*)

TABLE B-8 Food Composition Table

FOOD NO.	FOOD DESCRIPTION	MEASURE OF EDIBLE PORTION	WEIGHT (G)	WATER (%)	CALORIES (KCAL)	PROTEIN (G)	TOTAL FAT (G)	FATTY ACIDS SAT (G)	FATTY ACIDS MONO (G)	FATTY ACIDS POLY (G)	CHOLEST (MG)	CARB (G)	TOTAL D F(G)	CA (MG)	FE (MG)	K (MG)	NA (MG)	VITAMIN A (IU)	VITAMIN A (RE)	THMN (MG)	RIBOFL (MG)	NIACIN (MG)	ASCORBIC ACID (MG)
Grain Products (*Continued*)																							
	Pies (*Continued*)																						
622	Pecan	1 piece	122	20	503	6	27	4.9	13.6	7.0	106	64	2.2	39	1.8	162	320	410	109	0.23	0.22	1.0	Tr
623	Pumpkin	1 piece	155	59	316	7	14	4.9	5.7	2.8	65	41	2.9	146	2.0	288	349	11,833	1,212	0.14	0.31	1.2	3
624	Fried, cherry	1 pie	128	38	404	4	21	3.1	9.5	6.9	0	55	3.3	28	1.6	83	479	220	22	0.18	0.14	1.8	2
	Popcorn																						
625	Air popped, unsalted	1 cup	8	4	31	1	Tr	Tr	0.1	0.2	0	6	1.2	1	0.2	24	Tr	16	2	0.02	0.02	0.2	0
626	Oil popped, salted	1 cup	11	3	55	1	3	0.5	0.9	1.5	0	6	1.1	1	0.3	25	97	17	2	0.01	0.01	0.2	Tr
	Caramel coated																						
627	With peanuts	1 cup	42	3	168	3	3	0.4	1.1	1.4	0	34	1.6	28	1.6	149	124	27	3	0.02	0.05	0.8	0
628	Without peanuts	1 cup	35	3	152	1	5	1.3	1.0	1.6	2	28	1.8	15	0.6	38	73	18	4	0.02	0.02	0.8	0
629	Cheese flavor	1 cup	11	3	58	1	4	0.7	1.1	1.7	1	6	1.1	12	0.2	29	98	27	5	0.01	0.03	0.2	Tr
630	Popcorn cake	1 cake	10	5	38	1	Tr	Tr	0.1	0.1	0	8	0.3	1	0.2	33	29	7	1	0.01	0.02	0.6	0
	Pretzels, made with enriched flour																						
631	Stick, 2¼" long	10 pretzels	3	3	11	Tr	Tr	Tr	Tr	Tr	0	2	0.1	1	0.1	4	51	0	0	0.01	0.02	0.2	0
632	Twisted, regular	10 pretzels	60	3	229	5	2	0.5	0.8	0.7	0	48	1.9	22	2.6	88	1,029	0	0	0.28	0.37	3.2	0
633	Twisted, dutch, 2¾" × 2⅛"	1 pretzel	16	3	61	1	1	0.1	0.2	0.2	0	13	0.5	6	0.7	23	274	0	0	0.07	0.10	0.8	0
	Rice																						
634	Brown, long grain, cooked	1 cup	195	73	216	5	2	0.4	0.6	0.6	0	45	3.5	20	0.8	84	10	0	0	0.19	0.05	3.0	0
	White, long grain, enriched																						
	Regular																						
635	Raw	1 cup	185	12	675	13	1	0.3	0.4	0.3	0	148	2.4	52	8.0	213	9	0	0	1.07	0.09	7.8	0
636	Cooked	1 cup	158	68	205	4	Tr	0.1	0.1	0.1	0	45	0.6	16	1.9	55	2	0	0	0.26	0.02	2.3	0
637	Instant, prepared	1 cup	165	76	162	3	Tr	0.1	0.1	0.1	0	35	1.0	13	1.0	7	5	0	0	0.12	0.08	1.5	0
	Parboiled																						
638	Raw	1 cup	185	10	686	13	1	0.3	0.3	0.3	0	151	3.1	111	6.6	222	9	0	0	1.10	0.13	6.7	0
639	Cooked	1 cup	175	72	200	4	Tr	0.1	0.1	0.1	0	43	0.7	33	2.0	65	5	0	0	0.44	0.03	2.5	0
640	Wild, cooked	1 cup	164	74	166	7	1	0.1	0.1	0.3	0	35	3.0	5	1.0	166	5	0	0	0.09	0.14	2.1	0
641	Rice cake, brown rice, plain	1 cake	9	6	35	1	Tr	0.1	0.1	0.1	0	7	0.4	1	0.1	26	29	4	Tr	0.01	0.01	0.7	0
642	RICE KRISPIES Treat Squares	1 bar	22	6	91	1	2	0.3	0.6	1.1	0	18	0.1	1	0.5	9	77	200	60	0.15	0.18	2.0	0

(*Continues*)

TABLE B-8 Food Composition Table

FOOD NO.	FOOD DESCRIPTION	MEASURE OF EDIBLE PORTION	WEIGHT (G)	WATER (%)	CALORIES (KCAL)	PROTEIN (G)	TOTAL FAT (G)	FATTY ACIDS SAT (G)	MONO (G)	POLY (G)	CHOLEST (MG)	CARB (G)	TOTAL D F(G)	CA (MG)	FE (MG)	K (MG)	NA (MG)	VITAMIN A (IU)	(RE)	THMN (MG)	RIBOFL (MG)	NIACIN (MG)	ASCORBIC ACID (MG)
	Grain Products (*Continued*)																						
	Rolls																						
643	Dinner	1 roll	28	32	84	2	2	0.5	1.0	0.3	Tr	14	0.8	33	0.9	37	146	0	0	0.14	0.09	1.1	Tr
644	Hamburger or hotdog	1 roll	43	34	123	4	2	0.5	0.4	1.1	0	22	1.2	60	1.4	61	241	0	0	0.21	0.13	1.7	Tr
645	Hard, kaiser	1 roll	57	31	167	6	2	0.3	0.6	1.0	0	30	1.3	54	1.9	62	310	0	0	0.27	0.19	2.4	0
	Spaghetti, cooked																						
646	Enriched	1 cup	140	66	197	7	1	0.1	0.1	0.4	0	40	2.4	10	2.0	43	1	0	0	0.29	0.14	2.3	0
647	Whole wheat	1 cup	140	67	174	7	1	0.1	0.1	0.3	0	37	6.3	21	1.5	62	4	0	0	0.15	0.06	1.0	0
	Sweet rolls, cinnamon																						
648	Commercial, with raisins	1 roll	60	25	223	4	10	1.8	2.9	4.5	40	31	1.4	43	1.0	67	230	129	38	0.19	0.16	1.4	1
649	Refrigerated dough, baked, with frosting	1 roll	30	23	109	2	4	1.0	2.2	0.5	0	17	0.6	10	0.8	19	250	1	0	0.12	0.07	1.1	Tr
650	Taco shell, baked	1 medium	13	6	62	1	3	0.4	1.2	1.1	0	8	1.0	21	0.3	24	49	0	0	0.03	0.01	0.2	0
651	Tapioca, pearl, dry	1 cup	152	11	544	Tr	Tr	Tr	Tr	Tr	0	135	1.4	30	2.4	17	2	0	0	0.01	0.00	0.0	0
	Toaster pastries																						
652	Brown sugar cinnamon	1 pastry	50	11	206	3	7	1.8	4.0	0.9	0	34	0.5	17	2.0	57	212	493	112	0.19	0.29	2.3	Tr
653	Chocolate with frosting	1 pastry	52	13	201	3	5	1.0	2.7	1.1	0	37	0.6	20	1.8	82	203	500	NA	0.16	0.16	2.0	0
654	Fruit filled	1 pastry	52	12	204	2	5	0.8	2.2	2.0	0	37	1.1	14	1.8	58	218	501	2	0.15	0.19	2.0	Tr
655	Low fat	1 pastry	52	12	193	2	3	0.7	1.7	0.5	0	40	0.8	23	1.8	34	131	494	49	0.15	0.29	2.0	2
	Tortilla chips																						
	Plain																						
656	Regular	1 oz	28	2	142	2	7	1.4	4.4	1.0	0	18	1.8	44	0.4	56	150	56	6	0.02	0.05	0.4	0
657	Low fat, baked	10 chips	14	2	54	2	1	0.1	0.2	0.4	0	11	0.7	22	0.2	37	57	52	6	0.03	0.04	0.1	Tr
	Nacho flavor																						
658	Regular	1 oz	28	2	141	2	7	1.4	4.3	1.0	1	18	1.5	42	0.4	61	201	105	12	0.04	0.05	0.4	1
659	Light, reduced fat	1 oz	28	1	126	2	4	0.8	2.5	0.6	1	20	1.4	45	0.5	77	284	108	12	0.06	0.08	0.1	Tr
	Tortillas, ready to cook (about 6" dia)																						
660	Corn	1 tortilla	26	44	58	1	1	0.1	0.2	0.3	0	12	1.4	46	0.4	40	42	0	0	0.03	0.02	0.4	0
661	Flour	1 tortilla	32	27	104	3	2	0.6	1.2	0.3	0	18	1.1	40	1.1	42	153	0	0	0.17	0.09	1.1	0
	Waffles, plain																						
662	Prepared from recipe, 7" dia	1 waffle	75	42	218	6	11	2.1	2.6	5.1	52	25	0.7	191	1.7	119	383	171	49	0.20	0.26	1.6	Tr
663	Frozen, toasted, 4" dia	1 waffle	33	42	87	2	3	0.5	1.1	0.9	8	13	0.8	77	1.5	42	260	400	120	0.13	0.16	1.5	0

(*Continues*)

TABLE B-8 Food Composition Table

FOOD NO.	FOOD DESCRIPTION	MEASURE OF EDIBLE PORTION	WEIGHT (G)	WATER (%)	CALORIES (KCAL)	PROTEIN (G)	TOTAL FAT (G)	FATTY ACIDS SAT (G)	MONO (G)	POLY (G)	CHOLEST (MG)	CARB (G)	TOTAL D F(G)	CA (MG)	FE (MG)	K (MG)	NA (MG)	VITAMIN A (IU)	(RE)	THMN (MG)	RIBOFL (MG)	NIACIN (MG)	ASCORBIC ACID (MG)
Grain Products (*Continued*)																							
	Waffles, plain (*Continued*)																						
664	Low fat, 4" dia	1 waffle	35	43	83	2	1	0.3	0.4	0.4	9	15	0.4	20	1.9	50	155	506	NA	0.31	0.26	2.6	0
	Wheat flours																						
	All purpose, enriched																						
665	Sifted, spooned	1 cup	115	12	419	12	1	0.2	0.1	0.5	0	88	3.1	17	5.3	123	2	0	0	0.90	0.57	6.8	0
666	Unsifted, spooned	1 cup	125	12	455	13	1	0.2	0.1	0.5	0	95	3.4	19	5.8	134	3	0	0	0.98	0.62	7.4	0
667	Bread, enriched	1 cup	137	13	495	16	2	0.3	0.2	1.0	0	99	3.3	21	6.0	137	3	0	0	1.11	0.70	10.3	0
668	Cake or pastry flour, enriched, unsifted, spooned	1 cup	137	13	496	11	1	0.2	0.1	0.5	0	107	2.3	19	10.0	144	3	0	0	1.22	0.59	9.3	0
669	Self rising, enriched, unsifted, spooned	1 cup	125	11	443	12	1	0.2	0.1	0.5	0	93	3.4	423	5.8	155	1,588	0	0	0.84	0.52	7.3	0
670	Whole wheat, from hard wheats, stirred, spooned	1 cup	120	10	407	16	2	0.4	0.3	0.9	0	87	14.6	41	4.7	486	6	0	0	0.54	0.26	7.6	0
671	Wheat germ, toasted, plain	1 tbsp	7	6	27	2	1	0.1	0.1	0.5	0	3	0.9	3	0.6	66	Tr	0	0	0.12	0.06	0.4	Tr
Legumes, Nuts, and Seeds																							
	Almonds, shelled																						
672	Sliced	1 cup	95	5	549	20	48	3.7	30.5	11.6	0	19	11.2	236	4.1	692	1	10	1	0.23	0.77	3.7	0
673	Whole	1 oz (24 nuts)	28	5	164	6	14	1.1	9.1	3.5	0	6	3.3	70	1.2	206	Tr	3	Tr	0.07	0.23	1.1	0
	Beans, dry																						
	Cooked																						
674	Black	1 cup	172	66	227	15	1	0.2	0.1	0.4	0	41	15.0	46	3.6	611	2	10	2	0.42	0.10	0.9	0
675	Great Northern	1 cup	177	69	209	15	1	0.2	Tr	0.3	0	37	12.4	120	3.8	692	4	2	0	0.28	0.10	1.2	2
676	Kidney, red	1 cup	177	67	225	15	1	0.1	0.1	0.5	0	40	13.1	50	5.2	713	4	0	0	0.28	0.10	1.0	2
677	Lima, large	1 cup	188	70	216	15	1	0.2	0.1	0.3	0	39	13.2	32	4.5	955	4	0	0	0.30	0.10	0.8	0
678	Pea (navy)	1 cup	182	63	258	16	1	0.3	0.1	0.4	0	48	11.6	127	4.5	670	2	4	0	0.37	0.11	1.0	2
679	Pinto	1 cup	171	64	234	14	1	0.2	0.2	0.3	0	44	14.7	82	4.5	800	3	3	0	0.32	0.16	0.7	4
	Canned, solids and liquid																						
	Baked beans																						
680	Plain or vegetarian	1 cup	254	73	236	12	1	0.3	0.1	0.5	0	52	12.7	127	0.7	752	1,008	434	43	0.39	0.15	1.1	8
681	With frankfurters	1 cup	259	69	368	17	17	6.1	7.3	2.2	16	40	17.9	124	4.5	609	1,114	399	39	0.15	0.15	2.3	6

(*Continues*)

TABLE B-8 Food Composition Table

FOOD NO.	FOOD DESCRIPTION	MEASURE OF EDIBLE PORTION	WEIGHT (G)	WATER (%)	CALORIES (KCAL)	PROTEIN (G)	TOTAL FAT (G)	FATTY ACIDS SAT (G)	MONO (G)	POLY (G)	CHOLEST (MG)	CARB (G)	TOTAL D F(G)	CA (MG)	FE (MG)	K (MG)	NA (MG)	VITAMIN A (IU)	(RE)	THMN (MG)	RIBOFL (MG)	NIACIN (MG)	ASCORBIC ACID (MG)
Legumes, Nuts, and Seeds (*Continued*)																							
	Beans, dry (*Continued*)																						
682	With pork in tomato sauce	1 cup	253	73	248	13	3	1.0	1.1	0.3	18	49	12.1	142	8.3	759	1,113	314	30	0.13	0.12	1.3	8
683	With pork in sweet sauce	1 cup	253	71	281	13	4	1.4	1.6	0.5	18	53	13.2	154	4.2	673	850	288	28	0.12	0.15	0.9	8
684	Kidney, red	1 cup	256	77	218	13	1	0.1	0.1	0.5	0	40	16.4	61	3.2	658	873	0	0	0.27	0.23	1.2	3
685	Lima, large	1 cup	241	77	190	12	Tr	0.1	Tr	0.2	0	36	11.6	51	4.4	530	810	0	0	0.13	0.08	0.6	0
686	White	1 cup	262	70	307	19	1	0.2	0.1	0.3	0	57	12.6	191	7.8	1,189	13	0	0	0.25	0.10	0.3	0
	Black eyed peas, dry																						
687	Cooked	1 cup	172	70	200	13	1	0.2	0.1	0.4	0	36	11.2	41	4.3	478	7	26	3	0.35	0.09	0.9	1
688	Canned, solids and liquid	1 cup	240	80	185	11	1	0.3	0.1	0.6	0	33	7.9	48	2.3	413	718	31	2	0.18	0.18	0.8	6
689	Brazil nuts, shelled	1 oz (6-8 nuts)	28	3	186	4	19	4.6	6.5	6.8	0	4	1.5	50	1.0	170	1	0	0	0.28	0.03	0.5	Tr
690	Carob flour	1 cup	103	4	229	5	1	0.1	0.2	0.2	0	92	41.0	358	3.0	852	36	14	1	0.05	0.47	2.0	Tr
	Cashews, salted																						
691	Dry roasted	1 oz	28	2	163	4	13	2.6	7.7	2.2	0	9	0.9	13	1.7	160	181	0	0	0.06	0.06	0.4	0
692	Oil roasted	1 cup	130	4	749	21	63	12.4	36.9	10.6	0	37	4.9	53	5.3	689	814	0	0	0.55	0.23	2.3	0
693		1 oz (18 nuts)	28	4	163	5	14	2.7	8.1	2.3	0	8	1.1	12	1.2	150	177	0	0	0.12	0.05	0.5	0
694	Chestnuts, European, roasted, shelled	1 cup	143	40	350	5	3	0.6	1.1	1.2	0	76	7.3	41	1.3	847	3	34	3	0.35	0.25	1.9	37
	Chickpeas, dry																						
695	Cooked	1 cup	164	60	269	15	4	0.4	1.0	1.9	0	45	12.5	80	4.7	477	11	44	5	0.19	0.10	0.9	2
696	Canned, solids and liquid	1 cup	240	70	286	12	3	0.3	0.6	1.2	0	54	10.6	77	3.2	413	718	58	5	0.07	0.08	0.3	9
	Coconut																						
	Raw																						
697	Piece, about 2" × 2" × ½"	1 piece	45	47	159	1	15	13.4	0.6	0.2	0	7	4.1	6	1.1	160	9	0	0	0.03	0.01	0.2	1
698	Shredded, not packed	1 cup	80	47	283	3	27	23.8	1.1	0.3	0	12	7.2	11	1.9	285	16	0	0	0.05	0.02	0.4	3
699	Dried, sweetened, shredded	1 cup	93	13	466	3	33	29.3	1.4	0.4	0	44	4.2	14	1.8	313	244	0	0	0.03	0.02	0.4	1
700	Hazelnuts (filberts), chopped	1 cup	115	5	722	17	70	5.1	52.5	9.1	0	19	11.2	131	5.4	782	0	46	5	0.74	0.13	2.1	7
701		1 oz	28	5	178	4	17	1.3	12.9	2.2	0	5	2.7	32	1.3	193	0	11	1	0.18	0.03	0.5	2
702	Hummus, commercial	1 tbsp	14	67	23	1	1	0.2	0.6	0.5	0	2	0.8	5	0.3	32	53	4	Tr	0.03	0.01	0.1	0
703	Lentils, dry, cooked	1 cup	198	70	230	18	1	0.1	0.1	0.3	0	40	15.6	38	6.6	731	4	16	2	0.33	0.14	2.1	3

(*Continues*)

TABLE B-8 Food Composition Table

FOOD NO.	FOOD DESCRIPTION	MEASURE OF EDIBLE PORTION	WEIGHT (G)	WATER (%)	CALORIES (KCAL)	PROTEIN (G)	TOTAL FAT (G)	FATTY ACIDS SAT (G)	MONO (G)	POLY (G)	CHOLEST (MG)	CARB (G)	TOTAL D F(G)	CA (MG)	FE (MG)	K (MG)	NA (MG)	VITAMIN A (IU)	(RE)	THMN (MG)	RIBOFL (MG)	NIACIN (MG)	ASCORBIC ACID (MG)
Legumes, Nuts, and Seeds (*Continued*)																							
704	Macadamia nuts, dry roasted, salted	1 cup	134	2	959	10	102	16.0	79.4	2.0	0	17	10.7	94	3.6	486	355	0	0	0.95	0.12	3.0	1
705		1 oz (10-12 nuts)	28	2	203	2	22	3.4	16.8	0.4	0	4	2.3	20	0.8	103	75	0	0	0.20	0.02	0.6	Tr
	Mixed nuts, with peanuts, salted																						
706	Dry roasted	1 oz	28	2	168	5	15	2.0	8.9	3.1	0	7	2.6	20	1.0	169	190	4	Tr	0.06	0.06	1.3	Tr
707	Oil roasted	1 oz	28	2	175	5	16	2.5	9.0	3.8	0	6	2.6	31	0.9	165	185	5	1	0.14	0.06	1.4	Tr
	Peanuts																						
	Dry roasted																						
708	Salted	1 oz (about 28)	28	2	166	7	14	2.0	7.0	4.4	0	6	2.3	15	0.6	187	230	0	0	0.12	0.03	3.8	0
709	Unsalted	1 cup	146	2	854	35	73	10.1	36.0	22.9	0	31	11.7	79	3.3	961	9	0	0	0.64	0.14	19.7	0
710		1 oz (about 28)	28	2	166	7	14	2.0	7.0	4.4	0	6	2.3	15	0.6	187	2	0	0	0.12	0.03	3.8	0
711	Oil roasted, salted	1 cup	144	2	837	38	71	9.9	35.2	22.4	0	27	13.2	127	2.6	982	624	0	0	0.36	0.16	20.6	0
712		1 oz	28	2	165	7	14	1.9	6.9	4.4	0	5	2.6	25	0.5	193	123	0	0	0.07	0.03	4.0	0
	Peanut butter																						
	Regular																						
713	Smooth style	1 tbsp	16	1	95	4	8	1.7	3.9	2.2	0	3	0.9	6	0.3	107	75	0	0	0.01	0.02	2.1	0
714	Chunk style	1 tbsp	16	1	94	4	8	1.5	3.8	2.3	0	3	1.1	7	0.3	120	78	0	0	0.02	0.02	2.2	0
715	Reduced fat, smooth	1 tbsp	18	1	94	5	6	1.3	2.9	1.8	0	6	0.9	6	0.3	120	97	0	0	0.05	0.01	2.6	0
716	Peas, split, dry, cooked	1 cup	196	69	231	16	1	0.1	0.2	0.3	0	41	16.3	27	2.5	710	4	14	2	0.37	0.11	1.7	1
717	Pecans, halves	1 cup	108	4	746	10	78	6.7	44.0	23.3	0	15	10.4	76	2.7	443	0	83	9	0.71	0.14	1.3	1
718		1 oz (20 halves)	28	4	196	3	20	1.8	11.6	6.1	0	4	2.7	20	0.7	116	0	22	2	0.19	0.04	0.3	Tr
719	Pine nuts (pignolia), shelled	1 oz	28	7	160	7	14	2.2	5.4	6.1	0	4	1.3	7	2.6	170	1	8	1	0.23	0.05	1.0	1
720		1 tbsp	9	7	49	2	4	0.7	1.6	1.8	0	1	0.4	2	0.8	52	Tr	2	Tr	0.07	0.02	0.3	Tr
721	Pistachio nuts, dry roasted, with salt, shelled	1 oz (47 nuts)	28	2	161	6	13	1.6	6.8	3.9	0	8	2.9	31	1.2	293	121	151	15	0.24	0.04	0.4	1
722	Pumpkin and squash kernels, roasted, with salt	1 oz (142 seeds)	28	7	148	9	12	2.3	3.7	5.4	0	4	1.1	12	4.2	229	163	108	11	0.06	0.09	0.5	1

(Continues)

TABLE B-8 Food Composition Table

FOOD NO.	FOOD DESCRIPTION	MEASURE OF EDIBLE PORTION	WEIGHT (G)	WATER (%)	CALORIES (KCAL)	PROTEIN (G)	TOTAL FAT (G)	FATTY ACIDS			CHOLEST (MG)	CARB (G)	TOTAL D F(G)	CA (MG)	FE (MG)	K (MG)	NA (MG)	VITAMIN A		THMN (MG)	RIBOFL (MG)	NIACIN (MG)	ASCORBIC ACID (MG)
								SAT (G)	MONO (G)	POLY (G)								(IU)	(RE)				
Legumes, Nuts, and Seeds (*Continued*)																							
723	Refried beans, canned	1 cup	252	76	237	14	3	1.2	1.4	0.4	20	39	13.4	88	4.2	673	753	0	0	0.07	0.04	0.8	15
724	Sesame seeds	1 tbsp	8	5	47	2	4	0.6	1.7	1.9	0	1	0.9	10	0.6	33	3	5	1	0.06	0.01	0.4	0
725	Soybeans, dry, cooked	1 cup	172	63	298	29	15	2.2	3.4	8.7	0	17	10.3	175	8.8	886	2	15	2	0.27	0.49	0.7	3
	Soy products																						
726	Miso	1 cup	275	41	567	32	17	2.4	3.7	9.4	0	77	14.9	182	7.5	451	10,029	239	25	0.27	0.69	2.4	0
727	Soy milk	1 cup	245	93	81	7	5	0.5	0.8	2.0	0	4	3.2	10	1.4	345	29	78	7	0.39	0.17	0.4	0
	Tofu																						
728	Firm	¼ block	81	84	62	7	4	0.5	0.8	2.0	0	2	0.3	131	1.2	143	6	6	1	0.08	0.08	Tr	Tr
729	Soft, piece 2½" × 2¾" × 1	1 piece	120	87	73	8	4	0.6	1.0	2.5	0	2	0.2	133	1.3	144	10	8	1	0.06	0.04	0.6	Tr
730	Sunflower seed kernels, dry roasted, with salt	¼ cup	32	1	186	6	16	1.7	3.0	10.5	0	8	2.9	22	1.2	272	250	0	0	0.03	0.08	2.3	Tr
731		1 oz	28	1	165	5	14	1.5	2.7	9.3	0	7	2.6	20	1.1	241	221	0	0	0.03	0.07	2.0	Tr
732	Tahini	1 tbsp	15	3	89	3	8	1.1	3.0	3.5	0	3	1.4	64	1.3	62	17	10	1	0.18	0.07	0.8	0
733	Walnuts, English	1 cup, chopped	120	4	785	18	78	7.4	10.7	56.6	0	16	8.0	125	3.5	529	2	49	5	0.41	0.18	2.3	2
734		1 oz (14 halves)	28	4	185	4	18	1.7	2.5	13.4	0	4	1.9	29	0.8	125	1	12	1	0.10	0.04	0.5	Tr
Meat and Meat Products																							
	Beef, cooked																						
	Cuts braised, simmered, or pot roasted																						
	Relatively fat, such as chuck blade, piece, 2½" × 2½" × ¾"																						
735	Lean and fat	3 oz	85	47	293	23	22	8.7	9.4	0.8	88	0	0.0	11	2.6	196	54	0	0	0.06	0.20	2.1	0
736	Lean only	3 oz	85	55	213	26	11	4.3	4.8	0.4	90	0	0.0	11	3.1	224	60	0	0	0.07	0.24	2.3	0
	Relatively lean, such as bottom round, piece, 4¹/₈," × 2¼" × ½"																						
737	Lean and fat	3 oz	85	52	234	24	14	5.4	6.2	0.5	82	0	0.0	5	2.7	240	43	0	0	0.06	0.20	3.2	0
738	Lean only	3 oz	85	58	178	27	7	2.4	3.1	0.3	82	0	0.0	4	2.9	262	43	0	0	0.06	0.22	3.5	0
	Ground beef, broiled																						
739	83% lean	3 oz	85	57	218	22	14	5.5	6.1	0.5	71	0	0.0	6	2.0	266	60	0	0	0.05	0.23	4.2	0

(*Continues*)

TABLE B-8 Food Composition Table

FOOD NO.	FOOD DESCRIPTION	MEASURE OF EDIBLE PORTION	WEIGHT (G)	WATER (%)	CALORIES (KCAL)	PROTEIN (G)	TOTAL FAT (G)	FATTY ACIDS			CHOLEST (MG)	CARB (G)	TOTAL D F(G)	CA (MG)	FE (MG)	K (MG)	NA (MG)	VITAMIN A		THMN (MG)	RIBOFL (MG)	NIACIN (MG)	ASCORBIC ACID (MG)
								SAT (G)	MONO (G)	POLY (G)								(IU)	(RE)				
	Meat and Meat Products (*Continued*)																						
	Beef, cooked (*Continued*)																						
740	79% lean	3 oz	85	56	231	21	16	6.2	6.9	0.6	74	0	0.0	9	1.8	256	65	0	0	0.04	0.18	4.4	0
741	73% lean	3 oz	85	54	246	20	18	6.9	7.7	0.7	77	0	0.0	9	2.1	248	71	0	0	0.03	0.16	4.9	0
742	Liver, fried, slice, 6½" × 2³/₈" × ³/₈"	3 oz	85	56	184	23	7	2.3	1.4	1.5	410	7	0.0	9	5.3	309	90	30,689	9,120	0.18	3.52	12.3	20
	Roast, oven cooked, no liquid added																						
	Relatively fat, such as rib, 2 pieces, 4¹/₈" × 2¼" × ¼"																						
743	Lean and fat	3 oz	85	47	304	19	25	9.9	10.6	0.9	71	0	0.0	9	2.0	256	54	0	0	0.06	0.14	2.9	0
744	Lean only	3 oz	85	59	195	23	11	4.2	4.5	0.3	68	0	0.0	9	2.4	318	61	0	0	0.07	0.18	3.5	0
	Relatively lean, such as eye of round, two pieces, 2½" × 2½" × ³/₈"																						
745	Lean and fat	3 oz	85	59	195	23	11	4.2	4.7	0.4	61	0	0.0	5	1.6	308	50	0	0	0.07	0.14	3.0	0
746	Lean only	3 oz	85	65	143	25	4	1.5	1.8	0.1	59	0	0.0	4	1.7	336	53	0	0	0.08	0.14	3.2	0
	Steak, sirloin, broiled, piece, 2½" × 2½" × ³/₈"																						
747	Lean and fat	3 oz	85	57	219	24	13	5.2	5.6	0.5	77	0	0.0	9	2.6	311	54	0	0	0.09	0.23	3.3	0
748	Lean only	3 oz	85	62	166	26	6	2.4	2.6	0.2	76	0	0.0	9	2.9	343	56	0	0	0.11	0.25	3.6	0
749	Beef, canned, corned	3 oz	85	58	213	23	13	5.3	5.1	0.5	73	0	0.0	10	1.8	116	855	0	0	0.02	0.12	2.1	0
750	Beef, dried, chipped	1 oz	28	57	47	8	1	0.5	0.5	0.1	12	Tr	0.0	2	1.3	126	984	0	0	0.02	0.06	1.5	0
	Lamb, cooked																						
	Chops																						
	Arm, braised																						
751	Lean and fat	3 oz	85	44	294	26	20	8.4	8.7	1.5	102	0	0.0	21	2.0	260	61	0	0	0.06	0.21	5.7	0
752	Lean only	3 oz	85	49	237	30	12	4.3	5.2	0.8	103	0	0.0	22	2.3	287	65	0	0	0.06	0.23	5.4	0
	Loin, broiled																						
753	Lean and fat	3 oz	85	52	269	21	20	8.4	8.2	1.4	85	0	0.0	17	1.5	278	65	0	0	0.09	0.21	6.0	0
754	Lean only	3 oz	85	61	184	25	8	3.0	3.6	0.5	81	0	0.0	16	1.7	320	71	0	0	0.09	0.24	5.8	0
	Leg, roasted, two pieces, 4¹/₈" × 2¼" × ¼"																						
755	Lean and fat	3 oz	85	57	219	22	14	5.9	5.9	1.0	79	0	0.0	9	1.7	266	56	0	0	0.09	0.23	5.6	0

(*Continues*)

TABLE B-8 Food Composition Table

FOOD NO.	FOOD DESCRIPTION	MEASURE OF EDIBLE PORTION	WEIGHT (G)	WATER (%)	CALORIES (KCAL)	PROTEIN (G)	TOTAL FAT (G)	FATTY ACIDS SAT (G)	MONO (G)	POLY (G)	CHOLEST (MG)	CARB (G)	TOTAL D F(G)	CA (MG)	FE (MG)	K (MG)	NA (MG)	VITAMIN A (IU)	(RE)	THMN (MG)	RIBOFL (MG)	NIACIN (MG)	ASCORBIC ACID (MG)
Meat and Meat Products (*Continued*)																							
	Lamb, cooked (*Continued*)																						
756	Lean only	3 oz	85	64	162	24	7	2.3	2.9	0.4	76	0	0.0	7	1.8	287	58	0	0	0.09	0.25	5.4	0
	Rib, roasted, three pieces, 2½" × 2½" × ¼"																						
757	Lean and fat	3 oz	85	48	305	18	25	10.9	10.6	1.8	82	0	0.0	19	1.4	230	62	0	0	0.08	0.18	5.7	0
758	Lean only	3 oz	85	60	197	22	11	4.0	5.0	0.7	75	0	0.0	18	1.5	268	69	0	0	0.08	0.20	5.2	0
	Pork, cured, cooked																						
	Bacon																						
759	Regular	3 medium slices	19	13	109	6	9	3.3	4.5	1.1	16	Tr	0.0	2	0.3	92	303	0	0	0.13	0.05	1.4	0
760	Canadian style (6slices per 6-oz pkg)	2 slices	47	62	86	11	4	1.3	1.9	0.4	27	1	0.0	5	0.4	181	719	0	0	0.38	0.09	3.2	0
	Ham, light cure, roasted, two pieces, 4⅛" × 2¼" × ¼"																						
761	Lean and fat	3 oz	85	58	207	18	14	5.1	6.7	1.5	53	0	0.0	6	0.7	243	1,009	0	0	0.51	0.19	3.8	0
762	Lean only	3 oz	85	66	133	21	5	1.6	2.2	0.5	47	0	0.0	6	0.8	269	1,128	0	0	0.58	0.22	4.3	0
763	Ham, canned, roasted, two pieces, 4⅛" × 2¼" × ¼"	3 oz	85	67	142	18	7	2.4	3.5	0.8	35	Tr	0.0	6	0.9	298	908	0	0	0.82	0.21	4.3	0
	Pork, fresh, cooked																						
	Chop, loin (cut 3 per lb with bone)																						
	Broiled																						
764	Lean and fat	3 oz	85	58	204	24	11	4.1	5.0	0.8	70	0	0.0	28	0.7	304	49	8	3	0.91	0.24	4.5	Tr
765	Lean only	3 oz	85	61	172	26	7	2.5	3.1	0.5	70	0	0.0	26	0.7	319	51	7	2	0.98	0.26	4.7	Tr
	Pan fried																						
766	Lean and fat	3 oz	85	53	235	25	14	5.1	6.0	1.6	78	0	0.0	23	0.8	361	68	7	2	0.97	0.26	4.8	1
767	Lean only	3 oz	85	57	197	27	9	3.1	3.8	1.1	78	0	0.0	20	0.8	382	73	7	2	1.06	0.28	5.1	1
	Ham (leg), roasted, piece, 2½" × 2½" × ¾"																						
768	Lean and fat	3 oz	85	55	232	23	15	5.5	6.7	1.4	80	0	0.0	12	0.9	299	51	9	3	0.54	0.27	3.9	Tr
769	Lean only	3 oz	85	61	179	25	8	2.8	3.8	0.7	80	0	0.0	6	1.0	317	54	8	3	0.59	0.30	4.2	Tr
	Rib roast, piece, 2½" × 2½" × ¾"																						

(*Continues*)

TABLE B-8 Food Composition Table

FOOD NO.	FOOD DESCRIPTION	MEASURE OF EDIBLE PORTION	WEIGHT (G)	WATER (%)	CALORIES (KCAL)	PROTEIN (G)	TOTAL FAT (G)	FATTY ACIDS SAT (G)	FATTY ACIDS MONO (G)	FATTY ACIDS POLY (G)	CHOLEST (MG)	CARB (G)	TOTAL D F(G)	CA (MG)	FE (MG)	K (MG)	NA (MG)	VITAMIN A (IU)	VITAMIN A (RE)	THMN (MG)	RIBOFL (MG)	NIACIN (MG)	ASCORBIC ACID (MG)
	Meat and Meat Products (*Continued*)																						
	Pork, fresh, cooked (*Continued*)																						
770	Lean and fat	3 oz	85	56	217	23	13	5.0	5.9	1.1	62	0	0.0	24	0.8	358	39	5	2	0.62	0.26	5.2	Tr
771	Lean only	3 oz	85	59	190	24	9	3.7	4.5	0.7	60	0	0.0	22	0.8	371	40	5	2	0.64	0.27	5.5	Tr
	Ribs, lean and fat, cooked																						
772	Backribs, roasted	3 oz	85	45	315	21	25	9.3	11.4	2.0	100	0	0.0	38	1.2	268	86	8	3	0.36	0.17	3.0	Tr
773	Country style, braised	3 oz	85	54	252	20	18	6.8	7.9	1.6	74	0	0.0	25	1.0	279	50	7	2	0.43	0.22	3.3	1
774	Spareribs, braised	3 oz	85	40	337	25	26	9.5	11.5	2.3	103	0	0.0	40	1.6	272	79	9	3	0.35	0.32	4.7	0
	Shoulder cut, braised, three pieces, 2½" × 2½" × ¼"																						
775	Lean and fat	3 oz	85	48	280	24	20	7.2	8.8	1.9	93	0	0.0	15	1.4	314	75	8	3	0.46	0.26	4.4	Tr
776	Lean only	3 oz	85	54	211	27	10	3.5	4.9	1.0	97	0	0.0	7	1.7	344	87	7	2	0.51	0.31	5.0	Tr
	Sausages and luncheon meats																						
777	Bologna, beef and pork (8 slices per 8-oz pkg)	2 slices	57	54	180	7	16	6.1	7.6	1.4	31	2	0.0	7	0.9	103	581	0	0	0.10	0.08	1.5	0
778	Braunschweiger (6 slices per 6-oz pkg)	2 slices	57	48	205	8	18	6.2	8.5	2.1	89	2	0.0	5	5.3	113	652	8,009	2,405	0.14	0.87	4.8	0
779	Brown and serve, cooked, link, 4" × ⅞," raw	2 links	26	45	103	4	9	3.4	4.5	1.0	18	1	0.0	3	0.3	49	209	0	0	0.09	0.04	0.9	0
	Canned, minced luncheon meat																						
780	Pork, ham, and chicken, reduced sodium (7 slices per 7-oz can)	2 slices	57	56	172	7	15	5.1	7.1	1.5	43	1	0.0	0	0.4	321	539	0	0	0.15	0.10	1.8	18
781	Pork with ham (12 slices per 12-oz can)	2 slices	57	52	188	8	17	5.7	7.7	1.2	40	1	0.0	0	0.4	233	758	0	0	0.18	0.10	2.0	0
782	Pork and chicken (12 slices per 12-ozcan)	2 slices	57	64	117	9	8	2.7	3.8	0.8	43	1	0.0	0	0.7	352	539	0	0	0.10	0.12	2.0	18
783	Chopped ham (8 slices per 6-oz pkg)	2 slices	21	64	48	4	4	1.2	1.7	0.4	11	0	0.0	1	0.2	67	288	0	0	0.13	0.04	0.8	0
	Cooked ham (8 slices per 8-oz pkg)																						
784	Regular	2 slices	57	65	104	10	6	1.9	2.8	0.7	32	2	0.0	4	0.6	189	751	0	0	0.49	0.14	3.0	0
785	Extra lean	2 slices	57	71	75	11	3	0.9	1.3	0.3	27	1	0.0	4	0.4	200	815	0	0	0.53	0.13	2.8	0

(*Continues*)

TABLE B-8 Food Composition Table

								FATTY ACIDS										VITAMIN A					
FOOD NO.	FOOD DESCRIPTION	MEASURE OF EDIBLE PORTION	WEIGHT (G)	WATER (%)	CALORIES (KCAL)	PROTEIN (G)	TOTAL FAT (G)	SAT (G)	MONO (G)	POLY (G)	CHOLEST (MG)	CARB (G)	TOTAL D F(G)	CA (MG)	FE (MG)	K (MG)	NA (MG)	(IU)	(RE)	THMN (MG)	RIBOFL (MG)	NIACIN (MG)	ASCORBIC ACID (MG)
Meat and Meat Products (*Continued*)																							
	Frankfurter (10 per 1-lb pkg), heated																						
786	Beef and pork	1 frank	45	54	144	5	13	4.8	6.2	1.2	23	1	0.0	5	0.5	75	504	0	0	0.09	0.05	1.2	0
787	Beef	1 frank	45	55	142	5	13	5.4	6.1	0.6	27	1	0.0	9	0.6	75	462	0	0	0.02	0.05	1.1	0
	Pork sausage, fresh, cooked																						
788	Link (4" × 7/8" raw)	2 links	26	45	96	5	8	2.8	3.6	1.0	22	Tr	0.0	8	0.3	94	336	0	0	0.19	0.07	1.2	1
789	Patty (3 7/8" × ¼" raw)	1 patty	27	45	100	5	8	2.9	3.8	1.0	22	Tr	0.0	9	0.3	97	349	0	0	0.20	0.07	1.2	1
	Salami, beef and pork																						
790	Cooked type (8 slices per 8-oz pkg)	2 slices	57	60	143	8	11	4.6	5.2	1.2	37	1	0.0	7	1.5	113	607	0	0	0.14	0.21	2.0	0
791	Dry type, slice, 3 1/8" × 1/16"	2 slices	20	35	84	5	7	2.4	3.4	0.6	16	1	0.0	2	0.3	76	372	0	0	0.12	0.06	1.0	0
792	Sandwich spread (pork, beef)	1 tbsp	15	60	35	1	3	0.9	1.1	0.4	6	2	Tr	2	0.1	17	152	13	1	0.03	0.02	0.3	0
793	Vienna sausage (7 per 4-oz can)	1 sausage	16	60	45	2	4	1.5	2.0	0.3	8	Tr	0.0	2	0.1	16	152	0	0	0.01	0.02	0.3	0
	Veal, lean and fat, cooked																						
794	Cutlet, braised, 4 1/8" × 2¼" × ½"	3 oz	85	55	179	31	5	2.2	2.0	0.4	114	0	0.0	7	1.1	326	57	0	0	0.05	0.30	9.0	0
795	Rib, roasted, 2 pieces, 4 1/8" × 2¼" × ¼"	3 oz	85	60	194	20	12	4.6	4.6	0.8	94	0	0.0	9	0.8	251	78	0	0	0.04	0.23	5.9	0
Mixed Dishes and Fast Foods																							
	Mixed dishes																						
796	Beef macaroni, frozen, HEALTHY CHOICE	1 package	240	78	211	14	2	0.7	1.2	0.3	14	33	4.6	46	2.7	365	444	514	50	0.28	0.16	3.1	58
797	Beef stew, canned	1 cup	232	82	218	11	12	5.2	5.5	0.5	37	16	3.5	28	1.6	404	947	3,860	494	0.17	0.14	2.9	10
798	Chicken pot pie, frozen	1 small pie	217	60	484	13	29	9.7	12.5	4.5	41	43	1.7	33	2.1	256	857	2,285	343	0.25	0.36	4.1	2
799	Chili con carne with beans, canned	1 cup	222	74	255	20	8	2.1	2.2	1.4	24	24	8.2	67	3.3	608	1,032	884	93	0.15	0.15	2.1	1
800	Macaroni and cheese, canned, made with corn oil	1 cup	252	82	199	8	6	3.0	NA	1.3	8	29	3.0	113	2.0	123	1,058	713	NA	0.28	0.25	2.5	0
801	Meatless burger crumbles, MORNINGSTAR FARMS	1 cup	110	60	231	22	13	3.3	4.6	4.9	0	7	5.1	79	6.4	178	476	0	0	9.92	0.35	3.0	0
802	Meatless burger patty, frozen, MORNINGSTAR FARMS	1 patty	85	71	91	14	1	0.1	0.3	0.2	0	8	4.3	87	2.9	434	383	0	0	0.26	0.55	4.1	0

(Continues)

TABLE B-8 Food Composition Table

FOOD NO.	FOOD DESCRIPTION	MEASURE OF EDIBLE PORTION	WEIGHT (G)	WATER (%)	CALORIES (KCAL)	PROTEIN (G)	TOTAL FAT (G)	FATTY ACIDS SAT (G)	FATTY ACIDS MONO (G)	FATTY ACIDS POLY (G)	CHOLEST (MG)	CARB (G)	TOTAL D F(G)	CA (MG)	FE (MG)	K (MG)	NA (MG)	VITAMIN A (IU)	VITAMIN A (RE)	THMN (MG)	RIBOFL (MG)	NIACIN (MG)	ASCORBIC ACID (MG)
Mixed Dishes and Fast Foods (*Continued*)																							
	Mixed dishes (*Continued*)																						
803	Pasta with meatballs in tomato sauce, canned	1 cup	252	78	260	11	10	4.0	4.2	0.6	20	31	6.8	28	2.3	416	1,053	920	93	0.19	0.16	3.3	8
804	Spaghetti bolognese (meat sauce), frozen, HEALTHY CHOICE	1 package	283	78	255	14	3	1.0	0.9	0.9	17	43	5.1	51	3.5	408	473	492	48	0.35	3.77	0.5	15
805	Spaghetti in tomato sauce with cheese, canned	1 cup	252	80	192	6	2	0.7	0.3	0.3	8	39	7.8	40	2.8	305	963	932	58	0.35	0.28	4.5	10
806	Spinach souffle, home-prepared	1 cup	136	74	219	11	18	7.1	6.8	3.1	184	3	NA	230	1.3	201	763	3,461	675	0.09	0.30	0.5	3
807	Tortellini, pasta with cheese filling, frozen	¾ cup (yields 1 cup cooked)	81	31	249	11	6	2.9	1.7	0.4	34	38	1.5	123	1.2	72	279	50	13	0.25	0.25	2.2	0
	Fast foods																						
	Breakfast items																						
808	Biscuit with egg and sausage	1 biscuit	180	43	581	19	39	15.0	16.4	4.4	302	41	0.9	155	4.0	320	1,141	635	164	0.50	0.45	3.6	0
809	Croissant with egg, cheese, bacon	1 croissant	129	44	413	16	28	15.4	9.2	1.8	215	24	NA	151	2.2	201	889	472	120	0.35	0.34	2.2	2
	Danish pastry																						
810	Cheese filled	1 pastry	91	34	353	6	25	5.1	15.6	2.4	20	29	NA	70	1.8	116	319	155	43	0.26	0.21	2.5	3
811	Fruit filled	1 pastry	94	29	335	5	16	3.3	10.1	1.6	19	45	NA	22	1.4	110	333	86	24	0.29	0.21	1.8	2
812	English muffin with egg, cheese, Canadian bacon	1 muffin	137	57	289	17	13	4.7	4.7	1.6	234	27	1.5	151	2.4	199	729	586	156	0.49	0.45	3.3	2
813	French toast with butter	2 slices	135	51	356	10	19	7.7	7.1	2.4	116	36	NA	73	1.9	177	513	473	146	0.58	0.50	3.9	Tr
814	French toast sticks	5 sticks	141	30	513	8	29	4.7	12.6	9.9	75	58	2.7	78	3.0	127	499	45	13	0.23	0.25	3.0	0
815	Hashed brown potatoes	½ cup	72	60	151	2	9	4.3	3.9	0.5	9	16	NA	7	0.5	267	290	18	3	0.08	0.01	1.1	5
816	Pancakes with butter, syrup	2 pancakes	232	50	520	8	14	5.9	5.3	2.0	58	91	NA	128	2.6	251	1,104	281	70	0.39	0.56	3.4	3
	Burrito																						
817	With beans and cheese	1 burrito	93	54	189	8	6	3.4	1.2	0.9	14	27	NA	107	1.1	248	583	625	119	0.11	0.35	1.8	1
818	With beans and meat	1 burrito	116	52	255	11	9	4.2	3.5	0.6	24	33	NA	53	2.5	329	670	319	32	0.27	0.42	2.7	1
	Cheeseburger																						
	Regular size, with condiments																						

(Continues)

TABLE B-8 Food Composition Table

FOOD NO.	FOOD DESCRIPTION	MEASURE OF EDIBLE PORTION	WEIGHT (G)	WATER (%)	CALORIES (KCAL)	PROTEIN (G)	TOTAL FAT (G)	FATTY ACIDS SAT (G)	MONO (G)	POLY (G)	CHOLEST (MG)	CARB (G)	TOTAL D F(G)	CA (MG)	FE (MG)	K (MG)	NA (MG)	VITAMIN A (IU)	(RE)	THMN (MG)	RIBOFL (MG)	NIACIN (MG)	ASCORBIC ACID (MG)
Mixed Dishes and Fast Foods (*Continued*)																							
	Fast foods (*Continued*)																						
819	Double patty with mayo type dressing, vegetables	1 sandwich	166	51	417	21	21	8.7	7.8	2.7	60	35	NA	171	3.4	335	1,051	398	65	0.35	0.28	8.1	2
820	Single patty	1 sandwich	113	48	295	16	14	6.3	5.3	1.1	37	27	NA	111	2.4	223	616	462	94	0.25	0.23	3.7	2
	Regular size, plain																						
821	Double patty	1 sandwich	155	42	457	28	28	13.0	11.0	1.9	110	22	NA	233	3.4	308	636	332	79	0.25	0.37	6.0	0
822	Double patty with three-piece bun	1 sandwich	160	43	461	22	22	9.5	8.3	1.8	80	44	NA	224	3.7	285	891	277	66	0.34	0.38	6.0	0
823	Single patty	1 sandwich	102	37	319	15	15	6.5	5.8	1.5	50	32	NA	141	2.4	164	500	153	37	0.40	0.40	3.7	0
	Large, with condiments																						
824	Single patty with mayo type dressing, vegetables	1 sandwich	219	53	563	28	33	15.0	12.6	2.0	88	38	NA	206	4.7	445	1,108	613	129	0.39	0.46	7.4	8
825	Single patty with bacon	1 sandwich	195	44	608	32	37	16.2	14.5	2.7	111	37	NA	162	4.7	332	1,043	406	80	0.31	0.41	6.6	2
826	Chicken fillet (breaded and fried) sandwich, plain	1 sandwich	182	47	515	24	29	8.5	10.4	8.4	60	39	NA	60	4.7	353	957	100	31	0.33	0.24	6.8	9
	Chicken, fried. See Poultry and Poultry Products.																						
827	Chicken pieces, boneless, breaded and fried, plain	6 pieces	106	47	319	18	21	4.7	10.5	4.6	61	15	0.0	14	0.9	305	513	0	0	0.12	0.16	7.5	0
828	Chili con carne	1 cup	253	77	256	25	8	3.4	3.4	0.5	134	22	NA	68	5.2	691	1,007	1,662	167	0.13	1.14	2.5	2
829	Chimichanga with beef	1 chimi-changa	174	51	425	20	20	8.5	8.1	1.1	9	43	NA	63	4.5	586	910	146	16	0.49	0.64	5.8	5
830	Coleslaw	¾ cup	99	74	147	1	11	1.6	2.4	6.4	5	13	NA	34	0.7	177	267	338	50	0.04	0.03	0.1	8
	Desserts																						
831	Ice milk, soft, vanilla, in cone	1 cone	103	65	164	4	6	3.5	1.8	0.4	28	24	0.1	153	0.2	169	92	211	52	0.05	0.26	0.3	1
832	Pie, fried, with fruit filling (5" × 3¾")	1 pie	128	38	404	4	21	3.1	9.5	6.9	0	55	3.3	28	1.6	83	479	35	4	0.18	0.14	1.8	2
833	Sundae, hot fudge	1 sundae	158	60	284	6	9	5.0	2.3	0.8	21	48	0.0	207	0.6	395	182	221	57	0.06	0.30	1.1	2
834	Enchilada with cheese	1 enchi-lada	163	63	319	10	19	10.6	6.3	0.8	44	29	NA	324	1.3	240	784	1,161	186	0.08	0.42	1.9	1
835	Fish sandwich, with tartar sauce and cheese	1 sandwich	183	45	523	21	29	8.1	8.9	9.4	68	48	NA	185	3.5	353	939	432	97	0.46	0.42	4.2	3
836	French fries	1 small	85	35	291	4	16	3.3	9.0	2.7	0	34	3.0	12	0.7	586	168	0	0	0.07	0.03	2.4	10

(*Continues*)

TABLE B-8 Food Composition Table

FOOD NO.	FOOD DESCRIPTION	MEASURE OF EDIBLE PORTION	WEIGHT (G)	WATER (%)	CALORIES (KCAL)	PROTEIN (G)	TOTAL FAT (G)	FATTY ACIDS SAT (G)	MONO (G)	POLY (G)	CHOLEST (MG)	CARB (G)	TOTAL D F(G)	CA (MG)	FE (MG)	K (MG)	NA (MG)	VITAMIN A (IU)	(RE)	THMN (MG)	RIBOFL (MG)	NIACIN (MG)	ASCORBIC ACID (MG)
Mixed Dishes and Fast Foods (*Continued*)																							
	Fast foods (*Continued*)																						
837		1 medium	134	35	458	6	25	5.2	14.3	4.2	0	53	4.7	19	1.0	923	265	0	0	0.11	0.05	3.8	16
838		1 large	169	35	578	7	31	6.5	18.0	5.3	0	67	5.9	24	1.3	1,164	335	0	0	0.14	0.07	4.8	20
839	Frijoles (refried beans, chili sauce, cheese)	1 cup	167	69	225	11	8	4.1	2.6	0.7	37	29	NA	189	2.2	605	882	456	70	0.13	0.33	1.5	2
	Hamburger																						
	Regular size, with condiments																						
840	Double patty	1 sandwich	215	51	576	32	32	12.0	14.1	2.8	103	39	NA	92	5.5	527	742	54	4	0.34	0.41	6.7	1
841	Single patty	1 sandwich	106	45	272	12	10	3.6	3.4	1.0	30	34	2.3	126	2.7	251	534	74	10	0.29	0.24	3.9	2
	Large, with condiments, mayo type dressing, and vegetables																						
842	Double patty	1 sandwich	226	54	540	34	27	10.5	10.3	2.8	122	40	NA	102	5.9	570	791	102	11	0.36	0.38	7.6	1
843	Single patty	1 sandwich	218	56	512	26	27	10.4	11.4	2.2	87	40	NA	96	4.9	480	824	312	33	0.41	0.37	7.3	3
	Hot dog																						
844	Plain	1 sandwich	98	54	242	10	15	5.1	6.9	1.7	44	18	NA	24	2.3	143	670	0	0	0.24	0.27	3.6	Tr
845	With chili	1 sandwich	114	48	296	14	13	4.9	6.6	1.2	51	31	NA	19	3.3	166	480	58	6	0.22	0.40	3.7	3
846	With corn flour coating (corndog)	1 corndog	175	47	460	17	19	5.2	9.1	3.5	79	56	NA	102	6.2	263	973	207	37	0.28	0.70	4.2	0
847	Hush puppies	5 pieces	78	32	257	5	12	2.7	7.8	0.4	135	35	NA	69	1.4	188	965	94	27	0.00	0.02	2.0	0
848	Mashed potatoes	⅓ cup	80	79	66	2	1	0.4	0.3	0.2	2	13	NA	17	0.4	235	182	33	8	0.07	0.04	1.0	Tr
849	Nachos, with cheese sauce	6-8 nachos	113	40	346	9	19	7.8	8.0	2.2	18	36	NA	272	1.3	172	816	559	92	0.19	0.37	1.5	1
850	Onion rings, breaded and fried	8-9 rings	83	37	276	4	16	7.0	6.7	0.7	14	31	NA	73	0.8	129	430	8	1	0.08	0.10	0.9	1
	Pizza (slice = ⅛ of 12" pizza)																						
851	Cheese	1 slice	63	48	140	8	3	1.5	1.0	0.5	9	21	NA	117	0.6	110	336	382	74	0.18	0.16	2.5	1
852	Meat and vegetables	1 slice	79	48	184	13	5	1.5	2.5	0.9	21	21	NA	101	1.5	179	382	524	101	0.21	0.17	2.0	2
853	Pepperoni	1 slice	71	47	181	10	7	2.2	3.1	1.2	14	20	NA	65	0.9	153	267	282	55	0.13	0.23	3.0	2
854	Roast beef sandwich, plain	1 sandwich	139	49	346	22	14	3.6	6.8	1.7	51	33	NA	54	4.2	316	792	210	21	0.38	0.31	5.9	2
855	Salad, tossed, with chicken, no dressing	1½ cups	218	87	105	17	2	0.6	0.7	0.6	72	4	NA	37	1.1	447	209	935	96	0.11	0.13	5.9	17
856	Salad, tossed, with egg, cheese, no dressing	1½ cups	217	90	102	9	6	3.0	1.8	0.5	98	5	NA	100	0.7	371	119	822	115	0.09	0.17	1.0	10

(*Continues*)

TABLE B-8 Food Composition Table

FOOD NO.	FOOD DESCRIPTION	MEASURE OF EDIBLE PORTION	WEIGHT (G)	WATER (%)	CALORIES (KCAL)	PROTEIN (G)	TOTAL FAT (G)	FATTY ACIDS SAT (G)	MONO (G)	POLY (G)	CHOLEST (MG)	CARB (G)	TOTAL D F(G)	CA (MG)	FE (MG)	K (MG)	NA (MG)	VITAMIN A (IU)	(RE)	THMN (MG)	RIBOFL (MG)	NIACIN (MG)	ASCORBIC ACID (MG)
Mixed Dishes and Fast Foods (*Continued*)																							
	Fast foods (*Continued*)																						
	Shake																						
857	Chocolate	16 fl oz	333	72	423	11	12	7.7	3.6	0.5	43	68	2.7	376	1.0	666	323	310	77	0.19	0.82	0.5	1
858	Vanilla	16 fl oz	333	75	370	12	10	6.2	2.9	0.4	37	60	1.3	406	0.3	579	273	433	107	0.15	0.61	0.6	3
859	Shrimp, breaded and fried	6-8 shrimp	164	48	454	19	25	5.4	17.4	0.6	200	40	NA	84	3.0	184	1,446	120	36	0.21	0.90	0.0	0
	Submarine sandwich (6" long), with oil and vinegar																						
860	Cold cuts (with lettuce, cheese, salami, ham, tomato, onion)	1 sandwich	228	58	456	22	19	6.8	8.2	2.3	36	51	NA	189	2.5	394	1,651	424	80	1.00	0.80	5.5	12
861	Roast beef (with tomato, lettuce, mayo)	1 sandwich	216	59	410	29	13	7.1	1.8	2.6	73	44	NA	41	2.8	330	845	413	50	0.41	0.41	6.0	6
862	Tuna salad (with mayo, lettuce)	1 sandwich	256	54	584	30	28	5.3	13.4	7.3	49	55	NA	74	2.6	335	1,293	187	41	0.46	0.33	11.3	4
863	Taco, beef	1 small	171	58	369	21	21	11.4	6.6	1.0	56	27	NA	221	2.4	474	802	855	147	0.15	0.44	3.2	2
864		1 large	263	58	568	32	32	17.5	10.1	1.5	87	41	NA	339	3.7	729	1,233	1,315	226	0.24	0.68	4.9	3
865	Taco salad (with ground beef, cheese, taco shell)	1½ cups	198	72	279	13	15	6.8	5.2	1.7	44	24	NA	192	2.3	416	762	588	77	0.10	0.36	2.5	4
	Tostada (with cheese, tomato, lettuce)																						
866	With beans and beef	1 tostada	225	70	333	16	17	11.5	3.5	0.6	74	30	NA	189	2.5	491	871	1,276	173	0.09	0.50	2.9	4
867	With guacamole	1 tostada	131	73	181	6	12	5.0	4.3	1.5	20	16	NA	212	0.8	326	401	879	109	0.07	0.29	1.0	2
Poultry and Poultry Products																							
	Chicken																						
	Fried in vegetable shortening, meat with skin																						
	Batter dipped																						
868	Breast, ½ breast (5.6 oz with bones)	½ breast	140	52	364	35	18	4.9	7.6	4.3	119	13	0.4	28	1.8	281	385	94	28	0.16	0.20	14.7	0
869	Drumstick (3.4 oz with bones)	1 drumstick	72	53	193	16	11	3.0	4.6	2.7	62	6	0.2	12	1.0	134	194	62	19	0.08	0.15	3.7	0
870	Thigh	1 thigh	86	52	238	19	14	3.8	5.8	3.4	80	8	0.3	15	1.2	165	248	82	25	0.10	0.20	4.9	0
871	Wing	1 wing	49	46	159	10	11	2.9	4.4	2.5	39	5	0.1	10	0.6	68	157	55	17	0.05	0.07	2.6	0

(*Continues*)

TABLE B-8 Food Composition Table

FOOD NO.	FOOD DESCRIPTION	MEASURE OF EDIBLE PORTION	WEIGHT (G)	WATER (%)	CALORIES (KCAL)	PROTEIN (G)	TOTAL FAT (G)	FATTY ACIDS SAT (G)	MONO (G)	POLY (G)	CHOLEST (MG)	CARB (G)	TOTAL D F(G)	CA (MG)	FE (MG)	K (MG)	NA (MG)	VITAMIN A (IU)	(RE)	THMN (MG)	RIBOFL (MG)	NIACIN (MG)	ASCORBIC ACID (MG)
	Poultry and Poultry Products (*Continued*)																						
	Chicken (*Continued*)																						
	Flour coated																						
872	Breast, ½ breast (4.2 oz with bones)	½ breast	98	57	218	31	9	2.4	3.4	1.9	87	2	0.1	16	1.2	254	74	49	15	0.08	0.13	13.5	0
873	Drumstick (2.6 oz with bones)	1 drumstick	49	57	120	13	7	1.8	2.7	1.6	44	1	Tr	6	0.7	112	44	41	12	0.04	0.11	3.0	0
	Fried, meat only																						
874	Dark meat	3 oz	85	56	203	25	10	2.7	3.7	2.4	82	2	0.0	15	1.3	215	82	67	20	0.08	0.21	6.0	0
875	Light meat	3 oz	85	60	163	28	5	1.3	1.7	1.1	77	Tr	0.0	14	1.0	224	69	26	8	0.06	0.11	11.4	0
	Roasted, meat only																						
876	Breast, ½ breast (4.2 oz with bone and skin)	½ breast	86	65	142	27	3	0.9	1.1	0.7	73	0	0.0	13	0.9	220	64	18	5	0.06	0.10	11.8	0
877	Drumstick (2.9 oz with bone and skin)	1 drumstick	44	67	76	12	2	0.7	0.8	0.6	41	0	0.0	5	0.6	108	42	26	8	0.03	0.10	2.7	0
878	Thigh	1 thigh	52	63	109	13	6	1.6	2.2	1.3	49	0	0.0	6	0.7	124	46	34	10	0.04	0.12	3.4	0
879	Stewed, meat only, light and dark meat, chopped or diced	1 cup	140	56	332	43	17	4.3	5.7	4.0	116	0	0.0	18	2.0	283	109	157	46	0.16	0.39	9.0	0
880	Chicken giblets, simmered, chopped	1 cup	145	68	228	37	7	2.2	1.7	1.6	570	1	0.0	17	9.3	229	84	10,775	3,232	0.13	1.38	5.9	12
881	Chicken liver, simmered	1 liver	20	68	31	5	1	0.4	0.3	0.2	126	Tr	0.0	3	1.7	28	10	3,275	983	0.03	0.35	0.9	3
882	Chicken neck, meat only, simmered	1 neck	18	67	32	4	1	0.4	0.5	0.4	14	0	0.0	8	0.5	25	12	22	6	0.01	0.05	0.7	0
883	Duck, roasted, flesh only	½ duck	221	64	444	52	25	9.2	8.2	3.2	197	0	0.0	27	6.0	557	144	170	51	0.57	1.04	11.3	0
	Turkey																						
	Roasted, meat only																						
884	Dark meat	3 oz	85	63	159	24	6	2.1	1.4	1.8	72	0	0.0	27	2.0	247	67	0	0	0.05	0.21	3.1	0
885	Light meat	3 oz	85	66	133	25	3	0.9	0.5	0.7	59	0	0.0	16	1.1	259	54	0	0	0.05	0.11	5.8	0
886	Light and dark meat, chopped or diced	1 cup	140	65	238	41	7	2.3	1.4	2.0	106	0	0.0	35	2.5	417	98	0	0	0.09	0.25	7.6	0
	Ground, cooked																						
887	Patty, from 4 oz raw	1 patty	82	59	193	22	11	2.8	4.0	2.6	84	0	0.0	21	1.6	221	88	0	0	0.04	0.14	4.0	0
888	Crumbled	1 cup	127	59	298	35	17	4.3	6.2	4.1	130	0	0.0	32	2.5	343	136	0	0	0.07	0.21	6.1	0

(*Continues*)

TABLE B-8 Food Composition Table

FOOD NO.	FOOD DESCRIPTION	MEASURE OF EDIBLE PORTION	WEIGHT (G)	WATER (%)	CALORIES (KCAL)	PROTEIN (G)	TOTAL FAT (G)	FATTY ACIDS SAT (G)	MONO (G)	POLY (G)	CHOLEST (MG)	CARB (G)	TOTAL D F(G)	CA (MG)	FE (MG)	K (MG)	NA (MG)	VITAMIN A (IU)	(RE)	THMN (MG)	RIBOFL (MG)	NIACIN (MG)	ASCORBIC ACID (MG)
Poultry and Poultry Products (*Continued*)																							
889	Turkey giblets, simmered, chopped	1 cup	145	65	242	39	7	2.2	1.7	1.7	606	3	0.0	19	9.7	290	86	8,752	2,603	0.07	1.31	6.5	2
890	Turkey neck, meat only, simmered	1 neck	152	65	274	41	11	3.7	2.5	3.3	185	0	0.0	56	3.5	226	85	0	0	0.05	0.29	2.6	0
	Poultry food products																						
	Chicken																						
891	Canned, boneless	5 oz	142	69	234	31	11	3.1	4.5	2.5	88	0	0.0	20	2.2	196	714	166	48	0.02	0.18	9.0	3
892	Frankfurter (10 per 1 lb pkg)	1 frank	45	58	116	6	9	2.5	3.8	1.8	45	3	0.0	43	0.9	38	617	59	17	0.03	0.05	1.4	0
893	Roll, light meat (6 slices per 6-oz pkg)	2 slices	57	69	90	11	4	1.1	1.7	0.9	28	1	0.0	24	0.5	129	331	46	14	0.04	0.07	3.0	0
	Turkey																						
894	Gravy and turkey, frozen	5-oz package	142	85	95	8	4	1.2	1.4	0.7	26	7	0.0	20	1.3	87	787	60	18	0.03	0.18	2.6	0
895	Patties, breaded or battered, fried (2.25 oz)	1 patty	64	50	181	9	12	3.0	4.8	3.0	40	10	0.3	9	1.4	176	512	24	7	0.06	0.12	1.5	0
896	Roast, boneless, frozen, seasoned, light and dark meat, cooked	3 oz	85	68	132	18	5	1.6	1.0	1.4	45	3	0.0	4	1.4	253	578	0	0	0.04	0.14	5.3	0
Soups, Sauces, and Gravies																							
	Soups																						
	Canned, condensed																						
	Prepared with equal volume of whole milk																						
897	Clam chowder, New England	1 cup	248	85	164	9	7	3.0	2.3	1.1	22	17	1.5	186	1.5	300	992	164	40	0.07	0.24	1.0	3
898	Cream of chicken	1 cup	248	85	191	7	11	4.6	4.5	1.6	27	15	0.2	181	0.7	273	1,047	714	94	0.07	0.26	0.9	1
899	Cream of mushroom	1 cup	248	85	203	6	14	5.1	3.0	4.6	20	15	0.5	179	0.6	270	918	154	37	0.08	0.28	0.9	2
900	Tomato	1 cup	248	85	161	6	6	2.9	1.6	1.1	17	22	2.7	159	1.8	449	744	848	109	0.13	0.25	1.5	68
	Prepared with equal volume of water																						
901	Bean with pork	1 cup	253	84	172	8	6	1.5	2.2	1.8	3	23	8.6	81	2.0	402	951	888	89	0.09	0.03	0.6	2
902	Beef broth, bouillon, consomme	1 cup	241	96	29	5	0	0.0	0.0	0.0	0	2	0.0	10	0.5	154	636	0	0	0.02	0.03	0.7	1

(Continues)

TABLE B-8 Food Composition Table

FOOD NO.	FOOD DESCRIPTION	MEASURE OF EDIBLE PORTION	WEIGHT (G)	WATER (%)	CALORIES (KCAL)	PROTEIN (G)	TOTAL FAT (G)	FATTY ACIDS SAT (G)	MONO (G)	POLY (G)	CHOLEST (MG)	CARB (G)	TOTAL D F(G)	CA (MG)	FE (MG)	K (MG)	NA (MG)	VITAMIN A (IU)	(RE)	THMN (MG)	RIBOFL (MG)	NIACIN (MG)	ASCORBIC ACID (MG)
Soups, Sauces, and Gravies (*Continued*)																							
	Soups (*Continued*)																						
903	Beef noodle	1 cup	244	92	83	5	3	1.1	1.2	0.5	5	9	0.7	15	1.1	100	952	630	63	0.07	0.06	1.1	Tr
904	Chicken noodle	1 cup	241	92	75	4	2	0.7	1.1	0.6	7	9	0.7	17	0.8	55	1,106	711	72	0.05	0.06	1.4	Tr
905	Chicken and rice	1 cup	241	94	60	4	2	0.5	0.9	0.4	7	7	0.7	17	0.7	101	815	660	65	0.02	0.02	1.1	Tr
906	Clam chowder, Manhattan	1 cup	244	92	78	2	2	0.4	0.4	1.3	2	12	1.5	27	1.6	188	578	964	98	0.03	0.04	0.8	4
907	Cream of chicken	1 cup	244	91	117	3	7	2.1	3.3	1.5	10	9	0.2	34	0.6	88	986	561	56	0.03	0.06	0.8	Tr
908	Cream of mushroom	1 cup	244	90	129	2	9	2.4	1.7	4.2	2	9	0.5	46	0.5	100	881	0	0	0.05	0.09	0.7	1
909	Minestrone	1 cup	241	91	82	4	3	0.6	0.7	1.1	2	11	1.0	34	0.9	313	911	2,338	234	0.05	0.04	0.9	1
910	Pea, green	1 cup	250	83	165	9	3	1.4	1.0	0.4	0	27	2.8	28	2.0	190	918	203	20	0.11	0.07	1.2	2
911	Tomato	1 cup	244	90	85	2	2	0.4	0.4	1.0	0	17	0.5	12	1.8	264	695	688	68	0.09	0.05	1.4	66
912	Vegetable beef	1 cup	244	92	78	6	2	0.9	0.8	0.1	5	10	0.5	17	1.1	173	791	1,891	190	0.04	0.05	1.0	2
913	Vegetarian vegetable	1 cup	241	92	72	2	2	0.3	0.8	0.7	0	12	0.5	22	1.1	210	822	3,005	301	0.05	0.05	0.9	1
	Canned, ready to serve, chunky																						
914	Bean with ham	1 cup	243	79	231	13	9	3.3	3.8	0.9	22	27	11.2	78	3.2	425	972	3,951	396	0.15	0.15	1.7	4
915	Chicken noodle	1 cup	240	84	175	13	6	1.4	2.7	1.5	19	17	3.8	24	1.4	108	850	1,222	122	0.07	0.17	4.3	0
916	Chicken and vegetable	1 cup	240	83	166	12	5	1.4	2.2	1.0	17	19	NA	26	1.5	367	1,068	5,990	600	0.04	0.17	3.3	6
917	Vegetable	1 cup	240	88	122	4	4	0.6	1.6	1.4	0	19	1.2	55	1.6	396	1,010	5,878	588	0.07	0.06	1.2	6
	Canned, ready to serve, low fat, reduced sodium																						
918	Chicken broth	1 cup	240	97	17	3	0	0.0	0.0	0.0	0	1	0.0	19	0.6	204	554	0	0	Tr	0.03	1.6	1
919	Chicken noodle	1 cup	237	92	76	6	2	0.4	0.6	0.4	19	9	1.2	19	1.1	209	460	920	95	0.11	0.11	3.4	1
920	Chicken and rice	1 cup	241	88	116	7	3	0.9	1.3	0.7	14	14	0.7	22	1.0	422	482	2,010	202	0.05	0.13	5.0	2
921	Chicken and rice with vegetables	1 cup	239	91	88	6	1	0.4	0.5	0.5	17	12	0.7	24	1.2	275	459	1,644	165	0.12	0.07	2.6	1
922	Clam chowder, New England	1 cup	244	89	117	5	2	0.5	0.7	0.4	5	20	1.2	17	0.9	283	529	244	59	0.05	0.09	0.9	5
923	Lentil	1 cup	242	88	126	8	2	0.3	0.8	0.2	0	20	5.6	41	2.7	336	443	951	94	0.11	0.09	0.7	1
924	Minestrone	1 cup	241	87	123	5	3	0.4	0.9	1.0	0	20	1.2	39	1.7	306	470	1,357	135	0.15	0.08	1.0	1
925	Vegetable	1 cup	238	91	81	4	1	0.3	0.4	0.3	5	13	1.4	31	1.5	290	466	3,196	319	0.08	0.07	1.8	1
	Dehydrated																						
	Unprepared																						
926	Beef bouillon	1 packet	6	3	14	1	1	0.3	0.2	Tr	1	1	0.0	4	0.1	27	1,019	3	Tr	Tr	0.01	0.3	0

(Continues)

TABLE B-8 Food Composition Table

FOOD NO.	FOOD DESCRIPTION	MEASURE OF EDIBLE PORTION	WEIGHT (G)	WATER (%)	CALORIES (KCAL)	PROTEIN (G)	TOTAL FAT (G)	FATTY ACIDS			CHOLEST (MG)	CARB (G)	TOTAL D F(G)	CA (MG)	FE (MG)	K (MG)	NA (MG)	VITAMIN A		THMN (MG)	RIBOFL (MG)	NIACIN (MG)	ASCORBIC ACID (MG)
								SAT (G)	MONO (G)	POLY (G)								(IU)	(RE)				
Soups, Sauces, and Gravies (*Continued*)																							
	Soups (*Continued*)																						
927	Onion	1 packet	39	4	115	5	2	0.5	1.4	0.3	2	21	4.1	55	0.6	260	3,493	8	1	0.11	0.24	2.0	1
	Prepared with water																						
928	Chicken noodle	1 cup	252	94	58	2	1	0.3	0.5	0.4	10	9	0.3	5	0.5	33	578	15	5	0.20	0.08	1.1	0
929	Onion	1 cup	246	96	27	1	1	0.1	0.3	0.1	0	5	1.0	12	0.1	64	849	2	0	0.03	0.06	0.5	Tr
	Home prepared, stock																						
930	Beef	1 cup	240	96	31	5	Tr	0.1	0.1	Tr	0	3	0.0	19	0.6	444	475	0	0	0.08	0.22	2.1	0
931	Chicken	1 cup	240	92	86	6	3	0.8	1.4	0.5	7	8	0.0	7	0.5	252	343	0	0	0.08	0.20	3.8	Tr
932	Fish	1 cup	233	97	40	5	2	0.5	0.5	0.3	2	0	0.0	7	Tr	336	363	0	0	0.08	0.18	2.8	Tr
	Sauces																						
	Home recipe																						
933	Cheese	1 cup	243	67	479	25	36	19.5	11.5	3.4	92	13	0.2	756	0.9	345	1,198	1,473	389	0.11	0.59	0.5	1
934	White, medium, made with whole milk	1 cup	250	75	368	10	27	7.1	11.1	7.2	18	23	0.5	295	0.8	390	885	1,383	138	0.17	0.46	1.0	2
	Ready to serve																						
935	Barbecue	1 tbsp	16	81	12	Tr	Tr	Tr	0.1	0.1	0	2	0.2	3	0.1	28	130	139	14	Tr	Tr	0.1	1
936	Cheese	¼ cup	63	71	110	4	8	3.8	2.4	1.6	18	4	0.3	116	0.1	19	522	199	40	Tr	0.07	Tr	Tr
937	Hoisin	1 tbsp	16	44	35	1	1	0.1	0.2	0.3	Tr	7	0.4	5	0.2	19	258	2	Tr	Tr	0.03	0.2	Tr
938	Nacho cheese	¼ cup	63	70	119	5	10	4.2	3.1	2.1	20	3	0.5	118	0.2	20	492	128	32	Tr	0.08	Tr	Tr
939	Pepper or hot	1 tsp	5	90	1	Tr	Tr	Tr	Tr	Tr	0	Tr	0.1	Tr	Tr	7	124	14	1	Tr	Tr	Tr	4
940	Salsa	1 tbsp	16	90	4	Tr	Tr	Tr	Tr	Tr	0	1	0.3	5	0.2	34	69	96	10	0.01	0.01	0.1	2
941	Soy	1 tbsp	16	69	9	1	Tr	Tr	Tr	Tr	0	1	0.1	3	0.3	64	871	0	0	0.01	0.03	0.4	0
942	Spaghetti/marinara/pasta	1 cup	250	87	143	4	5	0.7	2.2	1.8	0	21	4.0	55	1.8	738	1,030	938	95	0.14	0.10	2.7	20
943	Teriyaki	1 tbsp	18	68	15	1	0	0.0	0.0	0.0	0	3	Tr	5	0.3	41	690	0	0	0.01	0.01	0.2	0
944	Tomato chili	¼ cup	68	68	71	2	Tr	Tr	Tr	0.1	0	17	4.0	14	0.5	252	910	462	46	0.06	0.05	1.1	11
945	Worcestershire	1 tbsp	17	70	11	0	0	0.0	0.0	0.0	0	3	0.0	18	0.9	136	167	18	2	0.01	0.02	0.1	2
	Gravies, canned																						
946	Beef	¼ cup	58	87	31	2	1	0.7	0.6	Tr	2	3	0.2	3	0.4	47	325	0	0	0.02	0.02	0.4	0
947	Chicken	¼ cup	60	85	47	1	3	0.8	1.5	0.9	1	3	0.2	12	0.3	65	346	221	67	0.01	0.03	0.3	0
948	Country sausage	¼ cup	62	75	96	3	8	2.0	2.9	2.2	13	4	0.4	4	0.3	48	236	0	0	0.10	0.04	0.7	Tr
949	Mushroom	¼ cup	60	89	30	1	2	0.2	0.7	0.6	0	3	0.2	4	0.4	64	342	0	0	0.02	0.04	0.4	0
950	Turkey	¼ cup	60	89	31	2	1	0.4	0.5	0.3	1	3	0.2	2	0.4	65	346	0	0	0.01	0.05	0.8	0

(*Continues*)

TABLE B-8 Food Composition Table

FOOD NO.	FOOD DESCRIPTION	MEASURE OF EDIBLE PORTION	WEIGHT (G)	WATER (%)	CALORIES (KCAL)	PROTEIN (G)	TOTAL FAT (G)	FATTY ACIDS SAT (G)	MONO (G)	POLY (G)	CHOLEST (MG)	CARB (G)	TOTAL D F(G)	CA (MG)	FE (MG)	K (MG)	NA (MG)	VITAMIN A (IU)	(RE)	THMN (MG)	RIBOFL (MG)	NIACIN (MG)	ASCORBIC ACID (MG)
Sugars and Sweets																							
	Candy																						
951	BUTTERFINGER (NESTLE)	1 fun size bar	7	2	34	1	1	0.7	0.4	0.2	Tr	5	0.2	2	0.1	27	14	0	0	0.01	Tr	0.2	0
	Caramel																						
952	Plain	1 piece	10	9	39	Tr	1	0.7	0.1	Tr	1	8	0.1	14	Tr	22	25	3	1	Tr	0.02	Tr	Tr
953	Chocolate flavored roll	1 piece	7	7	25	Tr	Tr	Tr	0.1	0.1	0	6	Tr	2	Tr	7	6	1	Tr	Tr	0.01	Tr	Tr
954	Carob	1 oz	28	2	153	2	9	8.2	0.1	0.1	1	16	1.1	86	0.4	179	30	7	2	0.03	0.05	0.3	Tr
	Chocolate, milk																						
955	Plain	1 bar (1.55 oz)	44	1	226	3	14	8.1	4.4	0.5	10	26	1.5	84	0.6	169	36	81	24	0.03	0.13	0.1	Tr
956	With almonds	1 bar (1.45 oz)	41	2	216	4	14	7.0	5.5	0.9	8	22	2.5	92	0.7	182	30	30	6	0.02	0.18	0.3	Tr
957	With peanuts, MR. GOODBAR (HERSHEY)	1 bar (1.75 oz)	49	1	267	5	17	7.3	5.7	2.4	4	25	1.7	53	0.6	219	73	70	18	0.08	0.12	1.6	Tr
958	With rice cereal, NESTLE CRUNCH	1 bar (1.55 oz)	44	1	230	3	12	6.7	3.8	0.4	6	29	1.1	74	0.2	151	59	30	9	0.15	0.25	1.7	Tr
	Chocolate chips																						
959	Milk	1 cup	168	1	862	12	52	31.0	16.7	1.8	37	99	5.7	321	2.3	647	138	311	92	0.13	0.51	0.5	1
960	Semisweet	1 cup	168	1	805	7	50	29.8	16.7	1.6	0	106	9.9	54	5.3	613	18	35	3	0.09	0.15	0.7	0
961	White	1 cup	170	1	916	10	55	33.0	15.5	1.7	36	101	0.0	338	0.4	486	153	60	2	0.11	0.48	1.3	1
962	Chocolate coated peanuts	10 pieces	40	2	208	5	13	5.8	5.2	1.7	4	20	1.9	42	0.5	201	16	0	0	0.05	0.07	1.7	0
963	Chocolate coated raisins	10 pieces	10	11	39	Tr	1	0.9	0.5	0.1	Tr	7	0.4	9	0.2	51	4	4	1	0.01	0.02	Tr	Tr
964	Fruit leather, pieces	1 oz	28	12	97	Tr	2	0.3	0.9	0.8	0	22	1.0	5	0.2	46	114	33	3	0.01	0.03	Tr	16
965	Fruit leather, rolls	1 large	21	11	74	Tr	1	0.1	0.3	0.1	0	18	0.8	7	0.2	62	13	24	3	0.01	Tr	Tr	1
966		1 small	14	11	49	Tr	Tr	0.1	0.2	0.1	0	12	0.5	4	0.1	41	9	16	2	0.01	Tr	Tr	1
	Fudge, prepared from recipe																						
	Chocolate																						
967	Plain	1 piece	17	10	65	Tr	1	0.9	0.4	0.1	2	14	0.1	7	0.1	18	11	32	8	Tr	0.01	Tr	Tr
968	With nuts	1 piece	19	7	81	1	3	1.1	0.8	1.0	3	14	0.2	10	0.1	30	11	38	9	0.01	0.02	Tr	Tr
	Vanilla																						
969	Plain	1 piece	16	11	59	Tr	1	0.5	0.2	Tr	3	13	0.0	6	Tr	8	11	33	8	Tr	0.01	Tr	Tr

(Continues)

TABLE B-8 Food Composition Table

FOOD NO.	FOOD DESCRIPTION	MEASURE OF EDIBLE PORTION	WEIGHT (G)	WATER (%)	CALORIES (KCAL)	PROTEIN (G)	TOTAL FAT (G)	FATTY ACIDS			CHOLEST (MG)	CARB (G)	TOTAL D F(G)	CA (MG)	FE (MG)	K (MG)	NA (MG)	VITAMIN A		THMN (MG)	RIBOFL (MG)	NIACIN (MG)	ASCORBIC ACID (MG)
								SAT (G)	MONO (G)	POLY (G)								(IU)	(RE)				
Sugars and Sweets (*Continued*)																							
	Candy (*Continued*)																						
970	With nuts	1 piece	15	8	62	Tr	2	0.6	0.5	0.8	2	11	0.1	7	0.1	17	9	30	7	0.01	0.01	Tr	Tr
	Gumdrops/gummy candies																						
971	Gumdrops (¾" dia)	1 cup	182	1	703	0	0	0.0	0.0	0.0	0	180	0.0	5	0.7	9	80	0	0	0.00	Tr	Tr	0
972		1 medium	4	1	16	0	0	0.0	0.0	0.0	0	4	0.0	Tr	Tr	Tr	2	0	0	0.00	Tr	Tr	0
973	Gummy bears	10 bears	22	1	85	0	0	0.0	0.0	0.0	0	22	0.0	1	0.1	1	10	0	0	0.00	Tr	Tr	0
974	Gummy worms	10 worms	74	1	286	0	0	0.0	0.0	0.0	0	73	0.0	2	0.3	4	33	0	0	0.00	Tr	Tr	0
975	Hard candy	1 piece	6	1	24	0	Tr	0.0	0.0	0.0	0	6	0.0	Tr	Tr	Tr	2	0	0	Tr	Tr	Tr	0
976		1 small piece	3	1	12	0	Tr	0.0	0.0	0.0	0	3	0.0	Tr	Tr	Tr	1	0	0	Tr	Tr	Tr	0
977	Jelly beans	10 large	28	6	104	0	Tr	Tr	0.1	Tr	0	26	0.0	1	0.3	10	7	0	0	0.00	0.00	0.0	0
978		10 small	11	6	40	0	Tr	Tr	Tr	Tr	0	10	0.0	Tr	0.1	4	3	0	0	0.00	0.00	0.0	0
979	KIT KAT (HERSHEY)	1 bar (1.5 oz)	42	2	216	3	11	6.8	3.1	0.3	3	27	0.8	69	0.4	122	32	68	20	0.07	0.23	1.1	Tr
	Marshmallows																						
980	Miniature	1 cup	50	16	159	1	Tr	Tr	Tr	Tr	0	41	0.1	2	0.1	3	24	1	0	Tr	Tr	Tr	0
981	Regular	1 regular	7	16	23	Tr	Tr	Tr	Tr	Tr	0	6	Tr	Tr	Tr	Tr	3	Tr	0	Tr	Tr	Tr	0
	M&M's (M&M MARS)																						
982	Peanut	¼ cup	43	2	222	4	11	4.4	4.7	1.8	4	26	1.5	43	0.5	149	21	40	10	0.04	0.07	1.6	Tr
983		10 pieces	20	2	103	2	5	2.1	2.2	0.8	2	12	0.7	20	0.2	69	10	19	5	0.02	0.03	0.7	Tr
984	Plain	¼ cup	52	2	256	2	11	6.8	3.6	0.3	7	37	1.3	55	0.6	138	32	106	28	0.03	0.11	0.1	Tr
985		10 pieces	7	2	34	Tr	1	0.9	0.5	Tr	1	5	0.2	7	0.1	19	4	14	4	Tr	0.01	Tr	Tr
986	MILKY WAY (M&M MARS)	1 fun size bar	18	6	76	1	3	1.4	1.1	0.1	3	13	0.3	23	0.1	43	43	19	6	0.01	0.04	0.1	Tr
987		1 bar (2.15 oz)	61	6	258	3	10	4.8	3.7	0.4	9	44	1.0	79	0.5	147	146	66	20	0.02	0.14	0.2	1
988	REESE'S Peanut butter cup (HERSHEY)	1 miniature cup	7	2	38	1	2	0.8	0.9	0.4	Tr	4	0.2	5	0.1	25	22	5	1	0.02	0.01	0.3	Tr
989		1 package (contains 2)	45	2	243	5	14	5.0	5.9	2.5	2	25	1.4	35	0.5	158	143	33	9	0.11	0.08	2.1	Tr
990	SNICKERS bar (M&M MARS)	1 fun size bar	15	5	72	1	4	1.3	1.6	0.7	2	9	0.4	14	0.1	49	40	23	6	0.01	0.02	0.6	Tr

(*Continues*)

TABLE B-8 Food Composition Table

FOOD NO.	FOOD DESCRIPTION	MEASURE OF EDIBLE PORTION	WEIGHT (G)	WATER (%)	CALORIES (KCAL)	PROTEIN (G)	TOTAL FAT (G)	FATTY ACIDS SAT (G)	MONO (G)	POLY (G)	CHOLEST (MG)	CARB (G)	TOTAL D F(G)	CA (MG)	FE (MG)	K (MG)	NA (MG)	VITAMIN A (IU)	(RE)	THMN (MG)	RIBOFL (MG)	NIACIN (MG)	ASCORBIC ACID (MG)
Sugars and Sweets (*Continued*)																							
	Candy (*Continued*)																						
991		1 king size bar (4 oz)	113	5	541	9	28	10.2	11.8	5.6	15	67	2.8	106	0.9	366	301	172	44	0.11	0.17	4.7	1
992		1 bar (2 oz)	57	5	273	5	14	5.1	6.0	2.8	7	34	1.4	54	0.4	185	152	87	22	0.06	0.09	2.4	Tr
993	SPECIAL DARK sweet chocolate (HERSHEY)	1 miniature	8	1	46	Tr	3	1.7	0.9	0.1	Tr	5	0.4	2	0.2	25	1	3	Tr	Tr	0.01	Tr	0
994	STARBURST fruit chews (M&M MARS)	1 piece	5	7	20	Tr	Tr	0.1	0.2	0.2	0	4	0.0	Tr	Tr	Tr	3	0	0	Tr	Tr	Tr	3
995		1 package (2.07 oz)	59	7	234	Tr	5	0.7	2.1	1.8	0	50	0.0	2	0.1	1	33	0	0	Tr	Tr	Tr	31
	Frosting, ready to eat																						
996	Chocolate	1/12 package	38	17	151	Tr	7	2.1	3.4	0.8	0	24	0.2	3	0.5	74	70	249	75	Tr	0.01	Tr	0
997	Vanilla	1/12 package	38	13	159	Tr	6	1.9	3.3	0.9	0	26	Tr	1	Tr	14	34	283	86	0.00	Tr	Tr	0
	Frozen desserts (nondairy)																						
998	Fruit and juice bar	1 bar (2.5 fl oz)	77	78	63	1	Tr	0.0	0.0	Tr	0	16	0.0	4	0.1	41	3	22	2	0.01	0.01	0.1	7
999	Ice pop	1 bar (2 fl oz)	59	80	42	0	0	0.0	0.0	0.0	0	11	0.0	0	0.0	2	7	0	0	0.00	0.00	0.0	0
1000	Italian ices	½ cup	116	86	61	Tr	Tr	0.0	0.0	0.0	0	16	0.0	1	0.1	7	5	194	0	0.01	0.01	0.8	1
1001	Fruit butter, apple	1 tbsp	17	56	29	Tr	0	0.0	0.0	0.0	0	7	0.3	2	0.1	15	1	20	2	Tr	Tr	Tr	Tr
	Gelatin dessert, prepared with gelatin dessert powder and water																						
1002	Regular	½ cup	135	85	80	2	0	0.0	0.0	0.0	0	19	0.0	3	Tr	1	57	0	0	0.00	Tr	Tr	0
1003	Reduced calorie (with aspartame)	½ cup	117	98	8	1	0	0.0	0.0	0.0	0	1	0.0	2	Tr	0	56	0	0	0.00	Tr	Tr	0
1004	Honey, strained or extracted	1 tbsp	21	17	64	Tr	0	0.0	0.0	0.0	0	17	Tr	1	0.1	11	1	0	0	0.00	0.01	Tr	Tr
1005		1 cup	339	17	1,031	1	0	0.0	0.0	0.0	0	279	0.7	20	1.4	176	14	0	0	0.00	0.13	0.4	2
1006	Jams and preserves	1 tbsp	20	30	56	Tr	Tr	Tr	Tr	0.0	0	14	0.2	4	0.1	15	6	2	Tr	0.00	Tr	Tr	2
1007		1 packet (0.5 oz)	14	30	39	Tr	Tr	Tr	Tr	0.0	0	10	0.2	3	0.1	11	4	2	Tr	0.00	Tr	Tr	1

(*Continues*)

TABLE B-8 Food Composition Table

FOOD NO.	FOOD DESCRIPTION	MEASURE OF EDIBLE PORTION	WEIGHT (G)	WATER (%)	CALORIES (KCAL)	PROTEIN (G)	TOTAL FAT (G)	FATTY ACIDS			CHOLEST (MG)	CARB (G)	TOTAL D F(G)	CA (MG)	FE (MG)	K (MG)	NA (MG)	VITAMIN A		THMN (MG)	RIBOFL (MG)	NIACIN (MG)	ASCORBIC ACID (MG)
								SAT (G)	MONO (G)	POLY (G)								(IU)	(RE)				
Sugars and Sweets (*Continued*)																							
1008	Jellies	1 tbsp	19	29	54	Tr	Tr	Tr	Tr	Tr	0	13	0.2	2	Tr	12	5	3	Tr	Tr	Tr	Tr	Tr
1009		1 packet (0.5 oz)	14	29	40	Tr	Tr	Tr	Tr	Tr	0	10	0.1	1	Tr	9	4	2	Tr	Tr	Tr	Tr	Tr
	Puddings																						
	Prepared with dry mix and 2% milk																						
	Chocolate																						
1010	Instant	½ cup	147	75	150	5	3	1.6	0.9	0.2	9	28	0.6	153	0.4	247	417	253	56	0.05	0.21	0.1	1
1011	Regular (cooked)	½ cup	142	74	151	5	3	1.8	0.8	0.1	10	28	0.4	160	0.5	240	149	253	68	0.05	0.21	0.2	1
	Vanilla																						
1012	Instant	½ cup	142	75	148	4	2	1.4	0.7	0.1	9	28	0.0	146	0.1	185	406	241	64	0.05	0.20	0.1	1
1013	Regular (cooked)	½ cup	140	76	141	4	2	1.5	0.7	0.1	10	26	0.0	153	0.1	193	224	252	70	0.04	0.20	0.1	1
	Ready to eat																						
	Regular																						
1014	Chocolate	4 oz	113	69	150	3	5	0.8	1.9	1.6	3	26	1.1	102	0.6	203	146	41	12	0.03	0.18	0.4	2
1015	Rice	4 oz	113	68	184	2	8	1.3	3.6	3.2	1	25	0.1	59	0.3	68	96	129	40	0.02	0.08	0.2	1
1016	Tapioca	4 oz	113	74	134	2	4	0.7	1.8	1.5	1	22	0.1	95	0.3	110	180	0	0	0.02	0.11	0.4	1
1017	Vanilla	4 oz	113	71	147	3	4	0.6	1.7	1.5	8	25	0.1	99	0.1	128	153	24	7	0.02	0.16	0.3	0
	Fat free																						
1018	Chocolate	4 oz	113	76	107	3	Tr	0.3	0.1	Tr	2	23	0.9	89	0.6	235	192	174	52	0.02	0.12	0.1	Tr
1019	Tapioca	4 oz	113	77	98	2	Tr	0.1	Tr	Tr	1	23	0.1	76	0.2	99	251	121	36	0.02	0.09	0.1	Tr
1020	Vanilla	4 oz	113	76	105	2	Tr	0.1	Tr	Tr	1	24	0.1	86	Tr	123	241	174	52	0.02	0.10	0.1	Tr
	Sugar																						
	Brown																						
1021	Packed	1 cup	220	2	827	0	0	0.0	0.0	0.0	0	214	0.0	187	4.2	761	86	0	0	0.02	0.02	0.2	0
1022	Unpacked	1 cup	145	2	545	0	0	0.0	0.0	0.0	0	141	0.0	123	2.8	502	57	0	0	0.01	0.01	0.1	0
1023		1 tbsp	9	2	34	0	0	0.0	0.0	0.0	0	9	0.0	8	0.2	31	4	0	0	Tr	Tr	Tr	0
	White																						
1024	Granulated	1 packet	6	0	23	0	0	0.0	0.0	0.0	0	6	0.0	Tr	Tr	Tr	Tr	0	0	0.00	Tr	0.0	0
1025		1 tsp	4	0	16	0	0	0.0	0.0	0.0	0	4	0.0	Tr	Tr	Tr	Tr	0	0	0.00	Tr	0.0	0
1026		1 cup	200	0	774	0	0	0.0	0.0	0.0	0	200	0.0	2	0.1	4	2	0	0	0.00	0.04	0.0	0

(*Continues*)

TABLE B-8 Food Composition Table

FOOD NO.	FOOD DESCRIPTION	MEASURE OF EDIBLE PORTION	WEIGHT (G)	WATER (%)	CALORIES (KCAL)	PROTEIN (G)	TOTAL FAT (G)	FATTY ACIDS SAT (G)	MONO (G)	POLY (G)	CHOLEST (MG)	CARB (G)	TOTAL D F(G)	CA (MG)	FE (MG)	K (MG)	NA (MG)	VITAMIN A (IU)	(RE)	THMN (MG)	RIBOFL (MG)	NIACIN (MG)	ASCORBIC ACID (MG)
Sugars and Sweets (*Continued*)																							
	Sugar (*Continued*)																						
1027	Powdered, unsifted	1 tbsp	8	Tr	31	0	Tr	Tr	Tr	Tr	0	8	0.0	Tr	Tr	Tr	Tr	0	0	0.00	0.00	0.0	0
1028		1 cup	120	Tr	467	0	Tr	Tr	Tr	0.1	0	119	0.0	1	0.1	2	1	0	0	0.00	0.00	0.0	0
	Syrup																						
	Chocolate flavored syrup or topping																						
1029	Thin type	1 tbsp	19	31	53	Tr	Tr	0.1	0.1	Tr	0	12	0.3	3	0.4	43	14	6	1	Tr	0.01	0.1	Tr
1030	Fudge type	1 tbsp	19	22	67	1	2	0.8	0.7	0.1	Tr	12	0.5	15	0.2	69	66	3	1	0.01	0.04	0.1	Tr
1031	Corn, light	1 tbsp	20	23	56	0	0	0.0	0.0	0.0	0	15	0.0	1	Tr	1	24	0	0	Tr	Tr	Tr	0
1032	Maple	1 tbsp	20	32	52	0	Tr	Tr	Tr	Tr	0	13	0.0	13	0.2	41	2	0	0	Tr	Tr	Tr	0
1033	Molasses, blackstrap	1 tbsp	20	29	47	0	0	0.0	0.0	0.0	0	12	0.0	172	3.5	498	11	0	0	0.01	0.01	0.2	0
1034		1 cup	328	29	771	0	0	0.0	0.0	0.0	0	199	0.0	2,821	57.4	8,174	180	0	0	0.11	0.17	3.5	0
	Table blend, pancake																						
1035	Regular	1 tbsp	20	24	57	0	0	0.0	0.0	0.0	0	15	0.0	Tr	Tr	Tr	17	0	0	Tr	Tr	Tr	0
1036	Reduced calorie	1 tbsp	15	55	25	0	0	0.0	0.0	0.0	0	7	0.0	Tr	Tr	Tr	30	0	0	Tr	Tr	Tr	0
Vegetables and Vegetable Products																							
1037	Alfalfa sprouts, raw	1 cup	33	91	10	1	Tr	Tr	Tr	0.1	0	1	0.8	11	0.3	26	2	51	5	0.03	0.04	0.2	3
1038	Artichokes, globe or French, cooked, drained	1 cup	168	84	84	6	Tr	0.1	Tr	0.1	0	19	9.1	76	2.2	595	160	297	30	0.11	0.11	1.7	17
1039		1 medium	120	84	60	4	Tr	Tr	Tr	0.1	0	13	6.5	54	1.5	425	114	212	22	0.08	0.08	1.2	12
	Asparagus, green Cooked, drained																						
1040	From raw	1 cup	180	92	43	5	1	0.1	Tr	0.2	0	8	2.9	36	1.3	288	20	970	97	0.22	0.23	1.9	19
1041		4 spears	60	92	14	2	Tr	Tr	Tr	0.1	0	3	1.0	12	0.4	96	7	323	32	0.07	0.08	0.6	6
1042	From frozen	1 cup	180	91	50	5	1	0.2	Tr	0.3	0	9	2.9	41	1.2	392	7	1,472	148	0.12	0.19	1.9	44
1043		4 spears	60	91	17	2	Tr	0.1	Tr	0.1	0	3	1.0	14	0.4	131	2	491	49	0.04	0.06	0.6	15
1044	Canned, spears, about 5" long, drained	1 cup	242	94	46	5	2	0.4	0.1	0.7	0	6	3.9	39	4.4	416	695	1,285	128	0.15	0.24	2.3	45
1045		4 spears	72	94	14	2	Tr	0.1	Tr	0.2	0	2	1.2	12	1.3	124	207	382	38	0.04	0.07	0.7	13
1046	Bamboo shoots, canned, drained	1 cup	131	94	25	2	1	0.1	Tr	0.2	0	4	1.8	10	0.4	105	9	10	1	0.03	0.03	0.2	1

(Continues)

TABLE B-8 Food Composition Table

FOOD NO.	FOOD DESCRIPTION	MEASURE OF EDIBLE PORTION	WEIGHT (G)	WATER (%)	CALORIES (KCAL)	PROTEIN (G)	TOTAL FAT (G)	FATTY ACIDS SAT (G)	MONO (G)	POLY (G)	CHOLEST (MG)	CARB (G)	TOTAL D F(G)	CA (MG)	FE (MG)	K (MG)	NA (MG)	VITAMIN A (IU)	(RE)	THMN (MG)	RIBOFL (MG)	NIACIN (MG)	ASCORBIC ACID (MG)
Vegetables and Vegetable Products (*Continued*)																							
	Beans																						
	Lima, immature seeds, frozen, cooked, drained																						
1047	Ford hooks	1 cup	170	74	170	10	1	0.1	Tr	0.3	0	32	9.9	37	2.3	694	90	323	32	0.13	0.10	1.8	22
1048	Baby limas	1 cup	180	72	189	12	1	0.1	Tr	0.3	0	35	10.8	50	3.5	740	52	301	31	0.13	0.10	1.4	10
	Snap, cut																						
	Cooked, drained																						
	From raw																						
1049	Green	1 cup	125	89	44	2	Tr	0.1	Tr	0.2	0	10	4.0	58	1.6	374	4	833	84	0.09	0.12	0.8	12
1050	Yellow	1 cup	125	89	44	2	Tr	0.1	Tr	0.2	0	10	4.1	58	1.6	374	4	101	10	0.09	0.12	0.8	12
	From frozen																						
1051	Green	1 cup	135	91	38	2	Tr	0.1	Tr	0.1	0	9	4.1	66	1.2	170	12	541	54	0.05	0.12	0.5	6
1052	Yellow	1 cup	135	91	38	2	Tr	0.1	Tr	0.1	0	9	4.1	66	1.2	170	12	151	15	0.05	0.12	0.5	6
	Canned, drained																						
1053	Green	1 cup	135	93	27	2	Tr	Tr	Tr	0.1	0	6	2.6	35	1.2	147	354	471	47	0.02	0.08	0.3	6
1054	Yellow	1 cup	135	93	27	2	Tr	Tr	Tr	0.1	0	6	1.8	35	1.2	147	339	142	15	0.02	0.08	0.3	6
	Beans, dry. See Legumes.																						
	Bean sprouts (mung)																						
1055	Raw	1 cup	104	90	31	3	Tr	Tr	Tr	0.1	0	6	1.9	14	0.9	155	6	22	2	0.09	0.13	0.8	14
1056	Cooked, drained	1 cup	124	93	26	3	Tr	Tr	Tr	Tr	0	5	1.5	15	0.8	125	12	17	1	0.06	0.13	1.0	14
	Beets																						
	Cooked, drained																						
1057	Slices	1 cup	170	87	75	3	Tr	Tr	0.1	0.1	0	17	3.4	27	1.3	519	131	60	7	0.05	0.07	0.6	6
1058	Whole beet, 2" dia	1 beet	50	87	22	1	Tr	Tr	Tr	Tr	0	5	1.0	8	0.4	153	39	18	2	0.01	0.02	0.2	2
	Canned, drained																						
1059	Slices	1 cup	170	91	53	2	Tr	Tr	Tr	0.1	0	12	2.9	26	3.1	252	330	19	2	0.02	0.07	0.3	7
1060	Whole beet	1 beet	24	91	7	Tr	Tr	Tr	Tr	Tr	0	2	0.4	4	0.4	36	47	3	Tr	Tr	0.01	Tr	1
1061	Beet greens, leaves and stems, cooked, drained, 1" pieces	1 cup	144	89	39	4	Tr	Tr	0.1	0.1	0	8	4.2	164	2.7	1,309	347	7,344	734	0.17	0.42	0.7	36
	Black eyed peas, immature seeds, cooked, drained																						
1062	From raw	1 cup	165	75	160	5	1	0.2	0.1	0.3	0	34	8.3	211	1.8	690	7	1,305	130	0.17	0.24	2.3	4

(Continues)

TABLE B-8 Food Composition Table

FOOD NO.	FOOD DESCRIPTION	MEASURE OF EDIBLE PORTION	WEIGHT (G)	WATER (%)	CALORIES (KCAL)	PROTEIN (G)	TOTAL FAT (G)	FATTY ACIDS			CHOLEST (MG)	CARB (G)	TOTAL D F(G)	CA (MG)	FE (MG)	K (MG)	NA (MG)	VITAMIN A		THMN (MG)	RIBOFL (MG)	NIACIN (MG)	ASCORBIC ACID (MG)
								SAT (G)	MONO (G)	POLY (G)								(IU)	(RE)				
Vegetables and Vegetable Products (*Continued*)																							
	Black eyed peas, immature seeds, cooked, drained (*Continued*)																						
1063	From frozen	1 cup	170	66	224	14	1	0.3	0.1	0.5	0	40	10.9	39	3.6	638	9	128	14	0.44	0.11	1.2	4
	Broccoli																						
	Raw																						
1064	Chopped or diced	1 cup	88	91	25	3	Tr	Tr	Tr	0.1	0	5	2.6	42	0.8	286	24	1,357	136	0.06	0.10	0.6	82
1065	Spear, about 5" long	1 spear	31	91	9	1	Tr	Tr	Tr	0.1	0	2	0.9	15	0.3	101	8	478	48	0.02	0.04	0.2	29
1066	Flower cluster	1 floweret	11	91	3	Tr	Tr	Tr	Tr	Tr	0	1	0.3	5	0.1	36	3	330	33	0.01	0.01	0.1	10
	Cooked, drained																						
	From raw																						
1067	Chopped	1 cup	156	91	44	5	1	0.1	Tr	0.3	0	8	4.5	72	1.3	456	41	2,165	217	0.09	0.18	0.9	116
1068	Spear, about 5" long	1 spear	37	91	10	1	Tr	Tr	Tr	0.1	0	2	1.1	17	0.3	108	10	514	51	0.02	0.04	0.2	28
1069	From frozen, chopped	1 cup	184	91	52	6	Tr	Tr	Tr	0.1	0	10	5.5	94	1.1	331	44	3,481	348	0.10	0.15	0.8	74
	Brussels sprouts, cooked, drained																						
1070	From raw	1 cup	156	87	61	4	1	0.2	0.1	0.4	0	14	4.1	56	1.9	495	33	1,122	112	0.17	0.12	0.9	97
1071	From frozen	1 cup	155	87	65	6	1	0.1	Tr	0.3	0	13	6.4	37	1.1	504	36	913	91	0.16	0.18	0.8	71
	Cabbage, common varieties, shredded																						
1072	Raw	1 cup	70	92	18	1	Tr	Tr	Tr	0.1	0	4	1.6	33	0.4	172	13	93	9	0.04	0.03	0.2	23
1073	Cooked, drained	1 cup	150	94	33	2	1	0.1	Tr	0.3	0	7	3.5	47	0.3	146	12	198	20	0.09	0.08	0.4	30
	Cabbage, Chinese, shredded, cooked, drained																						
1074	Pak choi or bok choy	1 cup	170	96	20	3	Tr	Tr	Tr	0.1	0	3	2.7	158	1.8	631	58	4,366	437	0.05	0.11	0.7	44
1075	Pe tsai	1 cup	119	95	17	2	Tr	Tr	Tr	0.1	0	3	3.2	38	0.4	268	11	1,151	115	0.05	0.05	0.6	19
1076	Cabbage, red, raw, shredded	1 cup	70	92	19	1	Tr	Tr	Tr	0.1	0	4	1.4	36	0.3	144	8	28	3	0.04	0.02	0.2	40
1077	Cabbage, savoy, raw, shredded	1 cup	70	91	19	1	Tr	Tr	Tr	Tr	0	4	2.2	25	0.3	161	20	700	70	0.05	0.02	0.2	22
1078	Carrot juice, canned	1 cup	236	89	94	2	Tr	0.1	Tr	0.2	0	22	1.9	57	1.1	689	68	25,833	2,584	0.22	0.13	0.9	20

(*Continues*)

TABLE B-8 Food Composition Table

FOOD NO.	FOOD DESCRIPTION	MEASURE OF EDIBLE PORTION	WEIGHT (G)	WATER (%)	CALORIES (KCAL)	PROTEIN (G)	TOTAL FAT (G)	FATTY ACIDS SAT (G)	MONO (G)	POLY (G)	CHOLEST (MG)	CARB (G)	TOTAL D F(G)	CA (MG)	FE (MG)	K (MG)	NA (MG)	VITAMIN A (IU)	(RE)	THMN (MG)	RIBOFL (MG)	NIACIN (MG)	ASCORBIC ACID (MG)
Vegetables and Vegetable Products (*Continued*)																							
	Carrots																						
	Raw																						
1079	Whole, 7½" long	1 carrot	72	88	31	1	Tr	Tr	Tr	0.1	0	7	2.2	19	0.4	233	25	20,253	2,025	0.07	0.04	0.7	7
1080	Grated	1 cup	110	88	47	1	Tr	Tr	Tr	0.1	0	11	3.3	30	0.6	355	39	30,942	3,094	0.11	0.06	1.0	10
1081	Baby	1 medium	10	90	4	Tr	Tr	Tr	Tr	Tr	0	1	0.2	2	0.1	28	4	1,501	150	Tr	0.01	0.1	1
	Cooked, sliced, drained																						
1082	From raw	1 cup	156	87	70	2	Tr	0.1	Tr	0.1	0	16	5.1	48	1.0	354	103	38,304	3,830	0.05	0.09	0.8	4
1083	From frozen	1 cup	146	90	53	2	Tr	Tr	Tr	0.1	0	12	5.1	41	0.7	231	86	25,845	2,584	0.04	0.05	0.6	4
1084	Canned, sliced, drained	1 cup	146	93	37	1	Tr	0.1	Tr	0.1	0	8	2.2	37	0.9	261	353	20,110	2,010	0.03	0.04	0.8	4
	Cauliflower																						
1085	Raw	1 floweret	13	92	3	Tr	Tr	Tr	Tr	Tr	0	1	0.3	3	0.1	39	4	2	Tr	0.01	0.01	0.1	6
1086		1 cup	100	92	25	2	Tr	Tr	Tr	0.1	0	5	2.5	22	0.4	303	30	19	2	0.06	0.06	0.5	46
	Cooked, drained, 1" pieces																						
1087	From raw	1 cup	124	93	29	2	1	0.1	Tr	0.3	0	5	3.3	20	0.4	176	19	21	2	0.05	0.06	0.5	55
1088		3 flowerets	54	93	12	1	Tr	Tr	Tr	0.1	0	2	1.5	9	0.2	77	8	9	1	0.02	0.03	0.2	24
1089	From frozen	1 cup	180	94	34	3	Tr	0.1	Tr	0.2	0	7	4.9	31	0.7	250	32	40	4	0.07	0.10	0.6	56
	Celery																						
	Raw																						
1090	Stalk, 7 ½ to 8" long	1 stalk	40	95	6	Tr	Tr	Tr	Tr	Tr	0	1	0.7	16	0.2	115	35	54	5	0.02	0.02	0.1	3
1091	Pieces, diced	1 cup	120	95	19	1	Tr	Tr	Tr	0.1	0	4	2.0	48	0.5	344	104	161	16	0.06	0.05	0.4	8
	Cooked, drained																						
1092	Stalk, medium	1 stalk	38	94	7	Tr	Tr	Tr	Tr	Tr	0	2	0.6	16	0.2	108	35	50	5	0.02	0.02	0.1	2
1093	Pieces, diced	1 cup	150	94	27	1	Tr	0.1	Tr	0.1	0	6	2.4	63	0.6	426	137	198	20	0.06	0.07	0.5	9
1094	Chives, raw, chopped	1 tbsp	3	91	1	Tr	Tr	Tr	Tr	Tr	0	Tr	0.1	3	Tr	9	Tr	131	13	Tr	Tr	Tr	2
1095	Cilantro, raw	1 tsp	2	92	Tr	Tr	Tr	Tr	Tr	Tr	0	Tr	Tr	1	Tr	8	1	98	10	Tr	Tr	Tr	1
1096	Coleslaw, home prepared	1 cup	120	82	83	2	3	0.5	0.8	1.6	10	15	1.8	54	0.7	217	28	762	98	0.08	0.07	0.3	39
	Collards, cooked, drained, chopped																						
1097	From raw	1 cup	190	92	49	4	1	0.1	Tr	0.3	0	9	5.3	226	0.9	494	17	5,945	595	0.08	0.20	1.1	35
1098	From frozen	1 cup	170	88	61	5	1	0.1	Tr	0.4	0	12	4.8	357	1.9	427	85	10,168	1,017	0.08	0.20	1.1	45

(*Continues*)

TABLE B-8 Food Composition Table

FOOD NO.	FOOD DESCRIPTION	MEASURE OF EDIBLE PORTION	WEIGHT (G)	WATER (%)	CALORIES (KCAL)	PROTEIN (G)	TOTAL FAT (G)	FATTY ACIDS			CHOLEST (MG)	CARB (G)	TOTAL D F(G)	CA (MG)	FE (MG)	K (MG)	NA (MG)	VITAMIN A		THMN (MG)	RIBOFL (MG)	NIACIN (MG)	ASCORBIC ACID (MG)
								SAT (G)	MONO (G)	POLY (G)								(IU)	(RE)				
	Vegetables and Vegetable Products (*Continued*)																						
	Corn, sweet, yellow																						
	Cooked, drained																						
1099	From raw, kernels on cob	1 ear	77	70	83	3	1	0.2	0.3	0.5	0	19	2.2	2	0.5	192	13	167	17	0.17	0.06	1.2	5
	From frozen																						
1100	Kernels on cob	1 ear	63	73	59	2	Tr	0.1	0.1	0.2	0	14	1.8	2	0.4	158	3	133*	13*	0.11	0.04	1.0	3
1101	Kernels	1 cup	164	77	131	5	1	0.1	0.2	0.3	0	32	3.9	7	0.6	241	8	361*	36*	0.14	0.12	2.1	5
	Canned																						
1102	Cream style	1 cup	256	79	184	4	1	0.2	0.3	0.5	0	46	3.1	8	1.0	343	730	248*	26*	0.06	0.14	2.5	12
1103	Whole kernel, vacuum pack	1 cup	210	77	166	5	1	0.2	0.3	0.5	0	41	4.2	11	0.9	391	571	506*	50*	0.09	0.15	2.5	17
1104	Corn, sweet, white, cooked, drained	1 ear	77	70	83	3	1	0.2	0.3	0.5	0	19	2.1	2	0.5	192	13	0	0	0.17	0.06	1.2	5
	Cucumber																						
	Peeled																						
1105	Sliced	1 cup	119	96	14	1	Tr	Tr	Tr	0.1	0	3	0.8	17	0.2	176	2	88	8	0.02	0.01	0.1	3
1106	Whole, 8¼" long	1 large	280	96	34	2	Tr	0.1	Tr	0.2	0	7	2.0	39	0.4	414	6	207	20	0.06	0.03	0.3	8
	Unpeeled																						
1107	Sliced	1 cup	104	96	14	1	Tr	Tr	Tr	0.1	0	3	0.8	15	0.3	150	2	224	22	0.02	0.02	0.2	6
1108	Whole, 8¼" long	1 large	301	96	39	2	Tr	0.1	Tr	0.2	0	8	2.4	42	0.8	433	6	647	63	0.07	0.07	0.7	16
1109	Dandelion greens, cooked, drained	1 cup	105	90	35	2	1	0.2	Tr	0.3	0	7	3.0	147	1.9	244	46	12,285	1,229	0.14	0.18	0.5	19
1110	Dill weed, raw	5 sprigs	1	86	Tr	Tr	Tr	Tr	Tr	Tr	0	Tr	Tr	2	0.1	7	1	77	8	Tr	Tr	Tr	1
1111	Eggplant, cooked, drained	1 cup	99	92	28	1	Tr	Tr	Tr	0.1	0	7	2.5	6	0.3	246	3	63	6	0.08	0.02	0.6	1
1112	Endive, curly (including escarole), raw, small pieces	1 cup	50	94	9	1	Tr	Tr	Tr	Tr	0	2	1.6	26	0.4	157	11	1,025	103	0.04	0.04	0.2	3
1113	Garlic, raw	1 clove	3	59	4	Tr	Tr	Tr	Tr	Tr	0	1	0.1	5	0.1	12	1	0	0	0.01	Tr	Tr	1
1114	Hearts of palm, canned	1 piece	33	90	9	1	Tr	Tr	Tr	0.1	0	2	0.8	19	1.0	58	141	0	0	Tr	0.02	0.1	3
1115	Jerusalem artichoke, raw, sliced	1 cup	150	78	114	3	Tr	0.0	Tr	Tr	0	26	2.4	21	5.1	644	6	30	3	0.30	0.09	2.0	6
	Kale, cooked, drained, chopped																						
1116	From raw	1 cup	130	91	36	2	1	0.1	Tr	0.3	0	7	2.6	94	1.2	296	30	9,620	962	0.07	0.09	0.7	53
1117	From frozen	1 cup	130	91	39	4	1	0.1	Tr	0.3	0	7	2.6	179	1.2	417	20	8,260	826	0.06	0.15	0.9	33

*White varieties contain only a trace amount of vitamin A; other nutrients are the same.

(Continues)

TABLE B-8 Food Composition Table

FOOD NO.	FOOD DESCRIPTION	MEASURE OF EDIBLE PORTION	WEIGHT (G)	WATER (%)	CALORIES (KCAL)	PROTEIN (G)	TOTAL FAT (G)	FATTY ACIDS SAT (G)	MONO (G)	POLY (G)	CHOLEST (MG)	CARB (G)	TOTAL D F(G)	CA (MG)	FE (MG)	K (MG)	NA (MG)	VITAMIN A (IU)	(RE)	THMN (MG)	RIBOFL (MG)	NIACIN (MG)	ASCORBIC ACID (MG)
Vegetables and Vegetable Products (*Continued*)																							
1118	Kohlrabi, cooked, drained, slices	1 cup	165	90	48	3	Tr	Tr	Tr	0.1	0	11	1.8	41	0.7	561	35	58	7	0.07	0.03	0.6	89
1119	Leeks, bulb and lower leaf portion, chopped or diced, cooked, drained	1 cup	104	91	32	1	Tr	Tr	Tr	0.1	0	8	1.0	31	1.1	90	10	48	5	0.03	0.02	0.2	4
	Lettuce, raw																						
	Butterhead, as Boston types																						
1120	Leaf	1 medium leaf	8	96	1	Tr	Tr	Tr	Tr	Tr	0	Tr	0.1	2	Tr	19	Tr	73	7	Tr	Tr	Tr	1
1121	Head, 5" dia	1 head	163	96	21	2	Tr	Tr	Tr	0.2	0	4	1.6	52	0.5	419	8	1,581	158	0.10	0.10	0.5	13
	Crisphead, as iceberg																						
1122	Leaf	1 medium	8	96	1	Tr	Tr	Tr	Tr	Tr	0	Tr	0.1	2	Tr	13	1	26	3	Tr	Tr	Tr	Tr
1123	Head, 6" dia	1 head	539	96	65	5	1	0.1	Tr	0.5	0	11	7.5	102	2.7	852	49	1,779	178	0.25	0.16	1.0	21
1124	Pieces, shredded or chopped	1 cup	55	96	7	1	Tr	Tr	Tr	0.1	0	1	0.8	10	0.3	87	5	182	18	0.03	0.02	0.1	2
	Looseleaf																						
1125	Leaf	1 leaf	10	94	2	Tr	Tr	Tr	Tr	Tr	0	Tr	0.2	7	0.1	26	1	190	19	0.01	0.01	Tr	2
1126	Pieces, shredded	1 cup	56	94	10	1	Tr	Tr	Tr	0.1	0	2	1.1	38	0.8	148	5	1,064	106	0.03	0.04	0.2	10
	Romaine or cos																						
1127	Innerleaf	1 leaf	10	95	1	Tr	Tr	Tr	Tr	Tr	0	Tr	0.2	4	0.1	29	1	260	26	0.01	0.01	0.1	2
1128	Pieces, shredded	1 cup	56	95	8	1	Tr	Tr	Tr	0.1	0	1	1.0	20	0.6	162	4	1,456	146	0.06	0.06	0.3	13
	Mushrooms																						
1129	Raw, pieces or slices	1 cup	70	92	18	2	Tr	Tr	Tr	0.1	0	3	0.8	4	0.7	259	3	0	0	0.06	0.30	2.8	2
1130	Cooked, drained, pieces	1 cup	156	91	42	3	1	0.1	Tr	0.3	0	8	3.4	9	2.7	555	3	0	0	0.11	0.47	7.0	6
1131	Canned, drained, pieces	1 cup	156	91	37	3	Tr	0.1	Tr	0.2	0	8	3.7	17	1.2	201	663	0	0	0.13	0.03	2.5	0
	Mushrooms, shiitake																						
1132	Cooked pieces	1 cup	145	83	80	2	Tr	0.1	0.1	Tr	0	21	3.0	4	0.6	170	6	0	0	0.05	0.25	2.2	Tr
1133	Dried	1 mush-room	4	10	11	Tr	Tr	Tr	Tr	Tr	0	3	0.4	Tr	0.1	55	Tr	0	0	0.01	0.05	0.5	Tr
1134	Mustard greens, cooked, drained	1 cup	140	94	21	3	Tr	Tr	0.2	0.1	0	3	2.8	104	1.0	283	22	4,243	424	0.06	0.09	0.6	35
	Okra, sliced, cooked, drained																						
1135	From raw	1 cup	160	90	51	3	Tr	0.1	Tr	0.1	0	12	4.0	101	0.7	515	8	920	93	0.21	0.09	1.4	26
1136	From frozen	1 cup	184	91	52	4	1	0.1	0.1	0.1	0	11	5.2	177	1.2	431	6	946	94	0.18	0.23	1.4	22

(Continues)

TABLE B-8 Food Composition Table

FOOD NO.	FOOD DESCRIPTION	MEASURE OF EDIBLE PORTION	WEIGHT (G)	WATER (%)	CALORIES (KCAL)	PROTEIN (G)	TOTAL FAT (G)	FATTY ACIDS SAT (G)	MONO (G)	POLY (G)	CHOLEST (MG)	CARB (G)	TOTAL D F(G)	CA (MG)	FE (MG)	K (MG)	NA (MG)	VITAMIN A (IU)	(RE)	THMN (MG)	RIBOFL (MG)	NIACIN (MG)	ASCORBIC ACID (MG)
Vegetables and Vegetable Products (*Continued*)																							
	Onions																						
	Raw																						
1137	Chopped	1 cup	160	90	61	2	Tr	Tr	Tr	0.1	0	14	2.9	32	0.4	251	5	0	0	0.07	0.03	0.2	10
1138	Whole, medium, 2½" dia	1 whole	110	90	42	1	Tr	Tr	Tr	0.1	0	9	2.0	22	0.2	173	3	0	0	0.05	0.02	0.2	7
1139	Slice, ⅛" thick	1 slice	14	90	5	Tr	Tr	Tr	Tr	Tr	0	1	0.3	3	Tr	22	Tr	0	0	0.01	Tr	Tr	1
1140	Cooked (whole or sliced), drained	1 cup	210	88	92	3	Tr	0.1	0.1	0.2	0	21	2.9	46	0.5	349	6	0	0	0.09	0.05	0.3	11
1141		1 medium	94	88	41	1	Tr	Tr	Tr	0.1	0	10	1.3	21	0.2	156	3	0	0	0.04	0.02	0.2	5
1142	Dehydrated flakes	1 tbsp	5	4	17	Tr	Tr	Tr	Tr	Tr	0	4	0.5	13	0.1	81	1	0	0	0.03	0.01	Tr	4
	Onions, spring, raw, top and bulb																						
1143	Chopped	1 cup	100	90	32	2	Tr	Tr	Tr	0.1	0	7	2.6	72	1.5	276	16	385	39	0.06	0.08	0.5	19
1144	Whole, medium, 4⅛" long	1 whole	15	90	5	Tr	Tr	Tr	Tr	Tr	0	1	0.4	11	0.2	41	2	58	6	0.01	0.01	0.1	3
1145	Onion rings, 2"-3" dia, breaded, par fried, frozen, oven heated	10 rings	60	29	244	3	16	5.2	6.5	3.1	0	23	0.8	19	1.0	77	225	135	14	0.17	0.08	2.2	1
1146	Parsley, raw	10 sprigs	10	88	4	Tr	Tr	Tr	Tr	Tr	0	1	0.3	14	0.6	55	6	520	52	0.01	0.01	0.1	13
1147	Parsnips, sliced, cooked, drained	1 cup	156	78	126	2	Tr	0.1	0.2	0.1	0	30	6.2	58	0.9	573	16	0	0	0.13	0.08	1.1	20
	Peas, edible pod, cooked, drained																						
1148	From raw	1 cup	160	89	67	5	Tr	0.1	Tr	0.2	0	11	4.5	67	3.2	384	6	210	21	0.20	0.12	0.9	77
1149	From frozen	1 cup	160	87	83	6	1	0.1	0.1	0.3	0	14	5.0	94	3.8	347	8	267	27	0.10	0.19	0.9	35
	Peas, green																						
1150	Canned, drained	1 cup	170	82	117	8	1	0.1	0.1	0.3	0	21	7.0	34	1.6	294	428	1,306	131	0.21	0.13	1.2	16
1151	Frozen, boiled, drained	1 cup	160	80	125	8	Tr	0.1	Tr	0.2	0	23	8.8	38	2.5	269	139	1,069	107	0.45	0.16	2.4	16
	Peppers																						
	Hot chili, raw																						
1152	Green	1 pepper	45	88	18	1	Tr	Tr	Tr	Tr	0	4	0.7	8	0.5	153	3	347	35	0.04	0.04	0.4	109
1153	Red	1 pepper	45	88	18	1	Tr	Tr	Tr	Tr	0	4	0.7	8	0.5	153	3	4,838	484	0.04	0.04	0.4	109
1154	Jalapeno, canned, sliced, solids and liquids	¼ cup	26	89	7	Tr	Tr	Tr	Tr	0.1	0	1	0.7	6	0.5	50	434	442	44	0.01	0.01	0.1	3
	Sweet (2¾" long, 2½" dia)																						

(Continues)

TABLE B-8 Food Composition Table

FOOD NO.	FOOD DESCRIPTION	MEASURE OF EDIBLE PORTION	WEIGHT (G)	WATER (%)	CALORIES (KCAL)	PROTEIN (G)	TOTAL FAT (G)	FATTY ACIDS SAT (G)	MONO (G)	POLY (G)	CHOLEST (MG)	CARB (G)	TOTAL D F(G)	CA (MG)	FE (MG)	K (MG)	NA (MG)	VITAMIN A (IU)	(RE)	THMN (MG)	RIBOFL (MG)	NIACIN (MG)	ASCORBIC ACID (MG)
Vegetables and Vegetable Products (*Continued*)																							
	Peppers (*Continued*)																						
	Raw																						
	Green																						
1155	Chopped	1 cup	149	92	40	1	Tr	Tr	Tr	0.2	0	10	2.7	13	0.7	264	3	942	94	0.10	0.04	0.8	133
1156	Ring (¼" thick)	1 ring	10	92	3	Tr	Tr	Tr	Tr	Tr	0	1	0.2	1	Tr	18	Tr	63	6	0.01	Tr	0.1	9
1157	Whole (2¾" × 2½")	1 pepper	119	92	32	1	Tr	Tr	Tr	0.1	0	8	2.1	11	0.5	211	2	752	75	0.08	0.04	0.6	106
	Red																						
1158	Chopped	1 cup	149	92	40	1	Tr	Tr	Tr	0.2	0	10	3.0	13	0.7	264	3	8,493	849	0.10	0.04	0.8	283
1159	Whole (2¾" × 2½")	1 pepper	119	92	32	1	Tr	Tr	Tr	0.1	0	8	2.4	11	0.5	211	2	6,783	678	0.08	0.04	0.6	226
	Cooked, drained, chopped																						
1160	Green	1 cup	136	92	38	1	Tr	Tr	Tr	0.1	0	9	1.6	12	0.6	226	3	805	80	0.08	0.04	0.6	101
1161	Red	1 cup	136	92	38	1	Tr	Tr	Tr	0.1	0	9	1.6	12	0.6	226	3	5,114	511	0.08	0.04	0.6	233
1162	Pimento, canned	1 tbsp	12	93	3	Tr	Tr	Tr	Tr	Tr	0	1	0.2	1	0.2	19	2	319	32	Tr	0.01	0.1	10
	Potatoes																						
	Baked (2⅓" × 4¾")																						
1163	With skin	1 potato	202	71	220	5	Tr	0.1	Tr	0.1	0	51	4.8	20	2.7	844	16	0	0	0.22	0.07	3.3	26
1164	Flesh only	1 potato	156	75	145	3	Tr	Tr	Tr	0.1	0	34	2.3	8	0.5	610	8	0	0	0.16	0.03	2.2	20
1165	Skin only	1 skin	58	47	115	2	Tr	Tr	Tr	Tr	0	27	4.6	20	4.1	332	12	0	0	0.07	0.06	1.8	8
	Boiled (2½" dia)																						
1166	Peeled after boiling	1 potato	136	77	118	3	Tr	Tr	Tr	0.1	0	27	2.4	7	0.4	515	5	0	0	0.14	0.03	2.0	18
1167	Peeled before boiling	1 potato	135	77	116	2	Tr	Tr	Tr	0.1	0	27	2.4	11	0.4	443	7	0	0	0.13	0.03	1.8	10
1168		1 cup	156	77	134	3	Tr	Tr	Tr	0.1	0	31	2.8	12	0.5	512	8	0	0	0.15	0.03	2.0	12
	Potato products, prepared																						
	Au gratin																						
1169	From dry mix, with whole milk, butter	1 cup	245	79	228	6	10	6.3	2.9	0.3	37	31	2.2	203	0.8	537	1,076	522	76	0.05	0.20	2.3	8
1170	From home recipe, with butter	1 cup	245	74	323	12	19	11.6	5.3	0.7	56	28	4.4	292	1.6	970	1,061	647	93	0.16	0.28	2.4	24
1171	French fried, frozen, oven heated	10 strips	50	57	100	2	4	0.6	2.4	0.4	0	16	1.6	4	0.6	209	15	0	0	0.06	0.01	1.0	5
	Hashed brown																						

(*Continues*)

TABLE B-8 Food Composition Table

FOOD NO.	FOOD DESCRIPTION	MEASURE OF EDIBLE PORTION	WEIGHT (G)	WATER (%)	CALORIES (KCAL)	PROTEIN (G)	TOTAL FAT (G)	FATTY ACIDS SAT (G)	MONO (G)	POLY (G)	CHOLEST (MG)	CARB (G)	TOTAL D F(G)	CA (MG)	FE (MG)	K (MG)	NA (MG)	VITAMIN A (IU)	(RE)	THMN (MG)	RIBOFL (MG)	NIACIN (MG)	ASCORBIC ACID (MG)
Vegetables and Vegetable Products *(Continued)*																							
	Potato products, prepared *(Continued)*																						
1172	From frozen (about 3" × 1½" × ½")	1 patty	29	56	63	1	3	1.3	1.5	0.4	0	8	0.6	4	0.4	126	10	0	0	0.03	0.01	0.7	2
1173	From home recipe	1 cup	156	62	326	4	22	8.5	9.7	2.5	0	33	3.1	12	1.3	501	37	0	0	0.12	0.03	3.1	9
	Mashed																						
1174	From dehydrated flakes (without milk); whole milk, butter, and salt added	1 cup	210	76	237	4	12	7.2	3.3	0.5	29	32	4.8	103	0.5	489	697	378	44	0.23	0.11	1.4	20
	From home recipe																						
1175	With whole milk	1 cup	210	78	162	4	1	0.7	0.3	0.1	4	37	4.2	55	0.6	628	636	40	13	0.18	0.08	2.3	14
1176	With whole milk and margarine	1 cup	210	76	223	4	9	2.2	3.7	2.5	4	35	4.2	55	0.5	607	620	355	42	0.18	0.08	2.3	13
1177	Potato pancakes, home prepared	1 pancake	76	47	207	5	12	2.3	3.5	5.0	73	22	1.5	18	1.2	597	386	109	11	0.10	0.13	1.6	17
1178	Potato puffs, from frozen	10 puffs	79	53	175	3	8	4.0	3.4	0.6	0	24	2.5	24	1.2	300	589	13	2	0.15	0.06	1.7	5
1179	Potato salad, home prepared	1 cup	250	76	358	7	21	3.6	6.2	9.3	170	28	3.3	48	1.6	635	1,323	523	83	0.19	0.15	2.2	25
	Scalloped																						
1180	From dry mix, with whole milk, butter	1 cup	245	79	228	5	11	6.5	3.0	0.5	27	31	2.7	88	0.9	497	835	363	51	0.05	0.14	2.5	8
1181	From home recipe, with butter	1 cup	245	81	211	7	9	5.5	2.5	0.4	29	26	4.7	140	1.4	926	821	331	47	0.17	0.23	2.6	26
	Pumpkin																						
1182	Cooked, mashed	1 cup	245	94	49	2	Tr	0.1	Tr	Tr	0	12	2.7	37	1.4	564	2	2,651	265	0.08	0.19	1.0	12
1183	Canned	1 cup	245	90	83	3	1	0.4	0.1	Tr	0	20	7.1	64	3.4	505	12	54,037	5,405	0.06	0.13	0.9	10
1184	Radishes, raw (¾" to 1" dia)	1 radish	5	95	1	Tr	Tr	Tr	Tr	Tr	0	Tr	0.1	1	Tr	10	1	Tr	Tr	Tr	Tr	Tr	1
1185	Rutabagas, cooked, drained, cubes	1 cup	170	89	66	2	Tr	Tr	Tr	0.2	0	15	3.1	82	0.9	554	34	954	95	0.14	0.07	1.2	32
1186	Sauerkraut, canned, solids and liquid	1 cup	236	93	45	2	Tr	0.1	Tr	0.1	0	10	5.9	71	3.5	401	1,560	42	5	0.05	0.05	0.3	35
	Seaweed																						
1187	Kelp, raw	2 tbsp	10	82	4	Tr	Tr	Tr	Tr	Tr	0	1	0.1	17	0.3	9	23	12	1	0.01	0.02	Tr	Tr
1188	Spirulina, dried	1 tbsp	1	5	3	1	Tr	Tr	Tr	Tr	0	Tr	Tr	1	0.3	14	10	6	1	0.02	0.04	0.1	Tr

(Continues)

TABLE B-8 Food Composition Table

FOOD NO.	FOOD DESCRIPTION	MEASURE OF EDIBLE PORTION	WEIGHT (G)	WATER (%)	CALORIES (KCAL)	PROTEIN (G)	TOTAL FAT (G)	FATTY ACIDS SAT (G)	MONO (G)	POLY (G)	CHOLEST (MG)	CARB (G)	TOTAL D F(G)	CA (MG)	FE (MG)	K (MG)	NA (MG)	VITAMIN A (IU)	(RE)	THMN (MG)	RIBOFL (MG)	NIACIN (MG)	ASCORBIC ACID (MG)
Vegetables and Vegetable Products (*Continued*)																							
1189	Shallots, raw, chopped	1 tbsp	10	80	7	Tr	Tr	Tr	Tr	Tr	0	2	0.2	4	0.1	33	1	119	12	0.01	Tr	Tr	1
1190	Soybeans, green, cooked, drained	1 cup	180	69	254	22	12	1.3	2.2	5.4	0	20	7.6	261	4.5	970	25	281	29	0.47	0.28	2.3	31
	Spinach																						
	Raw																						
1191	Chopped	1 cup	30	92	7	1	Tr	Tr	Tr	Tr	0	1	0.8	30	0.8	167	24	2,015	202	0.02	0.06	0.2	8
1192	Leaf	1 leaf	10	92	2	Tr	Tr	Tr	Tr	Tr	0	Tr	0.3	10	0.3	56	3	672	67	0.01	0.02	0.1	3
	Cooked, drained																						
1193	From raw	1 cup	180	91	41	5	Tr	0.1	Tr	0.2	0	7	4.3	245	6.4	839	126	14,742	1,474	0.17	0.42	0.9	18
1194	From frozen (chopped or leaf)	1 cup	190	90	53	6	Tr	0.1	Tr	0.2	0	10	5.7	277	2.9	566	163	14,790	1,478	0.11	0.32	0.8	23
1195	Canned, drained	1 cup	214	92	49	6	1	0.2	Tr	0.4	0	7	5.1	272	4.9	740	58	18,781	1,879	0.03	0.30	0.8	31
	Squash																						
	Summer (all varieties), sliced																						
1196	Raw	1 cup	113	94	23	1	Tr	Tr	Tr	0.1	0	5	2.1	23	0.5	220	2	221	23	0.07	0.04	0.6	17
1197	Cooked, drained	1 cup	180	94	36	2	1	0.1	Tr	0.2	0	8	2.5	49	0.6	346	2	517	52	0.08	0.07	0.9	10
1198	Winter (all varieties), baked, cubes	1 cup	205	89	80	2	1	0.3	0.1	0.5	0	18	5.7	29	0.7	896	2	7,292	730	0.17	0.05	1.4	20
1199	Winter, butternut, frozen, cooked, mashed	1 cup	240	88	94	3	Tr	Tr	Tr	0.1	0	24	2.2	46	1.4	319	5	8,014	802	0.12	0.09	1.1	8
	Sweetpotatoes																						
	Cooked (2" dia, 5" long raw)																						
1200	Baked, with skin	1 potato	146	73	150	3	Tr	Tr	Tr	0.1	0	35	4.4	41	0.7	508	15	31,860	3,186	0.11	0.19	0.9	36
1201	Boiled, without skin	1 potato	156	73	164	3	Tr	0.1	Tr	0.2	0	38	2.8	33	0.9	287	20	26,604	2,660	0.08	0.22	1.0	27
1202	Candied (2½" × 2" piece)	1 piece	105	67	144	1	3	1.4	0.7	0.2	8	29	2.5	27	1.2	198	74	4,398	440	0.02	0.04	0.4	7
	Canned																						
1203	Syrup pack, drained	1 cup	196	72	212	3	1	0.1	Tr	0.3	0	50	5.9	33	1.9	378	76	14,028	1,403	0.05	0.07	0.7	21
1204	Vacuum pack, mashed	1 cup	255	76	232	4	1	0.1	Tr	0.2	0	54	4.6	56	2.3	796	135	20,357	2,035	0.09	0.15	1.9	67
1205	Tomatillos, raw	1 medium	34	92	11	Tr	Tr	Tr	0.1	0.1	0	2	0.6	2	0.2	91	Tr	39	4	0.01	0.01	0.6	4
	Tomatoes																						
	Raw, year round average																						
1206	Chopped or sliced	1 cup	180	94	38	2	1	0.1	0.1	0.2	0	8	2.0	9	0.8	400	16	1,121	112	0.11	0.09	1.1	34

(Continues)

TABLE B-8 Food Composition Table

FOOD NO.	FOOD DESCRIPTION	MEASURE OF EDIBLE PORTION	WEIGHT (G)	WATER (%)	CALORIES (KCAL)	PROTEIN (G)	TOTAL FAT (G)	FATTY ACIDS SAT (G)	MONO (G)	POLY (G)	CHOLEST (MG)	CARB (G)	TOTAL D F(G)	CA (MG)	FE (MG)	K (MG)	NA (MG)	VITAMIN A (IU)	(RE)	THMN (MG)	RIBOFL (MG)	NIACIN (MG)	ASCORBIC ACID (MG)
Vegetables and Vegetable Products (*Continued*)																							
	Tomatoes (*Continued*)																						
1207	Slice, medium, ¼" thick	1 slice	20	94	4	Tr	Tr	Tr	Tr	Tr	0	1	0.2	1	0.1	44	2	125	12	0.01	0.01	0.1	4
	Whole																						
1208	Cherry	1 cherry	17	94	4	Tr	Tr	Tr	Tr	Tr	0	1	0.2	1	0.1	38	2	106	11	0.01	0.01	0.1	3
1209	Medium, 2⅗" dia	1 tomato	123	94	26	1	Tr	0.1	0.1	0.2	0	6	1.4	6	0.6	273	11	766	76	0.07	0.06	0.8	23
1210	Canned, solids and liquid	1 cup	240	94	46	2	Tr	Tr	Tr	0.1	0	10	2.4	72	1.3	530	355	1,428	144	0.11	0.07	1.8	34
	Sun dried																						
1211	Plain	1 piece	2	15	5	Tr	Tr	Tr	Tr	Tr	0	1	0.2	2	0.2	69	42	17	2	0.01	0.01	0.2	1
1212	Packed in oil, drained	1 piece	3	54	6	Tr	Tr	0.1	0.3	0.1	0	1	0.2	1	0.1	47	8	39	4	0.01	0.01	0.1	3
1213	Tomato juice, canned, with salt added	1 cup	243	94	41	2	Tr	Tr	Tr	0.1	0	10	1.0	22	1.4	535	877	1,351	136	0.11	0.08	1.6	44
	Tomato products, canned																						
1214	Paste	1 cup	262	74	215	10	1	0.2	0.2	0.6	0	51	10.7	92	5.1	2,455	231	6,406	639	0.41	0.50	8.4	111
1215	Puree	1 cup	250	87	100	4	Tr	0.1	0.1	0.2	0	24	5.0	43	3.1	1,065	85*	3,188	320	0.18	0.14	4.3	26
1216	Sauce	1 cup	245	89	74	3	Tr	0.1	0.1	0.2	0	18	3.4	34	1.9	909	1,482	2,399	240	0.16	0.14	2.8	32
	Spaghetti/marinara/pasta sauce. See Soups, Sauces, and Gravies.																						
1217	Stewed	1 cup	255	91	71	2	Tr	Tr	0.1	0.1	0	17	2.6	84	1.9	607	564	1,380	138	0.12	0.09	1.8	29
1218	Turnips, cooked, cubes	1 cup	156	94	33	1	Tr	Tr	Tr	0.1	0	8	3.1	34	0.3	211	78	0	0	0.04	0.04	0.5	18
	Turnip greens, cooked, drained																						
1219	From raw (leaves and stems)	1 cup	144	93	29	2	Tr	0.1	Tr	0.1	0	6	5.0	197	1.2	292	42	7,917	792	0.06	0.10	0.6	39
1220	From frozen (chopped)	1 cup	164	90	49	5	1	0.2	Tr	0.3	0	8	5.6	249	3.2	367	25	13,079	1,309	0.09	0.12	0.8	36
1221	Vegetable juice cocktail, canned	1 cup	242	94	46	2	Tr	Tr	Tr	0.1	0	11	1.9	27	1.0	467	653	2,831	283	0.10	0.07	1.8	67
	Vegetables, mixed																						
1222	Canned, drained	1 cup	163	87	77	4	Tr	0.1	Tr	0.2	0	15	4.9	44	1.7	474	243	18,985	1,899	0.07	0.08	0.9	8
1223	Frozen, cooked, drained	1 cup	182	83	107	5	Tr	0.1	Tr	0.1	0	24	8.0	46	1.5	308	64	7,784	779	0.13	0.22	1.5	6
1224	Waterchestnuts, canned, slices, solids and liquids	1 cup	140	86	70	1	Tr	Tr	Tr	Tr	0	17	3.5	6	1.2	165	11	6	0	0.02	0.03	0.5	2

*For product with no salt added: If salt added, consult the nutrition label for sodium value.

(*Continues*)

TABLE B-8 Food Composition Table

FOOD NO.	FOOD DESCRIPTION	MEASURE OF EDIBLE PORTION	WEIGHT (G)	WATER (%)	CALORIES (KCAL)	PROTEIN (G)	TOTAL FAT (G)	FATTY ACIDS SAT (G)	FATTY ACIDS MONO (G)	FATTY ACIDS POLY (G)	CHOLEST (MG)	CARB (G)	TOTAL D F(G)	CA (MG)	FE (MG)	K (MG)	NA (MG)	VITAMIN A (IU)	VITAMIN A (RE)	THMN (MG)	RIBOFL (MG)	NIACIN (MG)	ASCORBIC ACID (MG)
Miscellaneous Items																							
1225	Bacon bits, meatless	1 tbsp	7	8	31	2	2	0.3	0.4	0.9	0	2	0.7	7	0.1	10	124	0	0	0.04	Tr	0.1	Tr
	Baking powders for home use																						
	Double acting																						
1226	Sodium aluminum sulfate	1 tsp	5	5	2	0	0	0.0	0.0	0.0	0	1	Tr	270	0.5	1	488	0	0	0.00	0.00	0.0	0
1227	Straight phosphate	1 tsp	5	4	2	Tr	0	0.0	0.0	0.0	0	1	Tr	339	0.5	Tr	363	0	0	0.00	0.00	0.0	0
1228	Low sodium	1 tsp	5	6	5	Tr	Tr	Tr	Tr	Tr	0	2	0.1	217	0.4	505	5	0	0	0.00	0.00	0.0	0
1229	Baking soda	1 tsp	5	Tr	0	0	0	0.0	0.0	0.0	0	0	0.0	0	0.0	0	1,259	0	0	0.00	0.00	0.0	0
1230	Beef jerky	1 large piece	20	23	81	7	5	2.1	2.2	0.2	10	2	0.4	4	1.1	118	438	0	0	0.03	0.03	0.3	0
1231	Catsup	1 cup	240	67	250	4	1	0.1	0.1	0.4	0	65	3.1	46	1.7	1,154	2,846	2,438	245	0.21	0.18	3.3	36
1232		1 tbsp	15	67	16	Tr	Tr	Tr	Tr	Tr	0	4	0.2	3	0.1	72	178	152	15	0.01	0.01	0.2	2
1233		1 packet	6	67	6	Tr	Tr	Tr	Tr	Tr	0	2	0.1	1	Tr	29	71	61	6	0.01	Tr	0.1	1
1234	Celery seed	1 tsp	2	6	8	Tr	1	Tr	0.3	0.1	0	1	0.2	35	0.9	28	3	1	Tr	0.01	0.01	0.1	Tr
1235	Chili powder	1 tsp	3	8	8	Tr	Tr	0.1	0.1	0.2	0	1	0.9	7	0.4	50	26	908	91	0.01	0.02	0.2	2
	Chocolate, unsweetened, baking																						
1236	Solid	1 square	28	1	148	3	16	9.2	5.2	0.5	0	8	4.4	21	1.8	236	4	28	3	0.02	0.05	0.3	0
1237	Liquid	1 oz	28	1	134	3	14	7.2	2.6	3.0	0	10	5.1	15	1.2	331	3	3	Tr	0.01	0.08	0.6	0
1238	Cinnamon	1 tsp	2	10	6	Tr	Tr	Tr	Tr	Tr	0	2	1.2	28	0.9	11	1	6	1	Tr	Tr	Tr	1
1239	Cocoa powder, unsweetened	1 cup	86	3	197	17	12	6.9	3.9	0.4	0	47	28.6	110	11.9	1,311	18	17	2	0.07	0.21	1.9	0
1240		1 tbsp	5	3	12	1	1	0.4	0.2	Tr	0	3	1.8	7	0.7	82	1	1	Tr	Tr	0.01	0.1	0
1241	Cream of tartar	1 tsp	3	2	8	0	0	0.0	0.0	0.0	0	2	Tr	Tr	0.1	495	2	0	0	0.00	0.00	0.0	0
1242	Curry powder	1 tsp	2	10	7	Tr	Tr	Tr	0.1	0.1	0	1	0.7	10	0.6	31	1	20	2	0.01	0.01	0.1	Tr
1243	Garlic powder	1 tsp	3	6	9	Tr	Tr	Tr	Tr	Tr	0	2	0.3	2	0.1	31	1	0	0	0.01	Tr	Tr	1
1244	Horseradish, prepared	1 tsp	5	85	2	Tr	Tr	Tr	Tr	Tr	0	1	0.2	3	Tr	12	16	Tr	0	Tr	Tr	Tr	1
1245	Mustard, prepared, yellow	1 tsp or 1 packet	5	82	3	Tr	Tr	Tr	0.1	Tr	0	Tr	0.2	4	0.1	8	56	7	1	Tr	Tr	Tr	Tr
	Olives, canned																						
1246	Pickled, green	5 medium	17	78	20	Tr	2	0.3	1.6	0.2	0	Tr	0.2	10	0.3	9	408	51	5	0.00	0.00	Tr	0
1247	Ripe, black	5 large	22	80	25	Tr	2	0.3	1.7	0.2	0	1	0.7	19	0.7	2	192	89	9	Tr	0.00	Tr	Tr
1248	Onion powder	1 tsp	2	5	7	Tr	Tr	Tr	Tr	Tr	0	2	0.1	8	0.1	20	1	0	0	0.01	Tr	Tr	Tr

(Continues)

TABLE B-8 Food Composition Table

FOOD NO.	FOOD DESCRIPTION	MEASURE OF EDIBLE PORTION	WEIGHT (G)	WATER (%)	CALORIES (KCAL)	PROTEIN (G)	TOTAL FAT (G)	FATTY ACIDS			CHOLEST (MG)	CARB (G)	TOTAL D F(G)	CA (MG)	FE (MG)	K (MG)	NA (MG)	VITAMIN A		THMN (MG)	RIBOFL (MG)	NIACIN (MG)	ASCORBIC ACID (MG)
								SAT (G)	MONO (G)	POLY (G)								(IU)	(RE)				
Miscellaneous Items (*Continued*)																							
1249	Oregano, ground	1 tsp	2	7	5	Tr	Tr	Tr	Tr	0.1	0	1	0.6	24	0.7	25	Tr	104	10	0.01	Tr	0.1	1
1250	Paprika	1 tsp	2	10	6	Tr	Tr	Tr	Tr	0.2	0	1	0.4	4	0.5	49	1	1,273	127	0.01	0.04	0.3	1
1251	Parsley, dried	1 tbsp	1	9	4	Tr	Tr	Tr	Tr	Tr	0	1	0.4	19	1.3	49	6	303	30	Tr	0.02	0.1	2
1252	Pepper, black	1 tsp	2	11	5	Tr	Tr	Tr	Tr	Tr	0	1	0.6	9	0.6	26	1	4	Tr	Tr	0.01	Tr	Tr
	Pickles, cucumber																						
1253	Dill, whole, medium (3¾" long)	1 pickle	65	92	12	Tr	Tr	Tr	Tr	0.1	0	3	0.8	6	0.3	75	833	214	21	0.01	0.02	Tr	1
1254	Fresh (bread and butter pickles), slices 1½" dia, ¼" thick	3 slices	24	79	18	Tr	Tr	Tr	Tr	Tr	0	4	0.4	8	0.1	48	162	34	3	0.00	0.01	0.0	2
1255	Pickle relish, sweet	1 tbsp	15	62	20	Tr	Tr	Tr	Tr	Tr	0	5	0.2	Tr	0.1	4	122	23	2	0.00	Tr	Tr	Tr
1256	Pork skins/rinds, plain	1 oz	28	2	155	17	9	3.2	4.2	1.0	27	0	0.0	9	0.2	36	521	37	11	0.03	0.08	0.4	Tr
	Potato chips																						
	Regular																						
	Plain																						
1257	Salted	1 oz	28	2	152	2	10	3.1	2.8	3.5	0	15	1.3	7	0.5	361	168	0	0	0.05	0.06	1.1	9
1258	Unsalted	1 oz	28	2	152	2	10	3.1	2.8	3.5	0	15	1.4	7	0.5	361	2	0	0	0.05	0.06	1.1	9
1259	Barbecue flavor	1 oz	28	2	139	2	9	2.3	1.9	4.6	0	15	1.2	14	0.5	357	213	62	6	0.06	0.06	1.3	10
1260	Sour cream and onion flavor	1 oz	28	2	151	2	10	2.5	1.7	4.9	2	15	1.5	20	0.5	377	177	48	6	0.05	0.06	1.1	11
1261	Reduced fat	1 oz	28	1	134	2	6	1.2	1.4	3.1	0	19	1.7	6	0.4	494	139	0	0	0.06	0.08	2.0	7
1262	Fat free, made with olestra	1 oz	28	2	75	2	Tr	Tr	0.1	0.1	0	17	1.1	10	0.4	366	185	1,469	441	0.10	0.02	1.3	8
	Made from dried potatoes																						
1263	Plain	1 oz	28	1	158	2	11	2.7	2.1	5.7	0	14	1.0	7	0.4	286	186	0	0	0.06	0.03	0.9	2
1264	Sour cream and onion flavor	1 oz	28	2	155	2	10	2.7	2.0	5.3	1	15	0.3	18	0.4	141	204	214	28	0.05	0.03	0.7	3
1265	Reduced fat	1 oz	28	1	142	2	7	1.5	1.7	3.8	0	18	1.0	10	0.4	285	121	0	0	0.05	0.02	1.2	3
1266	Salt	1 tsp	6	Tr	0	0	0	0.0	0.0	0.0	0	0	0.0	1	Tr	Tr	2,325	0	0	0.00	0.00	0.0	0
	Trail mix																						
1267	Regular, with raisins, chocolate chips, salted nuts and seeds	1 cup	146	7	707	21	47	8.9	19.8	16.5	6	66	8.8	159	4.9	946	177	64	7	0.60	0.33	6.4	2
1268	Tropical	1 cup	140	9	570	9	24	11.9	3.5	7.2	0	92	10.6	80	3.7	993	14	69	7	0.63	0.16	2.1	11
1269	Vanilla extract	1 tsp	4	53	12	Tr	Tr	Tr	Tr	Tr	0	1	0.0	Tr	Tr	6	Tr	0	0	Tr	Tr	Tr	0

(*Continues*)

TABLE B-8 Food Composition Table

FOOD NO.	FOOD DESCRIPTION	MEASURE OF EDIBLE PORTION	WEIGHT (G)	WATER (%)	CALORIES (KCAL)	PROTEIN (G)	TOTAL FAT (G)	FATTY ACIDS SAT (G)	FATTY ACIDS MONO (G)	FATTY ACIDS POLY (G)	CHOLEST (MG)	CARB (G)	TOTAL D F(G)	CA (MG)	FE (MG)	K (MG)	NA (MG)	VITAMIN A (IU)	VITAMIN A (RE)	THMN (MG)	RIBOFL (MG)	NIACIN (MG)	ASCORBIC ACID (MG)
Miscellaneous Items (*Continued*)																							
	Vinegar																						
1270	Cider	1 tbsp	15	94	2	0	0	0.0	0.0	0.0	0	1	0.0	1	0.1	15	Tr	0	0	0.00	0.00	0.0	0
1271	Distilled	1 tbsp	17	95	2	0	0	0.0	0.0	0.0	0	1	0.0	0	0.0	2	Tr	0	0	0.00	0.00	0.0	0
	Yeast, baker's																						
1272	Dry, active	1 pkg	7	8	21	3	Tr	Tr	0.2	Tr	0	3	1.5	4	1.2	140	4	Tr	0	0.17	0.38	2.8	Tr
1273		1 tsp	4	8	12	2	Tr	Tr	0.1	Tr	0	2	0.8	3	0.7	80	2	Tr	0	0.09	0.22	1.6	Tr
1274	Compressed	1 cake	17	69	18	1	Tr	Tr	0.2	Tr	0	3	1.4	3	0.6	102	5	0	0	0.32	0.19	2.1	Tr

APPENDIX C

Harvesting and Storing Fruits, Nuts, and Vegetables

HARVESTING AND STORING VEGETABLES

Harvesting

For the best possible quality, vegetables should be harvested at peak maturity. This is possible when the vegetables will be used for home consumption, sold at a local market, or processed. The best harvesting time varies with each vegetable crop. Some will hold their quality for only a few days, while others can maintain quality over a period of several weeks. Frequent and timely harvest of crops is needed if the vegetable grower wishes to supply the market or the kitchen table with high-quality produce over a period of time. Specifications for harvesting each kind of vegetable crop should be consulted for the best results.

Harvesting is done by hand and through the use of machines. Mechanical harvesting is especially useful for the commercial grower. When harvesting a crop, care must be taken to prevent injuring the crop by both machinery and hand (see Figure C-1).

© B Brown/Shutterstock.com

FIGURE C-1 Mechanical harvesting of vegetable crops such as potatoes requires special machinery that is designed to separate the crop from leaves, stems, and soil.

Storing

When storing vegetables, ensuring they are in the proper stage of maturity is essential. There are specific temperatures and humidity levels for storing vegetable crops. Most crops need 90% to 95% humidity. Most homeowners find it difficult to reproduce the exact temperatures and humidity levels, but commercial growers can accomplish the correct storage conditions through specialized facilities.

Vegetables differ in the amount of time that they can be stored (see Table C-1). Some may be stored for several months, others for only a few days (Table C-2). For vegetables that store well, storage is essential to prolonging their marketing periods.

Fresh vegetables used for storage should be free of skin breaks, bruises, decay, and disease. Such damage or disease will decrease the vegetable's storage life.

Refrigeration storage or controlled atmosphere storage is recommended (Figure C-2). This type of storage reduces respiration and other metabolic

TABLE C-1
Temperatures and Humidities for Storing Vegetables

CROP	TEMPERATURE °F	RELATIVE HUMIDITY PERCENT
Asparagus	32	85-90
Beans, snap	45-50	85-90
Beans, lima	32	85-90
Beets	32	90-95
Broccoli	32	90-95
Brussels sprouts	32	90-95
Cabbage	32	90-95
Carrots	32	90-95
Cauliflower	32	85-90
Corn	31-32	85-90
Cucumbers	45-50	85-95
Eggplants	45-50	85-90
Lettuce	32	90-95
Cantaloupes	40-45	85-90
Onions	32	70-75
Parsnips	32	90-95
Peas, green	32	85-90
Potatoes	38-40	85-90

activities. It slows ripening, moisture loss, and wilting; spoilage from bacteria, fungus, and yeast; and undesirable growths such as that frequently seen with potato sprouts.

One activity that can increase the successful storage of vegetables is precooling. Precooling is the process of rapid removal of the heat from the crop before storage or shipment. One method of precooling is termed *hydrocooling*. In this method, vegetables are immersed in cold water long enough to lower their temperature to a desired level.

HARVEST AND STORAGE OF FRUITS AND NUTS

Harvesting

Harvesting fruits can be a daily practice. The advantage of home-grown fruit is that it can be picked at its peak of ripeness. However, fruits for commercial use must be picked several days ahead so they will withstand shipping and handling (Figure C-3).

For the best quality of fruit, apples, pears, and quince should be harvested when they begin to drop, soften, and become fully colored (Figure C-4). Some varieties will ripen over a 2-week period, requiring picking every day. Other varieties will ripen all at once.

Peaches, plums, apricots, and cherries should be harvested when the green disappears from the surface skin of the fruit. A yellow undercolor should be developed by this time. The fruit should be soft when pressed lightly in a cupped hand.

The nut varieties ripen from August to November. Nuts should be harvested immediately after they fall from the tree. Most nuts that do not fall can be knocked off tree with poles or mechanical harvesters.

FIGURE C-2 Some vegetables can be stored for extended periods of time in refrigerated storage or climate-controlled units such as this large potato storage unit.

TABLE C-2 Temperatures and Humidities for Storage and the Storage Life of Vegetables

STORAGE CONDITIONS							
VEGETABLE	TEMPERATURE [°F]	RELATIVE HUMIDITY (%)	STORAGE LIFE	VEGETABLE	TEMPERATURE [°F]	RELATIVE HUMIDITY (%)	STORAGE LIFE
Pea, English	32	95-98	1-2 weeks	Squash, winter	50	50-70	____[4]
Pea, southern	40-41	95	6-8 days	Strawberry	32	90-95	5-7 days
Pepper, chili (dry)	32-50	60-70	6 months	Sweet corn	32	95-98	5-8 days
Pepper, sweet	45-55	90-95	2-3 weeks	Sweet potato	55-60[3]	85-90	4-7 months
Potato, early	____[1]	90-95	____[1]	Tamarillo	37-40	85-95	10 weeks
Potato, late	____[2]	90-95	5-10 months	Taro	45-50	85-90	4-5 months
Pumpkin	50-55	50-70	2-3 months	Tomato, mature green	55-70	90-95	1-3 weeks
Radish, spring	32	95-100	3-4 weeks				
Radish, winter	32	95-100	2-4 months	Tomato, firm ripe	46-50	90-95	4-7 days
Rhubarb	32	95-100	2-4 weeks	Turnip	32	95	4-5 months
Rutabaga	32	98-100	4-6 months	Turnip greens	32	95-100	10-14 days
Salsify	32	95-98	2-4 months	Water chestnut	32-36	98-100	1-2 months
Spinach	32	95-100	10-14 days	Watercress	32	95-100	2-3 weeks
Squash, summer	41-50	95	1-2 weeks	Yam	61	70-80	6-7 months

Adapted from R. E. Hardenburg, A. E. Watada, and C. Y. Wang. *The Commercial Storage of Fruits, Vegetables, and Florist and Nursery Stocks*, USDA Agriculture Handbook 66 (1986).

[1]Spring- or summer-harvested potatoes are usually not stored. However, they can be held 4–5 months at 40 °F if cured 4 or more days at 60–70 °F before storage. Potatoes for chips should be held at 70 °F or conditioned for best chip quality.

[2]Fall-harvested potatoes should be cured at 50–60 °F and high relative humidity for 10–14 days. Storage temperatures for table stock or seed should be lowered gradually to 38–40 °F. Potatoes intended for processing should be stored at 50–55 °F; those stored at lower temperatures or with a high reducing sugar content should be conditioned at 70 °F for 1–4 weeks, or until cooking tests are satisfactory.

[3]Sweet potatoes should be cured immediately after harvest by holding at 85 °F and 90–95% relative humidity for 4–7 days.

[4]Winter squash varieties differ in storage life.

© Vitpho/Shutterstock.com

FIGURE C-3 Large trucks and modern highways move agricultural commodities quickly from farm to processor to market.

Pecan, hickory nuts, chestnuts, and Persian walnuts will lose their husks when ripe. However, the husks of black walnuts and other types of walnuts will need to be removed.

Grapes should be harvested only when fully ripe. The best way to judge when grapes are at full maturity is to sample an occasional grape.

Small bush and cane fruits such as strawberries and raspberries should be harvested when they are fully ripe. The best indicator is when the fruit is fully colored (Figure C-4). When these fruits and berries reach maturity, they should be picked every day. When harvesting, the fruit should be picked when dry to avoid mildew and mold damage.

© idiz/Shutterstock.com

FIGURE C-4 Fruit should be harvested for home use when it acquires the color of ripe fruit and begins to be soft to the touch.

Storage

Fruit storage is limited to varieties that mature late in the fall or that can be purchased at grocery markets during winter months (Figure C-5).

© Slavica Stajic/Shutterstock.com

FIGURE C-5A Apples ready for harvest.

© Zbynek Burival/Shutterstock.com

FIGURE C-5B Harvested apples waiting for sorting, packing, or storage.

The length of time fruits can be stored depends on their variety, stage of maturity, and soundness of the fruit at harvest (Table C-3).

To store apples long term, the temperature should be as close to 32 °F as possible. Apples can be stored in many ways, but they should be protected from freezing. A cellar or other area below ground level that is cooled by night air is a good place for apple storage. There should be moderate humidity in the storage area. Pears have the same storage requirements as apples.

Nuts should be air dried before they are stored. Nuts, especially pecans, keep longer if they are left in the shell and refrigerated at 35 °F. They can also be frozen if they are kept in their shells. Chestnuts have special requirements. They should be stored at 35 °F to 40 °F and high humidity shortly after harvest.

Small bush and cane fruits, including grapes, are not suitable for storage. These fruits are too perishable to keep long and are best used as soon as they are harvested.

TABLE C-3 Fruit Storage Temperatures as Adapted from USDA

		STORAGE CONDITIONS		LENGTH OF STORAGE PERIOD
FRUITS	FREEZING POINT	TEMPERATURE	HUMIDITY	
	°F	°F		
Apples	29°	32°	Moderate Moisture	Fall/Winter
Grapefruit	29.8°	32°	Moderate Moisture	4 to 6 weeks
Grapes	28.1°	32°	Moderate Moisture	1 to 2 months
Oranges	30.5°	32°	Moderate Moisture	4 to 6 weeks
Pears	29.2°	32°	Moderate Moisture	4 to 6 weeks

Adpated: USDA

GLOSSARY/GLOSARIO

A

absorption (1) Penetration of liquid into a solid that contains a porous structure; (2) adherence of molecules of liquid, gas, or solid to the surface of a solid.

absorción (1) La penetración de un líquido adentro de un sólido que contiene una estructura porosa; (2) la adhesión de las moléculas del líquido, gas o sólido a la superficie de un sólido.

acidulants Make a food acid or sour; added to foods primarily to change the taste and to control microbial growth.

acidulantes Hace un alimento ácidico o amargo; se agrega a los alimentos principalmente para cambiar el sabor y para controlar el crecimiento de microbios.

agglomeration Gathering into a cluster, mass, or ball.

aglomeración Recopilación de aglomeración en un clúster, masa o bola.

agglutination Sticking together as with glue.

aglutinación Pegadas como con pegamento.

aggregation Clumping together.

agregación Agregados juntos.

aging Holding beef in a cooler or refrigerator to tenderize its meat; commonly referred to as the "aging period."

envejecimiento Guardando la carne de res en un refrigerador o nevera para ablandar la carne; se refiere comúnmente a como el "período de envejecimiento."

albumen Also known as egg white; contains about 75 calories (kcal) of energy; provides humans with a high-quality protein containing every essential amino acid.

albúmina También conocida como la clara de huevo; contiene aproximadamente 75 calorías (kcal) de energía; proporciona a los seres humanos con una proteína de alta calidad que contiene todos los aminoácidos esenciales.

alcohols Chemical compounds characterized by an OH group.

alcoholes Compuestos químicos caracterizados por un grupo OH.

aldehydes Class of organic compounds characterized by the presence of the unsaturated carbonyl group (H–C=O) and a hydrogen atom attached to the carbon represented by (R–C=O).

aldehídos Una clase de compuestos orgánicos que se caracteriza por la presencia del grupo carbonilo insaturados (H–C=O) y un átomo de hidrógeno adjunta al carbono representada por (R–C=O).

alkaline Basic pH.

alcalino La base de la pH.

alkaloid Nitrogenous heterocyclic compounds product of plant metabolism many of which are poisonous.

alcaloide Compuestos heterocíclicos nitrogenados producto del metabolismo de las plantas, muchas de las ellos son venenosas.

all-purpose flour A blend of soft and hard wheat flours with a medium amount of gluten suitable for most baking purposes, including conventional handmade yeast breads.

harina para todo propósito Una mezcla de harinas de trigo blando y duro, con un importe medio de gluten apto para la mayoría de propósitos, incluyendo la cocción convencional de panes artesanales con levadura.

allied industry Supporting industry associated with food.

la industria aliada Apoyo del sector aliado asociados con los alimentos.

alpha-tocopherol A chemical with vitamin E activity represented by $C_{29}H_5O_2$.

el alfa-tocoferol La vitamina E un producto químico con actividad representada por $C_{29}H_5O_2$.

amino acid A basic building block of protein containing at least one amino group (NH_2) and at least one carboxyl group (–COOH) or acid group of small molecules, each having both an organic acid group (–COOH) and an amino acid group (–NH_2), which are the building units for protein molecules.

aminoácido Un bloque básico de la construcción de proteína que contiene al menos un grupo amino (NH_2) y al menos un grupo carboxilo (–COOH) o grupo de pequeñas moléculas de ácido, teniendo cada una de ellas tanto un grupo orgánico ácido (–COOH) y un grupo ácido amino (–NH_2), que son las unidades de construcción para moléculas de proteínas.

amino group A chemical group (NH_2) characteristic of all amino acids.

grupo amino Un grupo químico del grupo amino (NH_2) característico de todos los aminoácidos.

amylase An enzyme that hydrolyzes starch to produce dextrins, maltose, and glucose.

amilasa Una enzima que hidroliza el almidón para producir dextrinas, maltose, y glucosa.

amylopectin Long-chain branched fraction of starch.

amilopectina Ramificados de cadena larga fracción de almidón.

amylose Long chain or linear fraction of starch.

amilosa La cadena larga lineal o fracción de almidón.

anabolism Reactions involving the synthesis of compounds.

anabolismo Las reacciones relacionadas con la síntesis de compuestos.

anaerobic Without atmospheric oxygen.

anaerobio Sin oxígeno atmosférico.

anaerobic digestion Processes by which microorganisms break down biodegradable material in the absence of oxygen.

la digestión anaerobia Los procesos por los cuales los microorganismos descomponen material biodegradable en ausencia de oxígeno.

anorexia nervosa Serious eating disorder primarily of young women that is characterized by a pathological fear of weight gain; usually leads to harmful eating patterns, malnutrition, and excessive weight loss.

anorexia nerviosa Una grave trastorno alimentario, principalmente de mujeres jóvenes, que se caracteriza por un miedo patológico a aumentar de peso; conduce generalmente a patrones alimentarios perjudiciales, desnutrición y pérdida de peso excesiva.

antemortem Before slaughter or death.

antemortem Antes de la canicería o la muerte.

antimicrobial agents Substances that prevent or inhibit the growth of microorganisms.

los agentes antimicrobianos Sustancias que impiden o inhiben el crecimiento de microorganismos.

antioxidant A substance that can stop an oxidation reaction; a substance that slows down or interferes with the deterioration of fats through oxidation.

antioxidants Una sustancia antioxidante que puede detener una reacción de oxidación; una sustancia que ralentiza o interfiere con el deterioro de las grasas a través de la oxidación.

aquaculture The art, science, and business of cultivating plants and animals in water.

acuicultura El arte, la ciencia, y el cultivo de plantas y animales en el agua.

aquaponics System combining conventional aquaculture (raising animals such as fish or crayfish in tanks) with hydroponics (growing plants in water) in a symbiotic environment.

acuaponía Combinando el sistema convencional de la acuicultura (cría de animales como el pescado o el cangrejo de río en tanques) con hidroponía (cultivo de plantas en agua) en un entorno simbiótico.

aroma An odor detected by the olfactory sense.

aroma Olor detectado por el sentido del olfato.

ascorbic acid Vitamin C.

ácido ascórbico Vitamina C.

aseptic packaging Filling a container previously sterilized without recontaminating either the product or the container.

el envasado aséptico Llenar un recipiente previamente esterilizado sin recontaminando el producto o del contenedor.

aseptically Free from disease-producing microorganisms.

asépticamente Libre de microorganismos patógenos.

astringency The puckering, drawing, or shrinking sensation produced by certain compounds in food.

la astringencia La sensación de frunciendo, recogiendo, o reduciendo producida por ciertos compuestos en los alimentos.

atmospheric pressure Force per unit area exerted against a surface by the weight of the air above that surface.

presión atmosférica Fuerza por unidad de área ejercida sobre una superficie por el peso del aire por encima de la superficie.

atomic number An experimentally determined number typical of a chemical element that represents the number of protons in the nucleus, which in a neutral atom equals the number of electrons outside the nucleus, and that determines the place of the element in the periodic table.

número atómico Un número típico determinado experimentalmente de un elemento químico que representa el número de protones en el núcleo, el cual en un átomo neutro es igual al número de electrones fuera del núcleo, y que determina el lugar del elemento de la tabla periódica.

A_w Vapor pressure of food product at a specified temperature; abbreviation for *water activity.*

A_w La Presión de vapor de productos alimenticios a una temperatura especificada; abreviatura de *actividad de agua.*

B

bacteria Microorganisms usually consisting of a single cell composed of proteinaceous substances; some cause disease, others are used in food processing.

las bacterias Microorganismos que normalmente consisten de una sola célula compuesta de sustancias proteínicas; algunos provocan enfermedades, otros son utilizados en el procesamiento de alimentos.

bake To cook covered or uncovered in an oven, usually by dry heat; can also be done under coals, in ashes, or on heated stones or metals.

hornear Cocinar cubierto o descubierto en un horno, generalmente por calor seco; también puede realizarse bajo los carbones, en cenizas, o en piedras calentadas o metales.

baker's yeast Yeast used for raising bread, typically from the taxonomic group *Saccharomyces cerevisiae.*

la levadura de panadero Levadura usada para elevar el pan, normalmente desde el grupo taxonómico *Saccharomyces cerevisiae.*

bar International unit of pressure equal to 29.531 in. of mercury at 32 °F.

bar Unidad internacional de presión igual a 29.531. de mercurio a 32 °F.

barley A cereal of the genus *Hordeum*, a member of the Gramineae or grass family of plants. The two varieties of barley are two and six-rowed barley.

cebada Un cereal del género *Hordeum*, un miembro de la familia de las plantas Gramineae o hierba. Las dos variedades de cebada son la cebada de dos y de seis hileras.

baste To pour liquid composed of (1) drippings, fat, and water or (2) sugar and water over a food while it cooks.

hilvanara Verter líquido compuesto de (1) que los goteos, al grasa y el agua, o (2) el azúcar y el agua sobre una comida mientras cocina.

batch The amount of material prepared or required for one operation.

un lote La cantidad de lotes de material preparado o requerido para una operación.

beet sugar Sugar (sucrose) processed from the sugar beet plant.

azúcar de remolacha Azúcar (sacarosa) procesados a partir de la planta de remolacha azucarera.

beneficials Organisms that provide benefits for other species or ecosystems.

los organismos beneficiosos Organismos que proporcionan beneficios para otras especies o ecosistemas.

BHA *See* butylated hydroxyanisole.

BHA *Véase* butilhidroxianisol.

BHT *See* butylated hydroxytoluene.

BHT *Véase* hidroxitolueno butilado.

binge-eating disorder Eating disorder characterized by episodic consumption of large quantities of food (quickly and to discomfort); often purging follows out of guilt.

trastorno de alimentación compulsiva Trastorno alimentario caracterizado por el consumo episódico de grandes cantidades de alimentos (rápidamente y al malestar); a menudo la purga sigue fuera de la culpabilidad.

bioavailability In a form that can be used by the body.

biodisponibilidad En una forma que puede ser utilizado por el cuerpo.

biofilms Films formed by organisms.

biofilms (biopelículas) Películas formadas por organismos.

bio-intensive Organic agricultural system that achieves maximum yields from minimal land areas while increasing biodiversity and sustaining soil fertility.

bio-intensiva El sistema agrícola orgánica que logra rendimientos máximos desde el mínimo de áreas terrestres mientras aumenta la biodiversidad y mantener la fertilidad del suelo.

biological oxygen demand (BOD) A measure of water quality.

la demanda biológica de oxígeno (DBO) Un indicador de la calidad del agua.

bioproducts Designates a wide variety of corn-refining products made from natural, renewable raw materials that replace products made from nonrenewable resources; items produced by chemical synthesis.

bioproductos Designa una amplia variedad de maíz refinado de productos fabricados a partir de recursos naturales, materias primas renovables que sustituya productos fabricados a partir de recursos no renovables; los artículos producidos por síntesis química.

biosecurity The use of preventive measures to reduce the risk of transmitting infectious diseases in crops and livestock.

la bioseguridad La utilización de medidas preventivas para reducir el riesgo de transmisión de enfermedades infecciosas en los cultivos y el ganado.

biosensor (1) A device consisting of a biological component such as an enzyme or bacterium, that reacts with a target substance, and a signal-generating electrochemical component detecting the target substance; (2) a device that senses and transmits information about a biological process.

biosensor (1) Un dispositivo que consta de un componente biológico como una enzima o una bacteria, que reacciona con una sustancia de destino, y un generador de señales electroquímicas para detectar la sustancia componente de destino; (2) un dispositivo que detecta y transmite información acerca de un proceso biológico.

biotechnology Collection of industrial processes and tools that involve the use of biological systems such as plants, animals, and microorganisms.

la biotecnología Colección de herramientas y procesos industriales que impliquen el uso de sistemas biológicos tales como plantas, animals, y microorganismos.

biotin A water-soluble vitamin; functions in fatty acid synthesis.

la biotina Una vitamina soluble en agua; funciona en la síntesis de ácidos grasos.

birefringence The ability of a substance to refract light in two slightly different directions to form two rays; this produces a dark cross on each starch granule when viewed with a polarizing microscope.

birrefringencia La capacidad de una sustancia para refractar la luz en dos direcciones ligeramente diferentes para formar dos rayos; esto produce una cruz negra en cada gránulo de almidón cuando se ven con un microscopio de polarización.

black tea Type of tea made by tea leaves, then breaking and spreading them out in a fermentation room to oxidize, which turns the leaves to a copper color. The leaves are finally hot-air dried in a process that stops fermentation and turns the leaves black.

té negro Tipo de té hecho con hojas de té, rompiendo y distribuyéndolos en una sala de fermentación para oxidar, que convierte las hojas a un color cobre. Las hojas son finalmente aire caliente de secado en el proceso de fermentación que se detiene y gira las hojas de color negro.

blanching Pretreating with steam or boiling water to (1) partially inactivate enzymes and shrink food before canning, freezing, or drying, by heating with steam or boiling water; (2) aid in removal of skins from nuts and fruits by dipping into boiling water from 1 to 5 minutes; and (3) reduce strong flavors or set color of food by plunging into boiling water.

el escaldado El escaldado con vapor o agua hirviendo para (1) parcialmente inactivar enzimas y encoger los alimentos antes de enlatado, congelado, o secado por calentamiento con vapor o agua hirviendo; (2) ayudar en la extracción de pieles de nueces y frutas sumergiéndolo en agua hirviendo de 1 a 5 minutos; y (3) reducen los sabores fuertes o establecer el color de los alimentos al sumergirse en agua hirviendo.

bleached Flour processed with a "bleaching agent." Freshly ground wheat flour does not result in consistently good products. Over time, flour ages and whitens, and within several months it produces a better product. To hasten the improvement process, modern flour mills bleach and age flour chemically through the addition of a bleaching agent.

blanqueado Harina blanqueadas procesadas con un "agente blanqueador." Harina de trigo recién molida no resulta en productos consistentemente buenos. Con el pasar de tiempo, la harina se añeja y blanquea, y dentro de varios meses, produce un mejor producto. Para acelerar el proceso de mejora, molinos de harina modernos se añeja la harina químicamente mediante la adición de un agente blanqueador.

bleaching To make whiter or lighter, especially by physical or chemical removal of color.

blanqueo Para hacer más blanca o clara, especialmente por extracción física o química de color.

blend To mix two or more ingredients until they are well combined.

mezcla Para mezclar dos o más de los ingredientes hasta que estén bien combinados.

blood spot Also called *meat spots*; occasionally found on an egg yolk.

mancha de sangre También llamado *manchas de carne*; encuentra ocasionalmente en una yema de huevo.

BMI *See* body mass index.

BMI (IMC) Véase *el artículo sobre* Índice de Masa Corporal.

BOD *See* biological oxygen demand.

BOD *Véase* la demanda biológica de oxígeno.

body mass index (BMI) A formula that determines a person's relative health risks from obesity.

el índice de masa corporal (IMC) Una fórmula que determina una persona relativa de los riesgos para la salud derivadas de la obesidad.

boil To cook in boiling liquid. A liquid is boiling when bubbles break on the surface and steam is released. In a slowly boiling liquid, bubbles are small; in a rapidly boiling liquid, bubbles are large. As boiling changes from slow to rapid, more steam is formed but there is no increase in temperature. Boiling temperature of water at sea level is 100 °C or 212 °F. Boiling temperature is reduced by approximately 1 °C for every increase of 970 feet of elevation.

hervir Cocinar en líquido en ebullición. Un líquido hierve cuando las burbujas se rompen en la superficie y se libera el vapor. En un líquido en ebullición lentamente, las burbujas son pequeñas; en un líquido en ebullición rápidamente, las burbujas son grandes. Como la ebullición cambia de lento a rápido, más vapor está formado, pero no hay un aumento de la temperatura. La temperatura de ebullición del agua al nivel del mar es de 100 °C o 212 °F. La temperatura de ebullición se redujo en aproximadamente 1 °C por cada incremento de 970 metros de altitud.

boiling point The temperature at which a liquid vaporizes.

el punto de ebullición La temperatura a la cual un líquido se evapora.

bomb calorimeter Instrument for measuring the energy content of food.

calorímetro de bomba Un instrumento para medir el contenido de energía del alimento.

braise To cook in a covered utensil with a small amount of liquid; a moist heat method of cooking.

estofar Cocinar en un utensilio cubiertos con una pequeña cantidad de líquido; un método de cocción de calor húmedo.

bran Outer coat of a grain seed that is chiefly cellulose and contains minerals and some vitamins.

salvado De una capa exterior de la semilla del grano que es principalmente de celulosa y contiene minerales y algunas vitaminas.

breaded Coated with bread crumbs or similar flour product.

empanados Recubierto con migas de pan o harina similar producto.

bread flour Special flour that is higher in gluten; used for making yeast breads by hand and recommended for use in bread machines.

pan de harina Harina especial que es superior en el gluten; se utiliza para hacer panes de levadura a mano y se recomienda para su uso en máquinas de pan.

brewer's yeast An inactive yeast product that is a by-product of beer making and specially processed to be a nutritional supplement for humans.

la levadura de la cerveza Un producto de levadura inactiva que es un subproducto de la fabricación de cerveza y especialmente procesados para ser un suplemento nutricional para los seres humanos.

brewing Process of making beer, ale, or other similar cereal beverages that are fermented but not distilled.

proceso de fermentación Proceso de elaboración de cerveza, ale, u otras bebidas de cereales similares que son fermentados pero no destilada.

brine A salt solution.

salmuera Una solución salina.

broiler Type of chicken bred and raised specifically for meat production; younger and smaller than a roaster.

pollo de engorde Tipo de pollo de engorde y criados específicamente para la producción de carne; más joven y más pequeña que una para freír.

broiling Dry-heat method of cooking, usually by radiation but otherwise by direct exposure to heat source.

asar en el horno Un método de cocción de calor seco, generalmente por radiación, sino de lo contrario por la exposición directa a la fuente de calor.

bromelin A proteolytic enzyme found in pineapple; used in meat tenderization.

la bromelina Una enzima proteolítica bromelina encontrada en la piña; se utiliza en la carne tenderization.

brood To incubate eggs to hatch.

críar Para incubar los huevos para que empollen.

buckwheat Seed of a small plant that is ground into light or dark flour with greater fiber and a stronger flavor.

el alforfón Las semillas de una planta pequeña que es molinado en harina clara u oscura con mayor fibra y un sabor más fuerte.

bulimia nervosa Serious eating disorder occurring chiefly in females and characterized by compulsive overeating; self-induced vomiting or laxative or abuse often follows.

bulimia nerviosa Grave trastorno alimentario que ocurren principalmente en las mujeres y se caracteriza por el exceso de alimentación compulsiva; vómito provocado o abuso de laxantes a menudo sigue.

buttermilk Liquid by-product of butter making. Churning breaks the fat globule membrane so the emulsion breaks, fat coalesces, and water (buttermilk) escapes.

mazada Sub-producto líquido de la fabricación de mantequilla. Batiendo rompe la membrana de glóbulos de grasa de manera que la emulsión se rompe, la grasa se agrega y el agua (mazada) se escapa.

butylated hydroxyanisole (BHA) An antioxidant.

el butilhidroxianisol (BHA) Un antioxidante.

butylated hydroxytoluene (BHT) An antioxidant.

hidroxitolueno butilado (BHT) Un antioxidante.

by-products Secondary or incidental products from a manufacturing process.

los sub-productos Productos secundarias o accesorias de un proceso de fabricación.

C

caffeine Stimulant that is a plant alkaloid in coffee, tea, and selected carbonated beverages.

la cafeína Estimulante que la cafeína es un alcaloide de la planta de café, té y bebidas carbonatadas seleccionada.

calorie A unit of heat measurement; the small calorie used in chemistry. The kilocalorie is used in nutrition. One kilocalorie is equal to 1,000 small calories.

caloría Unidad de medición de calor; la pequeña caloría utilizada en química. La kilocaloría se utiliza en nutrición. Una kilocaloría es igual a 1,000 calorías pequeñas.

cane sugar A sucrose product processed from sugarcane.

azúcar de caña Un producto procesado de sacarosa de la caña de azúcar.

capon Male chicken castrated before 6 weeks of age and less than 8 months old.

capón Pollo macho castrado antes de 6 semanas de edad y menos de 8 meses de edad.

caramel Product formed by sugar decomposition because of the heating of sucrose; may also be made into a confectionery product because of the Maillard reaction.

caramel Producto formado por la descomposición del azúcar debido a la calefacción de sacarosa; también pueden hacerse en una confitería de producto debido a la reacción de Maillard.

caramelization Sucrose heated past the molten point so that it dehydrates and decomposes; the development of brown color and caramel flavor as dry sugar is heated to a high temperature and chemical decomposition occurs in the sugar.

caramelización Sacarosa calientada pasado el punto de fundido de modo que se deshidrata y se descompone; el desarrollo de color marrón y sabor de caramelo, como azúcar seca se calienta a una temperatura alta y descomposición química se produce en el azúcar.

carbohydrate A category of organic compounds with carbon, hydrogen, and oxygen; in sugars, the ratio is an approximate C:H:O [1:2:1] ratio.

carbohidratos Una categoría de compuestos orgánicos con carbono, hidrógeno, y oxígeno; en azúcares, la proporción es de un aproximado relación de C:H:O [1:2:1].

carbonated (1) Having carbon dioxide gas injected or dissolved in a liquid, creating an effervescence of pleasant taste and texture. (2) A category of organic compounds with carbon, hydrogen, and oxygen; in sugars, the ratio is an approximate C:H:O [1:2:1] ratio.

carbonatado (1) Con gas de dióxido de carbono inyectado o disuelta en un líquido, creando una efervescencia de sabor y textura agradable. (2) una categoría de compuestos orgánicos con carbono, hidrógeno, y oxígeno; en azúcares, la proporción es de un aproximado relación de C:H:O [1:2:1].

carbonator Equipment used to add carbon dioxide to a liquid.

carbonador Equipo utilizado para agregar dióxido de carbono a un líquido.

carbon dioxide (CO_2) Leavening agent for baked products that is made by chemical or biological means.

el dióxido de carbono (CO_2) Agentes de levadura para productos horneados que se hizo por medios químicos o biológicos.

carcinogen Cancer-causing substance.

carcinógeno Una sustancia que causa cáncer.

carotenoids Fat-soluble, yellow-orange pigments that are produced by plants; may be stored in the fatty tissues of animals.

carotenoides Solubles en grasa, pigmentos de color amarillo-anaranjado que son producidos por las plantas; puede almacenarse en los tejidos grasos de los animales.

case hardening During the drying process, food cooks on the outside before it dries on the inside.

endurecimiento superficial Durante el proceso de secado, cocineros de comida en el exterior antes de que se seque el interior.

casein Major protein found in milk.

caseína Proteína principal que se encuentra en la leche.

casings Tubular intestinal membrane of sheep, cattle, or hogs, or a synthetic facsimile used for sausage, salami, and the like.

tripas Membrana intestinal tubular de ovejas, vacas, o cerdo, o un facsímil sintéticos utilizados para salchichas, salami, y similares.

catabolism Breaking down of complex substances into simpler ones with the release of energy; the opposite of anabolism.

el catabolismo Rompiendo de las sustancias complejas en otras más simples con la liberación de energía; el opuesto de anabolismo.

catalyst A substance that changes the rate of a chemical reaction without being used up in the reaction; enzymes are catalysts.

catalizador Una sustancia que cambia la velocidad de una reacción química sin ser utilizado en la reacción; las enzimas son catalizadores.

celiac disease Chronic intestinal disorder in which the consumption of gluten results in an immune response that damages the intestinal mucosa.

la enfermedad celíaca Desorden crónico intestinal en el cual el consumo de gluten se traduce en una respuesta inmune que daña la mucosa intestinal.

caustic Capable of destroying living tissue.

cáustico Capaz de destruir el tejido vivo.

cellulose A plant carbohydrate of long chains of glucose; indigestible to humans.

la celulosa Una planta de celulosa carbohidrato de largas cadenas de glucosa; indigesta para los seres humanos.

centrifuge Equipment that uses centrifugal force to separate solids and liquids; which spinning speeds are used depends on separation needs.

centrifugar Equipo que utiliza la fuerza centrífuga para separar sólidos y líquidos, los cuales se utilizan velocidades de hilado depende de necesidades de separación.

chalaza Ropey strands of egg white that anchor the yolk in place in the center of the thick white.

chalaza Chungo, hebras de huevo que anclan la yema en su lugar en el centro de la blanco espeso.

chelating agent Substance that binds strongly to multivalent cations, by virtue of a number of anionic groups acting like pincers, for example, EDTA (ethylene diamine tetraacetic acid).

agente quelante Sustancia quelante que une fuertemente a los cationes multivalentes, en virtud de una serie de grupos aniónicos actuando como pinzas, por ejemplo, EDTA (ácido etilendiaminotetraacético).

chitin Water-insoluble polysaccharide containing amine groups; exoskeleton of insects and crustaceans.

quitina Polisacárido insoluble en agua que contienen grupos amino; cutícula de insectos y crustáceos.

Choice One of the quality grades.

Choice Uno de los grados de calidad.

choline A dietary component of many foods, is part of several major phospholipids (including phosphatidylcholine, which is also called lecithin) that are critical for normal membrane structure and function.

la colina Un componente dietético de muchos alimentos, forma parte de varios de los principales fosfolípidos (incluyendo fosfatidilcolina, que también se denomina lecitina) que son fundamentales para la estructura y la función de las membranas normales.

cholesterol Sterol with the formula $C_{27}H_{46}O$ abundant in animal fat, brain and nervous tissue, and eggs; functions in the body as a part of membranes and as a precursor of steroid hormones and bile acids.

colesterol Esteroles con la fórmula $C_{27}H_{46}O$ abundante en grasas animales, el cerebro, y el tejido nervioso, y huevos; funciones en el organismo como parte de membranas y como un precursor de las hormonas esteroides y ácidos biliares.

chroma Intensity or purity of a color.

croma Intensidad o pureza de un color.

chromatography A process in which a chemical mixture carried by a liquid or gas is separated into components as a result of differential distribution of the solutes as they flow around or over a stationary liquid or solid phase.

cromatografía Un proceso en el cual una mezcla química llevada por un líquido o un gas se divide en componentes como resultado de la distribución diferencial de los solutos como fluyen alrededor o a través de una fase estacionaria líquida o sólida.

churning The process that breaks the fat globule membrane so the emulsion breaks, fat coalesces, and water (buttermilk) escapes.

batiendo El proceso que rompe la membrana de glóbulos de grasa de manera que la emulsión se rompe, la grasa y el agua (coalesces mazada) se escapa.

cis Configuration has the hydrogen atoms on the same side of the double bond, particularly with unsaturated fatty acids.

cis Configuración cis tiene los átomos de hidrógeno en el mismo lado del doble enlace, especialmente con los ácidos grasos insaturados.

clean out of place (COP) System in which parts can be taken apart and cleaned away from the unit to which they belong.

limpiar fuera de lugar (CDP) En la que los componentes del sistema puede desmontarse y limpiarse fuera de la unidad a la que pertenecen.

climacteric Fruits that produce ethylene gas during ripening; ethylene sensitive.

climaterio Frutos que producen gas de etileno durante la maduración; sensible al etileno.

coagulate To form a clot, a semisolid mass, or a gel, after initial denaturation of a protein; to produce a firm mass or gel by denaturation of protein molecules followed by formation of new crosslinks.

coagular Para formar un coágulo, una masa semisólida, o un gel, después de una proteína desnaturalización inicial; para producir una masa firme o gel por la desnaturalización de las moléculas de proteínas, seguida por la formación de nuevos vínculos cruzados.

coagulation Aggregation of protein macromolecules into clumps or aggregates of semisolid material.

la coagulación La agregación de proteínas de coagulación macromoléculas en terrones o agregados de material semisólido.

coalesce To grow together or to unite into a whole.

se unen Para crecer juntos o a unirse en un todo.

coenzyme Compound required so an enzyme can function.

co-enzima Compuesto tan necesaria una enzima puede funcionar.

cold shortening A carcass that is chilled too rapidly; causes subsequent toughness.

el acortamiento en frío Una canal que está refrigerada con demasiada rapidez; causa dureza de carne después.

cold storage The use of low temperatures to preserve foods for extended periods of time, especially meats.

almencenamiento en frío El uso de almacenamiento en frío a bajas temperaturas para conservar alimentos durante largos períodos de tiempo, especialmente las carnes.

coliform Relating to, resembling, or being *E. coli.*

bacterias coliformes Relacionadas con coliformes, semejando, o siendo *E. coli.*

commercial sterility The condition in which not only all pathogenic and toxin-forming organisms have been destroyed but also other organisms capable of growth and spoilage under normal handling and storage conditions.

esterilidad commercial La condición en la que no sólo todos los patógenos y toxinas formando organismos han sido destruidos, pero también otros organismos capaces de crecimiento y descomposición en condiciones normales de manipulación y condiciones de almacenamiento.

competencies Abilities or capabilities of employees.

competencias Habilidades o capacidades de los empleados.

complex carbohydrates Carbohydrates made up of many small sugar units joined together—for example, starch and cellulose.

los carbohidratos complejos Carbohidratos compuesta de muchos pequeños se unieron unidades de azúcar, por ejemplo, el almidón, y la celulosa.

compliance Conformity in fulfilling official requirements.

en conformidad En el cumplimiento de los requisitos oficiales.

concentration Reducing the weight and volume of a product.

la concentración Reduce el peso y el volumen de un producto.

conching A flavor-development process that puts the chocolate through a "kneading" action and takes its name from the shell-like shape of the containers originally employed.

proceso de homogeneización de chocolate Un proceso de desarrollo de sabor que pone el chocolate a través de una acción de "amasar" y toma su nombre del shell-como la forma de los recipientes empleados originalmente.

conduction Heating transfers heat by direct contact of the heated molecules to those at a lower energy level.

la conducción La calefacción transfere calor por contacto directo de las moléculas calentadas a aquellos con un menor nivel de energía.

confectioners sugar A refined sugar product whose granule sizes range from coarse to powdered.

azúcar pastelero Un producto de azúcar refinado cuyo rango de tamaños de gránulo grueso para polvo.

constipation Fewer than three bowel movements a week; bowel movements produce stools that are hard, dry, and small, making them painful or difficult to pass.

estreñimiento Menos de tres evacuaciones a la semana; los movimientos intestinales producen las heces son duras, secas y pequeños, haciéndolos dolor o dificultad para pasar.

consumer Person or organization purchasing or using a service or commodity.

consumidor Persona u organización de comprar o utilizar un servicio o producto.

convection Motion of liquids and gases that helps distribute heat throughout a closed system. As liquids and gases are heated, they become lighter (less dense) and rise, whereas cooler molecules of the liquid or gas move to the bottom of a container or closed compartment.

la convección Los movimientos de convección de líquidos y gases que ayuda a distribuir el calor en un sistema cerrado. Como los líquidos y gases se calientan, se vuelven más ligeros (menos densa) y enfriador aumentando, mientras que las moléculas del líquido o gas se desplaza hacia la parte inferior de un contenedor o un compartimento cerrado.

convection oven Oven with a built-in fan that circulates the air and cooks food more evenly than a conventional oven.

horno de convección Horno con un ventilador incorporado que circula el aire y prepara la comida de forma más uniforme que un horno convencional.

cool storage Storage in which temperature ranges from 28 °F to 68 °F (−2 °C to 16 °F).

almacenamiento fresco En el que la temperatura oscila entre los 28 °F a 68°F (−2 °C a 16 °F).

copolymer Product of chemical reaction in which two molecules combine to form larger molecules that contain repeating structural units.

copolímero Producto de copolímero de reacción química en la que dos moléculas se combinan para formar moléculas más grandes que contienen unidades estructurales de repetición.

COP *See* clean out of place.

COP *Véase* limpio fuera de lugar.

corn sugar Processed sugar products from acid- or enzyme-hydrolyzed cornstarch.

azúcar de maíz Azúcar procesado de productos ácidos o enzimas hidrolizado de almidón de maíz.

corn syrup (1) Syrup made by partial hydrolysis of cornstarch to dextrose, maltose, and dextrins; (2) purified concentrated aqueous solution of nutritive saccharides obtained from edible starch.

el almíbar de maíz (1) Almíbar formado por hidrólisis parcial del almidón de maíz de dextrosa, maltosa, y dextrinas; (2) solución acuosa concentrada purificado de sacáridos nutritivos, obtenido a partir de almidón comestible.

covalent bond Strong chemical bond that joins two atoms together.

enlace covalente Un vínculo químico fuerte que une dos átomos juntos.

cream (1) Butterfat of milk; (2) to work one or more foods until soft and creamy by hand, spoon, electric mixer, or other implement.

la crema dulce (la nata dulce) (1) crema de materia grasa butírica de la leche; (2) para trabajar uno o más alimentos hasta el suave y cremoso a Mano, cuchara, batidora eléctrica, o cualquier otro accesorio.

creative thinking Ability to generate new ideas by making nonlinear or unusual connections or by changing or reshaping goals to imagine new possibilities; using imagination freely, combining ideas and information in new ways.

pensamiento creativo Capacidad para generar nuevas ideas, haciendo conexiones inusuales o no lineales o cambiando o reformulación de objetivos para imaginar nuevas posibilidades; usando la imaginación libremente, combinar las ideas y la información de nuevas maneras.

critical control point Any point in the process where loss of control may result in a health risk. *See also* Hazard Analysis and Critical Control Point (HACCP).

punto crítico de control Cualquier punto en el proceso donde la pérdida de control puede resultar en un riesgo para la salud. *Véase también el* Análisis de Peligros y Puntos Críticos de Control (HACCP).

critical temperature Temperature above which a gas can exist only as a gas, regardless of the pressure, because the motion of the molecules is so violent.

temperatura crítica La temperatura por encima de la cual un gas sólo puede existir como un gas, independientemente de la presión, porque el movimiento de las moléculas es tan violentos.

crohn's disease Chronic and long-lasting disease that causes inflammation, irritation, or swelling in the gastrointestinal tract; also called inflammatory bowel disease.

enfermedad de crohn Crónica y enfermedad prolongada que provoca inflamación, irritación o hinchazón en el tracto gastrointestinal; también se conoce como enfermedad inflamatoria intestinal.

croissant Classic crescent-shaped roll made from buttered layers of yeast dough much like a puff pastry.

croissant Rollo clásico hecho con forma de media luna de capas con mantequilla de la masa de levadura mucho como un hojaldre.

cross-contamination Contamination of one substance by another; for example, cooked chicken will become contaminated with salmonella organisms if it is cut on the same board used to cut raw chicken.

contaminación cruzada La contaminación de una sustancia por otra; por ejemplo, el pollo cocinado se contaminará con salmonelosis si se cortan en la misma placa utilizada para cortar pollo crudo.

crustaceans Shellfish with segmented, crusted shells and jointed appendages.

crustáceos Moluscos, crusted segmentada con conchas y apéndices articulados.

cryogenic Being or relating to very low temperatures.

criogénico Siendo criogénicos o relativos a temperaturas muy bajas.

crystalline Aggregation of molecules of a substance in a set, ordered pattern, forming individual crystals.

cristalina Agregación de las moléculas de una sustancia en un conjunto ordenado, patrón, formando cristales individuales.

crystallization The formation of crystals from the solidification of dispersed elements in a precise orderly structure.

cristalización La formación de cristales de la solidificación de elementos dispersos en una precisa estructura ordenada.

crystallize To form crystals from the solidification of dispersed elements in a precise orderly structure.

cristalizar Formar cristales a partir de la solidificación de elementos dispersos en una precisa estructura ordenada.

curd Substance consisting mainly of casein; obtained by coagulation of milk and used as food or made into cheese.

cuajada Sustancia formada principalmente de la caseína; obtenido por coagulación de la leche y utilizados como alimentos o hecho en el queso.

curing A preservative method; more often used for flavor and color enhancement.

curar Un conservante; el método más frecuentemente utilizado para la mejora del color y sabor.

Current Good Manufacturing Practices (CGMPS) Guidelines that a company uses to evaluate the design and construction of food-processing plants and equipment.

Actual de las Buenas Prácticas de Manufactura (CGMPS) Las directrices que una empresa utiliza para evaluar el diseño y construcción de plantas procesadoras de alimentos y equipo.

cut (1) To divide food material with knife or scissors; (2) to incorporate fat into dry ingredients with a pastry blender or two knives with the least possible amount of blending.

cortar (1) para dividir los alimentos material con un cuchillo o tijeras; (2) para incorporar la grasa en los ingredientes secos con una batidora de repostería o dos cuchillas con la menor cantidad posible de la mezcla.

cutability The percentage of closely trimmed, boneless, retail cuts from the four major beef wholesale cuts (round, loin, rib and chuck); it is an estimate of the relative amount of lean, edible meat from a carcass.

la cortabilidad El porcentaje de cortabilidad estrechamente recortado, deshuesados, cortes minoristas de los cuatro principales mayoristas de carne de vacuno cortes (ronda, lomo, costilla y Chuck); es una estimación de la cantidad relativa de carne magra, carne comestible de un cadáver.

cultural diversity Describes the American workplace representing people from different backgrounds.

la diversidad cultural Describe el lugar de trabajo estadounidense que representa a personas de orígenes diferentes

D

data sheet Similar to a résumé; contains pertinent information about potential employees.

hoja de datos Similar a un currículum vitae; contiene información pertinente acerca de los potenciales empleados.

deboning Removal of bone from meat.

el deshuesado Extracción del hueso de la carne.

degumming The first step in the oil-refining process, oils are mixed with water, which removes valuable emulsifiers such as lecithin; enhanced by adding phosphoric or citric acid or silica gel.

degumming El primer paso en el proceso de refinación de petróleo, los aceites se mezclan con el agua, el cual elimina la valiosa emulsionantes como la lecitina; mejorada mediante la adición de ácido cítrico o fosfórico o gel de sílice.

dehydrated food A food dried by artificial means to less than 5% moisture.

alimentos deshidratados Alimentos secado por medios artificiales a menos de 5% de humedad.

dehydration The almost complete removal of water from a product.

deshidratación La eliminación casi completa del agua de un producto.

dehydrogenase An enzyme that catalyzes a chemical reaction in which hydrogen is removed; similar to an oxidation reaction.

deshidrogenasa Una enzima que cataliza una reacción química en la que el hidrógeno es eliminado; similar a una reacción de oxidación.

dehydrogenation Removal of hydrogen from a compound.

la deshidrogenación La extracción de hidrógeno a partir de un compuesto.

Delaney clause Government action enforced by the FDA that basically says the food industry cannot add any substance to food if it induces cancer when ingested by humans or animals.

cláusula Delaney Acción gubernamental impuesta por la FDA que básicamente dice que la industria alimentaria no puede agregar cualquier sustancia a los alimentos si se induce el cáncer cuando es ingerido por los seres humanos o animales.

demographic Having to do with vital and social statistics.

demográfico Que tienen que ver con la demografía y las estadísticas sociales vitales.

denature A change in the molecular structure from the native structure of a protein.

la desnaturalización Un cambio en la estructura molecular de la estructura nativa de una proteína.

dental caries Cavity formation in teeth caused by bacteria that attach to the teeth and form acids in the presence of sucrose and other sugars; tooth decay.

caries dental La formación de cavidades de en los dientes causado por bacterias que se adhieren a los dientes y formar ácidos en presencia de sacarosa y otros azúcares; el decaimiento dental.

deodorization Removal of odor or smell.

desodorización Extracción de olor o el peste.

dermal Layer of protective tissue.

dérmica La capa de tejido protector.

deterioration The process of food progressively declining in quality and eventually becoming inedible.

deterioro El proceso de alimentos progresivamente disminuyendo en calidad y llegando a ser no comestible.

dextrinization Addition of various water-soluble gummy polysaccharides obtained from starch by the action of heat, acids, or enzymes; used as adhesives, sizes for paper and textiles, and thickening agents and in beer.

la dextrinization Adicción de diversos polisacáridos gomosos hidrosolubles obtenidos de almidón por la acción del calor, ácidos o enzimas; utilizado como adhesivos, tamaños de papel y textiles, y de agentes de espesamiento y en la cerveza.

dextrins Polysaccharides composed of many glucose units; produced at the beginning stages of starch hydrolysis (breakdown); somewhat smaller than starch molecules.

dextrinas Polisacáridos compuestos de muchas unidades de glucosa; producidos en las primeras etapas del proceso de hidrólisis de almidón (descomposición); algo más pequeña que las moléculas de almidón.

dextrose An alternate name for glucose, a monosaccharide having the chemical formula $C_6H_{12}O_6$.

dextrosa Un nombre alternativo para la glucosa, un monosacárido habiendo la fórmula química $C_6H_{12}O_6$.

diabetes Metabolic disorder that affects how the body uses blood sugar (glucose); has three main types—types 1 and 2 and gestational diabetes. In all types, the result is too much glucose in the blood.

diabetes Trastorno metabólico que afecta la forma como el cuerpo utiliza el azúcar (glucosa) en la sangre; tiene tres tipos principales tipos 1 y 2 y la diabetes gestacional. En todos los tipos, el resultado es una cantidad excesiva de glucosa en la sangre.

diarrhea Production of loose, watery stools three or more times per day. Diarrhea lasting more than two days could indicate more serious problems, and diarrhea lasting 4 or more weeks may indicate chronic disease.

la diarrea Producción de acuosa, heces sueltas tres o más veces al día. La diarrea que dura más de dos días podría indicar problemas más graves, y la diarrea dura 4 o más semanas puede indicar una enfermedad crónica.

Dietary Reference Intake (DRI) System of nutrition recommendations from the Institute of Medicine.

Ingesta Dietética de Referencia (DRI) Sistema de las recomendaciones de nutrición del Instituto de Medicina.

diffusion The movement of a substance from an area of higher concentration to an area of lower concentration.

la difusión El movimiento de una sustancia desde una zona de mayor concentración a una zona de menor concentración.

digestion The process that beaks down food into molecules small enough to be absorbed by the body.

la digestion El proceso de que descomposita los alimentos en moléculas lo suficientemente pequeñas para ser absorbidas por el cuerpo.

diglyceride Glycerol combined with two fatty acids.

diglyceride Glicerol combinado con dos ácidos grasos.

disaccharide Sugar composed of two simple sugars or monosaccharides; two monosaccharides linked together; simple sugars with two basic units.

disacárido Azúcar compuesto de dos azúcares simples o monosacáridos; dos monosacáridos entrelazados; azúcares simples con dos unidades básicas.

disinfecting The use of chemical products that destroy or inactivate microorganisms and prevent them from growing.

desinfección El uso de productos químicos que destruyen o inactivan los microorganismos y les impiden crecer.

dissolve To break into parts or to pass into solution.

disolvera Romper en partes o para pasar a la solución.

distribution Division and classification; deals with those aspects favorable to product sales, including product form, weight and bulk, storage requirements, and storage stability.

la distribución División y clasificación; se ocupa de los aspectos favorables para las ventas del producto, incluyendo la forma del producto, peso y volumen, los requisitos de almacenamiento y estabilidad en almacenamiento.

diverticular disease Condition occurring when small pouches or sacs form and pushed outward through weak spots in the colon wall. Most common in lower part of the colon; problems include diverticulitis (inflammation) and diverticular bleeding.

enfermedad diverticular Ocurre cuando pequeñas bolsas o sacos en forma y empujada hacia afuera a través de puntos débiles en la pared del colon. Más común en la parte inferior del colon; problemas incluyen la diverticulitis (inflamación) y sangrado diverticular.

double bonds Two bonds between atoms; in food science, typically carbon atoms (-C=C-).

dos enlaces Dobles enlaces entre los átomos; en ciencia de los alimentos, normalmente átomos de carbono (-C=C-).

DRI *See* Dietary Reference Intake.

DRI *Véase la* ingesta dietética de referencia.

dutch processing A mild alkali treatment of chocolate to change and darken color and improve flavor.

procesamiento holandés Un leve tratamiento alcalino de chocolate para cambiar y oscurecer el color y mejorar el sabor.

D value The time in minutes at a specified temperature to reduce the number of microorganisms by one log cycle.

valor D El tiempo en minutos a una temperatura especificada para reducir el número de microorganismos por un ciclo de registro.

E

egg The ova or female reproductive cell.

óvulo (HUEVO) El óvulos o célula reproductora femenina.

electrical stimulation The use of high-voltage electrical current to improve tenderness in many cuts of the beef carcass; used before slaughter of animals to render unconscious.

la estimulación eléctrica La utilización de corriente eléctrica de alto voltaje para mejorar la sensibilidad en muchos cortes de la carne en canal utilizado; antes del sacrificio de los animales para representar inconsciente.

electromagnetic energy Energy that has an electric and a magnetic component; for example, microwaves are electromagnetic energy.

la energía electromagnética Energía que tiene un componente eléctrico y magnético; por ejemplo, las microondas son energía electromagnética.

electron Chemical properties of an element are determined by the number of electrons in the outermost energy level of an atom; in its elemental state, the number of electrons of an atom equals the number of protons making the atom electrically neutral. Electrons travel around the nucleus at extremely high speeds.

electrón Las propiedades químicas de electrones de un elemento están determinadas por el número de electrones en el nivel energético más externo de un átomo; en su estado elemental, el número de electrones de un átomo es igual al número de protones haciendo el átomo eléctricamente neutro. Los electrones viajan alrededor del núcleo a velocidades extremadamente altas.

element One of a limited number of substances, such as hydrogen and carbon; composed of atoms; listed in the Periodic Table of Elements.

elemento Uno de un número limitado de sustancias, tales como el hidrógeno y el carbono; compuesto de átomos; enumeradas en la Tabla Periódica de Elementos.

emulsifier A substance that acts as a bridge at the interface between two immiscible liquids and allows the formation of an emulsion; a substance that aids in producing a fine division of fat globules; in ice cream, it also stabilizes the dispersion of air in the foam structure. Eggs contain the natural emulsifier lecithin.

emulsionante Una sustancia que actúa como un puente en la interfase entre dos líquidos inmiscibles y permite la formación de una emulsión; una sustancia que ayuda a producir una división de finos glóbulos de grasa; en helados, también estabiliza la dispersión del aire en la estructura de espuma. Los huevos contienen la natural emulsionante lecitina.

emulsion A system consisting of a liquid dispersed in an immiscible liquid usually in droplets of larger than colloidal size—for example, fat in milk.

emulsión Un sistema compuesto de un líquido disperso en un líquidos inmiscibles generalmente en gotas de mayor de tamaño coloidal-por ejemplo, la grasa en la leche.

endosperm Seed tissue surrounding the embryo that contains food reserves.

endosperma Tejido que rodea el embrión de la semilla que contiene reservas de alimentos.

energy Ability to do work; measured in food in terms of calories.

energía Capacidad de energía para hacer el trabajo; se mide en términos de calorías en los alimentos.

enhancers A substance added to a food to improve the quality or taste.

potenciadores Una sustancia añadida a un alimento para mejorar la calidad o sabor.

enriched flour Flour with added niacin, thiamin, riboflavin, folic acid, and iron to compensate for some of the nutrients lost during the milling process.

harina enriquecida Harina con el agregado de niacina, tiamina, riboflavina, ácido fólico y hierro para compensar algunos de los nutrientes perdidos durante el proceso de molienda.

enrober Covers and surrounds each candy center (nuts, nougats, fruit, and so on) with a blanket of chocolate.

bañadora Cubre y rodea a cada centro de caramelo (nueces, turrones, frutas, etc.) con una capa de chocolate.

entomophagy The practice of eating insects.

entomofagia La práctica de comer insectos.

entrepreneur One who starts and conducts a business assuming full control and risk.

un empresario El que inicia y dirige un negocio asume plenamente el control y el riesgo.

enzymatic browning Coloring of food caused by enzymes and prevented by blanching a food before drying.

pardeamiento enzimático Coloreando de alimentos causada por enzimas y prevenirse mediante escaldado comida antes del secado.

enzymatic reactions Those catalyzed by enzymes, special proteins produced by living cells.

reacciones enzimáticas Esas catalizadas por enzimas, proteínas especiales producidas por las células vivas.

enzyme Organic catalyst produced by living cells that changes the rate of a reaction without being used up in the reaction.

enzima Catalizador orgánicos producidos por células vivas que cambia la velocidad de una reacción sin ser utilizado en la reacción.

essential amino acid One that is required in the diet.

aminoácido esencial Uno de los aminoácidos esenciales que se requieren en la dieta.

essential oil Concentrated flavoring oil extracted from a food substance—for example, oil of orange and oil of peppermint.

aceite esencial Concentrado aceite aromatizante extraído de un alimento, por ejemplo, aceite de naranja y aceite de menta.

ester Type of chemical compound that results from combination of an organic acid (-COOH) with an alcohol (-OH) with the removal of one molecule of water.

ester Tipo de compuesto químico que resulta de la combinación de un ácido orgánico (-COOH) con un alcohol (-OH) con la eliminación de una molécula de agua.

ether extract The part of a complex organic material that is soluble in ether; consists chiefly of fats and fatty acids.

extracto etéreo La forma parte de un complejo material orgánico soluble en éter; se compone principalmente de grasas y ácidos grasos.

ethylene A small gaseous molecule (C_2H_4) produced by fruits and vegetables that initiates the ripening process.

etileno Una pequeña molécula gaseosa de etileno (C_2H_4) producido por frutas y verduras que inicia el proceso de maduración.

evaporation Removal of water, generally as a vapor either because of reaching the boiling point or at a lower temperature in a vacuum chamber.

la evaporación Extracción del agua, generalmente en forma de vapor o bien porque de llegar al punto de ebullición o a una temperatura inferior en una cámara de vacío.

eviscerate Removing the internal organs.

eviscerar Extracción de los órganos internos.

expenditures Expenses of money, time, or energy.

los gastos Gastos de dinero, tiempo o energía.

extraction Drawing out, pulling out, or removing.

la extracción La extracción, tirando o removimiento.

extratries Chemical interaction of packaging materials with foods.

extratries Interacción química de materiales envasado con alimentos.

extrusion Shaping through force.

la extrusión Conformación mediante la fuerza.

F

facultative Microorganisms that are both aerobic and anaerobic.

los facultativos Microorganismos que son tanto aerobia y anaerobia.

fahrenheit Thermometer scale in which the freezing point of water is 32 °F and the boiling point is 212 °F.

fahrenheit Una escala del termómetro en que el punto de congelación del agua es de 32 °F y el punto de ebullición es 212 °F.

famine A great shortage of food.

hambre Una gran escasez de alimentos.

FAO *See* Food and Agricultural Organization.

FAO *Véase* Organización alimentaria y agrícola.

fats Ester of glycerol and three fatty acids. Fats add richness, tenderness, calories, and flavor to many products.

las grasas Éster de glicerol y tres ácidos grasos. Añade riqueza, grasas, calorías, ternura y sabor a muchos productos.

fatty acid A chemical molecule consisting of carbon and hydrogen atoms bonded in a chainlike structure; combined through its acid group (–COOH) with the alcohol glycerol to form triglycerides.

ácidos grasos Una molécula química de ácidos grasos compuesto de carbono y átomos de hidrógeno enlazados en una estructura chainlike; combinado a través de su grupo ácido (–COOH) con el alcohol glicerol para formar triglicéridos.

fermentation Enzymatic decomposition of carbohydrates under anaerobic conditions.

la fermentación La descomposición enzimática de la fermentación de carbohidratos en condiciones anaeróbicas.

fiber Indigestible substances, including cellulose, hemicelluloses, pectin (all polysaccharides), and lignin, which is a noncarbohydrate material found particularly in woody parts of a vegetable.

la fibra Sustancias indigestible, incluyendo celulosa, hemicelluloses, pectina (todos los polisacáridos) y lignina, que es un material encontrado noncarbohydrate especialmente en las partes leñosas de una verdura.

ficin Used as a meat tenderizer.

ficin Usado como un ablandador de carne.

filtration Process of separating a solid from a liquid by applying a force to move the liquid through a barrier while retaining the solid. The force may be gravity.

filtración Proceso de separar un sólido de un líquido mediante la aplicación de una fuerza para mover el líquido a través de una barrera conservando el sólido. Puede ser la fuerza de la gravedad.

flatulence Accumulation of gas in the alimentary canal; intestinal gas.

la flatulancia Acumulación de gas en el tubo digestivo; gases intestinales.

flavor A blend of taste, smell, and general touch sensations evoked by the presence of a substance in the mouth.

sabor Una mezcla de sabores, sabores, olores y sensaciones táctiles generales evocados por la presencia de una sustancia en la boca.

flocculation Aggregation into a mass.

la floculación Agregación en una masa.

follow-up letter Letter written immediately after an interview.

carta de seguimiento Una carta escrita inmediatamente después de la entrevista.

food additive A substance, other than usual ingredients, that is added to a food product for a specific purpose such as flavoring, preserving, stabilizing, or thickening.

aditivo alimentario Una sustancia, excepto los ingredientes habituales, que se agrega a un alimento con un propósito específico , como aromatizante, preservando, estabilización o engrosamiento.

food allergy Abnormal response to food triggered by the body's immune system.

la alergia alimentaria Respuesta anormal a los alimentos desencadenada por el sistema inmunológico del cuerpo.

Food and Agricultural Organization (FAO) An agency of the United Nations that conducts research, provides technical assistance, conducts education programs, maintains statistics on world food, and publishes reports with the World Health Organization.

Organización de Alimentos y Agricultura (FAO) Un organismo de las Naciones Unidas que lleva a cabo la investigación, proporciona asistencia técnica, lleva a cabo programas de educación, mantiene estadísticas sobre el mundo de los alimentos, y publica informes con la Organización Mundial de la salud.

Food, Drug, and Cosmetic Act Regulates the labeling for all foods except meat and poultry.

ley de Alimentos, Medicamentos y Cosméticos Regula el etiquetado de todos los alimentos, excepto la carne y las aves de corral.

food-borne illness Any sickness resulting from the food spoilage of contaminated food, pathogenic bacteria, viruses, or parasites that contaminate food, as well as chemical or natural toxins.

las enfermedades transmitidas por los alimentos cualquier enfermedad resultante del deterioro de los alimentos de la contaminación de los alimentos, las bacterias patógenas, virus o parásitos que contaminan los alimentos, así como sustancias químicas o toxinas naturales.

food infection Illness produced by the presence and growth of pathogenic microorganisms in the gastrointestinal tract; they are often but not always present in large numbers.

infección alimentaria Enfermedad producida por la presencia y el crecimiento de microorganismos patógenos en el tracto gastrointestinal; son a menudo, pero no siempre está presente en grandes cantidades.

food insecure When a household is uncertain whether it can acquire enough food to meet the needs of all members because they do not have enough money or other resources for food.

inseguridad alimentaria Cuando un hogar es incierto si puede adquirir alimentos en cantidad suficiente para satisfacer las necesidades de todos los miembros porque no tienen suficiente dinero u otros recursos para la alimentación.

food irradiation Technology for controlling spoilage and eliminating food-borne pathogens; similar to conventional pasteurization and often called *cold pasteurization* or *irradiation pasteurization*.

la irradiación de alimentos La tecnología de irradiación de alimentos para el control de la corrupción y la eliminación de los agentes patógenos transmitidos por los alimentos; similar a la pasteurización convencional y a menudo llamado *la pasteurización fría* o *irradiación pasteurización*.

food labeling Labels show a product's name, the manufacturer's name and address, the amount of the product in the package, and the product ingredients.

etiquetado de alimentos Las etiquetas muestran el nombre del producto, el nombre y dirección del fabricante, la cantidad de producto en el paquete, y los ingredientes del producto.

food processing Procedure of taking raw materials and preparing them so that they become foods for human consumption.

el procesamiento de alimentos Procedimiento de tomar materias primas y prepararlos para que se conviertan en los alimentos para el consumo humano.

food safety Judgment of the acceptability of the risk involved in eating a food; if risk is relatively low, a food substance may be considered.

seguridad alimentaria La aceptabilidad de los riesgos implicados en la ingestión de un alimento; si el riesgo es relativamente bajo, un alimento puede ser considerado.

food system The totality of processes and infrastructures involved in feeding a population.

sistema alimentario La totalidad de infraestructuras y procesos implicados en la alimentación de una población.

food security When all people, at all times, have physical and economic access to sufficient, safe, and nutritious food to meet their dietary needs and food preferences for an active and healthy life.

la seguridad alimentaria Cuando todas las personas tienen en todo momento acceso físico y económico a alimentos suficientes, seguros y nutritivos para satisfacer sus necesidades dietéticas y preferencias alimentarias para una vida activa y sana.

food soil Unwanted matter on food-contact surfaces.

suelo de alimentos Materia no deseados en las superficies en contacto con los alimentos.

foreign aid Food, supplies, and money sent to people in need in foreign countries.

la ayuda extranjera Suministros, alimentos y dinero enviado a personas necesitadas en países extranjeros.

FPC Fish protein concentrate.

FPC Concentrado de proteína de pescado.

free radical Highly reactive atomic particle that has the potential to harm cells; created when an atom or molecule gains or loses an electron.

radicales libres Altamente reactivos de partículas atómicas que tiene el potencial de dañar células; creado cuando un átomo o molécula gana o pierde un electrón.

freeze-drying A drying process that involves first freezing the product and then placing it in a vacuum chamber, the ice sublimes (goes from solid to vapor phase without going through the liquid phase). The dried food is more flavorful and fresher in appearance because it does not become hot in the drying process.

liofilización Un proceso de secado que supone, en primer lugar, congelar el producto y luego colocarlo en una cámara de vacío, el hielo sublima (pasa de sólido a la fase de vapor sin pasar por la fase líquida). Los alimentos secos es más sabroso y fresco en apariencia porque no se calientan en el proceso de secado.

freezer burn Drying out while stored in a freezer.

quemaduras de congelador Secado fuera mientras se almacenan en un congelador.

fructose A sugar sometimes called *levulose* or *fruit sugar*; a monosaccharide with the chemical formula $C_6H_{12}O_6$.

fructosa Azúcar a veces llama *levulosa* o *azúcar de frutas*; un monosacárido con la fórmula química $C_6H_{12}O_6$.

fry (1) To cook in fat deep enough to float the food; also called *deep-fat fry* or *french fry*; (2) to cook in small amount of hot fat or drippings; also called *pan fry* or *sauté.*

freír (1) para cocinar en grasa lo suficientemente profundo para flotar los alimentos; también llamada *deep-fat* o como se frie las *papas fritas*; (2) para cocinar en una pequeña cantidad de grasa caliente o goteos; también llamado *pan freír* o *saltear.*

gastroenteritis Inflammation of the gastrointestinal tract.

gastroenteritis Inflamación del tracto gastrointestinal.

gastroesophogeal reflux disease (GERD) Regularly occurring heartburn produced when the lower esophageal sphincter becomes weak or relaxes, when allowing stomach contents and acid to enter the esophagus.

gastroesophogeal enfermedad por reflujo gastroesofágico (ERGE) Que ocurre regularmente cuando acidez producida el esfínter esofágico inferior se debilita o se relaja, al permitir que los contenidos del estómago y ácido para entrar en el esófago.

gelatinization Changes that occur in the first stages of heating starch granules in a moist environment; includes swelling of granules as water is absorbed and disruption of the organized granule structure.

el gelatinization Los cambios que ocurren en las primeras etapas de calefacción gránulos de almidón en un ambiente húmedo; incluye hinchazón de gránulos como el agua es absorbida y la desorganización de la estructura organizada de gránulo.

gelation The process of gelling.

la gelificación El proceso de gelificación de la gelatinización.

Generally Recognized as Safe (GRAS) A process or procedure that has been deemed safe by a panel of experts; the list that is maintained and periodically reevaluated by the FDA.

Generalmente Reconocidos como Seguros (GRAS) Un proceso o procedimiento que ha sido considerada segura por un panel de expertos; la lista que es mantenido y reevaluados periódicamente por la FDA.

generation time Time it takes for microorganisms to reproduce.

el tiempo de generación El Tiempo necesario para que los microorganismos se reproduzcan.

genetically modified organism (GMO) An organism whose genes have been artificially altered through genetic engineering.

los organismos genéticamente modificados (OGM) Un organismo cuyos genes han sido alterado artificialmente mediante la ingeniería genética.

genomics Field in genetics that uses recombinant DNA, DNA sequencing methods, and bioinformatics to determine the structures and analyze the functions of genomes, the complete sets of DNA within an organism.

la genómica El campo de la genética que utiliza ADN recombinante, los métodos de la secuenciación de ADN, y la bioinformática para determinar las estructuras y analizar las funciones de los genomas, los juegos completos de ADN dentro de un organismo.

GERD *See* gastroesophogeal reflux disease.

GERD *Véase* una enfermedad con el reflujo gastroesofogeal.

germ Small structure at the lower end of a kernel; rich in fat, protein, and mineral; contains most of the riboflavin content of the kernel.

pequeña Estructura germinal en el extremo inferior de un kernel; ricos en grasas, proteínas y minerales; contiene la mayoría de la riboflavina contenido del núcleo.

glazing Dipping a fish in cold water and then freezing a layer before dipping the fish again.

acristalamiento Inmersiando un pez en agua fría y luego congelando una capa antes de mojar el pescado de nuevo.

gliadin One of the wheat proteins that makes up one portion of gluten, the primary structural component.

la gliadina Una de las proteínas del trigo que conforma una parte de gluten, el principal componente estructural.

glucose A sugar sometimes called dextrose or blood sugar; a monosaccharide with the chemical formula $C_6H_{12}O_6$. This is the basic building block of starch.

glucosa Un azúcar a veces llamada dextrosa o azúcar en la sangre; un monosacárido con la fórmula química $C_6H_{12}O_6$. Este es el bloque de construcción básico de almidón.

gluten A protein in wheat and a limited number of other cereals that is formed when water is added to flour and, with kneading, gives structure to baked products.

gluten Una proteína presente en el trigo y un número limitado de otros cereales que se forma cuando el agua se añade a la harina y, con amasar, da estructura a los productos horneados.

glycerol Colorless liquid with chemical formula $C_3H_8O_3$; used as sweetener and preservative.

el glycerol Líquido incoloro de fórmula química $C_3H_8O_3$; se utiliza como conservante y edulcorante.

glycogen A complex carbohydrate—a polysaccharide—used for carbohydrate storage in the liver and muscles of the body; sometimes called *animal starch*.

el glucógeno Un carbohidrato complejo-un polisacárido-utilizados para el almacenamiento de carbohidratos en el hígado y los músculos del cuerpo; a veces llamado *almidón animal*.

Good Manufacturing Practices (GMPS) Guidelines a company uses to evaluate the design and construction of food-processing plants and equipment.

Buenas Prácticas de Fabricación (BPF) Directrices que utiliza una empresa para evaluar el diseño y construcción de plantas procesadoras de alimentos y equipo.

grades Positions in a scale of ranks or qualities.

los grados Posiciones en una escala de grados y calidades.

grain mill Machine designed to grind wheat and other grains to make flour.

molino de grano Máquina diseñada para moler el trigo y otros cereales para hacer harina.

granulated sugar White crystalline sugar or sucrose; the sugar referred to in most recipes that call for "sugar."

el azúcar granulado Azúcar o sacarosa blanca cristalina; azúcar contemplada en la mayoría de las recetas que llaman para "azúcar."

GRAS *See* Generally Recognized as Safe.

Grass *Véase* Generalmente Reconocido como Seguro.

gravity flow Movement of a liquid pulled by gravity.

flujo por gravedad El movimiento de un líquido tirado por gravedad.

green tea Unfermented tea in which leaves are heated before rolling in order to destroy enzymes; the leaf remains green throughout processing.

el té verde Hojas de té sin fermentar que se calientan antes de rodar en orden de destruir enzimas; la hoja permanece verde durante todo el proceso.

grill To cook by direct heat.

asar Para cocinar a fuego directo.

grind To put through a food chopper.

esmerilar Para pasar un picador de alimentos.

gum Any of several colloidal polysaccharide substances of plant origin that are gelatinous when moist but harden on drying and are salts of complex organic acids.

la encía Polisacárido coloidal cualquiera de varias sustancias de origen vegetal que son gelatinosos cuando húmedo pero se endurecen en la desecación y son sales de ácidos orgánicos complejos.

H

HACCP *See* Hazard Analysis and Critical Control Point.

HACCP *Véase* Análisis de Peligros y Puntos de Control Crítico.

hard wheat Wheat generally grown in northern climates; especially suited to bread making because of a high level of the gluten-forming wheat protein; a specific genus of wheat.

trigo duro Trigo cultivado generalmente en climas norteños; especialmente adecuado para hacer pan debido a un alto nivel del gluten de trigo formando proteínas; un género específico de trigo.

hazard A source of danger, long- or short-term, such as microbial food poisoning, cancer, birth defects, and so on.

peligro Una fuente de peligro, a corto o largo plazo, tales como la intoxicación alimentaria microbiana, cáncer, defectos de nacimiento, y así sucesivamente.

Hazard Analysis and Critical Control Point (HACCP) A food-safety system that aims to prevent food-borne illnesses.

Análisis de Peligros y Puntos Críticos de Control (HACCP) Un sistema de seguridad alimentaria que apunta a prevenir las enfermedades transmitidas por alimentos.

heating, ventilation, and air-conditioning (HVAC) System that modulates air temperature inside a facility such as a food-processing plant.

calefacción, ventilación y aire acondicionado (HVAC) Sistema que modula la temperatura del aire en el interior de una planta, como una planta procesadora de alimentos.

heat transfer The process by which energy in the form of heat is exchanged between two bodies.

la transferencia de calor Del proceso por el cual la energía en forma de calor se intercambian entre dos cuerpos.

helicobacter pylori Spiral-shaped bacteria that can damage the lining of the stomach and duodenum and cause peptic ulcer disease.

helicobacter pylori Bacteria en forma de espiral que puede dañar el revestimiento del estómago y del duodeno y causar la enfermedad de úlcera péptica.

herbal supplements Nonpharmaceutical, nonfood substances that are marketed to improve health; claims may or may not be true.

suplementos herbals Nonpharmaceutical, las sustancias no alimenticias que se comercializan para mejorar la salud; reclamaciones puede o no ser cierto.

hermetically Made airtight by fusion or sealing.

herméticamente Hecho sellado hermético o por fusión.

high-fructose corn syrup (HFCS) Invert sugar made from cornstarch.

el jarabe de maíz de alta fructosa (HFCS) El azúcar invertido fabricadas con almidón de maíz.

high hydrostatic pressure (HHP) Method of preserving and sterilizing food in which a product is processed under extremely high pressure, leading to the inactivation certain microorganisms and enzymes in the food.

alta presión hidrostática (HHP) Método de conservación y esterilización de alimentos en la cual un producto es procesado bajo presión extremadamente alta, conduciendo a la inactivación de ciertos microorganismos y enzimas en los alimentos.

high temperature, short time (HTST) Pasteurization method in which milk is heated to at least 161 °F for at least 15 seconds. The milk is immediately cooled to below 40 °F and packaged.

alta temperatura, corto tiempo (HTST) Método de pasteurización en el que se calienta la leche hasta al menos 161 °F durante al menos 15 segundos. La leche se enfría inmediatamente por debajo de los 40 °F y envasado.

homeostasis Tendency of a system or organism to maintain internal stability.

la homeostasis Tendencia de un sistema o un organismo para mantener la estabilidad interna.

homogenization Process in which whole milk is forced under pressure through very small openings, dividing the fat globules into minute particles.

la homogeneización Proceso en el que la leche entera es forzado bajo presión a través de aberturas muy pequeñas, dividiendo los glóbulos de grasa en las partículas diminutas.

hops The dried ripe cones of the female flowers of a plant used in brewing and medicine.

los conos De lúpulo desecado maduras de las flores femeninas de una planta utilizada en la industria cervecera y la medicina.

hot-fill *See* hot-pack.

hot-fill *Véase llenado en caliente* hot-pack.

hot-pack Filling unsterilized containers with sterilized food that is still hot enough to render the package commercially sterile.

hot-pack El llenado de recipientes no esterilizados con comida esterilizada que todavía está lo suficientemente caliente como para procesar el paquete comercialmente estéril.

HTST *See* High temperature, short time.

HTST *Véase* Alta temperatura, tiempo corto.

hue Property of light by which an object is classified in reference to the color spectrum—red, blue, green, or yellow.

tono La propiedad de luz por la que un objeto se clasifica en referencia al espectro de color—rojo, azul, verde o amarillo.

hunger Lack of food; desire or need for food.

el hambre La falta de alimentos; deseo o necesidad de alimentos.

HVAC *See* heating, ventilation, and air-conditioning.

HVAC *Véase* calefacción, ventilación y aire acondicionado.

hydrocooling Cooling with water.

hydrocooling Enfriamiento con agua.

hydrogenation Addition of hydrogen to oil; a selective process that can be controlled to produce various levels of hardening, from slight to almost solid.

hidrogenación La adición de hidrógeno al aceite; un proceso selectivo que puede ser controlada para producir distintos niveles de endurecimiento, desde leve hasta casi sólida.

hydrogen bond Relatively weak chemical bond that forms between a hydrogen atom and another atom with a slight negative charge, such as an oxygen or a nitrogen atom; each atom in this case is already covalently bonded to other atoms in the molecule of which it is part.

hidrógeno Enlace químico relativamente débil que se forma entre un átomo de hidrógeno y otro átomo con una ligera carga negativa, tales como oxígeno o un átomo de nitrógeno; cada átomo en este caso ya está covalentemente pegado a otros átomos en la molécula de la que es parte.

hydrolysis A chemical reaction in which a linkage between subunits of a large molecule is broken; a molecule of water enters the reaction and becomes part of the end products.

la hidrólisis Una reacción química en la que un vínculo entre las subunidades de una gran molécula está rota, una molécula de agua entra la reacción y pasa a formar parte de los productos finales.

hydrolysates Products of hydrolysis, the chemical process of decomposition involving the splitting of a bond and the addition of the hydrogen cation and the hydroxide anion of water.

hidrolizados Los productos de la hidrólisis, el proceso químico de descomposición que implica la división de un enlace y la adición del catión de hidrógeno y el anión hidróxido de agua.

hydrolyze To break a molecular linkage using a molecule of water; to break chemical linkages, by the addition of water, to yield smaller molecules.

hidrolizar Para romper un enlace molecular utilizando una molécula de agua; para romper los vínculos químicos, mediante la adición de agua, para producir moléculas más pequeñas.

hydrostatic retort Cans flow continuously. *See also* **retort**.

la retorta hidrostática Latas flujo continuo. *Véase también la* **retorte**.

hygroscopic Absorbing or attracting moisture from the air.

higroscópico Absorbiendo o atraer la humedad del aire.

hypobaric Reduced pressure; used along with low temperatures and humidity for cold storage.

hipobárico Reducción de la presión hipobárica; se utiliza junto con las bajas temperaturas y la humedad durante el almacenamiento en frío.

I

immersion freezing Intimate contact occurs between the food or package and the refrigerant.

congelación de inmersión Se produce por contacto íntimo entre la comida o el paquete y el refrigerante.

impeller A rotor for transmitting motion.

impulsor Un rotor para transmitir el movimiento.

Infant Health Formula Act Provides that manufactured formulas contain the known essential nutrients at the correct levels.

el Acto de la Fórmula de la Salud Infantil Dispone que fabrican fórmulas contienen los nutrientes esenciales conocidos en los niveles correctos.

inoculation Introducing a microorganism into surroundings suited to its growth.

la inoculación La introducción de un microorganismo en el entorno adecuado para su crecimiento.

insoluble Does not readily dissolve in water.

insoluble No insoluble fácilmente solubles en agua.

insoluble fiber Fiber that does not dissolve in water, allowing it to help move food particles through the digestive system.

la fibra insoluble La fibra que no se disuelven en el agua, lo que le permite ayudar a mover las partículas de alimentos a través del sistema digestivo.

inspection (1) Examining of food products or processes carefully and critically to ensure proper sanitary practices, labeling, or safety of the consumer; (2) organizing of food products and classifying them according to quality, such as grade A, B, or C; based on defined standards.

la inspección (1) inspección de productos alimenticios o de examinar cuidadosamente los procesos para garantizar que son correctos y críticamente las prácticas sanitarias, el etiquetado, o la seguridad de los consumidores; (2) la organización de productos alimenticios y clasificarlas de acuerdo a la calidad, tales como el grado A, B o C; basado en estándares definidos.

insulin Protein hormone synthesized in the pancreas that regulates blood glucose levels.

insulina Hormona proteica sintetizada en el páncreas que regula los niveles de glucosa en la sangre.

integrated System in which components are interconnected to perform a function.

integrado Sistema en el cual los componentes están interconectados para realizar una función.

integrated pest management (IPM) Control of one or more pests by a broad spectrum of techniques ranging from biological means to pesticides.

manejo integrado de plagas (MIP) El control de uno o más plagas insectiles por un amplio espectro de técnicas que van desde medios biológicos para plaguicidas.

interfering agent Substance used in candy making that hinders formation or reaction.

agente interfiriendo Sustancia utilizada en las golosinas que obstaculiza la formación o reacción.

international unit Standard unit of potency for a vitamin.

unidad internacional La unidad estándar de potencia de una vitamina.

inversion Reaction of the hydrolysis of sucrose to yield an equal mixture of glucose and fructose.

la inversión Reacción de la hidrólisis de sacarosa para producir una misma mezcla de glucosa y fructosa.

invert sugar The product of the hydrolysis of sucrose, yielding an equal mixture of glucose and fructose.

el azúcar invertido El producto de la hidrólisis de la sacarosa, dando igual mezcla de glucosa y fructosa.

invertase Enzyme that catalyzes reaction of the hydrolysis of sucrose to yield an equal mixture of glucose and fructose.

la invertasa Enzima que cataliza la reacción de la hidrólisis de sacarosa para producir una misma mezcla de glucosa y fructosa.

invisible fat Fat that occurs naturally in food products such as meats, dairy products, nuts, and seeds.

grasa invisible Grasa Que ocurre naturalmente en productos alimenticios como carnes, productos lácteos, nueces, y semillas.

iodine value Chemical test to determine how unsaturated the fatty acids are in a fat; the test is based on the amount of iodine absorbed by a fat per 100 grams. The higher the iodine value, the greater the degree of unsaturation.

valor de yodo Prueba química para determinar cómo son los ácidos grasos insaturados en grasa; la prueba se basa en la cantidad de yodo absorbido por una grasa por cada 100 gramos. Cuanto mayor sea el valor de yodo, mayor es el grado de saturación.

ion Electrically charged (+ or −) atom or group of atoms.

ion Cargado eléctricamente (+ o −), átomo o grupo de átomos.

ion exchange Process that uses specially fabricated porous beads that are chemically modified to exchange one ion for another as a solution is passed through them. Different types of ion-exchange columns exist.

intercambio iónico Un proceso que utiliza fabricado especialmente perlas porosas que son químicamente modificado para intercambiar un ion de otro como solución pasa a través de ellos. Los diferentes tipos de columnas de intercambio de iones que existen.

ionic bond A chemical bond formed when a complete transfer of electrons from one atom to another occurs.

enlace iónico Un enlace químico formado cuando una transferencia completa de electrones de un átomo a otro ocurre.

ionization Converted into ions.

la ionización Convertidos en iones.

ionomer Type of plastic material formed by ionic bonds.

ionómero Tipo de material plástico formado por enlaces iónicos.

irradiation Energy moving through space in invisible waves.

la irradición Energía en movimiento a través del espacio de irradiación en oleadas invisibles.

J

Joule heating Also known as ohmic heating and resistive heating; is the process by which the passage of an electric current through a conductor releases heat.

Calentamiento Joule También conocido como calentamiento óhmico y calentamiento resistivo; es el proceso por el cual el paso de una corriente eléctrica a través de un conductor libera calor.

juice Fluid expressed from a food product. In sugar, the juice with dissolved sucrose pressed out of sugar beet roots or sugarcane stalks.

jugo Líquido expresado desde un producto alimenticio. En el azúcar, el zumo con sacarosa disuelta se extrae de la remolacha azucarera, raíces o tallos de caña de azúcar.

Julian date A number—1 through 365—that indicates the day of the year.

fecha Juliana Un número—de 1 a 365—que indica el día del año.

K

kilocalorie Equal to 1,000 small calories; the small calorie is used in chemistry, and the kilocalorie is used in nutrition.

kilocaloría Igual a 1,000 calorías pequeñas; la pequeña caloría es usado en la química, y la kilocaloría es utilizado en la nutrición.

kimchi A spicy Korean pickled or fermented mixture containing cabbage and other seasonings.

el kimchi Un picante coreano encurtido o mezcla que contenga repoyo fermentado y otros condimentos.

knead The action used to manipulate bread dough that forms the gluten network in dough.

amasar La acción utilizada para manipular la masa de pan que forma la red de gluten en la masa.

L

lactase Enzyme that hydrolyzes lactose, a milk sugar, to glucose and galactose; found primarily in the intestines and in yeasts.

lactasa La enzima que hidroliza la lactosa, un azúcar de la leche, de glucosa y galactosa; encontrado principalmente en los intestinos y en levaduras.

lactose Disaccharide ($C_{12}H_{22}O_{11}$) present in milk.

lactosa Disacárido ($C_{12}H_{22}O_{11}$) presentes en la leche.

lactose intolerance Condition in which someone has digestive symptoms such as bloating, diarrhea, and gas after eating or drinking milk or milk products; caused by deficiency of lactase, the enzyme needed to break down lactose.

la intolerancia a la lactosa Una condición en la cual una persona tiene síntomas digestivos tales como distensión abdominal, diarrea y gas después de comer o beber leche o productos lácteos; causada por la deficiencia de lactasa, la enzima necesaria para descomponer la lactosa.

lagering Brewing by slow fermentation and maturing under refrigeration

lagering La elaboración de fermentación lenta y maduración bajo refrigeración

lag time The amount of time required for an organism to reach the log growth phase.

tiempo de retraso La cantidad de tiempo requerido para que un organismo para llegar a la fase de crecimiento.

lakes Any of various, usually bright, clear organic pigments composed basically of a soluble dye absorbed on or combined with an inorganic carrier.

lagos Cualquiera de los diversos lagos, generalmente claras y brillantes pigmentos orgánicos compuesto esencialmente de un tinte soluble absorbida o combinado con un transportista inorgánicos.

laminar Particles of fluid move in parallel or in adjacent layers; each layer has a constant velocity but this is relative to neighboring layers.

laminares Las partículas laminares de fluido se mueven en paralelo o en capas adyacentes; cada capa tiene una velocidad constante, pero esto es relativo a las capas vecinas.

landfilling Disposal of food-processing wastes in a landfill.

proceso de vertedero Eliminación de vertederos de desechos de procesamiento de comida en un vertedero.

leavening Substance that makes baked products lighter by helping them rise; yeast, baking powder, and baking soda are common leavening agents.

levadura Una sustancia que hace que los productos horneados encendedor ayudándoles alza; levadura, polvo de hornear y bicarbonato de soda son comunes los agentes leudantes.

legume Any of a large family of plants characterized by true pods enclosing seeds; for example, dried beans and peas.

legumbre Cualquiera de una gran familia de plantas caracterizado por una verdadera vainas que encierra semillas; por ejemplo, frijoles y guisantes secos.

letter of application Sent with a résumé or data sheet when applying for a job.

carta de solicitud Enviada con un curriculum vitae o hoja de datos al aplicar para un trabajo.

letter of inquiry Sent to a potential employee requesting possibility of employment.

carta de solicitud Enviada a un potencial empleado que solicita la posibilidad de empleo.

levulose Monosaccharide with the chemical formula $C_6H_{12}O_6$; also called *fructose* or *fruit sugar.*

levulosa Monosacárido con la fórmula química $C_6H_{12}O_6$; también se denomina *fructosa* o *azúcar de frutas.*

limiting amino acid Amino acid that is present in the lowest quantity compared to need; part of a complete diet.

aminoácido limitante Aminoácido que está presente en la cantidad más baja en comparación con la necesidad; como parte de una dieta completa.

linolenic acid Polyunsaturated fatty acid with 18 carbon atoms and 3 double bonds between carbon atoms; omega-3 fatty acid.

ácido linolénico Graso poliinsaturado ácido graso de 18 átomos de carbono y 3 dobles enlaces entre átomos de carbono; ácidos grasos omega-3.

lipase Enzyme that catalyzes the hydrolysis of triglycerides to yield glycerol and fatty acids.

la lipase Una enzima que cataliza la hidrólisis de triglicéridos para producir glicerol y ácidos grasos.

lipids A broad group of fatlike substances with similar properties.

los lípidos Un amplio grupo de sustancias similar a las grasas con propiedades similares.

lipolysis Breakdown of lipids.

lipólisis Desglose de lípidos.

lipoproteins Proteins combined with lipid or fatty material such as phospholipids.

las lipoproteínas Las proteínas combinadas con lípidos o material graso como los fosfolípidos.

liquid Ingredient that has flow or viscosity; may be milk, water, juices or other liquids; will dissolve and disperse ingredients.

líquido Ingrediente que tiene flujo o viscosidad; puede ser leche, agua, jugos u otros líquidos; disolver y dispersar los ingredientes.

logarithmic Expression of numbers using exponents.

logarítmico Expresión logarítmica de Números con exponentes.

low-temperature longer time (LTLT) Pasteurization method in which milk is heated to 145 °F for at least 30 minutes.

más tiempo de baja temperatura (LTLT) Un método de pasteurización de la leche que se calienta a 145 °F durante al menos 30 minutos.

M

macromineral Minerals in the diet that are used in larger amounts; includes calcium, phosphorus, potassium, sodium, chloride, magnesium, and sulfur.

macromineral Minerales en la dieta que se utilizan en grandes cantidades; incluye calcio, fósforo, potasio, sodio, cloro, magnesio y azufre.

Maillard reaction Special type of browning reaction involving a combination of proteins and sugars as a first step; may occur in relatively dry foods on long storage as well as in foods heated to high temperatures; the reaction between the amino group of an amino acid and protein and the reducing sugar to cause a brown color.

reacción de Maillard Un tipo especial de reacción de tostado que involucra una combinación de proteínas y azúcares como un primer paso; puede ocurrir en alimentos relativamente seco en almacenamiento prolongado, así como en los alimentos calentados a altas temperaturas; la reacción entre el grupo amino de un aminoácido y la proteína y el azúcar reductor para causar un color marrón.

malnutrition When a person eats but does not receive enough needed nutrients to keep his or her body healthy.

la malnutrición Cuando una persona come pero no recibe suficientes nutrientes necesarios para mantener su cuerpo sano.

malt Processed barley steeped in water, germinated on malting floors or in germination boxes or drums, and later dried in kilns for the purpose of converting insoluble starches in the grain to soluble substances and sugars in malt.

la malta La cebada procesada sumergidos en agua, germinadas en plantas de malteado o en cajas o bidones de germinación, y luego secados en hornos para el propósito de convertir los almidones insolubles en el grano para sustancias solubles y azúcares de la malta.

maltose A disaccharide made up of two glucose units.

la maltosa La disacárido formado por dos unidades de glucosa.

mammary gland Milk-producing gland found in female mammals.

la glándula mamaria La producción de leche de la glándula mamaria que se encuentra en hembras de mamíferos.

manufacturing Converts raw agricultural products to more refined or finished products.

la fabricación La fabricación de Productos agrícolas crudos convierte a más refinadas o productos acabados.

marbling Distribution of fat throughout the muscles of meat animals.

veteado La distribución de la grasa a través de los músculos de la carne de los animales.

marinating Soaking in a prepared liquid (oil and acid mixture) for a time to tenderize and season.

el marinado La inmersión en un preparado líquido (aceite y ácido mezcla) durante un tiempo para ablandar la carne y la temporada.

marketing The selling of foods—wholesale, retail, institutions, restaurants, and consumers.

la comercialización La venta de alimentos, mayoristas, minoristas, instituciones, restaurantes y consumidores.

market niches Specialized but limited market sectors.

nichos de mercado Especializado, pero limita los sectores del mercado.

mashing One of the steps of brewing; the infusion of malt, water, and crushed cereal grains at temperatures that encourage the complete conversion of cereal starches into sugars.

la maceración Uno de los pasos de la elaboración de la cerveza; la infusión de malta, agua y granos de cereales triturados a temperaturas que fomenten la plena conversión de almidones de cereales en azúcares.

mastitis Inflammation of one or more mammary glands within a cow's udder caused by an infection or a damaged teat.

mastitis La inflamación de una o más glándulas mamarias dentro de la ubre de una vaca por causa de una infección o un pezón dañado.

Meat Inspection Act Federal act authorized in 1906 and administered by the Food Safety and Inspection Service (FSIS) of the Department of Agriculture (USDA).

Ley Inespection Carne Ley Federal autorizó en 1906 y administrado por el Servicio de Inspección y Seguridad de Alimentos (FSIS) del Departamento de Agricultura de Estados Unidos (USDA).

mechanically separated Separation of bone from meat using automated equipment.

la separación mecánicamente Separada de los huesos de la carne mediante un equipo automatizado.

melt To liquefy by heat.

fundir Licuar por el calor.

melting point Temperature at which a solid fat becomes a liquid oil.

punto de fusion Temperatura en la cual una grasa sólida se convierte en un aceite líquido.

mesophilic Microorganism that prefers normal temperatures.

mesofílico El microorganismo que prefiere las temperaturas normales.

mesophilic bacteria Bacteria that grow best at moderate temperatures.

bacterias mesófilas Las bacterias que crecen mejor en temperaturas moderadas.

metabolism General term for all of the chemical reactions that occur in a living system. Metabolism can be divided into two processes: (1) anabolism, or reactions involving the synthesis of compounds, and (2) catabolism, or reactions involving the breakdown of compounds.

metabolismo Un término general para todas las reacciones químicas que ocurren en un sistema de vida. Metabolismo pueden dividirse en dos procesos: (1) anabolismo, o reacciones de síntesis de compuestos, y (2) el catabolismo, o reacciones que involucran la ruptura de compuestos.

microfiltration A membrane process that filters out or separates particles of extremely small size (0.02 to 2.00 microns), including starch, emulsified oils, and bacteria.

la microfiltración Una membrana proceso que filtra o separa las partículas de muy pequeño tamaño (0.02 a 2.00 micrones), tales como el almidón, aceites emulsionados y bacterias.

microminerals Minerals in the diet that are used in smaller amounts—chromium, cobalt, copper, fluorine, iodine, iron, manganese, molybdenum, nickel, selenium, silicon, tin, vanadium, and zinc.

microminerales Minerales en la dieta que se utilizan en cantidades pequeñas-cromo, cobalto, cobre, flúor, yodo, hierro, manganeso, molibdeno, níquel, selenio, silicio, el estaño, el vanadio y el zinc.

micron One millionth of a meter.

micron Millonésima parte de un metro.

microsensors Miniaturized, and possibly biological, indicators capable of accurately measuring the physiological state of plants, indicating temperature abuse for refrigerated foods, or to monitor the shelf life of food.

microsensores Miniaturizado y posiblemente, indicadores biológicos capaces de medir con precisión el estado fisiológico de las plantas, indicando la temperatura abuso de alimentos refrigerados, o supervisar la vida útil de los alimentos.

microwaves Electromagnetic energy that has an electric and a magnetic component.

microondas Energía electromagnética que tiene un componente eléctrico y magnético.

middlings Inner portion of a grain kernel.

medio grano Interior harinillas parte de un grano .

mill Grinding of grain to produce flour.

molino El moler el grano para producir harina.

millet Small yellow seed that lends texture and flavor to breads.

el mijo Pequeño semilla amarilla que da textura y sabor al pan.

milling Separating bran covering, germ, and endosperm from a grant to some desired extent.

el fresado La separación de la molienda bran cubriendo, el germen y endospermo de una donación a alguna medida deseada.

mill starch A starch-gluten suspension in the process of germ separation.

almidón molinero Almidón molino de suspensión de gluten en el proceso de separación del germen.

mince To cut or chop into tiny pieces.

carne picada Para cortar o picar en pequeños trozos.

mitochondria Microscopic sausage-shaped bodies in the cell cytoplasm that contain the enzymes necessary for energy metabolism.

las mitocondrias Organismos microscópicos en forma de salchicha en el citoplasma de la célula que contienen las enzimas necesarias para el metabolismo de la energía.

modified starches Natural starches that have been treated chemically to create some specific change in chemical structure, such as linking parts of the molecules together or adding some new chemical groups to the molecules; the chemical charges create new physical properties that improve starches in food preparation.

almidones modificados Almidones naturales que han sido tratadas químicamente para crear algún cambio específico en su estructura química, como la vinculación de partes de las moléculas juntos o agregar algunos nuevos grupos químicos a las moléculas; las cargas químicas crear nuevas propiedades físicas que mejoran los almidones en la preparación de alimentos.

molecule The smallest identifiable unit into which a pure substance can be divided and still retain the composition and chemical properties of that substance.

la molécula La más pequeña unidad identificable en la cual una sustancia pura puede dividirse y seguir manteniendo la composición y propiedades químicas de esa sustancia.

mollusk An invertebrate with a calcareous shell that is one or more pieces enclosing a soft, unsegmented body.

molusco Un invertebrado con una concha calcárea que es uno o más fragmentos adjuntando un suave, sin segmentar el cuerpo.

monoglyceride Glycerol combined with only one fatty acid

monoglicéridos Glicerol combinado con solamente un ácido graso

monosaccharide A simple sugar unit such as glucose, fructose, and galactose; a simple sugar with a single basic unit.

monosacárido Un azúcar simple unidad monosacáridos como la glucosa, la fructosa y la galactosa, un azúcar simple con una única unidad básica.

monounsaturated fatty acid (MUFA) Fat lacking a hydrogen bond on the carbon chain.

ácido graso monoinsaturadas (MUFA) Que carecen de una grasa de hidrógeno en la cadena de carbono.

mycotoxins Toxins produced by molds.

micotoxinas Las toxinas producidas por mohos.

myoglobin The name of the protein that is the primary color pigment of meat.

la mioglobina El nombre de la proteína que es el principal pigmento de color carne.

N

neutron Particle found in atomic nuclei having no charge.

neutrón Encontrado partículas de neutrones en el núcleo atómico que no tienen ningún cargo.

niacin One of the B vitamins.

niacina Una de las vitaminas del complejo B.

nib The roasted ground kernel of the cacao bean.

plumín La tierra asadas kernel del cacao en grano.

nonclimacteric Fruits that do not produce ethylene gas during ripening.

no climácterico Frutas que no producen gas de etileno durante la maduración.

noncrystalline Types of sugar candy such as hard candies, brittles, chewy candies, and gummy candies.

no cristalina Tipos de azúcar no cristalina dulces como caramelos, brittles, caramelos masticables y pegajosos de caramelos.

nonnutritive Term to describe a sweet product that contains little, if any, nutritive value; also called *artificial sweetener.*

no nutritivo Término para describir un dulce producto que contiene poco o ningún valor nutritivo; también se conoce como *edulcorante artificial.*

nonvolatile Lacking the ability to readily change to a vapor or to evaporate; not able to vaporize or form a gas at ordinary temperatures.

no volátil Faltan la capacidad de cambiar fácilmente a un vapor o se evaporen; no pueden vaporizar o formar un gas a temperaturas normales.

nutraceutical A food containing health-giving additives and having medicinal benefits.

la nutracéutica Los alimentos que contienen aditivos saludables y tener beneficios medicinales.

nutrigenomics Area of research examining how different foods interact with specific genes with the goal of modifying risks from common chronic diseases such as type 2 diabetes, obesity, heart disease, stroke, and certain cancers.

nutrigenómicos Área de investigación examinando cómo los diferentes alimentos interactúan con genes específicos con el objetivo de modificar los riesgos comunes de enfermedades crónicas tales como la diabetes tipo 2, obesidad, enfermedades del corazón, derrames cerebrales y algunos tipos de cáncer.

nutritionally enhanced Processed foods with added nutrient or nutrients such as vitamin C, B vitamins, iodine, iron, and so forth.

mejorados nutricionalmente Los alimentos procesados con el agregado de nutrientes o nutrientes como la vitamina C, vitaminas del complejo B, hierro, yodo, y así sucesivamente.

nutrition labeling Expression of the nutrient and caloric content of food products.

el etiquetado nutricional Expresión de los nutrientes y el contenido calórico de los alimentos.

Nutrition Labeling and Education Act Protects consumers against partial truths, mixed messages, and fraudulent nutrition information.

Etiquetado Nutricional y Educación Act Protege a los consumidores contra verdades parciales, mensajes contradictorios, y nutrición información fraudulenta.

nutritive sweetener Sweetener that has a caloric value.

edulcorante nutritivo Edulcorante que tiene un valor calórico.

O

obesity Condition characterized by excessive accumulation and storage of fat; in an adult, typically indicated by a BMI of 30 or more.

la obesidad Condición que se caracteriza por la acumulación excesiva y el almacenamiento de grasa; en un adulto, normalmente indicado por un IMC de 30 o más.

obligative Restricted to a particular condition such as organisms that can only survive in the absence of oxygen.

restringido Restringida a una condición particular como organismos que sólo puede sobrevivir en la ausencia de oxígeno.

offal Parts of a carcass that are considered by-products, such as feet, tongue, hide, and so on.

los despojos Las partes de un cuerpo que son considerados subproductos, tales como los pies, la lengua, ocultar y así sucesivamente.

ohmic heating Heating a food product by using an alternating current flowing between two electrodes.

calefacción óhmica Calefacción un producto alimentario mediante una corriente alterna que fluye entre dos electrodos.

oil Fat that is liquid at room temperature.

aciete Grasa que es líquido a temperatura ambiente.

olfactory Having to do with the sense of smell.

olfativas Que tienen que ver con el sentido del olfato.

oligosaccharide General term for a sugar composed of a few (often between 3 and 10) simple sugars or monosaccharides; a carbohydrate containing 2 to 20 sugar residues (the upper limit is not well defined); intermediate-size molecules containing approximately 10 or fewer basic units.

oligosacáridos Término general para un azúcar compuesto de unos pocos (a menudo entre 3 y 10) los azúcares simples o monosacáridos; un carbohidrato que contiene residuos de azúcar de 2 a 20 (el límite superior no está bien definido); las moléculas de tamaño intermedio que contiene aproximadamente 10 o menos unidades básicas.

oolong Type of tea that begins like black tea; when the leaf is fired or dried, a coppery color forms around the edge of the leaf while the center remains green; the oolong flavor is fruity and pungent.

té oolong Tipo de té oolong que comienza como el té negro; cuando la hoja se dispararon o secos, un color cobrizos formas alrededor del borde de la hoja, mientras que el centro permanece verde; el oolong sabor afrutado y picante.

organic acid Acid containing carbon atoms—for example, citric acid and acetic acid; generally, weak acids characterized by a carboxyl (-COOH) group.

ácido orgánico Que contiene átomos de carbono (por ejemplo, ácido cítrico y ácido acético; por lo general, ácidos débiles caracterizado por un extremo carboxilo (-COOH) del grupo.

organic compound Compound that has carbon included in its chemical formula; can be natural or human-made material.

compuesto orgánico Un compuesto de carbono que ha incluido en su fórmula química; pueden ser de origen natural o humano-hechas de material.

organic foods Foods grown or produced under conditions that supposedly replenish and maintain soil fertility, use only nationally approved materials in their production, and have verifiable records of the production system.

los alimentos orgánicos Alimentos cultivados o producidos bajo condiciones que supuestamente reponer y mantener la fertilidad del suelo, utilice sólo materiales aprobados a nivel nacional en su producción, y verificable de los registros en el sistema de producción.

organoleptic Perceived by any sense organ.

organolépticas Ninguna sensación percibida por el órgano.

osmosis Movement of water through a semipermeable membrane into a solution where the solvent concentration is higher, thus equalizing the concentration of water on both sides of the membrane.

osmosis El movimiento de agua a través de una membrana semipermeable en una solución donde la concentración de vapor del disolvente es mayor, por lo tanto igualando la concentración de agua en ambos lados de la membrana.

osmotic pressure The force that a dissolved substance exerts on a semipermeable membrane.

la presión osmótica La fuerza que una sustancia disuelta ejerce sobre una membrana semipermeable.

oxidase Enzyme that catalyzes oxidation reactions.

oxidasa Cataliza las reacciones de oxidación.

oxidation Chemical change that involves adding oxygen; for example, polyphenols are oxidized to produce different flavor and color compounds; a chemical reaction in which oxygen is added; addition of oxygen to carotenoid pigments to lighten color; chemical reactions in which oxygen is added, hydrogen is removed, or electrons are lost; gain in oxygen or loss of electrons.

la oxidación Un cambio químico que implica la adición de oxígeno; por ejemplo, los polifenoles se oxidan para producir compuestos de sabor y color diferente; una reacción química en la cual el oxígeno es añadido; adición de oxígeno de pigmentos carotenoides para aligerar el color; las reacciones químicas en las cuales se agrega oxígeno, el hidrógeno es eliminado, o los electrones se pierden; ganancia de oxígeno o pérdida de electrones.

oxidation-reduction reaction Reaction in which the loss of electrons from one atom produces a positive oxidation state and the gain of electrons by another atom results in a negative oxidation or reduction.

reacción de oxidación-reducción Una reacción en la que la pérdida de electrones de un átomo produce un estado de oxidación positivo y la ganancia de electrones por otro átomo resultados negativos en una oxidación o reducción.

P

palatability Acceptable taste and tenderness of a food product.

palatabilidad Sabor aceptable y la ternura de un producto alimenticio.

pan broil To cook uncovered on hot metal such as a grill or frying pan; the utensil may be oiled just enough to prevent sticking.

pan asar Cocinar destapado en un superficie de metal caliente como una plancha o sartén; el utensilio pueden lubricarse lo suficiente para evitar que se pegue.

pancreatitis Inflammation of the pancreas.

pancreatitis Una inflamación del páncreas.

pan fry To cook in a small amount of fat.

sartén sofreír Cocinar en una pequeña cantidad de grasa.

papain A vegetable enzyme used to tenderize meat.

la papaína Una enzima vegetal utilizado para ablandar la carne, la carne.

parboil To boil until partially cooked. Foods with strong or salt flavor are often parboiled, as are tough foods that are to be roasted or cooked in hot fat.

cueza parcialmente Hervir hasta parcialmente cocidos. Los alimentos con un fuerte sabor a sal o son a menudo escaldado, como son los alimentos que son difíciles para ser tostado o cocinado en grasa caliente.

pasteurization Process of heating a food to a specified temperature for a set period of time to destroy pathogenic organisms.

la pasteurización Proceso de calentar un alimento a una temperatura específica durante un período de tiempo establecido para destruir los organismos patógenos.

pasteurize To treat with mold heat to destroy pathogens—but not all microorganisms—present in a food product.

pasteurizar Para tratar el molde con calor para destruir los patógenos, pero no todos los microorganismos presentes en un producto alimenticio.

pathogenic Capable of causing disease.

patógeno Capaz de causar la enfermedad.

peptic ulcer A sore on the lining of the stomach or duodenum; causes include long-term use of certain drugs (NSAIDs) such as aspirin and ibuprofen, infection with *Helicobacter pylori* bacteria, and rare tumors in the stomach, duodenum, or pancreas.

úlcera péptica Una úlcera en el revestimiento del estómago o del duodeno; causas incluyen el uso a largo plazo de ciertos fármacos (AINES), como aspirina e ibuprofeno, la infección con la bacteria Helicobacter pylori, y raros tumores en el estómago, el duodeno o el páncreas.

pectin Gel-forming polysaccharide (polygalacturonic acid) found in plant tissue; a complex carbohydrate (polysaccharide) composed of galacturonic acid subunits, partially esterified with methyl alcohol and capable of forming a gel.

pectina Polisacárido formador de gel (ácido polygalacturonic) encontrada en el tejido vegetal; un carbohidrato complejo (polisacárido ácido galacturonic) compuesta de subunidades, parcialmente esterificados con alcohol metílico y capaces de formar un gel.

pectinase Enzyme that hydrolyzes the linkages that hold the small building blocks of galacturonic acid together in the pectic substances, producing smaller molecules.

actividades de pectinase Enzima que hidroliza los vínculos que mantienen los pequeños bloques de construcción del ácido galacturonic juntos en las substancias pécticas, produciendo moléculas más pequeñas.

pentose Simple sugar or monosaccharide with five carbon atoms.

la pentose Azúcar simple o monosacárido de cinco átomos de carbono.

peptide Variable number of amino acids joined together.

péptido Número variable de aminoácidos se unen.

peptide bond Bond formed by the condensation of the amino group (-NH_2) of one amino acid with the acid group (-COOH) of another amino acid, resulting in the loss of water.

vínculo del péptido Vínculo formado por la condensación del grupo amino (-NH_2) de un aminoácido con el grupo ácido (-COOH) de otro aminoácido, resultando en la pérdida de agua.

PER *See* protein efficiency ratio.

PER *Vease* coeficiente de eficiencia proteica.

per capita Per person.

per cápita Por persona.

permeate To penetrate through the pores; magnesium and iron sulfates that do not precipitate when boiled.

impregnar De penetrar a través de los poros; magnesio e hierro sulfatos que no precipitar al hervido.

peroxide value Indicates the degree of oxidation that has taken place in a fat or oil. The test is based on the amount of peroxides that form at the site of double bonds. These peroxides release iodine from potassium iodide when it is added to the system.

valor de peróxido Indica el grado de oxidación que ha tenido lugar en una grasa o aceite. La prueba se basa en la cantidad de peróxidos que se forman en el sitio de enlaces dobles. Estos peróxidos liberan yodo de yoduro de potasio cuando se agregan al sistema.

polyethylene terephithalate (PET) The plastic used for large 2-liter soft-drink bottles.

polietileno terephithalate (PET) El plástico utilizado para grandes botellas de refrescos de 2 litros.

pH Scale using numbers 1 to 14 to indicate how acidic or alkaline a substance is; 1 is the most acid, 7 is neutral, and 14 is the most alkaline.

pH La escala de pH con los números 1 al 14 para indicar cuán ácida o alcalina es una sustancia; 1 es el más ácido, 7 es neutro, y 14 es el más alcalino.

phenolic compound Organic compound that includes in its chemical structure an unsaturated ring with -OH groups on it; polyphenols have more than one -OH group; organic compounds that include in their chemical structures unsaturated rings with -OH groups. These compounds are easily oxidized, producing a brownish discoloration.

compuesto fenólico Compuesto orgánico que incluye en su estructura química un anillo insaturado con -grupos OH en ella; polifenoles tiene más de un grupo -OH; compuestos orgánicos que incluyen en sus estructuras químicas insaturados con anillos -grupos OH. Estos compuestos son oxidados fácilmente, produciendo una coloración pardusca.

phospholipid Type of lipid characterized chemically by glycerol combined with two fatty acids; phosphoric acid and a nitrogen-containing base—for example, lecithin.

fosfolípido Tipo de lípidos caracterizado químicamente por glicerol combinado con dos ácidos grasos; ácido fosfórico y una base que contiene nitrógeno, por ejemplo, la lecitina.

photosynthesis The conversion of carbon dioxide and water to sugars in the cells of plants.

la fotosíntesis La conversión de dióxido de carbono y el agua en azúcares en las células de plantas.

phytochemical Chemical from a plant.

fitoquímicos De una planta química.

phytonutrient Chemical compounds that help protect a plant; believed to be beneficial to human health and help prevent various diseases.

fitonutrientes Compuestos químicos que ayudan a proteger una planta; cree que es beneficiosa para la salud humana y ayuda a prevenir diversas enfermedades.

pigment Any biological substance that produces color in tissues.

pigmento Cualquier sustancia biológica que produce el color en los tejidos.

poach To cook in a hot liquid, carefully handling the food to retain its form.

escalfar Para cocinar en un líquido caliente, la manipulación de los alimentos con cuidado para conservar su forma.

polar Chemical molecules that have electrical charges (positive or negative); tend to be soluble in water; having two opposite natures such as both positive and negative charges.

polar Moléculas químicas que tienen cargas eléctricas (positivas o negativas); tienden a ser solubles en agua; tener dos naturalezas opuestas como tanto las cargas positivas y negativas.

polymer Giant molecule formed from smaller molecules that are chemically linked; a large molecule formed by linking together many smaller molecules of a similar kind; molecules of relatively high molecular weight that are composed of many small molecules acting as building blocks.

polímero Molécula gigante formado a partir de moléculas más pequeñas que son químicamente ligados; una gran molécula formada por la vinculación de muchas moléculas más pequeñas similares; de moléculas de peso molecular relativamente alto que están compuestos de muchas moléculas pequeñas que actúan como bloques de construcción.

polymerize To form large molecules by combining smaller chemical units.

se polimerizan Para formar moléculas grandes combinando unidades químicas más pequeñas.

polysaccharide Complex carbohydrate made of many simple sugar (monosaccharide) units linked together; in the case of starch, the simple sugars are all glucose; carbohydrate polymer consisting of at least 20 monosaccharides or monosaccharide derivatives; complex carbohydrates with many basic units (up to thousands).

polisacárido Complejo polisacárido hecha de muchos carbohidratos azúcar simple (monosacárido) unidades vinculadas entre sí; en el caso de almidón, azúcares simples son la glucosa; carbohidratos polímero compuesto de por lo menos 20 monosacáridos o monosacáridos derivados; carbohidratos complejos con muchas unidades básicas (miles).

polyunsaturated fatty acid (PUFA) Fatty acid with two or more double bonds between carbon atoms; for example, linoleic acid with two double bonds.

ácidos grasos poliinsaturados (PUFA) deácidos grasos con dos o más dobles enlaces entre átomos de carbono; por ejemplo, el ácido linoleico, con dos dobles enlaces.

postmortem After death.

postmortem Después de la muerte.

potable Drinkable.

agua potable Agua bebible.

poult A young turkey

los polluelos de pavo Un pavo joven

powdered sugar Sugar product produced by grinding a mixture of granulated sugar and corn starch.

azúcar en polvo Un producto producido por una mezcla de molienda de azúcar granulada y fécula de maíz.

pretzel Yeast dough typically rolled into a long rope and often knotted; can be crisp or soft and chewy.

pretzel Masa de levadura normalmente enrollada en una cuerda larga y a menudo anudadas; puede ser nítida y suave y masticable.

primal cut Larger section of a carcass from which retail cuts are made, including the chuck, shank, rib, plate, round, short loin, flank, sirloin, tenderloin, top sirloin, and bottom sirloin.

corte primigenia Sección mayor de un canal desde el que se realizan los cortes minoristas, incluyendo el chuck, Shank, costilla, placa redonda, Lomo corto, el flanco, solomillo, solomillo, solomillo superior e inferior el solomillo.

primary Containers that come into direct contact with food.

primarios Contenedores que entran en contacto directo con alimentos.

printability Ability to take on ink for printing.

impermeabilidad La habilidad para llevar a cabo la tinta para la impresión.

probiotics Microorganisms believed to provide health benefits when consumed.

los probióticos Microorganismos que se cree proporcionan beneficios para la salud cuando se consume.

processed meats Meat by-products combined with spices and additives to form a new product such as hot dogs, sausage, bologna, and jerky.

las carnes procesadas Subproductos cárnicos combinadas con especias y aditivos para formar un nuevo producto como perros calientes, salchichas, mortadela, y carne seca.

production In the food industry, includes such industries as farming, ranching, orchard management, fishing, and aquaculture.

la producción En la industria alimentaria, incluye industrias tales como la agricultura, la ganadería, la gestión de huertos, la pesca y la acuicultura.

protease Enzyme that breaks down or digests proteins.

proteasa Enzima que descompone o digiere proteínas.

protein Large molecules of long chains of amino acids.

las proteínas Grandes moléculas largas cadenas de aminoácidos.

protein efficiency ratio (PER) Measure of protein quality assessed by determining the extent of weight gain in experimental animals when fed the test item.

coeficiente de eficiencia proteica (PER) Medida de la calidad de la proteína evaluada por determinar el alcance de la ganancia de peso en los animales de experimentación cuando se alimenta el elemento de prueba.

protein quality Refers to amino acid content of a protein.

la calidad de la proteína Se refiere a contenido de aminoácidos de una proteína.

proteinase Enzyme that hydrolyzes protein to smaller fragments, eventually producing amino acids.

proteinasa Enzima que hidroliza la proteína para los fragmentos más pequeños, eventualmente producir aminoácidos.

protein Substance made of long chains of amino acids.

la proteína Sustancia proteica hechas de largas cadenas de aminoácidos.

protozoan One-celled animal.

el protozoario Un animal unicelular.

proximate analysis Approximate composition of food products consisting of proportions of water, carbohydrate, protein, fat, and ash.

análisis proximal La composición aproximada de los productos alimenticios consistente en proporciones de agua, carbohidratos, proteína, grasa y cenizas.

pounds per square inch (PSI) Measure of pressure as referenced to atmospheric pressure.

libras por pulgada cuadrada (PSI) Medida de presión como el referido a la presión atmosférica.

pyschrophilic Organisms capable of growth and reproduction in cold temperatures, ranging from 14 °F to 50 °F (−20 °C to +10 °C).

psicrófilo Los organismos capaces de vivir a temperaturas por debajo de los 14 °F hasta 50 °F (−20 °C to +10 °C).

psychrotrophic Bacteria that grow best at cold temperatures (cold-loving bacteria).

psicrotróficas Bacterias que crecen mejor en temperaturas frías (frío-amante de las bacterias).

purity Degree of singularity of a constituent in a product, generally expressed as a percentage. For example, in sugar it indicates the degree of sucrose and other extraneous products.

la pureza Grado de la singularidad de un componente de un producto, generalmente expresado como un porcentaje. Por ejemplo, en azúcar que indica el grado de sacarosa y otros productos superfluos.

quality assurance (QA) Continual monitoring of incoming raw and finished products to ensure compliance with compositional standards, microbiological standards, and various government regulations; requires many diverse technical and analytical skills.

garantía de calidad (QA) Monitoreo continuo de entrada de materias y productos acabados para garantizar el cumplimiento de normas de composición, las normas microbiológicas y diversas regulaciones gubernamentales; requiere muchas diversas técnicas y habilidades analíticas.

quality grades The results of meat and poultry grading; provide a labeling system that shows the quality of specific cuts.

grados de calidad Los resultados de la carne de aves de corral y clasificación; proporcionar un sistema de etiquetado que muestra la calidad de cortes específicos.

R

radiation Transfer from electromagnetic radiation of a body because of the vibration of its molecules.

la radiación Transferencia de las radiaciones electromagnéticas de un cuerpo debido a la vibración de sus moléculas.

radioisotope Chemical element that spontaneously emits radiation (such as electrons or alpha particles) and exhibits the same atomic number and nearly identical chemical behavior of an element but with a differing atomic mass and different physical properties.

la radioisótopos Elemento químico que espontáneamente emite radiación (como los electrones o partículas alfa) y exhibe el mismo número atómico y casi idéntico comportamiento químico de un elemento, pero con diferente masa atómica y diferentes propiedades físicas.

rancidity Special type of spoilage in fats that involves oxidation of unsaturated fatty acids; the deterioration of fats, usually by an oxidation process, resulting in objectionable flavors and odors.

la rancidez Un tipo especial de la descomposición de las grasas que involucra la oxidación de ácidos grasos insaturados; el deterioro de las grasas, generalmente por un proceso de oxidación, resultando en sabores y olores desagradables.

raw sugar Intermediate crystalline product of cane sugar factories resulting from the evaporation of water from sugarcane stalk juice; sometimes called *turbinado sugar*. True raw sugar cannot be sold in the United States because it contains too many impurities.

el azúcar en bruto Producto cristalino intermedio de caña azucareras resultante de la evaporación de agua de jugo de tallos de caña de azúcar; en ocasiones se denomina *azúcar turbinado*. Verdadero azúcar en bruto no puede ser vendido en los Estados Unidos porque contiene demasiadas impurezas.

RDA *See* Recommended Daily Allowances.

RDA *Véase* Dosis diaria recomendada.

reciprocating Moving forward and backward alternately.

reciprocar El Movimiento hacia adelante y hacia atrás alternativamente.

Recommended Daily Allowances (RDA) Average daily level of intake sufficient to meet the nutrient requirements of nearly all healthy people.

Allowences Diarias Recomendadas (RDA) Nivel de ingesta diaria promedio suficiente para satisfacer los requerimientos nutricionales de casi todas las personas sanas.

recycle To obtain, treat, or process used materials for reuse.

reciclar Para obtener, tratar o procesar materiales usados para su reutilización.

reduction reaction Chemical reaction in which there is a gain in hydrogen or electrons.

la reacción de reducción La reacción química en la cual existe una ganancia de hidrógeno o electrones.

refining Separating corn seed and soybeans into their component parts and converting these to high-value products.

la refinación Separando las semillas de maíz y soya en sus partes componentes y convertirlos en productos de alto valor.

refrigeration Holding a food product at temperatures that range from 40 °F to 45 °F (4.5 °C to 7 °C).

la refrigeración El reposo de un producto alimentario a temperaturas que oscilan entre 40 °F a 45 °F (4.5 °C a 7 °C).

rehydration Adding water to replace that lost during drying.

la rehidratación La adición de agua para sustituir el que perdió durante el secado.

rehydrated Dried or dehydrated products to which the lost water or fluid has been restored.

rehidratado Los productos desecados o deshidratados para que la perdida de agua o líquido ha sido restaurada.

rendering Freeing fat from connective tissue by means of heat.

fundir grasa Liberar grasa de tejido conectivo por medio de calor.

rennet Used for enzyme coagulation of milk.

cuajo Empleado por enzima de la coagulación de la leche.

respiration Metabolic process by which cells consume oxygen and give off carbon dioxide; process continues after harvest.

la respiración Proceso metabólico por el cual células consumen oxígeno y desprenden dióxido de carbono; el proceso continúa después de la cosecha.

résumé Written information for a prospective employer that may include any of the following: career objectives, work experience, education background, accomplishments, awards, or skills; also called a *data sheet*.

curriculum (résumé) Información escrita para un posible empleador, que puede incluir cualquiera de los siguientes: objetivos de carrera, experiencia laboral, educación Antecedentes, logros, premios o habilidades, también denominada *hoja de datos*.

retail cuts Family-sized or single-serving cuts of meat purchased at a market.

cortes minoristas De tamaño familiar o solo sirve cortes de carne comprada en un mercado.

retentate A concentrated fluid.

la reducción Un concentrado líquido.

retort Container in which a product is heated.

contenedor calentado El contenedor en el que un producto se calienta.

retrogradation Process in which starch molecules, particularly the amylose fraction, reassociate or bond together in an ordered structure after disruption by gelatinization.

la retrogradación Proceso en el cual las moléculas de almidón, especialmente la fracción amilosa, reassociate o bond juntos en una estructura ordenada tras la interrupción por la gelatinización.

reverse osmosis (RO) Process of "dewatering" in which water passes through a membrane but not ions and small molecules.

ósmosis inversa (OI) Proceso de "desecación" en el cual el agua pasa a través de una membrana, pero no iones y moléculas pequeñas.

rheology The study of the science of deformation of matter.

reología El estudio de la ciencia de la deformación de la materia.

riboflavin One of the B vitamins.

riboflavina Una de las vitaminas del complejo B.

rigor mortis Contraction and stiffening of muscles after slaughter.

rigor mortis Contracción y rigidez de los músculos después de la matanza.

ripening To age or cure to develop characteristic flavor, odor, body, texture, and color.

la maduración Dejar madurar o curar para desarrollar características de sabor, olor, textura, cuerpo y color.

rise Stage in the process of making yeast breads where the dough is set in a warm, draft-free place for a period of time (usually an hour or so) while the yeast ferments some of the sugars in the dough, forming carbon dioxide. This causes the bread to increase in size. A rising period usually lasts until the dough doubles in size.

la subida Una Etapa del proceso de hacer panes de levadura cuando la masa está establecido en un ambiente cálido, proyecto de espacio durante un período de tiempo (generalmente una hora o así) mientras que la levadura fermenta algunos de los azúcares en la masa, formando dióxido de carbono. Esto hace que el pan aumente de tamaño. Un aumento del periodo generalmente dura hasta que la masa se duplica en tamaño.

risk Measure of the probability and severity of harm to human health.

riesgo Medida de la probabilidad y la gravedad de los daños a la salud humana.

risk assessment The process of determining the likelihood and severity of loss from a possible risk and deciding how to best manage it.

evaluación del riesgo El proceso de determinar la probabilidad y severidad de la pérdida de un posible riesgo y decidir cómo administrar mejor.

RO *See* reverse osmosis.

RO *Véase* ósmosis inversa.

roast To bake; applied to certain foods such as meats.

asar Cocinar; se aplica a ciertos alimentos como la carne.

roe Fish eggs.

las huevas Los Huevos de peces.

ropey Capable of being drawn into a thread or tending to adhere in stringy masses.

chungo Capaz de ser arrastrado en una rosca o tienden a adherirse en masas fibrosas.

S

salmonella Bacteria species that may cause food poisoning.

la salmonela Las especies de bacterias de Salmonella que pueden causar intoxicaciones alimentarias.

salmonellosis Illness produced by ingestion of *Salmonella* organisms.

salmonelosis Una enfermedad producida por la ingestión de la *salmonelosis.*

salt Chemical compound derived from an acid by replacement of the hydrogen (H^+), wholly or in part, with a metal or electrically positive ion—for example, sodium citrate; sodium chloride crystals used as a flavoring.

sal Compuesto químico derivado de un ácido por la sustitución del hidrógeno (H^+), en su totalidad o en parte, con un metal o eléctricamente de iones positivos, por ejemplo, el citrato de sodio; los cristales de cloruro de sodio utilizado como aromatizante.

sanitization Removing or neutralizing elements that may present health dangers.

la higienización La retirar o neutralizar los elementos que pueden presentar peligros para la salud.

sanitizing The use of chemicals to reduce microorganisms from surfaces but not entirely remove them; enough organisms are removed to make a surface safe.

la desinfección El uso de productos químicos para reducir los microorganismos de las superficies, pero no enteramente eliminarlos; suficiente microorganismos se quita para hacer una superficie segura.

saponification value Indicates the average molecular weight of fatty acids in a fat. The value represents the number of milligrams of potassium hydroxide needed to saponify (convert to soap) 1 gram of fat. The value increases and decreases inversely (opposite of) with average molecular weight.

el valor de saponificación Indica el peso molecular promedio de ácidos grasos en la grasa. El valor representa el número de miligramos de hidróxido de potasio necesarios para saponificar (convertir en jabón) 1 gramo de grasa. El valor aumenta y disminuye inversamente (enfrente de) con peso molecular promedio.

saturated When the carbons in a fatty acid share are completely connected through single bonds.

saturados Cuando los carbones en un ácido graso comparte están totalmente conectadas mediante enlaces simples.

saturated fatty acid Fatty acid with no double bonds between carbon atoms; it holds all of the hydrogen that can be attached to the carbon atoms.

ácidos grasos saturados ácidos grasos sin dobles enlaces entre átomos de carbono; contiene todo el hidrógeno que se pueden adjuntar a los átomos de carbono.

sauté To cook in a small amount of fat.

saltear Cocinar en una pequeña cantidad de grasa.

scald To heat a liquid, usually milk, until bubbles appear around the edge, approximately 198 ° to 203 °F (92 ° to 95 °C); to blanch, as when preparing vegetables for freezing.

la escaldadura Para calentar un líquido, normalmente la leche, hasta que aparezcan burbujas alrededor del borde, aproximadamente 198 ° a 203 ° F (92 ° a 95 °C); escalfar, como al preparar verduras para congelar.

SCF *See* supercritical fluid.

SCF *Véase* fluido supercrítico.

science The general body of research that uses the scientific process or method to use observations to produce hypotheses and theorems.

la ciencia El cuerpo general de la investigación que utiliza el proceso o método científico para utilizar sus observaciones para elaborar hipótesis y teoremas.

SCP *See* single-celled protein.

SCP Véase proteína unicelular.

sear To brown the surface quickly by intense heat in order to develop color and flavor and to improve appearance; usually applied to meat.

dorear Pasar a la placha hasta que esté marrón la superficie rápidamente por el intenso calor para desarrollar sabor y color para mejorar la apariencia; generalmente aplicado a la carne

secondary Container that holds several primary containers together.

Contenedor secundario que contiene varios contenedores primarios juntos.

sensory evaluation The process of human analysis of the look, taste, smell, sound, and feel of food.

la evaluación sensorial El proceso de evaluación sensorial análisis humano de la apariencia, sabor, olor, sonido, y el tacto de los alimentos.

sensory analysis Human analysis of the look, taste, smell, sound, and feel of food.

análisis sensorial El análisis humano de la apariencia, sabor, olor, sonido, y el tacto de los alimentos.

separator Machine that separates cream and skim portions of milk.

el separador Máquina que separa la crema dulce y leche descremada porciones de leche.

septicemia Presence of pathogenic microorganisms in the blood.

septicemia Presencia de microorganismos patógenos en la sangre.

sequestrant Substance that binds or isolates other substances; for example, some molecules can tie up trace amounts of minerals that may have unwanted effects in a food product.

secuestrante Sustancia que une o aísla otras sustancias; por ejemplo, algunas moléculas pueden atar trazas de minerales que pueden tener efectos no deseados en los productos alimentarios.

shelf life Time required for a food product to reach an unacceptable quality.

el tiempo de vida útil El tiempo de vida útil de un producto alimenticio, necesarios para llegar a una calidad inaceptable.

shorts By-product of flour milling; small particles of bran, germ, aleurone layer, and coarse flour.

subproducto de harina Subproducto de la molinería; pequeñas partículas de salvado, germen, capa y aleurone harina gruesa.

shucked Removed from the shell.

debullado Quitado de la concha.

simmer To cook in liquid below the boiling point; a liquid is simmering when bubbles form slowly and break just below the surface, about 185 °F (85 °C).

hervir lentamente Cocinado en un líquido por debajo del punto de ebullición; un líquido está hirviendo cuando las burbujas se forman lentamente y romper justo debajo de la superficie, aproximadamente 185 °F (85 °C).

single-celled protein (SCP) Protein obtained from single-celled organisms such as yeast, bacteria, and algae grown on specifically prepared media.

proteína unicelular (SCP) Proteicos obtenidos a partir de organismos unicelulares como las levaduras, las bacterias y las algas cultivadas en medios preparados específicamente.

sleep apnea Common and often undiagnosed disorder in which an individual has pauses in breathing or shallow breaths while sleeping.

apnea del sueño Común y a menudo sin diagnosticar trastorno en el cual una persona tiene pausas en la respiración o respiraciones poco profundas mientras duerme.

smoking Preservation method for meat that inhibits microbial growth, protects fat from rancidity, contributes to characteristic color, and creates unique flavors in processed meats.

humear Método de conservación para la carne que inhibe el crecimiento de microbios, protege la grasa de la rancidez, contribuye a su color característico, y crea sabores únicos en las carnes procesadas.

SNF *See* solids-not-fat.

SNF *Véase* sólidos-no-grasa.

soft wheat General term for varieties of wheat that contain relatively small amounts of gluten.

el trigo blando Término general para el trigo blando, las variedades de trigo que contienen cantidades relativamente pequeñas de gluten.

solids-not-fat (SNF) Includes the carbohydrates, lactose, protein, and minerals of milk.

los sólidos-no-grasa (SNF) Incluye los hidratos de carbono, proteínas, lactosa y minerales de la leche.

soluble Able to be dissolved or liquefied.

soluble Capaz de ser soluble disuelto o licuado.

soluble fiber Fiber that dissolves in water; can help lower blood-level cholesterol and glucose levels.

fibra soluble Fibra que se disuelve en agua; pueden ayudar a bajar el nivel de colesterol de la sangre y los niveles de glucosa.

solute Dissolved or dispersed substance.

soluto Una sustancia disuelta o dispersa.

solution (1) Mixture resulting from the dispersion of small molecules or ions (the solute) in a liquid such as water (the solvent); (2) the resulting mixture of a solute dissolved in a solvent.

solución (1) la mezcla resultante de la dispersión de pequeñas moléculas o iones (soluto) en un líquido como el agua (disolvente); (2) la mezcla resultante de un soluto disuelto en un disolvente.

solvent Liquid in which other substances may be dissolved.

disolvente Líquido en el cual otras sustancias pueden ser disueltas.

somatic cell count (SCC) A main indicator of milk quality; the majority of somatic cells are leukocytes (white blood cells).

el recuento de células somáticas (SCC) Un indicador principal de la calidad de la leche; la mayoría de las células somáticas son los leucocitos (glóbulos blancos).

SOP *See* standard operating procedure.

SOP *Véase* procedimiento operativo estándar.

sorbitol Sugar alcohol similar to glucose in chemical structure but with an alcohol group (-C-OH) replacing the aldehyde group (H-C=O) of glucose; occurs naturally in fruit and berries. It is slowly absorbed by the body, and it is a caloric sweetener.

soluble Alcohol de azúcar similar a la glucosa en su estructura química, pero con un grupo alcohol (-C-OH) Sustitución del grupo aldehído (H-C=O) de la glucosa; se produce naturalmente en las frutas y bayas. Lentamente es absorbido por el cuerpo, y es un edulcorante calórico.

SPC *See* Standard plate count.

SPC *Véase la* norma SPC recuento en placa.

species Group of taxonomic classification consisting of organisms that can breed together.

especies Grupo de clasificación taxonómica compuesta de organismos que pueden criar juntos.

specific gravity Weight of a volume of material divided by the weight of an equal volume of water.

gravedad específica El peso de un volumen de material dividido por el peso de un volumen igual de agua.

specific heat The number of calories needed to raise the temperature of 1 gram of a given substance 1 degree Celsius; the specific heat of water has been set at 1.0; fats and sugars have lower specific heats, thus requiring less heat than water needs to raise their temperature an equal number of degrees.

calor específico El número de calorías necesarias para elevar la temperatura de un gramo de una sustancia dada 1 grado Celsius; el calor específico del agua se ha fijado en 1,0; grasas y azúcares han específico inferior se calienta, por lo que requieren menos calor que las necesidades de agua para elevar su temperatura un número igual de grados.

spectrophotometry A method of chemical analysis based on the transmission or absorption of light.

espectrofotometría Un método de análisis químico basado en la transmisión o absorción de la luz.

spore Microorganism in a dormant state or a one-celled reproductive organ of a fungus. Spores may be activated by appropriate environmental conditions.

espora Microorganismo en un estado de latencia o un uno-celled el órgano reproductor de un hongo. Las esporas pueden ser activados por condiciones ambientales apropiadas.

stability Resistance to chemical change, disintegration, or degradation.

la estabilidad La resistencia a los cambios químicos, la desintegración o degradación.

stabilizer A water-holding substance, such as a vegetable gum, that interferes with ice crystal formation and contributes to a smooth texture in frozen desserts.

un estabilizador Una sustancia agua-sosteniendo , tales como un vegetal, goma de mascar, que interfiere con la formación de cristales de hielo y contribuye a una textura suave en postres congelados.

standardized To have a rule or principle used for the basis of judgment applied.

estandarizado Para tener una regla o principio que se utiliza para la base de la sentencia aplicada.

standard operating procedure (SOP) Established procedures to be followed in carrying out a given operation or in a given situation.

procedimiento operativo estándar (SOP) Estableció los procedimientos que deben seguirse en la realización de una operación determinada o en una situación dada.

standard plate count (SPC) A test that determines the presence of microbiological organisms in a food.

recuento en placa estándar (SPC) Una prueba que determina la presencia de organismos microbiológicos en los alimentos.

standards Set up and established by authority as a rule for the measure of quantity, weight, extent, value, or quality. Set by the USDA to specifically describe a food; to be labeled as such, a food must meet these specifications.

estándares Establecer estándares y establecida por la autoridad como una regla para la medida de la cantidad, peso, medida, valor o calidad. Establecido por el USDA para describir específicamente a los alimentos; a ser etiquetada como tal, un alimento debe cumplir estas especificaciones.

starch A carbohydrate polymer made up of many units of glucose.

almidón Un carbohidrato polímero compuesto de muchas unidades de glucosa.

starch granule Millions of starch molecules laid down in a highly organized manner; the shape of a granule is typical for the plant species from which it comes.

gránulo de almidón Millones de moléculas de almidón establecidas en forma muy organizada; la forma de gránulo es típica de las especies de la planta de la que procede.

starter culture A concentrated number of the organisms desired to start the fermentation process.

el cultivo iniciador Un número concentrado de los organismos deseada para iniciar el proceso de fermentación.

steam To cook in steam, with or without pressure.

vapor Para cocinar al vapor de agua, con o sin presión.

steeping Soaking; also extracting flavor, color, or other qualities from a substance by allowing it to stand in liquid below the boiling point.

remojo Remojo; también la extracción de sabor, color u otras cualidades de una sustancia, permitiendo que permanezca en estado líquido por debajo del punto de ebullición.

sterilization The process of heating a material sufficiently to destroy essentially all micro-organisms.

la esterilización El proceso de esterilización de calentar un material suficientemente para destruir esencialmente todos los microorganismos.

sterilize To destroy essentially all microorganisms.

esterilizar Destruir esencialmente todos los microorganismos.

stew To simmer in a small to moderate quantity of liquid.

guisar De cocinar en una pequeña cantidad de líquido.

still retort *See* retort.

aún retorta *Véase* retorta.

stir To mix with circular motion, to blend food materials, or to obtain a uniform consistency as in sauces.

agitar (batir) Para mezclar con movimiento circular, para batir alimentos materiales, o para obtener una consistencia uniforme como en salsas.

stoma Small opening in the epidermis of leaves and some stems that opens to permit gas exchange and closes in conditions of water stress; flanked by stomatal guard cells, which regulate the opening of the stoma. Plural is *stomata*.

estoma Abertura pequeña en la epidermis de las hojas y algunos tallos que se abre para permitir el intercambio de gases y cierra en condiciones de estrés hídrico; flanqueado por estomas células de guarda, que regulan la apertura del estoma. plural es *estomas*.

storage tissue Located in the cytoplasm in leucoplasts; dominant in roots, tubers, bulbs, and seeds.

tejido de almacenamiento Ubicada en el citoplasma en leucoplasts; dominante en raíces, tubérculos, bulbos y semillas.

straight grade Flour that should contain all flour streams resulting from the milling process; most patent flours on the market include approximately 85% of the straight flour.

la harina de calidad recta La harina que debe contener toda la harina arroyos resultantes del proceso de molienda; más patentes las harinas en el mercado incluyen aproximadamente el 85% de la recta de harina.

stunting Reduced growth in children who have inadequate nutrition.

atrofia La reducción del crecimiento en los niños que tienen una nutrición inadecuada.

sublimation When water goes from a solid to a gas without passing through its liquid phase.

sublimación Cuando el agua pasa del estado sólido al estado gaseoso sin pasar por la fase líquida.

substrate Substance acted on by an enzyme.

el sustrato Actuó en la sustancia por una enzima.

sucrose The disaccharide with the chemical formula $C_{12}H_{22}O_{11}$.

sacarosa El disacárido con la fórmula química $C_{12}H_{22}O_{11}$.

sugar Often refers to sucrose, the disaccharide with the chemical formula $C_{12}H_{22}O_{11}$, but can also refer to other sugars like glucose, which is often called blood sugar.

azúcar A menudo se refiere a la sacarosa, el disacárido sacarosa, disacárido con la fórmula química $C_{12}H_{22}O_{11}$, pero también puede referirse a otros azúcares como la glucosa, que a menudo se llama azúcar en la sangre.

supercritical fluid (SCF) Process of separating one component (the extractant) from another (the matrix) using supercritical fluids as the extracting solvent. Carbon dioxide is the most common supercritical fluid.

fluidos supercríticos (SCF) El proceso de separar un componente (el extractante) de otro (la matriz) utilizando fluidos supercríticos como solvente de extracción. El dióxido de carbono es el más común de fluido supercrítico.

supercritical fluid (SCF) extraction Process of separating an extractant substance from another (called the matrix) using supercritical fluids as the extracting solvent.

fluido supercrítico (SCF) de extracción El proceso de separar una de otra sustancia extractante (llamada la matriz) utilizando fluidos supercríticos como solvente de extracción.

surface active agents Emulsifiers that improve the uniformity of a food—the fineness of grain—the smoothness and body of foods such as bakery goods, ice creams, and confectionery products.

agentes tensoactivos Emulsionantes que mejoran la uniformidad de una comida-la finura de grano-la suavidad y el cuerpo de alimentos tales como productos de panadería, helados y productos de confitería.

surface area Measure of exposed surface; measured in square inches, meters, centimeters, millimeters and so on.

área de superficie La medida de superficie expuesta; se mide en pulgadas cuadradas, metros, centímetros, milímetros, y así sucesivamente.

surimi A minced fish flesh washed to remove solubles including pigments (color) and flavors, leaving an odorless, flavorless, high-protein product.

surimi Una carne de pescado picada lavada para remover solutos incluidos los pigmentos (color) y sabores, no dejando olor, sin sbor, producto de alta proteína.

sustainable Enduring without giving away or yielding.

sostenible Duradero sin regalar o ceder.

sweeteners Any food that adds a sweet flavor to foods.

edulcorantes Cualquier alimento que añade un sabor dulce a los alimentos.

symbiotic Interdependent or mutually beneficially relationship.

simbiótico Una relación que es interdependiente o simbióticamente y mutuamente provechosa.

synchrometer Metering device that measures syrup and water in fixed proportion to the carbonator.

sincrónmetero Dispositivo de medición que mide el jarabe y el agua en una proporción fija para el carbonator.

syneresis The oozing of liquid from a rigid gel; sometimes called *weeping.*

sinergia La exudación de líquido desde un gel rígido; en ocasiones se denomina *llorando.*

synergism Interaction in which the effects or benefits of the mixture are greater than the effects of the sums of component parts; two or more factors acting cooperatively so that their combined effects when acting together exceed the sum of their effects when each acts alone.

el sinergismo La Interacción en el cual los efectos o los beneficios de la mezcla son mayores que los efectos de la suma de los componentes; dos o más factores que actúan de forma cooperativa, de modo que sus efectos combinados cuando actúan conjuntamente exceden la suma de sus efectos cuando cada uno actúa por su cuenta.

synthetic Compound produced by chemically combining two or more simple compounds or elements in the laboratory.

sintético Compuesto producido por químicamente combinando dos o más compuestos simples o elementos en el laboratorio.

syrup Viscous, concentrated sugar solution that occurs because of the evaporation of liquid.

jarabe Solución viscosa de azúcar concentrada que se produce a causa de la evaporación de un líquido.

T

tariffs A schedule of duties imposed by a government on imports or exports.

los aranceles (tarifas) Un calendario de obligaciones impuestas por el gobierno sobre las importaciones o exportaciones.

teff Smallest of grains and therefore with a high ratio of bran and germ.

teff Los granos más pequeños y, por tanto, con un alto coeficiente de salvado y germen.

tertiary A container that groups several secondary holders together into shipping units.

el nivel terciario Un contenedor que agrupa varios titulares de secundaria junto a unidades de envío.

texture (1) Arrangement of the parts of a material showing the structure; for example, the texture of baked flour products such as a slice of bread may be fine and even or coarse and open; or the texture of a cream sauce may be smooth or lumpy; (2) the arrangement of the particles or constituent parts of a material that provides its characteristic structure.

textura (1) Disposición de las piezas de un material que muestra la estructura; por ejemplo, la textura de la harina cocida productos como una rebanada de pan pueden ser finas o gruesas e incluso y abierto; o la textura de una salsa de crema pueden ser lisas o abultadas; (2) la disposición de las partículas o partes constitutivas de un material que proporciona su estructura característica.

textured protein Food products that are usually at least 50% protein and contain all eight essential amino acids and the vitamins and minerals found in meats; soybean protein is most commonly used. Other plant proteins—wheat gluten, yeast protein, and most other edible proteins—can be used singly or in combination.

proteína texturizada Productos alimenticios que son generalmente por lo menos el 50% de proteínas y contienen los ocho aminoácidos esenciales y las vitaminas y minerales que se encuentran en carnes; proteína de la soja es más comúnmente utilizado. Otras proteínas vegetales gluten de trigo, proteína de levadura, y la mayoría de las otras proteínas comestibles pueden ser utilizadas individualmente o en combinación.

thermization To heat.

la termización El calor.

thermophilic Preferring hot temperatures.

termófilas Prefiriendo a altas temperaturas.

thermotrophic Can tolerate high temperature.

termotrófico Puede tolerar la temperatura alta.

thiamin One of the B vitamins.

tiamina Una de las vitaminas del complejo B.

tom Young male turkey.

tom Pavos machos jóvenes.

Total Quality Management Management philosophy in which companies strive to provide customers with products and services that satisfy their needs.

la Gestión de la Calidad Total La filosofía de gestión en el que las empresas se esfuerzan por ofrecer a los clientes productos y servicios que satisfagan sus necesidades.

total sugars Measurement of the total amount of saccharides.

los azúcares totales La medición de la cantidad total de azúcares.

toxins Substances that are poisonous to humans.

toxinas Sustancias que son venenosas para el ser humano.

TQM See Total Quality Management.

TQM Véase Gestión de Calidad Total.

tramp material Soil and extraneous material.

material residual El suelo y el material ajeno.

trans Configuration in which the hydrogen atoms are on opposite sides of the double bond, particularly with unsaturated fatty acids.

trans Configuración en la cual los átomos de hidrógeno están en lados opuestos del doble enlace, especialmente con los ácidos grasos insaturados.

trans-fatty acid Unsaturated fatty acid formed especially during hydrogenation of vegetable oils; linked to an increase in blood cholesterol; chemically, a trans arrangement of the carbon atoms adjacent to its double bonds.

trans-ácidos grasos Ácidos grasos insaturados formado especialmente durante la hidrogenación de aceites vegetales; vinculado a un aumento en el nivel de colesterol en sangre; químicamente un ordenamiento trans de los átomos de carbono adyacentes a sus enlaces dobles.

transgenic crop *See* genetically modified organism.

cultivos transgénicos *Véase* organismo modificado genéticamente.

transgenic organism *See* genetically modified organism.

organismo transgénico *Véase* organismo modificado genéticamente.

translucency Partially transparent.

la translucidez Parcialmente transparente.

trends General direction of a market.

las tendencias Dirección General del mercado.

Trichinella Spiralis Tiny parasite that may be present in some fresh pork; if not destroyed by cooking, causes a disease called *trichinosis*.

Trichinella Spiralis Parásito pequeño que pueden estar presentes en algunos de carne de cerdo fresca; si no es destruida por la cocción, causa una enfermedad llamada *triquinosis*.

triglycerides Neutral fat molecule made up of three fatty acids joined to one glycerol molecule through a special chemical linkage called an *ester*. A type of lipid consisting chemically of one molecule of glycerol combined with three fatty acids.

los triglicéridos Molécula neutral de grasa compuesta de tres ácidos grasos se unió a una molécula de glicerol a través de un varillaje de químicos especiales llamados un éster. Un tipo de lípido compuesto químicamente de una molécula de glicerol combinado con tres ácidos grasos.

tuber A short, thickened, fleshy part of an underground stem such as a potato; new plants develop from the buds or eyes; an enlarged underground stem (for example, the potato).

tubérculo Un corto, grueso, parte carnosa de un tallo subterráneo como una patata; desarrollar nuevas plantas a partir de las yemas o los ojos; agrandamiento de tallo subterráneo (por ejemplo, la patata).

turbidity Cloudiness in a fluid; the opposite of translucent.

la turbidez La nubosidad de un fluido; lo contrario de traslúcido.

turbinado Semirefined light brown crystalline sugar. It is a steam-cleaned, partially refined sugar. In the United States it is sold as raw sugar.

turbinado Azúcar cristalino semirefinado de color marrón claro. Es un limpiado con vapor, parcialmente el azúcar refinado. En los Estados Unidos se vende como azúcar en bruto.

U

UHT See Ultrahigh temperature pasteurization.

UHT Véase temperatura ultraelevada pasteurización.

ultrafiltration A membrane process that filters out or separates particles of extremely small size (0.02 to 0.2 microns), including proteins, gums, glucose, and pigments; filtration through an extremely fine filter.

la ultrafiltración Una proceso de membrana que filtra o separa las partículas de muy pequeño tamaño (0,02 a 0,2 micrones), incluidas las proteínas, las encías, glucosa y pigmentos; filtración a través de un filtro extremadamente finos.

ultrahigh temperature (UHT) Sterilizes food by heating it above 135 °C (275 °F)—the temperature required to kill spores in milk—for 1 to 2 seconds; with this treatment, the sterilized milk is aseptically packaged, and the milk does not require refrigeration until it is opened.

temperatura ultra alta (UHT) Esteriliza los alimentos por calentamiento por más de 135 °C (275 °F)—la temperatura necesaria para matar las esporas en la leche—durante 1 a 2 segundos; con este tratamiento, la leche esterilizada es asépticamente empaquetado, y la leche no requieren refrigeración hasta su apertura.

ultrahigh temperature pasteurization Sterilizing food by heating it above 135 °C (275 °F).

altísima temperatura de pasteurización La esterilización de alimentos por calentamiento por encima de 135 °C (275 °F).

ultrapasteurization Heating milk to 138 °F or higher for 2 seconds, followed by rapid cooling to 45 °F or lower.

la ultrapasteurization La calefacción de leche a 138 °F o superior durante 2 segundos, seguido de un enfriamiento rápido a 45 °F o menos.

undernutrition When someone does not get enough food to live a healthy life.

la desnutrición Cuando alguien no recibe suficiente alimento para vivir una vida sana.

underweight Means of defining malnutrition; below weight for age in children.

insuficiencia de peso El medio de definir la desnutrición; por debajo de peso para la edad en los niños.

unleavened Bread or dough product containing no yeast or chemical (baking soda, baking powder) leavener.

los panes sin levadura Masa de levadura o productos que no contienen ningún producto de levadura química (bicarbonato de soda, el polvo de hornear).

unsaturated General term for any fatty acid with one or more double bonds between carbon atoms.

insaturado Término general para designar cualquier ácido graso con uno o más dobles enlaces entre átomos de carbono.

unsaturated fatty acid General term used to refer to any fatty acid with one or more double bonds between carbon atoms; capable of binding more hydrogen at these points of unsaturation.

ácidos grasos insaturados Un término general utilizado para referirse a cualquier ácido graso con uno o más dobles enlaces entre átomos de carbono; capaz de enlazar más hidrógeno en estos puntos de saturación.

urbanization Shift of populations from rural areas to denser urban areas.

la urbanización Un cambio de las poblaciones de las zonas rurales a las zonas urbanas más densas.

U.S. fancy Grade designations for vegetables; more uniform in shape and have fewer defects.

EE. UU. fancy Grado de designaciones de hortalizas; en forma más uniforme y tienen menos defectos.

U.S. grade a, b, c Quality grades for fruits and vegetables.

EE. UU. grado a, b, c Grados de calidad de las frutas y verduras.

U.S. no. 1, 2, 3 Quality grades for fruits, vegetables, and pigs.

EE. UU. no. 1, 2, 3 Grados de calidad de las frutas, hortalizas y cerdos.

V

vacuum A space partially or nearly completely exhausted of air.

un vacío Un espacio parcial o casi completamente agotado de aire.

vacuum cooling Cooling under conditions of no air (atmosphere).

enfriamiento de vacío Bajo condiciones de refrigeración sin aire (atmósfera).

vacuum drying Drying a product in a vacuum chamber; water vaporizes at a lower temperature than at atmospheric pressure.

secado al vacío Un producto de secado en una cámara de vacío; el agua se evapora a una temperatura inferior a la presión atmosférica.

vacuum evaporation Removal of water under conditions of no air (atmosphere).

la evaporación al vacío Extracción de agua bajo condiciones de ausencia de aire (atmósfera).

vacuum packed Airtight packaging in heavy foil container to minimize exposure to oxygen.

envasados al vacío En envases herméticos lámina pesada contenedor para minimizar la exposición al oxígeno.

value Magnitude, relative worth; estimated or assigned worth.

el valor La magnitud, valor relativo; estimados o asignado valor.

van der waals bond Extremely weak bonds formed between nonpolar molecules or nonpolar parts of a molecule.

los enlaces van der waals Los enlaces extremadamente débil formado entre moléculas no polar o no polar partes de una molécula.

vascular Includes the xylem and phloem of plants.

vascular Incluye del xilema y floema de plantas.

veal Young calf or meat from a calf 1 to 4 months old.

la ternera Los terneros de carne o de un ternero de 1 a 4 meses de edad.

vegan A person who only eats food of plant origin.

vegan Una persona que solo come alimentos de origen vegetal.

vegetable fruit Botanically, a fruit is the ovary and surrounding tissues, including the seeds, of a plant; fruit is the fruit part of a plant that is not sweet and is usually served with the main course of a meal.

frutas verduras Botánicamente, un fruto es el ovario y los tejidos circundantes, incluyendo las semillas, fruto de una planta; es el fruto de una planta que no es dulce y se sirve generalmente con el plato principal de una comida.

vegetable gums Polysaccharide substances derived from plants, including seaweed and various shrubs or trees; have the ability to hold water and often act as thickeners, stabilizers, or gelling agents in various food products—for example, algin, carrageenan, and gum arabic.

las encías vegetales Sustancias polisacárida derivadas de plantas, incluidas las algas y diversos arbustos o árboles; tienen la capacidad para retener el agua y a menudo actúan como espesantes, estabilizantes, o gelificantes en diversos productos alimenticios, por ejemplo, algin, carragenano y goma arábiga.

vinegar Sour liquid consisting of dilute acetic acid obtained from the fermentation of wine, beer, cider, or similar products.

vinagre Líquido amargo compuesto de ácido acético diluido obtenido de la fermentación del vino, cerveza, sidra, o productos similares.

vinification Conversion of fruit juices such as grape juice into wine by fermentation.

la vinificación La conversión de jugos de frutas como el jugo de uva en vino por fermentación.

viscosity Resistance to flow; increase in thickness or consistency.

viscosidad La resistencia al flujo; aumento de espesor o consistencia.

visible fat (1) Refined fats and oils used in food preparation, including edible oils, margarine, butter, lard, and shortenings; (2) fats seen and easily trimmed from meat.

grasa visible (1) Refinado las grasas y aceites utilizados en la preparación de alimentos, incluyendo los aceites comestibles, margarina, mantequilla, manteca de cerdo y grasas; (2) las grasas visto y fáciles de recortar de la carne.

vitamin fortified Vitamins added to processed foods.

fortificadas con vitaminas Vitaminas añadidos a los alimentos procesados.

vitamins Chemical compounds in food that are needed in minute amounts (in milligrams and micrograms) to regulate the chemical reactions in the human body. May be fat soluble or water soluble.

vitaminas Compuestos químicos en los alimentos que se necesitan en cantidades diminutas (en miligramos y microgramos) para regular las reacciones químicas en el cuerpo humano. Pueden ser liposolubles o solubles en agua.

vitelline membrane The covering of the yolk of an egg.

membrana vitelina El cubrimiento de la yema de un huevo.

volatile Evaporates readily.

volátil Se evapora fácilmente.

W

wastewater Water left over after being used in food processing.

las aguas residuales El agua remanente después de ser utilizados en el procesamiento de alimentos.

wasting Gradually reducing the fullness and strength of the body.

la emaciación Reduciendo gradualmente la plenitud y la fuerza del cuerpo.

water activity (A_w) Water in food which is not bound to food molecules can support the growth of bacteria, yeasts and molds (fungi).

la actividad de agua (A_w) El agua en los alimentos que no se une a moléculas de los alimentos puede apoyar el crecimiento de bacterias, levaduras y mohos (hongos).

waxy Starch granules devoid of amylose.

ceroso Los gránulos de almidón desprovisto de amilosa.

weeping When starch gives up water while cooling or during storage.

lloro Cuando el almidón da para arriba el agua mientras la refrigeración o durante el almacenamiento.

wet scrubber Device used to remove pollutants.

depurador húmedo Dispositivo utilizado para eliminar contaminantes.

whey Liquid by-product of cheese making.

el suero el Líquido subproducto de la fabricación de quesos.

whip To beat rapidly and produce expansion by incorporating air; applied to cream, eggs, and gelatin dishes.

batir rápidamente Batir rápidamente y producir la expansión mediante la incorporación de aire; aplicado a la crema, huevos y platos de gelatina.

white blood cell A cell that helps fight infection.

una celda de glóbulos blancos Que ayudan a combatir la infección.

wholesale cuts Larger primal cuts of meat that are shipped to grocery stores and meat markets.

cortes mayorista Primal grandes cortes de carne que son enviados a las tiendas de comestibles y mercados de la carne.

whole wheat Wheat flour milled using the entire wheat berry (germ, endosperm, bran).

harina de trigo entero La harina de trigo molida utilizando todo la semilla de trigo (germen, endospermo, afrecho).

winterization Process that keeps oils from becoming cloudy when chilled.

el acondicionamiento para el invierno Proceso de que mantiene los aceites de ser nublado cuando refrigeradas.

wort The clear filtrate from enzymatic action on the malt during the mashing process of beer making.

el mosto El filtrado claro de acción enzimática sobre la malta durante el proceso de maceración de la fabricación de cerveza.

X

xanthan gum Microbial produced from the fermentation of corn sugar; used as a thickener, emulsifier, and stabilizer in foods such as dairy products and salad dressings.

la goma xantan Microbial producidos a partir de la fermentación del azúcar de maíz; se usa como espesante, emulsionante y estabilizador en alimentos tales como productos lácteos y aderezos para ensaladas.

Y

yeast Single-celled fungus; one species—*Saccharomyces cerevisiae*—ferments sugars; its by-products are principally carbon dioxide and alcohol. Carbon dioxide raises breads.

la levadura Hongo unicelular; una especie-*Saccharomyces cerevisiae*-fermenta los azúcares; sus productos son principalmente el dióxido de carbono y alcohol. El dióxido de carbono plantea panes.

yeast bread Any bread whose primary leavening action results from the fermentation of sugar by yeast.

pan de levadura Pan leudado, cuyas principales medidas resultan de la fermentación del azúcar por la levadura.

yeast fermentation Process in which enzymes produced by the yeast break down sugars to carbon dioxide and alcohol; also produce some flavor substances.

la fermentación de la levadura El proceso en la que las enzimas producidas por las levaduras descomponer los azúcares en alcohol y dióxido de carbono; también producen algunas sustancias aromatizantes.

yield grade Classifies carcasses on the basis of the proportion of usable meat to bone and fat; used in conjunction with quality.

el grado de rendimiento Clasifica los cadáveres sobre la base de la proporción de carne utilizable a los huesos y grasa; se utiliza en conjunción con la calidad.

Z

z value The temperature required to decrease the time needed to obtain a one log reduction in cell numbers to 1/10th of the original value.

el valor z La temperatura necesaria para reducir el tiempo necesario para obtener una reducción de registro en los números de celda a 1/10 de su valor original.

INDEX

NOTE: Page numbers in *italics* refer to figures and page numbers in **boldface** refer to tables

A

B

C

D

E

G

H

M

N

O

P

R

T